Recursive Estimation and Control for Stochastic Systems

Probability and Mathematical Statistics (Continued)

ROHATGI • Statistical Inference
ROSS • Stochastic Processes
RUBINSTEIN • Simulation and The Monte Carlo Method
SCHEFFE • The Analysis of Variance
SEBER • Linear Regression Analysis
SEBER • Multivariate Observations
SEN • Sequential Nonparametrics: Invariance Principles and Statistical Inference
SERFLING • Approximation Theorems of Mathematical Statistics
TJUR • Probability Based on Radon Measures
WILLIAMS • Diffusions, Markov Processes, and Martingales, Volume I: Foundations
ZACKS • Theory of Statistical Inference

Applied Probability and Statistics

ABRAHAM and LEDOLTER • Statistical Methods for Forecasting
AGRESTI • Analysis of Ordinal Categorical Data
AICKIN • Linear Statistical Analysis of Discrete Data
ANDERSON, AUQUIER, HAUCK, OAKES, VANDAELE, and WEISBERG • Statistical Methods for Comparative Studies
ARTHANARI and DODGE • Mathematical Programming in Statistics
BAILEY • The Elements of Stochastic Processes with Applications to the Natural Sciences
BAILEY • Mathematics, Statistics and Systems for Health
BARNETT • Interpreting Multivariate Data
BARNETT and LEWIS • Outliers in Statistical Data, *Second Edition*
BARTHOLOMEW • Stochastic Models for Social Processes, *Third Edition*
BARTHOLOMEW and FORBES • Statistical Techniques for Manpower Planning
BECK and ARNOLD • Parameter Estimation in Engineering and Science
BELSLEY, KUH, and WELSCH • Regression Diagnostics: Identifying Influential Data and Sources of Collinearity
BHAT • Elements of Applied Stochastic Processes, *Second Edition*
BLOOMFIELD • Fourier Analysis of Time Series: An Introduction
BOX • R. A. Fisher, The Life of a Scientist
BOX and DRAPER • Evolutionary Operation: A Statistical Method for Process Improvement
BOX, HUNTER, and HUNTER • Statistics for Experimenters: An Introduction to Design, Data Analysis, and Model Building
BROWN and HOLLANDER • Statistics: A Biomedical Introduction
BROWNLEE • Statistical Theory and Methodology in Science and Engineering, *Second Edition*
BUNKE and BUNKE • Statistical Inference in Linear Models, Volume I
CHAMBERS • Computational Methods for Data Analysis
CHATTERJEE and PRICE • Regression Analysis by Example
CHOW • Analysis and Control of Dynamic Economic Systems
CHOW • Econometric Analysis by Control Methods
CLARKE and DISNEY • Probability and Random Processes: A First Course with Applications
COCHRAN • Sampling Techniques, *Third Edition*
COCHRAN and COX • Experimental Designs, *Second Edition*
CONOVER • Practical Nonparametric Statistics, *Second Edition*
CONOVER and IMAN • Introduction to Modern Business Statistics
CORNELL • Experiments with Mixtures: Designs, Models and The Analysis of Mixture Data
COX • Planning of Experiments
DANIEL • Biostatistics: A Foundation for Analysis in the Health Sciences, *Third Edition*
DANIEL • Applications of Statistics to Industrial Experimentation
DANIEL and WOOD • Fitting Equations to Data: Computer Analysis of Multifactor Data, *Second Edition*
DAVID • Order Statistics, *Second Edition*
DAVISON • Multidimensional Scaling
DEMING • Sample Design in Business Research

continued on back

Recursive Estimation and Control for Stochastic Systems

HAN-FU CHEN
Academia Sinica
Institute of Systems Science
Beijing, People's Republic of China

JOHN WILEY & SONS

New York Chichester Brisbane Toronto Singapore

Library of Congress Cataloging in Publication Data:

Chen, Han-Fu.
Recursive estimation and control for stochastic systems.

(Wiley series in probability and mathematical statistics, 0271-6232. Probability and mathematical statistics)
Includes index.
1. Stochastic systems. 2. Estimation theory. 3. Control theory. I. Title. II. Series.

QA402.C447 1985 519.2 84-20907
ISBN 0-471-81566-7

Printed in the United States of America

10 9 8 7 6 5 4 3 2 1

Preface

Almost without exception, real dynamical systems are subject to random disturbances. In some circumstances such systems can be approximated by deterministic ones by neglecting the random effects. However, to consider them as truly stochastic systems is not only very attractive from a theoretical point of view but it is in fact necessary in order to improve the performance of a system in an engineering context.

In order to control a stochastic system, one first has to construct its mathematical model; this is known as *system identification*. Further, in order to control a system once its mathematical model has been established, one needs to acquire information about the system in real time; to be specific, one needs to observe the system state. But since the observations on the behavior of a system are often incomplete (i.e., only partial observations of the system state are available) and since they are usually corrupted by random noise, the techniques of filtering and prediction are necessary. Having obtained a state estimate, one is then in a position to solve the optimal stochastic control problem with respect to any given performance index. Finally, a certain class of tracking, regulation, and adaptive control problems can be solved by the techniques of stochastic approximation; hence the convergence analysis of such algorithms is very important for the related control problems. These areas are the topics of this book.

The aims of the book are (1) to give a convergence analysis of recursive estimation algorithms arising in system identification and stochastic approximation with correlated noise, (2) to consider the recursive Gauss–Markov estimates in connection with the Kalman filter for both discrete- and continuous-time systems, and (3) to analyze some control and adaptive control problems closely related to recursive estimation (this will include the class of so-called singular problems).

This book summarizes the author's work in these fields but it also includes considerable other material in order to give a systematic and self-contained treatment; it is aimed at professional research workers and engineers in the areas of system and control theory, mathematical statistics, and the related application areas; in addition, it is intended to serve graduate students as a self-contained reference of the theory.

This book contains eight chapters. Chapter 1, Main Concepts of Probability Theory, gives without proof some basic facts from probability theory; they are restricted to lie within the limits of the material used later. Proofs are readily available in the literature cited in the references [1–6].

After the fundamental work in Refs. 7 and 8, stochastic approximation has continually attracted attention because of the intrinsic interest of its theory and its importance for applications [9–13]. Before the 1970s the main type of measurement errors considered in this field for the analysis of convergence and asymptotic behavior were independent random sequences or their continuous-time analogue; the theory of Markov processes was the main tool of analysis. This direction of stochastic approximation theory is well summarized in Ref. 14, which contains a great number of references, while its applications are presented in Refs. 12 and 13. In the mid-1970s there appeared an ordinary differential equation approach to treat the convergence analysis of algorithms [15–17]. This method suits more general measurement errors in comparison with the probabilistic method but needs much more complicated conditions to guarantee the convergence of the algorithms. In Chapter 2, Stochastic Approximation Algorithms, we combine both methods, based on the following approach: First we establish, by probabilistic methods, the boundedness of estimates generated by the algorithms and the convergence of a certain stochastic Liapounov function along the trajectories defined by the algorithms, and then, by the ordinary differential equation method, we prove the convergence of the estimates. This combined method not only successfully treats the measurement errors of ARMA (autoregressive moving average) type, but it also improves the results in the classical independent case. Both the discrete-time and continuous-time cases are considered in this chapter, the main material being taken from Refs. 18–21.

Least-squares estimation is an old statistical method, but it is still investigated by statisticians [22–24] because of its simplicity and practical importance. But the convergence results obtained within mathematical statistics cannot be directly applied to system identification, since the design matrix is no longer deterministic in contrast to the classical situation. However, the least-squares method has been analyzed and applied to the system identification problem for a long time. Clearly, the problem of the convergence of least-squares estimates in the system identification context must be treated specifically. The consistency of least-squares estimates for uncorrelated noise is discussed in Refs. 22, 23, and 25–27 but, in general, it is inconsistent for correlated noise, as can be easily shown. In Chapter 3, Strong Consistency of Least-Squares Identification, we consider measurement errors of ARMA type; we modify the design matrix used in the least-squares estimation method but leave the algorithm invariant, then we demonstrate the consistency of the estimates and, moreover, give convergence rates for the special uncorrelated noise case. For continuous-time systems, the concept of least-squares identification is also introduced and results parallel to those for the discrete time case are obtained. References 25 and 28 cover this material; see also Refs. 29–31 for applications to system identification and other related problems.

In order to treat the correlated noise case, there are many recursive algorithms other than least squares in existence (see, e.g., Refs. 32–36). The strong consistency of various identification algorithms is the topic of Chapter 4, Identification Algorithms of Stochastic Approximation Type and Adaptive Control. The combined method developed in Chapter 2 is applied in the analysis of the strong consistency of identification algorithms for the correlated noise case, when the positive real condition [17, 37, 38] is satisfied. We consider both discrete- and continuous-time systems [32, 33, 39, 40]. If the system not only has to be identified but also controlled, then the use of a self-timing regulator is called for. These have proved very successful in engineering applications [41–44], but until recently the algorithms lacked a convergence analysis. This is the second topic treated in Chapter 4. For the adaptive tracking problem [45–47], we give conditions under which the system is stabilized and the tracking error asymptotically tends to its minimum value. Further, for the more general quadratic index, we give an algorithm to compute adaptive control laws in real time and examine the question of the suboptimality [48, 49] of this scheme.

It is clear that the results obtained in Chapters 3 and 4 are also applicable to time series analysis [50–51], where the discrete-time system considered is called an ARMAX model.

Readers interested only in filtering and stochastic control problems may start directly from Chapter 5, Recursive Estimation and Control for Discrete-Time Systems, where first the filtering, prediction, and interpolation equations for conditionally Gaussian systems are derived [2], then the Gauss–Markov estimate and its recursive formulas are obtained and its connection with the estimates given by the Kalman filter is elucidated [52–55]. Further, in this chapter we introduce a concept of stochastic observability that differs from other definitions [56–58] and is a natural extension of the complete observability of deterministic systems; this notion is closely connected with the properties of the Gauss–Markov estimate. Several criteria are given for the stochastic observability of a system. Finally, the stochastic control problem is considered with quadratic cost indices [2, 59].

Chapter 6, Linear Unbiased Minimum Variance Estimates for Continuous-Time Systems, covers the well-known Kalman–Bucy filter, including the prediction and interpolation equations, stochastic control, and stochastic differential game problems. These are well-known results (see e.g., Refs. 59–64), but the proofs given here are simple and rigorous and are different from those in the aforementioned references.

Chapter 7, Singular Problems, in contrast to Chapter 6, considers estimation and control problems when the control weighting as well as the measurement noise covariance may be degenerate. With possibly degenerate measurement noise, the filtering value may not satisfy any stochastic differential equation since it may even be discontinuous, but one can derive the stochastic differential equations satisfied by the suboptimal filtering values that converge to the optimal ones as a small parameter goes to zero [2]. When the control weighting is degenerate (i.e., for singular control problems [65]), the extremum of the performance index may not be attained in the set of admissible controls. In this chapter, we describe a control sequence that can be applied for both the singular and nonsingular cases and for which the quadratic index converges to its minimum. Similar results are also shown for the singular stochastic differential game problem [66–70].

The last chapter, Gauss–Markov Estimates for Continuous-Time Systems, is the continuous analogue of the main part of Chapter 5, but the situation here is more complicated because the singular case is covered; hence we cannot write down the stochastic differential equations for the Gauss–Markov estimates. However, the stochastic differential equations for the suboptimal Gauss–Markov estimates are derived and their convergence to the optimal estimates is analyzed. A stochastic control problem related to the Gauss–Markov estimate is also discussed in this chapter [52, 55, 71].

This book was written while the author was a visiting professor during 1982–83 in the Department of Electrical Engineering of McGill University, Montreal, and he would like to express his appreciation to the department, the engineering faculty, the university, and the Natural Sciences and Engineering Research Council for the support he received during this visit.

The author would like to thank Professor P. E. Caines, who constantly supported the idea of publishing this book in English, read the manuscript, gave useful suggestions, and made many improvements in language. He would also like to thank Mindle Levitt, who typed this book with consummate skill, continual good humor, and great stamina.

HAN-FU CHEN

Beijing, People's Republic of China
January 1985

Contents

Recursive Estimation and Control for Stochastic Systems

CHAPTER 1

Main Concepts of Probability Theory

In this chapter we give some basic facts from probability theory and random processes. We do not give any proofs here, instead we refer the reader to the standard texts [1–6].

1.1. PROBABILITY SPACE, RANDOM VARIABLES AND MATHEMATICAL EXPECTATION

Let $(\Omega, \mathscr{F}, P)$ denote a probability space and ω denote a point of Ω which is also called an elementary event. $\mathscr{F}$ is the σ-algebra of subsets in Ω (i.e., $\mathscr{F}$ has the following properties):

1. $\Omega \in \mathscr{F}$.
2. The complementary set A^c of A belongs to $\mathscr{F}$, if $A \in \mathscr{F}$.
3. $\bigcup_{i=1}^{\infty} A_i \in \mathscr{F}$ if $A_i \in \mathscr{F}$, $i = 1, 2, \ldots$.

From here it follows immediately that

$$\bigcap_{i=1}^{\infty} A_i \in \mathscr{F}$$

if we notice that $\bigcap_{i=1}^{\infty} A_i = \bigcup_{i=1}^{\infty} A_i^c$.

A set $A \in \mathscr{F}$ is called a random event. P is called the probability measure on $(\Omega, \mathscr{F})$. It is a function defined on $\mathscr{F}$ with the properties

1. $P(A) \geq 0$, $\forall A \in \mathscr{F}$.
2. $P(\Omega) = 1$.
3. $P(\bigcup_{i=1}^{\infty} A_i) = \sum_{i=1}^{\infty} P(A_i)$, if $A_i \in \mathscr{F}$, and $A_i \cap A_j = \phi$, $\forall i \neq j$.

$P(A)$ is called the probability of the random event A.

Let B be any subset of a set $A \in \mathscr{F}$, where A is of probability zero. Then we assume that $B \in \mathscr{F}$ and that it also has probability zero. The probability space with such an extended σ-algebra is called a complete probability space. In the sequel we shall only consider complete probability spaces.

We shall always denote the l-dimensional Euclidean space by R^l and its Borel σ-algebra by B^l. By a Borel σ-algebra on a topological space we mean the smallest σ-algebra containing all the open sets of the topology. A measurable function $\boldsymbol{\xi} = \boldsymbol{\xi}(\omega)$ defined on $(\Omega, \mathscr{F})$ and valued in (R^l, B^l) is called the l-dimensional random vector.

Let $\boldsymbol{\xi}, \boldsymbol{\eta}$ be two l-dimensional random vectors. We say that $\boldsymbol{\xi}$ is equal to $\boldsymbol{\eta}$ with probability one, or almost surely, and denote this by

$$\boldsymbol{\xi} = \boldsymbol{\eta} \quad \text{a.s.}$$

if

$$P(\boldsymbol{\xi} \neq \boldsymbol{\eta}) = 0.$$

Let ξ be a one-dimensional nonnegative random variable and set

$$A_{ni} = \{\omega \colon i2^{-n} < \xi \leq (i+1)2^{-n}\}.$$

The mathematical expectation $E\xi$ of nonnegative random variable ξ is defined as the integral

$$E\xi = \int_\Omega \xi \, dP = \lim_{n \to \infty} \left[\sum_{i=1}^{n2^n} i2^{-n} PA_{ni} + nP(\xi > n) \right],$$

which may be infinite.

For an arbitrary random variable ξ, define

$$\xi^+ = \max(\xi, 0), \qquad \xi^- = \max(-\xi, 0).$$

These are both nonnegative and hence $E\xi^+$ and $E\xi^-$ are well defined. Notice that

$$\xi = \xi^+ - \xi^-,$$

so it is natural to define

$$E\xi = E\xi^+ - E\xi^-$$

if at least one of $E\xi^+$ and $E\xi^-$ is finite.

If $E|\xi| = E\xi^{+} + E\xi^{-} < \infty$, then ξ is said to be integrable or to have finite expectation.

Let

$$\boldsymbol{\xi} = [\xi^1, \ldots, \xi^l]^T$$

be an l-dimensional random vector. By its distribution function, we mean the function defined by

$$F_{\boldsymbol{\xi}}(x^1, \ldots, x^l) = P[\xi^1 < x^1, \ldots, \xi^l < x^l].$$

If there is a function $f_{\boldsymbol{\xi}}(x^1, \ldots, x^l)$ such that

$$F_{\boldsymbol{\xi}}(x^1, \ldots, x^l) = \int_{-\infty}^{x^1} \cdots \int_{-\infty}^{x^l} f_{\boldsymbol{\xi}}(\lambda^1, \ldots, \lambda^l)\, d\lambda^1 \cdots d\lambda^l,$$

then $f_{\boldsymbol{\xi}}(x^1, \ldots, x^l)$ is called the density of the distribution of $\boldsymbol{\xi}$ or simply the density of $\boldsymbol{\xi}$.

When ξ is one-dimensional, then its distribution function and density are denoted by $F_\xi(x)$ and $f_\xi(x)$, respectively. Notice that the mathematical expectation of a random variable ξ can be written as a Lebesgue–Stieltjes integral with respect to its distribution function:

$$E\xi = \int_\Omega \xi\, dP = \int_{-\infty}^{\infty} x\, dF_\xi(x).$$

1.2. CONVERGENCE THEOREMS

The convergence of a sequence of random variables ξ_n to its limit ξ can take place in several different ways:

1. Convergence with probability one or almost surely means that, with the possible exception of a set of probability zero, for any $\omega \in \Omega$ $\xi_n(\omega) \to \xi(\omega)$, that is,

$$P(\xi_n \to \xi) = 1.$$

For this type of convergence, we often write

$$\xi_n \to \xi \quad \text{a.s.}$$

2. Convergence in probability means that for any $\varepsilon > 0$

$$P[|\xi_n - \xi| > \varepsilon] \underset{n \to \infty}{\to} 0$$

and this is denoted by

$$\xi_n \xrightarrow{P} \xi.$$

3. Convergence in distribution or weak convergence means that for any x where $F_\xi(x)$ is continuous

$$F_{\xi_n}(x) \underset{n\to\infty}{\to} F_\xi(x),$$

and it is denoted by

$$\xi_n \underset{n\to\infty}{\xrightarrow{w}} \xi.$$

4. Convergence in the mean square sense means that

$$E|\xi_n - \xi|^2 \underset{n\to\infty}{\to} 0.$$

The following diagram explains the relationship of these convergence types

convergence a.s. ⇒	convergence in probability	⇒ convergence in distribution
	⇑	
	convergence in the mean square sense	

Let $E|\xi_n| < \infty$. We now give conditions for $E\xi_n \underset{n\to\infty}{\to} E\xi$.

Theorem 1.1 (Monotone Convergence Theorem). *If* $\xi_n \uparrow \xi (\xi_n \downarrow \xi)$ a.s. *and* $E\xi_1^- < \infty$ $(E\xi_1^+ < \infty)$, *then* $E\xi_n \uparrow E\xi (E\xi_n \downarrow E\xi)$.

Theorem 1.2 (Fatou Lemma). *If there exists an integrable random variable* η *such that* $\eta \le \xi_n$ $(\xi_n \le \eta)$, *then*

$$E \varliminf_{n\to\infty} \xi_n \le \varliminf_{n\to\infty} E\xi_n \left(\varlimsup_{n\to\infty} E\xi_n \le E \varlimsup_{n\to\infty} \xi_n \right).$$

Theorem 1.3 (Dominated Convergence Theorem). *If* $\xi_n \underset{n\to\infty}{\to} \xi$ a.s. *and there exists an integrable random variable* η *such that* $|\xi| \le \eta$, *then*

$$E|\xi_n - \xi| \underset{n\to\infty}{\to} 0.$$

1.3. INDEPENDENCE

Let $A_i \in \mathscr{F}$, $i = 1, 2, \ldots$. If for any set of indices $\{i_1, \ldots, i_k\}$

$$P \bigcap_{j=1}^{k} A_{i_j} = \prod_{j=1}^{k} PA_{i_j},$$

then random events are called mutually independent.

We say that $\mathscr{F}_1$ is a sub-σ-algebra of $\mathscr{F}$ if $A \in \mathscr{F}_1$ implies $A \in \mathscr{F}$ and $\mathscr{F}_1$ itself is a σ-algebra of sets in Ω. Sub-σ-algebras $\mathscr{F}_i$, $i = 1, 2, \ldots$ of $\mathscr{F}$ are called mutually independent if for any index set $\{i_1, \ldots, i_k\}$, the random events $A_1, \ldots, A_k$ are mutually independent when $A_j \in \mathscr{F}_{i_j}$ for $j = 1, \ldots, k$.

Let $\boldsymbol{\eta}$ be an l-dimensional random vector. Denote by $\mathscr{F}^{\boldsymbol{\eta}}$ the smallest σ-algebra containing all sets of form

$$\{\omega = \boldsymbol{\eta}^{-1}(B),\ B \in \mathscr{B}^l\}$$

and call it the σ-algebra generated by $\boldsymbol{\eta}$.

Random vectors $\boldsymbol{\eta}_i$, $i = 1, 2, \ldots$ are called mutually independent if σ-algebras $\mathscr{F}^{\boldsymbol{\eta}_i}$ are mutually independent.

If $\{\boldsymbol{\eta}_i\}$ are mutually independent and identically distributed with $E\|\boldsymbol{\eta}_i\| < \infty$, then

$$\frac{1}{n} \sum_{i=1}^{n} \boldsymbol{\eta}_i \underset{n \to \infty}{\to} E\boldsymbol{\eta}_i \quad \text{a.s.} \tag{1.1}$$

This is called the strong law of large numbers.

Theorem 1.4 (Borel–Cantelli Lemma). *Let $A_1, A_2, \ldots$ be random events.*

1. *If $\sum_{i=1}^{\infty} PA_i < \infty$, then $P\bigcap_{i=1}^{\infty}\bigcup_{j=i}^{\infty} A_j = 0$.*

2. *If the events $\{A_i\}$ are mutually independent and $\sum_{i=1}^{\infty} PA_i = \infty$, then $P\bigcap_{i=1}^{\infty}\bigcup_{j=i}^{\infty} A_j = 1$.*

The set $\bigcap_{i=1}^{\infty}\bigcup_{j=i}^{\infty} A_j$ is usually denoted by $\overline{\lim}_{j \to \infty} A_j$ and it consists of all ω which appear in an infinite number of A_j.

For the l-dimensional random vector $\boldsymbol{\eta}$ define a function $\psi(\boldsymbol{\lambda})$, called the characteristic function of $\boldsymbol{\eta}$, by

$$\psi(\boldsymbol{\lambda}) = Ee^{i\boldsymbol{\lambda}^T\boldsymbol{\eta}}, \qquad \boldsymbol{\lambda} \in R^l$$

This is clearly the Fourier–Stieltjes transformation of the distribution function $F_{\boldsymbol{\eta}}(x)$ of $\boldsymbol{\eta}$, that is,

$$\psi(\boldsymbol{\lambda}) = \int_{-\infty}^{\infty} \cdots \int_{-\infty}^{\infty} e^{i\boldsymbol{\lambda}^T\mathbf{x}}\, dF_{\boldsymbol{\eta}}(\mathbf{x}).$$

It is clear that the characteristic function is uniquely defined by its corresponding distribution function, and it can be shown that the converse is also true. Consequently, the characteristic function and the distribution function are in one-to-one correspondence.

Let $\mathbf{w} = [\mathbf{x}^\tau \mathbf{y}^\tau]^\tau$ be a random vector. Then it is a basic fact that the random vectors $\mathbf{x}$ and $\mathbf{y}$ are independent if and only if the characteristic function of $\mathbf{w}$ equals the product of characteristic functions of $\mathbf{x}$ and $\mathbf{y}$.

1.4. CONDITIONAL EXPECTATION

The relationships between any two quantities in the sequel always permits the failure of that relationship on a set with probability zero. This point will not be mentioned every time. For example, $\xi_n \xrightarrow[n\to\infty]{} \xi$ means convergence with probability one, but we shall frequently omit to write a.s., similarly, the quantifier $\forall \omega \in J$ corresponds to "all ω in J with the possible exception of a set in J with zero probability."

We always denote by I_A the indicator of a set A:

$$I_A = \begin{cases} 1, & \omega \in A \\ 0, & \omega \notin A \end{cases}$$

and define

$$\int_A \xi \, dP = E\xi I_A.$$

If on $\mathscr{F}$, besides the probability measure P, there is another measure Q such that for any $A \in \mathscr{F}$, $PA = 0$ implies $QA = 0$, then Q is called absolutely continuous with respect to P and this fact is denoted by $Q \ll P$.

Theorem 1.5 (Radon–Nikodym). *If $Q \ll P$, then there exists a nonnegative random variable ξ such that for any $A \in \mathscr{F}$*

$$Q(A) = \int_A \xi \, dP$$

and ξ is defined uniquely in the sense that if there is another nonnegative random variable η with the property

$$Q(A) = \int_A \eta \, dP \qquad \forall A \in \mathscr{F},$$

then $P(\xi \neq \eta) = 0$.

This kind of uniqueness is called uniqueness to within stochastic equivalence, and ξ is the Radon–Nikodym derivative (or the density of one measure (P) with respect to the other (Q), denoted by

$$\xi = \frac{dQ}{dP}.$$

Let $\mathscr{F}_1 \subset \mathscr{F}$ be a sub-σ-algebra and $P^{\mathscr{F}_1}$ be a probability measure on $\mathscr{F}_1$ defined simply by setting

$$P^{\mathscr{F}_1}(A) = P(A), \qquad \forall A \in \mathscr{F}_1.$$

Let η be a nonnegative random variable and define

$$Q(A) = \int_A \eta \, dP = \int_A \eta \, dP^{\mathscr{F}_1} \qquad \forall A \in \mathscr{F}_1. \tag{1.2}$$

Clearly, Q is a measure on $\mathscr{F}_1$ and is absolutely continuous with respect to $P^{\mathscr{F}_1}$, hence by Theorem 1.5 there exists an $\mathscr{F}_1$-measurable nonnegative random variable ξ such that

$$Q(A) = \int_A \xi \, dP^{\mathscr{F}_1} = \int_A \xi \, dP. \tag{1.3}$$

Comparing (1.3) with (1.2), we find that there is a nonnegative random variable ξ such that

$$\int_A \eta \, dP = \int_A \xi \, dP, \qquad \forall A \in \mathscr{F}_1.$$

ξ is called conditional expectation of ξ given $\mathscr{F}_1$ and is denoted by

$$\xi = E\left(\frac{\eta}{\mathscr{F}_1}\right) \quad \text{or} \quad E^{\mathscr{F}_1}\eta.$$

By Theorem 1.5, $E(\eta/\mathscr{F}_1)$ is unique to within stochastic equivalence.

For the general random variable η (not necessarily nonnegative) if $E\eta$ exists (i.e., at least one of $E\eta^+$ and $E\eta^-$ is finite), then define

$$E\left(\frac{\eta}{\mathscr{F}_1}\right) = E\left(\frac{\eta^+}{\mathscr{F}_1}\right) - E\left(\frac{\eta^-}{\mathscr{F}_1}\right).$$

The conditional expectation $E(\xi/\boldsymbol{\eta})$ of ξ given a random vector $\boldsymbol{\eta}$ is defined by

$$E\left(\frac{\xi}{\boldsymbol{\eta}}\right) = E\left(\frac{\xi}{\mathscr{F}^{\boldsymbol{\eta}}}\right).$$

Assume $E\|\boldsymbol{\xi}\| < \infty$, $E\|\boldsymbol{\eta}\| < \infty$. The conditional expectation has the following properties:

1. $E^{\mathscr{F}_1}(a\boldsymbol{\xi} + b\boldsymbol{\eta}) = aE^{\mathscr{F}_1}\boldsymbol{\xi} + bE^{\mathscr{F}_1}\boldsymbol{\eta}$, (1.4)
   where a, b are constants.
2. Let $\boldsymbol{\zeta}$ be a random vector. There exists a Borel measurable function $f(\cdot)$ such that

$$E\left(\frac{\boldsymbol{\xi}}{\boldsymbol{\zeta}}\right) = f(\boldsymbol{\zeta}). \tag{1.5}$$

3. $EE^{\mathscr{F}_1}\boldsymbol{\xi} = E\boldsymbol{\xi}$. (1.6)
4. $E^{\mathscr{F}_1}\boldsymbol{\xi} = \boldsymbol{\xi}$ if $\boldsymbol{\xi}$ is $\mathscr{F}_1$-measurable. (1.7)
5. $E^{\mathscr{F}_1}\zeta^T\xi = \zeta^T E^{\mathscr{F}_1}\xi$, (1.8)
   if ζ is $\mathscr{F}_1$-measurable and $E\|\zeta^T\xi\| < \infty$.
6. If $\mathscr{F}_1$ and $\mathscr{F}_2$ are sub-σ-algebras with $\mathscr{F}_1 \subset \mathscr{F}_2 \subset \mathscr{F}$, then

$$E^{\mathscr{F}_1}E^{\mathscr{F}_2}\xi = E^{\mathscr{F}_1}\xi. \tag{1.9}$$

7. If ξ and ζ are independent, then

$$E\left(\frac{\xi}{\zeta}\right) = E\xi \tag{1.10}$$

8. If $\mathscr{F}_1 = (\Omega, \phi)$, then

$$E^{\mathscr{F}_1}\xi = E\xi. \tag{1.11}$$

The conditional probability $P^{\mathscr{F}_1}A$ or $P(A/\mathscr{F}_1)$ of $A \in \mathscr{F}$ given $\mathscr{F}_1$ is defined by

$$P^{\mathscr{F}_1}(A) = E\left(\frac{I_A}{\mathscr{F}_1}\right),$$

and if $\mathscr{F}_1$ is the σ-algebra $\mathscr{F}^{\eta}$ generated by η, then $P(A/\mathscr{F}^{\eta})$ is called the conditional probability of A given η and is denoted by $P^{\eta}A$ or $P(A/\eta)$. Clearly,

$$P^{\mathscr{F}_1}(A) \geq 0, \qquad P^{\mathscr{F}_1}(\Omega) = 1,$$

and

$$P^{\mathscr{F}_1}\left(\bigcup_{i=1}^{\infty} A_i\right) = \sum_{i=1}^{\infty} P^{\mathscr{F}_1}(A_i)$$

if $A_i \cap A_j = \phi \; \forall i \neq j$.

Theorems 1.1–1.3 can be extended from expectation to the conditional expectation.

Theorem 1.6. *If $\xi_n \uparrow \xi (\xi_n \downarrow \xi)$ a.s. and $E\xi_1^- < \infty (E\xi_1^+ < \infty)$, then*

$$E^{\mathscr{F}_1}\xi_n \uparrow E^{\mathscr{F}_1}\xi \left(E^{\mathscr{F}_1}\xi_n \downarrow E^{\mathscr{F}_1}\xi\right) \text{ a.s.}$$

Theorem 1.7. *Suppose that η is an integrable random variable.*

1. *If $\eta \leq \xi_n$ $(\xi_n \leq \eta)$, then*

$$E^{\mathscr{F}_1} \varliminf_{n\to\infty} \xi_n \leq \varliminf_{n\to\infty} E\xi_n \left(\varlimsup_{n\to\infty} E^{\mathscr{F}_1}\xi_n \leq E^{\mathscr{F}_1} \varlimsup_{n\to\infty} \xi_n \right) \text{ a.s.}$$

2. *If $|\xi_n| \leq \eta$ and $\xi_n \underset{n\to\infty}{\to} \xi$ a.s., then*

$$E^{\mathscr{F}_1}|\xi_n - \xi| \underset{n\to\infty}{\to} 0 \text{ a.s.}$$

Random vectors $\boldsymbol{\xi}$ and $\boldsymbol{\eta}$ are called conditionally independent given ζ if

$$P^{\zeta}(\boldsymbol{\xi} < \mathbf{x}, \boldsymbol{\eta} < \mathbf{y}) = P^{\zeta}(\boldsymbol{\xi} < \mathbf{x}) P^{\zeta}(\boldsymbol{\eta} < \mathbf{y}), \qquad \forall \mathbf{x}, \mathbf{y},$$

where the inequality $\boldsymbol{\xi} < \mathbf{x}$ between vectors should be understood as inequalities between their components. Let

$$f_{\boldsymbol{\xi}}^{\zeta}(\boldsymbol{\lambda}) \triangleq E\left(\frac{e^{i\boldsymbol{\lambda}^{\tau}\boldsymbol{\xi}}}{\zeta}\right), \qquad f_{\boldsymbol{\eta}}^{\zeta}(\boldsymbol{\mu}) \triangleq E\left(\frac{e^{i\boldsymbol{\mu}^{\tau}\boldsymbol{\eta}}}{\zeta}\right), \qquad f_{\boldsymbol{\pi}}^{\zeta}(\boldsymbol{\gamma}) \triangleq E\left(\frac{e^{i\boldsymbol{\gamma}^{\tau}\boldsymbol{\pi}}}{\zeta}\right)$$

be conditional characteristic functions given ζ of $\boldsymbol{\xi}$, $\boldsymbol{\eta}$ and $\boldsymbol{\pi}$ respectively, where

$$\boldsymbol{\pi} = [\boldsymbol{\xi}^{\tau}\boldsymbol{\eta}^{\tau}]^{\tau}, \qquad \boldsymbol{\gamma} = [\boldsymbol{\lambda}^{\tau}, \boldsymbol{\mu}^{\tau}]^{\tau}.$$

Theorem 1.8. (1) *Given* ζ, ξ *and* η *are conditionally independent if and only if*

$$f_{\pi}^{\zeta}(\gamma) = f_{\xi}^{\zeta}(\lambda) f_{\eta}^{\zeta}(\mu).$$

(2) *If* η *and* $(\xi^T \zeta^T)^T$ *are independent then* ξ *and* η *are conditionally independent given* ζ. (3) *If* ξ *and* η *are conditionally independent given* ζ *then for any Borel set B*

$$P\left(\frac{[\xi \in B]}{\zeta, \eta}\right) = P\left(\frac{[\xi \in B]}{\zeta}\right)$$

and

$$E\left(\frac{\xi}{\zeta, \eta}\right) = E\left(\frac{\xi}{\zeta}\right).$$

Theorem 1.9. *Let* $f(\lambda, \mu)$ *be a measurable function defined on* $(R^l \times R^m, \mathscr{B}^l \times \mathscr{B}^m)$. *If the l-dimensional random vector* ξ *is independent of the m-dimensional random variable* η *then*

$$E\left(\frac{f(\xi, \eta)}{\xi}\right) = g(\xi)$$

and for any Borel set B

$$\left| P\left\{\frac{[f(\xi, \eta) \in B]}{\xi}\right\}\right|_{\xi = \lambda} = P[f(\lambda, \eta) \in B],$$

where

$$g(\lambda) = Ef(\lambda, \eta).$$

1.5. RANDOM PROCESSES

Let $T = [0, \infty)$ and let $\mathscr{B}(T)$ be the σ-algebra of Borel sets on T. A function $\xi_t(\omega)$ defined on $(\Omega \times T, \mathscr{F} \times \mathscr{B}(T))$ and taking values in $(R^l, \mathscr{B}^l)$ is called an l-dimensional continuous time stochastic process. If $\xi_t(\omega)$ is only defined at discrete times $t = 0, 1, 2, \ldots,$ then it is called a discrete time (parameter) stochastic process or a random sequence.

For fixed ω, $\xi_t(\omega)$ is a function of ω and is called a trajectory of the stochastic process.

If for any Borel set B

$$\{(\omega, t): \xi_t(\omega) \in B\} \in \mathscr{F} \times \mathscr{B}(T),$$

then $\xi_t(\omega)$ is called a measurable stochastic process.

We often omit ω and denote a process simply by ξ_t.

Theorem 1.10 (Fubini). *If ξ_t is a measurable stochastic process, then almost all of its trajectories are Borel measurable functions of t. In addition, if $E\xi_t$ exists $\forall t \in T$, then it is also a measurable function. Further, if*

$$\int_S E\|\xi_t\|\, dt < \infty$$

then

$$\int_S \|\xi_t\|\, dt < \infty \text{ a.s.}$$

and

$$E\int_S \xi_t\, dt = \int_S E\xi_t\, dt,$$

where S is any measurable set in T.

Two stochastic processes ξ_t and η_t are called stochastically equivalent if

$$P(\xi_t \neq \eta_t) = 0 \qquad \forall t \in T,$$

and in this case $\xi_t(\eta_t)$ is called a modification of $\eta_t(\xi_t)$.

If for all $\omega \in \Omega$, with the possible exception of a set of zero probability, the trajectories of ξ_t are continuous (left-continuous or right-continuous), then ξ_t is called continuous (left-continuous or right-continuous, respectively) process. A left- or right-continuous process is measurable.

Let $\{\mathscr{F}_t\}$ be a family of nondecreasing σ-algebras (i.e., $\mathscr{F}_s \subseteq \mathscr{F}_t \forall s \leq t$). If ξ_t is $\mathscr{F}_t$-measurable for any $t \in T$, then we say that ξ_t is $\mathscr{F}_t$-adapted and write $(\xi_t, \mathscr{F}_t)$.

If ξ_t is a measurable process, $E\|\xi_t\| < \infty$, $\forall t \in T$ and $\{\mathscr{F}_t\}$ is a nondecreasing family of σ-algebras, then in the equivalence class $E(\xi_t/\mathscr{F}_t)$ a modification can be chosen to be $\mathscr{F}_t$ measurable. In the sequel, we always assume that $E(\xi_t/\mathscr{F}_t)$ is so chosen.

1.6. MARTINGALES

DEFINITION 1.1. Let ξ_t be adapted to a nondecreasing family of σ-algebras $\{\mathscr{F}_t\}$ with $E|\xi_t| < \infty$. $(\xi_t, \mathscr{F}_t)$ is called a martingale (supermartingale, submartingale) if

$$E\left(\frac{\xi_t}{\mathscr{F}_s}\right) = \xi_s (\leq \xi_s, \geq \xi_s, \text{ respectively}) \quad \text{for any } s \leq t,\ s, t \in T.$$

This definition also holds for a discrete-time process.

Example. Suppose η_i $i = 1, 2, \ldots$ to be a mutually independent random sequence with $E\eta_i = 0$ $\forall i$.

Denote

$$\xi_n = \sum_{i=1}^{n} \eta_i, \qquad \mathscr{F}_n = \mathscr{F}_n^{\eta}$$

where $\mathscr{F}_n^{\eta}$ denotes the σ-algebra generated by $\eta_1, \ldots, \eta_n$. We know $E(\xi_m/\mathscr{F}_m) = \xi_m$ by (1.7) and $E(\sum_{i=m+1}^{n} \eta_i/\mathscr{F}_m) = E\sum_{i=m+1}^{n} \eta_i = 0$ by (1.10) for any $m \leq n$. Hence

$$E\left(\frac{\xi_n}{\mathscr{F}_m}\right) = E\left(\xi_m + \sum_{i=m+1}^{n} \frac{\eta_i}{\mathscr{F}_m}\right) = \xi_m$$

and $(\xi_n, \mathscr{F}_n)$ is a martingale.

Theorem 1.11. **(1) (Discrete-time version).** *Assume* $(\xi_n, \mathscr{F}_n)$ *to be a submartingale (supermartingale) and* $\sup_n E\xi_n^+ < \infty (\sup_n E\xi_n^- < \infty)$. *Then* ξ_n *converges to a finite limit* ξ a.s. *as* $n \to \infty$ *and* $E\xi^+ < \infty (E\xi^- < \infty)$. **(2) (Continuous-time version).** *Let* $(\xi_t, \mathscr{F}_t)$ *be a right continuous submartingale (supermartingale) with* $\sup_t E\xi_t^+ < \infty (\sup_t E\xi_t^- < \infty)$. *Then* ξ_t *tends to a finite limit* ξ a.s. *as* $t \to \infty$ *and* $E\xi^+ < \infty (E\xi^- < \infty)$.

Corollary 1. *If* $(\xi_n, \mathscr{F}_n)$ *is a nonpositive (nonnegative) submartingale (supermartingale), then* ξ_n *converges to a finite limit as* $n \to \infty$.

Corollary 2. *If* $(\xi_n, \mathscr{F}_n)$ *is a martingale, then* $E|\xi_n| = E\xi_n^+ + E\xi_n^- = 2E\xi_n^+ - E\xi_n = 2E\xi_n^+ - E\xi_1$. *Hence for martingale* $\sup_n E\xi_n^+ < \infty$ *(or* $\sup_n E\xi_n^- < \infty$*) is equivalent to* $\sup_n E|\xi_n| < \infty$.

The continuous analogue of these corollaries is also valid.

If $(\xi_n, \mathscr{F}_n)$ is a martingale, then $\{x_n\}$ defined by $x_1 = \xi_1$, $x_n = \xi_n - \xi_{n-1}$ is called a martingale difference sequence. The following two theorems are concerned with local convergence of martingales.

Theorem 1.12. *Let* $x_n = \xi_n - \xi_{n-1}$, $x_1 = \xi_1$ *be a martingale difference sequence, then* ξ_n *converges* a.s. *to a finite limit on* A, *where*

$$A = \left\{ \sum_{i=2}^{\infty} E\left[\frac{\left(|x_i|^2 I_{[|x_i| \leq a_i]} + |x_i| I_{[|x_i| > a_i]}\right)}{\mathscr{F}_{i-1}}\right] < \infty \right\}$$

and a_i *are constants with* $a_i \geq c > 0$.

As a consequence of this theorem, we obtain:

Theorem 1.13. *ξ_n converges to a finite limit* a.s. *on A where*

$$A = \left\{ \sum_{i=2}^{\infty} E\left(\frac{|x_i|^\rho}{\mathscr{F}_{i-1}} \right) < \infty \right\}, \qquad 1 \le \rho \le 2.$$

1.7. WIENER PROCESS

The l-dimensional random vector $\boldsymbol{\xi}$ with $E\boldsymbol{\xi} = \boldsymbol{\mu}$,

$$E(\boldsymbol{\xi} - \boldsymbol{\mu})(\boldsymbol{\xi} - \boldsymbol{\mu})^\tau = \underset{\sim}{\mathbf{R}}$$

is said to be normally distributed $\boldsymbol{\xi} \in N(\boldsymbol{\mu}, \underset{\sim}{\mathbf{R}})$ if its characteristic function is expressed by

$$\psi(\boldsymbol{\lambda}) = E \exp i\boldsymbol{\lambda}^\tau\boldsymbol{\xi} = \exp\left(i\boldsymbol{\lambda}^\tau\boldsymbol{\mu} - \tfrac{1}{2}\boldsymbol{\lambda}^\tau\underset{\sim}{\mathbf{R}}\boldsymbol{\lambda}\right).$$

$\boldsymbol{\xi}$ is also called a normal or Gaussian random vector. Clearly, $\mathbf{a} + \underset{\sim}{\mathbf{A}}\boldsymbol{\xi}$ is also with normal distribution if $\boldsymbol{\xi}$ is and the vector $\mathbf{a}$ and matrix $\underset{\sim}{\mathbf{A}}$ are deterministic. If $\underset{\sim}{\mathbf{R}} > 0$, then the l-dimensional normal random vector $\boldsymbol{\xi} \in N(\boldsymbol{\mu}, \underset{\sim}{\mathbf{R}})$ has density $f_{\boldsymbol{\xi}}(\mathbf{x})$:

$$f_{\boldsymbol{\xi}}(\mathbf{x}) = (2\pi)^{-l/2}(\det \underset{\sim}{\mathbf{R}})^{-1/2}\exp\left[-\tfrac{1}{2}(\mathbf{x} - \boldsymbol{\mu})^\tau\underset{\sim}{\mathbf{R}}^{-1}(\mathbf{x} - \boldsymbol{\mu})\right]. \quad (1.13)$$

In the following, we always denote by $\underset{\sim}{\mathbf{I}}$ an identity matrix, but its dimension may change depending upon the context.

DEFINITION 1.2. An l-dimensional process w_t which is adapted to the nondecreasing σ-algebra $\mathscr{F}_t$, $t \ge 0$ is called Wiener process, if $\mathbf{w}_0 = 0$, $E\mathbf{w}_t \equiv 0$, $E\|\mathbf{w}_t\|^2 < \infty$, $\forall t \ge 0$,

$$E\left[\frac{(\mathbf{w}_t - \mathbf{w}_s)(\mathbf{w}_t - \mathbf{w}_s)^\tau}{\mathscr{F}_s}\right] = (t - s)\underset{\sim}{\mathbf{I}}, \qquad t \ge s \quad (1.14)$$

and if $(\mathbf{w}_t, \mathscr{F}_t)$ is a continuous martingale.

Theorem 1.14. *A Wiener process $(\mathbf{w}_t, \mathscr{F}_t)$ is a process of independent increments. That is, $\mathbf{w}_{t_1} - \mathbf{w}_{t_2}$ are independent of $\mathbf{w}_{t_3} - \mathbf{w}_{t_4}$ for $t_1 \ge t_2 \ge t_3 \ge t_4$. Further, any of its increment $\mathbf{w}_t - \mathbf{w}_s$ is normally distributed, and the* *iterated logarithm law holds*:

$$\overline{\lim_{t\to\infty}} \frac{\|\mathbf{w}_t\|^2}{2lt \ln\ln t} = 1 \text{ a.s.},$$

where l *is the dimension of* $\mathbf{w}_t$.

From (1.14) *it is intuitively clear that* $\|\mathbf{w}_t - \mathbf{w}_s\|^2$ *is of order* $t - s$ *but it is not of order* $(t-s)^2$. *More precisely, we have the following theorem.*

Theorem 1.15. *Let* $0 \equiv t_{n_1} < t_{n_2} < \cdots < t_{n_n} \equiv t$ *be the partitions of* $[0,t]$ *with* $\max_{1\le i\le n}|t_{n_{i+1}} - t_{n_i}| \underset{n\to\infty}{\to} 0$. *Then*

$$\underset{n\to\infty}{\text{l.i.m.}} \sum_{i=0}^{n-1} \left(\mathbf{w}_{t_{n_{i+1}}} - \mathbf{w}_{t_{n_i}}\right)\left(\mathbf{w}_{t_{n_{i+1}}} - \mathbf{w}_{t_{n_i}}\right)^{\tau} = t\underset{\sim}{\mathbf{I}} \tag{1.15}$$

$$\lim_{n\to\infty} \sum_{i=0}^{n-1} \left(\mathbf{w}_{t_{n_{i+1}}} - \mathbf{w}_{t_{n_i}}\right)\left(\mathbf{w}_{t_{n_{i+1}}} - \mathbf{w}_{t_{n_i}}\right)^{\tau} = t\underset{\sim}{\mathbf{I}}, \tag{1.16}$$

where $\text{l.i.m.}_{n\to\infty}\mathbf{x}_n = \mathbf{x}$ *means* $E\|\mathbf{x}_n - \mathbf{x}\|^2 \underset{n\to\infty}{\to} 0$.

According to this theorem, we can formally write

$$\int_0^t d\mathbf{w}_s\, d\mathbf{w}_s^{\tau} = \int_0^t ds\, \underset{\sim}{\mathbf{I}}.$$

This fact plays an important role in Ito's stochastic calculus.

1.8. STOCHASTIC INTEGRAL

In what follows, $\mathscr{F}_t$ always denotes the nondecreasing σ-algebras. We say that process $(\boldsymbol{\xi}, \mathscr{F}_t) \in \mathscr{M}_T$, if

$$E\int_0^T \|\boldsymbol{\xi}_t\|^2\, dt < \infty,$$

and $(\boldsymbol{\xi}_t, \mathscr{F}_t) \in \mathscr{P}_T$, if

$$P\left\{\int_0^T \|\boldsymbol{\xi}_t\|^2\, dt < \infty\right\} = 1.$$

Suppose that for $0 = t_0 < t_1 < \cdots < t_n = T$, the random vectors $\boldsymbol{\xi}_i$ are $\mathscr{F}_{t_i}$-measurable, and that $\boldsymbol{\xi}$ is $\mathscr{F}_0$-measurable. Then the stochastic process

$$\boldsymbol{\xi}_t = \boldsymbol{\xi} I_{[0]} + \sum_{i=0}^{n-1} \boldsymbol{\xi}_i I_{(t_i, t_{i+1}]}$$

is called simple, where $I_{(t_i, t_{i+1}]}$ denotes the indicator of $(t_i, t_{i+1}]$.

DEFINITION 1.3. The Ito stochastic integral for the simple process $\boldsymbol{\xi}_t \in \mathscr{M}_T$ is defined by

$$\int_0^t \boldsymbol{\xi}_s dw_s \triangleq \sum_{i=0}^{m} \boldsymbol{\xi}_i (w_{t_{i+1}} - w_{t_i}) + \boldsymbol{\xi}_{m+1}(w_t - w_{t_{m+1}}), \qquad t_{m+1} < t \le t_{m+2},$$

where $(w_t, \mathscr{F}_t)$ is a one-dimensional Wiener process.

Theorem 1.16. *If $\xi_t \in \mathscr{M}_T$, then there exist simple processes $\xi_t^n \in \mathscr{M}_T$ and a random variable η such that*

$$E\int_0^T |\xi_t - \xi_t^n|^2 \, dt \underset{n\to\infty}{\to} 0, \tag{1.17}$$

$$\lim_{n\to\infty} E\left(\eta - \int_0^T \xi_t^n \, dw_t\right)^2 = 0 \tag{1.18}$$

and η is independent of the selection of ξ_t^n.

DEFINITION 1.4. For $\xi_t \in \mathscr{M}_T$, the Ito stochastic integral is defined by

$$\int_0^T \xi_t \, dw_t \triangleq \underset{n\to\infty}{\text{l.i.m.}} \int_0^T \xi_t^n \, dw_t \tag{1.19}$$

$$\int_0^t \xi_s \, dw_s \triangleq \int_0^T I_{[0,t]} \xi_s \, dw_s, \tag{1.20}$$

where ξ_t^n are simple processes satisfying (1.17).

The stochastic integral has the following properties, which can be easily verified for simple processes ξ_t and then for $\xi_t \in \mathscr{M}_T$ by passing to the limit.

Let $\xi_t, \xi_t^1, \xi_t^2 \in \mathscr{M}_T$, $t \in [0, T]$, we have that

$$\int_0^t (a\xi_s^1 + b\xi_s^2) \, dw_s = a\int_0^t \xi_s^1 \, dw_s + b\int_0^t \xi_s^2 \, dw_s, \tag{1.21}$$

where a and b are constants,

$$\int_0^t \xi_s \, dw_s = \int_0^u \xi_s \, dw_s + \int_u^t \xi_s \, dw_s, \tag{1.22}$$

$$\int_0^t \xi_s \, dw_s \text{ is continuous in } t \in [0.T] \tag{1.23}$$

$$E \int_0^t \xi_s \, dw_s = 0 \tag{1.24}$$

$$E\left(\int_0^t \frac{\xi_\lambda \, dw_\lambda}{\mathscr{F}_s} \right) = \int_0^s \xi_\lambda \, dw_\lambda, \qquad t \geq s, \tag{1.25}$$

$$E\left(\int_0^t \xi_s^1 \, dw_s \right)\left(\int_0^t \xi_s^2 \, dw_s \right) = E \int_0^t \xi_s^1 \xi_s^2 \, ds. \tag{1.26}$$

The Ito stochastic integral can be extended to processes belonging to $\mathscr{P}_T$.

Theorem 1.17. *If $\xi_t \in \mathscr{P}_T$, then there exist $\xi_t^n \in \mathscr{M}_T$ and a random variable independent of the selection of ξ_t^n such that*

$$\int_0^T |\xi_t - \xi_t^n|^2 \, dt \underset{n \to \infty}{\overset{P}{\to}} 0 \tag{1.27}$$

and

$$\int_0^T \xi_t^n \, dt \underset{n \to \infty}{\overset{P}{\to}} \eta. \tag{1.28}$$

DEFINITION 1.5. For $\xi_t \in \mathscr{P}_T$, the Ito stochastic integral is defined as

$$\int_0^T \xi_t \, dw_t \triangleq \lim_{n \to \infty} \int_0^T \xi_t^n \, dw_t \text{ in probability,}$$

where ξ_t^n satisfies (1.27).

It is worth pointing out that $(\int_0^t \xi_s \, ds, \mathscr{F}_t)$ is a square integrable martingale for $\xi_t \in \mathscr{M}_T$, while for $\xi_t \in \mathscr{P}_T$ it is no longer a martingale and even the expectation for it may not exist. For the latter, (1.21)–(1.23) remain valid but (1.24)–(1.26) do not hold.

1.9. ITO'S FORMULA

DEFINITION 1.6. Assume $(a_t, \mathscr{F}_t), (b_t, \mathscr{F}_t)$ to be measurable processes and $|a_t|^{1/2} \in \mathscr{P}_T$, $b_t \in \mathscr{P}_T$. The process ξ_t defined by

$$\xi_t = \xi_0 + \int_0^t a_s\, ds + \int_0^t b_s\, dw_s \tag{1.29}$$

is called an Ito process, and if a_s and b_s are measurable with respect to $\mathscr{F}_s^{\xi} \triangleq \sigma\{\xi_\lambda, \lambda \le s\}$, $\forall s \in [0.T]$, then it is called a diffusion process.

For ξ_t defined by (1.29), we also say that it has the stochastic differential

$$d\xi_t = a_t\, dt + b_t\, dw_t. \tag{1.30}$$

Theorem 1.18. (Ito's formula). *If function $f(t, x)$ together with its partial derivatives $f_t'(t, x)$, $f_x'(t, x)$, $f_{xx}''(t, x)$ are continuous then process $f(t, \xi_t)$ has the following stochastic differential*

$$df(t, \xi_t) = \left[f_t'(t, \xi_t) + f_t'(t, \xi_t) a_t + \tfrac{1}{2} f_{xx}''(t, \xi_t) b_t^2 \right] dt$$
$$+ f_x'(t, \xi_t) b_t\, dw_t, \tag{1.31}$$

where ξ_t is defined by (1.29).

In comparison with the differential calculus of deterministic functions, we have here an additional term $\frac{1}{2} f_{xx}''(t, \xi_t)$, which appears because of the facts given in Theorem 1.15.

Now we give the multidimensional Ito formula.

Theorem 1.19. *Assume that $(\mathbf{a}_t, \mathscr{F}_t)$ is an l-dimensional process, $\|\mathbf{a}_t\|^{1/2} \in \mathscr{P}_T$, $(\underset{\sim}{\mathbf{B}}_t, \mathscr{F}_t)$ is an $l \times m$ matrix, $\|\underset{\sim}{\mathbf{B}}_t\| \in \mathscr{P}_T$, $(\mathbf{w}_t, \mathscr{F}_t)$ is a Wiener process,*

$$d\boldsymbol{\xi}_t = \mathbf{a}_t\, dt + \underset{\sim}{\mathbf{B}}_t\, d\mathbf{w}_t$$

and function $f(t, \mathbf{x})$ with its partial derivatives $f_{\mathbf{x}}(t, \mathbf{x})$, $f_{\mathbf{xx}}(t, \mathbf{x})$ are continuous, where

$$f_{\mathbf{x}}(t, \mathbf{x}) \triangleq \begin{bmatrix} \dfrac{\partial f}{\partial x^1}(t, \mathbf{x}) \\ \vdots \\ \dfrac{\partial f}{\partial x^l}(t, \mathbf{x}) \end{bmatrix}, \qquad f_{\mathbf{xx}}(t, \mathbf{x}) = \begin{bmatrix} \dfrac{\partial^2 f(t, \mathbf{x})}{\partial x^1\, \partial x^1} & \cdots & \dfrac{\partial^2 f(t, \mathbf{x})}{\partial x^1\, \partial x^l} \\ \cdots & \cdots & \cdots \\ \dfrac{\partial^2 f(t, \mathbf{x})}{\partial x^1\, \partial x^l} & \cdots & \dfrac{\partial^2 f(t, \mathbf{x})}{\partial x^l\, \partial x^l} \end{bmatrix}$$

Then

$$df(t, \boldsymbol{\xi}_t) = \left[f_t'(t, \boldsymbol{\xi}_t) + f_{\mathbf{x}}^T(t, \boldsymbol{\xi}_t)\mathbf{a}_t + \tfrac{1}{2}\operatorname{tr} f_{\mathbf{xx}}(t, \boldsymbol{\xi}_t)\underset{\sim}{\mathbf{B}}_t \underset{\sim}{\mathbf{B}}_t^T \right] dt$$

$$+ f_{\mathbf{x}}^T(t, \boldsymbol{\xi}_t)\underset{\sim}{\mathbf{B}}_t \, d\mathbf{w}_t. \tag{1.32}$$

Example. Take $f(t, \mathbf{x}) = \mathbf{x}^T \underset{\sim}{\mathbf{C}}_t \mathbf{x}$, then by (1.32),

$$d\boldsymbol{\xi}_t^T \underset{\sim}{\mathbf{C}}_t \boldsymbol{\xi}_t = \left[\boldsymbol{\xi}_t^T \dot{\underset{\sim}{\mathbf{C}}}_t \boldsymbol{\xi}_t + \boldsymbol{\xi}_t^T (\underset{\sim}{\mathbf{C}}_t + \underset{\sim}{\mathbf{C}}_t^T)\mathbf{a}_t + \operatorname{tr} \underset{\sim}{\mathbf{C}}_t \underset{\sim}{\mathbf{B}}_t \underset{\sim}{\mathbf{B}}_t^T \right] dt$$

$$+ \boldsymbol{\xi}_t^T \underset{\sim}{\mathbf{C}}_t \underset{\sim}{\mathbf{B}}_t \, d\mathbf{w}_t + \boldsymbol{\xi}_t^T \underset{\sim}{\mathbf{C}}_t^T \underset{\sim}{\mathbf{B}}_t \, d\mathbf{w}_t. \tag{1.33}$$

Using Ito's formula, we can derive the following estimate.

Theorem 1.20. *Assume $(\xi_t, \mathscr{F}_t)$ is a measurable process and*

$$\int_0^T E\xi_t^{2m} \, dt < \infty.$$

Then

$$E\left(\int_0^t \xi_s \, dw_s \right)^{2m} \le [m(2m-1)]^m t^{m-1} \int_0^t E\xi_s^{2m} \, ds. \tag{1.34}$$

1.10. STOCHASTIC DIFFERENTIAL EQUATION

Let $(C_1, \mathscr{B}_T)$ denote the measurable space of continuous functions x defined on $[0, T]$, and let $\mathscr{B}_t = \sigma\{x\colon x_s, s \le t\}$ be the sub-σ-algebra of $\mathscr{B}_T$.

DEFINITION 1.7. Assume $a_t(x)$ and $b_t(x)$ are $\mathscr{B}_t$-measurable for any $t \in [0, T]$. $\mathscr{F}_t$-adapted process $(\xi_t, \mathscr{F}_t)$ is called the strong solution of stochastic differential equation

$$d\xi_t = a_t(\xi) \, dt + b_t(\xi) \, dw_t \tag{1.35}$$

with initial value η ($\mathscr{F}_0$-measurable) if

$$|a_t(\xi)|^{1/2} \in \mathscr{P}_T, \qquad b_t(\xi) \in \mathscr{P}_T$$

and for any $t \in [0, T]$

$$\xi_t = \eta + \int_0^t a_s(\xi) \, ds + \int_0^t b_s(\xi) \, dw_s, \tag{1.36}$$

where $(w_t, \mathscr{F}_t)$ is a Wiener process.

Theorem 1.21. *Suppose that a_t, b_t in (1.35) satisfy Lipschitz conditions. That is, there exist constants k_1 and k_2 and a nonnegative right-continuous function $0 \le k(s) \le 1$ such that for any continuous functions $x = \{x_t\}$ and $y = \{y_t\}$ defined on $[0, T]$*

$$|a_t(x) - a_t(y)|^2 + |b_t(x) - b_t(y)|^2 \le k_1 \int_0^t |x_s - y_s|^2\, dk(s) + k_2 |x_t - y_t|^2 \tag{1.37}$$

$$a_t^2(x) + b_t^2(x) \le k_1 \int_0^t (1 + x_s^2)\, dk(s) + k_2(1 + x_t^2), \tag{1.38}$$

and assume ξ_0 is $\mathscr{F}_0$-measurable, $P\{|\xi_0| < \infty\} = 1$. Then equation (1.35) has a unique strong solution $(\xi_t, \mathscr{F}_t)$ satisfying initial value ξ_0. If $E\xi_0^{2m} < \infty$, $m \ge 1$, then there is a constant c_m such that

$$E\xi_t^{2m} \le (1 + E\xi_0^{2m})e^{C_m} - 1.$$

The multidimensional version of this theorem is

Theorem 1.22. *If ξ_t in (1.35) is l-dimensional $\xi_t = [\xi_t^1 \cdots \xi_t^l]^\tau$, $\underset{\sim}{\mathbf{b}}_t$ is an $l \times m$ matrix, $\mathbf{w}_t$ is an m-dimensional Wiener process, and all components of $\mathbf{a}_t$ and $\underset{\sim}{\mathbf{b}}_t$ satisfy conditions (1.37) and (1.38) with x_s^2 and $|x_s - y_s|^2$ replaced by $\Sigma_{i=1}^l x_s^{i2}$ and $\Sigma_{i=1}^l (x_s^i - y_s^i)^2$, respectively, where x_s^i and y_s^i denote the i-components of l-dimensional continuous functions $\mathbf{x}_s$ and $\mathbf{y}_s$, respectively, then Equation (1.35) has the unique strong solution satisfying the initial value ξ_0 if $P(\|\xi_0\| < \infty) = 1$. Further,*

$$E \sum_{i=1}^{l} (\xi_t^i)^{2m} \le \left[1 + E \sum_{i=1}^{l} (\xi_0^i)^{2m}\right] e^{C_m} - 1$$

if $E\Sigma_{i=1}^l |\xi_0^i|^{2m} < \infty$.

Now consider the linear stochastic differential equation

$$d\xi_t = \underset{\sim}{\mathbf{A}}_t \xi_t\, dt + \mathbf{b}_t\, dt + \underset{\sim}{\mathbf{D}}_t\, d\mathbf{w}_t, \tag{1.39}$$

where ξ_t is l-dimensional and $(\mathbf{w}_t, \mathscr{F}_t)$ is an m-dimensional Wiener process, $\underset{\sim}{\mathbf{A}}_t, \underset{\sim}{\mathbf{D}}_t$ are deterministic while $\mathbf{b}_t$ is a random process with

$$\|\mathbf{b}_t\| \in \mathscr{P}_T \quad \text{and} \quad \int_0^T \|\underset{\sim}{\mathbf{A}}_s\|\, ds < \infty, \qquad \int_0^T \|\underset{\sim}{\mathbf{D}}_t\|^2\, dt < \infty.$$

The equation means that

$$\boldsymbol{\xi}_t = \boldsymbol{\xi}_0 + \int_0^t \mathbf{A}_s \boldsymbol{\xi}_s \, ds + \int_0^t \mathbf{b}_s \, ds + \int_0^t \mathbf{D}_s \, d\mathbf{w}_s .$$

Suppose $\boldsymbol{\Phi}_{ts}$ is the fundamental solution matrix of the linear ordinary differential equation. That is

$$\frac{d}{dt}\boldsymbol{\Phi}_{ts} = \mathbf{A}_t \boldsymbol{\Phi}_{ts}, \qquad \boldsymbol{\Phi}_{ss} = \mathbf{I}. \tag{1.40}$$

Then the solution of (1.39) is expressed by

$$\boldsymbol{\xi}_t = \boldsymbol{\Phi}_{t0}\boldsymbol{\xi}_0 + \int_0^t \boldsymbol{\Phi}_{ts} b_s \, ds + \int_0^t \boldsymbol{\Phi}_{ts} \mathbf{D}_s \, d\mathbf{w}_s , \tag{1.41}$$

and it is a normal (Gaussian) process, which means that any finite-dimensional distribution of $\boldsymbol{\xi}_t$ is normal. Formula (1.41) can be easily verified by Ito's formula. For this we only need to set

$$\mathbf{x}_t = \int_0^t \boldsymbol{\Phi}_{0s} \mathbf{D}_s \, d\mathbf{w}_s .$$

Making a comparison with Theorem 1.19, we see that

$$f(t, \mathbf{x}) = \boldsymbol{\Phi}_{t0}\boldsymbol{\xi}_0 + \int_0^t \boldsymbol{\Phi}_{ts}\mathbf{b}_s \, ds + \boldsymbol{\Phi}_{t0}\mathbf{x}, \qquad f(t, \mathbf{x}_t) = \boldsymbol{\xi}_t .$$

By (1.32) and (1.40), we have

$$\begin{aligned} d\boldsymbol{\xi}_t &= \mathbf{A}_t \boldsymbol{\Phi}_{t0}\boldsymbol{\xi}_0 \, dt + \mathbf{A}_t \boldsymbol{\Phi}_{t0} \int_0^t \boldsymbol{\Phi}_{0s}\mathbf{b}_s \, ds \, dt + \mathbf{b}_t \, dt + \mathbf{A}_t \boldsymbol{\Phi}_{t0}\mathbf{x}_t + \boldsymbol{\Phi}_{t0}\boldsymbol{\Phi}_{0t}\mathbf{D}_t \, d\mathbf{w}_t \\ &= \mathbf{A}_t \left[\boldsymbol{\Phi}_{t0}\boldsymbol{\xi}_0 + \int_0^t \boldsymbol{\Phi}_{ts}\mathbf{b}_s \, ds + \boldsymbol{\Phi}_{t0}\mathbf{x}_t \right] dt + \mathbf{b}_t \, dt + \mathbf{D}_t \, d\mathbf{w}_t \\ &= \mathbf{A}_t \boldsymbol{\xi}_t \, dt + \mathbf{b}_t \, dt + \mathbf{D}_t \, d\mathbf{w}_t , \end{aligned}$$

and hence the stochastic differential of $\boldsymbol{\xi}_t$ given by (1.41) indeed coincides with (1.39).

If $\mathbf{w}_t$ is not a Wiener process, but is a process of orthogonal increments, more precisely if

$$\mathbf{w}_0 = \mathbf{0}, \qquad E\mathbf{w}_t \equiv \mathbf{0},$$

$$E(\mathbf{w}_t - \mathbf{w}_s)(\mathbf{w}_t - \mathbf{w}_s)^{\tau} = (t - s)\mathbf{I},$$

and

$$E(\mathbf{w}_{t_1} - \mathbf{w}_{t_2})(\mathbf{w}_{t_3} - \mathbf{w}_{t_4})^{\tau} = \mathbf{0}$$

for any $t_1 \geq t_2 \geq t_3 \geq t_4$, then it is called a generalized Wiener process.

If $\underset{\sim}{\mathbf{D}}_t$ is deterministic and of compatible dimensions with $\int_0^T \|\underset{\sim}{\mathbf{D}}_t\|^2\,dt < \infty$, then we can define the stochastic integral

$$\int_0^t \underset{\sim}{\mathbf{D}}_s\,d\mathbf{w}_s, \qquad \forall t \in [0, T]$$

by an approximation with simple functions just as we did in 1.8.

For the stochastic integral with respect to generalized Wiener process properties, (1.21)–(1.24) remain valid, and

$$E\left(\int_0^t \underset{\sim}{\mathbf{D}}_\lambda\,d\mathbf{w}_\lambda\right)\left(\int_0^s \underset{\sim}{\mathbf{F}}_\lambda\,d\mathbf{w}_\lambda\right)^{\tau} = E\left(\int_0^s \underset{\sim}{\mathbf{D}}_\lambda\,d\mathbf{w}_\lambda\right)\left(\int_0^s \underset{\sim}{\mathbf{F}}_\lambda\,d\mathbf{w}_\lambda\right)^{\tau} = \int_0^s \underset{\sim}{\mathbf{D}}_\lambda \underset{\sim}{\mathbf{F}}_\lambda^{\tau}\,d\lambda,$$

$$0 \leq s \leq t \leq T, \qquad (1.42)$$

if $\underset{\sim}{\mathbf{D}}_s$ and $\underset{\sim}{\mathbf{F}}_s$ are deterministic and $\|\underset{\sim}{\mathbf{D}}_s\| \in L^2_{[0,T]}$, $\|\underset{\sim}{\mathbf{F}}_s\| \in L^2_{[0,T]}$.

If $\mathbf{w}_t$ in (1.39) is a generalized Wiener process, then it is still meaningful and the solution is again given by (1.41).

Let $\underset{\sim}{\mathbf{D}}_s$ and $\underset{\sim}{\mathbf{F}}_s$ be deterministic matrices and $\|\underset{\sim}{\mathbf{D}}_s\| \in L^1_{[0,T]}$, $\|\underset{\sim}{\mathbf{F}}_s\| \in L^2_{[0,T]}$. Then we have the formula for exchange of integration order

$$\int_0^t \underset{\sim}{\mathbf{D}}_s \int_0^s \underset{\sim}{\mathbf{F}}_\lambda\,d\mathbf{w}_\lambda\,ds = \int_0^t \int_\lambda^t \underset{\sim}{\mathbf{D}}_s\,ds\,\underset{\sim}{\mathbf{F}}_\lambda\,d\mathbf{w}_\lambda. \qquad (1.43)$$

If $\mathbf{w}_t$ is a Wiener process, then by setting

$$\underset{\sim}{\mathbf{G}}_t = \int_0^t \underset{\sim}{\mathbf{D}}_s\,ds, \qquad \mathbf{y}_t = \int_0^t \underset{\sim}{\mathbf{F}}_s\,d\mathbf{w}_s$$

Ito's formula gives

$$\underset{\sim}{\mathbf{G}}_t \mathbf{y}_t = \int_0^t d\underset{\sim}{\mathbf{G}}_s\,\mathbf{y}_s = \int_0^t \underset{\sim}{\mathbf{D}}_s \int_0^s \underset{\sim}{\mathbf{F}}_\lambda\,d\mathbf{w}_\lambda\,ds + \int_0^t \int_0^s \underset{\sim}{\mathbf{D}}_\lambda\,d\lambda\,\underset{\sim}{\mathbf{F}}_s\,d\mathbf{w}_s,$$

which leads to (1.43). If $\mathbf{w}_t$ is a generalized Wiener process, then (1.43) can be verified by approximation of simple functions.

1.11. MINIMUM VARIANCE ESTIMATION

We first briefly introduce the concept of pseudo-inverse of an $n \times m$ matrix $\underset{\sim}{\mathbf{A}}$. It is obvious that there are full-rank $n \times r$ matrix $\underset{\sim}{\mathbf{B}}$ and $r \times m$

matrix $\mathbf{C}$ such that

$$\mathbf{A} = \mathbf{BC}$$

if $\mathbf{A}$ is of rank $r \leq n \Lambda m$ [$\equiv \min(n, m)$]. It can be verified directly that

$$\mathbf{A}^{+} = \mathbf{C}^{T}(\mathbf{CC}^{T})^{-1}(\mathbf{B}^{T}\mathbf{B})^{-1}\mathbf{B}^{T} \tag{1.44}$$

satisfies the following system of matrix equations

$$\mathbf{AXA} = \mathbf{A}, \qquad \mathbf{XAX} = \mathbf{X}, \qquad (\mathbf{AX})^{T} = \mathbf{AX}, \qquad (\mathbf{XA})^{T} = \mathbf{XA}. \tag{1.45}$$

If $\mathbf{X}$ and $\mathbf{Y}$ are two solutions of (1.45), then

$$\begin{aligned}\mathbf{X} &= \mathbf{X}(\mathbf{AX})^{T} = \mathbf{XX}^{T}\mathbf{A}^{T} = \mathbf{XX}^{T}(\mathbf{AYA})^{T} = \mathbf{X}(\mathbf{AX})^{T}(\mathbf{AY})^{T} = \mathbf{XAY} \\ &= \mathbf{XAYAY} = (\mathbf{XA})^{T}(\mathbf{YA})^{T}\mathbf{Y} = (\mathbf{YAXA})^{T}\mathbf{Y} = (\mathbf{YA})^{T}\mathbf{Y} \\ &= \mathbf{YAY} = \mathbf{Y}.\end{aligned}$$

This chain of equalities shows that the solution of (1.45) is unique; hence $\mathbf{A}^{+}$ is the unique solution of (1.45), and it is called the pseudo-inverse of matrix $\mathbf{A}$. It is easy to verify

$$(\mathbf{A}^{T})^{+} = (\mathbf{A}^{+})^{T}, \qquad (\mathbf{A}^{+})^{+} = \mathbf{A} \tag{1.46}$$

$$(\mathbf{A}^{T}\mathbf{A})^{+} = \mathbf{A}^{+}(\mathbf{A}^{T})^{+} = \mathbf{A}^{+}(\mathbf{A}^{+})^{T}, \tag{1.47}$$

$$\mathbf{A}^{+} = (\mathbf{A}^{T}\mathbf{A})^{+}\mathbf{A}^{T} = \mathbf{A}^{T}(\mathbf{AA}^{T})^{+} \tag{1.48}$$

$$\mathbf{A}^{+}\mathbf{AA}^{T} = \mathbf{A}^{T}\mathbf{AA}^{+} = \mathbf{A}^{T} \tag{1.49}$$

$$(\mathbf{UAV})^{+} = \mathbf{V}^{T}\mathbf{A}^{+}\mathbf{U}^{T}, \tag{1.50}$$

if $\mathbf{U}$ and $\mathbf{V}$ are orthogonal matrices of compatible dimensions.

It is clear that $\mathbf{A}^{+} = \mathbf{A}^{-1}$ if $\mathbf{A}$ is square and $\det \mathbf{A} > 0$.

Let $\mathbf{x}, \mathbf{y}$ be two random vectors with $E\|\mathbf{x}\|^2 < \infty$, $E\|\mathbf{y}\|^2 < \infty$ and denote

$$\mathbf{R}_{\mathbf{x}} = E(\mathbf{x} - E\mathbf{x})(\mathbf{x} - E\mathbf{x})^{T}, \qquad \mathbf{R}_{\mathbf{xy}} = E(\mathbf{x} - E\mathbf{x})(\mathbf{y} - E\mathbf{y})^{T}.$$

We hope to choose a vector $\mathbf{c}$ and matrix $\mathbf{C}$ of compatible dimensions such that

$$E(\mathbf{x} - \mathbf{c} - \mathbf{C}\mathbf{y})(\mathbf{x} - \mathbf{c} - \mathbf{C}\mathbf{y})^{\tau} = \min. \tag{1.51}$$

We note

$$(\mathbf{y} - E\mathbf{y})^{\tau} = (\mathbf{y} - E\mathbf{y})^{\tau}\mathbf{R}_{\mathbf{y}}^{+}\mathbf{R}_{\mathbf{y}} \text{ a.s.} \tag{1.52}$$

since

$$E\left[(\mathbf{I} - \mathbf{R}_{\mathbf{y}}^{+}\mathbf{R}_{\mathbf{y}})(\mathbf{y} - E\mathbf{y})\right]\left[(\mathbf{I} - \mathbf{R}_{\mathbf{y}}^{+}\mathbf{R}_{\mathbf{y}})(\mathbf{y} - E\mathbf{y})\right]^{\tau} = 0$$

Using this fact, we have

$$\begin{aligned}
&E(\mathbf{x} - \mathbf{c} - \mathbf{C}\mathbf{y})(\mathbf{x} - \mathbf{c} - \mathbf{C}\mathbf{y})^{\tau} \\
&\quad = E\left[(\mathbf{x} - E\mathbf{x}) + (E\mathbf{x} - \mathbf{C}E\mathbf{y} - \mathbf{c}) - \mathbf{C}(\mathbf{y} - E\mathbf{y})\right] \\
&\qquad \times \left[(\mathbf{x} - E\mathbf{x}) - \mathbf{C}(\mathbf{y} - E\mathbf{y}) + (E\mathbf{x} - \mathbf{C}E\mathbf{y} - \mathbf{c})\right]^{\tau} \\
&\quad = (E\mathbf{x} - \mathbf{C}E\mathbf{y} - \mathbf{c})(E\mathbf{x} - \mathbf{C}E\mathbf{y} - \mathbf{c})^{\tau} \\
&\qquad + \mathbf{R}_{\mathbf{x}} - \mathbf{C}\mathbf{R}_{\mathbf{yx}} - \mathbf{R}_{\mathbf{xy}}\mathbf{C}^{\tau} + \mathbf{C}\mathbf{R}_{\mathbf{y}}\mathbf{C}^{\tau} \\
&\quad = (E\mathbf{x} - \mathbf{C}E\mathbf{y} - \mathbf{c})(E\mathbf{x} - \mathbf{C}E\mathbf{y} - \mathbf{c})^{\tau} \\
&\qquad + (\mathbf{C} - \mathbf{R}_{\mathbf{xy}}\mathbf{R}_{\mathbf{y}}^{+})\mathbf{R}_{\mathbf{y}}(\mathbf{C} - \mathbf{R}_{\mathbf{xy}}\mathbf{R}_{\mathbf{y}}^{+})^{\tau} \\
&\qquad + \mathbf{R}_{\mathbf{x}} - \mathbf{R}_{\mathbf{xy}}\mathbf{R}_{\mathbf{y}}^{+}\mathbf{R}_{\mathbf{yx}} \geq \mathbf{R}_{\mathbf{x}} - \mathbf{R}_{\mathbf{xy}}\mathbf{R}_{\mathbf{y}}^{+}\mathbf{R}_{\mathbf{yx}},
\end{aligned}$$

which turns into an equality if

$$\mathbf{C} = \mathbf{R}_{\mathbf{xy}}\mathbf{R}_{\mathbf{y}}^{+}, \qquad \mathbf{c} = E\mathbf{x} - \mathbf{R}_{\mathbf{xy}}\mathbf{R}_{\mathbf{y}}^{+}E\mathbf{y}.$$

Thus we conclude that the linear unbiased minimum variance estimate (LUMVE) of $\mathbf{x}$ based on $\mathbf{y}$ is $\hat{\mathbf{x}}$ given by

$$\hat{\mathbf{x}} = E\mathbf{x} + \mathbf{R}_{\mathbf{xy}}\mathbf{R}_{\mathbf{y}}^{+}(\mathbf{y} - E\mathbf{y}) \tag{1.53}$$

with estimation error covariance matrix

$$E(\mathbf{x} - \hat{\mathbf{x}})(\mathbf{x} - \hat{\mathbf{x}})^{\tau} = \underset{\sim}{\mathbf{R}}_{\mathbf{x}} - \underset{\sim}{\mathbf{R}}_{\mathbf{xy}}\underset{\sim}{\mathbf{R}}_{\mathbf{y}}^{+}\underset{\sim}{\mathbf{R}}_{\mathbf{yx}} \tag{1.54}$$

If random vectors $\mathbf{x}$ and $\mathbf{y}$ are independent then they are clearly uncorrelated, that is $\underset{\sim}{\mathbf{R}}_{\mathbf{xy}} = \underset{\sim}{\mathbf{0}}$, but the converse, generally speaking, is not true. However, these two concepts are equivalent for normal vectors. Indeed, if $\mathbf{w} = \{\mathbf{x}^{\tau}, \mathbf{y}^{\tau}\}^{\tau}$ is a normal vector and $\mathbf{x}$ and $\mathbf{y}$ are uncorrelated, then

$$\underset{\sim}{\mathbf{R}}_{\mathbf{w}} = \begin{pmatrix} \underset{\sim}{\mathbf{R}}_{\mathbf{x}} & \underset{\sim}{\mathbf{0}} \\ \underset{\sim}{\mathbf{0}} & \underset{\sim}{\mathbf{R}}_{\mathbf{y}} \end{pmatrix}$$

and its characteristic function will be

$$\begin{aligned} Ee^{i\boldsymbol{\lambda}^{\tau}\mathbf{w}} &= e^{i\boldsymbol{\lambda}^{\tau}E\mathbf{w}}e^{-\boldsymbol{\lambda}^{\tau}\underset{\sim}{\mathbf{R}}_{\mathbf{w}}\boldsymbol{\lambda}/2} \\ &= e^{i\boldsymbol{\mu}^{\tau}E\mathbf{x}}e^{i\boldsymbol{\nu}^{\tau}E\mathbf{y}}e^{-\boldsymbol{\mu}^{\tau}\underset{\sim}{\mathbf{R}}_{\mathbf{x}}\boldsymbol{\mu}/2}e^{-\boldsymbol{\nu}^{\tau}\underset{\sim}{\mathbf{R}}_{\mathbf{y}}\boldsymbol{\nu}/2} \\ &= Ee^{i\boldsymbol{\mu}^{\tau}\mathbf{x}}Ee^{i\boldsymbol{\nu}^{\tau}\mathbf{y}}, \end{aligned}$$

where $\boldsymbol{\lambda}^{\tau} = [\boldsymbol{\mu}^{\tau}\boldsymbol{\nu}^{\tau}]$ (see 1.7).

Hence $\mathbf{x}$ and $\mathbf{y}$ are independent (see 1.3).

Now consider LUMVE $\hat{\mathbf{x}}$ when $\mathbf{w} = [\mathbf{x}^{\tau}\mathbf{y}^{\tau}]^{\tau}$ is normal. $[\hat{\mathbf{x}}^{\tau}\mathbf{x}^{\tau}\mathbf{y}^{\tau}]^{\tau}$ is still normal since the linear combination of normal vectors is normal. Then by (1.52)

$$E(\mathbf{x} - \hat{\mathbf{x}})(\mathbf{y} - E\mathbf{y})^{\tau} = E\big[(\mathbf{x} - E\mathbf{x}) - \underset{\sim}{\mathbf{R}}_{\mathbf{xy}}\underset{\sim}{\mathbf{R}}_{\mathbf{y}}^{+}(\mathbf{y} - E\mathbf{y})\big](\mathbf{y} - E\mathbf{y})^{\tau} = \underset{\sim}{\mathbf{0}},$$

and $\mathbf{x} - \hat{\mathbf{x}}$ and $\mathbf{y} - E\mathbf{y}$ are independent. Hence by (1.10)

$$E\left(\mathbf{x} - \frac{\hat{\mathbf{x}}}{\mathbf{y}}\right) = E(\mathbf{x} - \hat{\mathbf{x}}) = \mathbf{0}.$$

On the other hand, since $\hat{\mathbf{x}}$ is measurable with respect to $\mathbf{y}$,

$$E\left(\frac{\hat{\mathbf{x}}}{\mathbf{y}}\right) = \hat{\mathbf{x}},$$

then

$$\hat{\mathbf{x}} = E\left(\frac{\mathbf{x}}{\mathbf{y}}\right). \tag{1.55}$$

Now we clarify the statistical meaning of the conditional expectation.

Let L be the totality of all measurable functions defined on R^m and valued in R^n with $E\|\mathbf{f}(\mathbf{y})\|^2 < \infty$, where m and n are dimensions of $\mathbf{x}$ and $\mathbf{y}$ respectively. Noticing

$$EE\left[\frac{\left(\mathbf{x} - E\left(\frac{\mathbf{x}}{\mathbf{y}}\right)\right)\left(E\left(\frac{\mathbf{x}}{\mathbf{y}}\right) - \mathbf{f}(\mathbf{y})\right)^{\tau}}{\mathbf{y}}\right]$$

$$= E\left[\frac{E\left(\mathbf{x} - E\left(\frac{\mathbf{x}}{\mathbf{y}}\right)\right)}{\mathbf{y}}\right]\left[E\left(\frac{\mathbf{x}}{\mathbf{y}}\right) - \mathbf{f}(\mathbf{y})\right]^{\tau} = \underset{\sim}{0}$$

by (1.7), we have for any $\mathbf{f} \in L$

$$\begin{aligned} E[\mathbf{x} - \mathbf{f}(\mathbf{y})][\mathbf{x} - \mathbf{f}(\mathbf{y})]^{\tau} &= E\left(\mathbf{x} - E\left(\frac{\mathbf{x}}{\mathbf{y}}\right) + E\left(\frac{\mathbf{x}}{\mathbf{y}}\right) - \mathbf{f}(\mathbf{y})\right) \\ &\quad \times \left(\mathbf{x} - E\left(\frac{\mathbf{x}}{\mathbf{y}}\right) + E\left(\frac{\mathbf{x}}{\mathbf{y}}\right) - \mathbf{f}(\mathbf{y})\right)^{\tau} \\ &= E\left(\mathbf{x} - E\left(\frac{\mathbf{x}}{\mathbf{y}}\right)\right)\left(\mathbf{x} - E\left(\frac{\mathbf{x}}{\mathbf{y}}\right)\right)^{\tau} \\ &\quad + E\left[\left(E\left(\frac{\mathbf{x}}{\mathbf{y}}\right) - \mathbf{f}(\mathbf{y})\right)\left(E\left(\frac{\mathbf{x}}{\mathbf{y}}\right) - \mathbf{f}(\mathbf{y})\right)^{\tau}\right], \end{aligned} \tag{1.56}$$

which reaches its minimum when $\mathbf{f}(\mathbf{y}) = E(\mathbf{x}/\mathbf{y})$. This means that $E(\mathbf{x}/\mathbf{y})$ is unbiased minimum variance estimate of $\mathbf{x}$ based on $\mathbf{y}$ in L, and (1.55) tells us that for normal vector LUMVE is optimal, not only in the class of linear estimates but also in the more general class of estimates.

CHAPTER 2

Stochastic Approximation Algorithms

2.1. INTRODUCTION

Let $\mathbf{h}(\cdot)$ be a defined on R^l and valued in R^m unknown function with a unique unknown zero $\mathbf{x}^0$

$$\mathbf{h}(\mathbf{x}^0) = \mathbf{0} \tag{2.1}$$

Assume that $\mathbf{h}(\cdot)$ can be observed at any $\mathbf{x} \in R^l$. If the measurement is free of errors, then there are many well-known algorithms which recursively approach $\mathbf{x}^0$, for example, the gradient method. But here we are concerned with the case when the measurement is corrupted by random errors, and we still hope to approximate $\mathbf{x}^0$ by observed data. At the beginning of the 1950s, Robbins and Monro first proposed an algorithm: take any $\mathbf{x}_0$ as the initial approximation of $\mathbf{x}^0$, then at time i, after having obtained the ith approximation $\mathbf{x}_i$ of $\mathbf{x}^0$, observe $\mathbf{h}(\cdot)$ at $\mathbf{x}_i$. Let the observed value by $\mathbf{y}_{i+1}$ with random error $\boldsymbol{\varepsilon}_{i+1}$. Then

$$\mathbf{y}_{i+1} = \mathbf{h}(\mathbf{x}_i) + \boldsymbol{\varepsilon}_{i+1}. \tag{2.2}$$

They suggested for the next approximation of $\mathbf{x}^0$ calculating from

$$\mathbf{x}_{i+1} = \mathbf{x}_i - a_i \mathbf{y}_{i+1}, \tag{2.3}$$

where a_i are real numbers satisfying

$$a_i > 0, \qquad \sum_{i=0}^{\infty} a_i = \infty, \qquad \sum_{i=0}^{\infty} a_i^2 < \infty. \tag{2.4}$$

They proved that

$$E|x_n - x^0|^2 \underset{n\to\infty}{\longrightarrow} 0$$

for the case when ε_i are mutually independent, $l = m = 1$, and $h(\cdot)$ satisfies certain conditions. This was the first fundamental work on stochastic approximation, and their algorithm became known as the Robbins–Monro (RM) procedure.

The next piece of important work was that due to Kiefer–Wolfowitz (KW). The KW procedure was first published in 1952. It is concerned with finding the extremum of $h(\cdot)$ when only $h(\cdot)$ itself but not its derivative can be observed, otherwise the problem would be reduced to the previous case and would be solved by RM procedure.

The sequence of real numbers with (2.4) is typical for stochastic approximation algorithms: If $\sum_{i=0}^{\infty} a_i < \infty$, then the correction at each time would be too small, so that the algorithm would converge quickly but the limit would differ from $\mathbf{x}^0$. This fact is easy to see from the simple example with $\boldsymbol{\varepsilon}_i \equiv 0$, $\|\mathbf{h}(\cdot)\| \leq C$, for which

$$\sum_{i=0}^{\infty} \|\mathbf{x}_{i+1} - \mathbf{x}_i\| \leq \sum_{i=0}^{\infty} a_i \|\mathbf{h}(\mathbf{x}_i)\| \leq C \sum_{i=0}^{\infty} a_i.$$

It means that the bound of sum of difference $\|\mathbf{x}_{i+1} - \mathbf{x}_i\|$ is independent of $\mathbf{x}_0$, and so $\mathbf{x}_i$ cannot go to $\mathbf{x}^0$ if the initial value $\mathbf{x}_0$ is far from $\mathbf{x}^0$. Hence the correction at each time should not be too small and the requirement $\sum_{i=0}^{\infty} a_i = \infty$ is necessary. On the other hand, it should not be too big; actually it should be reduced from time to time so that the influence of the measurement errors will vanish (i.e., $a_i \boldsymbol{\varepsilon}_{i+1} \underset{i \to \infty}{\longrightarrow} 0$). This point is guaranteed by the condition $\sum_{i=0}^{\infty} a_i^2 < \infty$.

For about 20 years since its pioneer work in stochastic approximation, the measurement errors considered have been restricted to a martingale difference sequence, and the algorithm itself is a Markov process. The convergence and other asymptotic properties of algorithms are proved mainly by using the theory of Markov process. In this kind of work, in contrast to the one that appeared later, the error is rather restricted but other conditions imposed for convergence are simple. The following two theorems are typical. Their proofs will follow from a more general theorem given later on.

Let $\mathscr{F}_i$ be the nondecreasing σ-algebras and k_i be constants independent of ω. Now we need the following condition.

CONDITION A_1. There exists a function $V(\mathbf{x})$ with bounded continuous second-order derivative such that

$$V(\mathbf{x}) > 0, \qquad \forall \mathbf{x} \neq \mathbf{x}^0, \qquad V(\mathbf{x}^0) = 0, \qquad V(\mathbf{x}) \underset{\|\mathbf{x}\| \to \infty}{\longrightarrow} \infty$$

and

$$\mathbf{h}^\tau(\mathbf{x}) V_{\mathbf{x}}(\mathbf{x}) > 0, \qquad \forall \mathbf{x} \neq \mathbf{x}^0,$$

where

$$V_{\mathbf{x}}(\mathbf{x}) = \left[\frac{\partial V}{\partial x^1} \cdots \frac{\partial V}{\partial x^l} \right]^\tau,$$

and x^i denote the components of $\mathbf{x}$.

Theorem 2.1. *Suppose that $l = m$ and the measurement error in* (2.2) $\boldsymbol{\varepsilon}_{i+1} = \boldsymbol{\varepsilon}_{i+1}(\mathbf{x}_i, \mathbf{e}_{i+1})$, *where $\boldsymbol{\varepsilon}_i(\mathbf{x}, \mathbf{e})$ is a*

$$(R^l \times R^s, \mathscr{B}^l \times \mathscr{B}^s) \to (R^l, \mathscr{B}^l)$$

measurable function, $\mathbf{e}_i$ is $\mathscr{F}_i$-measurable, independent of $\mathscr{F}_{i-1}$ and

$$E\boldsymbol{\varepsilon}_i(\mathbf{x}, \mathbf{e}_i) = \mathbf{0}, \qquad \forall \mathbf{x} \in R^l.$$

Further assume that Condition A_1 *holds,* $\{a_i\}$ *satisfies* (2.4), *and that*

$$\|\mathbf{h}(\mathbf{x})\|^2 + E\|\boldsymbol{\varepsilon}_i(\mathbf{x}, \mathbf{e}_i)\|^2 \le k_1(1 + V(\mathbf{x})) + k_2 \mathbf{h}^\tau(\mathbf{x}) V_{\mathbf{x}}(\mathbf{x}). \tag{2.5}$$

Then

$$\mathbf{x}_i \underset{i \to \infty}{\to} \mathbf{x}^0 \text{ a.s.}$$

where $\mathbf{x}_i$ is defined by (2.3) *with any initial value $\mathbf{x}_0$.*

Remark. Put

$$\boldsymbol{\xi}_n = \sum_{i=0}^{n} \boldsymbol{\varepsilon}_i(\mathbf{x}_{i-1}, \mathbf{e}_i).$$

Under the conditions of the theorem, $(\boldsymbol{\xi}_n, \mathscr{F}_n)$ is a martingale and hence the measurement errors $\boldsymbol{\varepsilon}_i$ compose a martingale difference sequence. But $\boldsymbol{\varepsilon}_i(\mathbf{x}_{i-1}, \mathbf{e}_i)$ is not necessarily independent of $\mathscr{F}_{i-1}$ if we only require that the measurement error is a martingale difference sequence.

From Theorem 2.1, the following well-known Gladyshev theorem [10] follows immediately if we take

$$V(\mathbf{x}) = (\mathbf{x} - \mathbf{x}^0)^\tau \underset{\sim}{C} (\mathbf{x} - \mathbf{x}^0),$$

where $\underset{\sim}{C}$ is a positive definite matrix.

Theorem 2.2. *If the measurement error is the same as in Theorem* 2.1 *and if*

$$\mathbf{h}^\tau(\mathbf{x})\underset{\sim}{C}(\mathbf{x} - \mathbf{x}^0) > 0, \qquad \mathbf{x} \neq \mathbf{x}^0,$$

$$\|\mathbf{h}(\mathbf{x})\|^2 + E\|\boldsymbol{\varepsilon}_i(\mathbf{x}, \omega)\|^2 \leq k_3(1 + \|\mathbf{x}\|^2) \tag{2.6}$$

then

$$\mathbf{x}_i \underset{i\to\infty}{\longrightarrow} \mathbf{x}^0 \text{ a.s.}$$

where $\mathbf{x}_i$ *is defined by* (2.3) *with any initial value* $\mathbf{x}_0$.

We note not only (2.5) and (2.6) are conditions on increasing rate of $\|\mathbf{h}(\mathbf{x})\|$ and $E\|\boldsymbol{\varepsilon}_i(\mathbf{x}, \mathbf{e}_i)\|^2$ as $\|\mathbf{x}\| \to \infty$, but they require the uniform boundedness in i for fixed $\mathbf{x}$.

In the mid-1970s, there appeared a new method, called the ordinary differential equation (ODE) method, which connects the convergence of stochastic approximation algorithms with stability of a certain ODE. This method can treat a wide range of measurement errors and requires only a minimal use of probability theory; however, for this method, more complicated and restrictive conditions are needed, including that $\mathbf{x}_i$ is a priori required to be bounded or to return to a compact set infinitely often.

In practice, ARMA processes are very important and are much more general than the martingale difference sequence; they will be the main type of error process considered in our convergence analysis later on. We shall give simple conditions guaranteeing the convergence of such stochastic approximation algorithms by use of a technique combining probabilistic and ODE methods. It will be seen that this combined method deals quite successfully with errors of the ARMA type, and even in the special case of martingale difference sequences it will give results better than classical technique.

In the sequel, we shall need the following theorems from analysis (e.g., see [72, 73]).

Theorem 2.3 (Arzela–Ascoli). *If* $\{\mathbf{f}_\alpha(t)\}$, $t \in [0, \infty)$, $\alpha \in A$ *is a set of equicontinuous* (i.e., *for any* $t \in [0, \infty)$ *and* $\varepsilon > 0$ *there exists* $\delta > 0$ *such that* $\|\mathbf{f}_\alpha(t) - \mathbf{f}_\alpha(s)\| \leq \varepsilon$, $\forall \alpha \in A$ *if* $|t - s| < \delta$) *and uniformly bounded functions, then there exists a subsequence* $\{\mathbf{f}_{\alpha_k}(t)\}$ *of* $\{\mathbf{f}_\alpha(t)\}$ *and a continuous function* $\mathbf{f}(t)$ *which is the uniform limit of* $\{\mathbf{f}_{\alpha_k}(t)\}$ *in any finite interval of* t.

Theorem 2.4. *If there is a function* $V(\mathbf{x})$ *with a continuous derivative and* $V(\mathbf{x}) > 0$; $\forall \mathbf{x} \neq \mathbf{x}^0$; $V(\mathbf{x}^0) = 0$, $V(\mathbf{x}) \underset{\|\mathbf{x}\| \to \infty}{\longrightarrow} \infty$,

$$V_{\mathbf{x}}^{T}(\mathbf{x})\mathbf{h}(\mathbf{x}) < 0, \qquad \forall \mathbf{x} \neq \mathbf{x}^0 \tag{2.7}$$

then, for the l-dimensional ODE

$$\mathbf{x}_t = \mathbf{f}(\mathbf{x}_t), \qquad t \geq 0, \tag{2.8}$$

the trajectory $\mathbf{x}_t$ *starting from any* $\mathbf{x}_0$ *converges to* $\mathbf{x}^0$. *In other words,* $\mathbf{x}^0$ *is the global asymptotically stable solution of* (2.7).

2.2. ANALYSIS VIA TECHNIQUES OF PROBABILITY AND DIFFERENTIAL EQUATION THEORY

In this section, the convergence of RM procedures with measurement errors equal to a martingale difference sequence is analyzed by combining probability and ODE techniques. Clearly, the results obtained here are special cases of the ones to come later; for the moment we are interested only in introducing the method itself, since the essentials of the method can be seen more clearly when the process of measurement errors is simple. The main idea is that the convergence of a Lyapunov function along the trajectory determined by the algorithm is first established by martingale convergence theorems and thus the uniform boundedness of estimates for any fixed ω follows. Then the properties of the "tail function" defined by the algorithm are considered via the ODE method, and finally the convergence of $\mathbf{x}_i$ to $\mathbf{x}^0$ is proven.

Assume that $l \geq m$, $\mathscr{F}_n$, $n \geq 1$ is a nondecreasing family of σ-algebras in $(\Omega, \mathscr{F}, P)$; further assume that $\mathbf{w}_i(\mathbf{x}, \mathbf{e})$ is a

$$(R^l \times R^s, \mathscr{B}^l \times \mathscr{B}^s) \to (R^m, \mathscr{B}^m)$$

measurable function such that $\mathbf{e}_{i+1}$ is $\mathscr{F}_{i+1}$-measurable and

$$E\left(\frac{\mathbf{w}_{i+1}(\boldsymbol{\xi}, \mathbf{e}_{i+1})}{\mathscr{F}_i}\right) = \mathbf{0} \tag{2.9}$$

whenever l-dimensional random vector $\boldsymbol{\xi}$ is $\mathscr{F}_i$-measurable.

Let $\mathbf{x}_i$ be the estimate of $\mathbf{x}^0$ at time i and assume the measurement error in (2.2) is given by

$$\boldsymbol{\varepsilon}_{i+1} = \mathbf{w}_{i+1}(\mathbf{x}_i, \mathbf{e}_{i+1}). \tag{2.10}$$

The rate of increase of $E\|\mathbf{w}_{i+1}(\mathbf{x}, \mathbf{e}_{i+1})\|^2$ and $\|\mathbf{h}(\mathbf{x})\|^2$ as $\|\mathbf{x}\| \to \infty$ is no longer constrained as in (2.5) and (2.6), and $E\|\mathbf{w}_{i+1}(\mathbf{x}, \mathbf{e}_{i+1})\|^2$ is allowed to increase as $i \to \infty$. However, we have to impose the requirement that their dominating functions are known; that is it is assumed that there are known functions $r_1(i)$ and $r_2(s) \geq 1 + s$, $s \geq 0$ such that

$$E\left(\frac{\|\mathbf{w}_{i+1}(\boldsymbol{\xi}, \mathbf{e}_{i+1})\|^2}{\mathscr{F}_i}\right) \leq r_1(i) r_2^2(\|\boldsymbol{\xi}\|), \tag{2.11}$$

and

$$\|\mathbf{h}(\mathbf{x})\| \leq r_2(\|\mathbf{x}\|) \tag{2.12}$$

where, for simplicity, the dominating function for $\mathbf{h}(\mathbf{x})$, namely $r_2(\cdot)$, is taken to be the same as that appearing in (2.11), and clearly, it is without loss of generality.

Since we have weakened conditions on $\mathbf{h}(\mathbf{x})$ and $\boldsymbol{\varepsilon}_{i+1}$ in comparison with those for Theorems 2.1 and 2.2 and we have allowed different dimensions for $\mathbf{x}$ and $\mathbf{h}(\mathbf{x})$, we must modify the RM procedure given by (2.3) and must also replace Condition A_1 by Condition A_2.

CONDITION A_2. There exist an $m \times l$ matrix $\underset{\sim}{\mathbf{L}}$ and a function $V(\mathbf{x})$ with bounded continuous second derivative such that

$$V(\mathbf{x}) > 0, \qquad \forall \mathbf{x} \neq \mathbf{x}^0, \qquad V(\mathbf{x}^0) = 0, \qquad V(\mathbf{x}) \to \infty \quad \text{as } \|\mathbf{x}\| \to \infty$$

and

$$\mathbf{h}^\tau(\mathbf{x}) \underset{\sim}{\mathbf{L}} V_{\mathbf{x}}(\mathbf{x}) > 0, \qquad \forall \mathbf{x} \neq \mathbf{x}^0.$$

Now let $b_i > 0$ be $\mathscr{F}_i$-measurable random variables such that

$$\sum_{i=0}^{\infty} b_i = \infty \text{ a.s.}, \quad E\sum_{i=0}^{\infty} b_i^2 < \infty, \quad \text{and} \quad E\sum_{i=0}^{\infty} b_i^2 r_1(i) < \infty. \tag{2.13}$$

Algorithm (2.3) is then modified to read

$$\mathbf{x}_{i+1} = \mathbf{x}_i - a_i \underset{\sim}{\mathbf{L}}^\tau \mathbf{y}_{i+1}, \tag{2.14}$$

where

$$a_i = \frac{b_i}{z_i}, \qquad a_0 = b_0 \tag{2.15}$$

$$z_i = r_2(\|\mathbf{x}_i\|). \tag{2.16}$$

We shall refer to this algorithm as the modified Robbins–Monro procedure or algorithm (MRM).

Theorem 2.5. *Suppose that* $\mathbf{h}(\mathbf{x})$ *is continuous, the measurement error is given by* (2.10), *the dominating functions in* (2.11) *and* (2.12) *are known, and that* b_i, $i \geq 0$, *satisfy* (2.13). *Then, if Condition* A_2 *holds,*

$$\mathbf{x}_i \underset{i \to \infty}{\to} \mathbf{x}^0 \text{ a.s.,}$$

where $\mathbf{x}_i$ *is defined by* MRM *procedure.*

Before we prove this result, we compare the conditions imposed here with those in Theorem 2.1.

Notice that Conditions A_1 and A_2 are somewhat equivalent, but the other conditions differ. In Theorem 2.1, it is required that $l = m$ and, as $\|\mathbf{x}\| \to \infty$, $E\|\boldsymbol{\varepsilon}_i(\mathbf{x}, \mathbf{e}_i)\|^2 + \|\mathbf{h}(\mathbf{x})\|^2$ is bounded by $k_1(1 + V(\mathbf{x})) + k_2\mathbf{h}^{\tau}(\mathbf{x})V_{\mathbf{x}}(\mathbf{x})$ where this quantity is independent of i, while in Theorem 2.5 we permit $l \leq m$ and, as $\|\mathbf{x}\| \to \infty$, $E\|\boldsymbol{\varepsilon}_i(\mathbf{x}, \mathbf{e}_i)\|^2 + \|\mathbf{h}(\mathbf{x})\|^2$ is allowed to increase arbitrarily (subject to some known dominating function) and also to increase with i at a specified rate. Concerning the stochastic nature of ε_i, Theorem 2.1 requires that $E\boldsymbol{\varepsilon}_i(\mathbf{x}, \mathbf{e}_i) = 0$ and that $\mathbf{e}_i$ is independent of $\mathscr{F}_{i-1}$. From this, if we put $\boldsymbol{\varepsilon}_i, \mathbf{e}_i, \mathbf{0}, \mathbf{x}_{i-1}$ in correspondence with $\mathbf{f}, \boldsymbol{\eta}, \mathbf{g}(\boldsymbol{\lambda}), \boldsymbol{\xi}$ of Theorem 1.9, respectively, and notice the measurability of $\mathbf{x}_{i-1}$ with respect to $\mathscr{F}_{i-1}$, it follows that

$$E\left[\frac{\boldsymbol{\varepsilon}_i(\mathbf{x}_{i-1}, \mathbf{e}_i)}{\mathscr{F}_{i-1}}\right] = \mathbf{0}$$

which is the condition imposed in Theorem 2.5.

Thus, with the exception of the continuity of $\mathbf{h}(\mathbf{x})$, the conditions in Theorem 2.5 are much weaker than those in Theorem 2.1.

For the proof of Theorem 2.5, we first establish some lemmas. We denote $\mathbf{w}_i(\mathbf{x}_{i-1}, \mathbf{e}_i)$ by $\mathbf{w}_i$ for simplicity.

Lemma 2.1. *Under the conditions of Theorem* 2.5, $E\|\mathbf{x}_i\|^2 < \infty$, $EV(\mathbf{x}_i) < \infty$ $\forall i$ *and*

$$E\sum_{i=0}^{\infty} a_i^2\|\mathbf{w}_{i+1}\|^2 < \infty, \qquad E\sum_{i=0}^{\infty} a_i^2\|\mathbf{h}(\mathbf{x}_i)\|^2 < \infty.$$

Proof. $\mathbf{x}_0$ is $\mathscr{F}_0$-measurable since $\mathbf{x}_0$ is a constant. Assume $\mathbf{x}_i$ is $\mathscr{F}_i$-measurable for any $i \leq n$. Then z_n defined by (2.16) is $\mathscr{F}_n$-measurable and, since b_n is $\mathscr{F}_n$-measurable, it follows that a_n is also. Thus $\varepsilon_{n+1}(\mathbf{x}_n, \mathbf{e}_{n+1})$

is $\mathscr{F}_{n+1}$ measurable since $\boldsymbol{\varepsilon}_{i+1}(\mathbf{x}, \mathbf{e})$ is $\mathscr{B}^l \times \mathscr{B}^s$-measurable. From here it follows that $\mathbf{y}_{n+1}$ and hence $\mathbf{x}_{n+1}$ is $\mathscr{F}_{n+1}$-measurable. By induction, we conclude that $\mathbf{x}_n$ is $\mathscr{F}_n$-measurable for any n.

By (1.6)–(1.8), we have

$$Ea_i^2\|\mathbf{w}_{i+1}\|^2 = EE^{\mathscr{F}_i}a_i^2\|\mathbf{w}_{i+1}\|^2 = Ea_i^2E^{\mathscr{F}_i}\|\mathbf{w}_{i+1}\|^2$$

and then by (2.11), (2.13), and (2.15), we see

$$\begin{aligned} Ea_i^2\|\mathbf{w}_{i+1}\|^2 &\le Ea_i^2 r_1(i) r_2^2(\|x_i\|) \\ &= E\left[\frac{b_i^2 r_1(i) r_2^2(\|\mathbf{x}_i\|)}{r_2^2(\|\mathbf{x}_i\|)}\right] = Eb_i^2 r_1(i) < \infty \qquad \forall i, \end{aligned} \tag{2.17}$$

and

$$E\sum_{i=0}^{\infty} a_i^2\|\mathbf{w}_{i+1}\|^2 = \sum_{i=0}^{\infty} Ea_i^2\|\mathbf{w}_{i+1}\|^2 \le \sum_{i=0}^{\infty} Eb_i^2 r_1(i) < \infty.$$

By (2.12) and (2.15), it follows that

$$E\sum_{i=0}^{\infty} a_i^2\|\mathbf{h}(\mathbf{x}_i)\| \le E\sum_{i=0}^{\infty}\left[\frac{b_i^2 r_2(\|\mathbf{x}_i\|)}{r_2^2(\|\mathbf{x}_i\|)}\right] \le E\sum_{i=0}^{\infty} b_i^2 < \infty. \tag{2.18}$$

and hence by noticing

$$E\|\mathbf{x}_{i+1}\|^2 \le 3E\|\mathbf{x}_i\|^2 + 3\|\underset{\sim}{\mathbf{L}}\|^2 Ea_i^2\|\mathbf{h}(\mathbf{x}_i)\|^2 + 3\|\underset{\sim}{\mathbf{L}}\|^2 Ea_i^2\|\mathbf{w}_{i+1}\|^2$$

we conclude by induction that

$$E\|\mathbf{x}_i\|^2 < \infty \qquad \forall i$$

since $E\|\mathbf{x}_0\|^2 < \infty$.

Denote by $V_{\mathbf{xx}}(\mathbf{x})$ the $l \times l$ matrix with (i, j)th element $V_{\mathbf{x}^i\mathbf{x}^j}(\mathbf{x}) = \partial^2 V(\mathbf{x})/\partial x^i \partial x^j$. According to Condition A_2, $V(\mathbf{x})$ has a bounded second derivative. Hence there exists a constant k_1 such that

$$|V_{\mathbf{x}^i\mathbf{x}^j}(\mathbf{x})| \le k_1, \qquad \forall i, j = 1, \ldots, l \tag{2.19}$$

In the sequel, for a matrix $\underset{\sim}{\mathbf{A}}$,

$$\underset{\sim}{\mathbf{A}} = \begin{bmatrix} a_{11} & \cdots & a_{1n} \\ \cdots & \cdots & \cdots \\ a_{m1} & \cdots & a_{mn} \end{bmatrix},$$

$\|\underset{\sim}{\mathbf{A}}\|$ will always denote the square root of the maximum eigenvalue of $\underset{\sim}{\mathbf{A}}\underset{\sim}{\mathbf{A}}^{\tau}$. Hence

$$\|\underset{\sim}{\mathbf{A}}\| \leq \sqrt{\operatorname{tr} \underset{\sim}{\mathbf{A}}\underset{\sim}{\mathbf{A}}^{\tau}} = \left(\sum_{i=1}^{m} \sum_{j=1}^{n} a_{ij}^{2} \right)^{1/2} \leq \left(mn \max_{i,j} a_{ij}^{2} \right)^{1/2} = \sqrt{mn} \max_{i,j} |a_{ij}| \tag{2.20}$$

and, in particular,

$$\| V_{\mathbf{xx}}(\mathbf{x}) \| \leq l \max_{i,j} \left| \frac{\partial^2 V(\mathbf{x})}{\partial x^i \, \partial x^j} \right| \leq l k_1, \qquad \forall x. \tag{2.21}$$

By use of the Taylor expansion, we have

$$\| V_{\mathbf{x}}(\mathbf{x}_n) \| \leq \| V_{\mathbf{x}}(\mathbf{x}^0) \| + l k_1 \left(\|\mathbf{x}_n\| + \|\mathbf{x}^0\| \right),$$

and hence

$$E \| V_{\mathbf{x}}(\mathbf{x}_n) \|^2 < \infty, \qquad \forall n \tag{2.22}$$

since $E\|\mathbf{x}_n\|^2 < \infty$.

By the formula for the remainder term in the Taylor expansion, we know that for any fixed ω there exists an $l \times l$ matrix $\underset{\sim}{\mathbf{V}}(n, \omega)$, whose components consist of second derivatives of $V(\mathbf{x})$ such that

$$V(\mathbf{x}_{n+1}) = V(\mathbf{x}_n) - a_n \left[\mathbf{h}^{\tau}(\mathbf{x}_n) + \mathbf{w}_{n+1}^{\tau} \right] \underset{\sim}{\mathbf{L}} V_{\mathbf{x}}(\mathbf{x}_n)$$
$$+ \tfrac{1}{2} a_n^2 \left[\mathbf{h}^{\tau}(\mathbf{x}_n) + \mathbf{w}_{n+1}^{\tau} \right] \underset{\sim}{\mathbf{L}} \underset{\sim}{\mathbf{V}}(n, \omega) \underset{\sim}{\mathbf{L}}^{\tau} \left[\mathbf{h}(\mathbf{x}_n) + \mathbf{w}_{n+1} \right]. \tag{2.23}$$

As in (2.20) and (2.21), we have $\|\underset{\sim}{\mathbf{V}}(n, \omega)\| \leq l k_1$. Hence

$$V(\mathbf{x}_{n+1}) \leq V(\mathbf{x}_n) - a_n \left[\mathbf{h}^{\tau}(\mathbf{x}_n) + \mathbf{w}_{n+1}^{\tau} \right] \underset{\sim}{\mathbf{L}} V_{\mathbf{x}}(\mathbf{x}_n)$$
$$+ \frac{l k_1}{2} \|\underset{\sim}{\mathbf{L}}\|^2 a_n^2 \|\mathbf{h}(\mathbf{x}_n) + \mathbf{w}_{n+1}\|^2. \tag{2.24}$$

From (2.17), (2.18), (2.22), and the Schwarz inequality, it follows that

$$E \left\| a_n \left[\mathbf{h}^{\tau}(\mathbf{x}_n) + \mathbf{w}_{n+1}^{\tau} \right] \underset{\sim}{\mathbf{L}} V_{\mathbf{x}}(\mathbf{x}_n) \right\|$$
$$\leq \|\underset{\sim}{\mathbf{L}}\| \left[E \| V_{\mathbf{x}}(\mathbf{x}_n) \|^2 \right]^{1/2} \left[E a_n^2 \|\mathbf{h}(\mathbf{x}_n) + \mathbf{w}_{n+1}\|^2 \right]^{1/2} < \infty.$$

Hence $EV(\mathbf{x}_n) < \infty$ yields $EV(\mathbf{X}_{n+1}) < \infty$. But we have $EV(\mathbf{x}_0) < \infty$, hence $EV(\mathbf{x}_n) < \infty, \forall n$. □

Lemma 2.2. *Under the conditions of Theorem* 2.5, *there are random variables* α, β, γ *such that* $0 \le \alpha < \infty$, $0 < \beta < \infty$, $1 \le \gamma < \infty$ a.s.

$$V(\mathbf{x}_n) \underset{n\to\infty}{\to} \alpha, \quad \|\mathbf{x}_i\| \le \beta, \quad r_2(\|\mathbf{x}_i\|) \le \gamma, \qquad \forall i \quad \text{a.s.} \tag{2.25}$$

Proof. Since a_n and $V_{\mathbf{x}}(\mathbf{x}_n)$ are $\mathscr{F}_n$-measurable, it follows from (1.8) and (2.9) that

$$E^{\mathscr{F}_n} a_n \mathbf{w}_{n+1}^\tau V_{\mathbf{x}}(\mathbf{x}_n) = a_n\left(E^{\mathscr{F}_n}\mathbf{w}_{n+1}^\tau\right)V_{\mathbf{x}}(\mathbf{x}_n) = 0.$$

Noticing that by Condition A_2

$$\mathbf{h}^\tau(\mathbf{x}_n)\underset{\sim}{\mathbf{L}}V_{\mathbf{x}}(\mathbf{x}_n) \ge 0,$$

and using (2.11) and (2.24), we obtain

$$\begin{aligned} E\left[\frac{V(\mathbf{x}_{n+1})}{\mathscr{F}_n}\right] &\le V(\mathbf{x}_n) + \frac{lk_1}{2}\|\underset{\sim}{\mathbf{L}}\|^2 a_n^2 \|\mathbf{h}(\mathbf{x}_n)\|^2 + \frac{lk_1}{2}\|\underset{\sim}{\mathbf{L}}\|^2 a_n^2 E\left[\frac{\|\mathbf{w}_{n+1}\|^2}{\mathscr{F}_n}\right] \\ &\le V(\mathbf{x}_n) + \frac{lk_1}{2}\|\underset{\sim}{\mathbf{L}}\|^2 a_n^2 \|\mathbf{h}(\mathbf{x}_n)\|^2 + \frac{lk_1}{2}\|\underset{\sim}{\mathbf{L}}\|^2 a_n^2 r_1(n) r_2^2(\|\mathbf{x}_n\|) \\ &\le V(\mathbf{x}_n) + \frac{lk_1}{2}\|\underset{\sim}{\mathbf{L}}\|^2 a_n^2 \|\mathbf{h}(\mathbf{x}_n)\|^2 + \frac{lk_1}{2}\|\underset{\sim}{\mathbf{L}}\|^2 b_n^2 r_1(n). \end{aligned} \tag{2.26}$$

Set

$$\begin{aligned} m_n = V(\mathbf{x}_n) &+ \frac{lk_1}{2}\|\underset{\sim}{\mathbf{L}}\|^2 E\left[\frac{\sum_{i=0}^{\infty} a_i^2\|\mathbf{h}(\mathbf{x}_i)\|^2}{\mathscr{F}_n}\right] \\ &- \frac{lk_1}{2}\|\underset{\sim}{\mathbf{L}}\|^2 \sum_{i=0}^{n-1} a_i^2\|\mathbf{h}(\mathbf{x}_i)\|^2 + \frac{lk_1}{2}\|\underset{\sim}{\mathbf{L}}\|^2 E\left[\frac{\sum_{i=0}^{\infty} b_i^2 r_1(i)}{\mathscr{F}_n}\right] \\ &- \frac{lk_1}{2}\|\underset{\sim}{\mathbf{L}}\|^2 \sum_{i=0}^{n-1} b_i^2 r_1(i). \end{aligned} \tag{2.27}$$

The expectations of all terms on the right-hand side of (2.27) are finite by (2.13) and Lemma 2.1. Hence by (2.26), we find

$$E\left(\frac{m_{n+1}}{\mathscr{F}_n}\right) \le V(\mathbf{x}_n) + \frac{lk_1}{2}\|\underset{\sim}{\mathbf{L}}\|^2 a_n^2 \|\mathbf{h}(\mathbf{x}_n)\|^2 + \frac{lk_1}{2}\|\underset{\sim}{\mathbf{L}}\|^2 b_n^2 r_1(n)$$

$$+ \frac{lk_1}{2}\|\underset{\sim}{\mathbf{L}}\|^2 E\left[\sum_{i=0}^{\infty} \frac{a_i \|\mathbf{h}(\mathbf{x}_i)\|^2}{\mathscr{F}_n}\right] - \frac{lk_1}{2}\|\underset{\sim}{\mathbf{L}}\|^2 \sum_{i=0}^{n} a_i^2 \|\mathbf{h}(\mathbf{x}_i)\|^2$$

$$+ \frac{lk_1}{2}\|\underset{\sim}{\mathbf{L}}\|^2 E\left[\frac{\sum_{i=0}^{\infty} b_i^2 r_1(i)}{\mathscr{F}_n}\right] - \frac{lk_1}{2}\|\underset{\sim}{\mathbf{L}}\|^2 \sum_{i=0}^{n} b_i^2 r_1(i) = m_n. \tag{2.28}$$

Thus $(m_n, \mathscr{F}_n)$ is a supermartingale and it is nonnegative, then by Corollary 1 of Theorem 1.11, there is a finite random variable $\mu \ge 0$ such that

$$m_n \underset{n\to\infty}{\to} \mu \quad \text{a.s.} \tag{2.29}$$

By the same type of reasoning and by use of (2.18), we see that the nonnegative martingale

$$\left(E\left[\sum_{i=0}^{\infty} \frac{a_i \|\mathbf{h}(\mathbf{x}_i)\|^2}{\mathscr{F}_n}\right], \mathscr{F}_n\right)$$

converges to a finite limit as $n \to \infty$ and that

$$\sum_{i=0}^{\infty} a_i \|\mathbf{h}(\mathbf{x}_i)\|^2 < \infty \quad \text{a.s.}$$

But by (1.7)

$$E\left(\sum_{i=0}^{\infty} \frac{a_i^2 \|\mathbf{h}(\mathbf{x}_i)\|^2}{\mathscr{F}_n}\right) - \sum_{i=0}^{n-1} a_i^2 \|\mathbf{h}(\mathbf{x}_i)\|^2$$

$$= E\left[\left(\sum_{i=0}^{\infty} a_i^2 \|\mathbf{h}(\mathbf{x}_i)\|^2 - \sum_{i=0}^{n-1} a_i^2 \|\mathbf{h}(\mathbf{x}_i)\|^2\right) \Big/ \mathscr{F}_n\right] \ge 0, \qquad \forall n,$$

hence the sum of the second and third terms on the right-hand side of (2.27) has finite, nonnegative limit. Similarly, its last two terms also have a finite nonnegative limit. Then from (2.29), it follows that there is a random variable α such that $0 \le \alpha < \infty$ a.s. and

$$V(\mathbf{x}_n) \underset{n\to\infty}{\to} \alpha \quad \text{a.s.}$$

Notice $V(\mathbf{x}) \to \infty$ as $\|\mathbf{x}\| \to \infty$. Then the preceding convergence means that for any fixed ω, $\mathbf{x}_n$ is uniformly bounded in n with the possible exception of a set with probability zero. From here the last two assertions in (2.25) are immediate. □

Corollary. *From $r_2(\|\mathbf{x}_i\|) \le \gamma \ \forall i$, it follows that*

$$\sum_{i=0}^{\infty} a_i = \sum_{i=0}^{\infty} \frac{b_i}{z_i} = \sum_{i=0}^{\infty} \frac{b_i}{r_2(\|\mathbf{x}_i\|)} \ge \frac{1}{\gamma} \sum_{i=0}^{\infty} b_i = \infty. \tag{2.30}$$

Up to now, using probabilistic methods only, we have proved the a.s. uniform boundedness of $\mathbf{x}_i$, $i \ge 0$ defined by (2.14), the convergence of $V(\mathbf{x}_i)$, $i \ge 0$, and we have shown that the coefficients a_i used in algorithm (2.14) have the properties (2.4) typically required by the stochastic approximation method. Now we proceed to apply the ODE method. Denote

$$t_n = \sum_{i=0}^{n-1} a_i, \qquad t_0 = 0 \tag{2.31}$$

and define

$$m(t) = \max(n\colon t_n \le t), \qquad t \ge 0, \tag{2.32}$$

which means

$$t_{m(t)} = \sum_{i=0}^{m(t)-1} a_i \le t < \sum_{i=0}^{m(t)} a_i = t_{m(t)+1}. \tag{2.33}$$

We rewrite (2.14) as

$$\mathbf{x}_{n+1} = \mathbf{x}_0 - \sum_{i=0}^{n} a_i \underset{\sim}{\mathbf{L}}^{\tau} \mathbf{h}(\mathbf{x}_i) - \sum_{i=0}^{n} a_i \underset{\sim}{\mathbf{L}}^{\tau} \mathbf{w}_{i+1}. \tag{2.34}$$

From now on, we shall always denote by $\underset{\sim}{\mathbf{A}}_t^0$ and $\overline{\underset{\sim}{\mathbf{A}}}_t$, respectively, the linear interpolation and the constant interpolation of a matrix sequence

$\{\underset{\sim}{\mathbf{A}}_n\}$ with interpolating length $\{a_n\}$. Explicitly

$$\underset{\sim}{\mathbf{A}}_t^0 = \frac{t_{n+1} - t}{a_n}\underset{\sim}{\mathbf{A}}_n + \frac{t - t_n}{a_n}\underset{\sim}{\mathbf{A}}_{n+1}, \qquad t \in [t_n, t_{n+1}]$$

$$\underset{\sim}{\mathbf{A}}_{t_n}^0 = \underset{\sim}{\mathbf{A}}_n, \tag{2.35}$$

and

$$\overline{\underset{\sim}{\mathbf{A}}}_t = \underset{\sim}{\mathbf{A}}_n, \qquad t \in [t_n, t_{n+1}). \tag{2.36}$$

Denote

$$\mathbf{q}_n = \sum_{i=0}^{n-1} a_i \underset{\sim}{\mathbf{L}}^\tau \mathbf{w}_{i+1}, \tag{2.37}$$

$$\mathbf{x}_t^0 = \mathbf{x}_0^0 - \underset{\sim}{\mathbf{L}}^\tau \int_0^t \mathbf{h}(\overline{\mathbf{x}}_s)\, ds - \mathbf{q}_t^0, \tag{2.38}$$

and define

$$\mathbf{x}_n(t) = \mathbf{x}_{t+n}^0. \tag{2.39}$$

Lemma 2.3. *Under the conditions of Theorem 2.5, for any $\omega \in \Omega$ (with the possible exception of a set with probability zero) there is a subsequence $\{\mathbf{x}_{n_k}(t)\}$ of $\{\mathbf{x}_n(t)\}$ and a continuous function $\mathbf{x}(t)$ which is the uniform limit of $\mathbf{x}_{n_k}(t)$:*

$$\mathbf{x}_{n_k}(t) \underset{k\to\infty}{\to} \mathbf{x}(t) \tag{2.40}$$

over any finite interval.

Proof. Let $\Delta > 0$. We have

$$\|\mathbf{x}_n(t+\Delta) - \mathbf{x}_n(t)\| \le \|L^\tau \int_{t+n}^{t+n+\Delta} \mathbf{h}(\overline{\mathbf{x}}_s)\, ds\| + \|\mathbf{q}_{t+n+\Delta}^0 - \mathbf{q}_{t+n}^0\| \tag{2.41}$$

It is clear that

$$E\left(\frac{\mathbf{q}_{n+1}}{\mathscr{F}_n}\right) = \sum_{i=0}^{n-1} a_i \underset{\sim}{\mathbf{L}}^\tau \mathbf{w}_{i+1} = \mathbf{q}_n$$

and $(\mathbf{q}_n, \mathscr{F}_n)$ is a martingale. Using the same method as with (2.17), we obtain

$$\sum_{n=2}^{\infty} E\left(\frac{\|\mathbf{q}_n - \mathbf{q}_{n-1}\|^2}{\mathscr{F}_{n-1}}\right) \le \|\underset{\sim}{\mathbf{L}}\|^2 \sum_{n=2}^{\infty} E\left(\frac{a_{n-1}^2\|\mathbf{w}_n\|^2}{\mathscr{F}_{n-1}}\right)$$

$$\le \|\underset{\sim}{\mathbf{L}}\|^2 \sum_{n=2}^{\infty} \frac{b_{n-1}^2}{z_{n-1}^2} r_1(n-1) r_2^2(\|\mathbf{x}_{n-1}\|)$$

$$\le \|L\|^2 \sum_{i=1}^{\infty} b_i^2 r_1(n) < \infty \text{ a.s.} \tag{2.42}$$

Thus $\mathbf{q}_n$ converges to a finite limit as $n \to \infty$ by Theorem 1.13. Hence for any $\varepsilon > 0$, there exists $T > 0$,

$$\|\mathbf{q}_{t+s}^0 - \mathbf{q}_t^0\| < \frac{\varepsilon}{2}, \qquad \forall s \ge 0, \tag{2.43}$$

if $t > T$. But for fixed T, if Δ is small enough, then

$$\|\mathbf{q}_{t+\Delta}^0 - \mathbf{q}_t^0\| < \frac{\varepsilon}{2}, \qquad \forall t \in [0, T] \tag{2.44}$$

since $\mathbf{q}_t^0$ is continuous. Hence by (2.43) and (2.44) for any n

$$\|\mathbf{q}_{t+n+\Delta}^0 - \mathbf{q}_{t+n}^0\| = I_{[t+n \ge T]}\|\mathbf{q}_{t+n+\Delta}^0 - \mathbf{q}_{t+n}^0\| + I_{[t+n<T]}\|\mathbf{q}_{t+n+\Delta}^0 - \mathbf{q}_{t+n}^0\|$$

$$< \frac{\varepsilon}{2} + \frac{\varepsilon}{2} = \varepsilon \tag{2.45}$$

and from (2.41) for small enough Δ, we have

$$\|\mathbf{x}_n(t+\Delta) - \mathbf{x}_n(t)\| \le \varepsilon + \|\underset{\sim}{\mathbf{L}}\| \left\| \int_0^{t_{m(t+n+\Delta)}} \mathbf{h}(\overline{\mathbf{x}}_s)\, ds - \int_0^{t_{m(t+n)}} \mathbf{h}(\overline{\mathbf{x}}_s)\, ds \right\|$$

$$+ \|\underset{\sim}{\mathbf{L}}\| \left\| \int_0^{t+n+\Delta} \mathbf{h}(\overline{\mathbf{x}}_s)\, ds - \int_0^{t_{m(t+n+\Delta)}} \mathbf{h}(\overline{\mathbf{x}}_s)\, ds \right\|$$

$$+ \|\underset{\sim}{\mathbf{L}}\| \left\| \int_0^{t+n} \mathbf{h}(\overline{\mathbf{x}}_s)\, ds - \int_0^{t_{m(t+n)}} \mathbf{h}(\overline{\mathbf{x}}_s)\, ds \right\|. \tag{2.46}$$

By Lemma 2.2, the second term on the right-hand side of (2.46) can be estimated as

$$\|\underset{\sim}{\mathbf{L}}\| \left\| \sum_{i=0}^{m(t+n+\Delta)-1} a_i \mathbf{h}(\mathbf{x}_i) - \sum_{i=0}^{m(t+n)-1} a_i \mathbf{h}(\mathbf{x}_i) \right\|$$

$$= \|\underset{\sim}{\mathbf{L}}\| \sum_{i=m(t+n)}^{m(t+n+\Delta)-1} a_i \mathbf{h}(\mathbf{x}_i) \leq \gamma \|\underset{\sim}{\mathbf{L}}\| \sum_{i=m(t+n)}^{m(t+n+\Delta)-1} a_i \leq \gamma \|\underset{\sim}{\mathbf{L}}\| \Delta, \qquad \forall n. \tag{2.47}$$

Since

$$\mathbf{h}(\bar{\mathbf{x}}_s) = \mathbf{h}(\mathbf{x}_{m(t+n+\Delta)})$$

for $s \in [t_{m(t+n+\Delta)}, t + n + \Delta]$, the third term on the right-hand side of (2.46) is equal to

$$\|\underset{\sim}{\mathbf{L}}\| \, \|\mathbf{h}(\mathbf{x}_{m(t+n+\Delta)})\| \big(t + n - \Delta - t_{m(t+n+\Delta)}\big)$$

$$\leq \|\underset{\sim}{\mathbf{L}}\| a_{m(t+n+\Delta)} \|\mathbf{h}(\mathbf{x}_{m(t+n+\Delta)})\| \leq \|\mathbf{L}\| \gamma a_{m(t+n+\Delta)}. \tag{2.48}$$

Similarly, the last term of (2.46) is estimated by

$$\|\underset{\sim}{\mathbf{L}}\| \left\| \int_0^{t+n} \mathbf{h}(\bar{\mathbf{x}}_s)\, ds - \int_0^{t_{m(t+n)}} \mathbf{h}(\bar{\mathbf{x}}_s)\, ds \right\| \leq \|\underset{\sim}{\mathbf{L}}\| \gamma a_{m(t+n)}. \tag{2.49}$$

Now we note that $m(t) \underset{t\to\infty}{\to} \infty$ since $\sum_{i=0}^{\infty} a_i = \infty$, and that $a_{m(t)} \underset{t\to\infty}{\to} 0$ since $\sum_{i=0}^{\infty} a_i^2 < \infty$ and $a_i \underset{i\to\infty}{\to} 0$.

Thus, if ω and t are fixed, then for any $\varepsilon > 0$, there exists an N such that the sum of the last two terms on the right-hand side of (2.46) is less than ε if $n > N$; that is, for any $\varepsilon > 0$, there exists an N such that

$$\|\mathbf{x}_n(t + \Delta) = \mathbf{x}_n(t)\| \leq \varepsilon + \gamma \|\underset{\sim}{\mathbf{L}}\| \Delta + \varepsilon, \qquad \forall n \geq N, \tag{2.50}$$

if Δ is sufficiently small.

Since $\mathbf{x}_n(t)$ is continuous, we have

$$\|\mathbf{x}_n(t + \Delta) - \mathbf{x}_n(t)\| < \varepsilon, \qquad \forall n \leq N \tag{2.51}$$

for ω and t fixed, for Δ sufficiently small.

For a given $\varepsilon > 0$, taking $\Delta \leq \varepsilon/\gamma\|\underset{\sim}{\mathbf{L}}\|$ and combining (2.50) and (2.51), we find that, for fixed t and ω,

$$\|\mathbf{x}_n(t+\Delta) - \mathbf{x}_n(t)\| \leq 3\varepsilon, \qquad \forall n \geq 0 \tag{2.52}$$

when Δ is sufficiently small.

For $\Delta < 0$, an estimate similar to (2.52) can be obtained if $|\Delta|$ is small enough. Hence for fixed ω, $\{\mathbf{x}_n(t)\}$ is equicontinuous; it is also uniformly bounded by Lemma 2.2. Hence the conclusion of the lemma follows from Theorem 2.2.

□

Lemma 2.4. *Under conditions of Theorem* 2.5, *the limiting function appearing in* (2.40) *satisfies the following differential equation*

$$\dot{\mathbf{x}}(t) = -\underset{\sim}{\mathbf{L}}^{\tau}\mathbf{h}(\mathbf{x}(t)) \tag{2.53}$$

Proof. In what follows, we assume $\Delta > 0$; for the case where $\Delta < 0$, the analysis is completely analogous.

We have

$$\begin{aligned}
\lim_{\Delta\to 0} & \frac{\mathbf{x}(t+s) - \mathbf{x}(t)}{\Delta} \\
&= \lim_{\Delta\to 0}\lim_{k\to\infty} \frac{\mathbf{x}_{n_k}(t+\Delta) - \mathbf{x}_{n_k}(t)}{\Delta} \\
&= \lim_{\Delta\to 0}\lim_{k\to\infty} \frac{\mathbf{x}_{t+n_k+\Delta} - \mathbf{x}_{t+n_k}}{\Delta} \\
&= -\lim_{\Delta\to 0}\frac{1}{\Delta}\lim_{k\to\infty}\left\{\underset{\sim}{\mathbf{L}}^{\tau}\int_{t+n_k}^{t+n_k+\Delta}\mathbf{h}(\overline{\mathbf{x}}_s)\,ds + \underset{\sim}{\mathbf{L}}^{\tau}\left(\mathbf{q}^0_{t+n_k+\Delta} - \mathbf{q}^0_{t+n_k}\right)\right\} \\
&= -\lim_{\Delta\to 0}\frac{1}{\Delta}\lim_{k\to\infty}\underset{\sim}{\mathbf{L}}^{\tau}\left[\left(\int_{t_{m(t+n_k)}}^{t_{m(t+n_k+\Delta)}}\mathbf{h}(\overline{\mathbf{x}}_s)\,ds\right.\right. \\
&\qquad \left. + \int_{t_{m(t+n_k+\Delta)}}^{t+n_k+\Delta}\mathbf{h}(\overline{\mathbf{x}}_s)\,ds - \int_{t_{m(t+n_k)}}^{t+n_k}\mathbf{h}(\overline{\mathbf{x}}_s)\,ds\right],
\end{aligned} \tag{2.54}$$

where we have used the fact that $\mathbf{q}^0_t$ goes to a finite limit as $t \to \infty$ and hence

$$\mathbf{q}^0_{t+n_k+\Delta} - \mathbf{q}^0_{t+n_k} \underset{k\to\infty}{\to} 0. \tag{2.55}$$

But from (2.48) and (2.49) we have

$$\left\| \int_{t_{m(t+n_k+\Delta)}}^{t+n_k+\Delta} \mathbf{h}(\bar{\mathbf{x}}_s)\, ds \right\| + \left\| \int_{t_{m(t+n_k)}}^{t+n_k} \mathbf{h}(\bar{\mathbf{x}}_s)\, ds \right\| \underset{k\to\infty}{\rightarrow} 0. \tag{2.56}$$

Hence

$$\begin{aligned} \lim_{\Delta\to 0} \frac{\mathbf{x}(t+\Delta) - \mathbf{x}(t)}{\Delta} &= -\lim_{\Delta\to 0} \frac{1}{\Delta} \lim_{k\to\infty} \underset{\sim}{\mathbf{L}}^{T} \int_{t_{m(t+n_k)}}^{t_{m(t+n_k+\Delta)}} \mathbf{h}(\bar{\mathbf{x}}_s)\, ds \\ &= -\lim_{\Delta\to 0} \frac{1}{\Delta} \lim_{k\to\infty} \underset{\sim}{\mathbf{L}}^{T} \sum_{i=m(t+n_k)}^{m(t+n_k+\Delta)} a_i \left[\mathbf{h}(\mathbf{x}_i) - \mathbf{h}(\mathbf{x}(t)) + \mathbf{h}(\mathbf{x}(t)) \right] \end{aligned} \tag{2.57}$$

Denote

$$\delta_{ki} = \sum_{j=m(t+n_k)}^{m(t+n_k)+i-1} a_j \tag{2.58}$$

Then we have

$$\delta_{ki} \le \Delta, \qquad \forall i \in \{1, 2, \ldots, m(t+n_k+\Delta) - m(t+n_k)\}$$

and

$$t + n_k - a_{m(t+n_k)} + \delta_{ki} < t_{m(t+n_k)} + \delta_{ki} = t_{m(t+n_k)+i} \le t + n_k + \delta_{ki}, \tag{2.59}$$

since

$$t_{m(t+n_k)} + a_{m(t+n_k)} > t + n_k.$$

Hence there exists $\delta'_{ki} \in [0, \Delta]$ such that

$$t_{m(t+n_k)+i} = n_k + t + \delta'_{ki},$$

$$\forall i \in \{0, 1, \ldots, m(t+n_k+\Delta) - m(t+n_k)\}. \tag{2.60}$$

Then, from (2.35), it follows that

$$\frac{1}{\Delta}\sum_{i=m(t+n_k)}^{m(t+n_k+\Delta)-1} a_i[\mathbf{h}(\mathbf{x}_i)-\mathbf{h}(\mathbf{x}(t))]$$

$$=\frac{1}{\Delta}\sum_{i=0}^{m(t+n_k+\Delta)-m(t+n_k)-1} a_{m(t+n_k)+i}\Big[\mathbf{h}\big(\mathbf{x}_{m(t+n_k)+i}\big)-\mathbf{h}(\mathbf{x}(t))\Big]$$

$$=\frac{1}{\Delta}\sum_{i=0}^{m(t+n_k+\Delta)-m(t+n_k)-1} a_{m(t+n_k)+i}\Big[\mathbf{h}\Big(\mathbf{x}^0_{t_{m(t+n_k)+i}}\Big)-\mathbf{h}(\mathbf{x}(t))\Big]$$

$$=\frac{1}{\Delta}\sum_{i=0}^{m(t+n_k+\Delta)-m(t+n_k)-1} a_{m(t+n_k)+i}\Big[\mathbf{h}\big(\mathbf{x}^0_{n_k+t+\delta'_{ki}}\big)-\mathbf{h}(\mathbf{x}(t))\Big]$$

$$+\frac{1}{\Delta}\sum_{i=0}^{m(t+n_k+\Delta)-m(t+n_k)-1} a_{m(t+n_k)+i}\Big[\mathbf{h}\big(\mathbf{x}_{n_k}(t+\delta'_{ki})\big)-\mathbf{h}(\mathbf{x}(t))\Big]$$

$$+\frac{1}{\Delta}\sum_{i=0}^{m(t+n_k+\Delta)-m(t+n_k)-1} a_{m(t+n_k)+i}\Big[\mathbf{h}\big(\mathbf{x}_{n_k}(t+\delta'_{ki})\big)-\mathbf{h}\big(\mathbf{x}_{n_k}(t)\big)\Big]$$

$$+\frac{1}{\Delta}\sum_{i=0}^{m(t+n_k+\Delta)-m(t+n_k)-1} a_{m(t+n_k)+i}\Big[\mathbf{h}\big(\mathbf{x}_{n_k}(t)\big)-\mathbf{h}(\mathbf{x}(t))\Big]. \tag{2.61}$$

For any $\varepsilon > 0$, if Δ is small enough, the following estimate holds because $\mathbf{x}_n(t)$ is equicontinuous

$$\Big\|\mathbf{h}\big(\mathbf{x}_{n_k}(t+\delta'_{ki})\big)-\mathbf{h}\big(\mathbf{x}_{n_k}(t)\big)\Big\| \le \varepsilon, \qquad \forall k.$$

Consequently the first term on the right-hand side of (2.61) is dominated by

$$\frac{\varepsilon}{\Delta}\sum_{i=0}^{m(t+n_k+\Delta)-m(t+n_k)-1} a_{m(t+n_k)+i}=\frac{\varepsilon}{\Delta}\sum_{i=m(t+n_k)}^{m(t+n_k+\Delta)-1} a_i \le \varepsilon \underset{\Delta\to 0}{\to} 0,$$

while the second term tends to zero as $k\to\infty$ since $\mathbf{x}_{n_k}(t)\to\mathbf{x}(t)$ and $\mathbf{h}(\cdot)$ is continuous. Hence from (2.61) it follows that

$$\lim_{\Delta\to 0}\lim_{k\to\infty}\frac{1}{\Delta}\sum_{i=m(t+n_k)}^{m(t+n_k+\Delta)-1} a_i[\mathbf{h}(\mathbf{x}_i)-\mathbf{h}(\mathbf{x}(t))]=0 \tag{2.62}$$

and from (2.57) and (2.62)

$$\lim_{\Delta \to 0} \frac{\mathbf{x}(t+\Delta) - \mathbf{x}(t)}{\Delta} = - \lim_{\Delta \to 0} \lim_{k \to \infty} \frac{1}{\Delta} \underset{\sim}{\mathbf{L}}^{\tau} \sum_{i=m(t+n_k)}^{m(t+n_k+\Delta)-1} a_i \mathbf{h}(\mathbf{x}(t)) \tag{2.63}$$

which gives (2.53) since $a_i \underset{i \to \infty}{\to} 0$ and $n_k \underset{k \to \infty}{\to} \infty$ implies

$$\sum_{i=m(t+n_k)}^{m(t+n_k+\Delta)-1} a_i \underset{k \to \infty}{\to} \Delta. \qquad \square$$

Proof of Theorem 2.5. By Condition A_2, $V(\mathbf{x})$ is the Lyapunov function of (2.53), hence

$$\mathbf{x}(t) \underset{t \to \infty}{\to} \mathbf{x}^0 \text{ a.s.}$$

by Theorem 2.4.

For a fixed ω and $\varepsilon > 0$, we can find t_1 such that

$$\|\mathbf{x}(t) - \mathbf{x}^0\| < \varepsilon \quad \text{for} \quad t \geq t_1.$$

By virtue of the uniform convergence $\mathbf{x}_{n_k}(t) \to \mathbf{x}(t)$ in any finite interval, we have

$$\|\mathbf{x}_{n_k}(t) - \mathbf{x}(t)\| < \varepsilon, \qquad \forall t \in [t_1, 2t_1] \tag{2.64}$$

or, equivalently,

$$\|\mathbf{x}^0_{t+n_k} - \mathbf{x}(t)\| < \varepsilon, \qquad \forall t \in [t_1, 2t_1] \tag{2.65}$$

for sufficiently large k, say $k \geq K$. Since $t_n \underset{n \to \infty}{\to} \infty$ and $a_n \underset{n \to \infty}{\to} 0$, we can find

$$t_{m_k} \in [n_k + t_1, n_k + 2t_1]$$

or, equivalently,

$$t_{m_k} = n_k + s, \qquad s \in [t_1, 2t_1]$$

if k is chosen large enough. Then from (2.65) we have

$$\|\mathbf{x}^0_{t_{m_k}} - \mathbf{x}(s)\| = \|\mathbf{x}^0_{n_k+s} - \mathbf{x}(s)\| < \varepsilon. \tag{2.66}$$

Notice $\mathbf{x}_{t_{m_k}} = \mathbf{x}_{m_k}$. Then (2.64) and (2.66) yield

$$\|\mathbf{x}_{m_k} - \mathbf{x}^0\| \le \|\mathbf{x}_{m_k} - \mathbf{x}(s)\| + \|\mathbf{x}(s) - \mathbf{x}^0\|$$

$$= \|\mathbf{x}^0_{t_{m_k}} - \mathbf{x}(s)\| + \|\mathbf{x}(s) - \mathbf{x}^0\| < 2\varepsilon,$$

and this means that for any ω, with the aforementioned possible exception, we can select subsequence $\mathbf{x}_{m_k} \underset{k\to\infty}{\to} \mathbf{x}^0$. But by (2.25),

$$V(\mathbf{x}_n) \underset{n\to\infty}{\to} \alpha.$$

Hence

$$\lim_{n\to\infty} V(\mathbf{x}_n) = \lim_{k\to\infty} V(\mathbf{x}_{m_k}) = V(\mathbf{x}^0) = 0.$$

From here we can conclude that $\mathbf{x}_n \underset{n\to\infty}{\to} \mathbf{x}^0$ a.s., otherwise for fixed ω we could select subsequence $\mathbf{x}_{l_k} \underset{k\to\infty}{\to} \mathbf{x}^1 \ne \mathbf{x}^0$, but $\mathbf{x}^0$ is the unique zero of $V(\mathbf{x})$, hence we would have

$$\lim_{k\to\infty} V(\mathbf{x}_{l_k}) = V(\mathbf{x}^1) > 0,$$

which contradicts

$$\lim_{k\to\infty} V(\mathbf{x}_{l_k}) = \lim_{n\to\infty} V(\mathbf{x}_n) = 0. \qquad \square$$

Remark 1. If we take

$$V(\mathbf{x}) = (\mathbf{x} - \mathbf{x}^0)^\tau \underset{\sim}{C}(\mathbf{x} - \mathbf{x}^0), \qquad \underset{\sim}{C} > 0,$$

then Condition A_2 requires

$$\mathbf{h}^\tau(\mathbf{x})\underset{\sim}{L}\underset{\sim}{C}(\mathbf{x} - \mathbf{x}^0) > 0, \qquad \mathbf{x} \ne \mathbf{x}^0.$$

Remark 2. If in Condition A_2 the converse inequality holds, that is,

$$\mathbf{h}^\tau(\mathbf{x})\underset{\sim}{L}V_{\mathbf{x}}(\mathbf{x}) < 0, \qquad \mathbf{x} \ne \mathbf{x}^0,$$

then for $-\mathbf{h}(\mathbf{x})$ the original Condition A_2 is verified. Hence if (2.14) is replaced by

$$\mathbf{x}_{i+1} = \mathbf{x}_i + a_i \mathbf{y}_{i+1},$$

then Theorem 2.5 remains valid.

2.3. FINDING THE ZERO OF A REGRESSION FUNCTION: CORRELATED NOISE CASE

In Section 2.2, using a method combining probability theory and differential equation theory, we have considered the problem of finding a zero of a regression function in the uncorrelated measurement error case. In this section, those results will be extended to the case of correlated measurement error.

We retain all the notation adapted in the preceding section, but the measurement error $\boldsymbol{\varepsilon}_{i+1}$ (2.10) will now be an ARMA process satisfying a stochastic difference equation as follows

$$\begin{aligned}\boldsymbol{\varepsilon}_n + \underset{\sim}{\mathbf{D}}_1\boldsymbol{\varepsilon}_{n-1} + \underset{\sim}{\mathbf{D}}_2\boldsymbol{\varepsilon}_{n-2} + \cdots + \underset{\sim}{\mathbf{D}}_d\boldsymbol{\varepsilon}_{n-d} \\ = \mathbf{w}_n(\mathbf{x}_{n-1}, \mathbf{e}_n) + \underset{\sim}{\mathbf{C}}_1\mathbf{w}_{n-1}(\mathbf{x}_{n-2}, \mathbf{e}_{n-1}) + \cdots \\ + \underset{\sim}{\mathbf{C}}_r\mathbf{w}_{n-r}(\mathbf{x}_{n-r-1}, \mathbf{e}_{n-r}), r \geq 0, d \geq 0 \end{aligned} \tag{2.67}$$

with initial values $\boldsymbol{\varepsilon}_j = 0$, $j \leq 0$, $\mathbf{w}_i(\mathbf{x}_{i-1}, \mathbf{e}) = 0$, $i \leq 0$. If $r = d = 0$, then (2.67) coincide with (2.10).

Denote

$$\underset{\sim}{\mathbf{C}}(z) = \underset{\sim}{\mathbf{I}} + \underset{\sim}{\mathbf{C}}_1 z + \cdots + \underset{\sim}{\mathbf{C}}_r z^r, \qquad \underset{\sim}{\mathbf{D}}(z) = \underset{\sim}{\mathbf{I}} + \underset{\sim}{\mathbf{D}}_1 z + \cdots + \underset{\sim}{\mathbf{D}}_d z^d. \tag{2.68}$$

Then (2.67) can be rewritten as

$$\underset{\sim}{\mathbf{D}}(z)\boldsymbol{\varepsilon}_n = \underset{\sim}{\mathbf{C}}(z)\mathbf{w}_n, \tag{2.69}$$

where here and henceforth z denotes the shift-back operator and $\mathbf{w}_i = \mathbf{w}_i(\mathbf{x}_{i-1}, \mathbf{e}_i)$.

In addition to (2.13), we shall also require the sequence b_i to be monotonically nonincreasing. In the present case, we have to modify Condition A_2 on the noise to Condition A_3. Let $\mathbf{x}_n$ and $\mathbf{f}_n$ be the input and output of the following difference equation:

$$\underset{\sim}{\mathbf{C}}(z)\mathbf{f}_{n+1} = \underset{\sim}{\mathbf{D}}(z)\mathbf{h}(\mathbf{x}_n). \tag{2.70}$$

CONDITION A_3. There exists a function $V(\mathbf{x})$ with bounded continuous second derivative such that

$$V(\mathbf{x}) > 0, \quad \forall \mathbf{x} \neq \mathbf{x}^0, \qquad V(\mathbf{x}^0) = 0, \quad V(\mathbf{x}) \underset{\|\mathbf{x}\| \to \infty}{\to} \infty \tag{2.71}$$

and such that for some $l \times m$ matrix $\underset{\sim}{\mathbf{L}}$ and some number n_0

$$\mathbf{f}_{i+1}^{\tau}\underset{\sim}{\mathbf{L}}V_{\mathbf{x}}(\mathbf{x}_i) > 0, \qquad \forall \mathbf{x}_i \neq \mathbf{x}^0, \quad i \geq n_0 \tag{2.72}$$

where $\mathbf{f}_{i+1}$ is defined from (2.70) with arbitrary initial values $\mathbf{f}_0, \ldots, \mathbf{f}_{r-1}$ with both $\mathbf{x}_i \equiv \mathbf{x}$ and $\mathbf{x}_i$ defined by (2.79).

It is obvious that if $d = r = 0$, then $\mathbf{f}_{i+1} = \mathbf{h}(\mathbf{x}_i)$ and Condition A_3 is equivalent to A_2.

Denote

$$\underset{\sim}{\mathbf{F}} = \begin{cases} \begin{bmatrix} -\underset{\sim}{C}_1 & \underset{\sim}{\mathbf{I}} & & \underset{\sim}{\mathbf{0}} \cdots \underset{\sim}{\mathbf{0}} \\ -\underset{\sim}{C}_2 & \underset{\sim}{\mathbf{0}} & \ddots & \underset{\sim}{\mathbf{0}} \\ \vdots & \vdots & \ddots & \underset{\sim}{\mathbf{I}} \\ -\underset{\sim}{C}_r & \underset{\sim}{\mathbf{0}} & \cdots & \underset{\sim}{\mathbf{0}} \end{bmatrix} & \text{for } r > 0 \\ \underbrace{[\mathbf{0}]}_{m}\}\, m & \\ \text{with } \underset{\sim}{\mathbf{F}}^0 = \underset{\sim}{\mathbf{I}} \text{ for } r = 0 & \end{cases} \tag{2.73}$$

$$\underset{\sim}{\mathbf{G}} = \begin{cases} \underbrace{[\underset{\sim}{\mathbf{I}}\underset{\sim}{\mathbf{0}} \cdots \underset{\sim}{\mathbf{0}}]}_{mr}\}\, m & \text{for } r > 0 \\ \underset{\sim}{\mathbf{I}} & \text{for } r = 0. \end{cases} \tag{2.74}$$

We recall that $\underset{\sim}{\mathbf{I}}$ denotes an identity matrix of appropriate dimension.

Now we modify algorithm (2.14) to suit the present case of correlated measurement error. Instead of $\mathbf{y}_n$, we shall use ζ_n produced by $\{\mathbf{y}_n\}$, that is, ζ_n is defined recursively by

$$\zeta_{n+1} = \underset{\sim}{\mathbf{F}}\zeta_n + \underset{\sim}{\mathbf{G}}^{\tau}\mathbf{u}_{n+1}, \qquad n \geq 0, \tag{2.75}$$

with ζ_0 an arbitrary $m(r \vee 1)$-dimensional deterministic vector with

$$\mathbf{u}_n = \underset{\sim}{\mathbf{D}}(z)\mathbf{y}_n, \qquad \mathbf{y}_i = 0, \qquad i \leq 0, \tag{2.76}$$

where we agree that

$$\mathbf{h}(\mathbf{x}_i) = 0, \qquad i < 0.$$

Denote

$$z_n = \max_{0 \le i \le n} r_2(\|\mathbf{x}_i\|) \tag{2.77}$$

$$a_0 = b_0, \qquad a_i = \frac{b_i}{z_i}. \tag{2.78}$$

Notice that a_i is monotonely nonincreasing since b_i is.

Given any deterministic $\mathbf{x}_0$, we define the MRM procedure as follows

$$\mathbf{x}_{n+1} = \mathbf{x}_n - a_n \underset{\sim}{\mathbf{L}}^\tau \underset{\sim}{\mathbf{G}} \zeta_{n+1} \tag{2.79}$$

or

$$\mathbf{x}_{n+1} = \mathbf{x}_n + a_n \underset{\sim}{\mathbf{L}}^\tau \underset{\sim}{\mathbf{G}} \zeta_{n+1} \tag{2.80}$$

if the converse inequality holds in (2.72) (Condition A_3).

For the special case when $r = 0$, $d = 0$, then $\underset{\sim}{\mathbf{G}} = \underset{\sim}{\mathbf{I}}$, $\mathbf{u}_n = \mathbf{y}_n$, $\zeta_{n+1} = \mathbf{y}_{n+1}$, and (2.79) reduces to (2.14).

Lemma 2.5. *If $r > 0$ and $\lambda_i \neq 0$, $i = 1, \ldots, rm$ are zeros of* $\det \underset{\sim}{\mathbf{C}}(\lambda)$, *then $1/\lambda_i$, $i = 1, \ldots, rm$ are eigenvalues of $\underset{\sim}{\mathbf{F}}$, and*

$$\underset{\sim}{\mathbf{F}}^{-1} = \begin{bmatrix} \underset{\sim}{\mathbf{0}} & \cdots & \underset{\sim}{\mathbf{0}} & -\underset{\sim}{\mathbf{C}}^{-1} \\ \underset{\sim}{\mathbf{I}} & & & -\underset{\sim}{\mathbf{C}}_1 \underset{\sim}{\mathbf{C}}_r^{-1} \\ \mathbf{0} & \ddots & \vdots & \vdots \\ \vdots & & \mathbf{0} & \vdots \\ \underset{\sim}{\mathbf{0}} & \cdots & \mathbf{0I} & -\underset{\sim}{\mathbf{C}}_{r-1} \underset{\sim}{\mathbf{C}}_r^{-1} \end{bmatrix}. \tag{2.81}$$

If $r = 0$ or $r > 0$, but $|\lambda_i| > 1$, $\forall i$, then there exist constants k_2 and $\rho \in (0,1)$ such that

$$\|\underset{\sim}{\mathbf{F}}^k\| < k_2 \rho^k, \qquad \forall k = 0, 1, \ldots . \tag{2.82}$$

Proof. By use of the expression for the determinant of 2×2 block matrix,

$$\det \begin{bmatrix} \underset{\sim}{\mathbf{R}} & \underset{\sim}{\mathbf{S}} \\ \underset{\sim}{\mathbf{T}} & \underset{\sim}{\mathbf{Q}} \end{bmatrix} = \det \underset{\sim}{\mathbf{Q}} \det\left(\underset{\sim}{\mathbf{R}} - \underset{\sim}{\mathbf{S}} \underset{\sim}{\mathbf{Q}}^{-1} \underset{\sim}{\mathbf{T}}\right), \tag{2.83}$$

we have

$$\det(\lambda \mathbf{I} - \mathbf{F}) = \begin{bmatrix} \lambda\mathbf{I} + \mathbf{C}_1 & -\mathbf{I} & \mathbf{0} & \cdots & \mathbf{0} \\ & & \ddots & \ddots & \vdots \\ \mathbf{C}_2 & \lambda\mathbf{I} & & & \\ & \mathbf{0} & \ddots & \ddots & \mathbf{0} \\ \vdots & \vdots & & \ddots & -\mathbf{I} \\ \mathbf{C}_r & \mathbf{0} & \mathbf{0}\cdots & \mathbf{0} & \lambda\mathbf{I} \end{bmatrix}$$

$$= \det(\lambda\mathbf{I})\det\left\{ \begin{bmatrix} \lambda\mathbf{I} + \mathbf{C}_1 & -\mathbf{I} & \mathbf{0} & \cdots & \mathbf{0} \\ \mathbf{C}_2 & \lambda\mathbf{I} & \ddots & \ddots & \vdots \\ \vdots & \mathbf{0} & \ddots & \ddots & \mathbf{0} \\ \vdots & \vdots & & \ddots & -\mathbf{I} \\ \mathbf{C}_{r-1} & \mathbf{0} & \cdots & \mathbf{0} & \lambda\mathbf{I} \end{bmatrix} - \frac{1}{\lambda} \begin{bmatrix} \mathbf{0} \\ \vdots \\ \vdots \\ \mathbf{0} \\ -\mathbf{I} \end{bmatrix} [\mathbf{C}_r \mathbf{0} \cdots \mathbf{0}] \right\}$$

$$= \det(\lambda\mathbf{I})\det \begin{bmatrix} \lambda\mathbf{I} + \mathbf{C}_1 & -\mathbf{I} & \mathbf{0} & \cdots & \mathbf{0} \\ & & & \ddots & \vdots \\ \mathbf{C}_2 & \lambda\mathbf{I} & -\mathbf{I} & & \\ & \mathbf{0} & \ddots & \ddots & \mathbf{0} \\ \vdots & \vdots & \ddots & \ddots & -\mathbf{I} \\ \mathbf{C}_{r-1} + \frac{1}{\lambda}\mathbf{C}_r & \mathbf{0} & \cdots & \mathbf{0} & \lambda\mathbf{I} \end{bmatrix}$$

$$= \det(\lambda\mathbf{I})^{r-1}\det\left(\lambda\mathbf{I} + \mathbf{C}_1 + \frac{1}{\lambda}\mathbf{C}_2 + \cdots + \frac{1}{\lambda^{r-1}}\mathbf{C}_r\right)$$

(repeating the first step)

$$= \lambda^{rm}\det\left(\mathbf{I} + \lambda^{-1}\mathbf{C}_1 + \cdots + \lambda^{-r}\mathbf{C}_r\right) = \lambda^{rm}\det \mathbf{C}(\lambda^{-1}).$$

Hence the eigenvalues of $\mathbf{F}$ coincide with the reciprocals of zeros of $\det \mathbf{C}(\lambda)$. Since all $\lambda_i \neq 0$, $\mathbf{F}$ is degenerate, hence $\mathbf{C}_r$ is degenerate and the right-hand side of (2.81) is meaningful. Then (2.81) is verified directly.

If $r = 0$, then by definition $\mathbf{F} = 0$, $\mathbf{F}^0 = \mathbf{I}$, and (2.82) is trivial. For $r > 0$, there exists $\rho_1 \in (0,1)$ such that $|\lambda_i| > 1/\rho_1$, $\forall i = 1, \dots, rm$ since $|\lambda_i| > 1$, and so $|1/\lambda_i| < \rho_1$. By use of the Jordan canonical form, it is clear that we can find a constant k_2' such that

$$\|\mathbf{F}^k\| \leq k_2' k^{rm-1}\rho_1^k, \qquad \forall k.$$

Since $k^{rm-1}\rho_1^{k/2} \underset{k\to\infty}{\to} 0$ we conclude that (2.82) is true if we take $\rho = \rho_1^{1/2}$ and take k_2 sufficiently large. □

Let us express ζ_n defined by (2.76) in vector components as

$$\zeta_n = \left[\zeta_n^{1\tau} \cdots \zeta_n^{r\tau}\right]^\tau, \tag{2.84}$$

where ζ_n^i, $i = 1, \ldots, r$ are all m-dimensional.

Lemma 2.6. *$\zeta_n^1 = \underset{\sim}{\mathbf{G}}\zeta_n$ satisfies the following difference equation:*

$$\underset{\sim}{\mathbf{C}}(z)\zeta_{n+1}^1 = \underset{\sim}{\mathbf{D}}(z)\mathbf{y}_{n+1} \tag{2.85}$$

Proof. For $r = 0$, the lemma is trivial. Let $r > 0$. From (2.73)–(2.75), we have

$$\begin{bmatrix} \zeta_{n+1}^1 \\ \vdots \\ \zeta_{n+1}^r \end{bmatrix} = \begin{bmatrix} -\underset{\sim}{\mathbf{C}}_1 & \mathbf{I} & \underset{\sim}{\mathbf{0}} & \cdots & \underset{\sim}{\mathbf{0}} \\ \vdots & \underset{\sim}{\mathbf{0}} & & \ddots & \vdots \\ & \vdots & & & \underset{\sim}{\mathbf{0}} \\ \cdot & \cdot & & & \mathbf{I} \\ -\underset{\sim}{\mathbf{C}}_r & \underset{\sim}{\mathbf{0}} & & \cdots & \underset{\sim}{\mathbf{0}} \end{bmatrix} \begin{bmatrix} \zeta_n^1 \\ \vdots \\ \zeta_n^r \end{bmatrix} + \begin{bmatrix} \mathbf{u}_{n+1} \\ \mathbf{0} \\ \vdots \\ \mathbf{0} \end{bmatrix}, \tag{2.86}$$

and

$$\left.\begin{aligned} \zeta_{n+1}^1 &= -\underset{\sim}{\mathbf{C}}_1\zeta_n^1 + \zeta_n^2 + \mathbf{u}_{n+1}, \\ \zeta_{n+1}^i &= -\underset{\sim}{\mathbf{C}}_i\zeta_n^1 + \zeta_n^{i+1}, \qquad i = 2, \ldots, r-1 \\ \zeta_{n+1}^r &= -\underset{\sim}{\mathbf{C}}_r\zeta_n^1 \end{aligned}\right\}. \tag{2.87}$$

From here, we obtain

$$\underset{\sim}{\mathbf{C}}(z)\zeta_{n+1}^1 = \mathbf{u}_{n+1}$$

and hence (2.85). □

From the measurement equation (2.2), we have

$$\underset{\sim}{\mathbf{D}}(z)\mathbf{y}_{n+1} = \underset{\sim}{\mathbf{D}}(z)\mathbf{h}(\mathbf{x}_n) + \underset{\sim}{\mathbf{D}}(z)\boldsymbol{\varepsilon}_{n+1}$$

and by (2.69)

$$\underset{\sim}{\mathbf{D}}(z)\mathbf{y}_{n+1} = \underset{\sim}{\mathbf{D}}(z)\mathbf{h}(\mathbf{x}_n) + \underset{\sim}{\mathbf{C}}(z)\mathbf{w}_{n+1}. \tag{2.88}$$

Substituting (2.88) into (2.85), we find

$$\underset{\sim}{\mathbf{C}}(z)(\boldsymbol{\zeta}_{n+1}^{1} - \mathbf{w}_{n+1}) = \underset{\sim}{\mathbf{D}}(z)\mathbf{h}(\mathbf{x}_n). \tag{2.89}$$

Define the $m(r \vee 1)$-dimensional vector $\boldsymbol{\xi}_n^{\tau} = [\boldsymbol{\xi}_n^{1\tau} \cdots \boldsymbol{\xi}_n^{(r\vee 1)\tau}]$ with $\boldsymbol{\xi}_n^{i}$ m-dimensional, $i = 1, \ldots, (r \vee 1)$, as follows:

$$\underset{\sim}{\mathbf{G}}\boldsymbol{\xi}_i = \boldsymbol{\xi}_i^{1} = \underset{\sim}{\mathbf{G}}\boldsymbol{\zeta}_i - \mathbf{w}_i, \qquad i = 0, 1, \ldots, r-1 \tag{2.90}$$

$$\boldsymbol{\xi}_{n+1} = \mathbf{F}\boldsymbol{\xi}_n + \underset{\sim}{\mathbf{G}}^{\tau}\underset{\sim}{\mathbf{D}}(z)\mathbf{h}(\mathbf{x}_n), \qquad n \geq 0. \tag{2.91}$$

From (2.91), we see

$$\boldsymbol{\xi}_i^{2} = \boldsymbol{\xi}_{i+1}^{1} + \underset{\sim}{\mathbf{C}}_i\boldsymbol{\xi}_i^{1} - \underset{\sim}{\mathbf{D}}(z)\mathbf{h}(\mathbf{x}_i), \qquad i = 0, 1, \ldots, r-2 \tag{2.92}$$

$$\boldsymbol{\xi}_i^{j} = \boldsymbol{\xi}_{i+1}^{j-1} + \underset{\sim}{\mathbf{C}}_{j-1}\boldsymbol{\xi}_i^{1}, \qquad i = 0, 1, \ldots, r-j, \quad j = 3, 4, \ldots, r. \tag{2.93}$$

Having obtained $\boldsymbol{\xi}_i^{2}$, $i = 0, 1, \ldots, r-2$ from (2.92), we find

$$\boldsymbol{\xi}_i^{3} = \boldsymbol{\xi}_{i+1}^{2} + \underset{\sim}{\mathbf{C}}_2\boldsymbol{\xi}_i^{1}, \qquad i = 0, 1, \ldots, r-3$$

by (2.93), and hence we can define $\boldsymbol{\xi}_i^{j}$, $i = 0, 1, \ldots, r-j$ for any $j \leq r$. Thus the initial value $\boldsymbol{\xi}_0$ of (2.91) is completely determined by (2.90).

By comparing (2.75) with (2.91), it follows from Lemma 2.6 that

$$\underset{\sim}{\mathbf{C}}(z)\boldsymbol{\xi}_{n+1}^{1} = \underset{\sim}{\mathbf{D}}(z)\mathbf{h}(\mathbf{x}_n) \tag{2.94}$$

which has, as can be seen from (2.90), the same initial value as (2.89). Hence we have $\boldsymbol{\xi}_{n+1}^{1} = \boldsymbol{\zeta}_{n+1}^{1} - \mathbf{w}_{n+1}$, or equivalently,

$$\underset{\sim}{\mathbf{G}}\boldsymbol{\zeta}_{n+1} = \underset{\sim}{\mathbf{G}}\boldsymbol{\xi}_{n+1} + \mathbf{w}_{n+1}. \tag{2.95}$$

Now algorithm (2.79) can be rewritten as

$$\mathbf{x}_{n+1} = \mathbf{x}_n - a_n\underset{\sim}{\mathbf{L}}^{\tau}(\mathbf{f}_{n+1} + \mathbf{w}_{n+1}), \tag{2.96}$$

where

$$\mathbf{f}_{n+1} = \underset{\sim}{\mathbf{G}}\boldsymbol{\xi}_{n+1}. \tag{2.97}$$

From (2.91), we have

$$\xi_{n+1} = (\xi_{n+1} - \underset{\sim}{F}\xi_n) + (\underset{\sim}{F}\xi_n - \underset{\sim}{F}^2\xi_{n-1}) + \cdots$$

$$+ (\underset{\sim}{F}^n\xi_1 - \underset{\sim}{F}^{n+1}\xi_0) + \underset{\sim}{F}^{n+1}\xi_0$$

$$= \sum_{i=0}^{n} \underset{\sim}{F}^i(\xi_{n-i+1} - \underset{\sim}{F}\xi_{n-i}) + \underset{\sim}{F}^{n+1}\xi_0$$

$$= \sum_{j=0}^{n} \underset{\sim}{F}^{n-j}(\xi_{j+1} - \underset{\sim}{F}\xi_j) + \underset{\sim}{F}^{n+1}\xi_0$$

$$= \sum_{j=0}^{n} \underset{\sim}{F}^{n-j}\underset{\sim}{G}^{\tau}\underset{\sim}{D}(z)\mathbf{h}(\mathbf{x}_j) + \underset{\sim}{F}^{n+1}\xi_0 \tag{2.98}$$

and

$$\mathbf{f}_{n+1} = \underset{\sim}{G}\sum_{j=0}^{n} \underset{\sim}{F}^{n-j}\underset{\sim}{G}^{\tau}\underset{\sim}{D}(z)\mathbf{h}(\mathbf{x}_j) + \underset{\sim}{G}\underset{\sim}{F}^{n+1}\xi_0. \tag{2.99}$$

Now we proceed to prove the convergence of the MRM procedure in a similar way to that used for Theorem 2.5.

Lemma 2.7. *Suppose that $r = 0$ or $r > 0$ but all zeros of* $\det \underset{\sim}{C}(\lambda)$ *lie outside the closed unit disk. Then*

$$E\|\mathbf{x}_i\|^2 < \infty, \quad EV(\mathbf{x}_i) < \infty, \qquad \forall i,$$

$$E\sum_{i=0}^{\infty} a_i^2\|\mathbf{w}_{i+1}\|^2 < \infty, \qquad E\sum_{i=0}^{\infty} a_i^2\|\mathbf{f}_{i+1}\|^2 < \infty.$$

Proof. Here we have

$$a_i = \frac{b_i}{\max_{0\le j\le i} r_2(\|\mathbf{x}_j\|)},$$

instead of $a_i = b_i/r_2(\|\mathbf{x}_i\|)$ as in Lemma 2.1, and

$$E\sum_{i=0}^{\infty} a_i^2\|\mathbf{w}_{i+1}\|^2 < \infty$$

follows directly from (2.17).

Now for any $i \ge 0$, $0 \le s \le i$,

$$Ea_i^2\|\mathbf{w}_{s+1}\|^2 \le Ea_i^2 r_1(s) r_2^2(\|\mathbf{x}_s\|) \le Eb_i^2 r_1(s) \le Eb_i^2 r_1(i) < \infty$$

and $\boldsymbol{\varepsilon}_0 = \mathbf{w}_0$, $Ea_i^2\|\boldsymbol{\varepsilon}_0\|^2 < \infty$, so the summability property

$$Ea_i^2\|\boldsymbol{\varepsilon}_{s+1}\|^2 < \infty, \qquad 0 \le s \le i$$

is easily seen by induction from

$$Ea_i^2\|\boldsymbol{\varepsilon}_{s+1}\|^2 \le Ea_i^2[\|\underset{\sim}{\mathbf{D}}_1\|\,\|\boldsymbol{\varepsilon}_s\| + \cdots + \|\underset{\sim}{\mathbf{D}}_d\|\,\|\boldsymbol{\varepsilon}_{s-d+1}\| + \|\mathbf{w}_{s+1}\| + \|\underset{\sim}{\mathbf{C}}_1\|\,\|\mathbf{w}_s\| + \cdots + \|\underset{\sim}{\mathbf{C}}_r\|\,\|\mathbf{w}_{s-r+1}\|]^2. \qquad (2.100)$$

Hence we have

$$Ea_i^2\|\mathbf{y}_{s+1}\|^2 \le 2Ea_i^2\left(\|\mathbf{h}(\mathbf{x}_s)\|^2 + \|\boldsymbol{\varepsilon}_{s+1}\|^2\right)$$
$$\le 2Ea_i^2\|\boldsymbol{\varepsilon}_{s+1}\|^2 + 2E\left[\frac{b_i}{\max\limits_{0\le j\le i} r_2^2\left(\|\mathbf{x}_j\|\right)} r_2^2\left(\|\mathbf{x}_s\|\right)\right] < \infty \qquad (2.101)$$

and from here by (2.76)

$$Ea_i^2\|\mathbf{u}_{s+1}\|^2 < \infty. \qquad (2.102)$$

Finally, by (2.102) and (2.75), we have

$$Ea_i^2\|\boldsymbol{\zeta}_{s+1}\|^2 < \infty, \qquad \forall i \ge 0, \quad 0 \le s \le i. \qquad (2.103)$$

From (2.79) we obtain the trivial estimate

$$E\|\mathbf{x}_{n+1}\|^2 \le 2E\|\mathbf{x}_n\|^2 + 2\|\mathbf{L}\|^2 Ea_n^2\|\boldsymbol{\zeta}_{n+1}\|^2 \qquad (2.104)$$

from which by induction and (2.103) it follows that

$$E\|\mathbf{x}_i\|^2 < \infty, \qquad \forall i.$$

Equalities (2.90)–(2.93) show that components of $\boldsymbol{\xi}_0$ consist of linear combinations of $\underset{\sim}{\mathbf{G}}\boldsymbol{\zeta}_i$, $\mathbf{w}_i$, and $\underset{\sim}{\mathbf{D}}(z)\mathbf{h}(\mathbf{x}_i)$, $i = 0, 1, \ldots, r-1$. But from (2.12), we have

$$Ea_r^2\|\mathbf{h}(\mathbf{x}_s)\|^2 \le \frac{Eb_r^2 r_2^2(\|\mathbf{x}_s\|)}{z_r^2} \le Eb_r^2 < \infty, \qquad \forall s \le r.$$

Hence

$$Ea_r^2\|\underset{\sim}{\mathbf{D}}(z)\mathbf{h}(\mathbf{x}_s)\|^2 < \infty \qquad (2.105)$$

when we agree to set $\mathbf{h}(\mathbf{x}_i) = 0$, $i < 0$. Thus it follows that for $i \geq r$

$$Ea_i^2 \|\boldsymbol{\xi}_0\|^2 \leq Ea_r^2 \|\boldsymbol{\xi}_0\|^2 < \infty \tag{2.106}$$

by (2.101), (2.103), (2.105), and by the nonincreasing property of the a_i sequence. Hence from (2.95), (2.101), and (2.103), we know that

$$Ea_i^2 \|\mathbf{f}_{i+1}\|^2 \leq 2Ea_i^2 \|\boldsymbol{\zeta}_{i+1}\|^2 + 2Ea_i^2 \|\mathbf{w}_{i+1}\|^2 < \infty, \qquad \forall i, \tag{2.107}$$

and from (2.82) and (2.99) that

$$\begin{aligned}
E \sum_{i=0}^{\infty} a_i^2 \|\mathbf{f}_{i+1}\|^2
&\leq \sum_{i=0}^{r-1} Ea_i^2 \|\mathbf{f}_{i+1}\|^2 + E \sum_{i=r}^{\infty} a_i^2 \|\mathbf{f}_{i+1}\|^2 \\
&\leq \sum_{i=0}^{r-1} Ea_i^2 \|\mathbf{f}_{i+1}\|^2 + 2E \sum_{i=r}^{\infty} a_i^2 \left\| \underset{\sim}{\mathbf{G}} \sum_{j=0}^{i} \underset{\sim}{\mathbf{F}}^{i-j} \underset{\sim}{\mathbf{G}}^{\tau} \underset{\sim}{\mathbf{D}}(z) \mathbf{h}(\mathbf{x}_j) \right\|^2 \\
&\quad + 2E \sum_{i=r}^{\infty} a_i^2 \| \underset{\sim}{\mathbf{G}} \underset{\sim}{\mathbf{F}}^{i+1} \boldsymbol{\xi}_0 \|^2, \\
&\leq \sum_{i=0}^{r-1} Ea_i^2 \|\mathbf{f}_{i+1}\|^2 + 2k_2^2 E \sum_{i=r}^{\infty} \left[\frac{b_i}{z_i} \sum_{j=0}^{i} \rho^{i-j} \sum_{s=0}^{d} \|\underset{\sim}{\mathbf{D}}_s\| \, \|\mathbf{h}(\mathbf{x}_{j-s})\| \right]^2 \\
&\quad + 2k_2^2 \sum_{i=r}^{\infty} \rho^{i+1} Ea_i^2 \|\boldsymbol{\xi}_0\|^2 \\
&\leq \sum_{i=0}^{r-1} Ea_i^2 \|\mathbf{f}_{i+1}\|^2 + 2k_2^2 \left(\sum_{s=0}^{d} \|\underset{\sim}{\mathbf{D}}_s\| \right)^2 E \sum_{i=r}^{\infty} b_i^2 \left(\sum_{j=0}^{i} \rho^{i-j} \right)^2 \\
&\quad + 2k_2^2 Ea_r^2 \|\boldsymbol{\xi}_0\|^2 \sum_{i=r}^{\infty} \rho^{i+1} \\
&\leq \sum_{i=0}^{r-1} Ea_i^2 \|\mathbf{f}_{i+1}\|^2 + \frac{2k_2^2}{(1-\rho^2)} \left(\sum_{s=0}^{d} \|\underset{\sim}{\mathbf{D}}_s\| \right)^2 E \sum_{i=r}^{\infty} b_i^2 \\
&\quad + 2k_2^2 Ea_r^2 \|\boldsymbol{\xi}_0\|^2 \frac{1}{1-\rho} < \infty,
\end{aligned}$$

where $\underset{\sim}{\mathbf{D}}_0$ denotes $\underset{\sim}{\mathbf{I}}$ for simplicity of notation.

We note that (2.19)–(2.22) in Lemma 2.1 remain valid since $E\|\mathbf{x}_i\|^2 < \infty$ has been verified.

According to (2.96), $\mathbf{h}(\mathbf{x}_n)$ in (2.23) and (2.24) will be replaced by $\mathbf{f}_{n+1}$ and the inequality corresponding to (2.24) will be

$$V(\mathbf{x}_{n+1}) \le V(\mathbf{x}_n) - a_n(\mathbf{f}_{n+1}^\tau + \mathbf{w}_{n+1}^\tau)\underset{\sim}{\mathbf{L}}V_{\mathbf{x}}(\mathbf{x}_n) + \frac{lk_1}{2}\|\underset{\sim}{\mathbf{L}}\|^2 a_n^2\|\mathbf{f}_{n+1} + \mathbf{w}_{n+1}\|^2. \tag{2.108}$$

By (2.101) and (2.107), we have

$$E\|a_n(\mathbf{f}_{n+1}^\tau + \mathbf{w}_{n+1}^\tau)\underset{\sim}{\mathbf{L}}V_{\mathbf{x}}(\mathbf{x}_n)\| \le \|\underset{\sim}{\mathbf{L}}\|\left(E\|V_{\mathbf{x}}(\mathbf{x}_n)\|^2\right)^{1/2}$$
$$\times\left(Ea_n^2\|\mathbf{f}_{n+1} + \mathbf{w}_{n+1}\|^2\right)^{1/2} < \infty$$

and hence $EV(\mathbf{x}_n) < \infty$, $\forall n$ can be established in a completely similar manner to that used in Lemma 2.1. □

Lemma 2.8. *Assume the conditions of Lemma* 2.7 *and Condition* A_3 *are satisfied. Then there are random variables* α, β, γ *such that*

$$0 \le \alpha < \infty, \qquad 0 < \beta < \infty, \qquad 1 \le \gamma < \infty,$$

and

$$V(\mathbf{x}_n) \underset{n\to\infty}{\longrightarrow} \alpha, \quad \|\mathbf{x}_i\| \le \beta, \quad r_2(\|\mathbf{x}_i\|) \le \gamma, \qquad \forall i, \text{a.s.} \tag{2.109}$$

Proof. By (2.99), $\mathbf{f}_{n+1}$ is $\mathscr{F}_n$-measurable, and by Condition A_3 and (2.108) [similar to (2.26)], it may be verified that

$$E\left(\frac{V(\mathbf{x}_{n+1})}{\mathscr{F}_n}\right) \le V(\mathbf{x}_n) + \frac{lk_1\|\underset{\sim}{\mathbf{L}}\|^2}{2}a_n^2\|\mathbf{f}_{n+1}\|^2 + \frac{lk_1\|\underset{\sim}{\mathbf{L}}\|^2}{2}b_n^2 r_1(n). \tag{2.110}$$

Then by using Lemma 2.7, the desired conclusions can be proved by the same argument used in Lemma 2.2 with $\mathbf{h}(\mathbf{x}_i)$ replaced by $\mathbf{f}_{i+1}$. □

Corollary. *In analogy with* (2.30)

$$\sum_{i=0}^{\infty} a_i = \infty \quad \text{a.s.} \tag{2.111}$$

From (2.96) and (2.99), we have

$$\mathbf{x}_{n+1} = \mathbf{x}_n - a_n \underset{\sim}{\mathbf{L}}^\tau \left[\underset{\sim}{\mathbf{G}} \sum_{j=0}^{n} \underset{\sim}{\mathbf{F}}^{n-j} \underset{\sim}{\mathbf{G}}^\tau \underset{\sim}{\mathbf{D}}(z) \mathbf{h}(\mathbf{x}_j) + \underset{\sim}{\mathbf{G}}\underset{\sim}{\mathbf{F}}^{n+1} \boldsymbol{\xi}_0 + \mathbf{w}_{n+1} \right]. \tag{2.112}$$

Denote

$$\underset{\sim}{\mathbf{G}}_{n,n} = 0, \underset{\sim}{\mathbf{G}}_{n+1,i} = \frac{1}{a_i} \sum_{j=i}^{n} a_j \underset{\sim}{\mathbf{L}}^\tau \underset{\sim}{\mathbf{G}}\underset{\sim}{\mathbf{F}}^{j-i} \underset{\sim}{\mathbf{G}}^\tau \underset{\sim}{\mathbf{D}}(z) \mathbf{h}(\mathbf{x}_i), \tag{2.113}$$

$$\mathbf{p}_n = \sum_{j=0}^{n-1} a_j \underset{\sim}{\mathbf{L}}^\tau \underset{\sim}{\mathbf{G}}\underset{\sim}{\mathbf{F}}^{j+1} \boldsymbol{\xi}_0. \tag{2.114}$$

Then from (2.112) we have

$$\begin{aligned}
\mathbf{x}_{n+1} &= \mathbf{x}_0 - \sum_{j=0}^{n} a_j \underset{\sim}{\mathbf{L}}^\tau \underset{\sim}{\mathbf{G}} \sum_{i=0}^{j} \underset{\sim}{\mathbf{F}}^{j-i} \underset{\sim}{\mathbf{G}}^\tau \underset{\sim}{\mathbf{D}}(z) \mathbf{h}(\mathbf{x}_i) - \mathbf{p}_{n+1} - \mathbf{q}_{n+1} \\
&= \mathbf{x}_0 - \sum_{i=0}^{n} \sum_{j=i}^{n} a_j \underset{\sim}{\mathbf{L}}^\tau \underset{\sim}{\mathbf{G}}\underset{\sim}{\mathbf{F}}^{j-i} \underset{\sim}{\mathbf{G}}^\tau \underset{\sim}{\mathbf{D}}(z) \mathbf{h}(\mathbf{x}_i) - \mathbf{p}_{n+1} - \mathbf{q}_{n+1} \\
&= \mathbf{x}_0 - \sum_{i=0}^{n} a_i \underset{\sim}{\mathbf{G}}_{n+1,i} - \mathbf{p}_{n+1} - \mathbf{q}_{n+1},
\end{aligned} \tag{2.115}$$

where $\mathbf{q}_n$ is given by (2.37). For the interpolation of $\mathbf{p}_n, \mathbf{q}_n$ we retain the notation introduced in (2.31)–(2.36). For $\underset{\sim}{\mathbf{G}}_{n+1,i}$ which has two indices, we shall denote by $\underset{\sim}{\mathbf{G}}^0_{t,i}$, $t \geq t_i$ the linear interpolation of $\{\underset{\sim}{\mathbf{G}}_{n,i}\}$ for fixed i with interpolating length $\{a_n\}$. That is,

$$\underset{\sim}{\mathbf{G}}^0_{t_n,i} = \underset{\sim}{\mathbf{G}}_{n,i},$$

$$\underset{\sim}{\mathbf{G}}^0_{t,i} = \frac{t_{n+1} - t}{a_n} \underset{\sim}{\mathbf{G}}_{n,i} + \frac{t - t_n}{a_n} \underset{\sim}{\mathbf{G}}_{n+1,i}, \qquad t \in [t_n, t_{n+1}], \tag{2.116}$$

and, for fixed t, by $\overline{\underset{\sim}{\mathbf{G}}}^0_{t,s}$ the constant interpolation of $\{\underset{\sim}{\mathbf{G}}^0_{t,i}\}$ on $[0, t]$ with $\{a_i\}$ being the interpolation length. That is,

$$\overline{\underset{\sim}{\mathbf{G}}}^0_{t,s} = \underset{\sim}{\mathbf{G}}^0_{t,j}, \qquad s \in \left[t_j, t_{j+1}\right). \tag{2.117}$$

Define the interpolating function $\mathbf{x}^*(t)$:

$$\mathbf{x}^*(t) = \mathbf{x}_0 - \int_0^t \overline{\mathbf{G}}^0_{t,s}\, ds - \mathbf{p}^0_t - \mathbf{q}^0_t. \tag{2.118}$$

For any fixed i, $\overline{\mathbf{G}}^0_{t,i}$ is a continuous function in t, and $\overline{\mathbf{G}}^0_{t,s}$ takes only a finite number of values $\mathbf{G}^0_{t,0}, \mathbf{G}^0_{t,1}, \ldots, \mathbf{G}^0_{t,m(t)}$ as s varies in $[0, t]$; hence $\int_0^t \overline{\mathbf{G}}^0_{t,s}\, ds$ is continuous in t and so is $\mathbf{x}^*(t)$ which, however, differs from $\mathbf{x}^0_t$ defined by (2.35). But, by (2.117) and (2.118), we still have

$$\begin{aligned}\mathbf{x}^*(t_n) &= \mathbf{x}_0 - \int_0^{t_n} \overline{\mathbf{G}}^0_{t_n,s}\, ds - \mathbf{p}^0_{t_n} - \mathbf{q}^0_{t_n} = \mathbf{x}_0 - \sum_{i=0}^{n-1} \int_{t_i}^{t_{i+1}} \overline{\mathbf{G}}^0_{t_n,s}\, ds - \mathbf{p}_n - \mathbf{q}_n \\ &= \mathbf{x}_0 - \sum_{i=0}^{n-1} a_i \mathbf{G}^0_{t_n,i} - \mathbf{p}_n - \mathbf{q}_n = \mathbf{x}_0 - \sum_{i=0}^{n-1} a_i \mathbf{G}_{n,i} - \mathbf{p}_n - \mathbf{q}_n = \mathbf{x}_n.\end{aligned} \tag{2.119}$$

Define

$$\mathbf{x}_n(t) \triangleq \mathbf{x}^*(t+n). \tag{2.120}$$

Clearly, $\mathbf{x}_n(t)$ is continuous for any n, since $\mathbf{x}^*(t)$ is continuous.

Lemma 2.9. *Under conditions of Lemma* 2.8, *for almost all* $\omega \in \Omega$, *there exists a subsequence* $\{\mathbf{x}_{n_k}(t)\}$ *of* $\{\mathbf{x}_n(t)\}$ *and a continuous function* $\mathbf{x}(t)$ *which is the uniform limit of* $\mathbf{x}_{n_k}(t)$

$$\mathbf{x}_{n_k}(t) \underset{k\to\infty}{\to} \mathbf{x}(t) \tag{2.121}$$

in any finite interval.

Proof. First we establish the equicontinuity of the family $\{\mathbf{x}_n(t)\}$. According to Lemma 2.3, $\mathbf{q}_n$ tends to a finite limit as $n \to \infty$. From (2.43)–(2.45), it can be seen that for any $\varepsilon > 0$, we can find Δ small enough such that

$$\|\mathbf{q}^0_{t+n+\Delta} - \mathbf{q}^0_{t+n}\| \le \frac{\varepsilon}{5}. \tag{2.122}$$

By (2.82), we have

$$\|\mathbf{p}_{n+m} - \mathbf{p}_n\| \le k_2 a_0 \|\mathbf{L}\|\, \|\boldsymbol{\xi}_0\| \sum_{j=n}^{m+n-1} \rho^{j+1} \underset{n\to\infty}{\to} 0,$$

and hence $\mathbf{p}_n$ converges to a finite limit as $n \to \infty$. Similar to (2.43)–(2.45), given $\varepsilon > 0$, we can find Δ such that

$$\|\mathbf{p}^0_{t+n+\Delta} - \mathbf{p}^0_{t+n}\| \le \frac{\varepsilon}{5}. \tag{2.123}$$

Since for $\Delta < 0$, it can be proved analogously, in what follows we consider the $\Delta > 0$ case only. From (2.118) and (2.120), we have

$$\begin{aligned}\|\mathbf{x}_n(t+\Delta) - \mathbf{x}_n(t)\| \le &\left\| \int_0^{t_{m(t+n+\Delta)}} \overline{\underset{\sim}{\mathbf{G}}}{}^0_{t_{m(t+n+\Delta)},s}\, ds - \int_0^{t_{m(t+n)}} \overline{\underset{\sim}{\mathbf{G}}}{}^0_{t_{m(t+n)},s}\, ds \right\| \\ &+ \left\| \int_0^{t+n+\Delta} \overline{\underset{\sim}{\mathbf{G}}}{}^0_{t+n+\Delta,s}\, ds - \int_0^{t_{m(t+n+\Delta)}} \overline{\underset{\sim}{\mathbf{G}}}{}^0_{t_{m(t+n+\Delta)},s}\, ds \right\| \\ &+ \left\| \int_0^{t+n} \overline{\underset{\sim}{\mathbf{G}}}{}^0_{t+n,s}\, ds - \int_0^{t_{m(t+n)}} \overline{\underset{\sim}{\mathbf{G}}}{}^0_{t_{m(t+n)},s}\, ds \right\| \\ &+ \|\mathbf{p}^0_{t+n+\Delta} - \mathbf{p}^0_{t+n}\| \\ &+ \|\mathbf{q}^0_{t+n+\Delta} - \mathbf{q}^0_{t+n}\|.\end{aligned} \tag{2.124}$$

Let $t \in [t_n, t_{n+1})$. Then for $s \in [t_n, t)$ from (2.116) and (2.117) it follows that

$$\overline{\underset{\sim}{\mathbf{G}}}{}^0_{t,s} = \underset{\sim}{\mathbf{G}}{}^0_{t,n} = \frac{t-t_n}{a_n}\underset{\sim}{\mathbf{G}}_{n+1,n} + \frac{t_{n+1}-t}{a_n}\underset{\sim}{\mathbf{0}} = \frac{t-t_n}{a_n}\underset{\sim}{\mathbf{L}}^{\tau}\underset{\sim}{\mathbf{D}}(z)\mathbf{h}(\mathbf{x}_n).$$

Hence from (2.116) and (2.117), we have

$$\begin{aligned}\left\| \int_0^t \overline{\underset{\sim}{\mathbf{G}}}{}^0_{t,s}\, ds - \int_0^{t_n} \overline{\underset{\sim}{\mathbf{G}}}{}^0_{t_n,s}\, ds \right\| &\le \left\| \int_{t_n}^t \overline{\underset{\sim}{\mathbf{G}}}{}^0_{t,s}\, ds \right\| + \left\| \int_0^{t_n} \left(\overline{\underset{\sim}{\mathbf{G}}}{}^0_{t,s} - \overline{\underset{\sim}{\mathbf{G}}}{}^0_{t_n,s} \right) ds \right\| \\ &\le \|\underset{\sim}{\mathbf{L}}\| \int_{t_n}^t \|\underset{\sim}{\mathbf{D}}(z)\mathbf{h}(\mathbf{x}_n)\|\, ds + \sum_{i=0}^{n-1} a_i \|\underset{\sim}{\mathbf{G}}{}^0_{t_i} - \underset{\sim}{\mathbf{G}}{}^0_{t_n,i}\| \\ &\le \|\underset{\sim}{\mathbf{L}}\| \sum_{i=0}^{d} \|\mathbf{D}_i\| r_2(\|\mathbf{x}_{n-i}\|)(t_{n+1}-t_n) \\ &\quad + \sum_{i=0}^{n-1} a_i \left\| \frac{t_{n+1}-t}{a_n}\underset{\sim}{\mathbf{G}}_{n,i} + \frac{t-t_n}{a_n}\underset{\sim}{\mathbf{G}}_{n+1,i} - \underset{\sim}{\mathbf{G}}_{n,i} \right\|.\end{aligned} \tag{2.125}$$

From Lemma 2.8, we have $r_2(\|\mathbf{x}_{n-i}\|) \le \gamma$. Hence by (2.125), (2.112), and by noticing that $t_{n+1} - t_n = a_n$, we find

$$\left\| \int_0^t \overline{\mathbf{G}}^0_{t,s}\,ds - \int_0^{t_n} \overline{\mathbf{G}}^0_{t_n,s}\,ds \right\| \le \|\mathbf{L}\| \sum_{i=0}^{d} \|\mathbf{D}_i\| a_n \gamma + \sum_{i=0}^{n-1} a_i \left\| \frac{t-t_n}{a_n} (\mathbf{G}_{n+1,i} - \mathbf{G}_{n,i}) \right\|$$

$$\le \|\mathbf{L}\|\gamma \sum_{i=0}^{d} \|\mathbf{D}_i\| a_n + \sum_{i=0}^{n-1} a_i \left\| \frac{t-t_n}{a_n} \left(\frac{a_n}{a_i} \mathbf{L}^\tau \mathbf{G} \mathbf{F}^{n-i} \mathbf{G}^\tau \mathbf{D}(z) \mathbf{h}(\mathbf{x}_i) \right) \right\|$$

$$\le \|\mathbf{L}\|\gamma \sum_{i=0}^{d} \|\mathbf{D}_i\| a_n + (t-t_n)\|\mathbf{L}\| k_2 \sum_{i=0}^{n-1} \rho^{n-i} \sum_{j=0}^{d} \|\mathbf{D}_j\| r_2(\|\mathbf{x}_{i-j}\|)$$

$$\le \|\mathbf{L}\|\gamma \sum_{i=0}^{d} \|\mathbf{D}_i\| a_n + k_2 \|\mathbf{L}\|\gamma a_n \sum_{j=0}^{d} \|\mathbf{D}_j\| \sum_{i=0}^{n-1} \rho^{n-i}$$

$$\le \|\mathbf{L}\|\gamma \sum_{i=0}^{d} \|\mathbf{D}_i\| a_n \left(1 + \frac{\rho}{1-\rho}\right) = \frac{\|\mathbf{L}\|\gamma}{1-\rho} \sum_{i=0}^{d} \|\mathbf{D}_i\| a_n, \tag{2.126}$$

which can be made less than $\varepsilon/5$ for a given ε if $n > N$ and N is large enough, since $a_n \underset{n\to\infty}{\to} 0$. Hence for $\varepsilon > 0$, the second and the third terms on the right-hand side of (2.124) both can be made less than $\varepsilon/5$ $\forall n > N$, while its last two terms both will be less than $\varepsilon/5$ if Δ is sufficiently small by (2.122) and (2.123). Finally, for the first term, we have

$$\left\| \int_0^{t_{m(t+n+\Delta)}} \overline{\mathbf{G}}^0_{t_{m(t+n+\Delta)},s}\,ds - \int_0^{t_{m(t+n)}} \overline{\mathbf{G}}^0_{t_{m(t+n)},s}\,ds \right\|$$

$$= \left\| \sum_{i=0}^{m(t+n+\Delta)-1} \sum_{j=i}^{m(t+n+\Delta)-1} a_j \mathbf{L}^\tau \mathbf{G} \mathbf{F}^{j-i} \mathbf{G}^\tau \mathbf{D}(z) \mathbf{h}(\mathbf{x}_i) - \sum_{i=0}^{m(t+n)-1} \sum_{j=i}^{m(t+n)-1} a_j \mathbf{L}^\tau \mathbf{G} \mathbf{F}^{j-i} \mathbf{G}^\tau \mathbf{D}(z) \mathbf{h}(\mathbf{x}_i) \right\|$$

$$= \left\| \sum_{j=m(t+n)}^{m(t+n+\Delta)-1} \sum_{i=0}^{j} a_j \mathbf{L}^\tau \mathbf{G} \mathbf{F}^{j-i} \mathbf{G}^\tau \mathbf{D}(z) \mathbf{h}(\mathbf{x}_i) \right\|$$

$$\le k_2 \|\mathbf{L}\| \sum_{s=0}^{d} \|\mathbf{D}_s\| \gamma \sum_{j=m(t+n)}^{m(t+n+\Delta)-1} \sum_{i=0}^{j} a_j \rho^{j-i}$$

$$\le \frac{k_2 \|\mathbf{L}\|\gamma}{1-\rho} \sum_{s=0}^{d} \|\mathbf{D}_s\| \sum_{j=m(t+n)}^{m(t+n+\Delta)-1} a_j \le \frac{k_2 \|\mathbf{L}\|\gamma}{1-\rho} \sum_{s=0}^{d} \|\mathbf{D}_s\| \Delta \underset{\Delta\to 0}{\to} 0. \tag{2.127}$$

By noticing that the right-hand side of (2.127) is independent of n, we conclude that for $\varepsilon > 0$ we can find N, for any $n \geq N$

$$\|\mathbf{x}_n(t+\Delta) - \mathbf{x}_n(t)\| \leq \varepsilon \tag{2.128}$$

for Δ sufficiently small. Further $\mathbf{x}_n(t)$ is continuous in t and N is a finite number, hence for any fixed t (2.128) holds also for $n \leq N$ if Δ is small enough. Thus $\{\mathbf{x}_n(t)\}$ is equicontinuous.

Now we prove the boundedness of $\{\mathbf{x}_n(t)\}$; for this we only need to prove that $\mathbf{x}^*(t)$ is a bounded function for any ω.

First, from (2.119) and Lemma 2.8, we know

$$\|\mathbf{x}^*(t_n)\| \leq \beta < \infty, \qquad \forall n. \tag{2.129}$$

Now let $t \in [t_n, t_{n+1})$. If we can prove that $\|\mathbf{x}^*(t) - \mathbf{x}^*(t_n)\|$ is uniformly bounded for $t \in [t_n, t_{n+1})$ and for all n, then the uniform boundedness of $\mathbf{x}^*(t)$ will follow from (2.129) immediately. By (2.118) we have

$$\|\mathbf{x}^*(t) - \mathbf{x}^*(t_n)\| \leq \left\| \int_0^t \underset{\sim}{\mathbf{G}}{}^0_{t,s}\, ds - \int_0^{t_n} \underset{\sim}{\mathbf{G}}{}^0_{t_n,s}\, ds \right\| + \|\mathbf{p}^0_t - \mathbf{p}^0_{t_n}\| + \|\mathbf{q}^0_t - \mathbf{q}^0_{t_n}\|. \tag{2.130}$$

The last two terms of the preceding expression have an upper bound independent of n and t since $\mathbf{p}^0_t$ and $\mathbf{q}^0_t$ tend to finite limits as $t \to \infty$ and $t_n \underset{n\to\infty}{\to} \infty$, while the term with integral goes to zero as $n \to \infty$ by (2.126) and hence it also has a bound independent of n and t. Thus $\{\mathbf{x}_n(t)\}$ is uniformly bounded and equicontinuous, and the assertion of the lemma follows from Theorem 2.2. □

Lemma 2.10. *Under conditions of Lemma* 2.8 *and if* $\mathbf{h}(\cdot)$ *is continuous, then the limiting function* $\mathbf{x}(t)$ *given by* (2.121) *satisfies differential equation*:

$$\frac{d\mathbf{x}(t)}{dt} = -\underset{\sim}{\mathbf{L}}^{\tau}\left(\underset{\sim}{\mathbf{I}} + \sum_{j=1}^{r} \underset{\sim}{\mathbf{C}}_j\right)^{-1}\left(\underset{\sim}{\mathbf{I}} + \sum_{i=1}^{d} \underset{\sim}{\mathbf{D}}_i\right)\mathbf{h}(\mathbf{x}(t)). \tag{2.131}$$

Proof. Let $\Delta > 0$. Similar to (2.124), we have

$$\begin{aligned}
\frac{\mathbf{x}(t+\Delta) - \mathbf{x}(t)}{\Delta} &= \lim_{k\to\infty} \frac{1}{\Delta}\left[\mathbf{x}_{n_k}(t+\Delta) - \mathbf{x}_{n_k}(t)\right] \\
&= -\frac{1}{\Delta}\lim_{k\to\infty}\Bigg[\left(\int_0^{t_{m(t+n_k+\Delta)}} \underset{\sim}{\mathbf{G}}{}^0_{t_{m(t+n_k+\Delta)},s}\, ds - \int_0^{t_{m(t+n_k)}} \underset{\sim}{\mathbf{G}}{}^0_{t_{m(t+n_k)},s}\, ds\right) \\
&\quad + \left(\int_0^{t+n_k+\Delta} \underset{\sim}{\mathbf{G}}{}^0_{t+n_k+\Delta,s}\, ds - \int_0^{t_{m(t+n_k+\Delta)}} \underset{\sim}{\mathbf{G}}{}^0_{t_{m(t+n_k+\Delta)},s}\, ds\right) \\
&\quad - \left(\int_0^{t+n_k} \underset{\sim}{\mathbf{G}}{}^0_{t+n_k,s}\, ds - \int_0^{t_{m(t+n_k)}} \underset{\sim}{\mathbf{G}}{}^0_{t_{m(t+n_k)},s}\, ds\right) + \left(\mathbf{p}^0_{t+n_k+\Delta} - \mathbf{p}^0_{t+n_k}\right) \\
&\quad + \left(\mathbf{q}^0_{t+n_k+\Delta} - \mathbf{q}^0_{t+n_k}\right)\Bigg]
\end{aligned} \tag{2.132}$$

From this, by (2.126) and by convergence of $\mathbf{p}_t^0$ and $\mathbf{q}_t^0$ as $t \to \infty$, we obtain

$$\lim_{\Delta\to 0} \frac{\mathbf{x}(t+\Delta)-\mathbf{x}(t)}{\Delta}$$

$$= -\lim_{\Delta\to 0}\lim_{k\to\infty}\frac{1}{\Delta}\left(\int_0^{t_{m(t+n_k+\Delta)}}\overline{\mathbf{G}}^0_{t_{m(t+n_k+\Delta)},s}\,ds - \int_0^{t_{m(t+n_k)}}\overline{\mathbf{G}}^0_{t_{m(t+n_k)},s}\,ds\right)$$

$$= \lim_{\Delta\to 0}\lim_{k\to\infty}\frac{1}{\Delta}\sum_{j=m(t+n_k)}^{m(t+n_k+\Delta)-1}\sum_{i=0}^{j} a_j\mathbf{L}^\tau\mathbf{G}\mathbf{F}^{j-i}\mathbf{G}^\tau\mathbf{D}(z)\mathbf{h}(\mathbf{x}_i). \quad (2.133)$$

By Lemma 2.8, we have

$$\lim_{\Delta\to 0}\lim_{k\to\infty}\frac{1}{\Delta}\left\|\sum_{j=m(t+n_k)}^{m(t+n_k+\Delta)-1} a_j \sum_{i=0}^{m(t+n_k)-1}\mathbf{L}^\tau\mathbf{G}\mathbf{F}^{j-i}\mathbf{G}^\tau\mathbf{D}(z)\mathbf{h}(\mathbf{x}_i)\right\|$$

$$\le \lim_{\Delta\to 0}\lim_{k\to\infty}\frac{k_2\|\mathbf{L}\|}{\Delta}\sum_{j=m(t+n_k)}^{m(t+n_k+\Delta)-1} a_j \sum_{i=0}^{m(t+n_k)-1}\rho^{j-i}\sum_{s=0}^{d}\|\mathbf{D}_s\|\,\|\mathbf{h}(\mathbf{x}_{i-s})\|$$

$$\le \lim_{\Delta\to 0}\lim_{k\to\infty}\frac{k_2\|\mathbf{L}\|\gamma}{\Delta}\sum_{s=0}^{d}\|\mathbf{D}_s\|\sum_{j=m(t+n_k)}^{m(t+n_k+\Delta)-1} a_j\sum_{i=0}^{m(t+n_k)-1}\rho^{j-i}$$

$$= \lim_{\Delta\to 0}\lim_{k\to\infty}\frac{k_2\|\mathbf{L}\|\gamma}{\Delta}\sum_{s=0}^{d}\|\mathbf{D}_s\|\sum_{j=m(t+n_k)}^{m(t+n_k+\Delta)-1}\frac{a_j\left(\rho^{j-m(t+n_k)+1}-\rho^{j+1}\right)}{1-\rho}$$

$$= \lim_{\Delta\to 0}\lim_{k\to\infty}\frac{k_2\|\mathbf{L}\|\gamma}{\Delta}\sum_{s=0}^{d}\|\mathbf{D}_s\|$$

$$\times\left(\sum_{i=0}^{N-1}\frac{a_{i+m(t+n_k)}\rho^{i+1}}{1-\rho} + \sum_{i=N}^{m(t+n_k+\Delta)-m(t+n_k)-1}\frac{a_{i+m(t+n_k)}\rho^{i+1}}{1-\rho}\right)$$

$$\le \frac{k_2\|\mathbf{L}\|\gamma}{\Delta}\sum_{s=0}^{d}\|\mathbf{D}_s\|\left(\frac{\Delta\varepsilon}{2(1-\rho)}+\frac{\Delta\varepsilon}{2(1-\rho)}\right)$$

$$\le k_2\|\mathbf{L}\|\gamma\sum_{s=0}^{d}\|\mathbf{D}_s\|\frac{\varepsilon}{1-\rho}\underset{\varepsilon\to 0}{\to} 0,$$

where the last inequality holds because for fixed $\varepsilon > 0$ N can be chosen such that $\rho^N < \varepsilon/2$ and for fixed N

$$\sum_{i=0}^{N-1} a_{i+m(t+n_k)} < \frac{\Delta\varepsilon}{2},$$

if k is large enough.

Hence from (2.133) it follows that

$$\lim_{\Delta\to 0} \frac{\mathbf{x}(t+\Delta)-\mathbf{x}(t)}{\Delta}$$

$$= -\lim_{\Delta\to 0}\lim_{k\to\infty}\frac{1}{\Delta}\sum_{j=m(t+n_k)}^{m(t+n_k+\Delta)-1}\sum_{i=m(t+n_k)}^{j} a_j \underset{\sim}{\mathbf{L}}^{\tau}\underset{\sim}{\mathbf{G}}\underset{\sim}{\mathbf{F}}^{j-i}\underset{\sim}{\mathbf{G}}^{\tau}\underset{\sim}{\mathbf{D}}(z)\mathbf{h}(\mathbf{x}_i)$$

$$= -\lim_{\Delta\to 0}\lim_{k\to\infty}\frac{1}{\Delta}\sum_{j=m(t+n_k)}^{m(t+n_k+\Delta)-1}\sum_{i=m(t+n_k)}^{j} a_j \underset{\sim}{\mathbf{L}}^{\tau}\underset{\sim}{\mathbf{G}}\underset{\sim}{\mathbf{F}}^{j-i}\underset{\sim}{\mathbf{G}}^{\tau}\underset{\sim}{\mathbf{D}}(z)[\mathbf{h}(\mathbf{x}_i)-\mathbf{h}(\mathbf{x}(t))]$$

$$-\lim_{\Delta\to 0}\lim_{k\to\infty}\frac{1}{\Delta}\sum_{j=m(t+n_k)}^{m(t+n_k+\Delta)-1}\sum_{i=m(t+n_k)}^{j} a_j \underset{\sim}{\mathbf{L}}^{\tau}\underset{\sim}{\mathbf{G}}\underset{\sim}{\mathbf{F}}^{j-i}\underset{\sim}{\mathbf{G}}^{\tau}\underset{\sim}{\mathbf{D}}(z)\mathbf{h}(\mathbf{x}(t)) \qquad (2.134)$$

Let us adopt the notation

$$\delta_{ki} = \begin{cases} \displaystyle\sum_{j=m(t+n_k)}^{m(t+n_k)+i-d-1} a_j, & d+1 \le i \le d + m(t+n_k+\Delta) - m(t+n_k) \\ 0, & i = d \\ \displaystyle\sum_{j=m(t+n_k)-d+i}^{m(t+n_k)-1} a_j, & 0 \le i \le d-1. \end{cases} \qquad (2.135)$$

For sufficiently large k, $\delta_{ki} \in [-\Delta/2, \Delta]$, hence we have

$$t + n_k - a_{m(t+n_k)} + \delta_{ki} < t_{m(t+n_k)} + \delta_{ki} = t_{m(t+n_k)} - d + i$$
$$\le t + n_k + \delta_{ki}$$

since

$$t_{m(t+n_k)} + a_{m(t+n_k)} > t + n_k \qquad (2.136)$$

For sufficiently large k, $a_{m(t+n_k)} < \Delta/2$ and hence there exists $\delta'_{ki} \in [-\Delta, \Delta]$ such that

$$t_{m(t+n_k)-d+i} = n_k + t + \delta'_{ki} \qquad (2.137)$$

for any i, $0 \le i \le d + m(t + n_k + \Delta) - m(t + n_k)$. Then

$$\begin{aligned}\left\|\mathbf{h}\left(\mathbf{x}_{m(t+n_k)-d+i}\right) - \mathbf{h}(\mathbf{x}(t))\right\| &= \left\|\mathbf{h}\left(\mathbf{x}^*\left(t_{m(t+n_k)-d+i}\right)\right) - \mathbf{h}(\mathbf{x}(t))\right\| \\ &= \left\|\mathbf{h}\left(\mathbf{x}^*\left(n_k + t + \delta'_{ki}\right)\right) - \mathbf{h}(\mathbf{x}(t))\right\| \\ &\le \left\|\mathbf{h}\left(\mathbf{x}_{n_k}\left(t + \delta'_{ki}\right)\right) - \mathbf{h}\left(\mathbf{x}_{n_k}(t)\right)\right\| + \left\|\mathbf{h}\left(\mathbf{x}_{n_k}(t)\right) - \mathbf{h}(\mathbf{x}(t))\right\|.\end{aligned}$$

Fix ω and t, then by equicontinuity the first term on the right-hand side of the preceding inequality can be made less than $\varepsilon/2$ for sufficiently small Δ, while the second term also can be made less $\varepsilon/2$ if k is large enough, since $\mathbf{h}(\cdot)$ is continuous and $\mathbf{x}_{n_k}(t) \to \mathbf{x}(t)$ uniformly in any finite interval. Hence for all i, with $0 \le i \le d + m(t + n_k + \Delta) - m(t + n_k)$, we have

$$\left\|\mathbf{h}\left(\mathbf{x}_{m(t+n_k)-d+i}\right) - \mathbf{h}(\mathbf{x}(t))\right\| < \varepsilon$$

or, equivalently,

$$\lim_{\Delta \to 0} \lim_{k \to \infty} \max_{0 \le i \le d + m(t+n_k+\Delta) - m(t+n_k)} \left\|\mathbf{h}\left(\mathbf{x}_{m(t+n_k)-d+i} - \mathbf{h}(\mathbf{x}(t))\right\| = 0 \tag{2.138}$$

Hence on the right-hand side of (2.134), the first term has norm less than or equal to

$$\begin{aligned}&\|\mathbf{L}\| k_2 \overline{\lim_{\Delta \to 0}} \overline{\lim_{k \to \infty}} \frac{1}{\Delta} \sum_{j=m(t+n_k)}^{m(t+n_k+\Delta)-1} \sum_{i=m(t+n_k)}^{j} a_j \rho^{j-i} \sum_{s=0}^{d} \|\underset{\sim}{\mathbf{D}}_s\| \\ &\qquad \times \left\|\mathbf{h}(\mathbf{x}_{i-s}) - \mathbf{h}(\mathbf{x}(t))\right\| \\ &\quad \le k_2 \|\underset{\sim}{\mathbf{L}}\| \sum_{s=0}^{d} \|\underset{\sim}{\mathbf{D}}_s\| \overline{\lim_{\Delta \to 0}} \overline{\lim_{k \to \infty}} \max_{0 \le i \le d + m(t+n_k+\Delta) - m(t+n_k)} \\ &\qquad \times \left\|\mathbf{h}\left(\mathbf{x}_{m(t+n_k)-d+i}\right) - \mathbf{h}(\mathbf{x}(t))\right\| \\ &\qquad \cdot \frac{1}{\Delta} \sum_{j=m(t+n_k)}^{m(t+n_k+\Delta)-1} \sum_{i=m(t+n_k)}^{j} \rho^{j-i} \\ &\quad \le \frac{k_2 \|\underset{\sim}{L}\|}{1 - \rho} \sum_{s=0}^{d} \|\underset{\sim}{\mathbf{D}}_s\| \overline{\lim_{\Delta \to 0}} \overline{\lim_{k \to \infty}} \max_{0 \le i \le d + m(t+n_k+\Delta) - m(t+n_k)} \\ &\qquad \times \left\|\mathbf{h}\left(\mathbf{x}_{m(t+n_k)-d+i}\right) - \mathbf{h}(\mathbf{x}(t))\right\| \frac{\Delta + a_{m(t+n_k)}}{\Delta} \\ &\quad = 0\end{aligned} \tag{2.139}$$

where the following inequality, derived from (2.136), has been used

$$\sum_{j=m(t+n_k)}^{m(t+n_k+\Delta)-1} a_j = t_{m(t+n_k+\Delta)} - t_{m(t+n_k)} \le t + n_k + \Delta - t_{m(t+n_k)}$$

$$\le t + n_k + \Delta - \left(t + n_k - a_{m(t+n_k)}\right) = \Delta + a_{m(t+n_k)}. \tag{2.140}$$

It is known that 1 is not the zero of $\det \underset{\sim}{C}(\lambda)$ and hence is not an eigenvalue of $\underset{\sim}{F}$ by Lemma 2.5; hence $\det(\underset{\sim}{I} - \underset{\sim}{F}) \neq 0$ and $(\underset{\sim}{I} - \underset{\sim}{F})^{-1}$ is meaningful. Then from (2.134) and (2.139), we know that

$$\lim_{\Delta \to 0} \frac{\mathbf{x}(t+\Delta) - \mathbf{x}(t)}{\Delta}$$

$$= - \lim_{\Delta \to 0} \lim_{k \to \infty} \frac{1}{\Delta} \sum_{j=m(t+n_k)}^{m(t+n_k+\Delta)-1} a_j$$

$$\times \sum_{i=m(t+n_k)}^{j} \underset{\sim}{L}^{\tau} \underset{\sim}{G} \underset{\sim}{F}^{j-i} \underset{\sim}{G}^{\tau} \left(\underset{\sim}{I} + \sum_{s=1}^{d} \underset{\sim}{D}_s \right) \mathbf{h}(\mathbf{x}(t))$$

$$= - \lim_{\Delta \to 0} \lim_{k \to \infty} \frac{1}{\Delta} \sum_{j=m(t+n_k)}^{m(t+n_k+\Delta)-1} a_j \underset{\sim}{L}^{\tau} \underset{\sim}{G} \left(\underset{\sim}{I} - \underset{\sim}{F}^{j-m(t+n_k)+1} \right)$$

$$\times (\underset{\sim}{I} - \underset{\sim}{F})^{-1} \underset{\sim}{G}^{\tau} \left(\underset{\sim}{I} + \sum_{s=1}^{d} \underset{\sim}{D}_s \right) \mathbf{h}(\mathbf{x}(t)). \tag{2.141}$$

Given $\varepsilon > 0$, take N large enough such that $\rho^N < \varepsilon/2$, then

$$\lim_{\Delta \to 0} \lim_{k \to \infty} \frac{1}{\Delta} \left\| \sum_{j=m(t+n_k)}^{m(t+n_k+\Delta)-1} a_j \underset{\sim}{L}^{\tau} \underset{\sim}{G} \underset{\sim}{F}^{j-m(t+n_k)+1} \right\|$$

$$\le \|\underset{\sim}{L}\| k_2 \lim_{\Delta \to 0} \lim_{k \to \infty} \frac{1}{\Delta} \sum_{j-m(t+n_k)}^{m(t+n_k+\Delta)-1} a_j \rho^{j-m(t+n_k)+1}$$

$$= \|\underset{\sim}{L}\| k_2 \lim_{\Delta \to 0} \lim_{k \to \infty} \frac{1}{\Delta} \sum_{i=1}^{m(t+n_k+\Delta)-m(t+n_k)} a_{i+m(t+n_k)+1} \rho^i$$

$$= \|\underset{\sim}{L}\| k_2 \lim_{\Delta \to 0} \lim_{k \to \infty} \frac{1}{\Delta} \left(\sum_{i=1}^{N-1} a_{i+m(t+n_k)+1} \rho^i \quad \sum_{i=N}^{m(t+n_k+\Delta)-m(t+n_k)} a_{i+m(t+n_k)+1} \rho^i \right)$$

$$\le \|\underset{\sim}{L}\| k_2 \lim_{\Delta \to 0} \lim_{k \to \infty} \left(\frac{1}{\Delta} \sum_{i=1}^{N-1} a_{i+m(t+n_k)+1} + \frac{\rho^N \Delta}{\Delta} \right) = 0 \tag{2.142}$$

since for fixed N

$$\lim_{\Delta\to 0}\lim_{k\to\infty}\frac{1}{\Delta}\sum_{i=0}^{N-1} a_{i+m(t+n_k)+1}=0$$

by $a_i \underset{i\to\infty}{\to} 0$.

Thus, from (2.141) and (2.142), it follows

$$\dot{\mathbf{x}}(t) = -\mathbf{L}^{\tau}\mathbf{G}(\mathbf{I}-\mathbf{F})^{-1}\mathbf{G}^{\tau}\left(\mathbf{I}+\sum_{s=1}^{d}\mathbf{D}_s\right)\mathbf{h}(\mathbf{x}(t)). \tag{2.143}$$

It may directly be verified that

$$(\mathbf{I}-\mathbf{F})^{-1}$$

$$=\begin{bmatrix}
\left(\sum_{j=0}^{r}\mathbf{C}_j\right)^{-1} & \left(\sum_{j=0}^{r}\mathbf{C}_j\right)^{-1} & \cdots & \cdots & \left(\sum_{j=0}^{r}\mathbf{C}_j\right)^{-1}\\
\sum_{i=0}^{1}\mathbf{C}_i\left(\sum_{j=0}^{r}\mathbf{C}_j\right)^{-1}-\mathbf{I} & \sum_{i=0}^{1}\mathbf{C}_i\left(\sum_{j=0}^{r}\mathbf{C}_j\right)^{-1} & & & \vdots\\
\sum_{i=0}^{2}\mathbf{C}_i\left(\sum_{j=0}^{r}\mathbf{C}_j\right)^{-1}-\mathbf{I} & \sum_{i=0}^{2}\mathbf{C}_i\left(\sum_{j=0}^{r}\mathbf{C}_j\right)^{-1}-\mathbf{I} & \vdots & & \vdots\\
\vdots & \vdots & \ddots & \ddots & \sum_{i=0}^{r-2}\mathbf{C}_i\left(\sum_{j=0}^{r}\mathbf{C}_j\right)^{-1}\\
\sum_{i=0}^{r-1}\mathbf{C}_i\left(\sum_{j=0}^{r}\mathbf{C}_j\right)^{-1}-\mathbf{I} & \sum_{i=0}^{r-1}\mathbf{C}_i\left(\sum_{j=0}^{r}\mathbf{C}_j\right)^{-1}-\mathbf{I} & \cdots & \sum_{i=0}^{r-1}\mathbf{C}_i\left(\sum_{j=0}^{r}\mathbf{C}_j\right)^{-1}-\mathbf{I} & \sum_{i=0}^{r-1}\mathbf{C}_i\left(\sum_{j=0}^{r}\mathbf{C}_j\right)^{-1}
\end{bmatrix}$$

where $\mathbf{C}_0$ denotes $\mathbf{I}$, hence

$$\mathbf{G}(\mathbf{I}-\mathbf{F})^{-1}\mathbf{G}^{\tau} = \left(\mathbf{I}+\sum_{i=1}^{r}\mathbf{C}_i\right)^{-1}, \tag{2.144}$$

and from here and (2.143), Equation (2.131) follows. □

Now we give conditions for convergence of the MRM procedure.

Theorem 2.6. *Assume that the measurement error in* (2.2) *is given by* (2.67), *the dominating functions* r_1 *and* r_2 *in* (2.11) *and* (2.12) *are known and that* $\mathbf{x}_n$ *is defined by* (2.79). *If the following conditions are satisfied*:

1. $r = 0$, *or* $r > 0$ *but all zeros of* $\det \mathbf{C}(\lambda)$ *lie outside the closed unit disk.*
2. $\mathbf{h}(\mathbf{x})$ *is continuous and* $\mathbf{h}(\mathbf{x}^0) = 0$.
3. *Condition* A_3 *holds,*

then

$$\mathbf{x}_n \underset{n\to\infty}{\to} \mathbf{x}^0 \text{ a.s.}$$

Proof. Comparing (2.94) with (2.70), we find $\underset{\sim}{\mathbf{G}}\boldsymbol{\xi}_{n+1}(=\boldsymbol{\xi}^1_{n+1})$ satisfies (2.70). If $\boldsymbol{\xi}_0$ is given, then, as can be seen from (2.90)–(2.94), $\boldsymbol{\xi}^1_i$, $i = 0, \ldots, r-1$ are given. Take an arbitrary $\mathbf{x} \neq \mathbf{x}^0$ and put

$$\boldsymbol{\xi}_0 = (\underset{\sim}{\mathbf{I}} - \underset{\sim}{\mathbf{F}})^{-1}\underset{\sim}{\mathbf{G}}^\tau\left(\underset{\sim}{\mathbf{I}} + \sum_{s=1}^{d} \underset{\sim}{\mathbf{D}}_s\right)\mathbf{h}(\mathbf{x}), \qquad \mathbf{x}_i \equiv \mathbf{x}.$$

Then from (2.98) it is obtained that

$$\begin{aligned}\boldsymbol{\xi}_{n+1} &= \sum_{j=0}^{n} \underset{\sim}{\mathbf{F}}^{n-j}\underset{\sim}{\mathbf{G}}^\tau\left(\underset{\sim}{\mathbf{I}} + \sum_{s=1}^{d} \underset{\sim}{\mathbf{D}}_s\right)\mathbf{h}(\mathbf{x}) + \underset{\sim}{\mathbf{F}}^{n+1}(\underset{\sim}{\mathbf{I}} - \underset{\sim}{\mathbf{F}})^{-1}\underset{\sim}{\mathbf{G}}^\tau\left(\underset{\sim}{\mathbf{I}} + \sum_{s=1}^{d} \underset{\sim}{\mathbf{D}}_s\right)\mathbf{h}(\mathbf{x})\\ &= (\underset{\sim}{\mathbf{I}} - \underset{\sim}{\mathbf{F}})^{-1}\underset{\sim}{\mathbf{G}}^\tau\left(\underset{\sim}{\mathbf{I}} + \sum_{s=1}^{d} \underset{\sim}{\mathbf{D}}_s\right)\mathbf{h}(\mathbf{x}).\end{aligned}$$

In other words, the solution of (2.70) with $\mathbf{x}_i \equiv \mathbf{x}$ and initial value $\boldsymbol{\xi}_0$ is given by

$$\mathbf{f}_{n+1} \equiv \underset{\sim}{\mathbf{G}}(\underset{\sim}{\mathbf{I}} - \underset{\sim}{\mathbf{F}})^{-1}\underset{\sim}{\mathbf{G}}^\tau\left(\underset{\sim}{\mathbf{I}} + \sum_{s=1}^{d} \underset{\sim}{\mathbf{D}}_s\right)\mathbf{h}(\mathbf{x}) = \left(\underset{\sim}{\mathbf{I}} + \sum_{i=1}^{r} \underset{\sim}{\mathbf{C}}_i\right)^{-1}\left(\underset{\sim}{\mathbf{I}} + \sum_{s=1}^{d} \underset{\sim}{\mathbf{D}}_s\right)\mathbf{h}(\mathbf{x}) \tag{2.145}$$

By Condition $\mathbf{A}_3$, we have

$$V_{\mathbf{x}}^\tau(\mathbf{x})\underset{\sim}{\mathbf{L}}^\tau\left(\underset{\sim}{\mathbf{I}} + \sum_{i=1}^{r} \underset{\sim}{\mathbf{C}}_i\right)^{-1}\left(\underset{\sim}{\mathbf{I}} + \sum_{s=1}^{d} \underset{\sim}{\mathbf{D}}_s\right)\mathbf{h}(\mathbf{x}) > 0, \qquad \forall \mathbf{x} \neq \mathbf{x}^0, \tag{2.146}$$

and hence $V(\mathbf{x})$ is the Lyapunov function of Equation (2.131). The remaining part of the proof coincides with the one given in Theorem 2.5. □

2.4. FINDING THE EXTREMUM OF REGRESSION FUNCTION

In this section we consider the problem of finding the extremum $\mathbf{x}_0$ of a function $h(\cdot)$ $(R^l, \mathscr{B}^l) \to (R^1, \mathscr{B}^1)$. If $h(\cdot)$ is differentiable and its derivative $h_{\mathbf{x}}(\mathbf{x})$ can be observed, then the problem will be reduced to seeking a zero of $h_{\mathbf{x}}(\mathbf{x})$ and this is the topic of the last section. The KW procedure solves the problem when only $h(\mathbf{x})$ can be measured.

$(\Omega, \mathscr{F}, P)$, $\{\mathscr{F}_n\}$, and $\mathbf{w}_{i+1}(\mathbf{x}, \mathbf{e}_{i+1})$ all retain their meaning as in the above and (2.9) and (2.11) are supposed to be satisfied.

Assume $h(\cdot)$ is differentiable and denote

$$\mathbf{x} = [x^1, \dots, x^l]^\tau, \qquad h_{\mathbf{x}}(\mathbf{x}) = \left[\frac{\partial h}{\partial x^1} \cdots \frac{\partial h}{\partial x^l}\right]^\tau. \tag{2.147}$$

Similar to (2.12), assume

$$\|h_{\mathbf{x}}(\mathbf{x})\| \le r_2(\|\mathbf{x}\|). \tag{2.148}$$

Let $b_i > 0$ and $c_i > 0$ be $\mathscr{F}_i$-measurable random variables $\forall i$, and b_i be monotonely nonincreasing and $c_i \underset{i\to\infty}{\to} 0$ with properties:

$$\sum_{i=0}^{\infty} b_i = \infty \text{ a.s.}, \quad E\sum_{i=0}^{\infty} b_i^2 < \infty, \quad E\sum_{i=0}^{\infty} b_i c_i < \infty, \quad E\sum_{i=0}^{\infty} b_i^2 r_1(i) < \infty. \tag{2.149}$$

Denote

$$\mathbf{x}_n = \left[x_n^1 \dots x_n^l\right]^\tau, \qquad \mathbf{x}_n^{i+} = \left[x_n^1, \dots, x_n^{i-1}, x_n^i + c_n, x_n^{i+1}, \dots, x_n^l\right]^\tau$$

$$\mathbf{x}_n^{i-} = \left[x_n^1, \dots, x_n^{i-1}, x_n^i - c_n, x_n^{i+1}, \dots, x_n^l\right]^\tau.$$

Kiefer and Wolfowitz suggested that one replace the unobservable quantity $h_{\mathbf{x}}(\mathbf{x}_n)$ by

$$\nabla h(\mathbf{x}_n) \triangleq \frac{1}{2c_n}\left[h(\mathbf{x}_n^{1+}) - h(\mathbf{x}_n^{1-}), \dots, h(\mathbf{x}_n^{l+}) - h(\mathbf{x}_n^{l-})\right]^\tau. \tag{2.150}$$

Assume the measured value is

$$\mathbf{y}_{n+1} = \nabla h(\mathbf{x}_n) + \boldsymbol{\varepsilon}_{n+1}. \tag{2.151}$$

The classical KW procedure [7] considers

$$\mathbf{x}_{n+1} = \mathbf{x}_n - b_n \mathbf{y}_{n+1} \tag{2.152}$$

with the measurement error $\boldsymbol{\varepsilon}_{n+1}$ defined by (2.10) and with $\mathbf{w}_{i+1}(\mathbf{x}, \mathbf{e}_{i+1})$ independent of $\mathscr{F}_i$ $\forall i$. There are many variations of conditions for convergence of KW procedure, we list one of them.

PROPOSITION. Assume the following conditions are satisfied:

1. There is a function $V(\mathbf{x})$ with bounded second-order derivative such that $V(\mathbf{x}) > 0$, $h_{\mathbf{x}}^{\tau} V_{\mathbf{x}}(\mathbf{x}) > 0$, $\mathbf{x} \neq \mathbf{x}^0$, $V(\mathbf{x}^0) = 0$ and $\lim_{\|\mathbf{x}\| \to \infty} V(\mathbf{x}) = \infty$.
2. $\| V_{\mathbf{x}}(\mathbf{x})\| + \|h_{\mathbf{x}}(\mathbf{x})\|^2 + E\|\boldsymbol{\varepsilon}_i\|^2 \leq k_0(1 + V(\mathbf{x}))$.
3. $h_{\mathbf{x}}(\cdot)$ satisfies global Lipschitz condition.
4. $\sum_{i=0}^{\infty} b_i^2/c_i^2 < \infty$ in addition to (2.149).

Then $\mathbf{x}_n \underset{n\to\infty}{\to} \mathbf{x}^0$ a.s., where $\mathbf{x}_n$ is defined by (2.152).

This proposition is concerned with the search for a global minimum of $h(\mathbf{x})$. If $h(\mathbf{x})$ has a unique maximum, then the minus sign in (2.152) should be replaced by a plus sign and inequality $>$ in Condition 1 by $<$.

Here we consider a more general case, that is, $\boldsymbol{\varepsilon}_n$ in (2.151) is given by (2.69) rather than (2.10). The algorithm (2.152) should be correspondingly modified.

Instead of (2.70), we consider

$$\underset{\sim}{C}(z)\mathbf{g}_{n+1} = \underset{\sim}{D}(z)h_{\mathbf{x}}(\mathbf{x}_n) \tag{2.153}$$

and introduce Condition A_4, which is similar to Condition A_3.

CONDITION A_4. There exists a function $V(\mathbf{x})$ with bounded continuous second-derivative such that

$$V(\mathbf{x}) > 0, \quad \forall \mathbf{x} \neq \mathbf{x}^0, \qquad V(\mathbf{x}^0) = 0, \quad V(\mathbf{x}) \to \infty \quad \text{as } \|\mathbf{x}\| \to \infty$$

$$|V_{x^i x^j}(\mathbf{x})| \leq k_1, \qquad i, j = 1, \ldots, l$$

and, for some n_0,

$$\mathbf{g}_{i+1}^{\tau} V_{\mathbf{x}}(\mathbf{x}_i) > 0, \qquad \forall \mathbf{x}_i \neq \mathbf{x}^0, \qquad i \geq n_0, \tag{2.154}$$

where $\mathbf{g}_{i+1}$ is defined from (2.153) with arbitrary initial values $\mathbf{g}_0, \ldots, \mathbf{g}_{r-1}$ for both $\mathbf{x}_i \equiv \mathbf{x}$ and $\mathbf{x}_i$ defined by (2.159).

For the special case $d = r = 0$, Condition A_4 is nothing but Condition 1 of the preceding proposition.

In our analysis we do not require $h_{\mathbf{x}}(\cdot)$ to satisfy the Lipschitz condition, but we assume that there exists a continuous function $\Psi(\mathbf{x}) \geq 1$ such that

$$\| \nabla h(\mathbf{x}) - h_{\mathbf{x}}(\mathbf{x})\| \leq \Psi(\mathbf{x})c, \qquad \forall x, c > 0, \tag{2.155}$$

where $\nabla h(\mathbf{x})$ is still defined by (2.150) but with $\mathbf{x}_n$ and c_n replaced by $\mathbf{x}$ and c, respectively.

We set

$$h_{\mathbf{x}}(\mathbf{x}_i) = 0, \quad \nabla h(\mathbf{x}_i) = 0, \qquad i < 0,$$
$$\mathbf{y}_i = 0, \qquad i \leq 0 \tag{2.156}$$

which are only adopted to simplify the notation.

Let $\underset{\sim}{\mathbf{F}}$ and $\underset{\sim}{\mathbf{G}}$ be defined by (2.73) and (2.74), and let $\mathbf{u}_n$ and ζ_n be defined by (2.75) and (2.76). However, in the present case, $\mathbf{y}_n$ in (2.75) will be given by (2.151).

Denote

$$\bar{\mathbf{x}}_n = \left[x_n^1 + c_n \operatorname{sign} x_1^1, \ldots, x_n^l + c_n \operatorname{sign} x_n^l\right]^{\tau}, \tag{2.157}$$

$$z_n = \max_{0 \leq i \leq n} r_2(\|\bar{x}_i\|) \max_{0 \leq j \leq n} \Psi(x_j), \qquad a_n = \frac{b_n}{z_n}. \tag{2.158}$$

Given $\mathbf{x}_0$ as the zeroth approximation of $\mathbf{x}^0$, we modify KW procedure as follows:

$$\mathbf{x}_{n+1} = \mathbf{x}_n - a_n \underset{\sim}{\mathbf{G}} \zeta_{n+1} \tag{2.159}$$

or $\mathbf{x}_{n+1} = \mathbf{x}_n + a_n \underset{\sim}{\mathbf{G}} \zeta_{n+1}$, if the converse inequality in (2.154) takes place. We call this algorithm the modified KW (MKW) procedure. For the special case $\underset{\sim}{\mathbf{D}}(z) = \underset{\sim}{\mathbf{C}}(z) = \underset{\sim}{\mathbf{I}}$ with bounded z_n, algorithms given by (2.158) and (2.152) coincide with each other (i.e., in this case the MKW procedure reduces to the KW procedure).

The equation corresponding to (2.91) will be

$$\xi_{n+1} = \underset{\sim}{\mathbf{F}} \xi_n + \underset{\sim}{\mathbf{G}}^{\tau} \underset{\sim}{\mathbf{D}}(z) \nabla h(\mathbf{x}_n) \tag{2.160}$$

with ξ_0 defined by (2.90)–(2.93). We see that (2.160) is generated by $\nabla h(\mathbf{x}_n)$; we also need a similar equation generated by $h_{\mathbf{x}}(\mathbf{x}_n)$:

$$\eta_{n+1} = \underset{\sim}{\mathbf{F}} \eta_n + \underset{\sim}{\mathbf{G}}^{\tau} \underset{\sim}{\mathbf{D}}(z) h_{\mathbf{x}}(\mathbf{x}_n), \qquad \eta_0 = \xi_0. \tag{2.161}$$

Denote

$$\mathbf{f}_{n+1} = \underset{\sim}{\mathbf{G}} \xi_{n+1}, \qquad \mathbf{g}_{n+1} = \underset{\sim}{\mathbf{G}} \eta_{n+1}. \tag{2.162}$$

From (2.99), we have

$$\mathbf{f}_{n+1} = \underset{\sim}{\mathbf{G}} \sum_{j=0}^{n} \underset{\sim}{\mathbf{F}}^{n-j} \underset{\sim}{\mathbf{G}}^{\tau} \mathbf{D}(z) \nabla h(\mathbf{x}_j) + \underset{\sim}{\mathbf{G}} \underset{\sim}{\mathbf{F}}^{n+1} \xi_0, \tag{2.163}$$

$$\mathbf{g}_{n+1} = \underset{\sim}{\mathbf{G}} \sum_{j=0}^{n} \underset{\sim}{\mathbf{F}}^{n-j} \underset{\sim}{\mathbf{G}}^{\tau} \underset{\sim}{\mathbf{D}}(z) h_{\mathbf{x}}(\mathbf{x}_j) + \underset{\sim}{\mathbf{G}} \underset{\sim}{\mathbf{F}}^{n+1} \xi_0. \tag{2.164}$$

Since (2.95) remains valid, algorithm (2.159) can be rewritten as

$$\mathbf{x}_{n+1} = \mathbf{x}_n - a_n(\mathbf{f}_{n+1} + \mathbf{w}_{n+1}). \tag{2.165}$$

Lemma 2.11. *Suppose that $r = 0$, or $r > 0$ but all zeros of $\det \underset{\sim}{C}(\lambda)$ lie outside the closed unit disk. Then*

$$E\|\mathbf{x}_i\|^2 < \infty, \quad EV(\mathbf{x}_i) < \infty, \qquad \forall i,$$

$$E\sum_{i=0}^{\infty} a_i^2\|\mathbf{w}_{i+1}\|^2 < \infty, \qquad E\sum_{i=0}^{\infty} a_i^2\|\mathbf{f}_{i+1}\|^2 < \infty.$$

Proof. Since $r_2(\cdot)$ is nondecreasing and $\|\bar{\mathbf{x}}_s\| \geq \|\mathbf{x}_i\|$, we have $r_2(\|\bar{\mathbf{x}}_s\|) \geq r_2(\|\mathbf{x}_s\|)$, and for any ω there are numbers δ_n^i such that $|\delta_n^i| \leq c_n$, $1 \leq i \leq l$ and

$$\|\nabla h(\mathbf{x}_n)\|^2 = \sum_{i=1}^{l}\left(\left.\frac{\partial h(\mathbf{x})}{\partial x^i}\right|_{\mathbf{x}=[x_n^1,\ldots,x_n^i+\delta_n^i,\ldots,x_n^l]^\tau}\right)^2 \leq l r_2^2(\|\bar{\mathbf{x}}_n\|). \tag{2.166}$$

The assertions of the lemma can be proved by the same arguments as those used to prove Lemma 2.7 with $h(\mathbf{x}_n)$ replaced by $\nabla h(\mathbf{x}_n)$ and a_n defined by (2.158). □

Lemma 2.12. *Suppose Condition A_4 and conditions of Lemma 2.11 are satisfied. Then there are random variables α, β, γ such that $0 \leq \alpha < \infty$, $0 < \beta < \infty$, $1 \leq \gamma < \infty$ and*

$$V(\mathbf{x}_n) \underset{n\to\infty}{\longrightarrow} \alpha, \quad \|\mathbf{x}_i\| \leq \beta, \quad r_2(\|\mathbf{x}_i\|) \leq \gamma, \qquad \forall i, \text{ a.s.} \tag{2.167}$$

Proof. As in (2.108), we obtain

$$V(\mathbf{x}_{n+1}) \leq V(\mathbf{x}_n) - a_n(\mathbf{f}_{n+1}^\tau + \mathbf{w}_{n+1}^\tau)V_{\mathbf{x}}(\mathbf{x}_n) + \frac{lk_1}{2}a_n^2\|\mathbf{f}_{n+1} + \mathbf{w}_{n+1}\|^2, \tag{2.168}$$

and by Condition A_4

$$E\left(\frac{v(\mathbf{x}_{n+1})}{\mathscr{F}_n}\right) \leq V(\mathbf{x}_n) + a_n(\mathbf{g}_{n+1}^\tau + \mathbf{f}_{n+1}^\tau)V_{\mathbf{x}}(\mathbf{x}_n) + \frac{lk_1}{2}b_n^2 r_1(n) + \frac{lk_1}{2}a_n^2\|\mathbf{f}_{n+1}\|^2, \qquad n \geq n_0. \tag{2.169}$$

Denote

$$k_3 = k_2\big(lk_1 + lk_1\|\mathbf{x}^0\| + \|V_{\mathbf{x}}(\mathbf{x}^0)\|\big). \tag{2.170}$$

Paying attention to (2.21), we see that

$$a_i\|V_{\mathbf{x}}(\mathbf{x}_i)\| \le \frac{b_i}{z_i}\big(\|V_{\mathbf{x}}(\mathbf{x}^0)\| + lk_1\|\mathbf{x}_i\| + lk_1\|\mathbf{x}^0\|\big). \tag{2.171}$$

From (2.163) and (2.164) by (2.82), we obtain

$$\|\mathbf{f}_{i+1} - \mathbf{g}_{i+1}\| \le k_2 \sum_{j=0}^{i} \rho^{i-j} \sum_{s=0}^{d} \|\underset{\sim}{\mathbf{D}}_s\| \, \|\nabla h(\mathbf{x}_{j-s}) - h_{\mathbf{x}}(\mathbf{x}_{j-s})\|$$

$$\le k_2 \sum_{j=0}^{i} \rho^{i-j} \sum_{s=0}^{d} \|\underset{\sim}{\mathbf{D}}_s\| \Psi(\mathbf{x}_{j-s}) c_{j-s}, \tag{2.172}$$

where the last inequality comes from (2.155) and the agreement $c_i = 0$, $i < 0$. Then from (2.171) and (2.172) we have

$$E\sum_{i=0}^{\infty} a_i\|\mathbf{f}_{i+1} - \mathbf{g}_{i+1}\| \, \|V_{\mathbf{x}}(\mathbf{x}_i)\|$$

$$\le k_3 E \sum_{i=0}^{\infty} \frac{b_i}{z_i}(1 + \|\mathbf{x}_i\|) \sum_{j=0}^{i} \rho^{i-j} \sum_{s=0}^{d} \|\underset{\sim}{\mathbf{D}}_s\| c_{j-s} \max_{0\le k\le i} \Psi(\mathbf{x}_k)$$

$$\le k_3 E \sum_{i=0}^{\infty} b_i \sum_{j=0}^{i} \rho^{i-j} \sum_{s=0}^{d} \|\underset{\sim}{\mathbf{D}}_s\| c_{j-s}$$

$$= k_3 \sum_{s=0}^{d} \|\underset{\sim}{\mathbf{D}}_s\| E \sum_{i=0}^{\infty} \sum_{j=0}^{i} \rho^{i-j} b_i c_{j-s}.$$

Notice $c_i = 0$, $i < 0$, and $i \ge j - s$ in the preceding expression. Hence $b_{j-s} \ge b_i$. Then we also have the estimate

$$E\sum_{i=0}^{\infty} a_i\|\mathbf{f}_{i+1} - \mathbf{g}_{i+1}\| \, \|V_{\mathbf{x}}(\mathbf{x}_i)\| \le k_3 \sum_{s=0}^{d} \|\underset{\sim}{\mathbf{D}}_s\| E \sum_{j=0}^{\infty} \sum_{i=j}^{\infty} \rho^{i-j} b_{j-s} c_{j-s}$$

$$= \frac{k_3}{1-\rho} \sum_{s=0}^{d} \|\underset{\sim}{\mathbf{D}}_s\| E \sum_{j=0}^{\infty} b_{j-s} c_{j-s} = \frac{k_3}{1-\rho} \sum_{s=0}^{d} \|\underset{\sim}{\mathbf{D}}_s\| E \sum_{j=0}^{\infty} b_j c_j < \infty.$$

Denote

$$m_n = V(\mathbf{x}_n) + E\left(\sum_{i=0}^{\infty} \frac{a_i \|\mathbf{f}_{i+1} - \mathbf{g}_{i+1}\| \, \|V_{\mathbf{x}}(\mathbf{x}_i)\|}{\mathscr{F}_n}\right)$$

$$- \sum_{i=0}^{n-1} a_i \|\mathbf{f}_{i+1} - \mathbf{g}_{i+1}\| \, \|V_{\mathbf{x}}(\mathbf{x}_i)\| + \frac{lk_1}{2} E\left(\sum_{i=0}^{\infty} \frac{b_i^2 r_1(i)}{\mathscr{F}_n}\right)$$

$$- \frac{lk_1}{2} \sum_{i=0}^{n-1} b_i^2 r_1(i) + \frac{lk_1}{2} E\left(\sum_{i=0}^{\infty} \frac{a_i^2 \|\mathbf{f}_{i+1}\|^2}{\mathscr{F}_n}\right) - \frac{lk_1}{2} \sum_{i=0}^{n-1} a_i^2 \|\mathbf{f}_{i+1}\|^2. \tag{2.173}$$

We know that each term in (2.173) has finite expectation and by (2.169) $(m_n, \mathscr{F}_n)$ is a nonnegative supermartingale. The remaining part of proof is the same as that in Lemma 2.2. □

In (2.112)–(2.115), without a change of notation, we replace $h(\mathbf{x}_j)$ by $\nabla h(\mathbf{x}_j)$ and take $\underset{\sim}{\mathbf{L}} = \underset{\sim}{\mathbf{I}}$. In particular, (2.113) changes to

$$\underset{\sim}{\mathbf{G}}_{n,n} = 0, \qquad \underset{\sim}{\mathbf{G}}_{n+1,i} = \frac{1}{a_i} \sum_{j=i}^{n} a_j \underset{\sim}{\mathbf{G}}\underset{\sim}{\mathbf{F}}^{j-i} \underset{\sim}{\mathbf{G}}^{\tau} \underset{\sim}{\mathbf{D}}(z) \nabla h(\mathbf{x}_i). \tag{2.174}$$

Estimate (2.166) will play role of (2.12).

Under the conditions of Lemma 2.12 it can be shown, just as in Lemma 2.9, that for almost all ω there exist a subsequence $\{\mathbf{x}_{n_k}(t)\}$ of $\{\mathbf{x}_n(t)\}$ and a continuous function $\mathbf{x}(t)$ which is the uniform limit of $\mathbf{x}_{n_k}(t)$

$$\mathbf{x}_{n_k}(t) \underset{k \to \infty}{\to} \mathbf{x}(t) \tag{2.175}$$

over any finite interval.

Lemma 2.13. *Under conditions of Lemma* 2.12, $\mathbf{x}(t)$ *figured in* (2.175) *satisfies the differential equation*

$$\frac{d\mathbf{x}(t)}{dt} = -\left(\underset{\sim}{\mathbf{I}} + \sum_{i=1}^{r} \underset{\sim}{\mathbf{C}}_i\right)^{-1} \left(\underset{\sim}{\mathbf{I}} + \sum_{i=1}^{d} \underset{\sim}{\mathbf{D}}_i\right) h_{\mathbf{x}}(\mathbf{x}). \tag{2.176}$$

Proof. We go back to the proof of Lemma 2.10. (2.132)–(2.134) remain true if we apply (2.166) instead of (2.12), replace $h(\mathbf{x}_i)$ by $\nabla h(\mathbf{x}_i)$, and take

$\underset{\sim}{\mathbf{L}} = \underset{\sim}{\mathbf{I}}$. Now (2.134) takes form

$$\lim_{\Delta \to 0} \frac{\mathbf{x}(t+\Delta) - \mathbf{x}(t)}{\Delta}$$

$$= - \lim_{\Delta \to 0} \lim_{k \to \infty} \frac{1}{\Delta} \sum_{j=m(t+n_k)}^{m(t+n_k+\Delta)-1} \sum_{i=m(t+n_k)}^{j} a_j \underset{\sim}{\mathbf{G}} \underset{\sim}{\mathbf{F}}^{j-i} \underset{\sim}{\mathbf{G}}^{\tau} \underset{\sim}{\mathbf{D}}(z)$$

$$\times \left[\nabla h(\mathbf{x}_i) - h_{\mathbf{x}}(\mathbf{x}(t)) \right]$$

$$- \lim_{\Delta \to 0} \lim_{k \to \infty} \frac{1}{\Delta} \sum_{j-m(t+n_k)}^{m(t+n_k+\Delta)-1} \sum_{i=m(t+n_k)}^{j} a_j \underset{\sim}{\mathbf{G}} \underset{\sim}{\mathbf{F}}^{j-i} \underset{\sim}{\mathbf{G}}^{\tau} \underset{\sim}{\mathbf{D}}(z) h_{\mathbf{x}}(\mathbf{x}(t)). \tag{2.177}$$

We note that (2.135)–(2.137) are still true, and for any i, $0 \le i \le d + m(t + n_k + \Delta) - m(t + n_k)$, there exists Δ_{ki}^j such that

$$|\Delta_{ki}^j| \le c_{m(t+n_k)-d+i}$$

and

$$\frac{1}{2c_{m(t+n_k)-d+i}} \left[h\left(\mathbf{x}_{m(t+n_k)-d+i}^{j+}\right) - h\left(\mathbf{x}_{m(t+n_k)-d+i}^{j-}\right) \right]$$

$$= \frac{1}{2c_{m(t+n_k)-d+i}} \left\{ h\left(\left[x^{*1}\left(t_{m(t+n_k)-d+i}\right), \dots, x^{*j}\left(t_{m(t-n_k)-d+i}\right) \right.\right.\right.$$

$$\left.\left. + c_{m(t+n_k)-d+i}, \dots, x^{*l}\left(t_{m(t+n_k)-d+i}\right) \right]^{\tau} \right)$$

$$- h\left(\left[x^{*1}\left(t_{m(t+n_k)-d+i}\right), \dots, x^{*j}\left(t_{m(t+n_k)-d+i}\right) \right.\right.$$

$$\left.\left.\left. - c_{m(t+n_k)-d+i}, \dots, x^{*l}\left(t_{m(t+n_k)-d+i}\right) \right] \right)^{\tau} \right\}$$

$$= h_{x^j}\left(\left[x^{*1}(n_k + t + \delta'_{ki}), \dots, x^{*j}(n_k + t + \delta'_{ki} + \Delta_{ki}^j), \right.\right.$$

$$\left.\left. \dots, x^{*l}(n_k + t + \delta'_{ki}) \right]^{\tau} \right)$$

$$= h_{x^j}\left(\left[x_{n_k}^1(t + \delta'_{ki}), \dots, x_{n_k}^j(t + \delta'_{ki} + \Delta_{ki}^j), \dots, x_{n_k}^l(t + \delta'_{ki}) \right]^{\tau} \right),$$

where $x^{*i}(t)$ and $x_{n_k}^i(t)$ denote the ith component of $\mathbf{x}^*(t)$ and $\mathbf{x}_{n_k}(t)$, respectively.

Hence we have

$$\left\| \frac{1}{2c_{m(t+n_k)-d+i}} \left[h\left(\mathbf{x}^{j+}_{m(t+n_k)-d+i}\right) - h\left(\mathbf{x}^{j-}_{m(t+n_k)-d+i}\right) \right] - h_{x^j}(\mathbf{x}(t)) \right\|$$

$$\leq \left\| h_{x^j}\left(\left[x^1_{n_k}(t+\delta'_{ki}), \ldots, x^j_{n_k}(t+\delta'_{ki}+\Delta_{ki}), \ldots, x^l_{n_k}(t+\delta'_{ki}) \right]^\tau \right) - h_{x^j}\left(\mathbf{x}_{n_k}(t)\right) \right\|$$

$$+ \left\| h_{x^j}\left(\mathbf{x}_{n_k}(t)\right) - h_{x^j}(\mathbf{x}(t)) \right\|.$$

By the continuity of h_{x^j}, the equicontinuity of $\{\mathbf{x}_n(t)\}$ and the fact that $\Delta_{ki} \underset{k\to\infty}{\to} 0\ \forall i$, with $0 \leq i \leq d + m(t+n_k+\Delta) - m(t+n_k)$, we find that for fixed ω and t on the right-hand side of the preceding inequality the first term can be made arbitrarily small if Δ is small enough and k large enough, while the second term by (2.175) will be also arbitrarily small if k is sufficiently large. Hence we have

$$\overline{\lim_{\Delta\to 0}}\ \overline{\lim_{k\to\infty}} \max_{0\leq i\leq d+m(t+n_k+\Delta)-m(t+n_k)} \left\| \nabla h\left(\mathbf{x}_{m(t+n_k)-d+i}\right) - h_x(\mathbf{x}(t)) \right\| = 0,$$

and the first term on the right-hand side of (2.177) is dominated by

$$\overline{\lim_{\Delta\to 0}}\ \overline{\lim_{k\to\infty}} \frac{k_2}{\Delta} \sum_{j=m(t+n_k)}^{m(t+n_k+\Delta)-1} \sum_{i=m(t+n_k)}^{j}$$

$$\times a_j \rho^{j-i} \sum_{s=0}^{d} \|D_s\| \left\| \nabla h(\mathbf{x}_{i-s}) - h_{\mathbf{x}}(\mathbf{x}(t)) \right\|$$

$$\leq \overline{\lim_{\Delta\to 0}}\ \overline{\lim_{k\to\infty}} k_2 \sum_{s=0}^{d} \|\underset{\sim}{\mathbf{D}}_s\|$$

$$\times \max_{0\leq i\leq d+m(t+n_k+\Delta)-m(t+n_k)} \left\| \nabla h\left(\mathbf{x}_{m(t+n_k)-d+i}\right) - h_{\mathbf{x}}(\mathbf{x}(t)) \right\|$$

$$\cdot \frac{1}{\Delta} \sum_{j=m(t+n_k)}^{m(t+n_k+\Delta)-1} a_j \sum_{i=m(t+n_k)}^{j} \rho^{j-i}$$

$$\leq k_2 \sum_{s=0}^{d} \|D_s\| \overline{\lim_{\Delta\to 0}}\ \overline{\lim_{k\to\infty}}$$

$$\times \max_{0\leq i\leq d+m(t+n_k+\Delta)-m(t+n_k)} \left\| \nabla h\left(\mathbf{x}_{m(t+n_k)-d+i}\right) - h_{\mathbf{x}}(\mathbf{x}(t)) \right\|$$

$$\times \frac{\left(\Delta + a_{m(t+n_k)}\right)}{\Delta}$$

$$= 0.$$

Finally, (2.176) follows by a similar argument as (2.141)–(2.144). □

Theorem 2.7. *Suppose that the measurement is given by* (2.151), *the* $\mathscr{F}_i$*-measurable random variables* $b_i > 0$, $c_i > 0$ *satisfy* (2.149), *that* $h(\mathbf{x})$ *is continuously differentiable and that* (2.148) *and* (2.155) *are fulfilled for* $h(\mathbf{x})$. *Further, if* $r = 0$, *or* $r > 0$ *and* $\det \underset{\sim}{C}(\lambda)$ *has all zeros outside the closed unit disk and if Condition* A_4 *holds, then*

$$\mathbf{x}_n \underset{n\to\infty}{\to} \mathbf{x}^0 \text{ a.s.}$$

where $\mathbf{x}_n$ *is given by* (2.159) *and* $\mathbf{x}^0$ *is the minimum of* $h(\mathbf{x})$ [*in* (2.154) *the converse inequality takes place and the minus sign is replaced by the plus sign in* (2.159), *then* $\mathbf{x}^0$ *is the maximum of* $h(\mathbf{x})$].

Proof. The proof of Theorem 2.6 gives required conclusion if we replace $h(\mathbf{x})$, $\mathbf{f}_{n+1}$ and $\underset{\sim}{L}$ by $h_{\mathbf{x}}(\mathbf{x})$, $\mathbf{g}_{n+1}$, and $\underset{\sim}{I}$, respectively. □

If we restrict Theorem 2.7 to the special case $r = d = 0$, we find that a comparison shows the conditions imposed here to be much weaker than those in the proposition mentioned at the begining of the section.

In the preceding proposition it is required that $E\boldsymbol{\varepsilon}_i(\mathbf{x}, \mathbf{e}_i) = 0$ and $\mathbf{e}_i$ is independent of $\mathscr{F}_{i-1}$. These imply, as noted in 2.2 $E[\boldsymbol{\varepsilon}_i(\mathbf{x}_{i-1}, \mathbf{e}_i)/\mathscr{F}_{i-1}] = 0$ which is imposed here in Theorem 2.7 for the case of $d = r = 0$. There, $E\|\boldsymbol{\varepsilon}_i(\mathbf{x}, \mathbf{e}_i)\|^2$ as a function of i is uniformly bounded, here it is allowed to increase at a certain rate; there as a function of $\|\mathbf{x}\|$ it satisfies a Lipschitz condition and is dominated by $k_0(1 + V(\mathbf{x}))$, here it satisfies (2.155) and the dominating function can be arbitrary; there $V_{\mathbf{x}}(\mathbf{x})$ is bounded by $k_0(1 + V(\mathbf{x}))$, here it is arbitrary; finally, there b_i, c_i figured in (2.149) are deterministic and must satisfy $\sum_{i=0}^{\infty} b_i^2/c_i^2 < \infty$, while here they can be random without the last restriction.

2.5. CONTINUOUS-TIME STOCHASTIC APPROXIMATION

In this section we consider the problem of seeking for zero $\mathbf{x}^0 \in R^l$ of an m-dimensional $\mathbf{h}(\mathbf{x})$ and for extremum $\mathbf{x}^0 \in R^l$ of a scalar function $h(\mathbf{x})$ when the measurement process evolves in continuous time.

Similar to (2.12) and (2.148) for discrete-time measurement, we place no constraint on the rate of increase of $\|\mathbf{h}(\mathbf{x})\|$ and $\|h_{\mathbf{x}}(\mathbf{x})\|$ as $\|\mathbf{x}\| \to \infty$, but assume their dominating function $r_2(\cdot)$ is known:

$$\|h(\mathbf{x})\| \le r_2(\|\mathbf{x}\|) \quad \text{for the MRM procedure} \tag{2.178}$$

$$\|h_{\mathbf{x}}(\mathbf{x})\| \le r_2(\|\mathbf{x}\|) \quad \text{for the MKW procedure,} \tag{2.179}$$

where the MRM and MKW procedures will be described later for measurement error process with dependent increments.

Clearly, without loss of generality, we can assume that $r_2(\cdot)$ is continuous, nondecreasing, and is bounded from below by

$$r_2(\|\mathbf{x}\|) \geq 1 + \|\mathbf{x}\|^2.$$

Denote

$$\mathbf{x} = [x^1, \ldots, x^l]^\tau, \qquad \mathbf{x}_t = [x_t^1, \ldots, x_t^l]^\tau, \tag{2.180}$$

and in addition

$$h_{\mathbf{x}}(\mathbf{x}) = \left[\frac{\partial h}{\partial x^1}, \ldots, \frac{\partial h}{\partial x^l}\right]^\tau, \tag{2.181}$$

$$\mathbf{x}_t^{i\pm} = [x_t^1, \ldots, x_t^{i-1}, x_t^i \pm c_t, x_t^{i+1}, \ldots, x_t^l]^\tau, \tag{2.182}$$

$$\bar{\mathbf{x}}_t = [x_t^1 + c_t \operatorname{sign} x_t^1, \ldots, x_t^l + c_t \operatorname{sign} x_t^l]^\tau, \tag{2.183}$$

$$\nabla h(\mathbf{x}_t) = \frac{1}{2c_t}\left[h(\mathbf{x}_t^{1+}) - h(\mathbf{x}_t^{1-}), \ldots, h(\mathbf{x}_t^{l+}) - h(\mathbf{x}_t^{l-})\right]^\tau, \tag{2.184}$$

if the MKW procedure applies, and in this case we also assume (2.155) holds with $\nabla h(\mathbf{x})$ defined by (2.184), but with $\mathbf{x}_t$, c_t replaced by $\mathbf{x}$ and c, respectively.

Let $(\Omega, \mathscr{F}, P)$ be the basic probability space and $\{\mathscr{F}_t\}$ be the nondecreasing family of σ-algebras and $(\mathbf{w}_t, \mathscr{F}_t)$ be an m'-dimensional Wiener process. Assume that $m \times m'$ matrix $\underset{\sim}{\mathbf{F}}_t(\boldsymbol{\xi}_t, \omega)$ is $\mathscr{F}_t$-measurable for any m-dimensional measurable and $\mathscr{F}_t$-adapted process $\boldsymbol{\xi}_t$ and that $\|\underset{\sim}{\mathbf{F}}_t(\mathbf{x}, \omega)\|$ may increase with t and $\|\mathbf{x}\|$ but its dominated function is known as in

$$\|\underset{\sim}{\mathbf{F}}_t(\mathbf{x}, \omega)\|^2 \leq r_1(t) r_2(\|\mathbf{x}\|), \tag{2.185}$$

where, without loss of generality, $r_2(\cdot)$ can be taken the same as in (2.178) and (2.179).

Denote by S the integral operator

$$S\mathbf{x}_t = \int_0^t \mathbf{x}_s \, ds, \tag{2.186}$$

and by $\mathbf{x}_t$ the estimate of $\mathbf{x}^0$ at time t.

When we look for a zero of $\mathbf{h}(\mathbf{x})$, $\mathbf{h}(\mathbf{x}_t)$ is observed and while we search for the extremum of $h(\mathbf{x})$, $\nabla h(\mathbf{x}_t)$ is observed, but both measurements are assumed to be corrupted by random error $\boldsymbol{\varepsilon}_t$ which is expressed by

$$\boldsymbol{\varepsilon}_t = \underset{\sim}{\mathbf{C}}(S)\int_0^t \underset{\sim}{\mathbf{F}}_s(\mathbf{x}_s, \omega)\, d\mathbf{w}_s \tag{2.187}$$

where the polynomial $\underset{\sim}{\mathbf{C}}(\lambda)$ is given by (2.68) and S is defined by (2.186) with $S^0 = \underset{\sim}{\mathbf{I}}$.

Assume that the measurement $\mathbf{y}_t$ is given by

$$\mathbf{y}_t = \mathbf{y}_0 + S^{r+1}\mathbf{h}(\mathbf{x}_t) + \boldsymbol{\varepsilon}_t \tag{2.188}$$

for the MRM procedure and by

$$\mathbf{y}_t = \mathbf{y}_0 + S^{r+1}\nabla h(\mathbf{x}_t) + \boldsymbol{\varepsilon}_t \tag{2.189}$$

for the MKW procedure.

Suppose that the stochastic processes b_t and c_t are $\mathscr{F}_t$-adapted, that b_t is nonincreasing, that

$$b_t > 0, \qquad c_t > 0, \qquad c_0 = 0, \qquad c_t \underset{t\to\infty}{\to} 0$$

and that

$$\int_0^\infty b_t\, dt = \infty \text{ a.s.}, \qquad E\int_0^\infty b_t^2\, dt < \infty, \qquad E\int_0^\infty b_t c_t\, dt < \infty,$$

$$E\int_0^\infty b_t^2 r_1(t)\, dt < \infty \tag{2.190}$$

c_t is introduced here only for the MKW procedure and the conditions on it are not required for the MRM procedure.

If $r = 0$ and x_t is $\mathscr{F}_t$-adapted, then $\boldsymbol{\varepsilon}_t$ is a process of independent increments, and in this case the classical RM and KW procedures are described by

$$d\mathbf{x}_t = b_t\, d\mathbf{y}_t, \qquad m = l. \tag{2.191}$$

However, we consider the general $r \geq 0$ case. In addition to (2.73) and

(2.74), we denote

$$H^\tau = \begin{cases} \Big[\underbrace{0 \ldots 0 I}_{mr}\Big]\Big\} m & \text{for } r > 0 \\ I & \text{for } r = 0 \end{cases} \tag{2.192}$$

$$z_t = \begin{cases} \max\limits_{0 \le s \le t} r_2(\|\bar{\mathbf{x}}_s\| & \text{for the MRM procedure} \\ \max\limits_{0 \le s \le t} r_2(\|\bar{\mathbf{x}}_s\|) \max\limits_{0 \le s \le t} \Psi(\mathbf{x}_s) & \text{for the MKW procedure} \end{cases} \tag{2.193}$$

$$a_t = \frac{b_t}{z_t}. \tag{2.194}$$

Algorithm (2.191) is modified to

$$d\mathbf{x}_t = -a_t \underset{\sim}{\mathbf{L}} \left[d\mathbf{y}_t - \underset{\sim}{\mathbf{C}}_1(\mathbf{y}_t - \mathbf{y}_0)\, dt + \int_0^t \underset{\sim}{\mathbf{G}}\underset{\sim}{\mathbf{F}}^2 e^{\mathbf{F}(t-\lambda)} \underset{\sim}{\mathbf{G}}^\tau (\mathbf{y}_\lambda - \mathbf{y}_0)\, d\lambda\, dt \right], \tag{2.195}$$

and is called an MRM procedure if $\underset{\sim}{\mathbf{L}}$ is an $l \times m$ matrix and the measurement is given by (2.188); it is called an MKW procedure if the measurement is given by (2.189) and $\underset{\sim}{\mathbf{L}}$ is an $l \times l$ matrix. Of course, we have to impose the condition that the stochastic differential equations (2.188) (or (2.189)) and (2.195) have a strong solution $([\mathbf{x}_t^\tau \mathbf{y}_t^\tau]^\tau, \mathscr{F}_t)$.

Define

$$\mathbf{f}_t = \begin{cases} \mathbf{h}(\mathbf{x}_t), & r = 0 \\ \underset{\sim}{\mathbf{G}} \int_0^t e^{\mathbf{F}(t-\lambda)} \underset{\sim}{\mathbf{H}} \mathbf{h}(\mathbf{x}_\lambda)\, d\lambda, & r > 0 \end{cases} \quad \text{for MRM} \tag{2.196}$$

or

$$\mathbf{f}_t = \begin{cases} \nabla h(\mathbf{x}_t), & r = 0 \\ \underset{\sim}{\mathbf{G}} \int_0^t e^{\mathbf{F}(t-\lambda)} \underset{\sim}{\mathbf{H}} \nabla h(\mathbf{x}_\lambda)\, d\lambda, & r > 0 \end{cases} \quad \text{for MKW.} \tag{2.197}$$

In the sequel we denote $\underset{\sim}{\mathbf{F}}_t(\mathbf{x}_t, \omega)$ simply by $\underset{\sim}{\mathbf{F}}_t$.

Lemma 2.14. *Algorithm* (2.195) *can be rewritten as*

$$d\mathbf{x}_t = -a_t\mathbf{L}(\mathbf{f}_t\,dt + \mathbf{F}_t\,d\mathbf{w}_t). \tag{2.198}$$

Proof. For $r = 0$, the lemma is trivial. Now assume $r > 0$ and consider the mr-dimensional vector $\boldsymbol{\zeta}_t$ defined by

$$\boldsymbol{\zeta}_t = S\mathbf{F}\boldsymbol{\zeta}_t + \mathbf{G}^\tau(\mathbf{y}_t - \mathbf{y}_0). \tag{2.199}$$

Obviously, $\boldsymbol{\zeta}_t$ can be expressed as

$$\boldsymbol{\zeta}_t = \mathbf{G}^\tau(\mathbf{y}_t - \mathbf{y}_0) + \int_0^t \mathbf{F}e^{\mathbf{F}(t-\lambda)}\mathbf{G}^\tau(\mathbf{y}_\lambda - \mathbf{y}_0)\,d\lambda. \tag{2.200}$$

From the proof of Lemma 2.6, it is easy to see

$$\mathbf{C}(S)\mathbf{G}\boldsymbol{\zeta}_t = \mathbf{y}_t - \mathbf{y}_0. \tag{2.201}$$

Hence from (2.188), (2.189), (2.200), and (2.201), it follows that

$$\begin{aligned}
&\mathbf{C}(S)\left[\mathbf{y}_t - \mathbf{y}_0 + \mathbf{G}\int_0^t \mathbf{F}e^{\mathbf{F}(t-\lambda)}\mathbf{G}^\tau(\mathbf{y}_\lambda - \mathbf{y}_0)\,d\lambda - \int_0^t \mathbf{F}_s\,d\mathbf{w}_s\right]\\
&= \mathbf{y}_t - \mathbf{y}_0 - \mathbf{C}(S)\int_0^t \mathbf{F}_s\,d\mathbf{w}_s\\
&= \begin{cases} S^{r+1}\mathbf{h}(\mathbf{x}_t) & \text{for the MRM procedure}\\ S^{r+1}\nabla h(\mathbf{x}_t) & \text{for the MKW procedure.}\end{cases}
\end{aligned} \tag{2.202}$$

From the differential equation

$$\dot{\boldsymbol{\eta}}_t = \begin{cases} \mathbf{F}\boldsymbol{\eta}_t + \mathbf{H}S\mathbf{h}(\mathbf{x}_t) & \text{for MRM}\\ \mathbf{F}\boldsymbol{\eta}_t + \mathbf{H}\nabla h(\mathbf{x}_t) & \text{for MKW,}\end{cases} \tag{2.203}$$

and in analogy with Lemma 2.6, we obtain

$$\mathbf{C}(S)\mathbf{G}\boldsymbol{\eta}_t = \begin{cases} S^{r+1}\mathbf{h}(\mathbf{x}_t) & \text{for MRM}\\ S^{r+1}\nabla h(\mathbf{x}_t) & \text{for MKW}\end{cases} \tag{2.204}$$

From (2.202)–(2.204), we know that

$$\mathbf{y}_t - \mathbf{y}_0 + \underset{\sim}{\mathbf{G}} \int_0^t \underset{\sim}{\mathbf{F}} e^{\underset{\sim}{\mathbf{F}}(t-\lambda)} \underset{\sim}{\mathbf{G}}^\tau (\mathbf{y}_\lambda - \mathbf{y}_0)\, d\lambda - \int_0^t \underset{\sim}{\mathbf{F}}_s\, d\mathbf{w}_s$$

$$= \begin{cases} \underset{\sim}{\mathbf{G}} \int_0^t e^{\underset{\sim}{\mathbf{F}}(t-s)} \underset{\sim}{\mathbf{H}} \int_0^s \mathbf{h}(\mathbf{x}_\lambda)\, d\lambda\, ds & \text{for MRM} \\ \underset{\sim}{\mathbf{G}} \int_0^t e^{\underset{\sim}{\mathbf{F}}(t-s)} \underset{\sim}{\mathbf{H}} \int_0^s \nabla h(\mathbf{x}_\lambda)\, d\lambda\, ds & \text{for MKW.} \end{cases} \tag{2.205}$$

Hence (2.195) can be rewritten as

$$d\mathbf{x}_t = \begin{cases} -a_t \underset{\sim}{\mathbf{L}} \left[\underset{\sim}{\mathbf{G}}\, d \int_0^t e^{\underset{\sim}{\mathbf{F}}(t-s)} \underset{\sim}{\mathbf{H}} \int_0^s \mathbf{h}(\mathbf{x}_\lambda)\, d\lambda\, ds + \underset{\sim}{\mathbf{F}}_t\, d\mathbf{w}_t \right. \\ -a_t \underset{\sim}{\mathbf{L}} \left[\underset{\sim}{\mathbf{G}}\, d \int_0^t e^{\underset{\sim}{\mathbf{F}}(t-s)} \underset{\sim}{\mathbf{H}} \int_0^s \nabla h(\mathbf{x}_\lambda)\, d\lambda\, ds + \underset{\sim}{\mathbf{F}}_t\, d\mathbf{w}_t. \right. \end{cases} \tag{2.206}$$

Noticing that $\underset{\sim}{\mathbf{G}}\underset{\sim}{\mathbf{H}} = \underset{\sim}{\mathbf{0}}$, integration by parts yields

$$\underset{\sim}{\mathbf{G}}\, d \int_0^t e^{\underset{\sim}{\mathbf{F}}(t-s)} \underset{\sim}{\mathbf{H}} \int_0^s \mathbf{h}(\mathbf{x}_\lambda)\, d\lambda\, ds = \underset{\sim}{\mathbf{G}}\underset{\sim}{\mathbf{F}} \int_0^t e^{\underset{\sim}{\mathbf{F}}(t-s)} \underset{\sim}{\mathbf{H}} \int_0^s \mathbf{h}(\mathbf{x}_\lambda)\, d\lambda\, ds\, dt$$

$$= -\underset{\sim}{\mathbf{G}} \int_0^t \frac{d}{ds} e^{\underset{\sim}{\mathbf{F}}(t-s)} \underset{\sim}{\mathbf{H}} \int_0^s \mathbf{h}(\mathbf{x}_\lambda)\, d\lambda\, ds\, dt$$

$$= -\underset{\sim}{\mathbf{G}} e^{\underset{\sim}{\mathbf{F}}(t-s)} \underset{\sim}{\mathbf{H}} \int_0^s \mathbf{h}(\mathbf{x}_\lambda)\, d\lambda \Big|_0^t dt + \underset{\sim}{\mathbf{G}} \int_0^t e^{\underset{\sim}{\mathbf{F}}(t-s)} \underset{\sim}{\mathbf{H}} \mathbf{h}(\mathbf{x}_s)\, ds\, dt$$

$$= \underset{\sim}{\mathbf{G}} \int_0^t e^{\underset{\sim}{\mathbf{F}}(t-s)} \underset{\sim}{\mathbf{H}} \mathbf{h}(\mathbf{x}_s)\, ds\, dt,$$

and from here and (2.206) the desired expression (2.198) follows. □

Let $\mathbf{g}_t$ be the solution of the equations

$$\frac{d^r}{dt^r} \underset{\sim}{\mathbf{C}}(S)\mathbf{g}_t = h_{\mathbf{x}}(\mathbf{x}_t) \quad \text{for MKW} \tag{2.207}$$

$$\frac{d^r}{dt^r} \underset{\sim}{\mathbf{C}}(S)\mathbf{g}_t = \mathbf{h}(\mathbf{x}_t) \quad \text{for MRM} \tag{2.208}$$

with initial values $\mathbf{g}_0^{(i)}$, $i = 0, \ldots, r-1$.

We need the following conditions.

CONDITION A_5. There exists an $l \times l$ matrix $\underset{\sim}{L}$ and a function $V(\mathbf{x})$, with continuous second derivative, such that $V(\mathbf{x}) > \tilde{0}\ \forall \mathbf{x} \neq \mathbf{x}^0$, $V(\mathbf{x}^0) = 0$, $V(\mathbf{x}) \underset{\|\mathbf{x}\| \to \infty}{\to} \infty$, $|V_{x^i x^j}(\mathbf{x})| \le k_1$, $i, j = 1, \dots, l$, and for some t_0,

$$V_{\mathbf{x}}^T(\mathbf{x}_t)\underset{\sim}{L}\mathbf{g}_t > 0, \qquad \forall \mathbf{x}_t \neq \mathbf{x}^0,\ t \ge t_0,$$

where $\mathbf{g}_t$ is the solution of (2.207) for both $\mathbf{x}_t \equiv \mathbf{x}$, $\forall \mathbf{x} \neq \mathbf{x}^0$ and for $\mathbf{x}_t$ given by (2.195).

CONDITION A_6. The same condition as A_5 but with (2.207) replaced by (2.208) and with the dimension of $\underset{\sim}{L}$ being $l \times m$, $l \le m$.

Lemma 2.15. *Suppose that* $r = 0$, *or* $r > 0$ *and all zeros of* $\det \underset{\sim}{C}(\lambda)$ *have negative real parts. Then*

$$E\|\mathbf{x}_t\|^2 < \infty \quad \textit{and} \quad E\int_0^\infty a_t^2 \|\underset{\sim}{F}_t\|^2\, dt < \infty,$$

where $\mathbf{x}_t$ *is defined by* (2.198).

Proof. Since $r_2(\|\mathbf{x}_t\|) \le r_2(\|\bar{\mathbf{x}}_t\|)$, we have

$$E\int_0^\infty a_t^2 \|\underset{\sim}{F}_t\|^2\, dt \le E\int_0^\infty \frac{1}{\max\limits_{0\le s\le t} r_2^2(\|\bar{\mathbf{x}}_s\|)} b_t^2 r_1(t) r_2(\|\mathbf{x}_t\|)\, dt$$

$$\le E\int_0^\infty b_t^2 r_1(t)\, dt < \infty. \tag{2.209}$$

Let λ_k, $k = 1, \dots, mr$, denote the zeros of $\det \underset{\sim}{C}(\lambda)$

$$\lambda_k = a_k + ib_k, \quad a_k < 0, \qquad k = 1, \dots, mr.$$

By Lemma 2.5, the eigenvalues of $\underset{\sim}{F}$ are given by

$$\mu_k \triangleq \frac{1}{\lambda_k} = \frac{1}{a_k + ib_k} = \frac{a_k - ib_k}{a_k^2 + b_k^2}, \qquad k = 1, \dots, mr \tag{2.210}$$

and these have negative real parts. Hence there exists a constant $c > 0$ such that

$$\operatorname{Re}\mu_i < -2c, \qquad i = 1, \dots, mr.$$

Assume the Jordan block $\underset{\sim}{\mathbf{J}}_i$ corresponding to μ_i is a $\gamma_i \times \gamma_i$ matrix. Then we have

$$\exp(\underset{\sim}{\mathbf{J}}_i t) = \begin{bmatrix} \exp(\mu_i t) & t\exp(\mu_i t) & \cdots & \dfrac{t^{\gamma_{i-1}}}{(\gamma_{i-1})!}\exp(\mu_i t) \\ 0 & & & \vdots \\ \vdots & \ddots & \ddots & \\ 0 & \cdots & 0 & \exp(\mu_i t) \end{bmatrix} \tag{2.211}$$

and the nonzero elements in (2.211) are estimated by

$$\left| \frac{t^s}{s!} e^{\mu_i t} \right| \le t^s e^{-2ct} = e^{-ct} t^s e^{-ct} \le k_4 e^{-ct} \tag{2.212}$$

with a common constant k_4 for any s, $1 \le s \le \gamma_{i-1}$ any i, $i = 1, \ldots, mr$ and any $t \in [0, \infty)$ since $t^s e^{-ct} \underset{t\to\infty}{\to} 0$. Thus we have the estimate

$$\|e^{\underset{\sim}{\mathbf{F}}t}\| \le k_4 e^{-ct}, \qquad \forall t \in [0, \infty) \tag{2.213}$$

Just as in (2.166), for any ω, there exists δ_t^i, $|\delta_t^i| \le c_t$, $1 \le i \le l$, such that

$$\|\nabla h(\mathbf{x}_t)\|^2 = \sum_{i=1}^{l} \left(\frac{\partial h(\mathbf{x})}{\partial x^i} \bigg|_{\mathbf{x}=[x_t^1, \ldots, x_t^i + \delta_t^i, \ldots, x_t^l]^\tau} \right)^2 \le l r_2^2(\|\overline{\mathbf{x}}_t\|), \tag{2.214}$$

and hence, for $r > 0$, from (2.178), (2.196), (2.197), and (2.213), it follows that

$$a_t^2 \|\mathbf{f}_t\|^2 \le \begin{cases} k_4^2 \dfrac{b_t^2}{z_t^2} \left[\displaystyle\int_0^t e^{-c(t-\lambda)} \|\mathbf{h}(\mathbf{x}_\lambda)\| \, d\lambda \right]^2 \\ k_4^2 \dfrac{b_t^2}{z_t^2} \left[\displaystyle\int_0^t e^{-c(t-\lambda)} \|\nabla h(\mathbf{x}_\lambda)\| \, d\lambda \right]^2 \end{cases}$$

$$\le \begin{cases} k_4^2 b_t^2 \left[\displaystyle\int_0^t e^{-c(t-\lambda)} \, d\lambda \right]^2 & \text{for MRM} \\ k_4^2 b_t^2 \left[\displaystyle\int_0^t e^{-c(t-\lambda)} \, d\lambda \right]^2 & \text{for MKW} \end{cases}$$

and

$$E\int_0^\infty a_t^2 \|\mathbf{f}_t\|^2\,dt \le \frac{k_4^2}{c^2} E\int_0^\infty b_t^2\,dt < \infty. \tag{2.215}$$

Finally, for $r > 0$, from (2.198), (2.209), and (2.215), we have

$$E\|\mathbf{x}_t\|^2 \le 3E\|\mathbf{x}_0\|^2 + 3t\|\underset{\sim}{\mathbf{L}}\|^2 E\int_0^t a_t^2\|\mathbf{f}_t\|^2\,dt + 3\|\underset{\sim}{\mathbf{L}}\|^2 E\int_0^t a_s^2\|\underset{\sim}{\mathbf{F}}_s\|^2\,ds < \infty,$$

$$\forall t \in [0, \infty).$$

When $r = 0$, from the definition of $\mathbf{f}_t$ and (2.178) and (2.214), we know that

$$E\int_0^\infty a_t^2\|\mathbf{f}_t\|^2\,dt \le lE\int_0^\infty b_t^2\,dt < \infty$$

and by (2.198) and (2.209)

$$E\|\mathbf{x}_t\|^2 < \infty, \qquad \forall t \in [0, \infty). \qquad \square$$

Lemma 2.16. *Under Condition* A_5 *and the conditions of Lemma* 2.15, *there are random variables* α, β, γ *such that* $0 \le \alpha < \infty$, $0 < \beta < \infty$, $0 < \gamma < \infty$ *and*

$$V(\mathbf{x}_t) \underset{t\to\infty}{\longrightarrow} \alpha,\ \|\mathbf{x}_t\| \le \beta,\ r_2(\|\overline{\mathbf{x}_t}\|) \le \gamma \qquad \forall t \in [0, \infty) \text{ a.s.},$$

where $\mathbf{x}_t$ *is defined by the* MKW *procedure.*

Proof. Because of $|V_{x^i x^j}(\mathbf{x})| \le k_1$, we have

$$\|V_{\mathbf{x}}(\mathbf{x}_t)\| \le \|V_{\mathbf{x}}(\mathbf{x}^0)\| + lk_1(\|\mathbf{x}_t\| + \|\mathbf{x}^0\|), \tag{2.216}$$

and by Lemma 2.15 and continuity of $\mathbf{x}_t$

$$E\|V_{\mathbf{x}}(\mathbf{x}_t)\|^2 < \infty, \qquad E\int_0^t \|V_{\mathbf{x}}(\mathbf{x}_\lambda)\|^2\,d\lambda < \infty. \tag{2.217}$$

Noticing that

$$r_2(\|\mathbf{x}_t\|) \ge 1 + \|\mathbf{x}_t\|^2$$

we obtain

$$a_t\|V_{\mathbf{x}}(\mathbf{x}_t)\| \le \frac{b_t}{\max\limits_{0\le s\le t} r_2(\|\bar{\mathbf{x}}_s\|)\max\limits_{0\le s\le t}\Psi(\mathbf{x}_s)}\left[\|V_{\mathbf{x}}(\mathbf{x}^0) + k_1 l(\|\mathbf{x}_t\| + \|\mathbf{x}^0\|\right]$$

$$\le k_5 b_t / \max_{0\le s\le t}\Psi(\mathbf{x}_s),$$

where

$$k_5 = \|V_{\mathbf{x}}(\mathbf{x}^0)\| + k_1 l(1 + \|\mathbf{x}^0\|).$$

Now consider

$$\frac{d}{dt}\boldsymbol{\phi}_t = \underset{\sim}{\mathbf{F}}\boldsymbol{\phi}_t + \underset{\sim}{\mathbf{H}}h_{\mathbf{x}}(\mathbf{x}_t) \quad \text{with initial value } \boldsymbol{\phi}_0. \tag{2.218}$$

On the one hand,

$$\boldsymbol{\phi}_t = e^{\underset{\sim}{\mathbf{F}}t}\boldsymbol{\phi}_0 + \int_0^t e^{\underset{\sim}{\mathbf{F}}(t-\lambda)}\underset{\sim}{\mathbf{H}}h_{\mathbf{x}}(\mathbf{x}_\lambda)\,d\lambda, \qquad r > 0, \tag{2.219}$$

and on the other hand, a treatment similar to that given in (2.86) and (2.87) shows that the first m components $\underset{\sim}{\mathbf{G}}\boldsymbol{\phi}_t$ of $\boldsymbol{\phi}_t$ satisfy

$$\frac{d^r}{dt^r}\underset{\sim}{\mathbf{C}}(S)\underset{\sim}{\mathbf{G}}\boldsymbol{\phi}_t = h_{\mathbf{x}}(\mathbf{x}_t). \tag{2.220}$$

Take $\boldsymbol{\phi}_0 = 0$ and set

$$\mathbf{g}_t = \begin{cases} \underset{\sim}{\mathbf{G}}\int_0^t e^{\underset{\sim}{\mathbf{F}}(t-\lambda)}\underset{\sim}{\mathbf{H}}h_{\mathbf{x}}(\mathbf{x}_\lambda)\,d\lambda, & r > 0 \\ h_{\mathbf{x}}(\mathbf{x}_t), & r = 0. \end{cases} \tag{2.221}$$

From (2.220), it is known that $\mathbf{g}_t$ satisfies (2.207). Now by (2.155), (2.197), and (2.213), we have

$$\|\mathbf{f}_t - \mathbf{g}_t\| \le \begin{cases} k_4\int_0^t e^{-c(t-\lambda)}\|\nabla h(\mathbf{x}_\lambda) - h_{\mathbf{x}}(\mathbf{x}_\lambda)\|\,d\lambda & \\ \qquad \le k_4\int_0^t e^{-c(t-\lambda)}\Psi(\mathbf{x}_\lambda)c_\lambda\,d\lambda, & r > 0 \\ \Psi(\mathbf{x}_t)c_t, & r = 0. \end{cases} \tag{2.222}$$

Hence for $r > 0$,

$$E\int_0^\infty a_\lambda \| V_{\mathbf{x}}(\mathbf{x}_\lambda)\| \, \|\mathbf{f}_\lambda - \mathbf{g}_\lambda\| \, d\lambda$$

$$\leq k_4 k_5 E\int_0^\infty \frac{b_\lambda}{\max\limits_{0\leq s\leq \lambda} \Psi(\mathbf{x}_s)} \int_0^\lambda e^{-c(\lambda-\mu)} \Psi(\mathbf{x}_\mu) c_\mu \, d\mu \, d\lambda$$

$$\leq k_4 k_5 E\int_0^\infty b_\lambda \int_0^\lambda e^{-c(\lambda-\mu)} c_\mu \, d\mu \, d\lambda$$

$$= k_4 k_5 E\int_0^\infty \int_0^\infty b_\lambda e^{-c(\lambda-\mu)} c_\mu \, d\mu \, d\lambda$$

$$\leq k_4 k_5 E\int_0^\infty \int_\mu^\infty e^{-c(\lambda-\mu)} b_\mu c_\mu \, d\lambda \, d\mu$$

$$= \frac{k_4 k_5}{c} E\int_0^\infty b_\mu c_\mu \, d\mu < \infty, \tag{2.223}$$

and for $r = 0$

$$E\int_0^\infty a_\lambda \| V_{\mathbf{x}}(\mathbf{x}_\lambda)\| \, \|\mathbf{f}_\lambda - \mathbf{g}_\lambda\| \, d\lambda \leq k_5 E\int_0^\infty \frac{b_\lambda}{\max\limits_{0\leq s\leq \lambda} \Psi(\mathbf{x}_s)} \Psi(\mathbf{x}_\lambda) c_\lambda \, d\lambda$$

$$\leq k_5 E\int_0^\infty b_\lambda c_\lambda \, d\lambda < \infty. \tag{2.224}$$

By (2.185) and (2.216) and the fact that $r_2(\|\mathbf{x}\|) \geq 1 + \|\mathbf{x}\|^2$, it follows that there exists a constant k_6 such that

$$E\int_0^\infty a_t^2 \| V_{\mathbf{x}}(\mathbf{x}_t)\|^2 \|\underset{\sim}{\mathbf{F}}_t\|^2 \, dt \leq E\int_0^\infty \frac{b_t^2}{\max\limits_{0\leq s\leq t} r_2^2(\|\mathbf{x}_s\|)} \| V_{\mathbf{x}}(\mathbf{x}_t)\|^2 r_1(t) r_2(\|\mathbf{x}_t\|) \, dt$$

$$\leq k_6 E\int_0^\infty b_t^2 r_1(t) \, dt < \infty. \tag{2.225}$$

As we did earlier, we shall denote the matrix with elements $V_{x^i x^j}(x)$ by $V_{\mathbf{xx}}(\mathbf{x})$ and then we have

$$\| V_{\mathbf{xx}}(\mathbf{x})\| \leq l k_1. \tag{2.226}$$

Hence

$$E\int_0^\infty a_t^2 \operatorname{tr} V_{\mathbf{xx}}(\mathbf{x}_t)\underset{\sim}{\mathbf{F}}_t\underset{\sim}{\mathbf{F}}_t^\tau\, dt < \infty. \tag{2.227}$$

By Ito's formula (1.32) and Condition A_5, we have

$$V(\mathbf{x}_t) \le V(\mathbf{x}_s) - \int_s^t a_\lambda V_{\mathbf{x}}^\tau(\mathbf{x}_\lambda)\underset{\sim}{\mathbf{L}}(\mathbf{f}_\lambda - \mathbf{g}_\lambda)\, d\lambda - \int_s^t a_\lambda V_{\mathbf{x}}^\tau(\mathbf{x}_\lambda)\underset{\sim}{\mathbf{L}}\underset{\sim}{\mathbf{F}}_\lambda\, d\mathbf{w}_\lambda$$

$$+ \frac{1}{2}\int_s^t a_\lambda^2 \operatorname{tr} V_{\mathbf{xx}}(\mathbf{x}_\lambda)\underset{\sim}{\mathbf{L}}\underset{\sim}{\mathbf{F}}_\lambda\underset{\sim}{\mathbf{F}}_\lambda^\tau\underset{\sim}{\mathbf{L}}^\tau\, d\lambda, \qquad t \ge s \ge t_0. \tag{2.228}$$

Denote

$$m_t = V(\mathbf{x}_t) + \|\underset{\sim}{\mathbf{L}}\| E\left(\int_0^\infty \frac{a_\lambda \|V_{\mathbf{x}}(\mathbf{x}_\lambda)\|\, \|\mathbf{f}_\lambda - \mathbf{g}_\lambda\|\, d\lambda}{\mathscr{F}_t}\right)$$

$$- \|\underset{\sim}{\mathbf{L}}\| \int_0^t a_\lambda \|V_{\mathbf{x}}(\mathbf{x}_\lambda)\|\, \|\mathbf{f}_\lambda - \mathbf{g}_\lambda\|\, d\lambda + \frac{\|\underset{\sim}{\mathbf{L}}\| k_1}{2} E\left(\int_0^\infty \frac{a_\lambda^2 \operatorname{tr} \mathbf{F}_\lambda \mathbf{F}_\lambda^\tau\, d\lambda}{\mathscr{F}_t}\right)$$

$$- \frac{\|\underset{\sim}{\mathbf{L}}\|^2 k_1}{2}\int_0^t a_\lambda^2 \operatorname{tr} \underset{\sim}{\mathbf{F}}_\lambda\underset{\sim}{\mathbf{F}}_\lambda^\tau\, d\lambda, \tag{2.229}$$

each term of which has finite expectation as shown by (2.223)–(2.227).

By (2.228), it is easy to prove that $(m_t, \mathscr{F}_t)$ is a nonnegative supermartingale and hence, by Theorem 1.11, it tends to a finite limit as $t \to \infty$. Starting from this point, all of the conclusions of the lemma can be verified in analogy with the proof of Lemma 2.2. □

Lemma 2.17. *Assume Condition* A_6 *and all the conditions of Lemma* 2.15 *hold. Then there are random variables* α, β, γ *such that* $0 \le \alpha < \infty$, $0 < \beta < \infty$, $0 < \gamma < \infty$, *and*

$$V(\mathbf{x}_t) \underset{t\to\infty}{\to} \alpha, \ \|\mathbf{x}_t\| \le \beta, \ r_2(\|\mathbf{x}_t\|) \le \gamma, \qquad \forall t \in [0, \infty) \text{ a.s.},$$

where $\mathbf{x}_t$ *is given by the* MRM *procedure.*

Proof. The proof of Lemma 2.16 will establish all the desired assertions if we replace $h_{\mathbf{x}}(\mathbf{x}_t)$ in Lemma 2.16 by $\mathbf{h}(\mathbf{x}_t)$, and furthermore delete (2.222)–(2.224) as well as all terms containing $(\mathbf{f}_\lambda - \mathbf{g}_\lambda)$ in (2.228) and (2.229). □

Corollary. *For almost all* $\omega \in \Omega$ x_t *is bounded, hence*

$$\int_0^\infty a_s\, ds = \infty \text{ a.s.} \tag{2.230}$$

Define

$$\mu(t) = \int_0^t a_s\,ds, \qquad \nu(\mu(t)) = t. \tag{2.231}$$

$\mu(t)$ is continuous and its inverse function $\nu(t)$ is also continuous since $a_t > 0$ and, by (2.230),

$$V(t) \underset{t\to\infty}{\rightarrow} \infty \text{ a.s.}$$

Denote

$$\mathbf{x}_n(t) = \mathbf{x}_{\nu(t+n)}, \qquad \mathbf{p}_t = \int_0^t a_s \underset{\sim}{\mathbf{L}} \mathbf{f}_s\,ds, \qquad \mathbf{q}_t = \int_0^t a_s \underset{\sim}{\mathbf{L}} \underset{\sim}{\mathbf{F}}_s\,d\mathbf{w}_s. \tag{2.232}$$

Lemma 2.18. *Under the conditions of Lemma* 2.16 *or* 2.17, *for almost all ω, there exists a subsequence $\mathbf{x}_{n_k}(t)$ of $\mathbf{x}_n(t)$ and a continuous function $\mathbf{x}(t)$ which is the uniform limit of $\mathbf{x}_{n_k}(t)$ in any finite interval.*

Proof. Since by Lemma 2.15

$$E\int_0^\infty a_t^2 \|\underset{\sim}{\mathbf{F}}_t\|^2\,dt < \infty,$$

$(\mathbf{q}_t, \mathscr{F}_t)$ is a square integrable martingale and hence by Theorem 1.11 it tends almost surely to a finite limit as $t \to \infty$. Given $\varepsilon > 0$, we can find T such that

$$\|\mathbf{q}_{\nu(t+n+\Delta)} - \mathbf{q}_{\nu(t+n)}\| \le \varepsilon \qquad \forall n, t \ge T, \tag{2.233}$$

and, for $t \in [0, T]$, $\|\mathbf{q}_{\nu(t+n+\Delta)} - \mathbf{q}_{\nu(t+n)}\|$ can be also small uniformly in n if Δ is small enough because of the continuity of $\mathbf{q}_{\nu(t)}$. Thus for fixed t, (2.233) can be assumed to hold for any n if Δ is small.

From (2.178), (2.196), (2.213), and Lemma 2.17 for the MRM procedure, we have

$$a_t\|\mathbf{f}_t\| \le \begin{cases} r_2(\|\mathbf{x}_t\|)a_t \\ k_4 a_t \int_0^t e^{-c(t-\lambda)} r_2(\|\mathbf{x}_\lambda\|)\,d\lambda \end{cases} \le \begin{cases} \gamma a_t, & r = 0 \\ \dfrac{k_4 \gamma a_t}{c}, & r > 0 \end{cases} \tag{2.234}$$

while for the MKW procedure it follows from (2.197), (2.214), (2.213), and Lemma 2.16 that

$$a_t\|\mathbf{f}_t\| \le \begin{cases} \sqrt{l}\, r_2(\|\bar{\mathbf{x}}_t\|)a_t \\ \sqrt{l}\, k_4 a_t \int_0^t e^{-c(t-\lambda)} r_2(\|\bar{\mathbf{x}}_\lambda\|)\,d\lambda \end{cases} \le \begin{cases} \sqrt{l}\,\gamma a_t, & r = 0 \\ \dfrac{k_4\sqrt{l}\,\gamma a_t}{c}, & r > 0. \end{cases} \tag{2.235}$$

Denote

$$k_6 = \begin{cases} \gamma, & r = 0 \\ \dfrac{k_4\gamma}{c}, & r > 0 \end{cases} \quad \text{for MRM}$$

and

$$k_6 = \begin{cases} \sqrt{l}\,\gamma, & r = 0 \\ \dfrac{\sqrt{l}\,k_4\gamma}{c}, & r > 0 \end{cases} \quad \text{for MRW.}$$

Then we have

$$\|\mathbf{p}_{\nu(t+n+\Delta)} - \mathbf{p}_{\nu(t+n)}\| \le \|\underset{\sim}{\mathbf{L}}\| \int_{\nu(t+n)}^{\nu(t+n+\Delta)} a_s$$

$$\|\mathbf{f}_s\|\,ds \le k_6\|\underset{\sim}{\mathbf{L}}\| \int_{\nu(t+n)}^{\nu(t+n+\Delta)} a_s\,ds = k_6\|\underset{\sim}{\mathbf{L}}\|\Delta. \qquad (2.236)$$

Hence from (2.233), (2.236), and the inequality

$$\|\mathbf{x}_n(t+\Delta) - \mathbf{x}_n(t)\| \le \|\mathbf{p}_{\nu(t+n+\Delta)} - \mathbf{p}_{\nu(t+n)}\| + \|\mathbf{q}_{\nu(t+n+\Delta)} - \mathbf{q}_{\nu(t+n)}\|$$

we conclude that $\{\mathbf{x}_n(t)\}$ is equicontinuous. In addition, Lemmas 2.16 and 2.17 guarantee the uniform boundedness of $\mathbf{x}_n(t)$ for all t, n. So the lemma follows from Theorem 2.3. □

Lemma 2.19. *Under the conditions of Lemma* 2.16, $\mathbf{x}(t)$ *satisfies*

$$\dot{\mathbf{x}}(t) = \begin{cases} -\underset{\sim}{\mathbf{L}}h_{\mathbf{x}}(\mathbf{x}(t)), & r = 0 \\ -\underset{\sim}{\mathbf{L}}\underset{\sim}{\mathbf{C}}_r^{-1}h_{\mathbf{x}}(\mathbf{x}(t)), & r > 0 \end{cases} \quad \textit{for } \text{MKW} \qquad (2.237)$$

and under the conditions of Lemma 2.17 $\mathbf{x}(t)$ *satisfies*

$$\dot{\mathbf{x}}(t) = \begin{cases} -\underset{\sim}{\mathbf{L}}\mathbf{h}(\mathbf{x}(t)), & r = 0 \\ -\underset{\sim}{\mathbf{L}}\underset{\sim}{\mathbf{C}}_r^{-1}\mathbf{h}(\mathbf{x}(t)), & r > 0 \end{cases} \quad \textit{for } \text{MRM.} \qquad (2.238)$$

Proof. We first prove the lemma for the MKW procedure. Let $\Delta > 0$, $r > 0$. Since

$$\|\mathbf{q}_{\nu(t+n+\Delta)} - \mathbf{q}_{\nu(t+n)}\| \underset{n\to\infty}{\to} 0,$$

we have

$$\lim_{\Delta\to 0}\frac{\mathbf{x}(t+\Delta)-\mathbf{x}(t)}{\Delta} = -\lim_{\Delta\to 0}\lim_{k\to\infty}\frac{1}{\Delta}\int_{\nu(t+n_k)}^{\nu(t+n_k+\Delta)}$$

$$\times\int_0^s a_s\underset{\sim}{\mathbf{L}}\underset{\sim}{\mathbf{G}}e^{\mathbf{F}(s-\lambda)}\underset{\sim}{\mathbf{H}}\nabla\mathbf{h}(\mathbf{x}_\lambda)\,d\lambda\,ds. \quad (2.239)$$

Noticing that for any fixed $\Delta'\in(0,\Delta)$

$$\exp[-c(s-\nu(t+n_k))]\underset{k\to\infty}{\to}0 \quad (2.240)$$

uniformly in $s\geq\nu(t+n_k+\Delta')$, it follows by (2.213), (2.214), and Lemma 2.16 that

$$\overline{\lim_{k\to\infty}}\frac{1}{\Delta}\left\|\int_{\nu(t+n_k)}^{\nu(t+n_k+\Delta)}\int_0^{\nu(t+n_k)}a_s\underset{\sim}{\mathbf{L}}\underset{\sim}{\mathbf{G}}e^{\mathbf{F}(s-\lambda)}\underset{\sim}{\mathbf{H}}\nabla h(\mathbf{x}_s)\,d\lambda\,ds\right\|$$

$$\leq\overline{\lim_{k\to\infty}}\,k_4\sqrt{l}\,\gamma\frac{1}{\Delta}\int_{\nu(t+n_k)}^{\nu(t+n_k+\Delta)}\int_0^{\nu(t+n_k)}a_s e^{-c(s-\lambda)}\,d\lambda\,ds$$

$$\leq\overline{\lim_{k\to\infty}}\frac{k_4\sqrt{l}\,\gamma}{\Delta c}\left[\int_{\nu(t+n_k)}^{\nu(t+n_k+\Delta')}a_s e^{c(\nu(t+n_k)-s)}ds\right.$$

$$\left.+\int_{\nu(t+n_k+\Delta')}^{\nu(t+n_k+\Delta)}a_s e^{c(\nu(t+n_k)-s)}\,ds\right]$$

$$=\overline{\lim_{k\to\infty}}\frac{k_4\sqrt{l}\,\gamma}{\Delta c}\int_{\nu(t+n_k)}^{\nu(t+n_k+\Delta')}a_s e^{c(\nu(t+n_k)-s)}\,ds\underset{\Delta'\to 0}{\to}0. \quad (2.241)$$

Then from (2.239) we have

$$\lim_{\Delta\to 0}\frac{\mathbf{x}(t+\Delta)-\mathbf{x}(t)}{\Delta} = -\lim_{\Delta\to 0}\lim_{k\to\infty}\frac{1}{\Delta}\int_{\nu(t+n_k)}^{\nu(t+n_k+\Delta)}$$

$$\times\int_{\nu(t+n_k)}^{s}a_s\underset{\sim}{\mathbf{L}}\underset{\sim}{\mathbf{G}}e^{\mathbf{F}(s-\lambda)}\underset{\sim}{\mathbf{H}}\nabla h(\mathbf{x}_\lambda)\,d\lambda\,ds. \quad (2.242)$$

For both of the cases $r = 0$ and $r > 0$, and for any $\lambda \in [0, \nu(t + n_k + \Delta) - \nu(t + n_k)]$, there exists $\Delta(\lambda) \in [0, \Delta]$ such that

$$\nu(t + n_k) + \lambda = \nu(t + n_k + \Delta(\lambda))$$

and as in Lemma 2.13 there are numbers such that

$$|\Delta^j_{k\lambda}| < c_{\nu(t+n_k)+\lambda}$$

and

$$\frac{1}{2c_{\nu(t+n_k)+\lambda}}\left[h\left(\mathbf{x}^{j+}_{\nu(t+n_k)+\lambda}\right) - h\left(\mathbf{x}^{j-}_{\nu(t+n_k)+\lambda}\right)\right]$$

$$= h_{x^j}\Big(\Big[x^1_{\nu(t+n_k+\Delta(\lambda))}, \ldots, x^{j-1}_{\nu(t+n_k+\Delta(\lambda))}, x^j_{\nu(t+n_k+\Delta(\lambda)+\Delta^j_{k\lambda})},$$

$$\ldots, x^{j+1}_{\nu(t+n_k+\Delta(\lambda))}, \ldots, x^l_{\nu(t+n_k+\Delta(\lambda))}$$

$$= h_{x^j}\Big(\Big[x^1_{n_k}(t + \Delta(\lambda)), \ldots, x^{j-1}_{n_k}(t + \Delta(\lambda)), x^j_{n_k}(t + \Delta(\lambda) + \Delta^j_{k\lambda}),$$

$$\times x^{j+1}_{n_k}(t + \Delta(\lambda)),, \ldots, x^l_{n_k}(t + \Delta(\lambda))\Big]^\tau\Big).$$

Hence we have

$$\left\|\frac{1}{2c_{\nu(t+n_k)+\lambda}}\left[h\left(\mathbf{x}^{j+}_{\nu(t+n_k)+\lambda}\right) - h\left(\mathbf{x}^{j-}_{\nu(t+n_k)+\lambda}\right)\right] - h_{x^j}(\mathbf{x}(t))\right\|$$

$$\leq \Big\|h_{x^j}\Big(\Big[x^1_{n_k}(t + \Delta(\lambda)), \ldots, x^j_{n_k}(t + \Delta(\lambda) + \Delta^j_{k\lambda}),$$

$$\ldots, x^l_{n_k}(t + \Delta(\lambda))\Big]^\tau\Big) - h_{x^j}(\mathbf{x}_{n_k}(t))\Big\|$$

$$+ \left\|h_{x^j}(\mathbf{x}_{n_k}(t)) - h_{x^j}(\mathbf{x}(t))\right\|, \qquad \forall j = 1, \ldots, l.$$

Since $c_t \underset{t\to\infty}{\to} 0$, the convergence $\Delta^j_{k\lambda} \underset{k\to\infty}{\to} 0$ is uniform in j, $0 \leq j \leq l$ and in $\lambda \in [0, \nu(t + n_k + \Delta) - \nu(t + n_k)]$. For fixed t and a given $\varepsilon > 0$, the first term on the right-hand side of the preceding inequality can be made less than $\varepsilon/2$ if k is large enough and Δ is small enough since $h_{x^j}(\cdot)$ is continuous and $\{\mathbf{x}_n(t)\}$ is equicontinuous for fixed ω, while the second term will be also less than $\varepsilon/2$ if k is sufficiently large. Therefore,

$$\overline{\lim_{\Delta\to 0}}\ \overline{\lim_{k\to\infty}}\ \max_{0\le\lambda\le\nu(t+n_k+\Delta)-\nu(t+n_k)} \left\| \nabla h\left(\mathbf{x}_{\nu(t+n_k)+\lambda}\right) - h_{\mathbf{x}}(\mathbf{x}(t)) \right\| = 0, \tag{2.243}$$

and from (2.242), it follows that

$$\begin{aligned} \lim_{\Delta\to 0} \frac{\mathbf{x}(t+\Delta)-\mathbf{x}(t)}{\Delta} &= -\lim_{\Delta\to 0}\lim_{k\to\infty}\frac{1}{\Delta}\int_{\nu(t+n_k)}^{\nu(t+n_k+\Delta)} \\ &\quad\times \int_{\nu(t+n_k)}^{s} a_s \underset{\sim}{\mathbf{L}}\underset{\sim}{\mathbf{G}} e^{\underset{\sim}{\mathbf{F}}(s-\lambda)} \underset{\sim}{\mathbf{H}} h_{\mathbf{x}}(\mathbf{x}(t))\, d\lambda\, ds \\ &\quad - \lim_{\Delta\to 0}\lim_{k\to\infty}\frac{1}{\Delta}\int_{\nu(t+n_k)}^{\nu(t+n_k+\Delta)}\int_{\nu(t+n_k)}^{s} a_s \underset{\sim}{\mathbf{L}}\underset{\sim}{\mathbf{G}} e^{\underset{\sim}{\mathbf{F}}(s-\lambda)} \underset{\sim}{\mathbf{H}} \\ &\quad\times \left[\nabla h(\mathbf{x}_\lambda) - h_{\mathbf{x}}(\mathbf{x}(t))\right] d\lambda\, ds, \end{aligned} \tag{2.244}$$

whose second term is, in norm, less than or equal to

$$\begin{aligned} &\overline{\lim_{\Delta\to 0}}\ \overline{\lim_{k\to\infty}}\ \max_{0\le\lambda\le\nu(t+n_k+\Delta)-\nu(t+n_k)} \left\| \nabla h\left(\mathbf{x}_{\nu(t+n_k)+\lambda}\right) - h_{\mathbf{x}}(\mathbf{x}(t)) \right\| \\ &\quad\times \int_{\nu(t+n_k)}^{\nu(t+n_k+\Delta)}\int_{\nu(t+n_k)}^{s} a_s e^{-c(s-\lambda)}\, d\lambda\, ds \\ &= \overline{\lim_{\Delta\to 0}}\ \overline{\lim_{k\to\infty}}\ \max_{0\le\lambda\le\nu(t+n_k+\Delta)-\nu(t+n_k)} \left\| \nabla h\left(\mathbf{x}_{\nu(t+n_k)+\lambda}\right) - h_{\mathbf{x}}(\mathbf{x}(t)) \right\| \\ &\quad\cdot \frac{k_4\|\underset{\sim}{\mathbf{L}}\|}{\Delta c}\int_{\nu(t+n_k)}^{\nu(t+n_k+\Delta)} a_s\, ds \\ &= -\frac{k_4\|\underset{\sim}{\mathbf{L}}\|}{c}\ \overline{\lim_{\Delta\to 0}}\ \overline{\lim_{k\to\infty}}\ \max_{0\le\lambda\le\nu(t+n_k+\Delta)-\nu(t+n_k)} \\ &\quad\times \left\| \nabla h\left(\mathbf{x}_{\nu(t+n_k)}\right) - h_{\mathbf{x}}(\mathbf{x}(t)) \right\| = 0 \end{aligned}$$

by (2.243).

Hence from (2.244), we have

$$\lim_{\Delta\to 0}\frac{\mathbf{x}(t+\Delta)-\mathbf{x}(t)}{\Delta}$$

$$= -\lim_{\Delta\to 0}\lim_{k\to\infty}\frac{1}{\Delta}\int_{\nu(t+n_k)}^{\nu(t+n_k+\Delta)} a_s \underset{\sim}{\mathbf{L}}\underset{\sim}{\mathbf{G}}\underset{\sim}{\mathbf{F}}^{-1} e^{\mathbf{F}(s-\lambda)} \int_{\nu(t+n_k)}^{s} \underset{\sim}{\mathbf{H}} h_{\mathbf{x}}(\mathbf{x}(t))\, ds$$

$$= -\underset{\sim}{\mathbf{L}}\underset{\sim}{\mathbf{G}}\underset{\sim}{\mathbf{F}}^{-1}\underset{\sim}{\mathbf{H}} h_{\mathbf{x}}(\mathbf{x}(t))$$

$$+ \lim_{\Delta\to 0}\lim_{k\to\infty}\frac{1}{\Delta}\int_{\nu(t+n_k)}^{\nu(t+n_k+\Delta)} a_s \underset{\sim}{\mathbf{L}}\underset{\sim}{\mathbf{G}}\underset{\sim}{\mathbf{F}}^{-1} e^{\mathbf{F}(s-\nu(t+n_k))}\, ds\, \underset{\sim}{\mathbf{H}} h_{\mathbf{x}}(\mathbf{x}(t)), \tag{2.245}$$

whose second term, by a similar treatment to that used in (2.241), can be estimated in norm by

$$\lim_{\Delta\to 0}\lim_{k\to\infty}\frac{k_4\|\underset{\sim}{\mathbf{L}}\|\;\|\underset{\sim}{\mathbf{F}}^{-1}\|}{\Delta}\int_{\nu(t+n_k)}^{\nu(t+n_k+\Delta)} a_s e^{c(\nu(t+n_k)-s)}\, ds \|h_{\mathbf{x}}(\mathbf{x}(t))\| = 0.$$

For $\Delta < 0$ an analogous relationship can be obtained, and hence by (2.81) we have for the MKW procedure the ODE

$$\dot{\mathbf{x}}(t) = -\underset{\sim}{\mathbf{L}}\underset{\sim}{\mathbf{C}}_r^{-1} h_{\mathbf{x}}(\mathbf{x}(t)), \qquad r > 0. \tag{2.246}$$

For $r = 0$, the right-hand side of (2.239) is

$$-\lim_{\Delta\to 0}\lim_{k\to\infty}\frac{1}{\Delta}\int_{\nu(t+n_k)}^{\nu(t+n_k+\Delta)} a_\lambda \underset{\sim}{\mathbf{L}}\mathbf{f}_\lambda\, d\lambda$$

$$= -\lim_{\Delta\to 0}\lim_{k\to\infty}\frac{1}{\Delta}\int_{\nu(t+n_k)}^{\nu(t+n_k+\Delta)} a_\lambda \underset{\sim}{\mathbf{L}}\nabla h(\mathbf{x}_\lambda)\, d\lambda \tag{2.247}$$

and applying (2.243) to (2.247) leads directly to the conclusion of the lemma.

For the MRM *procedure we only need to replace* $\nabla h(\mathbf{x}_\lambda)$ *by* $h(\mathbf{x}_\lambda)$ *and* $h_{\mathbf{x}}(\mathbf{x}(t))$ *by* $\mathbf{h}(\mathbf{x}(t))$ [*in* (2.239)–(2.242) *and in* (2.244)–(2.247)] *and to change the quantities in the limit* (2.243) *to*

$$\overline{\lim_{\Delta\to 0}}\ \overline{\lim_{k\to\infty}} \max_{0\le\lambda\le\nu(t+n_k+\Delta)-\nu(t+n_k)} \|\mathbf{h}(\mathbf{x}_{\nu(t+n_k)+\lambda}) - \mathbf{h}(\mathbf{x}(t))\|$$

$$= \max_{0\le\lambda\le\nu(t+n_k+\Delta)-\nu(t+n_k)} \|\mathbf{h}(\mathbf{x}_{n_k}(t+\Delta(\lambda))) - \mathbf{h}(\mathbf{x}(t))\|$$

$$\le \overline{\lim_{\Delta\to 0}}\ \overline{\lim_{k\to\infty}} \max_{0\le\lambda\le\nu(t+n_k+\Delta)-\nu(t+n_k)} \|\mathbf{h}(\mathbf{x}_{n_k}(t+\Delta(\lambda))) - \mathbf{h}(\mathbf{x}_{n_k}(t))\|$$

$$+ \max_{0\le\lambda\le\nu(t+n_k+\Delta)-\nu(t+n_k)} \|\mathbf{h}(\mathbf{x}_{n_k}(t)) - \mathbf{h}(\mathbf{x}(t))\|,$$

which is equal to zero by the equicontinuity of $\{\mathbf{x}_n(t)\}$ *and continuity of* $\mathbf{h}(\cdot)$. *Thus* (2.238) *follows.* □

Theorem 2.8. *Assume that the measurement process is given by* (2.188), *that the algorithm* (2.195) *has a strong solution, that* b_t *satisfies* (2.190) ($c_t \equiv 0$), *and that* (2.178) *and* (2.185) *hold. Then for the* MRM *procedure*

$$\mathbf{x}_t \underset{t\to\infty}{\to} \mathbf{x}^0 \text{ a.s.}$$

whenever the following conditions hold:

1. $r = 0$, *or* $r > 0$ *but all zeros of* $\det \underset{\sim}{C}(\lambda)$ *have negative real parts.*
2. *Condition* A_6 *holds.*
3. $\mathbf{h}(\mathbf{x})$ *is continuous and* $\mathbf{h}(\mathbf{x}^0) = 0$.

Proof. From (2.218)–(2.220), we see that, for any ϕ_0,

$$\mathbf{g}_t \triangleq \begin{cases} \underset{\sim}{G}\phi_t = \underset{\sim}{G} e^{\mathbf{F}t}\phi_0 + \underset{\sim}{G}\int_0^t e^{\underset{\sim}{F}(t-\lambda)}\underset{\sim}{H}\mathbf{h}(\mathbf{x}_\lambda)\, d\lambda, & r > 0 \\ \mathbf{h}(\mathbf{x}_t), & r = 0 \end{cases} \tag{2.248}$$

satisfies (2.208) and that $\mathbf{g}_0^{(i)}$, $i = 0, \ldots, r-1$ are completely determined by ϕ_0.

Take an arbitrary $\mathbf{x} \neq \mathbf{x}^0$, and set $\mathbf{x}_s \equiv \mathbf{x}$ and $\phi_0 = -\underset{\sim}{F}^{-1}\underset{\sim}{H}\mathbf{h}(\mathbf{x})$, then

$$\mathbf{g}_t = \begin{cases} \underset{\sim}{G} e^{\underset{\sim}{F}t}\phi_0 + \underset{\sim}{G} e^{\mathbf{F}t}\underset{\sim}{F}^{-1}\underset{\sim}{H}\mathbf{h}(\mathbf{x}) - \underset{\sim}{G}\underset{\sim}{F}^{-1}\underset{\sim}{H}\mathbf{h}(\mathbf{x}) = -\underset{\sim}{G}\underset{\sim}{F}^{-1}\underset{\sim}{H}\mathbf{h}(\mathbf{x}), & r > 0, \\ \mathbf{h}(\mathbf{x}), & r = 0 \end{cases}$$

and by (2.81)

$$\mathbf{g}_t = \begin{cases} \underset{\sim}{\mathbf{C}}_r^{-1}\mathbf{h}(\mathbf{x}), & r > 0 \\ h(x), & r = 0. \end{cases}$$

By Condition A_6 we know that

$$V_{\mathbf{x}}^{\tau}(\mathbf{x})\underset{\sim}{\mathbf{L}}\mathbf{g}_t > 0, \qquad \forall \mathbf{x} \neq \mathbf{x}^0,$$

hence $V(\mathbf{x})$ is the Lyapunov function for (2.238) and

$$\mathbf{x}(t) \underset{t\to\infty}{\to} \mathbf{x}^0 \text{ a.s.}$$

by Theorem 2.4.

For fixed ω and any $\varepsilon > 0$ there exists t_1 such that $\|\mathbf{x}(t) - \mathbf{x}^0\| < \varepsilon$ for $\forall t \geq t_1$, and exists K such that

$$\|\mathbf{x}_{n_k}(t) - \mathbf{x}(t)\| < \varepsilon, \qquad \forall t \in [t_1, 2t_1], \quad \forall k \geq K$$

or

$$\|\mathbf{x}_{\nu(t+n_k)} - \mathbf{x}(t)\| < \varepsilon, \qquad \forall t \in [t_1, 2t_1], \quad \forall k \geq K$$

since the convergence $\mathbf{x}_{n_k}(t) \underset{k\to\infty}{\to} \mathbf{x}(t)$ is uniform over any finite interval. Hence there is a subsequence $\mathbf{x}_{t_k} \underset{k\to\infty}{\to} \mathbf{x}^0$ with $t_k \underset{k\to\infty}{\to} \infty$, but $V(\mathbf{x}_t) \underset{t\to\infty}{\to} \alpha$ according to Lemma 2.17. Then we have

$$\lim_{t\to\infty} V(\mathbf{x}_t) = \lim_{k\to\infty} V(\mathbf{x}_{t_k}) = V(\mathbf{x}^0) = 0.$$

From here we conclude $\mathbf{x}_t \underset{t\to\infty}{\to} \mathbf{x}^0$ a.s.; otherwise, for fixed ω, there would be a subsequence

$$\mathbf{x}_{s_i} \underset{i\to\infty}{\to} \mathbf{x}^1 \neq \mathbf{x}^0, \qquad s_i \underset{i\to\infty}{\to} \infty,$$

and

$$\lim_{i\to\infty} V(\mathbf{x}_{s_i}) = V(\mathbf{x}^1) > 0,$$

since $\mathbf{x}^0$ is the unique zero of $V(\mathbf{x})$. This contradicts the fact that $\lim_{i\to\infty} V(\mathbf{x}_{s_i}) = \lim_{t\to\infty} V(\mathbf{x}_t) = 0$. □

Theorem 2.9. *Suppose that equations given by* (2.189) *and* (2.195) *have a strong solution, that* b_t *and* c_t *fulfill* (2.190) *and that* (2.179), (2.185) *hold. Then for the* MKW *procedure*

$$\mathbf{x}_t \underset{t\to\infty}{\to} \mathbf{x}^0 \text{ a.s.}$$

if

1. $r = 0$ *or* $r > 0$ *but all zeros of* $\det \underset{\sim}{\mathbf{C}}(\lambda)$ *have negative real parts.*
2. *Condition* A_5 *is valid.*
3. $h_{\mathbf{x}}(\mathbf{x})$ *is continuously differentiable, reaches its minimum at* $\mathbf{x}^0$, *and* (2.155) *holds.*

Proof. If, in (2.248), $h(\mathbf{x}_\lambda)$ is replaced by $h_{\mathbf{x}}(\mathbf{x}_\lambda)$ then $\mathbf{g}_t$ satisfies (2.207). The proof of Theorem 2.8 can then be applied to the present case if we replace $\mathbf{h}(\cdot)$ and Condition A_6 by $h_{\mathbf{x}}(\cdot)$ and Condition A_5, respectively. □

2.6. AN EXAMPLE

Methods of stochastic approximation have been widely applied in regulation and tracking systems, adaptive control systems, in problems of optimization and in other related fields. Here we give an example of an application to the adaptive beam former. The problem is concerned with the determination of the azimuth of a target by using a matrix composed of sensors. The outputs $\mathbf{x}_k$ of sensors are weighted by a matrix $\underset{\sim}{\mathbf{W}}$ in order that the signal $\underset{\sim}{\mathbf{W}}\mathbf{x}_k$ approximates as well as possible the unknown target signal $\mathbf{y}_k$ in the mean square sense. To maintain an error-free signal, the following constraint on $\underset{\sim}{\mathbf{W}}$ is necessary:

$$\underset{\sim}{\mathbf{W}}^\tau \underset{\sim}{\mathbf{C}} = \underset{\sim}{\mathbf{\Phi}}, \tag{2.249}$$

where $\underset{\sim}{\mathbf{C}}$ and $\underset{\sim}{\mathbf{\Phi}}$ are known matrices. Then the problem is reduced to finding $\underset{\sim}{\mathbf{W}}$, constrained by (2.249), such that

$$E(\underset{\sim}{\mathbf{W}}^\tau \mathbf{x}_k - \mathbf{y}_k)(\underset{\sim}{\mathbf{W}}^\tau \mathbf{x}_k - \mathbf{y}_k)^\tau = \min. \tag{2.250}$$

Assume $\mathbf{z}_k = [\mathbf{x}_k^\tau \ \mathbf{y}_k^\tau]^\tau$ is an independent random sequence with second-order moment independent of k where $\mathbf{x}_k$ and $\mathbf{y}_k$ are assumed to be n- and m-dimensional, respectively. Diagonalizing $E\mathbf{z}_k\mathbf{z}_k^\tau$ by an orthogonal matrix one can easily express the matrices of second-order moments in the following way

$$E\mathbf{x}_k\mathbf{x}_k^\tau = \underset{\sim}{\mathbf{H}}\underset{\sim}{\mathbf{H}}^\tau, \qquad E\mathbf{x}_k\mathbf{y}_k^\tau = \underset{\sim}{\mathbf{H}}\underset{\sim}{\mathbf{L}}^\tau, \qquad E\mathbf{y}_k\mathbf{y}_k^\tau = \underset{\sim}{\mathbf{L}}\underset{\sim}{\mathbf{L}}^\tau$$

where, of course, $\mathbf{H}$ and $\mathbf{L}$ are unknown.

Suppose that $E\|\mathbf{z}_k\|^4 < \infty$, that $E\mathbf{x}_k\mathbf{x}_k^\tau$ and $E\mathbf{y}_k\mathbf{y}_k^\tau$ are unknown but $E\mathbf{x}_k\mathbf{y}_k^\tau$ is known.

Under (2.249), we have

$$\mathbf{\Phi} = \mathbf{\Phi}\mathbf{C}^+\mathbf{C} \tag{2.251}$$

since

$$\mathbf{W}\mathbf{C}(\mathbf{I} - \mathbf{C}^+\mathbf{C}) = \mathbf{0}.$$

Conversely, if (2.251) holds, then (2.249) is satisfied for $\mathbf{W}^\tau = \mathbf{\Phi}\mathbf{C}^+$. This means that (2.249) is solvable with respect to W if and only if (2.251) holds. Denote

$$\mathbf{P} = \mathbf{I} - \mathbf{C}\mathbf{C}^+. \tag{2.252}$$

For (2.250), the optimal weight $\mathbf{W}_{\text{opt}}$ satisfying (2.249) is expressed by (see Section 5.3)

$$\mathbf{W}_{\text{opt}} = \mathbf{C}^{+\tau}\mathbf{\Phi}^\tau + (\mathbf{P}\mathbf{H}\mathbf{H}^\tau\mathbf{P})^+(\mathbf{H}\mathbf{L}^\tau - \mathbf{H}\mathbf{H}^\tau\mathbf{C}^{+\tau}\mathbf{\Phi}). \tag{2.253}$$

But this is unavailable since $\mathbf{H}\mathbf{H}^\tau$ is unknown. On the other hand, it is time-consuming to compute the pseudo-inverse in (2.253) since n is large. Hence it is desirable to approximate $\mathbf{W}_{\text{opt}}$ by a matrix $\mathbf{W}_k$ which can be corrected on the basis of the measurements $\{\mathbf{x}_k\}$ and is such that

$$\mathbf{W}_k \underset{k\to\infty}{\to} \mathbf{W}_{\text{opt}}.$$

Such a weight $\mathbf{W}_k$ is called adaptive beam former, for which we suggest the algorithm as follows. Let

$$a_k > 0, \qquad \sum_{k=0}^{\infty} a_k = \infty, \qquad \sum_{k=0}^{\infty} a_k^2 < \infty,$$

$$\mathbf{W}_{k+1} = \mathbf{C}^{+\tau}\mathbf{\Phi}^\tau + \mathbf{P}\left[\mathbf{W}_k + a_k(\mathbf{H}\mathbf{L}^\tau - \mathbf{x}_k\mathbf{x}_k^\tau\mathbf{W}_k)\right] \tag{2.254}$$

$$W_0 = C^{+\tau}\Phi^\tau. \tag{2.255}$$

Now we prove $\mathbf{W}_k \underset{k\to\infty}{\to} \mathbf{W}_{\text{opt}}$ a.s. Since

$$\mathbf{W}_{k+1} = \mathbf{P}\mathbf{W}_{k+1} + \mathbf{C}\mathbf{C}^+\mathbf{W}_{k+1} = \mathbf{P}\mathbf{W}_{k+1} + \mathbf{C}^{+\tau}\mathbf{\Phi}^\tau, \tag{2.256}$$

convergence of $\underset{\sim}{\mathbf{W}}_k$ will follow from that of $\underset{\sim}{\mathbf{S}}_k \triangleq \underset{\sim}{\mathbf{P}}\underset{\sim}{\mathbf{W}}_k$. From (2.254) and (2.256), we have

$$\underset{\sim}{\mathbf{S}}_{k+1} = \underset{\sim}{\mathbf{S}}_k + a_k \underset{\sim}{\mathbf{P}}[\underset{\sim}{\mathbf{H}}\underset{\sim}{\mathbf{L}}^\tau - \mathbf{x}_k \mathbf{x}_k^\tau (\underset{\sim}{\mathbf{S}}_k + \underset{\sim}{\mathbf{C}}^{+\tau}\underset{\sim}{\mathbf{\Phi}}^\tau)], \qquad \underset{\sim}{\mathbf{S}}_0 = 0. \quad (2.257)$$

Let r be an arbitrary m-dimensional vector and denote

$$\underset{\sim}{\mathbf{T}} \triangleq (\underset{\sim}{\mathbf{P}}\underset{\sim}{\mathbf{H}}\underset{\sim}{\mathbf{H}}^\tau\underset{\sim}{\mathbf{P}})^+ (\underset{\sim}{\mathbf{H}}\underset{\sim}{\mathbf{L}}^\tau - \underset{\sim}{\mathbf{H}}\underset{\sim}{\mathbf{H}}^\tau\underset{\sim}{\mathbf{C}}^{+\tau}\underset{\sim}{\mathbf{\Phi}}^\tau) \quad (2.258)$$

$$\underset{\sim}{\mathbf{S}}_k\mathbf{r} \triangleq \mathbf{s}_k, \qquad \underset{\sim}{\mathbf{T}}\mathbf{r} = \mathbf{t}. \quad (2.259)$$

Clearly, the dimension of $\mathbf{s}_k$ coincides with the number of rows of matrix $\underset{\sim}{\mathbf{C}}$, let it be l. Then $\mathbf{s}_k \in R^l$. From (2.257), we have

$$\mathbf{s}_{k+1} = \mathbf{s}_k + a_k\{-\underset{\sim}{\mathbf{P}}\underset{\sim}{\mathbf{H}}\underset{\sim}{\mathbf{H}}^\tau\underset{\sim}{\mathbf{P}}(\mathbf{s}_k - \mathbf{t}) + \underset{\sim}{\mathbf{P}}\underset{\sim}{\mathbf{H}}\underset{\sim}{\mathbf{H}}^\tau\underset{\sim}{\mathbf{P}}(\mathbf{s}_k - \mathbf{t})$$

$$+\underset{\sim}{\mathbf{P}}[\underset{\sim}{\mathbf{H}}\underset{\sim}{\mathbf{L}}^\tau\mathbf{r} - \mathbf{x}_k\mathbf{x}_k^\tau(\mathbf{s}_k + \underset{\sim}{\mathbf{C}}^{+\tau}\underset{\sim}{\mathbf{\Phi}}^\tau\mathbf{r})]\}, \qquad \mathbf{s}_0 = \mathbf{0}. \quad (2.260)$$

Since $\mathbf{r}$ is arbitrary, the desired result will follow if we can prove $\mathbf{s}_k \underset{k\to\infty}{\to} \mathbf{t}$. There are two cases:

1. If $\underset{\sim}{\mathbf{P}}\underset{\sim}{\mathbf{H}} = \underset{\sim}{\mathbf{0}}$, then

$$E\underset{\sim}{\mathbf{P}}\mathbf{x}_k\mathbf{x}_k^\tau\underset{\sim}{\mathbf{P}} = \underset{\sim}{\mathbf{P}}\underset{\sim}{\mathbf{H}}\underset{\sim}{\mathbf{H}}^\tau\underset{\sim}{\mathbf{P}} = \underset{\sim}{\mathbf{0}},$$

which means $\underset{\sim}{\mathbf{P}}\mathbf{x}_k = \mathbf{0}$ a.s. Hence

$$\underset{\sim}{\mathbf{S}}_k = \mathbf{0}, \qquad \forall k \text{ a.s.}$$

and

$$\mathbf{s}_k = \mathbf{t} \text{ a.s.} \qquad \forall k.$$

2. Now let $\underset{\sim}{\mathbf{P}}\underset{\sim}{\mathbf{H}} \neq \underset{\sim}{\mathbf{0}}$. Denote by $\underset{\sim}{\mathbf{V}}_1$ the matrix whose columns constitute an orthonormal base of the linear space $\mathscr{L}(\underset{\sim}{\mathbf{C}})$ spanned by column-vectors of matrix $\underset{\sim}{\mathbf{C}}$, and by $\underset{\sim}{\mathbf{V}}_2$ the matrix corresponding to $\mathscr{L}(\underset{\sim}{\mathbf{P}}\underset{\sim}{\mathbf{H}})$ in the same sense. Let $\underset{\sim}{\mathbf{V}}_3$ be the matrix that makes $\underset{\sim}{\mathbf{V}} = [\underset{\sim}{\mathbf{V}}_1 \quad \underset{\sim}{\mathbf{V}}_2 \quad \underset{\sim}{\mathbf{V}}_3]$ an orthonormal matrix. Then there exists a matrix $\underset{\sim}{\mathbf{K}}$ such that

$$\underset{\sim}{\mathbf{P}}\underset{\sim}{\mathbf{H}} = \underset{\sim}{\mathbf{V}}_2\underset{\sim}{\mathbf{K}} \quad \text{or} \quad \underset{\sim}{\mathbf{P}}\underset{\sim}{\mathbf{H}}\underset{\sim}{\mathbf{H}}^\tau\underset{\sim}{\mathbf{P}} = \underset{\sim}{\mathbf{V}}_2\underset{\sim}{\mathbf{K}}\underset{\sim}{\mathbf{K}}^\tau\underset{\sim}{\mathbf{V}}_2^\tau > \underset{\sim}{\mathbf{0}}$$

and hence

$$E\mathbf{V}_3^\tau \mathbf{P}\mathbf{x}_k\mathbf{x}_k^\tau \mathbf{P}\mathbf{V}_3 = \mathbf{V}_3^\tau \mathbf{P}\mathbf{H}\mathbf{H}^\tau \mathbf{P}\mathbf{V}_3 = \mathbf{0}$$

$$\mathbf{V}_3^\tau \mathbf{P}\mathbf{x}_3 = \mathbf{0}, \qquad \forall k \text{ a.s.}$$

Since

$$\mathbf{V}_1^\tau \mathbf{P}\mathbf{x}_k = \mathbf{0}$$

we have $\mathbf{P}\mathbf{x}_k \in \mathscr{L}(\mathbf{P}\mathbf{H})$ a.s. for all k and from (2.260) it follows $\mathbf{s}_k \in \mathscr{L}(\mathbf{P}\mathbf{H})$ a.s. $\forall k$.

From the proof of Theorem 2.5, it is easy to see that if $\mathbf{x}_i$ defined by the procedure there, and the prescribed limit $\mathbf{x}^0$ belong to some subspace of R^l, then the conclusions of the theorem remain valid if the conditions of theorem are fulfilled only in this subspace. In the present case, this subspace is $\mathscr{L}(\mathbf{P}\mathbf{H})$. We now use Remark 1 of Theorem 2.5, where $\mathbf{h}(\mathbf{x})$, $\mathbf{x}^0$, $\mathbf{C}$, and $\boldsymbol{\varepsilon}_i(\mathbf{x}, \omega)$ correspond to

$$\mathbf{h}(\mathbf{x}) = -\mathbf{P}\mathbf{H}\mathbf{H}^\tau \mathbf{P}(\mathbf{x} - \mathbf{t}), \qquad \mathbf{x}^0 = \mathbf{t}, \qquad \mathbf{C} = \mathbf{I},$$

$$\boldsymbol{\varepsilon}_{k+1}(\mathbf{x}, \omega) = \mathbf{P}\mathbf{H}\mathbf{H}^\tau \mathbf{P}(\mathbf{x} - \mathbf{t}) - \mathbf{P}\left[\mathbf{H}\mathbf{L}^\tau \mathbf{r} - \mathbf{x}_k\mathbf{x}_k^\tau(\mathbf{x} + \mathbf{C}^{+\tau}\boldsymbol{\Phi}^\tau \mathbf{r})\right]$$

for the present case.

For $\mathbf{x} \in \mathscr{L}(\mathbf{P}\mathbf{H})$,

$$\mathbf{P}\mathbf{H}\mathbf{H}^\tau \mathbf{P}(\mathbf{P}\mathbf{H}\mathbf{H}^\tau \mathbf{P})^+ \mathbf{H} = \mathbf{V}_2\mathbf{V}_2^\tau \mathbf{H} = \mathbf{P}\mathbf{H}. \tag{2.261}$$

Hence

$$-E\boldsymbol{\varepsilon}_{k+1} = \mathbf{P}\mathbf{H}\mathbf{H}^\tau \mathbf{x} - \mathbf{P}(\mathbf{H}\mathbf{L}^\tau - \mathbf{H}\mathbf{H}^\tau \mathbf{C}^{+\tau}\boldsymbol{\Phi}^\tau)\mathbf{r}$$
$$+\mathbf{P}(\mathbf{H}\mathbf{L}^\tau \mathbf{r} - \mathbf{H}\mathbf{H}^\tau \mathbf{x} - \mathbf{H}\mathbf{H}^\tau \mathbf{C}^{+\tau}\boldsymbol{\Phi}^\tau \mathbf{r}) = \mathbf{0}$$

and

$$\mathbf{h}^\tau(\mathbf{x})(\mathbf{x} - \mathbf{x}^0) = (\mathbf{x} - \mathbf{t})^\tau \mathbf{P}\mathbf{H}\mathbf{H}^\tau \mathbf{P}(\mathbf{x} - \mathbf{t}) = (\mathbf{x} - \mathbf{t})^\tau \mathbf{V}_2\mathbf{K}\mathbf{K}^\tau \mathbf{V}_2^\tau(\mathbf{x} - \mathbf{t}).$$

Since $\mathbf{x}$ and $\mathbf{t}$ belong to the same $\mathscr{L}(\mathbf{PH})$ there exists a nonzero vector $\mathbf{u}$ such that

$$\mathbf{x} - \mathbf{t} = \mathbf{V}_2\mathbf{u} \neq \mathbf{0}$$

if $\mathbf{x} \neq \mathbf{t}$. Paying attention to

$$\mathbf{V}_2\mathbf{K}\mathbf{K}^T\mathbf{V}_2^T > \mathbf{0},$$

we have

$$(\mathbf{x} - \mathbf{t})^T\mathbf{V}_2\mathbf{K}\mathbf{K}^T\mathbf{V}_2^T(\mathbf{x} - \mathbf{t}) > 0.$$

By Remark 1 of Theorem 2.5, we find

$$\mathbf{s}_k \underset{k\to\infty}{\to} \mathbf{t} \text{ a.s.}.$$

Thus the convergence of the adaptive beam former is proved by using a convergence theorem of stochastic approximation.

CHAPTER 3

Strong Consistency of Least-Squares Identification

3.1. INTRODUCTION

Given a physical system S, in order to control it or to predict its evolution, one first has to construct its mathematical model. In some circumstances one can derive it theoretically starting from relationships provided by physics or mechanics; an example is the equations of motion of a satellite in its orbit. However, the mathematical model obtained in such a way may contain a certain number of unknown parameters, for example, the motion equations of a plane derived from the mechanical relationships may include some unknown dynamic coefficients. In many cases one cannot obtain a model of the system from physics and mechanics at all. Consider, for example, the process arising in a complicated chemical reaction. Hence it is of great importance to define the mathematical model for a system based on its inputs and outputs. For example, for an aircraft in flight the change of its rudder angle may be regarded as an input and the three coordinates of its position in space may be viewed as the output of the system; for a chemical reaction the product depends on the levels of, say, a temperature, pressure, and a catalytic agent, these can be viewed as system inputs and the product as the system output. The task of system identification is to find the equations of the system. Since the measured data are usually corrupted by random noise, the identified system is a system under random influences. Several aspects must be considered in identification of a stochastic system.

1. *Selection of Model Set $M(\theta)$.* $M(\theta)$ is parameterized by some parameter θ to be selected. The true system S may lie in $M(\theta)$, but for

most cases S does not belong to $M(\theta)$. Thus θ has to be chosen such that $M(\theta)$ approximates S as well as possible.

2. *Parameter Estimation.* With $M(\theta)$ having been selected and with input–output data having been obtained, the next step is to construct the estimate $\hat{\theta}$ of θ such that $M(\hat{\theta})$ is consistent with true data as well as possible.

3. *Design of Input Signal.* This problem arises when for identification the input of system can be given arbitrarily since the result of identification may depend on input signal.

4. *Properties of the Parameter Estimates.* Having specified the parameter estimate $\hat{\theta}$, one has to determine its properties. It is usually desirable that $\hat{\theta}$ has at least the following property: if $S \in M(\theta)$ then $\hat{\theta}$ asymptotically converges to the true parameter θ as the data increase. This is the so-called consistency problem. In addition, one also wishes to obtain the convergence rate of the estimate, its asymptotic distribution, the efficiency of the estimate, and other related properties.

We shall mainly discuss the consistency problem. First, for the case with the system noise being a martingale difference sequence for the least-squares identification, we give not only strong consistency theorems but also its convergence rate by using probabilistic methods. Then for the correlated noise case, we also prove a strong consistency theorem for the least-squares identification by the same method.

Results obtained in this chapter and in the next chapter are also suitable for time series analysis since the dynamic model considered here is nothing but the ARMAX model in the time series analysis.

3.2. LEAST-SQUARES IDENTIFICATION

Let $(\Omega, \mathscr{F}, P)$ be the basic probability space, and let the input $\{\mathbf{u}_n\}$, output $\{\mathbf{y}_n\}$ and dynamic noise $\{\boldsymbol{\varepsilon}_n\}$ processes be l-, m- and m'-dimensional, respectively. Consider the following discrete-time system with known orders p and q:

$$\mathbf{y}_n = \underset{\sim}{\mathbf{A}}_1 \mathbf{y}_{n-1} + \cdots + \underset{\sim}{\mathbf{A}}_p \mathbf{y}_{n-p} + \underset{\sim}{\mathbf{B}}_1 \mathbf{u}_{n-1} + \cdots + \underset{\sim}{\mathbf{B}}_q \mathbf{u}_{n-q} + \boldsymbol{\varepsilon}_n. \quad (3.1)$$

Let $\mathbf{w}_i$ be m-dimensional random vectors with

$$E\left(\frac{\mathbf{w}_n}{\mathscr{F}_{n-1}}\right) = 0, \qquad E\left(\frac{\mathbf{w}_n \mathbf{w}_n^{\tau}}{\mathscr{F}_{n-1}}\right) = \underset{\sim}{\mathbf{I}} \quad (3.2)$$

and let the noise ε_n be expressed by

$$\varepsilon_n = \underset{\sim}{F}_n \mathbf{w}_n, \tag{3.3}$$

where $\mathscr{F}_n$ is the σ-algebra generated by $\mathbf{w}_i$, that is, where

$$\mathscr{F}_n = \sigma\{\mathbf{w}_i, 0 \leq i \leq n\}, \tag{3.4}$$

and where $\underset{\sim}{F}_n$ are $m \times m'$ $\mathscr{F}_{n-1}$-measurable matrices.

It is worth noting that we do not require

$$E \operatorname{tr} \underset{\sim}{F}_n \underset{\sim}{F}_n^\tau < \infty,$$

and we do not even require the finiteness of expectation for ε_n.

In order to include the feedback controls, we shall require $\mathbf{u}_n$ to be $\mathscr{F}_n$-measurable.

Denote

$$\underset{\sim}{\theta}^\tau = \left[\underset{\sim}{A}_1 \cdots \underset{\sim}{A}_p \underset{\sim}{B}_1 \cdots \underset{\sim}{B}_q\right] \tag{3.5}$$

$$\phi_n^\tau = \left[\mathbf{y}_n^\tau \mathbf{y}_{n-1}^\tau \cdots \mathbf{y}_{n-p+1}^\tau \mathbf{u}_n^\tau \cdots \mathbf{u}_{n-q+1}^\tau\right], \qquad \phi_{-1}\text{-arbitrary.} \tag{3.6}$$

Then (3.1) is rewritten as

$$\mathbf{y}_n = \underset{\sim}{\theta}^\tau \phi_{n-1} + \varepsilon_n. \tag{3.7}$$

Further, denote

$$\underset{\sim}{Y}_{n+1} = \begin{bmatrix} \mathbf{y}_1^\tau \\ \vdots \\ \mathbf{y}_{n+1}^\tau \end{bmatrix}, \quad \underset{\sim}{\Psi}_n = \begin{bmatrix} \phi_0^\tau \\ \vdots \\ \phi_n^\tau \end{bmatrix}, \quad \underset{\sim}{H}_n = \begin{bmatrix} \varepsilon_1^\tau \\ \vdots \\ \varepsilon_{n+1}^\tau \end{bmatrix} \tag{3.8}$$

which leads to

$$\underset{\sim}{Y}_{n+1} = \underset{\sim}{\Psi}_n \underset{\sim}{\theta} + \underset{\sim}{H}_n \tag{3.9}$$

by (3.7).

In the present setup of the problem, the model set $M(\underset{\sim}{\theta})$ is given by (3.1), the unknown parameter $\underset{\sim}{\theta}$ is defined by (3.5), and we hope to give an estimate θ_n for θ based on Ψ_n and Y_{n+1}. The most intuitive motivated and frequently applied method is the least-squares identification technique, which requires one to select θ_{n+1} such that

$$(Y_{n+1} - \Psi_n\theta_{n+1})^\tau (Y_{n+1} - \Psi_n\theta_{n+1}) = \min_{\theta} (Y_{n+1} - \Psi_n\theta)^\tau (Y_{n+1} - \Psi_n\theta). \tag{3.10}$$

Assume

$$\Psi_n^\tau \Psi_n > 0,$$

that is,

$$\sum_{i=0}^{n} \phi_i \phi_i^\tau > 0. \tag{3.11}$$

It may be directly verified that

$$(Y_{n+1} - \Psi_n\theta)^\tau (Y_{n+1} - \Psi_n\theta) = \left[\theta - (\Psi_n^\tau \Psi_n)^{-1} \Psi_n^\tau Y_{n+1}\right]^\tau \Psi_n^\tau \Psi_n$$
$$\cdot \left[\theta - (\Psi_n^\tau \Psi_n)^{-1} \Psi_n^\tau Y_{n+1}\right] + Y_{n+1}^\tau \left[I - \Psi_n (\Psi_n^\tau \Psi_n)^{-1} \Psi_n^\tau\right] Y_{n+1} \tag{3.12}$$

where only the first term depends on θ. Hence

$$\theta_{n+1} = (\Psi_n^\tau \Psi_n)^{-1} \Psi_n^\tau Y_{n+1} \tag{3.13}$$

or, by (3.8),

$$\theta_{n+1} = \left(\sum_{i=0}^{n} \phi_i \phi_i^\tau\right)^{-1} \sum_{i=0}^{n} \phi_i y_{i+1}^\tau \tag{3.14}$$

makes (3.12) reach its minimum. θ_{n+1} given here is called the least-squares identification (or estimate).

In mathematical statistics for linear model (3.9), matrix Ψ_n, called the design matrix, is deterministic, while here it is random, and hence results obtained in mathematical statistics cannot be applied directly to system identification.

For on-line computing it is convenient to write θ_n in the recursive form.

Lemma 3.1. *The estimate* θ_n *given by least-squares identification satisfies recursive formulas*:

$$\theta_{n+1} = \theta_n + a_n P_n \phi_n (y_{n+1}^\tau - \phi_n^\tau \theta_n) = \theta_n + P_{n+1} \phi_n (y_{n+1}^\tau - \phi_n^\tau \theta_n), \quad (3.15)$$

$$P_{n+1} = P_n - a_n P_n \phi_n \phi_n^\tau P_n, \quad (3.16)$$

where

$$P_n = \left(\sum_{i=0}^{n-1} \phi_i \phi_i^\tau \right)^{-1}, \qquad a_n = (1 + \phi_n^\tau P_n \phi_n)^{-1}. \quad (3.17)$$

Proof. We first show a matrix identify

$$C - B^\tau (A + BC^{-1}B^\tau)^{-1} B = [C^{-1} + C^{-1}B^\tau A^{-1} BC^{-1}]^{-1}. \quad (3.18)$$

Since

$$0 = -(A + BC^{-1}B^\tau)^{-1} + A^{-1} - A^{-1}BC^{-1}B^\tau (A + BC^{-1}B^\tau)^{-1},$$

we have

$$I = I - C^{-1}B^\tau (A + BC^{-1}B^\tau)^{-1} B + C^{-1}B^\tau A^{-1} B$$
$$- C^{-1}B^\tau A^{-1}BC^{-1}B^\tau (A + BC^{-1}B^\tau)^{-1} B$$

or

$$I = (I + C^{-1}B^\tau A^{-1}B) - (C^{-1} + C^{-1}B^\tau A^{-1}BC^{-1})B^\tau (A + BC^{-1}B^\tau)^{-1} B$$
$$= (C^{-1} + C^{-1}B^\tau A^{-1}BC^{-1})C$$
$$- (C^{-1} + C^{-1}B^\tau A^{-1}BC^{-1})B^\tau (A + BC^{-1}B^\tau)^{-1} B$$
$$= (C^{-1} + C^{-1}B^\tau A^{-1}BC^{-1})\left[C - B^\tau (A + BC^{-1}B^\tau)^{-1} B\right],$$

from which (3.18) follows.

By definition

$$\mathbf{P}_{n+1} = \left[\sum_{i=0}^{n-1} \phi_i \phi_i^T + \phi_n \phi_n^T \right]^{-1} = \left[\mathbf{P}_n^{-1} + \phi_n \phi_n^T \right]^{-1}. \tag{3.19}$$

In (3.18), taking $\mathbf{C} = \mathbf{P}_n$, $\mathbf{B} = \phi_n^T \mathbf{P}_n$, $\mathbf{A} = 1$ gives

$$\begin{aligned} \left[\mathbf{C}^{-1} + \mathbf{C}^{-1}\mathbf{B}^T\mathbf{A}^{-1}\mathbf{B}\mathbf{C}^{-1} \right]^{-1} &= \left[\mathbf{P}_n^{-1} + \phi_n \phi_n^T \right]^{-1} \\ &= \mathbf{P}_n - \mathbf{P}_n \phi \left(1 + \phi_n^T \mathbf{P}_n \mathbf{P}_n^{-1} \mathbf{P}_n \phi_n \right)^{-1} \phi_n^T \mathbf{P}_n \\ &= \mathbf{P}_n - \mathbf{P}_n \phi_n \left(1 + \phi_n^T \mathbf{P}_n \phi_n \right)^{-1} \phi_n^T \mathbf{P}_n \\ &= \mathbf{P}_n - a_n \mathbf{P}_n \phi_n \phi_n^T \mathbf{P}_n, \end{aligned} \tag{3.20}$$

which together with (3.19) yields (3.16).

From (3.14) and (3.16), we have

$$\begin{aligned} \theta_{n+1} &= \mathbf{P}_{n+1} \sum_{i=0}^{n} \phi_i \mathbf{y}_{i+1}^T = \left(\mathbf{P}_n - a_n \mathbf{P}_n \phi_n \phi_n^T \mathbf{P}_n \right) \left(\sum_{i=0}^{n-1} \phi_i \mathbf{y}_{i+1}^T + \phi_n \mathbf{y}_{n+1}^T \right) \\ &= \theta_n - a_n \mathbf{P}_n \phi_n \phi_n^T \theta_n + \mathbf{P}_n \phi_n \mathbf{y}_{n+1}^T - a_n \mathbf{P}_n \phi_n \phi_n^T \mathbf{P}_n \phi_n \mathbf{y}_{n+1}^T) \\ &= \theta_n - a_n \mathbf{P}_n \phi_n \phi_n^T \theta_n + \mathbf{P}_n \phi_n \left(1 - a_n \phi_n^T \mathbf{P}_n \phi_n \right) \mathbf{y}_{n+1}^T \\ &= \theta_n - a_n \mathbf{P}_n \phi_n \phi_n^T \theta_n + a_n \mathbf{P}_n \phi_n \mathbf{y}_{n+1}^T = \theta_n + a_n \mathbf{P}_n \phi_n \left(\mathbf{y}_{n+1}^T - \phi_n^T \theta_n \right), \end{aligned}$$

which gives the first equality in (3.15). Further, by (3.17) and (3.16) we find

$$a_n \mathbf{P}_n \phi \left(1 - a_n \phi_n^T \mathbf{P}_n \phi_n \right) = a_n \mathbf{P}_{n+1} \phi_n$$

and

$$a_n \mathbf{P}_n \phi_n = \mathbf{P}_{n+1} \phi_n,$$

from which the second equality in (3.15) follows. □

We now give initial values for (3.15) and (3.16) as follows

$$\mathbf{P}_0 = (mp + lq)\mathbf{I},$$

$$\boldsymbol{\theta}_0 = \text{any deterministic matrix of compatible dimension.} \tag{3.21}$$

We still call their solution $\boldsymbol{\theta}_n$ the least-squares estimate. With such a initial value, we shall have no trouble when (3.11) is not fulfilled, since the solution of (3.16) with initial value (3.21) is nonsingular and is given by

$$\mathbf{P}_n = \left(\sum_{i=1}^{n-1} \boldsymbol{\phi}_i \boldsymbol{\phi}_i^\tau + \frac{1}{mp + lq}\mathbf{I} \right)^{-1}, \tag{3.22}$$

which is directly verified by (3.18) in a way similar to that used in Lemma 3.1. Analogously we can verify

$$\boldsymbol{\theta}_n = \mathbf{P}_n \sum_{i=0}^{n-1} \boldsymbol{\phi}_i \mathbf{y}_{i+1}^\tau + \mathbf{P}_n \mathbf{P}_0^{-1} \boldsymbol{\theta}_0. \tag{3.23}$$

Denote by $\lambda_{\max}^n$ and $\lambda_{\min}^n$ the maximum and minimum eigenvalues of $\mathbf{P}_n$, respectively, and denote

$$r_n \triangleq \operatorname{tr} \mathbf{P}_n^{-1} = 1 + \sum_{i=0}^{n-1} \|\boldsymbol{\phi}_i\|^2. \tag{3.24}$$

In what follows, $k_i > 0$ always denotes a constant independent of $\omega \in \Omega$ if no other explanation is given.

r_n is nondecreasing as $n \to \infty$, but if it has a finite limit, then the nondecreasing matrix $\sum_{i=0}^{n-1} \boldsymbol{\phi}_i \boldsymbol{\phi}_i^\tau$ will also converge to a finite limit, and from (3.22), $\mathbf{P}_n$ will go to a limit $\mathbf{P}$. In this case, as can be seen from (3.23), as $n \to \infty$, $\boldsymbol{\theta}_n$ tends to a finite limit, which, generally speaking, depends on $\boldsymbol{\theta}_0$ and is not the true parameter.

From now on we assume

$$r_n \underset{n\to\infty}{\longrightarrow} \infty \text{ a.s.} \tag{3.25}$$

We shall use the following theorem from analysis:

Lemma 3.2. *If* $a_i \geq 0$, $b_n = \sum_{i=0}^n a_i \underset{n\to\infty}{\longrightarrow} \infty$, *then*

$$\sum_{i=1}^{\infty} \frac{a_i}{b_i} = \infty, \qquad \sum_{i=1}^{\infty} \frac{a_i}{b_i^{1+\delta}} < \infty, \quad \delta > 0. \tag{3.26}$$

Further, if $\sum_{i=1}^{\infty}(1/b_i)\underset{\sim}{\mathbf{A}}_i < \infty$, *then*

$$\frac{1}{b_n}\sum_{i=1}^{n}\underset{\sim}{\mathbf{A}}_i \underset{n\to\infty}{\to} 0, \tag{3.27}$$

where $\underset{\sim}{\mathbf{A}}_i$ *are matrices.*

Proof. For fixed n

$$\frac{a_{n+1}}{b_{n+1}} + \frac{a_{n+2}}{b_{n+1}} + \cdots + \frac{a_{n+r}}{b_{n+r}} \geq \frac{b_{n+r} - b_n}{b_{n+r}} = 1 - \frac{b_n}{b_{n+r}} \underset{r\to\infty}{\to} 1.$$

Hence to any n there is a corresponding r so that the preceding sum is greater than $\frac{1}{2}$ and thus

$$\sum_{i=1}^{a}\frac{a_i}{b_i} = \infty.$$

We shall use the following inequality [74]

$$\delta x^{\delta-1}(x-y) \leq x^{\delta} - y^{\delta} \tag{3.28}$$

which is valid for any $x \geq 0$, $y \geq 0$, and $0 < \delta \leq 1$.

Setting $b_0 = b_1$ for $0 < \delta \leq 1$, we have

$$\sum_{i=1}^{\infty}\frac{a_i}{b_i^{1+\delta}} \leq \sum_{i=1}^{\infty}\frac{a_i}{b_i b_{i-1}^{\delta}} = \sum_{i=1}^{\infty}\frac{\delta(b_i - b_{i-1})b_i^{\delta-1}}{\delta b_{i-1}^{\delta} b_i^{1-\delta}} \leq \frac{1}{\delta}\sum_{i=1}^{\infty}\frac{b_i^{\delta} - b_{i-1}^{\delta}}{b_{i-1}^{\delta} b_i^{\delta}}$$

$$= \frac{1}{\delta}\sum_{i=1}^{\infty}\left(\frac{1}{b_{i-1}^{\delta}} - \frac{1}{b_i^{\delta}}\right) = \frac{1}{\delta}\frac{1}{b_1^{\delta}} < \infty,$$

which implies $\sum_{i=1}^{\infty}a_i/b_i^{1+\delta} < \infty$ for $\delta > 1$.

Denote

$$\underset{\sim}{\mathbf{B}}_n = \sum_{i=1}^{n}\frac{\underset{\sim}{\mathbf{A}}_i}{b_i}, \qquad \underset{\sim}{\mathbf{B}}_0 = 0.$$

By the condition of the lemma

$$\underset{\sim}{\mathbf{B}}_n \underset{n\to\infty}{\to} \underset{\sim}{\mathbf{B}} < \infty.$$

Hence given $\varepsilon > 0$, we can find N so that

$$\|\underset{\sim}{\mathbf{B}}_{i-1} - \underset{\sim}{\mathbf{B}}\| < \varepsilon, \qquad \forall i - 1 \geq N.$$

Then we have

$$\left\| \frac{1}{b_n} \sum_{i=1}^{n} \underset{\sim}{\mathbf{A}}_i \right\| = \left\| \frac{1}{b_n} \sum_{i=1}^{n} b_i (\underset{\sim}{\mathbf{B}}_i - \underset{\sim}{\mathbf{B}}_{i-1}) \right\| = \left\| \underset{\sim}{\mathbf{B}}_n - \frac{1}{b_n} \sum_{i=1}^{n} (b_i - b_{i-1}) \underset{\sim}{\mathbf{B}}_{i-1} \right\|$$

$$= \left\| \underset{\sim}{\mathbf{B}}_n - \frac{1}{b_n} \sum_{i=1}^{n} a_i \underset{\sim}{\mathbf{B}}_{i-1} \right\| = \left\| \underset{\sim}{\mathbf{B}}_n - \underset{\sim}{\mathbf{B}} - \frac{1}{b_n} \sum_{i=1}^{n} a_i (\underset{\sim}{\mathbf{B}}_{i-1} - \mathbf{B}) \right\|$$

$$\leq \| \underset{\sim}{\mathbf{B}}_n - \underset{\sim}{\mathbf{B}} \| + \frac{1}{b_n} \sum_{i=1}^{N-1} a_i \| \underset{\sim}{\mathbf{B}}_{i-1} - \underset{\sim}{\mathbf{B}} \| + \varepsilon \underset{\substack{n \to \infty \\ \varepsilon \to 0}}{\longrightarrow} 0. \qquad \square$$

Theorem 3.1. *Provided that* (3.25) *holds and there are constants* $k_0 > 0$, $\varepsilon \in [0, 1)$, $\alpha \in [0, \frac{1}{2})$ *and an* a.s. *finite random variable* γ *such that*

$$\operatorname{tr} \underset{\sim}{\mathbf{F}}_n \underset{\sim}{\mathbf{F}}_n^{\tau} \leq k_0 r_n^{\varepsilon} \text{ a.s.}, \qquad \forall n \tag{3.29}$$

$$\lambda_{\max}^n \leq \frac{\gamma}{r_n^{1-\alpha}}, \qquad \alpha + \frac{\varepsilon}{2} < \frac{1}{2}, \tag{3.30}$$

then $\underset{\sim}{\theta}_n$ *given by* (3.15) *and* (3.16) *with initial values* (3.21) *is strongly consistent. That is,*

$$\underset{\sim}{\theta}_n \underset{n \to \infty}{\longrightarrow} \underset{\sim}{\theta} \text{ a.s.} \tag{3.31}$$

with convergence rate

$$\| \underset{\sim}{\theta}_n - \underset{\sim}{\theta} \| = o\left(r_n^{\delta - 1/2} \right), \qquad \forall \delta \in \left(\alpha + \frac{\varepsilon}{2}, \frac{1}{2} \right]. \tag{3.32}$$

Proof. Putting (3.7) into (3.23) we see

$$\underset{\sim}{\theta}_n = \underset{\sim}{\mathbf{P}}_n \sum_{i=0}^{n-1} \phi_i \phi_i^{\tau} \underset{\sim}{\theta} + \underset{\sim}{\mathbf{P}}_n \sum_{i=0}^{n-1} \phi_i \mathbf{w}_{i+1}^{\tau} \underset{\sim}{\mathbf{F}}_{i+1}^{\tau} + \underset{\sim}{\mathbf{P}}_n \underset{\sim}{\mathbf{P}}_o^{-1} \underset{\sim}{\theta}_o$$

$$= \underset{\sim}{\theta} - \frac{1}{mp + lq} \underset{\sim}{\mathbf{P}}_n \underset{\sim}{\theta} + \underset{\sim}{\mathbf{P}}_n \sum_{i=0}^{n-1} \phi_i \mathbf{w}_{i+1}^{\tau} \underset{\sim}{\mathbf{F}}_{i+1}^{\tau} + \underset{\sim}{\mathbf{P}}_n \underset{\sim}{\mathbf{P}}_o^{-1} \underset{\sim}{\theta}_o \tag{3.33}$$

and

$$r_n^{1/2 - \delta} \| \underset{\sim}{\theta}_n - \underset{\sim}{\theta} \| \leq \frac{1}{mp + lq} r_n^{1/2 - \delta} \| \underset{\sim}{\mathbf{P}}_n \| \, \| \underset{\sim}{\theta} \| + r_n^{1/2 - \delta} \left\| \underset{\sim}{\mathbf{P}}_n \sum_{i=0}^{n-1} \phi_i \mathbf{w}_{i+1}^{\tau} \underset{\sim}{\mathbf{F}}_{i+1}^{\tau} \right\|$$

$$+ \| r_n^{1/2 - \delta} \underset{\sim}{\mathbf{P}}_n \underset{\sim}{\mathbf{P}}_o^{-1} \underset{\sim}{\theta}_o \|, \tag{3.34}$$

whose first and third terms go to zero a.s. as $n \to \infty$ since by (3.30) and (3.31),

$$r_n^{1/2-\delta}\|\underset{\sim}{\mathbf{P}}_n\| \le r_n^{1/2-\delta}\lambda_{\max}^n \le \gamma r_n^{1/2-\delta+\alpha-1} = \gamma r_n^{\alpha-1/2-\delta} \underset{n\to\infty}{\to} 0 \text{ a.s.} \tag{3.35}$$

By $\|\mathbf{ab}^\tau\|$, we mean the square root of the maximum eigenvalue of matrix $\mathbf{ab}^\tau(\mathbf{ab}^\tau)^\tau$, where $\mathbf{a}, \mathbf{b}$ are vectors. But $\mathbf{aa}^\tau$ has $\mathbf{a}^\tau\mathbf{a}$ as its unique nonzero eigenvalue, consequently, $\|\mathbf{aa}^\tau\| = \sqrt{\mathbf{a}^\tau\mathbf{a}}$ and

$$\|\mathbf{ab}^\tau\| = \sqrt{\mathbf{b}^\tau\mathbf{b}}\,\sqrt{\mathbf{a}^\tau\mathbf{a}} = \|\mathbf{a}\|\,\|\mathbf{b}\|. \tag{3.36}$$

By this fact and (1.8), (3.36), (3.29), and (3.31), we have

$$\begin{aligned}
E\left\|\frac{\boldsymbol{\phi}_{i-1}\mathbf{w}_i^\tau\underset{\sim}{\mathbf{F}}_i^\tau}{r_i^{1/2+\delta-\alpha}}\right\|^2 &= E\frac{\|\boldsymbol{\phi}_{i-1}\|^2\|\underset{\sim}{\mathbf{F}}_i\mathbf{w}_i\|^2}{r_i^{1+2\delta-2\alpha}} = E\frac{\|\boldsymbol{\phi}_{i-1}\|^2\operatorname{tr}\underset{\sim}{\mathbf{F}}_i\mathbf{w}_i\mathbf{w}_i^\tau\underset{\sim}{\mathbf{F}}_i^\tau}{r_i^{1+2\delta-2\alpha}}\\
&= E\left[\frac{\|\boldsymbol{\phi}_{i-1}\|^2}{r_i^{1+2\delta-2\alpha}}E\left(\operatorname{tr}\frac{\underset{\sim}{\mathbf{F}}_i\mathbf{w}_i\mathbf{w}_i^\tau\underset{\sim}{\mathbf{F}}_i^\tau}{\mathscr{F}_{i-1}}\right)\right]\\
&= E\frac{\|\boldsymbol{\phi}_{i-1}\|^2\operatorname{tr}\underset{\sim}{\mathbf{F}}_i\underset{\sim}{\mathbf{F}}_i^\tau}{r_i^{1+2\delta-2\alpha}} \le k_0E\frac{\|\boldsymbol{\phi}_{i-1}\|^2 r_i^\varepsilon}{r_i^{1+2\delta-2\alpha}} \le k_0E\frac{\|\boldsymbol{\phi}_{i-1}\|^2}{r_i} \le k_0,
\end{aligned} \tag{3.37}$$

where we have used the fact that r_i and $\underset{\sim}{\mathbf{F}}_i$ are $\mathscr{F}_{i-1}$-measurable.

Then by (3.2), we find

$$\begin{aligned}
E\left(\sum_{i=0}^n\frac{\boldsymbol{\phi}_{i-1}\mathbf{w}_i^\tau\underset{\sim}{\mathbf{F}}_i^\tau}{r_i^{1/2+\delta-\alpha}}\Big/\mathscr{F}_{n-1}\right) &= \sum_{i=0}^{n-1}\frac{\boldsymbol{\phi}_{i-1}\mathbf{w}_i^\tau\underset{\sim}{\mathbf{F}}_i^\tau}{r_i^{1/2+\delta-\alpha}} + E\left(\frac{\boldsymbol{\phi}_{n-1}\mathbf{w}_n^\tau\underset{\sim}{\mathbf{F}}_n^\tau}{r_n^{1/2+\delta-\alpha}}\Big/\mathscr{F}_{n-1}\right)\\
&= \sum_{i=0}^{n-1}\frac{\boldsymbol{\phi}_{i-1}\mathbf{w}_i^\tau\underset{\sim}{\mathbf{F}}_i^\tau}{r_i^{1/2+\delta-\alpha}} + \frac{\boldsymbol{\phi}_{n-1}}{r_n^{1/2+\delta-\alpha}}E\left(\frac{\mathbf{w}_n^\tau}{\mathscr{F}_{n-1}}\right)\underset{\sim}{\mathbf{F}}_n^\tau\\
&= \sum_{i=0}^{n-1}\frac{\boldsymbol{\phi}_{i-1}\mathbf{w}_i^\tau\underset{\sim}{\mathbf{F}}_i^\tau}{r_i^{1/2+\delta-\alpha}},
\end{aligned}$$

and thus

$$\left(\sum_{i=0}^n\frac{\boldsymbol{\phi}_{i-1}\mathbf{w}_i^\tau\underset{\sim}{\mathbf{F}}_i^\tau}{r_i^{1/2+\delta-\alpha}}, \mathscr{F}_n\right) \tag{3.38}$$

is a martingale with finite second moment.

Similar to (3.37), we see by (3.26)

$$\sum_{i=0}^{\infty} E\left[\frac{\left\|\frac{\phi_{i-1}\mathbf{w}_i^\tau \mathbf{F}_i^\tau}{r_i^{1/2+\delta-\alpha}}\right\|^2}{\mathscr{F}_{i-1}}\right] = \sum_{i=0}^{\infty} \frac{\|\phi_{i-1}\|^2}{r_i^{1+2\delta-2\alpha}} \operatorname{tr} \mathbf{F}_i \mathbf{F}_i^\tau \le k_0 \sum_{i=0}^{\infty} \frac{\|\phi_{i-1}\|^2}{r_i^{1+2\delta_0}} < \infty,$$

where by (3.32)

$$\delta_0 \triangleq \delta - \alpha - \frac{\varepsilon}{2} > 0. \tag{3.39}$$

By Theorem 1.13, we conclude that

$$\sum_{i=1}^{\infty} \frac{\phi_{i-1}\mathbf{w}_i^\tau \mathbf{F}_i^\tau}{r_i^{1/2+\delta-\alpha}} < \infty \text{ a.s.}$$

and by (3.27)

$$\frac{1}{r_n^{1/2+\delta-\alpha}} \sum_{i=1}^{n} \phi_{i-1}\mathbf{w}_i^\tau \mathbf{F}_i^\tau \underset{n\to\infty}{\longrightarrow} 0 \text{ a.s.}$$

By use of (3.30) the second term on the right-hand side of (3.34), is estimated by

$$\begin{aligned} r_n^{1/2-\delta}\left\|\mathbf{P}_n \sum_{i=0}^{n-1} \phi_i \mathbf{w}_{i+1}^\tau \mathbf{F}_{i+1}^\tau\right\| &\le r_n^{1/2-\delta}\lambda_{\max}^n \left\|\sum_{i=0}^{n-1} \phi_i \mathbf{w}_{i+1}^\tau \mathbf{F}_{i+1}^\tau\right\| \\ &\le \gamma r_n^{1/2-\delta} r_n^{\alpha-1}\left\|\sum_{i=0}^{n-1} \phi_i \mathbf{w}_{i+1}^\tau \mathbf{F}_{i+1}^\tau\right\| \\ &= \gamma r_n^{\alpha-\delta-1/2}\left\|\sum_{i=0}^{n-1} \phi_i \mathbf{w}_{i+1}^\tau \mathbf{F}_{i+1}^\tau\right\| \underset{n\to\infty}{\longrightarrow} 0 \text{ a.s.} \quad \square \end{aligned}$$

Remark 1. In this theorem we do not require $E\|\varepsilon_i\| < \infty$, and furthermore $\operatorname{tr} \mathbf{F}_n \mathbf{F}_n^\tau$ is allowed to increase as $n \to \infty$. Notice that

$$\frac{r_n}{mp+lq} = \frac{\operatorname{tr}(\mathbf{P}_n^{-1})}{mp+lq} \le \|\mathbf{P}_n^{-1}\| \le \operatorname{tr}(\mathbf{P}_n^{-1}) = r_n,$$

$$\frac{1}{r_n} \le \lambda_{\min}^n = \frac{1}{\|\mathbf{P}_n^{-1}\|} \le \frac{mp+lq}{r_n}, \tag{3.40}$$

and condition (3.30) is satisfied if

$$\lambda^n_{\max} = \frac{1}{r_n^{1-\alpha}}$$

But in this case

$$\frac{\lambda^n_{\max}}{\lambda^n_{\min}} \geq \frac{1}{r_n^{1-\alpha}}\left(\frac{r_n}{mp+lq}\right) = \frac{r_n^{\alpha}}{mp+lq} \underset{n\to\infty}{\to} \infty,$$

which means that for strong consistency of least-squares estimate the ratio $\lambda^n_{\max}/\lambda^n_{\min}$ is not necessarily bounded.

Remark 2. From (3.40), it is known that (3.30) is equivalent to

$$\frac{\lambda^n_{\max}}{\lambda^n_{\min}} \leq \gamma_1 r_n^{\alpha} \tag{3.41}$$

where γ_1 is some a.s. finite random variable.

From this remark and Theorem 3.1, the Theorem 3.2 follows.

Theorem 3.2. *If there is a constant* σ^2 *such that* $\operatorname{tr} \underset{\sim}{\mathbf{F}}_n \underset{\sim}{\mathbf{F}}_n^{\tau} \leq \sigma^2$, $\forall n$ *and if* (3.25) *and* (3.41) *for some* $\alpha \in [0, \frac{1}{2})$ *hold, then*

$$\|\underset{\sim}{\theta}_n - \underset{\sim}{\theta}\| = o\left(r_n^{\delta-1/2}\right), \qquad \forall \delta \in \left(\alpha, \tfrac{1}{2}\right] \tag{3.42}$$

Theorem 3.3. *Let* $E\|\underset{\sim}{\mathbf{F}}_{\lambda}\|^2 < \infty$, $\varepsilon \in (0,1)$, $\alpha \in [0, \frac{1}{2})$, $\varepsilon/2 + \alpha < \frac{1}{2}$, γ *be a finite random variable. Then*

$$\|\underset{\sim}{\theta}_n - \underset{\sim}{\theta}\| = o\left(r_n^{\delta-1/2}\right), \qquad \forall \delta \in \left(\alpha + \frac{\varepsilon}{2}, \tfrac{1}{2}\right]$$

for almost all $\omega \in A$, *where*

$$A = \left\{\omega\colon r_n \to \infty, \operatorname{tr} \underset{\sim}{\mathbf{F}}_n \underset{\sim}{\mathbf{F}}_n^{\tau} \leq k_o r_n^{\varepsilon}, \lambda^n_{\max}/\lambda^n_{\min} \leq \gamma r_n^{\alpha}\right\}.$$

Proof. $(\sum_{i=0}^{n} \phi_{i-1} \mathbf{w}_i^{\tau} \underset{\sim}{\mathbf{F}}_i^{\tau} / r_i^{1/2+\delta-\alpha}, \mathscr{F}_n)$ is a martingale whenever $\delta \geq \alpha$, since in this case

$$E\left\|\frac{\phi_{i-1}\mathbf{w}_i^{\tau}\underset{\sim}{\mathbf{F}}_i^{\tau}}{r_i^{1/2+\delta-\alpha}}\right\| \leq E\|\mathbf{w}_i\|\,\|\underset{\sim}{\mathbf{F}}_i\| \leq \sqrt{E\|\mathbf{w}_i\|^2 E\|\mathbf{F}_i\|^2} < \infty.$$

In the proof of Theorem 3.1, all formulas following (3.38) are verified on A. Thus by Theorem 1.13 the required assertion follows. □

Now we give a theorem with no condition imposed on eigenvalues of $\underset{\sim}{\mathbf{P}}_n$.

Theorem 3.4. *If* $\operatorname{tr} \underset{\sim}{\mathbf{F}}_n \underset{\sim}{\mathbf{F}}_n \leq \sigma^2 \forall n$ *with* σ^2 *a constant, then*

$$\underset{\sim}{\boldsymbol{\theta}}_n \underset{n\to\infty}{\longrightarrow} \underset{\sim}{\boldsymbol{\theta}}, \qquad \forall \omega \in J$$

where

$$J \triangleq \left\{ \omega : \underset{\sim}{\mathbf{P}}_n \underset{n\to\infty}{\longrightarrow} 0,\ \overline{\lim_{n\to\infty}} \frac{1}{\varepsilon} \left\| \underset{\sim}{\mathbf{P}}_n \sum_{i=0}^{n-1} \varepsilon_i \boldsymbol{\phi}_i \boldsymbol{\phi}_i^{\tau} \right\| < \infty, \qquad \forall \varepsilon > 0, \forall \varepsilon_i \in [0, \varepsilon] \right\}. \tag{3.43}$$

Proof. Since

$$(\underset{\sim}{\mathbf{P}}_j - \underset{\sim}{\mathbf{P}}_{j+1}) \underset{\sim}{\mathbf{P}}_j^{-1} (\underset{\sim}{\mathbf{P}}_j - \underset{\sim}{\mathbf{P}}_{j+1}) \geq 0$$

we have

$$\underset{\sim}{\mathbf{P}}_j - 2\underset{\sim}{\mathbf{P}}_{j+1} + \underset{\sim}{\mathbf{P}}_{j+1} \underset{\sim}{\mathbf{P}}_j^{-1} \underset{\sim}{\mathbf{P}}_{j+1} \geq 0,$$

or

$$\underset{\sim}{\mathbf{P}}_j - \underset{\sim}{\mathbf{P}}_{j+1} \geq \underset{\sim}{\mathbf{P}}_{j+1} - \underset{\sim}{\mathbf{P}}_{j+1} \underset{\sim}{\mathbf{P}}_j^{-1} \underset{\sim}{\mathbf{P}}_{j+1},$$

or

$$\underset{\sim}{\mathbf{P}}_j - \underset{\sim}{\mathbf{P}}_{j+1} \geq \underset{\sim}{\mathbf{P}}_{j+1} \left(\underset{\sim}{\mathbf{P}}_{j+1}^{-1} - \underset{\sim}{\mathbf{P}}_j^{-1} \right) \underset{\sim}{\mathbf{P}}_{j+1}, \tag{3.44}$$

from which along with (3.22) it follows

$$\begin{aligned} E \operatorname{tr} \mathbf{P}_{j+1} \boldsymbol{\phi}_j \mathbf{w}_{j+1}^{\tau} \mathbf{w}_{j+1} \boldsymbol{\phi}_j^{\tau} \underset{\sim}{\mathbf{P}}_{j+1} &\leq E \| \underset{\sim}{\mathbf{F}}_{j+1} \mathbf{w}_{j+1} \|^2 \operatorname{tr} \underset{\sim}{\mathbf{P}}_{j+1} \left(\underset{\sim}{\mathbf{P}}_{j+1}^{-1} - \underset{\sim}{\mathbf{P}}_j^{-1} \right) \underset{\sim}{\mathbf{P}}_{j+1} \\ &\leq E \| \underset{\sim}{\mathbf{F}}_{j+1} \mathbf{w}_{j+1} \|^2 \left(\underset{\sim}{\mathbf{P}}_j - \underset{\sim}{\mathbf{P}}_{j+1} \right) \leq E \| \underset{\sim}{\mathbf{F}}_{j+1} \mathbf{w}_{j+1} \|^2 \underset{\sim}{\mathbf{P}}_j \\ &\leq \sigma^2 E \| \mathbf{w}_{j+1} \|^2 (mp + lq) \leq \sigma^2 (mp + lq). \end{aligned} \tag{3.45}$$

Denote

$$\underset{\sim}{\mathbf{S}}_n = \sum_{i=0}^{n-1} \underset{\sim}{\mathbf{P}}_{i+1}\boldsymbol{\phi}_i \mathbf{w}_{i+1}^{\tau} \underset{\sim}{\mathbf{F}}_{i+1}^{\tau}, \qquad \underset{\sim}{\mathbf{S}}_o = 0.$$

It is obvious that $E\|\underset{\sim}{\mathbf{S}}_n\|^2 < \infty$ by (3.45). Hence $(\underset{\sim}{\mathbf{S}}_n, \mathscr{F}_n)$ is a martingale since $\underset{\sim}{\mathbf{P}}_n, \underset{\sim}{\mathbf{F}}_n$ are $\mathscr{F}_{n-1}$-measurable. Further, by (3.44)

$$\sum_{i=0}^{n-1} E\left[\frac{\underset{\sim}{\mathbf{P}}_{i+1}\boldsymbol{\phi}_i \mathbf{w}_{i+1}^{\tau}\underset{\sim}{\mathbf{F}}_{i+1}^{\tau}\underset{\sim}{\mathbf{F}}_{i+1}\mathbf{w}_{i+1}\boldsymbol{\phi}_i^{\tau}\underset{\sim}{\mathbf{P}}_{i+1}}{\mathscr{F}_i}\right]$$

$$\leq \sigma^2 E\left[\sum_{i=0}^{n-1} E\left(\frac{\mathbf{w}_{i+1}^{\tau}\mathbf{w}_{i+1}}{\mathscr{F}_i}\right)\underset{\sim}{\mathbf{P}}_{i+1}\boldsymbol{\phi}_i\boldsymbol{\phi}_i^{\tau}\underset{\sim}{\mathbf{P}}_{i+1}\right]$$

$$\leq m'\sigma^2 E\sum_{i=0}^{n-1}\underset{\sim}{\mathbf{P}}_{i+1}(\underset{\sim}{\mathbf{P}}_{i+1}^{-1} - \underset{\sim}{\mathbf{P}}_i^{-1})\underset{\sim}{\mathbf{P}}_{i+1} \leq m'\sigma^2 E\sum_{i=0}^{n-1}(\underset{\sim}{\mathbf{P}}_j - \underset{\sim}{\mathbf{P}}_{j+1})$$

$$\leq m'\sigma^2\underset{\sim}{\mathbf{P}}_0 = m'\sigma^2(mp + lq)I,$$

where m' is the dimension of $\mathbf{w}_{i+1}$.

By Theorem 1.13, $\underset{\sim}{\mathbf{S}}_n$ tends to a limit as $n \to \infty$

$$\underset{\sim}{\mathbf{S}}_n \underset{n\to\infty}{\to} \underset{\sim}{\mathbf{S}}, \qquad \|\underset{\sim}{\mathbf{S}}\| < \infty \text{ a.s.} \tag{3.46}$$

Now, from (3.33), we have

$$\|\underset{\sim}{\boldsymbol{\theta}}_n - \underset{\sim}{\boldsymbol{\theta}}\| \leq \left\|\underset{\sim}{\mathbf{P}}_n\left(\underset{\sim}{\mathbf{P}}_o^{-1}\underset{\sim}{\boldsymbol{\theta}}_o - \frac{1}{mp+lq}\underset{\sim}{\boldsymbol{\theta}}\right)\right\| + \left\|\underset{\sim}{\mathbf{P}}_n\sum_{i=0}^{n-1}\underset{\sim}{\mathbf{P}}_{i+1}^{-1}(\underset{\sim}{\mathbf{S}}_{i+1} - \underset{\sim}{\mathbf{S}}_i)\right\|$$

$$= \left\|\underset{\sim}{\mathbf{P}}_n\left(\underset{\sim}{\mathbf{P}}_o^{-1}\underset{\sim}{\boldsymbol{\theta}}_o - \frac{1}{mp+lq}\underset{\sim}{\boldsymbol{\theta}}\right)\right\| + \left\|\underset{\sim}{\mathbf{S}}_n - \underset{\sim}{\mathbf{P}}_n\sum_{i=0}^{n-1}(\underset{\sim}{\mathbf{P}}_{i+1}^{-1} - \underset{\sim}{\mathbf{P}}_i^{-1})\underset{\sim}{\mathbf{S}}_i\right\|$$

$$\leq \left\|\underset{\sim}{\mathbf{P}}_n\left(\underset{\sim}{\mathbf{P}}_o^{-1}\underset{\sim}{\boldsymbol{\theta}} - \frac{1}{mp+lq}\underset{\sim}{\boldsymbol{\theta}}\right)\right\| + \left\|\underset{\sim}{\mathbf{S}}_n - \underset{\sim}{\mathbf{P}}_n\sum_{i=0}^{n-1}\boldsymbol{\phi}_i\boldsymbol{\phi}_i^{\tau}\underset{\sim}{\mathbf{S}}\right\|$$

$$+ \left\|\underset{\sim}{\mathbf{P}}_n\sum_{i=0}^{n-1}\boldsymbol{\phi}_i\boldsymbol{\phi}_i^{\tau}(\underset{\sim}{\mathbf{S}}_i - \underset{\sim}{\mathbf{S}})\right\| \leq \left\|\underset{\sim}{\mathbf{P}}_n\left(\underset{\sim}{\mathbf{P}}_o^{-1}\underset{\sim}{\boldsymbol{\theta}} - \frac{1}{mp+lq}\underset{\sim}{\boldsymbol{\theta}}\right)\right\|$$

$$+ \left\|\underset{\sim}{\mathbf{P}}_n\left(\frac{1}{mp+lq}\right)\right\| + \|\underset{\sim}{\mathbf{S}}_n - \underset{\sim}{\mathbf{S}}\| + \left\|\underset{\sim}{\mathbf{P}}_n\sum_{i=0}^{n-1}\boldsymbol{\phi}_i\boldsymbol{\phi}_i^{\tau}(\underset{\sim}{\mathbf{S}}_i - \underset{\sim}{\mathbf{S}})\right\|, \tag{3.47}$$

for which the first and second terms go to zero since, for $\omega \in J$, $\underset{\sim}{\mathbf{P}}_n \underset{n\to\infty}{\to} 0$, while the third term vanishes as $n \to \infty$ by (3.46). Hence to prove the theorem we only need to verify that

$$\underset{\sim}{\mathbf{P}}_n \sum_{i=0}^{n-1} \phi_i \phi_i^\tau (\underset{\sim}{\mathbf{S}}_i - \underset{\sim}{\mathbf{S}}) \underset{n\to\infty}{\to} 0, \qquad \omega \in J. \tag{3.48}$$

Denote by s_i^{jk} the jth row and kth column element of matrix $\mathbf{S}_i - \mathbf{S}$. For any $\varepsilon > 0$, there exists an N so that

$$|s_i^{jk}| < \frac{\varepsilon}{2}, \qquad \forall j, k$$

whenever $i \geq N$.

Clearly, for (3.48) we only need to prove

$$\underset{\sim}{\mathbf{P}}_n \sum_{i=0}^{n-1} \phi_i \phi_i^\tau s_i^{jk} \underset{n\to\infty}{\to} 0. \tag{3.49}$$

Denote

$$\varepsilon_i = \begin{cases} \dfrac{\varepsilon}{2} - s_i^{jk}, & i > N \\ \dfrac{\varepsilon}{2}, & i \leq N. \end{cases}$$

Then $0 \leq \varepsilon_i \leq \varepsilon$, $\forall i$, and

$$\underset{\sim}{\mathbf{P}}_n \sum_{i=0}^{n-1} s_i^{jk} \phi_i \phi_i^\tau = \underset{\sim}{\mathbf{P}}_n \sum_{i=0}^{n-1} \frac{\varepsilon}{2} \phi_i \phi_i^\tau - \underset{\sim}{\mathbf{P}}_n \sum_{i=0}^{n} \left(\frac{\varepsilon}{2} - s_i^{jk} \right) \phi_i \phi_i^\tau$$

$$= \frac{\varepsilon}{2} \underset{\sim}{\mathbf{I}} - \frac{\varepsilon}{2(mp + lq)} \underset{\sim}{\mathbf{P}}_n + \underset{\sim}{\mathbf{P}}_n \sum_{i=0}^{n} s_i^{jk} \phi_i \phi_i^\tau - \underset{\sim}{\mathbf{P}}_n \sum_{i=0}^{n-1} \varepsilon_i \phi_i \phi_i^\tau.$$

Hence we have

$$\overline{\lim_{n\to\infty}} \left\| \underset{\sim}{\mathbf{P}}_n \sum_{i=0}^{n-1} s_i^{jk} \phi_i \phi_i^\tau \right\| \leq \frac{\varepsilon}{2} + + \varepsilon \overline{\lim_{n\to\infty}} \left\| \frac{1}{\varepsilon} \underset{\sim}{\mathbf{P}}_n \sum_{i=0}^{n-1} \varepsilon_i \phi_i \phi_i^\tau \right\| \underset{\varepsilon\to 0}{\to} 0 \quad \text{for } \omega \in J,$$

and the theorem is proved. □

Remark. Let $0 \le \varepsilon_i \le \varepsilon$. By the triangle inequality and Schwarz inequality, we have

$$\frac{1}{\varepsilon}\left\| \underset{\sim}{\mathbf{P}}_n \sum_{i=0}^{n-1} \varepsilon_i \phi_i \phi_i^T \right\| \le \frac{1}{\varepsilon} \sum_{i=0}^{n-1} \varepsilon_i \|\underset{\sim}{\mathbf{P}}_n \phi_i \phi_i^T\| \le \sum_{i=0}^{n-1} \|\underset{\sim}{\mathbf{P}}_n \phi_i \phi_i^T\|$$

$$\le \sum_{i=0}^{n-1} \|\underset{\sim}{\mathbf{P}}_n\| \, \|\phi_i\|^2 \le \lambda_{\max}^n \frac{1}{r_n} \le \frac{(mp + lq)\lambda_{\max}^n}{\lambda_{\min}^n}. \tag{3.50}$$

From this chain of inequalities, various conditions guaranteeing $\omega \in J$ can be proposed.

3.3. LEAST SQUARES UNDER CORRELATED NOISE

In Section 3.2 we discussed least-squares identification when the noise is the martingale difference sequence and this, of course, is uncorrelated. However, if the noise is correlated and the stochastic regressor ϕ_n is still defined by (3.6), then (3.14), generally speaking, will give inconsistent estimate. We give a simple example taken from Ref. 31.

Assume $m = l = p = q = 1$, $|a| < 1$ and

$$y_n = ay_{n-1} + bu_{n-1} + \varepsilon_n, \qquad y_i = u_i = 0, \quad i < 0, \tag{3.51}$$

where $\{\varepsilon_n\}$ and $\{u_n\}$ are supposed to be the wide sense stationary processes. That is,

$$E\varepsilon_n = m_\varepsilon, \qquad Eu_n = m_u, \qquad \forall n$$

$$E(\varepsilon_n - m_\varepsilon)(\varepsilon_{n+s} - m_\varepsilon) = R_{\varepsilon\varepsilon}(s), \qquad \forall n, s$$

$$E(u_n - m_u)(u_{n+s} - m_u) = R_{uu}(s).$$

For simplicity assume $m_\varepsilon = m_u = 0$, $R_{\varepsilon\varepsilon}(\tau) = 0$, $|\tau| \ge 2$ and $\{\varepsilon_n\}$ and $\{u_n\}$ are independent. By (3.14) and the ergodicity of stationary process we have

$$\theta_n \triangleq \begin{bmatrix} a_n \\ b_n \end{bmatrix} = \begin{bmatrix} \frac{1}{n}\sum_{i=0}^{n} y_i^2 & \frac{1}{n}\sum_{i=0}^{n-1} u_i y_i \\ \frac{1}{n}\sum_{i=1}^{n-1} u_i y_i & \frac{1}{n}\sum_{i=0}^{n-1} u_i^2 \end{bmatrix} \begin{bmatrix} \frac{1}{n}\sum_{i=0}^{n-1} y_i y_{i+1} \\ \frac{1}{n}\sum_{i=0}^{n-1} u_i y_{i+1} \end{bmatrix}$$

$$\underset{n\to\infty}{\overset{P}{\to}} \frac{1}{\Delta}\begin{bmatrix} R_{uu}(0)R_{yy}(1) - R_{uy}(0)R_{uy}(1) \\ -R_{uy}(0)R_{yy}(1) + R_{yy}(0)R_{uy}(1) \end{bmatrix},$$

where R_{uu}, R_{yy}, R_{uy} denote correlation functions and

$$\Delta = R_{yy}(0)R_{uu}(0) - R_{uy}^2(0).$$

From (3.51), it is easy to compute

$$R_{yy}(1) = aR_{yy}(0) + bR_{uy}(0) + R_{\varepsilon\varepsilon}(1)$$

$$R_{uy}(1) = aR_{uy}(0) + bR_{uu}(0),$$

hence

$$\boldsymbol{\theta}_n = \begin{bmatrix} a_n \\ b_n \end{bmatrix} \xrightarrow[n\to\infty]{P} \begin{bmatrix} a \\ b \end{bmatrix} + \frac{1}{\Delta}\begin{bmatrix} R_{\varepsilon\varepsilon}(1)R_{uu}(0) \\ -R_{\varepsilon\varepsilon}(1)R_{uy}(0) \end{bmatrix},$$

which shows that $\boldsymbol{\theta}_n$ is asymptotically biased.

The reason for this bias is that the method does not take the noise model into account and treats it simply as an error that is minimized in the mean square sense to obtain the estimate.

Now let $\mathbf{y}_n$, $\mathbf{u}_n$, $\boldsymbol{\varepsilon}_n$, $\underset{\sim}{\mathbf{F}}_n$, $\mathscr{F}_n$ have the same dimensions and meaning as in Section 3.2. Assume

$$E\left(\frac{\mathbf{w}_n\mathbf{w}_n^{\tau}}{\mathscr{F}_{n-1}}\right) = \underset{\sim}{\mathbf{I}}, \quad E\left(\frac{\mathbf{w}_n}{\mathscr{F}_{n-1}}\right) = \mathbf{0} \tag{3.52}$$

$$\mathbf{y}_n = \underset{\sim}{\mathbf{A}}_1\mathbf{y}_{n-1} + \cdots + \underset{\sim}{\mathbf{A}}_p\mathbf{y}_{n-p} + \underset{\sim}{\mathbf{B}}_1\mathbf{u}_{n-1} + \cdots + \underset{\sim}{\mathbf{B}}_q\mathbf{u}_{n-q} + \boldsymbol{\varepsilon}_n, \tag{3.53}$$

$$\mathbf{y}_n = \mathbf{0}, \qquad n < 0.$$

The noise $\boldsymbol{\varepsilon}_n$ now differs from that in (3.1) and is correlated. It is supposed to be an autoregressive moving average (ARMA) process as follows

$$\boldsymbol{\varepsilon}_n + \underset{\sim}{\mathbf{D}}_1\boldsymbol{\varepsilon}_{n-1} + \cdots + \underset{\sim}{\mathbf{D}}_d\boldsymbol{\varepsilon}_{n-d} = \mathbf{w}_n + \underset{\sim}{\mathbf{C}}_1\underset{\sim}{\mathbf{F}}_{n-1}\mathbf{w}_{n-1} + \cdots + \underset{\sim}{\mathbf{C}}_r\underset{\sim}{\mathbf{F}}_{n-r}\mathbf{w}_{n-r},$$

$$\mathbf{w}_i = \mathbf{0}, \qquad \boldsymbol{\varepsilon}_i = \mathbf{0}, \qquad i < 0. \tag{3.54}$$

The task of system identification here is to estimate unknown matrices $\underset{\sim}{\mathbf{A}}_i$, $\underset{\sim}{\mathbf{B}}_j$ in the dynamic model as well as $\underset{\sim}{\mathbf{D}}_k$, $\underset{\sim}{\mathbf{C}}_s$ in the noise model, $i = 1, \ldots, p$, $j = 1, \ldots, q$, $k = 1, \ldots, d$, $s = 1, \ldots, r$.

Introduce the following notations

$$\underset{\sim}{\mathbf{A}}(z) = \mathbf{I} - \underset{\sim}{\mathbf{A}}_1 z - \cdots - \underset{\sim}{\mathbf{A}}_p z^p \tag{3.55}$$

$$\underset{\sim}{\mathbf{B}}(z) = \underset{\sim}{\mathbf{B}}_1 + \underset{\sim}{\mathbf{B}}_2 z + \cdots + \underset{\sim}{\mathbf{B}}_q z^q \tag{3.56}$$

$$\underset{\sim}{\mathbf{C}}(z) = \underset{\sim}{\mathbf{I}} + \underset{\sim}{\mathbf{C}}_1 z + \cdots + \underset{\sim}{\mathbf{C}}_r z^r \tag{3.57}$$

$$\underset{\sim}{\mathbf{D}}(z) = \underset{\sim}{\mathbf{I}} + \underset{\sim}{\mathbf{D}}_1 z + \cdots + \underset{\sim}{\mathbf{D}}_d z^d \tag{3.58}$$

then (3.53) and (3.54) become

$$\underset{\sim}{\mathbf{A}}(z)\mathbf{y}_n = \underset{\sim}{\mathbf{B}}(z)\mathbf{u}_{n-1} + \boldsymbol{\varepsilon}_n, \qquad \underset{\sim}{\mathbf{D}}(z)\boldsymbol{\varepsilon}_n = \underset{\sim}{\mathbf{C}}(z)\underset{\sim}{\mathbf{F}}_n\mathbf{w}_n, \tag{3.59}$$

where z denotes the shift-back operator.

If we set

$$\underset{\sim}{\mathbf{A}}^1(z) \triangleq \underset{\sim}{\mathbf{D}}(z)\underset{\sim}{\mathbf{A}}(z), \qquad \underset{\sim}{\mathbf{B}}^1(z) \triangleq \underset{\sim}{\mathbf{D}}(z)\underset{\sim}{\mathbf{B}}(z) \tag{3.60}$$

then from (3.59)

$$\underset{\sim}{\mathbf{A}}^1(z)\mathbf{y}_n = \underset{\sim}{\mathbf{B}}^1(z)\mathbf{u}_{n-1} + \underset{\sim}{\mathbf{C}}(z)\underset{\sim}{\mathbf{F}}_n\mathbf{w}_n \tag{3.61}$$

and the problem is reduced to estimate matrix coefficients of polynomials $\underset{\sim}{\mathbf{A}}^1(z)$, $\underset{\sim}{\mathbf{B}}^1(z)$, and $\underset{\sim}{\mathbf{C}}(z)$.

Therefore, for parameter estimation of system (3.53), without loss of generality, we can assume $\underset{\sim}{\mathbf{D}}_i = 0$, $i = 1, \ldots, d$ in (3.54). Thus in what follows we always assume that $\boldsymbol{\varepsilon}_n$ is driven by a moving average (MA) model. That is,

$$\boldsymbol{\varepsilon}_n = \underset{\sim}{\mathbf{F}}_n\mathbf{w}_n + \underset{\sim}{\mathbf{C}}_1\underset{\sim}{\mathbf{F}}_{n-1}\mathbf{w}_{n-1} + \cdots + \underset{\sim}{\mathbf{C}}_r\underset{\sim}{\mathbf{F}}_{n-r}\mathbf{w}_{n-r}. \tag{3.62}$$

Denote

$$\underset{\sim}{\boldsymbol{\theta}}^\tau = \left[\underset{\sim}{\mathbf{A}}_1 \cdots \underset{\sim}{\mathbf{A}}_p \underset{\sim}{\mathbf{B}}_1 \cdots \underset{\sim}{\mathbf{B}}_q \underset{\sim}{\mathbf{C}}_1 \cdots \underset{\sim}{\mathbf{C}}_r\right], \tag{3.63}$$

$$\boldsymbol{\phi}_n^\tau = \begin{cases} \left[\mathbf{y}_n^\tau \cdots \mathbf{y}_{n-p+1}^\tau \mathbf{u}_n^\tau \cdots \mathbf{u}_{n-q+1}^\tau\right] \quad \text{for } r = 0 & (3.64) \\ \left[\mathbf{y}_n^\tau \cdots \mathbf{y}_{n-p+1}^\tau \mathbf{u}_n^\tau \cdots \mathbf{u}_{n-q+1}^\tau \mathbf{y}_n^\tau - \boldsymbol{\phi}_{n-1}^\tau \underset{\sim}{\boldsymbol{\theta}}_n \cdots \mathbf{y}_{n-r+1}^\tau \right. & \\ \quad \left. - \boldsymbol{\phi}_{n-r+1}^\tau - \boldsymbol{\phi}_{n-r}^\tau \underset{\sim}{\boldsymbol{\theta}}_{n-r+1}\right] \quad \text{for } r > 0, & (3.65) \end{cases}$$

where $\underset{\sim}{\boldsymbol{\theta}}_n$ is the estimate of $\underset{\sim}{\boldsymbol{\theta}}$ and is given later.

Comparing (3.64), (3.65), with (3.6) we see ϕ_n remains invariant for $r = 0$ but is extended by some additional terms for $r > 0$.

By the least-squares estimate we still mean the solution θ_n of (3.15)–(3.17) with an arbitrary initial θ_0 and

$$\mathbf{P}_0 = (mp + lq + mr)\mathbf{I}, \tag{3.66}$$

but ϕ_n is defined by (3.64) or (3.65) depending on whether $r = 0$ or $r > 0$.

Obviously, (3.15)–(3.17) and (3.65) [or (3.64)] form a recursive algorithm: Having obtained θ_n, we can form ϕ_n, and by (3.16) can compute $\mathbf{P}_{n+1}$. Then by using the new measurement $\mathbf{y}_{n+1}$ we can obtain θ_{n+1} according to (3.15) and so on.

Similar to (3.22) and (3.23), we have

$$\mathbf{P}_n = \left(\sum_{i=0}^{n-1} \phi_i \phi_i^\tau + \frac{1}{mp + lq + mr} \mathbf{I} \right)^{-1} \tag{3.67}$$

$$\theta_n = \mathbf{P}_n \sum_{i=0}^{n-1} \phi_i \mathbf{y}_{i+1}^\tau + \mathbf{P}_n \mathbf{P}_0^{-1} \theta_0 . \tag{3.68}$$

Let r_n, $\lambda_{\max}^n$ and $\lambda_{\min}^n$ have the same meaning as in Section 3.2 but with ϕ_n defined by (3.64) (or (3.64)). Denote

$$\beta_i = \frac{\|\phi_i\|^2}{r_{i+1}} . \tag{3.69}$$

Lemma 3.3. (1) $\sum_{i=0}^{\infty} \beta_i = \infty$ *if and only if* $r_n \underset{n \to \infty}{\rightarrow} \infty$, (2) *if* $\sum_{i=0}^{\infty} \|\mathbf{u}_i\|^2 = \infty$, *then* $\sum_{i=0}^{\infty} \beta_i = \infty$, (3) *if* $\overline{\lim}_{n \to \infty} 1/n \sum_{i=1}^{n} \|\mathbf{w}_i\|^2 > 0$ *and all zeros of* $\det \mathbf{C}(\lambda)$ *lie outside the closed unit disk, then* $\sum_{i=0}^{\infty} \beta_i = \infty$.

Proof. (1) By Lemma 3.2, $r_n \to \infty$ yields $\sum_{i=0}^{\infty} \beta_i = \infty$. Conversely, if $\sum_{i=0}^{\infty} \beta_i = \infty$ and the opposite were true, that is, $r_n \underset{n \to \infty}{\rightarrow} r' < \infty$, then

$$\sum_{i=0}^{\infty} \beta_i = \lim_{n \to \infty} \sum_{i=0}^{n} \frac{\|\phi_i\|^2}{r_{i+1}} \le \lim_{n \to \infty} \sum_{i=0}^{n} \frac{r'}{r_{n+1}} \frac{\|\phi_i\|^2}{r_{i+1}} \le \lim_{n \to \infty} \frac{r'}{r_1 r_{n+1}} \sum_{i=0}^{n} \|\phi_i\|^2$$

$$= \frac{r'}{r_1} < \infty,$$

which contradicts $\sum_{i=0}^{\infty} \beta_i = \infty$. Thus $r_n \underset{n \to \infty}{\rightarrow} \infty$. (2) Since $\|\phi_i\| \ge \|\mathbf{u}_i\|$,

$\Sigma_{i=0}^{\infty}\|\mathbf{u}_i\|^2 = \infty$ implies $r_n \to \infty$ and hence $\Sigma_{i=0}^{\infty}\beta_i = \infty$ by (1). (3) Suppose the converse were true, that is, $\Sigma_{i=0}^{\infty}\beta_i < \infty$. Then by (1) r_n would have a finite limit and hence $\|\boldsymbol{\phi}_i\| \underset{i\to\infty}{\to} 0$ from which $\mathbf{y}_n \underset{n\to\infty}{\to} 0$. This means for $\varepsilon > 0$, $\|\boldsymbol{\varepsilon}_n\| < \varepsilon$ whenever $n \geq N$ if N is large enough.

Consider $m(r \vee 1)$-dimensional vector $\boldsymbol{\zeta}_n$ defined by

$$\boldsymbol{\zeta}_{n+1} = \underset{\sim}{\mathbf{F}}\boldsymbol{\zeta}_n + \underset{\sim}{\mathbf{G}}^{\tau}\boldsymbol{\varepsilon}_{n+1}, \qquad \boldsymbol{\zeta}_{-1} = \mathbf{0},$$

where $\underset{\sim}{\mathbf{F}}, \underset{\sim}{\mathbf{G}}$ are given by (2.73) and (2.74).

By Lemma 2.6, the first components $\boldsymbol{\zeta}_n^1 = \underset{\sim}{\mathbf{G}}\boldsymbol{\zeta}_n$ of $\boldsymbol{\zeta}_n$ satisfy the equation:

$$\underset{\sim}{\mathbf{C}}(z)\boldsymbol{\zeta}_n^1 = \boldsymbol{\varepsilon}_n, \qquad \boldsymbol{\zeta}_n^1 = \mathbf{0}, \qquad \boldsymbol{\varepsilon}_n = \mathbf{0}, \qquad n < 0,$$

hence $\boldsymbol{\zeta}_n^1 = \underset{\sim}{\mathbf{F}}_n\mathbf{w}_n$.

Replacing $\underset{\sim}{\mathbf{D}}(z)\mathbf{h}(\mathbf{x}_n)$ in (2.91) and (2.98) by $\boldsymbol{\varepsilon}_{n+1}$ yields

$$\boldsymbol{\zeta}_{n+1} = \sum_{j=-1}^{n} \underset{\sim}{\mathbf{F}}^{n-j}\underset{\sim}{\mathbf{G}}^{\tau}\boldsymbol{\varepsilon}_{j+1} = \sum_{j=0}^{n+1} \underset{\sim}{\mathbf{F}}^{n-j+1}\underset{\sim}{\mathbf{G}}^{\tau}\boldsymbol{\varepsilon}_j$$

and

$$\underset{\sim}{\mathbf{F}}_n\mathbf{w}_n = \underset{\sim}{\mathbf{G}}\boldsymbol{\zeta}_n = \underset{\sim}{\mathbf{G}}\sum_{j=0}^{n} \underset{\sim}{\mathbf{F}}^{n-j}\underset{\sim}{\mathbf{G}}^{\tau}\boldsymbol{\varepsilon}_j.$$

Then by (2.82), it follows that

$$\|\underset{\sim}{\mathbf{F}}_n\mathbf{w}_n\| \leq k_2\sum_{j=0}^{n}\rho^{n-j}\|\boldsymbol{\varepsilon}_j\| \leq k_2\sum_{j=0}^{N}\rho^{n-j}\|\boldsymbol{\varepsilon}_j\| + k_2\sum_{j=N+1}^{n}\rho^{n-j}\varepsilon$$

$$\leq k_2\rho^n\sum_{j=0}^{n}\rho^{-j}\|\boldsymbol{\varepsilon}_j\| + \frac{k_2\varepsilon}{1-\rho} \underset{\substack{n\to\infty\\ \varepsilon\to 0}}{\to} 0$$

and

$$\frac{1}{n}\sum_{i=0}^{n}\|\underset{\sim}{\mathbf{F}}_i\mathbf{w}_i\|^2 \underset{n\to\infty}{\to} 0,$$

which contradicts our hypothesis, hence

$$\sum_{i=0}^{\infty}\beta_i = \infty. \qquad \square$$

Lemma 3.4. *Denote*

$$\xi_n = y_n - F_n w_n - \theta_n^\tau \phi_{n-1}. \tag{3.70}$$

Then

$$\theta_{n+1} = \theta_n + P_n \phi_n (\xi_{n+1}^\tau + w_{n+1}^\tau F_{n+1}^\tau), \tag{3.71}$$

$$\xi_{n+1} = (\theta - \theta_{n+1})^\tau \phi_n + G \sum_{j=0}^{n-1} F^{n-j} G^\tau (\theta - \theta_{j+1})^\tau \phi_j$$

$$+ G F^{n+1} G^\tau (\theta - \theta_0)^\tau \phi_{-1}, \tag{3.72}$$

where F *and* G *are given by* (2.73) (2.74).

Proof. By (3.15), we have

$$\begin{aligned} y_n - \theta_n^\tau \phi_{n-1} &= y_n - [\theta_{n-1} + a_{n-1} P_{n-1} \phi_{n-1} (y_n^\tau - \phi_{n-1}^\tau \theta_{n-1})]^\tau \phi_{n-1} \\ &= (y_n - \theta_{n-1}^\tau \phi_{n-1})(1 - a_{n-1} \phi_{n-1}^\tau P_{n-1} \phi_{n-1}) \\ &= a_{n-1}(y_n - \theta_{n-1}^\tau \phi_{n-1}), \end{aligned} \tag{3.73}$$

consequently,

$$\theta_{n+1} = \theta_n + P_n \phi_n (y_{n+1}^\tau - \phi_n^\tau \theta_{n+1}).$$

Then (3.71) follows from this and (3.70).

Denote

$$\tilde{\theta}_n = \theta - \theta_n, \tag{3.74}$$

then

$$\begin{aligned} &C(z)(y_n - \theta_n^\tau \phi_{n-1} - F_n w_n) \\ &= y_n + C_1 y_{n-1} + \cdots + C_r Y_{n-r} - C(z) F_n w_n - C(z) \theta_n^\tau \phi_{n-1} \\ &= A_1 y_{n-1} + \cdots + A_p y_{n-p} + B_1 u_{n-1} + \cdots + B_q u_{n-q} \\ &\quad + C_1 y_{n-1} + \cdots + C_r y_{n-r} - C(z) \theta_n^\tau \phi_{n-1} \\ &= \theta^\tau \phi_{n-1} - \theta_n^\tau \phi_{n-1} = \tilde{\theta}_n^\tau \phi_{n-1}, \end{aligned}$$

or

$$\underset{\sim}{C}(z)\xi_n = \tilde{\underset{\sim}{\theta}}_n^\tau \phi_{n-1}. \tag{3.75}$$

We have agreed $\underset{\sim}{\theta}_i = \underset{\sim}{0}$, $\mathbf{y}_i = \mathbf{0}$, $\mathbf{w}_i = \mathbf{0}$, $i < 0$. Therefore, $\xi_i = 0$, $i < 0$. Now let us consider

$$\eta_{n+1} = \underset{\sim}{F}\eta_n + \underset{\sim}{G}^\tau \tilde{\underset{\sim}{\theta}}_{n+1}^\tau \phi_n, \qquad \eta_0^\tau = \left[\phi_{-1}^\tau \tilde{\underset{\sim}{\theta}}_0 \mathbf{0} \cdots \mathbf{0}\right], \tag{3.76}$$

By Lemma 2.6, $\eta_n^1 \triangleq \underset{\sim}{G}\eta_n$ satisfies

$$\underset{\sim}{C}(z)\eta_n^1 = \tilde{\underset{\sim}{\theta}}_n^\tau \phi_{n-1} \tag{3.77}$$

with initial values $\eta_i^1 = \mathbf{0}$, $i < 0$, $\eta_0^1 = \tilde{\underset{\sim}{\theta}}_0^\tau \phi_{-1}$ which come from (3.76) and coincide with initial values of ξ_n given by (3.75). Hence

$$\xi_n = \eta_n^1 = \underset{\sim}{G}\eta_n.$$

Similar to (2.98), we may obtain

$$\eta_{n+1} = \sum_{j=0}^{n} \underset{\sim}{F}^{n-j} \underset{\sim}{G}^\tau \tilde{\underset{\sim}{\theta}}_{j+1}^\tau \phi_j + \underset{\sim}{F}^{n+1}\eta_0$$

$$= \underset{\sim}{G}^\tau \tilde{\underset{\sim}{\theta}}_{n+1}^\tau \phi_n + \sum_{j=0}^{n-1} \underset{\sim}{F}^{n-j} \underset{\sim}{G}^\tau \tilde{\underset{\sim}{\theta}}_{j+1}^\tau \phi_j + \underset{\sim}{F}^{n+1} \underset{\sim}{G}^\tau \tilde{\underset{\sim}{\theta}}_0^\tau \phi_{-1},$$

which yields (3.72). □

We introduce a concept from the theory of hyperstability.

DEFINITION 3.1. A rational matrix transfer function $\underset{\sim}{H}(z)$ is called strictly positive real if there are constants $k_0 > 0$, $k_1 > 0$ such that

$$\sum_{i=1}^{n} \mathbf{g}_i^\tau \mathbf{f}_i + k_1 \geq 0, \qquad \forall n, \tag{3.78}$$

where $\{\mathbf{g}_i\}$ and $\{\mathbf{f}_i\}$ are the output and input sequences of the system:

$$g_n = \left[H(z) - k_0 I\right] f_n.$$

Property (3.78) is equivalent to the following properties of the matrix $\underline{H}'(z) \triangleq \underline{H}(z) - k_0 \underline{I}$

1. All the elements of $\underline{H}'(z)$ are analytic in $|z| < 1$

2. The matrix

$$\underline{H}'(z) + \underline{H}''(z^*) \Big|_{z=e^{i\omega}} = \underline{H}'(e^{j\omega}) + \underline{H}''(e^{-j\omega})$$

is a positive definite Hermitian for all ω.

We refer the reader to Refs. 37 and 38 for more details on this subject.
Denote

$$\underline{\Pi}_n = \underline{P}_n \sum_{i=0}^{n-1} \phi_i \left(\underline{G} \sum_{j=0}^{i-1} \underline{F}^{i-j} \underline{G}^\tau \tilde{\underline{\theta}}_{j+1}^\tau \phi_j \right)^\tau. \tag{3.79}$$

We allow $\operatorname{tr} F_n F_n^\tau$ to increase at a rate of r_n^ε as $n \to \infty$. Suppose there is a $k_3 > 0$ and $\varepsilon \in [0, 1)$ such that

$$\operatorname{tr} \underline{F}_n \underline{F}_n^\tau \le k_3 r_n^\varepsilon \quad \text{and} \quad E \operatorname{tr} \underline{F}_n \underline{F}_n^\tau < \infty. \tag{3.80}$$

Theorem 3.5. *For system (3.53) and noise [(3.52), (3.62) and (3.80)], let the estimate $\underline{\theta}_n$ be given by (3.65) [or (3.64) (3.15) (3.16), and (3.66)] with arbitrary $\underline{\theta}_0$. Assume that*

A. $r = 0$, *or* $r > 0$ *but* $\underline{C}^{-1}(z) - \frac{1}{2}\underline{I}$ *is strictly positive real and all zeros of* $\det \underline{C}(\lambda)$ *lie outside the closed unit disk.*

B. $\lambda_{\max}^n / \lambda_{\min}^n \le k_4$, $\forall n$ *for some constant* k_4.

Then

$$\underline{\Pi}_n \underset{n \to \infty}{\to} \underline{0} \tag{3.81}$$

and

$$\|\tilde{\underline{\theta}}_n + \underline{\Pi}_n\| = o\left(r_n^{\delta - 1/2}\right), \qquad \forall \delta \in \left(\frac{\varepsilon}{2}, \frac{1}{2}\right] \tag{3.82}$$

for almost all $\omega \in J \triangleq \{\omega: r_n \to \infty\}$.

Remark. In Condition A_1 we have emphasized the zero location for $\det \underline{C}(z)$ but, in fact, this property is a consequence of the strictly positive realness of the transfer function (see [37], [38]).

We first prove some lemmas.

Lemma 3.5. *Under the conditions of Theorem* 3.5

$$E\left(\frac{v_{n+1}}{\mathscr{F}_n}\right) \le v_n - \frac{2}{r_{n+1}} E\left[\frac{\phi_n^\tau \tilde{\theta}_{n+1}\left(\xi_{n+1} - \frac{1}{2}\tilde{\theta}_{n+1}^\tau \phi_n\right)}{\mathscr{F}_n}\right]$$

$$+2\frac{\operatorname{tr} \mathbf{F}_{n+1}\mathbf{F}_{n+1}^\tau}{r_{n+1}} \frac{\phi_n^\tau \mathbf{P}_n \phi_n}{1 + \phi_n^\tau \mathbf{P}_n \phi_n}, \tag{3.83}$$

where

$$v_n = \frac{1}{r_n} \operatorname{tr} \tilde{\theta}_n^\tau \mathbf{P}_n^{-1} \tilde{\theta}_n. \tag{3.84}$$

Proof. From (3.71) and (3.74), we have

$$\tilde{\theta}_{n+1} = \tilde{\theta}_n - \mathbf{P}_n \phi_n \left(\xi_{n+1}^\tau + \mathbf{w}_{n+1}^\tau \mathbf{F}_{n+1}^\tau\right). \tag{3.85}$$

From here and (3.16), it follows that

$$\operatorname{tr} \tilde{\theta}_{n+1}^\tau \mathbf{P}_{n+1}^{-1} \tilde{\theta}_{n+1} = \operatorname{tr} \tilde{\theta}_n^\tau \mathbf{P}_n^{-1} \tilde{\theta}_n - 2\left(\xi_{n+1}^\tau + \mathbf{w}_{n+1}^\tau \mathbf{F}_{n+1}^\tau\right) \tilde{\theta}_n^\tau \phi_n$$

$$+ \phi_n^\tau \mathbf{P}_n \phi_n \left(\xi_{n+1}^\tau + \mathbf{w}_{n+1}^\tau \mathbf{F}_{n+1}^\tau\right)\left(\xi_{n+1} + \mathbf{F}_{n+1}\mathbf{w}_{n+1}\right) + \operatorname{tr} \tilde{\theta}_{n+1}^\tau \phi_n \phi_n^\tau \tilde{\theta}_{n+1}$$

$$= \operatorname{tr} \tilde{\theta}_n^\tau \mathbf{P}_n^{-1} \tilde{\theta}_n - 2\left(\xi_{n+1}^\tau + \mathbf{w}_{n+1}^\tau \mathbf{F}_{n+1}^\tau\right) \tilde{\theta}_{n+1}^\tau \phi_n$$

$$- \phi_n^\tau \mathbf{P}_n \phi_n \|\xi_{n+1} + \mathbf{F}_{n+1}\mathbf{w}_{n+1}\|^2 + \|\tilde{\theta}_{n+1}^\tau \phi_n\|^2$$

$$\le \operatorname{tr} \tilde{\theta}_n^\tau \mathbf{P}_n^{-1} \tilde{\theta}_n - 2\left[\phi_n^\tau \tilde{\theta}_{n+1}\left(\xi_{n+1} - \tfrac{1}{2}\tilde{\theta}_{n+1}^\tau \phi_n\right)\right] - 2\mathbf{w}_{n+1}^\tau \mathbf{F}_{n+1}^\tau \tilde{\theta}_{n+1}^\tau \phi_n$$

$$\tag{3.86}$$

Set $\lambda = mp + lq + mr$. Noticing

$$\|\mathbf{P}_n^{-1}\| \ge \lambda^{-1} \operatorname{tr} \mathbf{P}_n^{-1} = \lambda^{-1} r_n, \tag{3.87}$$

from (3.67) and Condition B_1 we have

$$\frac{1}{r_n}\mathbf{I} \le \frac{1}{\|\mathbf{P}_n^{-1}\|}\mathbf{I} = \lambda_{\min}^n \mathbf{I} \le \mathbf{P}_n \le k_4 \lambda_{\min}^n \mathbf{I} = \frac{k_4}{\|P_n^{-1}\|}\mathbf{I} \le \frac{k_4 \lambda}{r_n}\mathbf{I}. \tag{3.88}$$

Since

$$\frac{\|\phi_n\| \, \|\mathbf{y}_{n-i}\|}{r_{n+1}} \le 1, \qquad \frac{\|\phi_n\| \, \|\mathbf{u}_{n-j}\|}{r_{n+1}} \le 1,$$

$$i = 0, 1, \ldots, p-1, \; j = 0, 1, \ldots, q-1, \tag{3.89}$$

by (3.53) it follows that

$$E\left\| \frac{\phi_n \mathbf{y}_{n+1}^\tau}{r_{n+1}} \right\|^2 < \infty$$

and by (3.88)

$$E\|\underset{\sim}{\mathbf{P}}_{n+1}\phi_n \mathbf{y}_{n+1}^\tau\|^2 < \infty. \tag{3.90}$$

Further from (3.15) and (3.88), we know

$$\begin{aligned} \|\underset{\sim}{\theta}_{n+1}\|^2 &\le 3\|\underset{\sim}{\theta}_n\|^2 + 3\|\underset{\sim}{\mathbf{P}}_{n+1}\phi_n \mathbf{y}_{n+1}^\tau\|^2 + 3\|\underset{\sim}{\mathbf{P}}_{n+1}\phi_n\phi_n^\tau\underset{\sim}{\theta}_n\|^2 \\ &\le 3\|\underset{\sim}{\theta}_n\|^2 + 3\|\underset{\sim}{\mathbf{P}}_{n+1}\phi_n \mathbf{y}_{n+1}^\tau\|^2 + 3k_4^2\lambda^2\|\underset{\sim}{\theta}_n\|^2. \end{aligned} \tag{3.91}$$

From (3.90), (3.91) and $E\|\underset{\sim}{\theta}_0\|^2 < \infty$ we conclude by induction that

$$E\|\underset{\sim}{\theta}_n\|^2 < \infty, \qquad \forall n. \tag{3.92}$$

Similarly, by (3.53) and (3.89), we can prove

$$E\frac{\|\mathbf{y}_n\|^2}{r_n} < \infty.$$

Hence

$$E\frac{\|\mathbf{y}_n - \underset{\sim}{\theta}_n^\tau \phi_{n-1}\|^2}{r_n} \le 2E\frac{\|\mathbf{y}_n\|^2}{r_n} + 2E\|\underset{\sim}{\theta}_n\|^2 < \infty, \tag{3.93}$$

and by (3.70) and (3.92)

$$E\frac{\|\xi_n\|^2}{r_n} \le 2E\frac{\|\mathbf{y}_n - \underset{\sim}{\theta}_n^\tau \phi_{n-1}\|^2}{r_n} + 2E\frac{\|\underset{\sim}{\mathbf{F}}_n \mathbf{w}_n\|^2}{r_n} < \infty, \tag{3.94}$$

$$E\frac{\|\tilde{\underset{\sim}{\theta}}_n^\tau \phi_{n-1}\|^2}{r_n} \le E\|\tilde{\underset{\sim}{\theta}}_n\|^2 \le 2E\|\underset{\sim}{\theta}\|^2 + 2E\|\underset{\sim}{\theta}_n\|^2 < \infty. \tag{3.95}$$

Therefore, divided by r_{n+1}, each term on the right-hand side of (3.86) has finite expectation and consequently we may compute

$$E\left(\frac{v_{n+1}}{\mathscr{F}_n}\right) \le v_n - E\left[\frac{2}{r_{n+1}} \frac{\phi_n^\tau \tilde{\theta}_{n+1}\left(\xi_{n+1} - \frac{1}{2}\tilde{\theta}_{n+1}^\tau \phi_n\right)}{\mathscr{F}_n}\right]$$

$$-2E\left(\frac{\mathbf{w}_{n+1}^\tau \mathbf{F}_{n+1}^\tau \tilde{\theta}_{n+1}^\tau \phi_n}{r_{n+1}} / \mathscr{F}_n\right). \qquad (3.96)$$

From (3.70) and (3.73), we have

$$\xi_{n+1} + \mathbf{F}_{n+1}\mathbf{w}_{n+1} = a_n\left(\mathbf{y}_{n+1} - \theta_n^\tau \phi_n - \mathbf{F}_{n+1}\mathbf{w}_{n+1} + \mathbf{F}_{n+1}\mathbf{w}_{n+1}\right). \quad (3.97)$$

From (3.53) and (3.62), it is clear that $y_{n+1} - F_{n+1}w_{n+1}$ is $\mathscr{F}_n$-measurable and, in addition, r_{n+1} and $\mathbf{F}_{n+1}$ are also $\mathscr{F}_n$-measurable. Then by (3.85) and (3.97), we have

$$E\left(\frac{\mathbf{w}_{n+1}^\tau \mathbf{F}_{n+1}^\tau \tilde{\theta}_{n+1}^\tau \phi_n}{r_{n+1}} / \mathscr{F}_n\right)$$

$$= E\left[\frac{1}{r_{n+1}} \frac{\mathbf{w}_{n+1}^\tau \mathbf{F}_{n+1}^\tau\left(\tilde{\theta}_n^\tau \phi_n - \phi_n^\tau \mathbf{P}_n \phi_n\left(\xi_{n+1} + \mathbf{F}_{n+1}\mathbf{w}_{n+1}\right)\right)}{\mathscr{F}_n}\right]$$

$$= -E\left[\frac{1}{r_{n+1}} \frac{\mathbf{w}_{n+1}^\tau \mathbf{F}_{n+1}^\tau\left(\xi_{n+1} + \mathbf{F}_{n+1}\mathbf{w}_{n+1}\right)\phi_n^\tau \mathbf{P}_n \phi_n}{\mathscr{F}_n}\right]$$

$$= -E\left[\frac{a_n}{r_{n+1}} \frac{\phi_n^\tau \mathbf{P}_n \phi_n \mathbf{w}_{n+1}^\tau \mathbf{F}_{n+1}^\tau\left(\mathbf{y}_{n+1} - \mathbf{F}_{n+1}\mathbf{w}_{n+1} - \theta_n^\tau \phi_n + \mathbf{F}_{n+1}\mathbf{w}_{n+1}\right)}{\mathscr{F}_n}\right]$$

$$= -\frac{a_n \phi_n^\tau \mathbf{P}_n \phi_n}{r_{n+1}} E\left[\frac{\|\mathbf{F}_{n+1}\mathbf{w}_{n+1}\|^2}{\mathscr{F}_n}\right] = -\frac{a_n \phi_n^\tau \mathbf{P}_n \phi_n}{r_{n+1}} \operatorname{tr} \mathbf{F}_{n+1}\mathbf{F}_{n+1}^\tau, \qquad (3.98)$$

which together with (3.96) gives (3.83). □

Lemma 3.6. *Under the conditions of Theorem* 3.5

$$\sum_{i=0}^{\infty} \frac{\|\tilde{\theta}_i^\tau \phi_{i-1}\|^2}{r_i} < \infty \text{ a.s.} \qquad (3.99)$$

Proof. Because $\underset{\sim}{\mathbf{C}}^{-1}(z) - \frac{1}{2}\underset{\sim}{\mathbf{I}}$ is strictly positive real we can find $k_0 > 0$, $k_1 > 0$ such that

$$S_n \triangleq \sum_{i=1}^{n} \phi_{i-1}^T \underset{\sim}{\tilde{\theta}}_i \left[\underset{\sim}{\mathbf{C}}^{-1}(z) - \left(\frac{1+k_0}{2} \right) \underset{\sim}{\mathbf{I}} \right] \underset{\sim}{\tilde{\theta}}_i^T \phi_{i-1} + k_1 \geq 0, \qquad \forall n,$$

or, by (3.75),

$$S_n = \sum_{i=1}^{n} \phi_{i-1}^T \underset{\sim}{\tilde{\theta}}_i \xi_i - \frac{1+k_0}{2} \sum_{i=1}^{n} \| \underset{\sim}{\tilde{\theta}}_i^T \phi_{i-1} \|^2 + k_1 \geq 0, \qquad \forall n. \quad (3.100)$$

From (3.95) and (3.99), we have

$$\frac{ES_n}{r_n} < \infty \quad (3.101)$$

and by (3.80) and (3.88)

$$\sum_{i=0}^{\infty} \frac{\operatorname{tr} \underset{\sim}{\mathbf{F}}_{i+1} \underset{\sim}{\mathbf{F}}_{i+1}^T \phi_i^T \underset{\sim}{\mathbf{P}}_i \phi_i}{r_{i+1}(1 + \phi_i^T \underset{\sim}{\mathbf{P}}_i \phi_i)} \leq k_4 \lambda \sum_{i=0}^{\infty} \frac{\|\phi_i\|^2 \operatorname{tr} \underset{\sim}{\mathbf{F}}_{i+1} \underset{\sim}{\mathbf{F}}_{i+1}^T}{r_{i+1} r_i \left(1 + \dfrac{1}{r_i}\|\phi_i\|^2\right)}$$

$$\leq k_3 k_4 \lambda \sum_{i=0}^{\infty} \frac{\|\phi_i\|^2}{r_{i+1}^{2-\varepsilon}}. \quad (3.102)$$

Inequality (3.28) gives

$$(1-\varepsilon)\|\phi_i\|^2 r_{i+1}^{-\varepsilon} \leq r_{i+1}^{1-\varepsilon} - r_i^{1-\varepsilon},$$

and putting this into (3.102) leads to

$$\sum_{i=0}^{\infty} \frac{\operatorname{tr} \underset{\sim}{\mathbf{F}}_{i+1} \underset{\sim}{\mathbf{F}}_{i+1}^T \phi_i^T \underset{\sim}{\mathbf{P}}_i \phi_i}{r_{i+1}(1 + \phi_i^T \underset{\sim}{\mathbf{P}}_i \phi_i)} \leq \frac{k_3 k_4 \lambda}{(1-\varepsilon)} \sum_{i=0}^{\infty} \frac{r_{i+1}^{1-\varepsilon} - r_i^{1-\varepsilon}}{r_{i+1}^{2-2\varepsilon}}$$

$$\leq \frac{k_3 k_4 \lambda}{1-\varepsilon} \sum_{i=0}^{\infty} \left(\frac{1}{r_i^{1-\varepsilon}} - \frac{1}{r_{i+1}^{1-\varepsilon}} \right)$$

$$= \frac{k_3 k_4 \lambda}{r_0^{1-\varepsilon}(1-\varepsilon)} \leq \frac{k_3 k_4 \lambda}{1-\varepsilon}. \quad (3.103)$$

Equations (3.95), (3.101), and (3.103) make the quantity m_n defined below meaningful,

$$m_n = v_n + \frac{2S_n}{r_n} + E\left[\sum_{i=0}^{\infty} \frac{2\,\mathrm{tr}\,\underset{\sim}{\mathbf{F}}_{i+1}\underset{\sim}{\mathbf{F}}_{i+1}^{\tau}}{r_{i+1}} \frac{\boldsymbol{\phi}_i^{\tau}\underset{\sim}{\mathbf{P}}_i\boldsymbol{\phi}_i}{1+\boldsymbol{\phi}_i^{\tau}\underset{\sim}{\mathbf{P}}_i\boldsymbol{\phi}_i} \Big/ \mathscr{F}_n\right]$$

$$- \sum_{i=0}^{n-1} \frac{2\,\mathrm{tr}\,\underset{\sim}{\mathbf{F}}_{i+1}\underset{\sim}{\mathbf{F}}_{i+1}^{\tau}\boldsymbol{\phi}_i^{\tau}\underset{\sim}{\mathbf{P}}_i\boldsymbol{\phi}_i}{r_{i+1}(1+\boldsymbol{\phi}_i^{\tau}\underset{\sim}{\mathbf{P}}_i\boldsymbol{\phi}_i)} + k_0 \sum_{i=0}^{n} \frac{\|\tilde{\underset{\sim}{\boldsymbol{\theta}}}_i^{\tau}\boldsymbol{\phi}_{i-1}\|^2}{r_i}, \qquad (3.104)$$

since all its terms have finite expectation.

By Lemma 3.5, we have

$$E\left(\frac{m_{n+1}}{\mathscr{F}_n}\right) \le v_n - \frac{2}{r_{n+1}} E\left(\frac{\boldsymbol{\phi}_n^{\tau}\tilde{\underset{\sim}{\boldsymbol{\theta}}}_{n+1}\boldsymbol{\xi}_{n+1}}{\mathscr{F}_n}\right) + \frac{1}{r_{n+1}} E\left(\frac{\|\tilde{\underset{\sim}{\boldsymbol{\theta}}}_{n+1}^{\tau}\boldsymbol{\phi}_n\|^2}{\mathscr{F}_n}\right)$$

$$+2\frac{\mathrm{tr}\,\underset{\sim}{\mathbf{F}}_{n+1}\underset{\sim}{\mathbf{F}}_{n+1}^{\tau}\boldsymbol{\phi}_n^{\tau}\underset{\sim}{\mathbf{P}}_n\boldsymbol{\phi}_n}{r_{n+1}(1+\boldsymbol{\phi}_n^{\tau}\underset{\sim}{\mathbf{P}}_n\boldsymbol{\phi}_n)} + E\left[\sum_{i=0}^{\infty} \frac{2\,\mathrm{tr}\,\underset{\sim}{\mathbf{F}}_{i+1}\underset{\sim}{\mathbf{F}}_{i+1}^{\tau}\boldsymbol{\phi}_i^{\tau}\underset{\sim}{\mathbf{P}}_i\boldsymbol{\phi}_i}{r_{i+1}(1+\boldsymbol{\phi}_i^{\tau}\underset{\sim}{\mathbf{P}}_i\boldsymbol{\phi}_i)} \Big/ \mathscr{F}_n\right]$$

$$- \sum_{i=0}^{n} \frac{2\,\mathrm{tr}\,\underset{\sim}{\mathbf{F}}_{i+1}\underset{\sim}{\mathbf{F}}_{i+1}^{\tau}\boldsymbol{\phi}_i^{\tau}\underset{\sim}{\mathbf{P}}_i\boldsymbol{\phi}_i}{r_{i+1}(1+\boldsymbol{\phi}_i^{\tau}\underset{\sim}{\mathbf{P}}_i\boldsymbol{\phi}_i)} + k_0 \sum_{i=0}^{n} \frac{\|\tilde{\underset{\sim}{\boldsymbol{\theta}}}_i^{\tau}\boldsymbol{\phi}_{i-1}\|^2}{r_i}$$

$$+k_0 E\left(\frac{\|\underset{\sim}{\boldsymbol{\theta}}_{n+1}^{\tau}\boldsymbol{\phi}_n\|^2}{r_{n+1}} \Big/ \mathscr{F}_n\right) + 2E\left(\frac{\boldsymbol{\phi}_n^{\tau}\tilde{\underset{\sim}{\boldsymbol{\theta}}}_{n+1}^{\tau}\boldsymbol{\xi}_{n+1}}{r_{n+1}} \Big/ \mathscr{F}_n\right)$$

$$- \frac{(1+k_0)}{r_{n+1}} E\left(\frac{\|\tilde{\underset{\sim}{\boldsymbol{\theta}}}_{n+1}^{\tau}\boldsymbol{\phi}_n\|^2}{\mathscr{F}_n}\right) + \frac{2S_n}{r_{n+1}}$$

$$= v_n + E\left[\sum_{i=0}^{\infty} \frac{2\,\mathrm{tr}\,\underset{\sim}{\mathbf{F}}_{i+1}\underset{\sim}{\mathbf{F}}_{i+1}^{\tau}\boldsymbol{\phi}_i^{\tau}\underset{\sim}{\mathbf{P}}_i\boldsymbol{\phi}_i}{r_{i+1}(1+\boldsymbol{\phi}_i^{\tau}\underset{\sim}{\mathbf{P}}_i\boldsymbol{\phi}_i)} \Big/ \mathscr{F}_n\right] - 2\sum_{i=0}^{n-1} \frac{\mathrm{tr}\,\underset{\sim}{\mathbf{F}}_{i+1}\underset{\sim}{\mathbf{F}}_{i+1}^{\tau}\boldsymbol{\phi}_i^{\tau}\underset{\sim}{\mathbf{P}}_i\boldsymbol{\phi}_i}{r_{i+1}(1+\boldsymbol{\phi}_i^{\tau}\underset{\sim}{\mathbf{P}}_i\boldsymbol{\phi}_i)}$$

$$+ \frac{2S_n}{r_{n+1}} + k_0 \sum_{i=0}^{n} \frac{\|\tilde{\underset{\sim}{\boldsymbol{\theta}}}_i^{\tau}\boldsymbol{\phi}_{i-1}\|^2}{r_i} \le m_n.$$

Hence $(m_n, \mathscr{F}_n)$ is a nonnegative supermartingale and it converges to a finite limit as $n \to \infty$ by Theorem 1.11. In the expression (3.104) for m_n the sum of the third and fourth terms is nonnegative and all the other terms, are nonnegative. Hence (3.99) follows. □

Proof of Theorem 3.5. By (3.82) and (3.88), we have

$$
\begin{aligned}
\|\Pi_n\| &\le \frac{k_2k_4\lambda}{r_n}\sum_{i=0}^{n-1}\sum_{j=0}^{i-1}\|\phi_i\|\rho^{i-j}\|\tilde{\theta}_{j+1}^{\tau}\phi_j\| \\
&\le \frac{k_2k_4\lambda}{r_n}\left[\sum_{i=0}^{n-1}\sum_{j=0}^{i-1}\|\phi_i\|^2\rho^{i-j}\sum_{i=0}^{n-1}\sum_{j=0}^{i}\rho^{i-j}\|\tilde{\theta}_{j+1}^{\tau}\phi_j\|^2\right]^{1/2} \\
&\le k_2k_4\lambda\left[\frac{1}{r_n^2}\sum_{i=0}^{n-1}\|\phi_i\|^2\frac{\rho}{1-\rho}\sum_{j=0}^{n-1}\sum_{i=j}^{n-1}\rho^{i-j}\|\tilde{\theta}_{j+1}^{\tau}\phi_j\|^2\right]^{1/2} \\
&\le \frac{k_2k_4\lambda}{1-\rho}\left[\frac{1}{r_n}\sum_{i=0}^{n-1}\|\tilde{\theta}_{i+1}^{\tau}\phi_i\|^2\right]^{1/2},
\end{aligned}
$$

which goes to zero by (3.27) and Lemma 3.6. Thus (3.81) has been proved.

Now from (3.68), (3.70), and (3.72), we have

$$
\begin{aligned}
\theta_n &= P_n\sum_{i=0}^{n-1}\phi_i(\xi_{i+1}^{\tau}+\phi_i^{\tau}\theta_{i+1}+w_{i+1}^{\tau}F_{i+1}^{\tau})+P_nP_0^{-1}\theta_0 \\
&= P_n\sum_{i=0}^{n-1}\phi_i\Bigg(\phi_i^{\tau}\tilde{\theta}_{i+1}+\sum_{j=0}^{i-1}\phi_j^{\tau}\tilde{\theta}_{j+1}GF^{\tau(i-j)}G^{\tau}+\phi_{-1}^{\tau}\tilde{\theta}_0GF^{\tau(i+1)} \\
&\qquad +\phi_i^{\tau}\theta_{i+1}+w_{i+1}^{\tau}F_{i+1}^{\tau}\Bigg)+P_nP_0^{-1}\theta_0. \qquad (3.105)
\end{aligned}
$$

Therefore, for any $\delta \in (\varepsilon/2, \frac{1}{2}]$,

$$
\begin{aligned}
r_n^{1/2-\delta}(\tilde{\theta}_n+\Pi_n) &= r_n^{1/2-\delta}\left(I-P_n\sum_{i=0}^{n-1}\phi_i\phi_i^{\tau}\right)\theta - r_n^{1/2-\delta}P_n\sum_{i=0}^{n-1}\phi_i\phi_i^{\tau}\theta_0GF^{\tau(i+1)} \\
&\quad - r_n^{1/2-\delta}P_nP_0^{-1}\theta_0 - r_n^{1/2-\delta}P_n\sum_{i=0}^{n-1}\phi_iw_{i+1}^{\tau}F_{i+1}^{\tau}. \qquad (3.106)
\end{aligned}
$$

Now we proceed to prove that each term in (3.106) goes to zero as $n \to \infty$.

First, since $\omega \in J$, $r_n \to \infty$, by (3.88) we have

$$
r_n^{1/2-\delta}\|P_n\|\,\|P_0^{-1}\theta_0\| \le \frac{k_4\lambda}{r_n^{1/2+\delta}}\|P_0^{-1}\theta_0\| \underset{n\to\infty}{\longrightarrow} 0. \qquad (3.107)
$$

By (3.67), it is easy to verify that

$$\left(\underset{\sim}{\mathbf{I}} - \underset{\sim}{\mathbf{P}}_n \sum_{i=0}^{n-1} \phi_i \phi_i^\tau\right) = \frac{1}{\lambda} \underset{\sim}{\mathbf{P}}_n$$

and hence

$$r_n^{1/2-\delta} \left\| \left(\underset{\sim}{\mathbf{I}} - \underset{\sim}{\mathbf{P}}_n \sum_{i=0}^{n-1} \phi_i \phi_i^\tau\right) \underset{\sim}{\boldsymbol{\theta}} \right\| \le \frac{1}{\lambda} r_n^{1/2-\delta} \|\underset{\sim}{\mathbf{P}}_n\| \, \|\underset{\sim}{\boldsymbol{\theta}}\| \underset{n\to\infty}{\to} 0, \qquad \forall \omega \in J. \tag{3.108}$$

Then, by (2.82), (3.88), and the Schwarz inequality, it follows

$$\left\| r_n^{1/2-\delta} \underset{\sim}{\mathbf{P}}_n \sum_{i=0}^{n-1} \phi_i \phi_{-1}^\tau \underset{\sim}{\tilde{\boldsymbol{\theta}}}_0 \underset{\sim}{\mathbf{G}} \underset{\sim}{\mathbf{F}}^{\tau(i+1)} \right\| \le \frac{k_2 k_4 \lambda}{r_n^{1/2+\delta}} \|\phi_{-1}\| \, \|\underset{\sim}{\tilde{\boldsymbol{\theta}}}_0\|$$

$$\left[\sum_{i=0}^{n-1} \|\phi_i\|^2 \sum_{i=0}^{n-1} \rho^{2(i+1)}\right]^{1/2} \le \frac{k_2 k_4 \lambda \|\phi_{-1}\| \, \|\underset{\sim}{\tilde{\boldsymbol{\theta}}}_0\|}{\sqrt{1-\rho^2}\, r_n^{\delta}} \underset{n\to\infty}{\to} 0, \qquad \forall \omega \in J. \tag{3.109}$$

Thus, to complete the proof we only need to show that the last term on the right-hand side of (3.106) also tends to zero as $n \to \infty$.

Since $\delta \in (\varepsilon/2, \frac{1}{2}]$, by (3.80) we know

$$E\left\| \frac{\phi_i \mathbf{w}_{i+1}^\tau \underset{\sim}{\mathbf{F}}_{i+1}^\tau}{r_{i+1}^{1/2+\delta}} \right\|^2 \le E\left(\frac{\|\phi_i\|^2 \|\mathbf{w}_{i+1}\|^2 \operatorname{tr} \underset{\sim}{\mathbf{F}}_{i+1} \underset{\sim}{\mathbf{F}}_{i+1}^\tau}{r_{i+1}^{1+2\delta}} \right) \le k_3 E \frac{\|\phi_i\|^2 \|\mathbf{w}_{i+1}\|^2}{r_{i+1}^{1+2\delta-\varepsilon}}$$

$$\le k_3 E \|\mathbf{w}_{i+1}\|^2 < \infty.$$

Consequently,

$$\left(\sum_{i=0}^{n-1} \frac{\phi_i \mathbf{w}_{i+1}^\tau \underset{\sim}{\mathbf{F}}_{i+1}^\tau}{r_{i+1}^{1/2+\delta}}, \mathscr{F}_{n-1} \right)$$

is a martingale with finite second moment since $\underset{\sim}{\mathbf{F}}_{i+1}$, r_{i+1}, and ϕ_i are all $\mathscr{F}_i$-measurable.

By (3.52), we see that

$$\sum_{i=0}^{\infty} E\left(\frac{\|\phi_i\|^2 \operatorname{tr} \underset{\sim}{F}_{i+1} \mathbf{w}_{i+1} \mathbf{w}_{i+1}^{\tau} \underset{\sim}{F}_{i+1}^{\tau}}{r_{i+1}^{1+2\delta}} \Big/ \mathscr{F}_i \right) \le \sum_{i=0}^{\infty} \frac{\|\phi_i\|^2 \operatorname{tr} \underset{\sim}{F}_{i+1} \underset{\sim}{F}_{i+1}^{\tau}}{r_{i+1}^{1+2\delta}}$$

$$\le k_3 \sum_{i=0}^{\infty} \frac{\|\phi_i\|^2}{r_{i+1}^{1+2\delta-\varepsilon}},$$

with $2\delta > \varepsilon$, which is convergent by Lemma 3.2.

Then by Theorem 1.13, we find that

$$\sum_{i=0}^{n-1} \frac{\phi_i \mathbf{w}_{i+1}^{\tau} \underset{\sim}{F}_{i+1}^{\tau}}{r_i^{1/2+\delta}}$$

converges to a finite limit and by (3.27)

$$\frac{1}{r_n^{1/2+\delta}} \sum_{i=0}^{n-1} \phi_i \mathbf{w}_{i+1}^{\tau} \underset{\sim}{F}_{i+1}^{\tau} \underset{n\to\infty}{\to} 0 \qquad \forall \omega \in J.$$

By this fact and (3.88), we have estimate

$$\left\| r_n^{1/2-\delta} \underset{\sim}{P}_n \sum_{i=0}^{n-1} \phi_i \mathbf{w}_{i+1}^{\tau} \underset{\sim}{F}_{i+1}^{\tau} \right\| \le \frac{k_4 \lambda}{r_n^{1/2+\delta}} \left\| \sum_{i=0}^{n-1} \phi_i \mathbf{w}_{i+1}^{\tau} \underset{\sim}{F}_{i+1}^{\tau} \right\| \underset{n\to\infty}{\to} 0, \qquad \forall \omega \in J$$

which together with (3.107)–(3.109) complete the proof. □

This theorem tells us that under certain conditions the least-squares estimate is strongly consistent and the convergence rate is determined by the rate with which $\underset{\sim}{\Pi}_n$ tends to zero. When $r = 0$, then $\underset{\sim}{\Pi}_n \equiv 0$ and this theorem gives convergence a rate, but for this special case Theorem 3.1 has given better results.

3.4. CONTINUOUS-TIME LEAST-SQUARES IDENTIFICATION

Let $t \in [0, \infty)$ be the continuous time parameter, S be the integral operator as defined by (2.186), and $\mathbf{y}_t, \mathbf{u}_t, \boldsymbol{\varepsilon}_t$ be output, input, and noise processes of m, l, and m dimensions, respectively. Consider the continuous-time system

$$\mathbf{y}_t = \underset{\sim}{A}_1 S \mathbf{y}_t + \cdots + \underset{\sim}{A}_p S^p \mathbf{y}_t + \underset{\sim}{B}_1 S \mathbf{u}_t + \cdots + \underset{\sim}{B}_q S^q \mathbf{u}_t + \boldsymbol{\varepsilon}_t, \qquad \mathbf{y}_0 = \mathbf{0} \tag{3.110}$$

where the noise $\boldsymbol{\varepsilon}_t$ is driven by an m'-dimensional Wiener process $\mathbf{w}_t$ with σ-algebra

$$\mathscr{F}_t = \sigma\{\mathbf{w}_s, 0 \le s \le t\} \tag{3.111}$$

as follows

$$\boldsymbol{\varepsilon}_t = \underset{\sim}{\mathbf{C}}(S)\int_0^t \underset{\sim}{\mathbf{F}}_s \, d\mathbf{w}_s. \tag{3.112}$$

Here

$$\underset{\sim}{\mathbf{C}}(\lambda) = \underset{\sim}{\mathbf{I}} + \underset{\sim}{\mathbf{C}}_1\lambda + \cdots + \underset{\sim}{\mathbf{C}}_r\lambda^r, \tag{3.113}$$

the $m \times m'$ matrix $\underset{\sim}{\mathbf{F}}_t$ is a measurable process adapted to $\mathscr{F}_t$ and

$$E \operatorname{tr} \underset{\sim}{\mathbf{F}}_t \underset{\sim}{\mathbf{F}}_t^{\tau} < \infty, \qquad \int_0^t E \operatorname{tr} \underset{\sim}{\mathbf{F}}_s \underset{\sim}{\mathbf{F}}_s^{\tau}\, ds < \infty, \qquad \forall t. \tag{3.114}$$

Assume the input $\mathbf{u}_t$ is also a measurable process adapted to $\mathscr{F}_t$. Set

$$\underset{\sim}{\boldsymbol{\theta}}^{\tau} = \left[\underset{\sim}{\mathbf{A}}_1 \cdots \underset{\sim}{\mathbf{A}}_p \underset{\sim}{\mathbf{B}}_1 \cdots \underset{\sim}{\mathbf{B}}_q \underset{\sim}{\mathbf{C}}_1 \cdots \underset{\sim}{\mathbf{C}}_r\right], \tag{3.115}$$

$$\boldsymbol{\psi}_t^{\tau} = \begin{cases} \left[\mathbf{y}_t^{\tau}\ S\mathbf{y}_t^{\tau} \cdots S^{p-1}\mathbf{y}_t^{\tau}\ \mathbf{u}_t^{\tau}\ S\mathbf{u}_t^{\tau} \cdots S^{q-1}\mathbf{u}_t^{\tau}\ \mathbf{y}_t^{\tau} \right. \\ \qquad \left. -S(\boldsymbol{\psi}_t^{\tau}\underset{\sim}{\boldsymbol{\theta}}_t) \cdots S^{r-1}\mathbf{y}_t^{\tau} - S^r(\boldsymbol{\psi}_t^{\tau}\underset{\sim}{\boldsymbol{\theta}}_t)\right], & r > 0 \\ \left[\mathbf{y}_t^{\tau}\ S\mathbf{y}_t^{\tau} \cdots S^{p-1}\mathbf{y}_t^{\tau}\ \mathbf{u}_t^{\tau}\ S\mathbf{u}_t^{\tau} \cdots S^{q-1}\mathbf{u}_t^{\tau}\right], & r = 0, \end{cases} \tag{3.116}$$

where $\underset{\sim}{\boldsymbol{\theta}}_t$ is defined in analogue with (3.15) and (3.16):

$$d\underset{\sim}{\boldsymbol{\theta}}_t = \underset{\sim}{\mathbf{P}}_t\boldsymbol{\psi}_t(d\mathbf{y}_t^{\tau} - \boldsymbol{\psi}_t^{\tau}\underset{\sim}{\boldsymbol{\theta}}_t\, dt), \qquad \underset{\sim}{\boldsymbol{\theta}}_0\text{-arbitrary}, \tag{3.117}$$

$$d\underset{\sim}{\mathbf{P}}_t = -\underset{\sim}{\mathbf{P}}_t\boldsymbol{\psi}_t\boldsymbol{\psi}_t^{\tau}\underset{\sim}{\mathbf{P}}_t\, dt, \qquad \underset{\sim}{\mathbf{P}}_0 = \lambda\underset{\sim}{\mathbf{I}}, \quad \lambda = mp + lq + mr. \tag{3.118}$$

(3.116)–(3.118) constitute a system of nonlinear stochastic differential equations, for which we shall assume the existence of a strong solution $(\underset{\sim}{\boldsymbol{\theta}}_t, \mathscr{F}_t)$ called the least-squares estimate for $\underset{\sim}{\boldsymbol{\theta}}$. This situation differs from the discrete-time case where (3.15), (3.16), and (3.65) (or 3.64) can be solved recursively and the question of the solvability of the equations does not arise.

It is easy to verify that (3.117) and (3.118) are equivalent to

$$\underset{\sim}{\theta}_t = \underset{\sim}{P}_t \int_0^t \psi_s d\mathbf{y}_s^\tau + \underset{\sim}{P}_t \underset{\sim}{P}_0^{-1} \theta_0, \tag{3.119}$$

$$\underset{\sim}{P}_t = \left(\int_0^t \psi_s \psi_s^\tau \, ds + \frac{1}{\lambda} \underset{\sim}{I} \right)^{-1} \tag{3.120}$$

For this we only need to take the stochastic differential of (3.119) by using (3.118)

$$d\underset{\sim}{\theta}_t = \dot{\underset{\sim}{P}}_t \left(\int_0^t \psi_s \, d\mathbf{y}_s^\tau + \underset{\sim}{P}_0^{-1} \underset{\sim}{\theta}_0 \right) + \underset{\sim}{P}_t \psi_t \, d\mathbf{y}_t^\tau = \underset{\sim}{P}_t \psi_t \left(d\mathbf{y}_t^\tau - \psi_t^\tau \underset{\sim}{\theta}_t \, dt \right).$$

Denote by $\lambda_{\max}^t$ and $\lambda_{\min}^t$, the maximum and minimum eigenvalues of $\underset{\sim}{P}_t$, respectively, and denote

$$r_t \triangleq \operatorname{tr} \underset{\sim}{P}_t^{-1} = 1 + \int_0^t \|\psi_s\|^2 \, ds. \tag{3.121}$$

As in the discrete-time case we do not impose the boundedness of $\operatorname{tr} \underset{\sim}{F}_t \underset{\sim}{F}_t^\tau$, but assume there exist constants k_3 and $\varepsilon \in [0, 1)$ such that

$$\operatorname{tr} \underset{\sim}{F}_t \underset{\sim}{F}_t^\tau \le k_3 r_t^\varepsilon \text{ a.s.,} \qquad \forall t \in [0, \infty) \tag{3.122}$$

Similar to (3.69), denote

$$\beta_t = \frac{\|\psi_t\|^2}{r_t} \tag{3.123}$$

The analogue of Lemma 3.3 is

Lemma 3.7. (1) $\int_0^\infty \beta_t \, dt = \infty$ *if and only if* $r_t \underset{t\to\infty}{\rightarrow} \infty$. (2) *If* $\int_0^\infty \|\mathbf{u}_s\|^2 \, ds = \infty$, *then* $\int_0^\infty \beta_t \, dt = \infty$. (3) *If* $\underset{\sim}{F}_n \equiv \underset{\sim}{D}$, $\underset{\sim}{D}\underset{\sim}{D}^\tau \ge \sigma^2 \underset{\sim}{I}$ *and all zeros of* $\det \underset{\sim}{C}(\lambda)$ *have negative real parts, then* $\int_0^\infty \beta_t \, dt = \infty$, *where* $\underset{\sim}{D}$ *and* σ^2 *are deterministic.*

Proof. Conclusion 1 follows from

$$\int_0^\infty \beta_t \, dt = \int_0^\infty \frac{\|\psi_t\|^2}{r_t} \, dt = \int_0^\infty \frac{dr_t}{r_t} = \ln r_t;$$

Conclusion 2, from $\|\mathbf{u}_s\|^2 \le \|\psi_s\|^2$.

From (2.199)–(2.201), it is easy to see that the first m components $\zeta_t^1 = \underset{\sim}{G}\zeta_t$ of ζ_t defined by

$$\zeta_t = S\underset{\sim}{F}\zeta_t + \underset{\sim}{G}^\tau\varepsilon_t \tag{3.124}$$

satisfy the equation

$$\underset{\sim}{C}(S)\zeta_t^1 = \varepsilon_t, \tag{3.125}$$

where $\underset{\sim}{F}$ is given by (2.73). Comparing (3.112) with (3.125) gives

$$\zeta_t^1 = \int_0^t \underset{\sim}{D}\, d\mathbf{w}_s,$$

hence

$$\int_0^t \underset{\sim}{D}\, d\mathbf{w}_s = \varepsilon_t + \int_0^t \underset{\sim}{G}\underset{\sim}{F}e^{\mathbf{F}(t-\lambda)}\underset{\sim}{G}^\tau \varepsilon_\lambda\, d\lambda.$$

Suppose $\int_0^\infty \beta_t\, dt < \infty$, then r_t tends to a finite limit as $t \to \infty$, hence $\|\psi_t\| \underset{t\to\infty}{\to} 0$ and $\|\varepsilon_t\| \underset{t\to\infty}{\to} 0$. By (2.213), we have $\|\varepsilon_t\| < \varepsilon$ for $t \geq T$ and

$$\left\|\int_0^t \underset{\sim}{D}\, d\mathbf{w}_s\right\| \leq \|\varepsilon_t\| + k_4\|\underset{\sim}{F}\| \int_0^t e^{-c(t-\lambda)}\|\varepsilon_\lambda\|\, d\lambda$$

$$\leq \varepsilon + k_4\|\underset{\sim}{F}\|\left(e^{-ct}\varepsilon\int_T^t e^{c\lambda}\, d\lambda + e^{-ct}\int_0^T e^{c\lambda}\|\varepsilon_\lambda\|\, d\lambda\right)$$

$$\leq \varepsilon + k_4\|\underset{\sim}{F}\|\left[\frac{\varepsilon}{c}e^{-ct}(e^{ct} - e^{cT}) + e^{-ct}\int_0^T e^{c\lambda}\|\varepsilon_\lambda\|\, d\lambda\right] \underset{\substack{t\to\infty\\ \varepsilon\to 0}}{\to} 0,$$

which implies

$$\frac{1}{t}\left(\int_0^t \underset{\sim}{D}\, d\mathbf{w}_s\right)\left(\int_0^t \underset{\sim}{D}\, d\mathbf{w}_s\right)^\tau \underset{t\to\infty}{\to} \underset{\sim}{0} \text{ a.s.} \tag{3.126}$$

On the other hand, by Theorem 1.14,

$$\overline{\lim_{t\to\infty}} \frac{1}{t}\operatorname{tr}\left(\int_0^t \underset{\sim}{D}\, d\mathbf{w}_s\right)\left(\int_0^t \underset{\sim}{D}\, d\mathbf{w}_s^\tau\right) = \overline{\lim_{t\to\infty}} \frac{1}{t}\operatorname{tr} \underset{\sim}{D}\mathbf{w}_t\mathbf{w}_t^\tau \underset{\sim}{D}^\tau$$

$$= \overline{\lim_{t\to\infty}} \operatorname{tr}\frac{\mathbf{w}_t^\tau \underset{\sim}{D}^\tau \underset{\sim}{D}\mathbf{w}_t}{t} \geq \sigma^2 \overline{\lim_{t\to\infty}} \frac{\|\mathbf{w}_t\|^2}{t} = \sigma^2 \overline{\lim_{t\to\infty}} \frac{\|\mathbf{w}_t\|^2 \ln\ln t}{2m't\ln\ln t} = \infty,$$

which contradicts with (3.126). Thus

$$\int_0^\infty \beta_t \, dt = \infty. \qquad \square$$

Denote

$$\tilde{\theta}_t = \theta - \theta_t, \tag{3.127}$$

$$\xi_t = \frac{d}{dt}\left(\mathbf{y}_t - \int_0^t \mathbf{F}_s \, d\mathbf{w}_s\right) - \theta_t^\tau \psi_t, \qquad \xi_0 = \mathbf{B}_1 \mathbf{u}_0 - \theta_0^\tau \psi_0. \tag{3.128}$$

We note ξ_t is meaningful since $\mathbf{y}_t - \int_0^t \mathbf{F}_s \, d\mathbf{w}_s$ is differentiable.

It is clear that by (3.118) and (3.128)

$$d\theta_t = \mathbf{P}_t \psi_t (\xi_t^\tau \, dt + \mathbf{F}_t \, d\mathbf{w}_t) \tag{3.129}$$

and by (3.110)

$$\begin{aligned} C(S) S \xi_t &= C(S)\left(\mathbf{y}_t - \int_0^t \mathbf{F}_s \, d\mathbf{w}_s - S(\theta_t^\tau \psi_t)\right) \\ &= \mathbf{y}_t + (C(S) - \mathbf{I})\mathbf{y}_t - \varepsilon_t - C(S) S(\theta_t^\tau \psi_t) \\ &= \mathbf{A}_1 S \mathbf{y}_t + \cdots + \mathbf{A}_p S^p \mathbf{y}_t + \mathbf{B}_1 S \mathbf{u}_t + \cdots + \mathbf{B}_q S^q \mathbf{u}_t \\ &\quad + (C(S) - \mathbf{I})\mathbf{y}_t - C(S) S(\theta_t^\tau \psi_t). \end{aligned}$$

Taking derivatives of this expression leads to

$$C(S)\xi_t = \theta^\tau \psi_t - \theta_t^\tau \psi_t = \tilde{\theta}_t^\tau \psi_t. \tag{3.130}$$

Define the $m(r \vee 1)$-dimensional vector η_t

$$\eta_t = S \mathbf{F} \eta_t + G^\tau \tilde{\theta}_t^\tau \psi_t \tag{3.131}$$

with $G^\tau \eta_0 = \xi_0$ and the other components of η_0 being zero.

Similar to (3.124)–(3.126), the first m elements η_t^1 of η_t satisfy the equation as follows

$$C(s)\eta_t^1 = \tilde{\theta}_t^\tau \psi_t, \tag{3.132}$$

which yields by (3.130), (3.132), and (3.126)

$$\xi_t = \underset{\sim}{G}\eta_t = \tilde{\underset{\sim}{\theta}}_t^{\tau}\psi_t + \int_0^t \underset{\sim}{G}\underset{\sim}{F}e^{\underset{\sim}{F}(t-\lambda)}\underset{\sim}{G}^{\tau}\tilde{\underset{\sim}{\theta}}_\lambda^{\tau}\psi_\lambda \, d\lambda. \tag{3.133}$$

Let

$$\underset{\sim}{\Pi}_t = \underset{\sim}{P}_t\int_0^t \psi_s \int_0^s \left[\underset{\sim}{G}\underset{\sim}{F}e^{\underset{\sim}{F}(s-\lambda)}\underset{\sim}{G}^{\tau}\tilde{\underset{\sim}{\theta}}_\lambda^{\tau}\psi_\lambda \, d\lambda\right]^{\tau} ds \tag{3.134}$$

Similar to the discrete-time case we would expect the strictly positive realness of $\underset{\sim}{C}^{-1}(S) - \frac{1}{2}\underset{\sim}{I}$, but which, unfortunately, has no such a property for $r > 0$. However, it is still possible that for a given input $\tilde{\underset{\sim}{\theta}}_t^{\tau}\psi_t$, the following condition holds:

There exist constants k_0 and k_1 such that

$$S_t \triangleq \int_0^t \left\{ \xi_s^{\tau} - \left(\tfrac{1}{2} + k_0\right)\psi_s^{\tau}\tilde{\underset{\sim}{\theta}}_s \right\} \tilde{\underset{\sim}{\theta}}_s^{\tau}\psi_s \, ds + k_1 \geq 0, \qquad \forall t \in [0, \infty). \tag{3.135}$$

Theorem 3.6. *For the system and algorithm defined by* (3.110)–(3.122), *if there is a constant* $k_2 > 0$ *such that*

$$\frac{\|\psi_t\|^2}{r_t} \leq k_2, \qquad \forall t \in [0, \infty) \tag{3.136}$$

and if the following conditions are satisfied:

A. $r = 0$, *or* $r > 0$, *but all zeros of* $\det \underset{\sim}{C}(z)$ *have negative real parts, and Inequality* (3.135) *holds.*

B. *There is a constant* $k_5 > 0$ *such that*

$$\frac{\lambda_{\max}^t}{\lambda_{\min}^t} \leq k_5,$$

then $\underset{\sim}{\Pi}_t \underset{t\to\infty}{\longrightarrow} 0$ *and*

$$\|\tilde{\underset{\sim}{\theta}}_t + \underset{\sim}{\Pi}_t\| = o\left(r_t^{\delta - 1/2}\right), \qquad \forall \delta \in \left(\frac{\varepsilon}{2}, \frac{1}{2}\right]$$

for almost all $\omega \in J \triangleq \{\omega \colon r_t \to \infty\}$.

We first prove the lemmas.

Lemma 3.8. *Under the conditions of Theorem* 3.6,

$$E\left(\frac{\operatorname{tr}\tilde{\boldsymbol{\theta}}_t^\tau \mathbf{P}_t^{-1}\tilde{\boldsymbol{\theta}}_t}{r_t}\Big/\mathscr{F}_s\right) \le \frac{\operatorname{tr}\tilde{\boldsymbol{\theta}}_s^\tau \mathbf{P}_s^{-1}\tilde{\boldsymbol{\theta}}_s}{r_s} - 2E\left[\int_s^t \frac{1}{r_\lambda}\left(\boldsymbol{\xi}_\lambda^\tau - \frac{1}{2}\boldsymbol{\psi}_\lambda^\tau\tilde{\boldsymbol{\theta}}_\lambda\right)\frac{\tilde{\boldsymbol{\theta}}_\lambda^\tau\boldsymbol{\psi}_\lambda\, d\lambda}{\mathscr{F}_s}\right]$$

$$+E\left[\int_s^t\left(\frac{\operatorname{tr}\mathbf{F}_\lambda\mathbf{F}_\lambda^\tau\boldsymbol{\psi}_\lambda^\tau\mathbf{P}_\lambda\boldsymbol{\psi}_\lambda}{r_\lambda}\right)d\lambda/\mathscr{F}_s\right] \tag{3.137}$$

for $t \ge s$.

Proof. By Condition B_2, similar to (3.88), we have

$$\frac{1}{r_t}\mathbf{I} \le \mathbf{P}_t \le \frac{k_5\lambda}{r_t}\mathbf{I}, \tag{3.138}$$

By using Ito formula and by induction, it is easy to see that

$$S^k\int_0^t \mathbf{F}_\lambda\, d\mathbf{w}_\lambda = \int_0^t \frac{(t-\lambda)^k}{k!}\mathbf{F}_\lambda\, d\mathbf{w}_\lambda.$$

Hence for $k > 0$, we have

$$E\left\|\frac{1}{r_t}\int_0^t \boldsymbol{\psi}_s\left(S^k\int_0^s \mathbf{F}_\lambda\, d\mathbf{w}_\lambda\right)^\tau ds\right\|^2 \le E\int_0^t \frac{\|\boldsymbol{\psi}_s\|^2}{r_t^2}\, ds\, E\int_0^t\left\|\int_0^s \frac{(s-\lambda)^k}{k!}\mathbf{F}_\lambda\, d\mathbf{w}_\lambda\right\|^2 ds$$

$$\le \frac{t^{2k}}{(k!)^2}E\int_0^t\frac{\|\boldsymbol{\psi}_s\|^2}{r_t^2}\, ds\int_0^t\int_0^s \operatorname{tr}\mathbf{F}_\lambda\mathbf{F}_\lambda^\tau\, d\lambda\, ds$$

$$\le \frac{k_3 t^{2k+2}}{(k!)^2}E\int_0^t \frac{\dot{r}_s}{r_s^{2-\varepsilon}}\, ds\, d\lambda \le \frac{k_3 t^{2k+2}}{(k!)^2}. \tag{3.139}$$

Applying Ito's formula to $r_t\int_0^t(\boldsymbol{\psi}_s/r_s)(\mathbf{F}_s\, d\mathbf{w}_s)^\tau$ leads to

$$\frac{1}{r_t}\int_0^t \boldsymbol{\psi}_s(\mathbf{F}_s\, d\mathbf{w}_s)^\tau = \int_0^t \frac{\boldsymbol{\psi}_s}{r_s}(\mathbf{F}_s\, d\mathbf{w}_s)^\tau + \frac{1}{r_t}\int_0^t \|\boldsymbol{\psi}_s\|^2\int_0^s \frac{\boldsymbol{\psi}_\lambda}{r_\lambda}(\mathbf{F}_\lambda\, d\mathbf{w}_\lambda)^\tau\, ds.$$

Therefore, by (3.136),

$$E\left\|\frac{1}{r_t}\int_0^t \psi_s(\underset{\sim}{\mathbf{F}}_s\, d\mathbf{w}_s)^\tau\right\|^2 \le E\int_0^t \frac{\|\psi_s\|^2}{r_s^2}\operatorname{tr}\underset{\sim}{\mathbf{F}}_s\underset{\sim}{\mathbf{F}}_s^\tau\, ds + k_2 tE\int_0^t\left\|\int_0^s \frac{\psi_\lambda}{r_\lambda}(\underset{\sim}{\mathbf{F}}_\lambda\, d\mathbf{w}_\lambda)^\tau\right\|^2 ds$$

$$\le k_3\int_0^t \frac{\|\psi_s\|^2}{r_s^{2-\varepsilon}}\, ds + k_2k_3\int_0^t E\int_0^s \frac{\|\psi_\lambda\|^2}{r_\lambda^{2-\varepsilon}}\, d\lambda\, ds < \infty. \tag{3.140}$$

This together with (3.138) means that

$$E\left\|\underset{\sim}{\mathbf{P}}_t\int_0^t \psi_s(d\boldsymbol{\varepsilon}_s)^\tau\right\|^2 < \infty. \tag{3.141}$$

Notice that

$$\|\psi_t(d\mathbf{y}_t - d\boldsymbol{\varepsilon}_t)^\tau\| \le \text{const}\|\psi_t\|^2\, dt \tag{3.142}$$

and

$$\underset{\sim}{\boldsymbol{\theta}}_t = \underset{\sim}{\mathbf{P}}_t\int_0^t \psi_s(d\mathbf{y}_s^\tau - d\boldsymbol{\varepsilon}_s^\tau) + \underset{\sim}{\mathbf{P}}_t\int_0^t \psi_s\, d\boldsymbol{\varepsilon}_s^\tau + \underset{\sim}{\mathbf{P}}_t\underset{\sim}{\mathbf{P}}_0^{-1}\underset{\sim}{\boldsymbol{\theta}}_0.$$

Then from (3.138), (3.121), (3.141), and (3.142), it follows that

$$E\|\underset{\sim}{\boldsymbol{\theta}}_t\|^2 < \infty, \qquad \forall t. \tag{3.143}$$

Now by Ito's formula and (3.129), we have

$$\frac{\operatorname{tr}\underset{\sim}{\tilde{\boldsymbol{\theta}}}_t^\tau\underset{\sim}{\mathbf{P}}_t^{-1}\underset{\sim}{\tilde{\boldsymbol{\theta}}}_t}{r_t} = \frac{\operatorname{tr}\underset{\sim}{\tilde{\boldsymbol{\theta}}}_s^\tau\underset{\sim}{\mathbf{P}}_s^{-1}\underset{\sim}{\tilde{\boldsymbol{\theta}}}_s}{r_s} - 2\int_s^t \frac{1}{r_\lambda}(\boldsymbol{\xi}_\lambda^\tau - \tfrac{1}{2}\psi_\lambda^\tau\underset{\sim}{\tilde{\boldsymbol{\theta}}}_\lambda)\underset{\sim}{\tilde{\boldsymbol{\theta}}}_\lambda^\tau\psi_\lambda\, d\lambda$$

$$-2\int_s^t \frac{1}{r_\lambda}\psi_\lambda^\tau\underset{\sim}{\tilde{\boldsymbol{\theta}}}_\lambda\underset{\sim}{\mathbf{F}}_\lambda\, d\mathbf{w}_\lambda + \int_s^t \frac{1}{r_\lambda}\operatorname{tr}\underset{\sim}{\mathbf{F}}_\lambda\underset{\sim}{\mathbf{F}}_\lambda^\tau\psi_\lambda^\tau\underset{\sim}{\mathbf{P}}_\lambda\psi_\lambda\, d\lambda$$

$$-\int_s^t \operatorname{tr}\underset{\sim}{\tilde{\boldsymbol{\theta}}}_\lambda^\tau\frac{\|\psi_\lambda\|^2\underset{\sim}{\mathbf{P}}_\lambda^{-1}}{r_\lambda^2}\underset{\sim}{\tilde{\boldsymbol{\theta}}}_\lambda\, d\lambda$$

$$\le \frac{\operatorname{tr}\underset{\sim}{\tilde{\boldsymbol{\theta}}}_s^\tau\underset{\sim}{\mathbf{P}}_s^{-1}\underset{\sim}{\tilde{\boldsymbol{\theta}}}_s}{r_s} - 2\int_s^t \frac{1}{r_\lambda}(\boldsymbol{\xi}_\lambda^\tau - \tfrac{1}{2}\psi_\lambda^\tau\underset{\sim}{\tilde{\boldsymbol{\theta}}}_\lambda)\underset{\sim}{\tilde{\boldsymbol{\theta}}}_\lambda^\tau\psi_\lambda\, d\lambda$$

$$-2\int_s^t \frac{1}{r_\lambda}\psi_\lambda^\tau\underset{\sim}{\tilde{\boldsymbol{\theta}}}_\lambda\underset{\sim}{\mathbf{F}}_\lambda\, d\mathbf{w}_\lambda + \int_s^t \frac{\operatorname{tr}\underset{\sim}{\mathbf{F}}_\lambda\underset{\sim}{\mathbf{F}}_\lambda^\tau\psi_\lambda^\tau\underset{\sim}{\mathbf{P}}_\lambda\psi_\lambda}{r_\lambda}\, d\lambda, \tag{3.144}$$

where the left-hand side has finite expectation because of (3.138) and (3.143). Now $r_t \geq 1$ and $\boldsymbol{\xi}_t$ and r_t are continuous, so from (3.136), (3.133), and (3.143), it is clear that

$$\int_0^t E\frac{\|\boldsymbol{\xi}_s\|^2}{r_s}\,ds < \infty, \tag{3.145}$$

$$E\int_0^t \frac{\|\boldsymbol{\psi}_s\|^2\|\boldsymbol{\theta}_s\|^2}{r_s}\,ds \leq k_2\int_0^t E\|\tilde{\boldsymbol{\theta}}_s\|^2\,ds < \infty, \qquad \forall t \in [0,\infty). \tag{3.146}$$

Then by (3.122), we have

$$E\int_0^t \frac{\|\boldsymbol{\psi}_\lambda\|^2\|\tilde{\boldsymbol{\theta}}_\lambda\|^2\|\mathbf{F}_\lambda\|^2}{r_\lambda^2}\,d\lambda \leq k_3 E\int_0^t \frac{\|\boldsymbol{\psi}_\lambda\|^2\|\tilde{\boldsymbol{\theta}}_\lambda\|^2}{r_\lambda^{2-\varepsilon}}\,d\lambda$$

$$\leq k_2 k_3 \int_0^t E\|\tilde{\boldsymbol{\theta}}_\lambda\|^2\,d\lambda < \infty \tag{3.147}$$

and by (3.141) and (3.122)

$$E\int_0^\infty \frac{\operatorname{tr}\mathbf{F}_\lambda\mathbf{F}_\lambda^\tau\boldsymbol{\psi}_\lambda^\tau\mathbf{P}_\lambda\boldsymbol{\psi}_\lambda}{r_\lambda}\,d\lambda \leq k_3 E\int_0^\infty \frac{\boldsymbol{\psi}_\lambda^\tau\mathbf{P}_\lambda\boldsymbol{\psi}_\lambda}{r_\lambda^{1-\varepsilon}}\,d\lambda$$

$$\leq k_3 k_5 \lambda E\int_0^\infty \left(\frac{\|\boldsymbol{\psi}_\lambda\|^2}{r_\lambda^{2-\varepsilon}}\right)d\lambda$$

$$= k_3 k_5 \lambda E\int_0^\infty \frac{dr_\lambda}{r_\lambda^{2-\varepsilon}} = \frac{k_3 k_5 \lambda}{1-\varepsilon}. \tag{3.148}$$

Formulas (3.145)–(3.148) show that each term on the right-hand side of (3.144) has finite expectation, and (3.137) is obtained by taking the conditional expectation $E(\cdot/\mathscr{F}_s)$. □

Lemma 3.9. *Under the conditions of Theorem* 3.6

$$\int_0^\infty \frac{1}{r_s}\|\tilde{\boldsymbol{\theta}}_s^\tau\boldsymbol{\psi}_s\|^2\,ds < \infty. \tag{3.149}$$

Proof. By Condition A_2 we have that

$$S_t \triangleq \int_0^t [\boldsymbol{\xi}_s^\tau - (\tfrac{1}{2} + k_0)\boldsymbol{\psi}_s^\tau\tilde{\boldsymbol{\theta}}_s]\tilde{\boldsymbol{\theta}}_s^\tau\boldsymbol{\psi}_s\,ds + k_1 \geq 0, \qquad \forall t \in [0,\infty). \tag{3.150}$$

for which by (3.143) and (3.145)

$$E\left(\frac{S_t}{r_t}\right) < \infty, \qquad \forall t \in [0, \infty). \tag{3.151}$$

Set

$$m_t = \frac{\operatorname{tr}\tilde{\theta}_t^\tau P_t^{-1}\tilde{\theta}_t}{r_t} + 2\int_0^t S_\lambda \frac{\|\psi_\lambda\|^2}{r_\lambda^2}\,d\lambda + \frac{2S_t}{r_t} + 2k_0\int_0^t \frac{1}{r_s}\|\tilde{\theta}_s^\tau \psi_s\|^2\,ds$$
$$+ E\left[\int_0^\infty \frac{1}{r_s}\operatorname{tr} F_s F_s^\tau \psi_s^\tau P_s \psi_s\,ds/\mathscr{F}_t\right] - \int_0^t \frac{1}{r_s}\operatorname{tr} F_s F_s^\tau \psi_s^\tau P_s \psi_s\,ds, \tag{3.152}$$

each term of which has finite expectation by (3.143)–(3.148) and by the boundedness of $\|\psi_t\|^2/r_t$. Then by Lemma 3.8, we have

$$E\left(\frac{m_t}{\mathscr{F}_s}\right) \le \frac{\operatorname{tr}\tilde{\theta}_s^\tau P_s^{-1}\tilde{\theta}_s}{r_s} - 2E\left[\int_s^t \frac{1}{r_\lambda}\left(\xi_\lambda^\tau - \left(\tfrac{1}{2} + k_0\right)\psi_\lambda^\tau\tilde{\theta}_\lambda\right)\tilde{\theta}_\lambda^\tau\psi_\lambda\,d\lambda/\mathscr{F}_s\right]$$
$$-2k_0E\left[\int_s^t \frac{1}{r_\lambda}\|\tilde{\theta}_\lambda^\tau\psi_\lambda\|^2\,d\lambda/\mathscr{F}_s\right] + E\left[\int_s^t\left(\frac{\operatorname{tr} F_\lambda F_\lambda^\tau \psi_\lambda^\tau P_\lambda \psi_\lambda}{r_\lambda}\right)d\lambda/\mathscr{F}_s\right]$$
$$+2\int_0^s S_\lambda\frac{\|\psi_\lambda\|^2}{r_\lambda^2}\,d\lambda + 2E\left[\int_s^t S_\lambda \frac{\|\psi_\lambda\|^2}{r_\lambda^2}\,d\lambda/\mathscr{F}_s\right] + E\left(\frac{2S_t}{r_t}\Big/\mathscr{F}_s\right)$$
$$+2k_0\int_0^s \frac{1}{r_\lambda}\|\tilde{\theta}_\lambda^\tau\psi_\lambda\|^2\,d\lambda + 2k_0E\left[\int_s^t \frac{1}{r_\lambda}\|\tilde{\theta}_\lambda^\tau\psi_\lambda\|^2\,d\lambda/\mathscr{F}_s\right]$$
$$+E\left[\int_0^\infty \frac{1}{r_\lambda}\frac{\operatorname{tr} F_\lambda F_\lambda^\tau \psi_\lambda^\tau P_\lambda \psi_\lambda\,d\lambda}{\mathscr{F}_s}\right] - \int_0^s \frac{1}{r_\lambda}\operatorname{tr} F_\lambda F_\lambda^\tau \psi_\lambda^\tau P_\lambda \psi_\lambda\,d\lambda$$
$$-E\left[\int_s^t \frac{1}{r_\lambda}\frac{\operatorname{tr} F_\lambda F_\lambda^\tau \psi_\lambda^\tau P_\lambda \psi_\lambda\,d\lambda}{\mathscr{F}_s}\right], \qquad \forall t \ge s. \tag{3.153}$$

Notice

$$2\int_s^t \frac{1}{r_\lambda}\left[\xi_\lambda^\tau - \left(\frac{1}{2} + k_0\right)\psi_\lambda^\tau\tilde{\theta}_\lambda\right]\tilde{\theta}_\lambda^\tau\psi_\lambda\,d\lambda = 2\int_s^t \frac{1}{r_\lambda}\,dS_\lambda$$
$$= \frac{2S_t}{r_t} - \frac{2S_s}{r_s} + 2\int_s^t S_\lambda \frac{\|\psi_\lambda\|^2}{r_\lambda^2}\,d\lambda,$$

then $E(m_t/\mathscr{F}_s) \le m_s$ a.s. $t \ge s$, and by Theorem 1.11, as $t \to \infty$, m_t converges to a finite limit. But from the expression for m_t it is easy to see that its first four terms are nonnegative and the sum of the last two terms are also nonnegative. Hence (3.149) is valid. □

Proof of Theorem 3.6. Denoting $k_6 = k_4 k_5 \lambda$ and using (2.213), (3.138), we obtain

$$\|\underset{\sim}{\Pi}_t\| \le k_6 \frac{\|\underset{\sim}{\mathbf{F}}\|}{r_t} \int_0^t \|\psi_s\| \int_0^s e^{-c(s-\lambda)} \|\underset{\sim}{\tilde{\theta}}_\lambda^\tau \psi_\lambda\| \, d\lambda \, ds$$

$$\le k_6 \frac{\|\underset{\sim}{\mathbf{F}}\|}{r_t} \left[\int_0^t \int_0^s \|\psi_s\|^2 e^{-c(s-\lambda)} \, d\lambda \, ds \int_0^t \int_0^s e^{-c(s-\lambda)} \|\underset{\sim}{\tilde{\theta}}_\lambda^\tau \psi_\lambda\|^2 \, d\lambda \, ds \right]^{1/2}$$

$$\le k_6 \frac{\|\underset{\sim}{\mathbf{F}}\|}{c r_t} \left[\int_0^t \|\psi_s\|^2 \, ds \int_0^t \left(\int_\lambda^t - de^{-c(s-\lambda)} \right) \|\underset{\sim}{\tilde{\theta}}_\lambda^\tau \psi_\lambda\|^2 \, d\lambda \right]^{1/2}$$

$$\le \frac{k_6}{c} \|\underset{\sim}{\mathbf{F}}\| \left[\frac{1}{r_t} \int_0^t \|\underset{\sim}{\tilde{\theta}}_\lambda^\tau \psi_\lambda\|^2 \, d\lambda \right]^{1/2}$$

$$\le \frac{k_6 \|\underset{\sim}{\mathbf{F}}\|}{c} \left[\frac{1}{r_t} \int_0^t \|\underset{\sim}{\tilde{\theta}}_\lambda^\tau \psi_\lambda\|^2 \, d\lambda + \int_T^\infty \frac{1}{r_\lambda} \|\underset{\sim}{\tilde{\theta}}_\lambda^\tau \psi_\lambda\|^2 \, d\lambda \right]^{1/2} \underset{t \to \infty}{\to} 0$$

for all $\omega \in J$, because, for fixed T, the first term goes to zero as $t \to \infty$ while the second term tends to zero as $T \to \infty$ by Lemma 3.9.

From (3.119) and (3.128), we have

$$\begin{aligned} \underset{\sim}{\theta}_t &= \underset{\sim}{\mathbf{P}}_t \underset{\sim}{\mathbf{P}}_0^{-1} \underset{\sim}{\theta}_0 + \underset{\sim}{\mathbf{P}}_t \int_0^t \psi_s (\xi_s^\tau \, ds + \psi_s^\tau \underset{\sim}{\theta}_s \, ds + d\mathbf{w}_s^\tau \underset{\sim}{\mathbf{F}}_s^\tau) \\ &= \underset{\sim}{\mathbf{P}}_t \underset{\sim}{\mathbf{P}}_0^{-1} \underset{\sim}{\theta}_0 + \underset{\sim}{\mathbf{P}}_t \int_0^t \psi_s \left[\psi_s^\tau \underset{\sim}{\tilde{\theta}}_s + \left(\int_0^s \underset{\sim}{\mathbf{G}} \underset{\sim}{\mathbf{F}} e^{\mathbf{F}(t-\lambda)} \underset{\sim}{\mathbf{G}}^\tau \underset{\sim}{\tilde{\theta}}_\lambda^\tau \psi_\lambda \, d\lambda \right)^\tau ds \right. \\ &\quad \left. + \psi_s^\tau \underset{\sim}{\theta}_s \, ds + d\mathbf{w}_s^\tau \underset{\sim}{\mathbf{F}}_s^\tau \right]. \end{aligned} \tag{3.154}$$

Hence

$$\begin{aligned} r_t^{1/2-\delta} (\underset{\sim}{\tilde{\theta}}_t + \underset{\sim}{\Pi}_t) &= -r_t^{1/2-\delta} \underset{\sim}{\mathbf{P}}_t \underset{\sim}{\mathbf{P}}_0^{-1} \underset{\sim}{\theta}_0 + r_t^{1/2-\delta} \left(\underset{\sim}{\mathbf{I}} - \underset{\sim}{\mathbf{P}}_t \int_0^t \psi_s \psi_s^\tau \, ds \right) \underset{\sim}{\theta} \\ &\quad - r_t^{1/2-\delta} \underset{\sim}{\mathbf{P}}_t \int_0^t \psi_s \, d\mathbf{w}_s^\tau \underset{\sim}{\mathbf{F}}_s^\tau. \end{aligned} \tag{3.155}$$

We shall consider the term on the right-hand side of (3.155) separately. By (3.138), its first term tends to zero as $t \to \infty$ for $\omega \in J$, while the second term goes to zero for $\omega \in J$ because

$$\left\| \underset{\sim}{\mathbf{I}} - \underset{\sim}{\mathbf{P}}_t \int_0^t \psi_s \psi_s^{\tau} \, ds \right\| \leq \left\| \underset{\sim}{\mathbf{I}} - \underset{\sim}{\mathbf{P}}_t \left(\underset{\sim}{\mathbf{P}}_t^{-1} - \frac{1}{\lambda} \underset{\sim}{\mathbf{I}} \right) \right\| \leq \frac{k_4}{r_t}.$$

For the third term, we take

$$0 = t_0 < t_1 < t_2 < \cdots < t_n < \cdots, \qquad t_n \underset{n \to \infty}{\longrightarrow} \infty$$

and denote

$$\underset{\sim}{\mathbf{X}}_{t_n} = \int_0^{t_n} \frac{1}{r_s^{1/2+\delta}} \psi_s \, d\mathbf{w}_s^{\tau} \underset{\sim}{\mathbf{F}}_s^{\tau} = \sum_{i=0}^{n-1} \int_{t_i}^{t_{i+1}} \frac{1}{r_s^{1/2+\delta}} \psi_s \, d\mathbf{w}_s^{\tau} \underset{\sim}{\mathbf{F}}_s^{\tau}. \tag{3.156}$$

$(\underset{\sim}{\mathbf{X}}_{t_n}, \mathscr{F}_{t_n})$ is a martingale with finite second-order moment since by (3.122)

$$E \left\| \int_{t_i}^{t_{i+1}} \frac{1}{r_s^{1/2+\delta}} \psi_s \, d\mathbf{w}_s^{\tau} \underset{\sim}{\mathbf{F}}_s^{\tau} \right\|^2 \leq E \int_{t_i}^{t_{i+1}} \frac{\|\psi_s\|^2 \|\underset{\sim}{\mathbf{F}}_s\|^2}{r_s^{1+2\delta}} \, ds \leq k_3 E \int_{t_i}^{t_{i+1}} \frac{\|\psi_s\|^2}{r_s^{1+2\delta-\varepsilon}} \, ds$$

$$= \frac{k_3}{\varepsilon - 2\delta} E \frac{1}{r_s^{2\delta-\varepsilon}} \Bigg|_{t_i}^{t_{i+1}} = \frac{k_3}{2\delta - \varepsilon} E \left(\frac{1}{r_{t_i}^{2\delta-\varepsilon}} - \frac{1}{r_{t_{i+1}}^{2\delta-\varepsilon}} \right) \leq \frac{k_3}{2\delta - \varepsilon}, \tag{3.157}$$

and $\underset{\sim}{\mathbf{X}}_{t_n}$ converges to a finite limit as $n \to \infty$ by Theorem 1.13 because

$$\sum_{i=0}^{\infty} E \left(\left\| \int_{t_i}^{t_{i+1}} \frac{1}{r_s^{1/2+\delta}} \psi_s \, d\mathbf{w}_s^{\tau} \underset{\sim}{\mathbf{F}}_s^{\tau} \right\|^2 \Big/ \mathscr{F}_{t_i} \right)$$

$$\leq \frac{k_3}{2\delta - \varepsilon} \sum_{i=0}^{\infty} \left(\frac{1}{r_{t_i}^{2\delta-\varepsilon}} - \frac{1}{r_{t_{i+1}}^{2\delta-\varepsilon}} \right) \leq \frac{k_3}{2\delta - \varepsilon} \tag{3.158}$$

Since the t_i are arbitrarily chosen, convergence of $\underset{\sim}{\mathbf{X}}_{t_n}$ implies

$$\underset{\sim}{\mathbf{X}}_t \triangleq \int_0^t \frac{1}{r_s^{1/2+\delta}} \psi_s \, d\mathbf{w}_s^{\tau} \underset{\sim}{\mathbf{F}}_s^{\tau} \underset{t \to \infty}{\longrightarrow} \underset{\sim}{\mathbf{X}} < \infty. \tag{3.159}$$

Applying Ito's formula to $r_t^{1/2+\delta}\underset{\sim}{\mathbf{X}}_t$, we have

$$r_t^{1/2+\delta}\underset{\sim}{\mathbf{X}}_t = \int_0^t \underset{\sim}{\mathbf{X}}_s \, dr_t^{1/2+\delta} + \int_0^t \psi_s \, d\mathbf{w}_s^T \underset{\sim}{\mathbf{F}}_s^T$$

and

$$\frac{1}{r_t^{1/2+\delta}} \int_0^t \psi_s \, d\mathbf{w}_s^T \underset{\sim}{\mathbf{F}}_s^T = \underset{\sim}{\mathbf{X}}_t - \frac{1}{r_t^{1/2+\delta}} \int_0^t \underset{\sim}{\mathbf{X}}_s \, dr_s^{1/2+\delta}$$

$$= \underset{\sim}{\mathbf{X}}_t - \frac{1}{r_t^{1/2+\delta}} \int_0^t (\underset{\sim}{\mathbf{X}}_s - \underset{\sim}{\mathbf{X}}) \, dr_s^{1/2+\delta} - \frac{r_t^{1/2+\delta} - 1}{r_t^{1/2+\delta}} \underset{\sim}{\mathbf{X}}$$

$$= (\underset{\sim}{\mathbf{X}}_t - \underset{\sim}{\mathbf{X}}) + \frac{1}{r_t^{1/2+\delta}} \underset{\sim}{\mathbf{X}} + \frac{1}{r_t^{1/2+\delta}} \int_0^t (\underset{\sim}{\mathbf{X}}_s - \underset{\sim}{\mathbf{X}}) \, dr_s^{1/2+\delta}$$

$$+ \frac{1}{r_t^{1/2+\delta}} \int_T^t (\underset{\sim}{\mathbf{X}}_s - \underset{\sim}{\mathbf{X}}) \, dr_s^{1/2+\delta}. \tag{3.160}$$

Given $\varepsilon > 0$, we can take T sufficiently large that $\|\underset{\sim}{\mathbf{X}}_s - \underset{\sim}{\mathbf{X}}\| < \varepsilon \ \forall t \geq T$, hence on the right-hand side of (3.160) both the first and the last terms are less than ε for $t \geq T$, while for fixed T the remaining two terms go to zero as $t \to \infty$ for $\omega \in J$. Applying (3.160) and (3.138) to the last term of (3.155) leads to

$$\left\| r_t^{1/2-\delta} \underset{\sim}{\mathbf{P}}_t \int_0^t \psi_s \, d\mathbf{w}_s^T \underset{\sim}{\mathbf{F}}_s^T \right\| \leq k_5 \lambda \frac{1}{r_t^{1/2+\delta}} \left\| \int_0^t \psi_s \, d\mathbf{w}_s^T \underset{\sim}{\mathbf{F}}_s^T \right\| \underset{t\to\infty}{\longrightarrow} 0,$$

which completes the proof. □

Similar to the discrete-time case with $r = 0$, we can prove a result better than Theorem 3.6 without need of the conditions (3.136) and $E\int_0^t \operatorname{tr} \underset{\sim}{\mathbf{F}}_s \underset{\sim}{\mathbf{F}}_s^T \, ds < \infty$. Theorem 3.7 is the analogue of Theorem 3.1.

Theorem 3.7. *Assume $r = 0$ and there are constants $k_3 > 0$, $\varepsilon \in [0, 1)$, $\alpha \in [0, \frac{1}{2})$ and an* a.s. *finite random variable γ such that* (3.122) *holds,*

$$\lambda_{\max}^t \leq \frac{\gamma}{r_t^{1-\alpha}}, \tag{3.161}$$

and

$$\alpha + \frac{\varepsilon}{2} < \tfrac{1}{2}, \tag{3.162}$$

Then

$$\|\underset{\sim}{\theta} - \underset{\sim}{\theta}_t\| = o\left(r_t^{\delta - 1/2}\right), \qquad \forall \delta \in \left(\alpha + \frac{\varepsilon}{2}, \tfrac{1}{2}\right] \tag{3.163}$$

for almost all $\omega \in J$, *where*

$$J \triangleq \left\{\omega : r_t \underset{t\to\infty}{\to} \infty\right\},$$

$\underset{\sim}{\theta}_t$ *is the strong solution of* (3.117) *and* (3.118) *with arbitrary initial value* $\underset{\sim}{\theta}_0$.

Proof. In the proof of Theorem 3.6, Lemmas 3.8 and 3.9 are used to show $\underset{\sim}{\Pi}_t \underset{t\to\infty}{\to} 0$. But for the present case $\underset{\sim}{\Pi}_t \equiv \underset{\sim}{0}$, as can be seen from (3.133) and (3.134). Noticing that with $\mu = mp + lq$,

$$\int_0^t \psi_s \psi_s^\tau \, ds = \underset{\sim}{P}_t^{-1} - \frac{1}{\mu}\underset{\sim}{I},$$

by (3.155) we find that

$$r_t^{1/2-\delta}\|\underset{\sim}{\theta} - \underset{\sim}{\theta}_t\| \le \|r_t^{1/2-\delta}\underset{\sim}{P}_t\underset{\sim}{P}_0^{-1}\underset{\sim}{\theta}_0\| + \frac{1}{\mu}r_t^{1/2-\delta}\|\underset{\sim}{P}_t\|\,\|\underset{\sim}{\theta}\|$$

$$+ r_t^{1/2-\delta}\left\|\underset{\sim}{P}_t\int_0^t \psi_s\, d\mathbf{w}_s^\tau \underset{\sim}{F}_s^\tau\right\|. \tag{3.164}$$

The first and second terms of (3.164) go to zero for $\omega \in J$ as $t \to \infty$ by (3.162). Notice that in (3.156)–(3.160), the only condition used is (3.122), which remains valid for the present case. Hence we still have

$$\frac{1}{r_t^{1/2+\delta-\alpha}}\int_0^t \psi_s\, d\mathbf{w}_s^\tau \underset{\sim}{F}_s^\tau \underset{t\to\infty}{\to} 0, \qquad \omega \in J \tag{3.165}$$

which is the result of (3.156)–(3.160) with δ replaced by $\delta - \alpha$. Consequently, by (3.161) and (3.165) for $\omega \in J$

$$r_t^{1/2-\delta}\left\|\underset{\sim}{P}_t\int_0^t \psi_s\, d\mathbf{w}_s^\tau \underset{\sim}{F}_s^\tau\right\| \le \frac{\gamma}{r_t^{1/2+\delta-\alpha}}\left\|\int_0^t \psi_s\, d\mathbf{w}_s^\tau \underset{\sim}{F}_s^\tau\right\| \underset{t\to\infty}{\to} 0,$$

and thus all terms on the right-hand side of (3.164) go to zero as $t \to \infty$ for $\omega \in J$.

□

Similar to the discrete-time case for $r = 0$, we have

$$\frac{r_t}{\mu} = \frac{\operatorname{tr}(\underset{\sim}{\mathbf{P}}_t^{-1})}{\mu} \le \|\underset{\sim}{\mathbf{P}}_t^{-1}\| \le \operatorname{tr} \underset{\sim}{\mathbf{P}}_t^{-1} = r_t \tag{3.166}$$

and

$$\frac{1}{r_t} \le \lambda^t_{\min} = \frac{1}{\|\underset{\sim}{\mathbf{P}}_t^{-1}\|} \le \frac{\mu}{r_t}. \tag{3.167}$$

Hence Condition (3.161) is equivalent to

$$\frac{\lambda^t_{\max}}{\lambda^t_{\min}} \le \gamma_1 r_t^{\alpha} \tag{3.168}$$

with γ_1 a finite random variable.

Theorem 3.8. *If $r = 0$, $\operatorname{tr} \underset{\sim}{\mathbf{F}}_t \underset{\sim}{\mathbf{F}}_t^{\tau} \le \sigma^2$, $\forall t$ with σ^2 a constant, and if* (3.168) *holds for some $\alpha \in [0, \frac{1}{2})$, then*

$$\|\underset{\sim}{\theta} - \underset{\sim}{\theta}_t\| = o\left(r_t^{\delta - 1/2}\right), \qquad \forall \delta \in \left(\alpha, \tfrac{1}{2}\right].$$

The method of proof of this theorem can be seen from that for Theorem 3.7. *The following theorem corresponds to Theorem* 3.4.

Theorem 3.9. *Assume* $\operatorname{tr} \underset{\sim}{\mathbf{F}}_t \underset{\sim}{\mathbf{F}}_t^{\tau} \le \sigma^2$ $\forall t$ *for some constant* γ^2. *Then*

$$\underset{\sim}{\theta}_n \to \underset{\sim}{\theta}$$

on J_1, where

$$J_1 = \left\{\omega: \underset{\sim}{\mathbf{P}}_t \underset{t \to \infty}{\to} 0,\ \overline{\lim_{t \to \infty}} \frac{1}{\varepsilon} \left\| \underset{\sim}{\mathbf{P}}_t \int_0^t \varepsilon_s \psi_s \psi_s^{\tau}\, ds \right\| < \infty, \qquad \forall \varepsilon,\ \forall \varepsilon_s \in [0, \varepsilon] \right\}$$

Proof. From (3.164), we have

$$\|\underset{\sim}{\theta} - \underset{\sim}{\theta}_t\| \le \|\underset{\sim}{\mathbf{P}}_t \underset{\sim}{\mathbf{P}}_0^{-1} \underset{\sim}{\theta}_0\| + \frac{1}{\mu} \|\underset{\sim}{\mathbf{P}}_t\| \, \|\underset{\sim}{\theta}\| + \left\| \underset{\sim}{\mathbf{P}}_t \int_0^t \psi_s \, d\mathbf{w}_s^{\tau} \underset{\sim}{\mathbf{F}}_s^{\tau} \right\|.$$

Hence in order to prove the theorem we only need to show that

$$\underset{\sim}{\mathbf{P}}_t \int_0^t \psi_s \, d\mathbf{w}_s^{\tau} \underset{\sim}{\mathbf{F}}_s^{\tau} \underset{t \to \infty}{\to} 0. \tag{3.169}$$

We denote by $\mathbf{f}_s^i$ the ith column of $\underset{\sim}{\mathbf{F}}_s^\tau$ and

$$\mathbf{m}_t^i = \int_0^t \underset{\sim}{\mathbf{P}}_s \psi_s \mathbf{f}_s^{i\tau}\, d\mathbf{w}_s.$$

It is easy to see that $(\mathbf{m}_t^i, \mathscr{F}_t)$ is a martingale with bounded second-order moment and hence that

$$\underset{\sim}{\mathbf{M}}_t \underset{t\to\infty}{\to} \underset{\sim}{\mathbf{M}} < \infty, \tag{3.170}$$

where $\underset{\sim}{\mathbf{M}}_t$ and $\underset{\sim}{\mathbf{M}}$ are matrices

$$\underset{\sim}{\mathbf{M}}_t = \int_0^t \underset{\sim}{\mathbf{P}}_s \psi_s\, d\mathbf{w}_s^\tau \underset{\sim}{\mathbf{F}}_s^\tau.$$

Applying Ito's formula to $\underset{\sim}{\mathbf{P}}_t^{-1}\underset{\sim}{\mathbf{M}}_t$ yields

$$\underset{\sim}{\mathbf{P}}_t^{-1}\underset{\sim}{\mathbf{M}}_t = \int_0^t \psi_s\psi_s^\tau \underset{\sim}{\mathbf{M}}_s\, ds + \int_0^t \psi_s\, d\mathbf{w}_s^\tau \underset{\sim}{\mathbf{F}}_s^\tau$$

and

$$\begin{aligned}\underset{\sim}{\mathbf{P}}_t \int_0^t \psi_s\, d\mathbf{w}_s^\tau \underset{\sim}{\mathbf{F}}_s^\tau &= \underset{\sim}{\mathbf{M}}_t - \underset{\sim}{\mathbf{P}}_t \int_0^t \psi_s\psi_s^\tau \underset{\sim}{\mathbf{M}}_s\, ds \\ &= \underset{\sim}{\mathbf{M}}_t - \underset{\sim}{\mathbf{M}} + \frac{1}{\mu}\underset{\sim}{\mathbf{P}}_t\underset{\sim}{\mathbf{M}} - \underset{\sim}{\mathbf{P}}_t \int_0^t \psi_s\psi_s^\tau (\underset{\sim}{\mathbf{M}}_s - \underset{\sim}{\mathbf{M}})\, ds.\end{aligned} \tag{3.171}$$

Paying attention to (3.170) and $\omega \in J_1$ for (3.169), we need to show that

$$\underset{\sim}{\mathbf{P}}_t \int_0^t \psi_s\psi_s^\tau (\underset{\sim}{\mathbf{M}}_s - \underset{\sim}{\mathbf{M}})\, ds \underset{t\to\infty}{\to} 0, \quad \text{for } \omega \in J_1. \tag{3.172}$$

But this convergence can be proved completely similarly as for (3.48). □

Remark. Let $0 \le \varepsilon_s \le \varepsilon$. Then

$$\begin{aligned}\frac{1}{\varepsilon}\left\| \underset{\sim}{\mathbf{P}}_t \int_0^t \varepsilon_s \psi_s\psi_s^\tau\, ds \right\| &\le \frac{1}{\varepsilon}\int_0^t \varepsilon_s \|\underset{\sim}{\mathbf{P}}_t \psi_s\psi_s^\tau\|\, ds \le \int_0^t \|\underset{\sim}{\mathbf{P}}_t \psi_s\psi_s^\tau\|\, ds \\ &\le \int_0^t \|\underset{\sim}{\mathbf{P}}_t\|\, \|\underset{\sim}{\psi}_s\|^2\, ds \le \lambda_{\max}^t (r_t - 1) \le \frac{\mu\lambda_{\max}^t}{\lambda_{\min}^t},\end{aligned}$$

which gives various conditions for $\omega \in J_1$.

CHAPTER 4

Identification Algorithms of Stochastic Approximation Type and Adaptive Control

4.1. A CLASS OF IDENTIFICATION ALGORITHMS FOR DISCRETE TIME

In this section we continue to consider the system described by (3.52), (3.53), and (3.61) with $\mathbf{y}_n$, $\mathbf{u}_n$, $\boldsymbol{\varepsilon}_n$, $\underset{\sim}{\mathbf{F}}_n$, and $\mathscr{F}_n$ retaining their meaning. Notations (3.63) and (3.65) will remain unchanged, especially,

$$r_n = 1 + \sum_{i=0}^{n-1} \|\boldsymbol{\phi}_i\|^2, \qquad \beta_i = \frac{\|\boldsymbol{\phi}_i\|^2}{r_{i+1}}. \tag{4.1}$$

In Chapter 3, where least-squares identification was studied, the algorithms were characterized by the fact that the estimates $\underset{\sim}{\boldsymbol{\theta}}_n$ could be explicitly expressed in terms of the variables $\boldsymbol{\phi}_{i-1}$, $\mathbf{y}_i$, $i \le n$; this made probability methods available for consistency analysis. But in the least-squares case, for computing the new matrix gain, which is necessary for correcting previous estimates, one must solve equations for $\underset{\sim}{\mathbf{P}}_n$ and $\underset{\sim}{\boldsymbol{\theta}}_n$ simultaneously. In order to simplify the computation, we introduce here algorithms of stochastic approximation type.

In order to study their consistency, we shall again use a method combining probability and differential equation developed in Chapter 2.

Assume $\underset{\sim}{\mathbf{F}}_n$ is $\mathscr{F}_{n-1}$-measurable and

$$E \operatorname{tr} \underset{\sim}{\mathbf{F}}_n \underset{\sim}{\mathbf{F}}_n^{\tau} < \infty, \qquad \operatorname{tr} \underset{\sim}{\mathbf{F}}_n \underset{\sim}{\mathbf{F}}_n^{\tau} \le k_0^1 r_n^{\varepsilon}, \qquad \varepsilon \in [0,1) \tag{4.2}$$

with k_0^1 a constant.

Let $\underset{\sim}{\mathbf{R}}_{n+1}$ be a given $\mathscr{F}_n$-measurable $\lambda \times \lambda$ positive definite matrix, $n = 1, \ldots,$ with $\lambda = mp + lq + mr$.

Given an initial value $\underset{\sim}{\theta}_0$ let the estimate $\underset{\sim}{\theta}_n$ be defined recursively as follows:

$$\underset{\sim}{\theta}_{n+1} = \underset{\sim}{\theta}_n + \mathbf{R}_{n+1}\phi_n(y_{n+1}^\tau - \phi_n^\tau\underset{\sim}{\theta}_n) \tag{4.3}$$

As earlier, we consider only the case where

$$r_n \underset{n\to\infty}{\to} \infty, \qquad \sum_{i=0}^{\infty}\beta_i = \infty. \tag{4.4}$$

Denote

$$b_{n+1} = (1 - \phi_n^\tau\mathbf{R}_{n+1}\phi_n), \tag{4.5}$$

then

$$y_n - \underset{\sim}{\theta}_n^\tau\phi_{n-1} = y_n - [\underset{\sim}{\theta}_{n-1} + \mathbf{R}_n\phi_{n-1}(y_n^\tau - \phi_{n-1}\underset{\sim}{\theta}_{n-1})]^\tau\phi_{n-1}$$

$$= (y_n - \underset{\sim}{\theta}_{n-1}^\tau\phi_{n-1})(1 - \phi_{n-1}^\tau\mathbf{R}_n\phi_{n-1}) = b_n(y_n - \underset{\sim}{\theta}_{n-1}^\tau\phi_{n-1}).$$

$$\tag{4.6}$$

Hence for $b_n > 0$,

$$\tilde{\underset{\sim}{\theta}}_{n+1} \triangleq \underset{\sim}{\theta} - \underset{\sim}{\theta}_{n+1} = \tilde{\underset{\sim}{\theta}}_n - \frac{1}{b_{n+1}}\mathbf{R}_{n+1}\phi_n(y_{n+1} - \underset{\sim}{\theta}_{n+1}^\tau\phi_n)^\tau. \tag{4.7}$$

We retain the notation (3.70) and make an agreement that $\underset{\sim}{\theta}_i = 0$, $y_i = 0$, $\mathbf{w}_i = 0$, $i < 0$.

From the proof of Lemma 3.4, it is easy to see that we still have

$$\underset{\sim}{C}(z)\xi_n = \tilde{\underset{\sim}{\theta}}_n^\tau\phi_{n-1} \tag{4.8}$$

$$\xi_{n+1} = \underset{\sim}{G}\sum_{j=0}^{n}\mathbf{F}^{n-j}\underset{\sim}{G}\tilde{\underset{\sim}{\theta}}_{j+1}^\tau\phi_j + \underset{\sim}{G}\mathbf{F}^{n+1}\underset{\sim}{G}\tilde{\underset{\sim}{\theta}}_0^\tau\phi_{-1}. \tag{4.9}$$

Then by (4.7) and (3.70), it follows that

$$\tilde{\underset{\sim}{\theta}}_{n+1} = \tilde{\underset{\sim}{\theta}}_n - \frac{1}{b_{n+1}}\mathbf{R}_{n+1}\phi_n(\xi_{n+1} + \mathbf{F}_{n+1}\mathbf{w}_{n+1})^\tau \tag{4.10}$$

In (2.31), (2.32), and (2.35)–(2.37), we replace a_i by β_i but do not change the notation t_n, $\underset{\sim}{\mathbf{A}}_t^0$, $\overline{\underset{\sim}{\mathbf{A}}}_t$.

Now we have

$$t_n = \sum_{i=0}^{n-1} \beta_i$$

and $\underset{\sim}{\mathbf{A}}_t^0$ is the linear interpolation of $\{\underset{\sim}{\mathbf{A}}_n\}$ with interpolating length $\{\beta_n\}$.

Theorem 4.1. *Let the dynamic system be given by* (3.52), (3.53), (3.62), *and* (4.2) *and the algorithm by* (3.63), (3.65), *and* (4.3). *Assume that*:

A_1. *As $n \to \infty$, $1/b_n r_n$ is nonincreasing, matrix $r_n \underset{\sim}{\mathbf{R}}_n$ is nondecreasing, and there are constants $b > 0$, $k_0 > 0$, $k_1 > 0$ such that*

$$b < b_n, \quad \forall n, \qquad \frac{k_0}{r_n}\underset{\sim}{\mathbf{I}} \leq \underset{\sim}{\mathbf{R}}_n \leq \frac{k_1}{r_n}\underset{\sim}{\mathbf{I}}, \quad \forall n, \tag{4.11}$$

B_1. *$r = 0$, or $r > 0$ with the transfer function $\underset{\sim}{\mathbf{C}}^{-1}(z)$ strictly positive real and all zeros of* det $\underset{\sim}{\mathbf{C}}(z)$ *lie outside of the closed unit disk,*

C_1. *There are* a.s. *finite random variables $\alpha > 0$, $\beta > 0$, and $T > 0$ such that*

$$\sum_{i=m(t)}^{m(t+\alpha)} \frac{\phi_i \phi_i^\tau}{r_{i+1}} \geq \beta I, \qquad \forall t \geq T. \tag{4.12}$$

Then

$$\underset{\sim}{\boldsymbol{\theta}}_n \underset{n\to\infty}{\longrightarrow} \underset{\sim}{\boldsymbol{\theta}} \text{ a.s.}$$

To begin with we prove some lemmas, but we first note that (4.12) implies $r_n \to \infty$, since if r_n were bounded then $\sum_{i=0}^{\infty}\beta_i < \infty$, which contradicts with (4.12).

Lemma 4.1. *Under the conditions of Theorem* 4.1,

$$E\left(\frac{v_{n+1}}{\mathscr{F}_n}\right) \leq v_n - 2E\left[\frac{\phi_n^\tau \underset{\sim}{\tilde{\boldsymbol{\theta}}}_{n+1}\boldsymbol{\xi}_{n+1}}{b_{n+1}r_{n+1}}\Big/\mathscr{F}_n\right] + \frac{2}{b_{n+1}r_{n+1}}\phi_n^\tau \underset{\sim}{\mathbf{R}}_{n+1}\phi_n \operatorname{tr} \underset{\sim}{\mathbf{F}}_{n+1}\underset{\sim}{\mathbf{F}}_{n+1}^\tau,$$

where

$$v_n = \frac{1}{r_n}\operatorname{tr}\underset{\sim}{\tilde{\boldsymbol{\theta}}}_n^\tau \underset{\sim}{\mathbf{R}}_n^{-1}\underset{\sim}{\tilde{\boldsymbol{\theta}}}_n. \tag{4.13}$$

Proof. Noticing

$$\frac{\|\boldsymbol{\phi}_n\|}{r_{n+1}} \leq 1, \quad \frac{\|\boldsymbol{\phi}_n\| \, \|\mathbf{y}_{n-i}\|}{r_{n+1}} \leq 1,$$

$$\frac{\|\boldsymbol{\phi}_n\| \, \|\mathbf{u}_{n-j}\|}{r_{n+1}} \leq 1, \qquad i = 0, \ldots, p-1, \; j = 0, \ldots, q-1 \tag{4.14}$$

We find that

$$E\left\| \frac{\boldsymbol{\phi}_n \mathbf{y}_{n+1}^{\tau}}{r_{n+1}} \right\|^2 < \infty \tag{4.15}$$

if we substitute $\mathbf{y}_{n+1}$ by its expression given by (3.53). Then by (4.11) and (4.15) from (4.3), we can conclude that

$$E\|\underset{\sim}{\boldsymbol{\theta}}_n\|^2 < \infty, \quad E\|\tilde{\underset{\sim}{\boldsymbol{\theta}}}_n\|^2 < \infty, \qquad \forall n \tag{4.16}$$

by induction.

Hence

$$E\left\| \frac{\tilde{\underset{\sim}{\boldsymbol{\theta}}}_n^{\tau} \boldsymbol{\phi}_{n-1}}{r_n} \right\|^2 \leq E\|\tilde{\underset{\sim}{\boldsymbol{\theta}}}_n\|^2 < \infty \tag{4.17}$$

and, similar to (4.15),

$$E\frac{\|\mathbf{y}_{n+1}\|^2}{r_{n+1}} < \infty.$$

Hence

$$E\frac{\|\boldsymbol{\xi}_n\|^2}{r_n} \leq 2E\frac{\|\mathbf{y}_n - \underset{\sim}{\boldsymbol{\theta}}_n^{\tau} \boldsymbol{\phi}_{n-1}\|^2}{r_n} + 2E\frac{\|\underset{\sim}{\mathbf{F}}_n \mathbf{w}_n\|^2}{r_n} < \infty. \tag{4.18}$$

From (4.10), we have

$$\begin{aligned}
\frac{\operatorname{tr} \tilde{\underset{\sim}{\boldsymbol{\theta}}}_{n+1}^{\tau} \underset{\sim}{\mathbf{R}}_{n+1}^{-1} \tilde{\underset{\sim}{\boldsymbol{\theta}}}_{n+1}}{r_{n+1}} = \frac{1}{r_{n+1}} \Bigg[& \operatorname{tr} \tilde{\underset{\sim}{\boldsymbol{\theta}}}_n^{\tau} \underset{\sim}{\mathbf{R}}_{n+1}^{-1} \tilde{\underset{\sim}{\boldsymbol{\theta}}}_n - \frac{2\boldsymbol{\phi}_n^{\tau} \tilde{\underset{\sim}{\boldsymbol{\theta}}}_{n+1} \boldsymbol{\xi}_{n+1}}{b_{n+1}} - \frac{2\mathbf{w}_{n+1}^{\tau} \underset{\sim}{\mathbf{F}}_{n+1}^{\tau} \tilde{\underset{\sim}{\boldsymbol{\theta}}}_{n+1}^{\tau} \boldsymbol{\phi}_n}{b_{n+1}} \\
& - \frac{\boldsymbol{\phi}_n^{\tau} \underset{\sim}{\mathbf{R}}_{n+1} \boldsymbol{\phi}_n}{b_{n+1}^2} \|\boldsymbol{\xi}_{n+1} + \underset{\sim}{\mathbf{F}}_{n+1} \mathbf{w}_{n+1}\|^2 \Bigg] \\
\leq \frac{1}{r_{n+1}} \Bigg[& \operatorname{tr} \tilde{\underset{\sim}{\boldsymbol{\theta}}}_n^{\tau} \underset{\sim}{\mathbf{R}}_{n+1}^{-1} \tilde{\underset{\sim}{\boldsymbol{\theta}}}_n - \frac{2\boldsymbol{\phi}_n^{\tau} \underset{\sim}{\boldsymbol{\theta}}_{n+1}^{\tau} \boldsymbol{\xi}_{n+1}}{b_{n+1}} - 2\frac{\mathbf{w}_{n+1}^{\tau} \underset{\sim}{\mathbf{F}}_{n+1}^{\tau} \tilde{\underset{\sim}{\boldsymbol{\theta}}}_{n+1}^{\tau} \boldsymbol{\phi}_n}{b_{n+1}} \Bigg],
\end{aligned} \tag{4.19}$$

each term of which has finite expectation by (4.11) and (4.16)–(4.18). Taking account of $\mathscr{F}_n$-measurability of $\mathbf{y}_{n+1} - \underset{\sim}{\mathbf{F}}_{n+1}\mathbf{w}_{n+1}$, $\underset{\sim}{\mathbf{F}}_{n+1}$, and b_{n+1} by (4.6),

$$E\left(\frac{\mathbf{w}_{n+1}^{\tau}\underset{\sim}{\mathbf{F}}_{n+1}^{\tau}\tilde{\underset{\sim}{\boldsymbol{\theta}}}_{n+1}^{\tau}\boldsymbol{\phi}_n}{\mathscr{F}_n}\right)$$

$$= E\left[\mathbf{w}_{n+1}^{\tau}\underset{\sim}{\mathbf{F}}_{n+1}^{\tau}\left(\tilde{\underset{\sim}{\boldsymbol{\theta}}}_n^{\tau}\boldsymbol{\phi}_n - \frac{\boldsymbol{\phi}_n^{\tau}\underset{\sim}{\mathbf{R}}_{n+1}\boldsymbol{\phi}_n}{b_{n+1}}(\boldsymbol{\xi}_{n+1} + \underset{\sim}{\mathbf{F}}_{n+1}\mathbf{w}_{n+1})\right)/\mathscr{F}_n\right]$$

$$= -E\left[\frac{1}{b_{n+1}}\mathbf{w}_{n+1}^{\tau}\underset{\sim}{\mathbf{F}}_{n+1}^{\tau}\boldsymbol{\phi}_n^{\tau}\underset{\sim}{\mathbf{R}}_{n+1}\boldsymbol{\phi}_n(\boldsymbol{\xi}_{n+1} + \underset{\sim}{\mathbf{F}}_{n+1}\mathbf{w}_{n+1})/\mathscr{F}_n\right]$$

$$= -\frac{\boldsymbol{\phi}_n^{\tau}\underset{\sim}{\mathbf{R}}_{n+1}\boldsymbol{\phi}_n}{b_{n+1}}E\left[\frac{\mathbf{w}_{n+1}^{\tau}\underset{\sim}{\mathbf{F}}_{n+1}^{\tau}(\mathbf{y}_{n+1} - \underset{\sim}{\mathbf{F}}_{n+1}\mathbf{w}_{n+1} - \underset{\sim}{\boldsymbol{\theta}}_{n+1}^{\tau}\boldsymbol{\phi}_n + \underset{\sim}{\mathbf{F}}_{n+1}\mathbf{w}_{n+1})}{\mathscr{F}_n}\right]$$

$$= -\boldsymbol{\phi}_n^{\tau}\underset{\sim}{\mathbf{R}}_{n+1}\boldsymbol{\phi}_n E\left[\frac{\mathbf{w}_{n+1}^{\tau}\underset{\sim}{\mathbf{F}}_{n+1}^{\tau}(\mathbf{y}_{n+1} - \underset{\sim}{\mathbf{F}}_{n+1}\mathbf{w}_{n+1} - \underset{\sim}{\boldsymbol{\theta}}_n^{\tau}\boldsymbol{\phi}_n + \underset{\sim}{\mathbf{F}}_{n+1}\mathbf{w}_{n+1})}{\mathscr{F}_n}\right]$$

$$= -\boldsymbol{\phi}_n^{\tau}\underset{\sim}{\mathbf{R}}_{n+1}\boldsymbol{\phi}_n E\left(\frac{\|\underset{\sim}{\mathbf{F}}_{n+1}\mathbf{w}_{n+1}\|^2}{\mathscr{F}_n}\right)$$

$$= -\boldsymbol{\phi}_n^{\tau}\underset{\sim}{\mathbf{R}}_{n+1}\boldsymbol{\phi}_n E\left(\frac{\operatorname{tr}\underset{\sim}{\mathbf{F}}_{n+1}\mathbf{w}_{n+1}\mathbf{w}_{n+1}^{\tau}\underset{\sim}{\mathbf{F}}_{n+1}^{\tau}}{\mathscr{F}_n}\right)$$

$$= -\boldsymbol{\phi}_n^{\tau}\underset{\sim}{\mathbf{R}}_{n+1}\boldsymbol{\phi}_n \operatorname{tr}\underset{\sim}{\mathbf{F}}_{n+1}\underset{\sim}{\mathbf{F}}_{n+1}^{\tau}. \tag{4.20}$$

By Condition A_1, $\underset{\sim}{\mathbf{R}}_{n+1}^{-1}/r_{n+1}$ is nonincreasing as $n \to \infty$, then substituting (4.20) into (4.19) yields the assertion of the lemma. □

Lemma 4.2. *Under the conditions of Theorem* 4.1, *there exists an* a.s. *finite random variable* γ *such that*

$$\|\underset{\sim}{\boldsymbol{\theta}}_n\| \leq \gamma, \qquad \forall n \tag{4.21}$$

and

$$\sum_{i=0}^{\infty}\frac{\|\tilde{\underset{\sim}{\boldsymbol{\theta}}}_i^{\tau}\boldsymbol{\phi}_{i-1}\|}{r_i b_i} < \infty \text{ a.s.} \tag{4.22}$$

Proof. The transfer function $\underset{\sim}{\mathbf{C}}^{-1}(z)$ of (4.8) is strictly positive real. This guarantees the existence of constants $k_3 > 0$, $k_4 > 0$ such that

$$\sum_{i=1}^{n}\boldsymbol{\phi}_{i-1}^{\tau}\tilde{\underset{\sim}{\boldsymbol{\theta}}}_i[\underset{\sim}{\mathbf{C}}^{-1}(z) - k_3\underset{\sim}{\mathbf{I}}]\tilde{\underset{\sim}{\boldsymbol{\theta}}}_i^{\tau}\boldsymbol{\phi}_{i-1} + k_4 \geq 0, \qquad \forall n$$

or

$$\sum_{i=1}^{n} \left(\phi_{i-1}^{\tau} \tilde{\underset{\sim}{\theta}}_i \xi_i - k_3 \|\tilde{\underset{\sim}{\theta}}_i^{\tau} \phi_{i-1}\|^2 \right) + k_4 \geq 0, \qquad \forall n.$$

Denote

$$S_n = 2 \sum_{i=1}^{n} \left(\phi_{i-1}^{\tau} \tilde{\underset{\sim}{\theta}}_i \xi_i - k_3 \|\tilde{\underset{\sim}{\theta}}_i^{\tau} \phi_{i-1}\|^2 \right) + 2k_4. \tag{4.23}$$

From (4.2) and (4.11) and the definition of r_i, it follows that

$$\sum_{i=1}^{\infty} \left(\frac{\phi_{i-1}^{\tau} \underset{\sim}{\mathbf{R}}_i \phi_{i-1}}{b_i r_i} \operatorname{tr} \underset{\sim}{\mathbf{F}}_i \underset{\sim}{\mathbf{F}}_i^{\tau} \right) \leq \frac{k_0^1 k_1}{b} \sum_{i=1}^{\infty} \frac{\|\phi_{i-1}\|^2}{r_i^{2-\varepsilon}} \leq \frac{k_0^1 k_1}{b(1-\varepsilon)},$$

the last inequality of which comes from (3.102) and (3.103). This makes m_n, defined in the sequel, meaningful

$$m_n = v_n + \frac{S_n}{r_n b_n} + E\left[\sum_{i=0}^{\infty} \frac{2\phi_i^{\tau} \underset{\sim}{\mathbf{R}}_{i+1} \phi_i}{b_{i+1} r_{i+1}} \operatorname{tr} \underset{\sim}{\mathbf{F}}_{i+1} \underset{\sim}{\mathbf{F}}_{i+1}^{\tau} / \mathscr{F}_n \right]$$
$$- \sum_{i=0}^{n-1} \frac{2\phi_i^{\tau} \underset{\sim}{\mathbf{R}}_{i+1} \phi_i}{b_{i+1} r_{i+1}} \operatorname{tr} \underset{\sim}{\mathbf{F}}_{i+1} \underset{\sim}{\mathbf{F}}_{i+1}^{\tau} + 2k_3 \sum_{i=0}^{n} \frac{\|\tilde{\underset{\sim}{\theta}}_i^{\tau} \phi_{i-1}\|^2}{b_i r_i}. \tag{4.24}$$

By Lemma 4.1 and the fact $1/r_{n+1} b_{n+1} \leq 1/r_n b_n$, it is seen that

$$E\left(\frac{m_{n+1}}{\mathscr{F}_n} \right) \leq v_n - 2E\left(\frac{\phi_n^{\tau} \tilde{\underset{\sim}{\theta}}_{n+1} \xi_{n+1}}{b_{n+1} r_{n+1}} \Big/ \mathscr{F}_n \right) + \frac{2}{b_{n+1} r_{n+1}} \phi_n^{\tau} \underset{\sim}{\mathbf{R}}_{n+1} \phi_n \operatorname{tr} \underset{\sim}{\mathbf{F}}_{n+1} \underset{\sim}{\mathbf{F}}_{n+1}^{\tau}$$
$$+ \frac{S_n}{b_{n+1} r_{n+1}} + \frac{2}{r_{n+1} b_{n+1}} E\left[\frac{\left(\phi_n^{\tau} \tilde{\underset{\sim}{\theta}}_{n+1} \xi_{n+1} - k_3 \|\tilde{\underset{\sim}{\theta}}_{n+1}^{\tau} \phi_n\|^2 \right)}{\mathscr{F}_n} \right]$$
$$+ E\left[\sum_{i=0}^{\infty} \frac{2\phi_i^{\tau} \underset{\sim}{\mathbf{R}}_{i+1} \phi_i}{b_{i+1} r_{i+1}} \operatorname{tr} \underset{\sim}{\mathbf{F}}_{i+1} \underset{\sim}{\mathbf{F}}_{i+1}^{\tau} / \mathscr{F}_n \right]$$
$$- 2 \sum_{i=0}^{n} \frac{\phi_i^{\tau} \underset{\sim}{\mathbf{R}}_{i+1} \phi_i}{b_{i+1} r_{i+1}} \operatorname{tr} \underset{\sim}{\mathbf{F}}_{i+1} \underset{\sim}{\mathbf{F}}_{i+1}^{\tau}$$
$$+ 2k_3 \sum_{i=0}^{n+1} E\left(\frac{\|\tilde{\theta}_i \phi_{i-1}\|^2}{b_i r_i} / \mathscr{F}_n \right) \leq m_n.$$

Then (4.22) follows from Theorem 1.11, which also leads to the boundedness of v_n for fixed ω. But according to (4.11),

$$v_n = \frac{1}{r_n} \operatorname{tr} \underset{\sim}{\theta}_n^{\tau} \underset{\sim}{\mathbf{R}}_n^{-1} \tilde{\underset{\sim}{\theta}}_n \geq \frac{1}{k_1} \operatorname{tr} \tilde{\underset{\sim}{\theta}}_n^{\tau} \tilde{\underset{\sim}{\theta}}_n.$$

From here, (4.21) follows. □

Lemma 4.3. *Under conditions of Theorem* 4.1,

$$v_n \underset{n\to\infty}{\to} v \text{ a.s.,} \tag{4.25}$$

where v_n is defined in Lemma 4.1.

Proof. Since $m_n \underset{n\to\infty}{\to} M < \infty$, from (4.24) it is easy to see in order to prove the lemma we only need to show

$$\frac{S_n}{r_n b_n} \underset{n\to\infty}{\to} 0, \tag{4.26}$$

for which it is sufficient to prove that

$$\frac{1}{b_n r_n} \sum_{i=1}^{n} \left(\boldsymbol{\phi}_{i-1}^{\tau} \underset{\sim}{\tilde{\boldsymbol{\theta}}}_i \boldsymbol{\xi}_i - k_3 \|\underset{\sim}{\tilde{\boldsymbol{\theta}}}_i^{\tau} \boldsymbol{\phi}_{i-1}\|^2 \right) \underset{n\to\infty}{\to} 0, \tag{4.27}$$

because of (4.23), $b_n \geq b > 0$, and $r_n \to \infty$.

Applying Lemma 3.2 to (4.22) leads to

$$\frac{1}{r_n b_n} \sum_{i=0}^{n} \|\tilde{\boldsymbol{\theta}}_i^{\tau} \boldsymbol{\phi}_{i-1}\|^2 \underset{n\to\infty}{\to} 0 \tag{4.28}$$

hence by (4.27) and (4.28) for (4.26), it is enough to show

$$\frac{1}{b_n r_n} \sum_{i=1}^{n} \boldsymbol{\phi}_{i-1}^{\tau} \underset{\sim}{\tilde{\boldsymbol{\theta}}}_i \boldsymbol{\xi}_i \underset{n\to\infty}{\to} 0, \tag{4.29}$$

for which the sufficient condition is

$$\frac{1}{b_n r_n} \sum_{i=1}^{n} \boldsymbol{\phi}_{i-1}^{\tau} \underset{\sim}{\tilde{\boldsymbol{\theta}}}_i \left(\underset{\sim}{\mathbf{G}} \sum_{j=0}^{i-1} \underset{\sim}{\mathbf{F}}^{i-j-1} \underset{\sim}{\mathbf{G}}^{\tau} \underset{\sim}{\tilde{\boldsymbol{\theta}}}_{j+1}^{\tau} \boldsymbol{\phi}_j + \underset{\sim}{\mathbf{G}} \underset{\sim}{\mathbf{F}}^i \underset{\sim}{\mathbf{G}}^{\tau} \underset{\sim}{\tilde{\boldsymbol{\theta}}}_0^{\tau} \boldsymbol{\phi}_{-1} \right) \underset{n\to\infty}{\to} 0 \tag{4.30}$$

by (4.9). But by (2.82) and (4.28), we have

$$\begin{aligned}
\frac{1}{b_n r_n} \left\| \sum_{i=1}^{n} \boldsymbol{\phi}_{i-1}^{\tau} \underset{\sim}{\tilde{\boldsymbol{\theta}}}_i \underset{\sim}{\mathbf{G}} \underset{\sim}{\mathbf{F}}^i \underset{\sim}{\mathbf{G}}^{\tau} \underset{\sim}{\tilde{\boldsymbol{\theta}}}_0^{\tau} \boldsymbol{\phi}_{-1} \right\| &\leq \frac{k_2}{b_n r_n} \sum_{i=1}^{n} \|\boldsymbol{\phi}_{i-1}^{\tau} \underset{\sim}{\tilde{\boldsymbol{\theta}}}_i\| \rho^i \|\underset{\sim}{\tilde{\boldsymbol{\theta}}}_0^{\tau} \boldsymbol{\phi}_{-1}\| \\
&\leq \frac{k_2}{b_n r_n} \left(\sum_{i=1}^{n} \|\boldsymbol{\phi}_{i-1}^{\tau} \underset{\sim}{\tilde{\boldsymbol{\theta}}}_i\|^2 \sum_{i=1}^{n} \rho^{2i} \right)^{1/2} \\
&\leq \frac{k_2}{\sqrt{b_n r_n}} \left(\frac{1}{b_n r_n} \sum_{i=1}^{n} \|\boldsymbol{\phi}_{i-1}^{\tau} \underset{\sim}{\tilde{\boldsymbol{\theta}}}_i\|^2 \frac{1}{1-\rho^2} \right)^{1/2} \underset{n\to\infty}{\to} 0.
\end{aligned} \tag{4.31}$$

Hence for (4.30), it is enough to prove

$$\frac{1}{b_n r_n}\left\|\sum_{i=1}^{n}\phi_{i-1}^{\tau}\tilde{\theta}_i\left(\underset{\sim}{G}\sum_{j=0}^{i-1}\underset{\sim}{F}^{i-j-1}\underset{\sim}{G}^{\tau}\tilde{\theta}_{j+1}^{\tau}\phi_j\right)\right\|$$

$$\leq \frac{k_2}{b_n r_n}\sum_{i=1}^{n}\|\phi_{i-1}^{\tau}\tilde{\underset{\sim}{\theta}}_i\|\sum_{j=0}^{i-1}\rho^{i-j-1}\|\tilde{\underset{\sim}{\theta}}_{j+1}^{\tau}\phi_j\| \underset{n\to\infty}{\to} 0. \tag{4.32}$$

By Lemma 3.2, for this we only need to show

$$\sum_{i=1}^{\infty}\frac{1}{b_i r_i}\left(\|\phi_{i-1}^{\tau}\tilde{\underset{\sim}{\theta}}_i\|\sum_{j=0}^{i-1}\rho^{i-j-1}\|\tilde{\underset{\sim}{\theta}}_{j+1}^{\tau}\phi_j\|\right) < \infty, \tag{4.33}$$

but by Lemma (4.2) we know

$$\sum_{i=1}^{\infty}\frac{1}{b_i r_i}\left(\|\phi_{i-1}^{\tau}\tilde{\underset{\sim}{\theta}}_i\|\sum_{j=0}^{i-1}\rho^{i-j-1}\|\tilde{\underset{\sim}{\theta}}_{j+1}^{\tau}\phi_j\|\right)$$

$$= \sum_{i=1}^{\infty}\frac{1}{b_i r_i}\|\phi_{i-1}^{\tau}\tilde{\underset{\sim}{\theta}}_i\|\sum_{j=1}^{i}\rho^{i-j}\|\tilde{\underset{\sim}{\theta}}_j^{\tau}\phi_{j-1}\|$$

$$\leq \sum_{i=1}^{\infty}\frac{1}{\sqrt{b_i r_i}}\|\phi_{i-1}^{\tau}\tilde{\underset{\sim}{\theta}}_i\|\sum_{j=1}^{i}\frac{\rho^{i-j}}{\sqrt{b_j r_j}}\|\phi_{j-1}^{\tau}\tilde{\underset{\sim}{\theta}}_j\|$$

$$\leq \frac{1}{2}\sum_{i=1}^{\infty}\sum_{j=1}^{i}\frac{\|\phi_{i-1}^{\tau}\tilde{\underset{\sim}{\theta}}_i\|^2}{b_i r_i}\rho^{i-j} + \frac{1}{2}\sum_{i=1}^{\infty}\sum_{j=1}^{i}\frac{\|\phi_{j-1}^{\tau}\tilde{\underset{\sim}{\theta}}_j\|^2}{b_j r_j}\rho^{i-j}$$

$$\leq \frac{1}{2(1-\rho)}\sum_{i=1}^{\infty}\frac{\|\phi_{i-1}^{\tau}\tilde{\underset{\sim}{\theta}}_i\|^2}{b_i r_i} + \frac{1}{2}\sum_{j=1}^{\infty}\sum_{i=j}^{\infty}\frac{\|\phi_{j-1}^{\tau}\tilde{\underset{\sim}{\theta}}_j\|^2}{b_j r_j}\rho^{i-j}$$

$$\leq \frac{1}{1-\rho}\sum_{i=1}^{\infty}\frac{\|\phi_{i-1}^{\tau}\tilde{\underset{\sim}{\theta}}_i\|^2}{b_i r_i} < \infty \text{ a.s.} \tag{4.34}$$

This proves (4.33) and hence the lemma. □

By using (4.9) and (4.10), $\tilde{\theta}_{n+1}$ can be expressed by

$$\tilde{\theta}_{n+1} = \tilde{\theta}_0 - \sum_{i=0}^{n} \frac{1}{b_{i+1}} \mathbf{R}_{i+1}\phi_i \xi_{i+1}^{\tau} - \sum_{i=1}^{n+1} \frac{1}{b_i} \mathbf{R}_i \phi_{i-1} \mathbf{w}_i^{\tau} \mathbf{F}_i^{\tau}$$

$$= \tilde{\theta}_0 - \sum_{i=0}^{n} \frac{1}{b_{i+1}} \mathbf{R}_{i+1}\phi_i \sum_{j=0}^{i} \phi_j^{\tau} \tilde{\theta}_{j+1} \mathbf{G} \mathbf{F}^{\tau(i-j)} \mathbf{G}^{\tau}$$

$$- \sum_{i=1}^{n+1} \frac{1}{b_i} \mathbf{R}_i \phi_{i-1} \phi_{-1}^{\tau} \tilde{\theta}_0 \mathbf{G} \mathbf{F}^{\tau i} \mathbf{G}^{\tau} - \sum_{i=1}^{n+1} \frac{1}{b_i} \mathbf{R}_i \phi_{i-1} \mathbf{w}_i^{\tau} \mathbf{F}_i^{\tau}. \tag{4.35}$$

Denote

$$\mathbf{G}_{n+1,i} = -\frac{1}{\beta_i} \sum_{j=i}^{n} \frac{\mathbf{R}_{j+1}}{b_{j+1}} \phi_j \phi_i^{\tau} \tilde{\theta}_{i+1} \mathbf{G} \mathbf{F}^{\tau(j-i)} \mathbf{G}^{\tau}, \qquad \mathbf{G}_{nn} = 0,$$

and use $\mathbf{J}_{n+1}$ and $\mathbf{H}_{n+1}$ for the last two terms in (4.35). Then

$$\tilde{\theta}_{n+1} = \tilde{\theta}_0 - \sum_{j=0}^{n} \frac{1}{b_{i+1}} \mathbf{R}_{i+1}\phi_i \sum_{i=j}^{n} \phi_j^{\tau} \tilde{\theta}_{j+1} \mathbf{G} \mathbf{F}^{\tau(i-j)} \mathbf{G}^{\tau} + \mathbf{J}_{n+1} + \mathbf{H}_{n+1}$$

$$= \tilde{\theta}_0 + \sum_{i=0}^{n} \beta_i \mathbf{G}_{n+1,i} + \mathbf{J}_{n+1} + \mathbf{H}_{n+1}. \tag{4.36}$$

For fixed i by $\mathbf{G}_{t,i}^0$ denote the linear interpolation of $\{\mathbf{G}_{n,i}\}$ on $t \geq t_i$ with interpolating length $\{\beta_n\}$. By (2.36),

$$\mathbf{G}_{t_{k},i}^0 = \mathbf{G}_{k,i} \quad \text{for } t = t_k.$$

For fixed t by $\overline{\mathbf{G}}_{t,s}$ denote the constant interpolation of $\{\mathbf{G}_{t,i}^0\}$ on $[0, t]$ with interpolating length $\{\beta_i\}$. Then for $t = t_{n+1}$

$$\int_0^t \overline{\mathbf{G}}_{t,s}\, ds = -\sum_{i=0}^{n} \frac{1}{b_{j+1}} \mathbf{R}_{j+1}\phi_j \sum_{j=i}^{n} \phi_i^{\tau} \tilde{\theta}_{i+1} \mathbf{G} \mathbf{F}^{\tau(j-i)} \mathbf{G}^{\tau}. \tag{4.37}$$

For $t \geq 0$, let $\mathbf{J}_t^0$ and $\mathbf{H}_t^0$ be defined by (2.35) with a_n replaced by β_n. Now define the interpolating function $\tilde{\theta}(t)$

$$\tilde{\theta}(t) \triangleq \tilde{\theta}_0 + \int_0^t \overline{\mathbf{G}}_{t,s}^0\, ds + \mathbf{J}_t^0 + \mathbf{H}_t^0. \tag{4.38}$$

For fixed i, $\mathbf{G}_{t,i}^0$ is continuous in t and as s varies in $[0, t)$ $\overline{\mathbf{G}}_{t,s}^0$ takes only finite values $\mathbf{G}_{t0}^0, \ldots, \mathbf{G}_{t,m(t)}^0$. Hence $\int_0^t \overline{\mathbf{G}}_{t,s}^0\, ds$ is continuous function of t and so is $\tilde{\theta}(t)$.

Clearly,

$$\tilde{\theta}(t_n) = \tilde{\theta}_n. \tag{4.39}$$

Define

$$\tilde{\underset{\sim}{\theta}}_n(t) = \tilde{\underset{\sim}{\theta}}(t+n), \qquad t \geq 0. \tag{4.40}$$

Lemma 4.4. *Under the conditions of Theorem 4.1 for almost all fixed $\omega \in \Omega$, there is a subsequence $\tilde{\underset{\sim}{\theta}}_{n_k}(t)$ of $\{\tilde{\underset{\sim}{\theta}}_n(t)\}$ and a constant matrix $\underset{\sim}{\theta}^0$ which is the uniform limit of $\underset{\sim}{\theta}_{n_k}(t)$ on any finite interval.*

Proof. Let $t_n \leq t < t_{n+1}$. Then for $s \in [t_n, t]$

$$\begin{aligned}\overline{\underset{\sim}{\mathbf{G}}}{}^0_{t,s} = \underset{\sim}{\mathbf{G}}{}^0_{t,n} &= \frac{(t-t_n)}{\beta_n}\underset{\sim}{\mathbf{G}}_{n+1,n} + \frac{(t_{n+1}-t)}{\beta_n}\underset{\sim}{\mathbf{G}}_{n,n} \\ &= -\frac{(t-t_n)\underset{\sim}{\mathbf{R}}_{n+1}}{\beta_n^2 b_{n+1}}\phi_n\phi_n^\tau\tilde{\underset{\sim}{\theta}}_{n+1}.\end{aligned} \tag{4.41}$$

By (4.11), (4.21), (2.82) and definition of β_n for $t \in [t_n, t_{n+1})$ we have

$$\begin{aligned}
&\left\|\int_0^t \overline{\underset{\sim}{\mathbf{G}}}{}^0_{t,s}\,ds - \int_0^{t_n}\overline{\underset{\sim}{\mathbf{G}}}{}^0_{t_n,s}\,ds\right\| \\
&\leq \left\|\int_{t_n}^t \overline{\underset{\sim}{\mathbf{G}}}{}^0_{t,s}\,ds\right\| + \left\|\int_0^{t_n}\left(\overline{\underset{\sim}{\mathbf{G}}}{}^0_{t,s} - \overline{\underset{\sim}{\mathbf{G}}}{}^0_{t_n,s}\right)ds\right\| \\
&\leq \frac{k_1\|\phi_n^\tau\tilde{\underset{\sim}{\theta}}_{n+1}\|\,\|\phi_n\|}{b_{n+1}r_{n+1}} + \sum_{i=0}^{n-1}\beta_i\|\underset{\sim}{\mathbf{G}}{}^0_{t,i} - \underset{\sim}{\mathbf{G}}{}^0_{t_n,i}\| \\
&= \frac{k_1}{b_{n+1}r_{n+1}}\|\phi_n^\tau\tilde{\underset{\sim}{\theta}}_{n+1}\|\,\|\phi_n\| \\
&\quad + \sum_{i=0}^{n-1}\beta_i\left\|\frac{t_{n+1}-t}{\beta_n}\underset{\sim}{\mathbf{G}}_{n,i} + \frac{t-t_n}{\beta_n}\underset{\sim}{\mathbf{G}}_{n+1,i} - \underset{\sim}{\mathbf{G}}_{n,i}\right\| \\
&= \frac{k_1}{b_{n+1}r_{n+1}}\|\phi_n^\tau\tilde{\underset{\sim}{\theta}}_{n+1}\|\,\|\phi_n\| + \sum_{i=0}^{n-1}\beta_i\left\|\frac{t-t_n}{\beta_n}(\underset{\sim}{\mathbf{G}}_{n+1,i} - \underset{\sim}{\mathbf{G}}_{n,i})\right\| \\
&= \frac{k_1}{b_{n+1}r_{n+1}}\|\phi_n^\tau\tilde{\underset{\sim}{\theta}}_{n+1}\|\,\|\phi_n\| + \sum_{i=0}^{n-1}\frac{t-t_n}{\beta_n}\left\|\frac{\underset{\sim}{\mathbf{R}}_{n+1}}{b_{n+1}}\phi_n\phi_i^\tau\tilde{\underset{\sim}{\theta}}_{i+1}\underset{\sim}{\mathbf{G}}\underset{\sim}{\mathbf{F}}^{\tau(n-i)}\underset{\sim}{\mathbf{G}}^\tau\right\| \\
&\leq \frac{k_1}{\sqrt{bb_{n+1}r_{n+1}}}\|\phi_n^\tau\tilde{\underset{\sim}{\theta}}_{n+1}\| + k_1k_2\sum_{i=0}^{n-1}\frac{1}{r_{n+1}b_{n+1}}\|\phi_n\|\,\|\phi_i^\tau\tilde{\underset{\sim}{\theta}}_{i+1}\|\rho^{n-i} \\
&\leq \frac{k_1}{\sqrt{bb_{n+1}r_{n+1}}}\|\phi_n^\tau\tilde{\underset{\sim}{\theta}}_{n+1}\| \\
&\quad + k_1k_2\left[\frac{1}{r_{n+1}^2b_{n+1}^2}\sum_{i=1}^{n-1}\|\phi_n\|^2\rho^{2(n-i)}\sum_{i=0}^{n-1}\|\phi_i^\tau\tilde{\underset{\sim}{\theta}}_{i+1}\|^2\right]^{1/2} \underset{n\to\infty}{\longrightarrow} 0.
\end{aligned} \tag{4.42}$$

The last convergence holds because of (4.22) and (4.28).

Lemma 3.2 guarantees convergence of the following series:

$$\sum_{i=1}^{\infty} E\left(\frac{1}{b_{i+1}^2}\|\underset{\sim}{\mathbf{R}}_{i+1}\|^2\|\phi_i\|^2\mathbf{w}_{i+1}^{\tau}\underset{\sim}{\mathbf{F}}_{i+1}^{\tau}\underset{\sim}{\mathbf{F}}_{i+1}\mathbf{w}_{i+1}/\mathscr{F}_i\right)$$

$$= \sum_{i=1}^{\infty}\frac{1}{b_{i+1}^2}\|\underset{\sim}{\mathbf{R}}_{i+1}\|^2\|\phi_i\|^2 \operatorname{tr}\underset{\sim}{\mathbf{F}}_{i+1}\underset{\sim}{\mathbf{F}}_{i+1}^{\tau} \le \frac{k_0^1 k_1^2}{b^2}\sum_{i=1}^{\infty}\frac{\|\phi_{i+1}\|^2}{r_{i+1}^{2-\varepsilon}} < \infty.$$

Hence the martingale $(\underset{\sim}{\mathbf{H}}_n, \mathscr{F}_n)$ tends to a finite limit as $n \to \infty$ by Theorem 1.13 and for fixed ω, $\underset{\sim}{\mathbf{H}}_t^0$ is uniformly bounded in t.

Similar to (2.45), it can be shown that for fixed t and any $\varepsilon > 0$

$$\|\underset{\sim}{\mathbf{H}}_{t+n+\Delta}^0 - \underset{\sim}{\mathbf{H}}_{t+n}^0\| < \varepsilon, \qquad \forall n \tag{4.43}$$

if Δ is small enough.

Taking notice of

$$\sum_{i=1}^{\infty}\left\|\frac{1}{b_i}\underset{\sim}{\mathbf{R}}_i\phi_{i-1}\phi_{-1}^{\tau}\tilde{\underset{\sim}{\theta}}_0\underset{\sim}{\mathbf{G}}\mathbf{F}^{\tau i}\underset{\sim}{\mathbf{G}}^{\tau}\right\| \le \frac{k_1 k_2}{b}\sum_{i=1}^{\infty}\frac{\|\phi_{i-1}\|}{r_i}\|\phi_{-1}\tilde{\underset{\sim}{\theta}}_0\|\rho^i$$

$$\le \frac{k_1 k_2}{b}\|\phi_{-1}\tilde{\underset{\sim}{\theta}}_0\|\left(\sum_{i=1}^{\infty}\frac{\|\phi_{i-1}\|^2}{r_i^2}\sum_{i=1}^{\infty}\rho^{2i}\right)^{1/2}$$

$$< \infty,$$

we find that $\underset{\sim}{\mathbf{J}}_n$ also goes to a finite limit as $n \to \infty$ and hence $\underset{\sim}{\mathbf{J}}_t^0$ is bounded on $[0, \infty)$ and for fixed t and any $\varepsilon > 0$

$$\|\underset{\sim}{\mathbf{J}}_{t+n+\Delta}^0 - \underset{\sim}{\mathbf{J}}_{t+n}^0\| < \varepsilon, \qquad \forall n \tag{4.44}$$

if Δ is sufficiently small.

Since $r_n \underset{n\to\infty}{\to} \infty$, for any $t \in [0, \infty)$, there is a t_n such that

$$t_n \le t < t_{n+1}$$

and consequently $\sum_{i=0}^{\infty}\beta_i = \infty$. Hence

$$\|\tilde{\underset{\sim}{\theta}}(t) - \tilde{\underset{\sim}{\theta}}(t_n)\| \le \left\|\int_0^t \overline{\underset{\sim}{\mathbf{G}}}_{t,s}^0\, ds - \int_0^{t_n}\overline{\underset{\sim}{\mathbf{G}}}_{t_n,s}^0\, ds\right\| + \|\underset{\sim}{\mathbf{J}}_t^0 - \underset{\sim}{\mathbf{J}}_{t_n}^0\| + \|\underset{\sim}{\mathbf{H}}_t^0 - \underset{\sim}{\mathbf{H}}_{t_n}^0\|,$$

which is bounded in n because of (4.42) and the boundedness of $\underset{\sim}{\mathbf{J}}_t^0$ and $\underset{\sim}{\mathbf{H}}_t^0$. But we know that $\tilde{\underset{\sim}{\theta}}(t_n) = \tilde{\underset{\sim}{\theta}}_n$ and that for fixed ω $\tilde{\underset{\sim}{\theta}}_n$ is bounded in n by Lemma 4.2; hence for fixed ω, $\tilde{\underset{\sim}{\theta}}(t)$ is bounded in $t \in [0, \infty)$. Now let

$\Delta > 0$, $t \in [0, \infty)$). We have

$$\|\tilde{\underset{\sim}{\theta}}_n(t+\Delta) - \tilde{\underset{\sim}{\theta}}_n(t)\| \le \left\| \int_0^{t+n+\Delta} \overline{\underset{\sim}{\mathbf{G}}}{}^0_{t+n+\Delta, s}\, ds - \int_0^{t+n} \overline{\underset{\sim}{\mathbf{G}}}{}^0_{t+n, s}\, ds \right\|$$

$$+ \|\underset{\sim}{\mathbf{J}}{}^0_{t+n+\Delta} - \underset{\sim}{\mathbf{J}}{}^0_{t+n}\| + \|\underset{\sim}{\mathbf{H}}{}^0_{t+n+\Delta} - \underset{\sim}{\mathbf{H}}{}^0_{t+n}\|$$

$$\le \left\| \int_0^{t_{m(t+n+\Delta)}} \overline{\underset{\sim}{\mathbf{G}}}{}^0_{t_{m(t+n+\Delta)}, s}\, ds - \int_0^{t_{m(t+n)}} \overline{\underset{\sim}{\mathbf{G}}}{}^0_{t_{m(t+n)}, s}\, ds \right\|$$

$$+ \left\| \int_0^{t+n+\Delta} \overline{\underset{\sim}{\mathbf{G}}}{}^0_{t+n+\Delta}\, ds - \int_0^{t_{m(t+n+\Delta)}} \overline{\underset{\sim}{\mathbf{G}}}{}^0_{t_{m(t+n+\Delta)}, s}\, ds \right\|$$

$$+ \left\| \int_0^{t+n} \overline{\underset{\sim}{\mathbf{G}}}{}^0_{t+n, s}\, ds - \int_0^{t_{m(t+n)}} \overline{\underset{\sim}{\mathbf{G}}}{}^0_{t_{m(t+n)}, s}\, ds \right\|$$

$$+ \|\underset{\sim}{\mathbf{J}}{}^0_{t+n+\Delta} - \underset{\sim}{\mathbf{J}}{}^0_{t+n}\| + \|\underset{\sim}{\mathbf{H}}{}^0_{t+n+\Delta} - \underset{\sim}{\mathbf{H}}{}^0_{t+n}\|. \tag{4.45}$$

But by (4.37), we obtain

$$\left\| \int_0^{t_{m(t+n+\Delta)}} \overline{\underset{\sim}{\mathbf{G}}}{}^0_{t_{m(t+n+\Delta)}, s}\, ds - \int_0^{t_{m(t+n)}} \overline{\underset{\sim}{\mathbf{G}}}{}^0_{t_{m(t+n)}, s}\, ds \right\|$$

$$= \left\| \sum_{j=0}^{m(t+n+\Delta)-1} \sum_{i=0}^{j} \frac{1}{b_{j+1}} \underset{\sim}{\mathbf{R}}_{j+1} \phi_j \phi_i^{\tau} \tilde{\underset{\sim}{\theta}}_{i+1} \underset{\sim}{\mathbf{G}} \underset{\sim}{\mathbf{F}}^{\tau(j-i)} \underset{\sim}{\mathbf{G}}^{\tau} \right.$$

$$\left. - \sum_{j=0}^{m(t+n)-1} \sum_{i=0}^{j} \frac{1}{b_{j+1}} \underset{\sim}{\mathbf{R}}_{j+1} \phi_j \phi_i^{\tau} \tilde{\underset{\sim}{\theta}}_{i+1} \underset{\sim}{\mathbf{G}} \underset{\sim}{\mathbf{F}}^{\tau(j-i)} \underset{\sim}{\mathbf{G}}^{\tau} \right\|$$

$$\le k_1 k_2 \sum_{j=m(t+n)}^{m(t+n+\Delta)-1} \sum_{i=0}^{j} \frac{\|\phi_j\| \, \|\phi_i^{\tau} \tilde{\underset{\sim}{\theta}}_{i+1}\| \rho^{j-i}}{b_{j+1} r_{j+1}}$$

$$\le k_1 k_2 \left(\sum_{j=m(t+n)}^{m(t+n+\Delta)-1} \sum_{i=0}^{j} \frac{\|\phi_j\|^2}{b_{j+1} r_{j+1}} \rho^{j-i} \right.$$

$$\left. \times \sum_{j=m(t+n)}^{m(t+n+\Delta)-1} \sum_{i=0}^{j} \frac{\|\phi_i^{\tau} \tilde{\underset{\sim}{\theta}}_{i+1}\|^2}{b_{i+1} r_{i+1}} \rho^{j-i} \right)^{1/2}$$

$$\le k_1 k_2 \left[\sum_{j=m(t+n)}^{m(t+n+\Delta)-1} \frac{\beta_j}{b_{j+1}(1-\rho)} \right.$$

$$\times \left(\sum_{i=m(t+n)}^{m(t+n+\Delta)-1} \frac{\|\phi_i^{\tau}\tilde{\theta}_{i+1}\|^2}{b_{i+1}r_{i+1}} \sum_{j=i}^{m(t+n+\Delta)-1} \rho^{j-i} \right.$$

$$\left. + \sum_{i=0}^{m(t+n)-1} \sum_{j=m(t+n)}^{m(t+n+\Delta)-1} \rho^{j-i} \frac{\|\phi_i^{\tau}\tilde{\theta}_{i+1}\|^2}{b_{i+1}r_{i+1}} \right]^{1/2}$$

$$\le \frac{k_1 k_2}{1-\rho} \left[\frac{1+\Delta}{b} \left(\sum_{i=m(t+n)}^{m(t+n+\Delta)-1} \frac{\|\tilde{\theta}_{i+1}^{\tau}\phi_i\|^2}{r_i} \right. \right.$$

$$\left. \left. + \sum_{i=0}^{N} \frac{\|\tilde{\theta}_{i+1}^{\tau}\phi_i\|^2}{r_i} \rho^{[m(t+n)-i]/2} + \sum_{i=N+1}^{m(t+n)-1} \frac{\|\tilde{\theta}_{i+1}^{\tau}\phi_i\|^2}{r_i} \right) \right]^{1/2} \underset{\substack{n\to\infty \\ N\to\infty}}{\longrightarrow} 0. \tag{4.46}$$

Similar properties hold for $\Delta < 0$.

Given $\varepsilon > 0$, by (4.42) there exists an N such that each of the first three terms on the right-hand side of (4.45) will be less than $\varepsilon/5$ if $n \ge N$. Then, by (4.43) and (4.44) the remaining two terms can also be made less than $\varepsilon/5$ uniformly in $n \ge N$ if Δ is small enough. Hence, for sufficiently small Δ, for $n \ge N$

$$\|\tilde{\theta}_n(t+\Delta) - \tilde{\theta}_n(t)\| \le \varepsilon. \tag{4.47}$$

This is obviously also valid for $n < N$ if Δ is sufficiently small, since for any n $\tilde{\theta}_n(t)$ is continuous.

Thus the family $\{\tilde{\theta}_n(t)\}$ is equicontinuous. Therefore, by Theorem 2.3, for any fixed ω there is a subsequence $\tilde{\theta}_{n_k}(t)$ which converges to a continuous matrix function $\theta(t)$ uniformly on any finite interval.

It is clear that

$$\sup_{t\ge s}\|H_t^0 - H_s^0\| \underset{s\to\infty}{\longrightarrow} 0, \qquad \sup_{t\ge s}\|J_t^0 - J_s^0\| \underset{s\to\infty}{\longrightarrow} 0, \tag{4.48}$$

since H_n and J_n both tend to finite limits as $n \to \infty$.

Now we return to treat (4.45) via an analysis along subsequences and using (4.42), (4.46), and (4.48) to obtain:

$$\|\underset{\sim}{\theta}(t+\Delta)-\underset{\sim}{\theta}(t)\| = \lim_{k\to\infty}\|\tilde{\underset{\sim}{\theta}}_{n_k}(t+\Delta)-\tilde{\underset{\sim}{\theta}}_{n_k}(t)\|$$

$$\le \lim_{k\to\infty}\frac{k_1k_2}{1-\rho}\left(\frac{\Delta}{b}\sum_{i=m(t+n_k)}^{m(t+n_k+\Delta)-1}\frac{\|\phi_i^\tau\tilde{\underset{\sim}{\theta}}_{i-1}\|^2}{b_{i+1}r_{i+1}}\right)^{1/2} = 0$$

the last equality of which follows from Lemma 4.2. Hence $\underset{\sim}{\theta}(t)$ is a constant matrix.

□

Proof of Theorem 4.1. Since $0 \le \beta_i \le 1$, we have

$$\sum_{i=m(t+n_k)}^{m(t+n_k+\alpha)}\beta_i = \beta_{m(t+n_k+\alpha)} + \sum_{i=0}^{m(t+n_k+\alpha)-1}\beta_i - \sum_{i=0}^{m(t+n_k)}\beta_i + \beta_{m(t+n_k)}$$

$$\le 2 + (t+n_k+\alpha) - (t+n_k) = \alpha + 2.$$

By the definition of b_{n+1}, $b_{n+1} = 1 - \phi_n^\tau\underset{\sim}{R}_{n+1}\phi_n \le 1$. Hence using Lemma 4.2, we know that

$$\lim_{k\to\infty}\left\|\sum_{i=m(t+n_k)}^{m(t+n_k+\alpha)}\frac{\phi_i\phi_i^\tau\tilde{\underset{\sim}{\theta}}_{i+1}}{r_{i+1}}\right\| \le \lim_{k\to\infty}\sum_{i=m(t+n_k)}^{m(t+n_k+\alpha)}\beta_i^{1/2}\frac{\|\phi_i^\tau\tilde{\underset{\sim}{\theta}}_{i+1}\|}{\sqrt{b_{i+1}r_{i+1}}}$$

$$\le \lim_{k\to\infty}\left[\sum_{i=m(t+n_k)}^{m(t+n_k+\alpha)}\beta_i\sum_{i=m(t+n_k)}^{m(t+n_k+\alpha)}\frac{\|\phi_i^\tau\tilde{\underset{\sim}{\theta}}_{i+1}\|^2}{b_{i+1}r_{i+1}}\right]^{1/2}$$

$$\le \sqrt{\alpha+2}\lim_{k\to\infty}\sum_{i=m(t+n_k)}^{m(t+n_k+\alpha)}\frac{\|\phi_i^\tau\tilde{\underset{\sim}{\theta}}_{i+1}\|^2}{b_{i+1}r_{i+1}} = 0. \tag{4.49}$$

For any $i \in [0, 1, \ldots, m(t+n_k+\alpha) - m(t+n_k)]$, we have

$$t = t + n_k - n_k < t_{m(t+n_k)+1} - n_k \le t_{m(t+n_k)+1+i} - n_k$$

$$= t_{m(t+n_k)} + \sum_{j=m(t+n_k)}^{m(t+n_k)+i}\beta_j - n_k \le t_{m(t+n_k)} + \sum_{j=m(t+n_k)}^{m(t+n_k+\alpha)}\beta_j - n_k$$

$$\le t + n_k + \alpha + 2 - n_k = t + \alpha + 2,$$

which means that

$$t_{m(t+n_k)+i+1} - n_k \in [t, t+\alpha+2].$$

According to Lemma 4.4,

$$\tilde{\underset{\sim}{\theta}}_{n_k}\left(t_{m(t+n_k)+i+1} - n_k\right) \underset{k\to\infty}{\to} \underset{\sim}{\theta}^0$$

uniformly in i satisfying (4.50). But by (4.39) and (4.40), this convergence is equivalent to $\tilde{\underset{\sim}{\theta}}(t_{m(t+n_k)+i+1}) \underset{k\to\infty}{\to} \underset{\sim}{\theta}^0$ and $\tilde{\underset{\sim}{\theta}}_{m(t+n_k)+i+1} \underset{k\to\infty}{\to} \underset{\sim}{\theta}^0$. Hence for any i satisfying (4.50)

$$\lim_{k\to\infty} \sum_{i=m(t+n_k)}^{m(t+n_k+\alpha)} \beta_i \|\tilde{\underset{\sim}{\theta}}_{i+1} - \underset{\sim}{\theta}^0\| = \lim_{k\to\infty} \sum_{i=m(t+n_k)+1}^{m(t+n_k+\alpha)+1} \beta_{i-1} \|\tilde{\underset{\sim}{\theta}}_i - \underset{\sim}{\theta}^0\| = 0. \tag{4.51}$$

Then by (4.49) and (4.51), we have

$$\begin{aligned}
\underset{\sim}{\theta}^{0\tau} \lim_{k\to\infty} \sum_{i=m(t+n_k)}^{m(t+n_k+\alpha)} \frac{\phi_i\phi_i^\tau}{r_{i+1}} \underset{\sim}{\theta}^0 &\le \|\underset{\sim}{\theta}^0\| \lim_{k\to\infty} \left\| \sum_{i=m(t+n_k)}^{m(t+n_k+\alpha)} \frac{\phi_i\phi_i^\tau}{r_{i+1}} \left(\underset{\sim}{\theta}^0 - \tilde{\underset{\sim}{\theta}}_{i+1}\right) \right\| \\
&\quad + \|\underset{\sim}{\theta}^0\| \lim_{k\to\infty} \left\| \sum_{i=m(t+n_k)}^{m(t+n_k+\alpha)} \frac{\phi_i\phi_i^\tau \tilde{\underset{\sim}{\theta}}_{i+1}}{r_{i+1}} \right\| \\
&\le \|\underset{\sim}{\theta}^0\| \lim_{k\to\infty} \sum_{i=m(t+n_k)}^{m(t+n_k+\alpha)} \beta_i \|\underset{\sim}{\theta}^0 - \tilde{\underset{\sim}{\theta}}_{i+1}\| \\
&\quad + \|\underset{\sim}{\theta}^0\| \lim_{k\to\infty} \left\| \sum_{i=m(t+n_k)}^{m(t+n_k+\alpha)} \frac{\phi_i\phi_i^\tau \tilde{\underset{\sim}{\theta}}_{i+1}}{r_{i+1}} \right\| = 0.
\end{aligned}$$

On the other hand, by Condition (4.12), we know that

$$0 = \theta^{0\tau} \overline{\lim_{k\to\infty}} \sum_{i=m(t+n_k)}^{m(t+n_k+\alpha)} \frac{\phi_i\phi_i^\tau}{r_{i+1}} \underset{\sim}{\theta}^0 \ge \beta \underset{\sim}{\theta}^{0\tau} \underset{\sim}{\theta}^0.$$

Hence

$$\underset{\sim}{\theta}^0 = 0,$$

and

$$\tilde{\boldsymbol{\theta}}(t + n_k) \underset{k\to\infty}{\to} \mathbf{0}$$

uniformly in $t \in [\mu, \nu]$, with $[\mu, \nu]$ being any finite interval. But $\tilde{\boldsymbol{\theta}}(t_n) = \tilde{\boldsymbol{\theta}}_n$, hence for fixed ω, there exists a subsequence

$$\tilde{\boldsymbol{\theta}}_{m_k} \underset{k\to\infty}{\to} 0.$$

Finally, by (4.11) and Lemma 4.3

$$\overline{\lim_{n\to\infty}} \frac{1}{k_1} \operatorname{tr} \tilde{\boldsymbol{\theta}}_n^\tau \tilde{\boldsymbol{\theta}}_n \leq \overline{\lim_{n\to\infty}} \operatorname{tr} \frac{\tilde{\boldsymbol{\theta}}_n^\tau \mathbf{R}_n^{-1} \tilde{\boldsymbol{\theta}}_n}{r_n} = \lim_{k\to\infty} v_{m_k} = \lim_{k\to\infty} \tilde{\boldsymbol{\theta}}_{m_k}^\tau \mathbf{R}_{m_k}^{-1} \tilde{\boldsymbol{\theta}}_{m_k}$$

$$\leq \frac{1}{k_0} \overline{\lim_{k\to\infty}} \operatorname{tr} \tilde{\boldsymbol{\theta}}_{m_k}^\tau \tilde{\boldsymbol{\theta}}_{m_k} = 0$$

which implies $\tilde{\boldsymbol{\theta}}_n \underset{n\to\infty}{\to} 0$ a.s. □

For the consistency of least-squares identification (Theorem 3.5), we require that the ratio of the maximum and minimum eigenvalues of $\mathbf{P}_n$ is bounded. Then, in this case, from (3.89) it is seen that

$$\frac{1}{r_n}\mathbf{I} \leq \mathbf{P}_n \leq \frac{c}{r_n}\mathbf{I}$$

with c a constant. This motivates us to replace $a_n \mathbf{P}_n$ in the least-squares estimate simply by $1/r_{n+1}$ to get a simple algorithm. We call this the quasi-least-squares estimate:

$$\boldsymbol{\theta}_{n+1} = \boldsymbol{\theta}_n + \frac{1}{r_{n+1}} \boldsymbol{\phi}_n (\mathbf{y}_{n+1}^\tau - \boldsymbol{\phi}_n^\tau \boldsymbol{\theta}_n). \tag{4.52}$$

By comparing (4.52) with (4.3), we find that

$$\mathbf{R}_{n+1} = \frac{1}{r_{n+1}} \mathbf{I}, \tag{4.53}$$

which is convenient for computation since r_n is obtained via the recursive procedure. Clearly, (4.11) holds true and for the present case

$$b_{n+1} = \left(1 - \frac{\|\boldsymbol{\phi}_n\|^2}{r_{n+1}}\right), \qquad r_{n+1} b_{n+1} = r_{n+1} - \|\boldsymbol{\phi}_n\|^2 = r_n,$$

hence $1/b_n r_n$ is nonincreasing. If $\|\phi_n\|^2/r_{n+1} \leq k_0 < 1$ with k_0 a constant, then $b_{n+1} \geq 1 - k_0 \triangleq b > 0$. Then from Theorem 4.1, a strong consistency theorem follows for the quasi-least-squares estimate:

Theorem 4.2. *Assume that a dynamic system is given by* (3.52), (3.53), (3.62), *and* (4.2) *and the algorithm given by* (3.63), (3.65), *and* (4.52) *is employed. If Conditions B_1 and C_1 of Theorem* 4.1 *hold and* $\|\phi_n\|^2/r_{n+1} \leq k_0$, *for all n, for some constant $k_0 < 1$, then*

$$\underset{\sim}{\theta}_n \underset{n\to\infty}{\to} \underset{\sim}{\theta} \text{ a.s.}$$

4.2. MODIFIED LEAST-SQUARES IDENTIFICATION AND ESTIMATION FOR THE COVARIANCE MATRIX

For the strong consistency of least-squares identification, we have imposed the uniform boundedness of $\lambda^n_{\max}/\lambda^n_{\min}$, which guarantees that $\underset{\sim}{P}_n$ is well conditioned. Now we do not a priori require this condition, but calculate $\lambda^n_{\max}/\lambda^n_{\min}$ at each instant, and, roughly speaking, the estimation will be carried out as for least-squares identification while $\lambda^n_{\max}/\lambda^n_{\min}$ does not exceed a prescribed limit. Otherwise, the algorithm will be modified to keep this quantity at the same level. Such an algorithm is called a modified least-squares (MLS) estimate, and it is described as follows:

Let the system still be given by (3.52), (3.53), (3.62), (4.2), and let (3.63)–(3.65) be unchanged.

Denote

$$r_n = r_{n-1} + \|\phi_{n-1}\|^2, \qquad r_0 = 1, \tag{4.54}$$

$$\beta_i = \frac{\|\phi_i\|^2}{r_{i+1}}. \tag{4.55}$$

Given any θ_0 the estimate $\underset{\sim}{\theta}_n$ for $\underset{\sim}{\theta}$ is given by

$$\underset{\sim}{\theta}_{n+1} = \underset{\sim}{\theta}_n + a_n \underset{\sim}{R}_{n+1}\phi_n(\mathbf{y}^\tau_{n+1} - \phi^\tau_n \underset{\sim}{\theta}_n), \tag{4.56}$$

where a_n and $\underset{\sim}{R}_{n+1}$ are defined as follows.

We are given constants $k_0 \geq 1$, $k_1 \in (0,1)$ and the initial value $\underset{\sim}{R}_0 = \lambda \underset{\sim}{I}$. Suppose $\underset{\sim}{R}_n$ has been obtained. Then we calculate $\underset{\sim}{R}'_{n+1}$ and its maximum and minimum eigenvalues $\mu^{n+1}_{\max}$ and $\mu^{n+1}_{\min}$, where

$$\underset{\sim}{R}'_{n+1} = \underset{\sim}{R}_n - (1 + \phi^\tau_n \underset{\sim}{R}_n \phi_n)^{-1} \underset{\sim}{R}_n \phi_n \phi^\tau_n \underset{\sim}{R}_n. \tag{4.57}$$

If

$$\frac{\mu_{\max}^{n+1}}{\mu_{\min}^{n+1}} \leq k_0 \quad \text{and} \quad \phi_n^\tau \underset{\sim}{\mathbf{R}}'_{n+1}\phi_n \leq k_1, \tag{4.58}$$

then take $a_n = 1$ and $\underset{\sim}{\mathbf{R}}_{n+1} = \underset{\sim}{\mathbf{R}}'_{n+1}$.

If at least one of the inequalities in (4.58) does not hold, take

$$\underset{\sim}{\mathbf{R}}_{n+1} = \frac{r_n}{r_{n+1}}\underset{\sim}{\mathbf{R}}_n, \qquad a_n = (1 + \phi_n^\tau \underset{\sim}{\mathbf{R}}_{n+1}\phi_n)^{-1}. \tag{4.59}$$

Theorem 4.3. *Assume* $r = 0$, *or* $r \geq 1$ *with the transfer function* $\underset{\sim}{\mathbf{C}}^{-1}(z) - \frac{1}{2}\underset{\sim}{\mathbf{I}}$ *strictly positive real and all zeros of* $\det \underset{\sim}{\mathbf{C}}(\lambda)$ *outside the closed unit disk. If Condition* $\underset{\sim}{\mathbf{C}}_1$ *of Theorem* 4.1 *is fulfilled, then*

$$\underset{\sim}{\boldsymbol{\theta}}_n \underset{n\to\infty}{\longrightarrow} \underset{\sim}{\boldsymbol{\theta}} \text{ a.s.}$$

where $\underset{\sim}{\boldsymbol{\theta}}_n$ *is given by* MLS *algorithm* (4.56)–(4.59).

Proof. We first prove

$$r_n = \operatorname{tr} \underset{\sim}{\mathbf{R}}_n^{-1}, \tag{4.60}$$

which is trivial for $n = 0$. Now assuming (4.60) is true for n, then it is clearly true for $n + 1$ if (4.59) is invoked. When $\underset{\sim}{\mathbf{R}}_{n+1}$ is computed according to (4.57), then by (3.19) and (3.20), we have

$$\underset{\sim}{\mathbf{R}}_{n+1}^{-1} = \underset{\sim}{\mathbf{R}}_n^{-1} + \phi_n\phi_n^\tau$$

and hence (4.60) also holds for $n + 1$. Thus (4.60) is verified for any n.

From (4.56), we have

$$\begin{aligned} \mathbf{y}_n - \underset{\sim}{\boldsymbol{\theta}}_n^\tau\phi_{n-1} &= \mathbf{y}_n - [\underset{\sim}{\boldsymbol{\theta}}_{n-1} + a_{n-1}\underset{\sim}{\mathbf{R}}_n\phi_{n-1}(\mathbf{y}_n^\tau - \phi_{n-1}^\tau\underset{\sim}{\boldsymbol{\theta}}_{n-1})]^\tau\phi_{n-1} \\ &= (1 - a_{n-1}\phi_{n-1}^\tau\underset{\sim}{\mathbf{R}}_n\phi_{n-1})(\mathbf{y}_n - \underset{\sim}{\boldsymbol{\theta}}_{n-1}^\tau\phi_{n-1}). \end{aligned} \tag{4.61}$$

Analogously to (3.70), define $\boldsymbol{\xi}_n$ as follows:

$$\boldsymbol{\xi}_n = \mathbf{y}_n - \underset{\sim}{\mathbf{F}}_n\mathbf{w}_n - \underset{\sim}{\boldsymbol{\theta}}_n^\tau\phi_{n-1}. \tag{4.62}$$

Then from (4.56), (4.61), and (4.62)

$$\begin{aligned}\underset{\sim}{\boldsymbol{\theta}}_{n+1} &= \underset{\sim}{\boldsymbol{\theta}}_n + a_n\left(1 - a_n\boldsymbol{\phi}_n^\tau \underset{\sim}{\mathbf{R}}_{n+1}\boldsymbol{\phi}_n\right)^{-1}\underset{\sim}{\mathbf{R}}_{n+1}\boldsymbol{\phi}_n\left(\mathbf{y}_{n+1} - \underset{\sim}{\boldsymbol{\theta}}_{n+1}^\tau\boldsymbol{\phi}_n\right)^\tau \\ &= \underset{\sim}{\boldsymbol{\theta}}_n + a_n\left(1 - a_n\boldsymbol{\phi}_n^\tau \underset{\sim}{\mathbf{R}}_{n+1}\boldsymbol{\phi}_n\right)^{-1}\underset{\sim}{\mathbf{R}}_{n+1}\boldsymbol{\phi}_n\left(\boldsymbol{\xi}_{n+1} + \underset{\sim}{\mathbf{F}}_{n+1}\mathbf{w}_{n+1}\right)^\tau.\end{aligned} \tag{4.63}$$

If (4.58) holds, then $a_n = 1$, $\boldsymbol{\phi}_n^\tau \underset{\sim}{\mathbf{R}}_{n+1}\boldsymbol{\phi}_n \le k_1$, and

$$1 - a_n\boldsymbol{\phi}_n^\tau \underset{\sim}{\mathbf{R}}_{n+1}\boldsymbol{\phi}_n \ge 1 - k_1 > 0. \tag{4.64}$$

Otherwise,

$$\begin{aligned}1 - a_n\boldsymbol{\phi}_n^\tau \underset{\sim}{\mathbf{R}}_{n+1}\boldsymbol{\phi}_n &= 1 - \frac{\boldsymbol{\phi}_n^\tau \underset{\sim}{\mathbf{R}}_{n+1}\boldsymbol{\phi}_n}{1 + \boldsymbol{\phi}_n^\tau \underset{\sim}{\mathbf{R}}_{n+1}\boldsymbol{\phi}_n} = \frac{1}{1 + \boldsymbol{\phi}_n^\tau \underset{\sim}{\mathbf{R}}_{n+1}\boldsymbol{\phi}_n} \ge \frac{1}{1 + \mu_{\max}^{n+1}\|\boldsymbol{\phi}_n\|^2} \\ &\ge \left[1 + \frac{k_0\lambda}{r_{n+1}}\|\boldsymbol{\phi}_n\|^2\right]^{-1} \ge \frac{1}{1 + k_0\lambda} > 0,\end{aligned} \tag{4.65}$$

since

$$\mu_{\max}^n \le k_0\mu_{\min}^n \le \frac{k_0\lambda}{\operatorname{tr}\underset{\sim}{\mathbf{R}}_n^{-1}} = \frac{k_0\lambda}{r_n} \tag{4.66}$$

where we recall that $\lambda = mp + lq + mr$.

Hence for both cases there is a constant $k_3 > 0$ such that

$$1 - a_n\boldsymbol{\phi}_n^\tau \underset{\sim}{\mathbf{R}}_{n+1}\boldsymbol{\phi}_n \ge k_3, \tag{4.67}$$

which makes (4.63) meaningful.

Since $0 < a_n \le 1$ by (4.65) and (4.67), we have

$$\left\|a_n\left(1 - a_n\boldsymbol{\phi}_n^\tau \underset{\sim}{\mathbf{R}}_{n+1}\boldsymbol{\phi}_n\right)^{-1}\underset{\sim}{\mathbf{R}}_{n+1}\right\| \le \frac{k_0\lambda}{k_3 r_{n+1}}.$$

Then, by use of (4.15), it is easy to prove that

$$E\|\underset{\sim}{\boldsymbol{\theta}}_n\|^2 < \infty \tag{4.68}$$

in a proof similar to that for (4.16). Clearly, (4.17) and (4.18) remain valid.

Write

$$v_n = \frac{\operatorname{tr} \tilde{\theta}_n^\tau R_n^{-1} \tilde{\theta}_n}{r_n}.$$

If R_{n+1} is calculated according to (4.57), then

$$\begin{aligned} 1 - \phi_n^\tau R_{n+1} \phi_n &= 1 - \phi_n^\tau R_n \phi_n + (1 + \phi_n^\tau R_n \phi_n)^{-1} (\phi_n^\tau R_n \phi_n)^2 \\ &= (1 + \phi_n^\tau R_n \phi_n)^{-1}. \end{aligned} \tag{4.69}$$

Substituting this into (4.63) leads to

$$\begin{aligned} \theta_{n+1} &= \theta_n + (1 + \phi_n^\tau R_n \phi_n) R_{n+1} \phi_n (\xi_{n+1} + F_{n+1} w_{n+1})^\tau \\ &= \theta_n + (1 + \phi_n^\tau R_n \phi_n) \left[R_n - (1 + \phi_n^\tau R_n \phi_n)^{-1} R_n \phi_n \phi_n^\tau R_n \right] \\ &\quad \times \phi_n (\xi_{n+1} + F_{n+1} w_{n+1})^\tau \\ &= \theta_n + R_n \phi_n (\xi_{n+1} + F_{n+1} w_{n+1})^\tau, \end{aligned} \tag{4.70}$$

or

$$\tilde{\theta}_{n+1} = \tilde{\theta}_n - R_n \phi_n (\xi_{n+1} + F_{n+1} w_{n+1})^\tau. \tag{4.71}$$

Paying attention to the fact that $R_{n+1}^{-1} = R_n^{-1} + \phi_n \phi_n^\tau$, we have

$$\begin{aligned} \operatorname{tr} \tilde{\theta}_{n+1}^\tau R_{n+1}^{-1} \tilde{\theta}_{n+1} &= \operatorname{tr} \tilde{\theta}_n^\tau R_n^{-1} \tilde{\theta}_n - 2(\xi_{n+1} + F_{n+1} w_{n+1})^\tau \tilde{\theta}_n^\tau \phi_n \\ &\quad + \phi_n^\tau R_n \phi_n \|\xi_{n+1} + F_{n+1} w_{n+1}\|^2 + \|\tilde{\theta}_{n+1}^\tau \phi_n\|^2 \\ &= \operatorname{tr} \tilde{\theta}_n^\tau R_n^{-1} \tilde{\theta}_n - 2(\xi_{n+1} + F_{n+1} w_{n+1})^\tau \tilde{\theta}_{n+1}^\tau \phi_n \\ &\quad - \phi_n^\tau R_n \phi_n \|\xi_{n+1} + F_{n+1} w_{n+1}\|^2 + \|\tilde{\theta}_{n+1}^\tau \phi_n\|^2 \\ &\leq \operatorname{tr} \tilde{\theta}_n^\tau R_n^{-1} \tilde{\theta}_n - 2\phi_n^\tau \tilde{\theta}_{n+1} (\xi_{n+1} + F_{n+1} w_{n+1}) + \|\tilde{\theta}_{n+1}^\tau \phi_n\|^2. \end{aligned}$$

From here and the fact that $r_{n+1} \geq r_n$, it follows that

$$v_{n+1} \leq v_n - \frac{2}{r_{n+1}} \phi_n^\tau \tilde{\theta}_{n+1}(\xi_{n+1} + F_{n+1} w_{n+1}) + \frac{\|\tilde{\theta}_{n+1}^\tau \phi_n\|^2}{r_{n+1}}. \quad (4.72)$$

Now we show that (4.72) holds true if R_{n+1} is defined by (4.59). In this case,

$$\frac{R_{n+1}^{-1}}{r_{n+1}} = \frac{R_n^{-1}}{r_n}, \qquad a_n = (1 + \phi_n^\tau R_{n+1} \phi_n)^{-1}$$

and

$$a_n(1 - a_n \phi_n^\tau R_{n+1} \phi_n)^{-1} = 1. \quad (4.73)$$

From (4.63), it is known that

$$\theta_{n+1} = \theta_n + R_{n+1} \phi_n (\xi_{n+1} + F_{n+1} w_{n+1})^\tau$$

and

$$\begin{aligned} \operatorname{tr} \tilde{\theta}_n^\tau R_{n+1}^{-1} \tilde{\theta}_n &= \operatorname{tr} \tilde{\theta}_n^\tau R_{n+1}^{-1} \tilde{\theta}_n - 2 \operatorname{tr}(\xi_{n+1} + F_{n+1} w_{n+1}) \phi_n^\tau \tilde{\theta}_n \\ &\quad + \phi_n^\tau R_{n+1} \phi_n \|\xi_{n+1} + F_{n+1} w_{n+1}\|^2 \\ &= \operatorname{tr} \tilde{\theta}_n^\tau R_{n+1}^{-1} \tilde{\theta}_n - 2 \phi_n^\tau \tilde{\theta}_{n+1}(\xi_{n+1} + F_{n+1} w_{n+1}) \\ &\quad - \phi_n^\tau R_{n+1} \phi_n \|\xi_{n+1} + F_{n+1} w_{n+1}\|^2. \end{aligned}$$

Hence

$$\begin{aligned} v_{n+1} &\leq \frac{\operatorname{tr} \tilde{\theta}_n^\tau R_{n+1}^{-1} \tilde{\theta}_n}{r_{n+1}} - \frac{2 \phi_n^\tau \tilde{\theta}_{n+1}}{r_{n+1}} (\xi_{n+1} + F_{n+1} w_{n+1}) \\ &= v_n - \frac{2 \phi_n^\tau \tilde{\theta}_{n+1}}{r_{n+1}} (\xi_{n+1} + F_{n+1} w_{n+1}), \end{aligned}$$

and (4.72) is also verified.

Notice that $\mathbf{y}_{n+1} - \underset{\sim}{\mathbf{F}}_{n+1}\mathbf{w}_{n+1}$, $\underset{\sim}{\mathbf{F}}_{n+1}$, $\underset{\sim}{\mathbf{P}}_{n+1}$, $\tilde{\underset{\sim}{\theta}}_n$, and ϕ_n are all $\mathscr{F}_n$-measurable, and so

$$E\left[\frac{1}{r_{n+1}}(\underset{\sim}{\mathbf{F}}_{n+1}\mathbf{w}_{n+1})^\tau \tilde{\underset{\sim}{\theta}}_{n+1}\phi_n/\mathscr{F}_n\right]$$

$$= E\left\{\frac{1}{r_{n+1}}\mathbf{w}_{n+1}^\tau\underset{\sim}{\mathbf{F}}_{n+1}\left[\tilde{\underset{\sim}{\theta}}_n - a_n\underset{\sim}{\mathbf{R}}_{n+1}\phi_n\right.\right.$$

$$\left.\left.\cdot(\mathbf{y}_{n+1}^\tau - \mathbf{w}_{n+1}^\tau\underset{\sim}{\mathbf{F}}_{n+1}^\tau + \mathbf{w}_{n+1}^\tau\underset{\sim}{\mathbf{F}}_{n+1}^\tau - \phi_n^\tau\underset{\sim}{\theta}_n)\right]^\tau\phi_n/\mathscr{F}_n\right\}$$

$$= -\frac{1}{r_{n+1}}E(\mathbf{w}_{n+1}^\tau\underset{\sim}{\mathbf{F}}_{n+1}^\tau\underset{\sim}{\mathbf{F}}_{n+1}\mathbf{w}_{n+1}\phi_n^\tau\underset{\sim}{\mathbf{R}}_{n+1}\phi_n a_n/\mathscr{F}_n)$$

$$= -\frac{a_n}{r_{n+1}}\phi_n^\tau\underset{\sim}{\mathbf{R}}_{n+1}\phi_n \operatorname{tr} E(\underset{\sim}{\mathbf{F}}_{n+1}\mathbf{w}_{n+1}\mathbf{w}_{n+1}^\tau\underset{\sim}{\mathbf{F}}_{n+1}^\tau/\mathscr{F}_n)$$

$$= -\frac{a_n}{r_{n+1}}\phi_n^\tau\underset{\sim}{\mathbf{R}}_{n+1}\phi_n \operatorname{tr}\underset{\sim}{\mathbf{F}}_{n+1}\underset{\sim}{\mathbf{F}}_{n+1}^\tau.$$

Substituting this into (4.72) and noting (3.75) yields

$$E(v_{n+1}/\mathscr{F}_n) \le v_n - E\left\{\frac{2}{r_{n+1}}\left[\phi_n^\tau\tilde{\underset{\sim}{\theta}}_{n+1}(\xi_{n+1} - \tfrac{1}{2}\tilde{\underset{\sim}{\theta}}_{n+1}^\tau\phi_n)\right]/\mathscr{F}_n\right\}$$

$$+ \frac{2k_0'\phi_n^\tau\underset{\sim}{\mathbf{R}}_{n+1}\phi_n}{r_{n+1}^{1-\varepsilon}}$$

$$= v_n - \frac{2}{r_{n+1}}E\left[\phi_n^\tau\tilde{\underset{\sim}{\theta}}_{n+1}(\underset{\sim}{C}^{-1}(z) - \tfrac{1}{2}\underset{\sim}{\mathbf{I}})\tilde{\underset{\sim}{\theta}}_{n+1}^\tau\phi_n/\mathscr{F}_n\right]$$

$$+ \frac{2k_0'\phi_n^\tau\underset{\sim}{\mathbf{R}}_{n+1}\phi_n}{r_{n+1}^{1-\varepsilon}} \tag{4.74}$$

Let us write

$$S_n = 2\sum_{i=1}^{n}\phi_i^\tau\tilde{\underset{\sim}{\theta}}_{i+1}\left(\underset{\sim}{C}^{-1}(z) - \frac{1+k_5}{2}\underset{\sim}{\mathbf{I}}\right)\tilde{\underset{\sim}{\theta}}_{i+1}^\tau\phi_i + k_6 \tag{4.75}$$

and set

$$m_n = v_n + \frac{S_{n-1}}{r_n} + 2k_0' E\left(\sum_{i=0}^{\infty} \frac{\phi_i^\tau \underset{\sim}{\mathbf{R}}_{i+1}\phi_i}{r_{i+1}^{1-\varepsilon}} \Big/ \mathscr{F}_n \right) - 2k_0' \sum_{i=0}^{n-1} \frac{\phi_i^\tau \underset{\sim}{\mathbf{R}}_{i+1}\phi_i}{r_{i+1}^{1-\varepsilon}}$$

$$+ k_5 \sum_{i=0}^{n-1} \frac{\|\tilde{\underset{\sim}{\boldsymbol{\theta}}}_{i+1}^\tau \phi_i\|^2}{r_{i+1}}. \tag{4.76}$$

This is finite since, by (4.66), (3.102), and (3.103),

$$\sum_{i=0}^{\infty} \frac{\phi_i^\tau \underset{\sim}{\mathbf{R}}_{i+1}\phi_i}{r_{i+1}^{1-\varepsilon}} \le k_0 \lambda \sum_{i=0}^{\infty} \frac{\|\phi_i\|^2}{r_{i+1}^{2-\varepsilon}} = \frac{k_0 \lambda}{1-\varepsilon}.$$

From (4.74) and (4.75) we know that

$$E(m_{n+1}/\mathscr{F}_n) \le m_n,$$

and by Theorem 1.11 m_n converges to a finite limit as $n \to \infty$. Hence

$$\sum_{i=0}^{\infty} \frac{\|\tilde{\underset{\sim}{\boldsymbol{\theta}}}_{i+1}^\tau \phi_i\|^2}{r_{i+1}} < \infty \text{ a.s.} \tag{4.77}$$

Next we note that if $\underset{\sim}{\mathbf{R}}_{n+1}$ is given by (4.59), then

$$\mu_{\max}^{n+1}/\mu_{\min}^{n+1} = \mu_{\max}^{n}/\mu_{\min}^{n}.$$

Hence for the MLS algorithm, we have

$$\mu_{\max}^{n}/\mu_{\min}^{n} \le k_0, \qquad \forall n, \tag{4.78}$$

whichever of the alternative parts of the algorithm give $\underset{\sim}{\mathbf{R}}_{n+1}$.

It follows that

$$\frac{1}{r_n}\underset{\sim}{\mathbf{I}} = \frac{1}{\operatorname{tr} \underset{\sim}{\mathbf{R}}_n^{-1}}\underset{\sim}{\mathbf{I}} \le \mu_{\min}^n \underset{\sim}{\mathbf{I}} \le \underset{\sim}{\mathbf{R}}_n \le k_0 \mu_{\min}^n \underset{\sim}{\mathbf{I}} \le \frac{1}{\operatorname{tr} \underset{\sim}{\mathbf{R}}_n^{-1}} k_0 \lambda \underset{\sim}{\mathbf{I}}$$

$$= \frac{k_0 \lambda \underset{\sim}{\mathbf{I}}}{r_n}. \tag{4.79}$$

From here, we conclude that for fixed ω $\|\tilde{\theta}_n\|$ is uniformly bounded since v_n is uniformly bounded and

$$v_n = \frac{1}{r_n}\operatorname{tr}\tilde{\theta}_n^{\tau}\mathbf{R}_n^{-1}\tilde{\theta}_n \geq \operatorname{tr}\tilde{\theta}_n^{\tau}\tilde{\theta}_n / k_0\lambda.$$

Now we proceed along the lines of Theorem 4.1.

At this point we have already proved (4.21) and (4.22) for $b_i \equiv 1$. The difference betwen (4.75) and (4.23) is that k_3 in (4.23) is replaced by $(1 + k_5)/2$ in (4.75). Lemma 4.3 is also valid here if in its proof we replace b_i by 1. From (4.63), we see that we should replace $1/b_{i+1}$ by

$$a_n(1 - a_n\phi_n^{\tau}\mathbf{R}_{n+1}\phi_n)^{-1},$$

starting from (4.35) in the proof.

By (4.66) and (4.73), we know that

$$1 \leq a_n(1 - a_n\phi_n^{\tau}\mathbf{R}_{n+1}\phi_n)^{-1}$$

$$= \begin{cases} (1 - \phi_n^{\tau}\mathbf{R}_{n+1}\phi_n)^{-1} \leq \dfrac{1}{1 - k_1} & \text{if (4.58) holds} \\ 1 & \text{otherwise.} \end{cases}$$

Hence

$$1 \leq a_n(1 - a_n\phi_n^{\tau}\mathbf{R}_{n+1}\phi_n)^{-1} \leq \frac{1}{1 - k_1} \tag{4.80}$$

independently of whether (4.58) holds or not.

By (4.78), we have

$$\frac{1}{r_n}\mathbf{I} \leq \mathbf{R}_n \leq \frac{k_0\lambda}{r_n}\mathbf{I}, \tag{4.81}$$

which corresponds to Condition (4.11) in Theorem 4.1. By (4.80) and (4.81), it is easy to see that the proof of Theorem 4.1 starting from Lemma 4.3 is completely parallel to the present case, and this yields Theorem 4.3. □

In Theorems 4.1–4.3, Condition (4.12) has played an important role. However, it looks a little complicated. We shall now give some sufficient conditions for it to hold. As in Chapter 3, we denote by $\lambda_{\max}^n$ and $\lambda_{\min}^n$ the maximum and minimum eigenvalues of

$$\underset{\sim}{\mathbf{P}}_n \triangleq \left(\sum_{i=0}^{n-1} \phi_i \phi_i^\tau + \frac{1}{\lambda} \underset{\sim}{\mathbf{I}} \right)^{-1}$$

It is clear that the following condition is weaker than Condition B_1, which is crucial to Theorem 3.5.

There exists an a.s. finite random variable $c > 0$ such that

$$\lambda_{\max}^n / \lambda_{\min}^n \leq c \text{ a.s.} \tag{4.82}$$

In addition to (4.82), we introduce a slightly different condition by changing the definition of ϕ_n. Namely, we replace the last r terms of ϕ_n by the driving noise given by (3.65) and denote it by

$$\phi_n'^\tau = \left[\mathbf{y}_n^\tau \cdots \mathbf{y}_{n+p+1}^\tau \mathbf{u}_n^\tau \cdots \mathbf{u}_{n-q+1}^\tau (\underset{\sim}{\mathbf{F}}_n \mathbf{w}_n)^\tau \cdots (\underset{\sim}{\mathbf{F}}_{n-r+1} \mathbf{w}_{n-r+1})^\tau \right]. \tag{3.65$'$}$$

Correspondingly, we shall denote by $\lambda_{\min}'^n$ and $\lambda_{\max}'^n$ the minimum and maximum eigenvalues of

$$\underset{\sim}{\mathbf{P}}_n' \triangleq \left(\sum_{i=0}^{n-1} \phi_i' \phi_i'^\tau + \frac{1}{\lambda} \underset{\sim}{\mathbf{I}} \right)^{-1}$$

The condition we shall consider together with (4.82) is

$$\lambda_{\max}'^n / \lambda_{\min}'^n \leq c', \tag{4.82$'$}$$

where c' is some a.s. finite random variable.

Theorem 4.4. (1) *Condition* (4.82) *implies Condition* (4.12). (2) *If*

$$\frac{1}{n} \underset{\sim}{\mathbf{P}}_n^{-1} \to \underset{\sim}{\mathbf{P}} > 0, \tag{4.83}$$

then both (4.82) *and* (4.12) *are satisfied*. (3) *Under the conditions of Theorem* 4.1 *or* 4.2 *or* 4.3, *Condition* (4.82) *is equivalent to* (4.82′).

Proof. (1) Assume (4.82), then from

$$\underset{\sim}{\mathbf{P}}_n^{-1} \geq \frac{1}{\lambda_{\max}^n} \underset{\sim}{\mathbf{I}} \geq \frac{1}{c \lambda_{\min}^n} \underset{\sim}{\mathbf{I}} = \frac{\|\underset{\sim}{\mathbf{P}}_n^{-1}\|}{c} \underset{\sim}{\mathbf{I}} \geq \frac{r_n}{c\lambda} \underset{\sim}{\mathbf{I}},$$

it follows that

$$\sum_{i=m(t)}^{m(r+\alpha)} \frac{\phi_i\phi_i^\tau}{r_{i+1}} = \sum_{i=m(t)}^{m(t+\alpha)} \frac{1}{r_{i+1}}(\underset{\sim}{\mathbf{P}}_{i+1}^{-1} - \underset{\sim}{\mathbf{P}}_i^{-1})$$

$$= \frac{\underset{\sim}{\mathbf{P}}_{m(t+\alpha)+1}^{-1}}{r_{m(t+\alpha)+1}} - \frac{\underset{\sim}{\mathbf{P}}_{m(t)}^{-1}}{r_{m(t)+1}} + \sum_{i=m(t)+1}^{m(t+\alpha)} \underset{\sim}{\mathbf{P}}_i^{-1}\left(\frac{1}{r_i} - \frac{1}{r_{i+1}}\right)$$

$$\geq \sum_{i=m(t)+1}^{m(t+\alpha)} \frac{\underset{\sim}{\mathbf{P}}_i^{-1}\|\phi_i\|^2}{r_i r_{i+1}} - \underset{\sim}{\mathbf{I}}$$

$$\geq \frac{1}{c\lambda}\sum_{i=m(t)+1}^{m(t+\alpha)} \frac{\|\phi_i\|^2}{r_{i+1}}\underset{\sim}{\mathbf{I}} - \underset{\sim}{\mathbf{I}} \geq \frac{\alpha - 1 - c\lambda}{c\lambda}\underset{\sim}{\mathbf{I}},$$

which implies (4.12) if we take

$$\alpha > 1 + c\lambda.$$

(2) This is trivial since (4.82) follows from (4.83). (3) In Theorems 4.1, 4.2 and 4.3 we have used the single symbol ξ_n to denote the quantity appearing in (3.70) and for which (4.8) and (4.9) take place. By (2.82), we have

$$\frac{1}{r_n}\sum_{i=0}^{n-1}\|\xi_{i+1}\|^2 \leq \frac{2k_2}{r_n}\sum_{i=0}^{n-1}\left[\left(\sum_{j=0}^{i}\rho^{i-j}\|\underset{\sim}{\tilde{\theta}}_{j+1}^\tau\phi_j\|\right)^2 + \rho^{2(i+1)}\|\underset{\sim}{\tilde{\theta}}_0^\tau\phi_{-1}\|^2\right]$$

$$\leq \frac{2k_2}{r_n}\sum_{i=0}^{n-1}\left[\sum_{j=0}^{i}\sum_{s=0}^{i}\rho^{i-j}\rho^{i-s}\|\underset{\sim}{\tilde{\theta}}_{j+1}^\tau\phi_j\|\,\|\underset{\sim}{\tilde{\theta}}_{s+1}^\tau\phi_s\| + \rho^{2(i+1)}\|\underset{\sim}{\tilde{\theta}}_0^\tau\phi_{-1}\|^2\right]$$

$$\leq \frac{2k_2}{r_n}\sum_{i=0}^{n-1}\left(\frac{1}{2}\sum_{j=0}^{i}\sum_{s=0}^{i}\rho^{i-j}\rho^{i-s}\|\underset{\sim}{\tilde{\theta}}_{j+1}^\tau\phi_j\|^2 + \frac{1}{2}\sum_{j=0}^{i}\sum_{s=0}^{i}\rho^{i-s}\rho^{i-j}\|\underset{\sim}{\tilde{\theta}}_{s+1}^\tau\phi_s\|^2\right)$$

$$+ \frac{2k_2}{r_n}\sum_{i=0}^{n-1}\rho^{2(i+1)}\|\underset{\sim}{\tilde{\theta}}_0^\tau\phi_{-1}\|^2$$

$$\leq \frac{2k_2}{(1-\rho)r_n}\sum_{i=0}^{n-1}\sum_{j=0}^{i}\rho^{i-j}\|\underset{\sim}{\tilde{\theta}}_{j+1}^\tau\phi_j\|^2 + \frac{2k_2\|\underset{\sim}{\tilde{\theta}}_0^\tau\phi_{-1}\|^2}{r_n(1-\rho^2)}$$

$$= \frac{2k_2}{(1-\rho)r_n}\sum_{j=0}^{n-1}\sum_{i=j}^{n-1}\rho^{i-j}\|\underset{\sim}{\tilde{\theta}}_{j+1}^\tau\phi_j\|^2 + \frac{2k_2\|\underset{\sim}{\tilde{\theta}}_0^\tau\phi_{-1}\|^2}{r_n(1-\rho^2)}$$

$$\leq \frac{2k_2}{(1-\rho)r_n}\sum_{j=0}^{n-1}\|\underset{\sim}{\tilde{\theta}}_{j+1}^\tau\phi_j\|^2 + \frac{2k_2\|\underset{\sim}{\tilde{\theta}}_0^\tau\phi_{-1}\|^2}{r_n(1-\rho^2)} \underset{n\to\infty}{\to} 0 \text{ a.s.}, \tag{4.84}$$

in which the second term goes to zero since $r_n \underset{n\to\infty}{\to} \infty$, whereas the first term tends to zero by (4.22) and $b_i > b > 0\ \forall i$ for Theorems 4.1 and 4.2 and by (4.77) for Theorem 4.3.

Denote

$$\boldsymbol{\phi}_n^{\xi} = [\mathbf{0} \ldots \mathbf{0} \boldsymbol{\xi}_n^{\tau} \ldots \boldsymbol{\xi}_{n-r+1}^{\tau}]^{\tau}$$

and

$$r_n' = 1 + \sum_{i=0}^{n-1} \|\boldsymbol{\phi}_i'\|^2.$$

Then

$$\boldsymbol{\phi}_n' = \boldsymbol{\phi}_n - \boldsymbol{\phi}_n^{\xi}$$

and

$$\frac{r_n'}{r_n} = \frac{r_n - 2\sum_{i=0}^{n-1} \boldsymbol{\phi}_i^{\tau}\boldsymbol{\phi}_i^{\xi} + \sum_{i=0}^{n-1} \|\boldsymbol{\phi}_i^{\xi}\|^2}{r_n} \to 1.$$

Hence we have

$$\left\| \frac{1}{r_n'} \sum_{i=0}^{n-1} \boldsymbol{\phi}_i'\boldsymbol{\phi}_i'^{\tau} - \frac{1}{r_n} \sum_{i=0}^{n-1} \boldsymbol{\phi}_i\boldsymbol{\phi}_i^{\tau} \right\|$$

$$\leq \left\| \frac{1}{r_n} \sum_{i=0}^{n-1} \boldsymbol{\phi}_i'\boldsymbol{\phi}_i'^{\tau} - \frac{1}{r_n} \sum_{i=0}^{n-1} \boldsymbol{\phi}_i\boldsymbol{\phi}_i^{\tau} \right\| + \left\| \frac{1}{r_n} \sum_{i=0}^{n-1} \boldsymbol{\phi}_i'\boldsymbol{\phi}_i'^{\tau} - \frac{1}{r_n'} \sum_{i=0}^{n-1} \boldsymbol{\phi}_i'\boldsymbol{\phi}_i'^{\tau} \right\| \underset{n\to\infty}{\to} 0$$

since by (4.84) and the Schwarz inequality

$$\left\| \frac{1}{r_n} \left(\sum_{i=0}^{n-1} \boldsymbol{\phi}_i'\boldsymbol{\phi}_i'^{\tau} - \sum_{i=0}^{n-1} \boldsymbol{\phi}_i\boldsymbol{\phi}_i^{\tau} \right) \right\| \leq \frac{1}{r_n} \left\| \sum_{i=0}^{n-1} \boldsymbol{\phi}_i^{\xi}\boldsymbol{\phi}_i^{\xi\tau} - \boldsymbol{\phi}_i\boldsymbol{\phi}_i^{\xi\tau} - \boldsymbol{\phi}_i^{\xi}\boldsymbol{\phi}_i^{\tau} \right\| \underset{n\to\infty}{\to} 0$$

and

$$\left\| \frac{1}{r_n} \sum_{i=0}^{n-1} \boldsymbol{\phi}_i'\boldsymbol{\phi}_i'^{\tau} - \frac{1}{r_n'} \sum_{i=0}^{n-1} \boldsymbol{\phi}_i'\boldsymbol{\phi}_i'^{\tau} \right\| \leq \left(\frac{r_n'}{r_n} - 1 \right) \left\| \frac{1}{r_n'} \sum_{i=0}^{n-1} \boldsymbol{\phi}_i'\boldsymbol{\phi}_i'^{\tau} \right\| \underset{n\to\infty}{\to} 0. \qquad \square$$

Now assume $\underset{\sim}{\mathbf{F}}_n$ is a constant matrix $\underset{\sim}{\mathbf{D}}$. We are going to estimate the covariance matrix $\underset{\sim}{\mathbf{D}}\underset{\sim}{\mathbf{D}}^{\tau}$.

Theorem 4.5. *Suppose that* $\underset{\sim}{F}_n \equiv \underset{\sim}{D}$, $\{\mathbf{w}_i\}$ *are mutually independent and* $r_n/n \underset{n\to\infty}{\longrightarrow} 1$. *Then under the conditions of Theorem* 4.1, 4.2, *or* 4.3

$$\frac{1}{n}\sum_{i=1}^{n}(\mathbf{y}_i - \underset{\sim}{\theta}_i^{\tau}\boldsymbol{\phi}_{i-1})(\mathbf{y}_i - \underset{\sim}{\theta}_i^{\tau}\boldsymbol{\phi}_{i-1})^{\tau} \underset{n\to\infty}{\longrightarrow} \underset{\sim}{D}\underset{\sim}{D}^{\tau} \text{ a.s.,} \tag{4.85}$$

where $\underset{\sim}{\theta}_i$ *and* $\boldsymbol{\phi}_i$ *are given in the corresponding theorems.*

Proof. From (3.70) and (4.84), we have

$$\begin{aligned}
&\left\|\frac{1}{r_{n+1}}\sum_{i=1}^{n}(\mathbf{y}_i - \underset{\sim}{\theta}_i^{\tau}\boldsymbol{\phi}_{i-1})(\mathbf{y}_i - \underset{\sim}{\theta}_i^{\tau}\boldsymbol{\phi}_{i-1})^{\tau} - \frac{1}{r_{n+1}}\sum_{i=1}^{n}\mathbf{w}_i\mathbf{w}_i^{\tau}\right\| \\
&= \left\|\frac{1}{r_{n+1}}\sum_{i=1}^{n}(\boldsymbol{\xi}_i + \underset{\sim}{D}\mathbf{w}_i)(\boldsymbol{\xi}_i + \underset{\sim}{D}\mathbf{w}_i)^{\tau} - \frac{1}{r_{n+1}}\sum_{i=1}^{n}\underset{\sim}{D}\mathbf{w}_i\mathbf{w}_i^{\tau}\underset{\sim}{D}^{\tau}\right\| \\
&= \left\|\frac{1}{r_{n+1}}\sum_{i=1}^{n}\boldsymbol{\xi}_i\boldsymbol{\xi}_i^{\tau} + \frac{1}{r_{n+1}}\sum_{i=1}^{n}(\boldsymbol{\xi}_i\mathbf{w}_i^{\tau}\underset{\sim}{D}^{\tau} + \underset{\sim}{D}\mathbf{w}_i\boldsymbol{\xi}_i^{\tau})\right\| \\
&\le \frac{1}{r_{n+1}}\sum_{i=1}^{n}\|\boldsymbol{\xi}_i\|^2 + 2\left(\frac{1}{r_{n+1}}\sum_{i=1}^{n}\|\boldsymbol{\xi}_i\|^2\right)^{1/2}\left(\frac{1}{r_{n+1}}\sum_{i=1}^{n}\|\underset{\sim}{D}\mathbf{w}_i\|^2\right)^{1/2},
\end{aligned} \tag{4.86}$$

and

$$\|\underset{\sim}{D}\mathbf{w}_i\|^2 \le 3\|\mathbf{y}_i\|^2 + 3\|\underset{\sim}{\theta}_i^{\tau}\boldsymbol{\phi}_{i-1}\|^2 + 3\|\boldsymbol{\xi}_i\|^2 \le 3\|\boldsymbol{\phi}_i\|^2 + 3\|\boldsymbol{\xi}_i\|^2.$$

Hence

$$\frac{1}{r_{n+1}}\sum_{i=1}^{n}\|\underset{\sim}{D}\mathbf{w}_i\|^2 \le \frac{3}{r_{n+1}}\sum_{i=1}^{n}\|\boldsymbol{\phi}_i\|^2 + \frac{3}{r_{n+1}}\sum_{i=1}^{n}\|\boldsymbol{\xi}_i\|^2 \le 3 + \frac{3}{r_{n+1}}\sum_{i=1}^{n}\|\boldsymbol{\xi}_i\|^2.$$

Putting this estimate into (4.86) yields

$$\left\|\frac{1}{r_{n+1}}\sum_{i=1}^{n}(\mathbf{y}_i - \underset{\sim}{\theta}_i^{\tau}\boldsymbol{\phi}_{i-1})(\mathbf{y}_i - \underset{\sim}{\theta}_i^{\tau}\boldsymbol{\phi}_{i-1})^{\tau} - \frac{1}{r_{n+1}}\sum_{i=1}^{n}\underset{\sim}{D}\mathbf{w}_i\mathbf{w}_i^{\tau}\underset{\sim}{D}^{\tau}\right\| \underset{n\to\infty}{\longrightarrow} 0. \tag{4.87}$$

Then by (1.3), $r_n/n \to 1$ and $E\mathbf{w}_i\mathbf{w}_i^\tau = \underset{\sim}{\mathbf{I}}$, and this leads to

$$\frac{1}{r_{n+1}}\sum_{i=1}^{n}\underset{\sim}{\mathbf{D}}\mathbf{w}_i\mathbf{w}_i^\tau\underset{\sim}{\mathbf{D}}^\tau = \frac{n+1}{r_{n+1}}\frac{n}{n+1}\underset{\sim}{\mathbf{D}}\left(\frac{1}{n}\sum_{i=1}^{n}\mathbf{w}_i\mathbf{w}_i^\tau\right)\underset{\sim}{\mathbf{D}}^\tau \underset{n\to\infty}{\to} \underset{\sim}{\mathbf{D}}\underset{\sim}{\mathbf{D}}^\tau \text{ a.s.} \tag{4.88}$$

Finally, (4.85) follows from (4.87) and (4.88). □

4.3. CONTINUOUS-TIME ALGORITHMS

We return to the system and notations introduced by (3.110)–(3.116) and assume that

$$\operatorname{tr}\underset{\sim}{\mathbf{F}}_t\underset{\sim}{\mathbf{F}}_t^\tau \le k_0' r_t^\varepsilon, \qquad \varepsilon \in [0,1]. \tag{4.89}$$

As in Chapter 3, we put

$$r_t = 1 + \int_0^t \|\psi_s\|^2\,ds, \qquad \beta_t = \frac{\|\psi_t\|^2}{r_t} \tag{4.90}$$

$$\mu(t) = \int_0^t \beta_s\,ds, \qquad \nu(\mu(t)) = t \tag{4.91}$$

Given any deterministic $\underset{\sim}{\theta}_0$, we introduce an algorithm of the stochastic approximation type as follows:

$$d\underset{\sim}{\theta}_t = \underset{\sim}{\mathbf{R}}_t\psi_t(d\mathbf{y}_t^\tau - \psi_t^\tau\underset{\sim}{\theta}_t\,dt), \tag{4.92}$$

where it is assumed that the strong solution $(\underset{\sim}{\theta}_t, \mathscr{F}_t)$ to this equation exists. The following theorem is the continuous-time analogue of Theorem 4.1.

Theorem 4.6. *Suppose that*

$A_2 \cdot \beta_t \neq 0$, *the positive definite matrix* $\underset{\sim}{\mathbf{R}}_t$ *is an* $\mathscr{F}_t$*-adapted measurable process with* $(d/dt)(\underset{\sim}{\mathbf{R}}_t^{-1}/r_t)$ *being nonpositive definite, and there are constants* $k_0 > 0$, $k_1 > 0$ *and* $k_2 > 0$ *such that*

$$\frac{k_0}{r_t}\underset{\sim}{\mathbf{I}} \le \underset{\sim}{\mathbf{R}}_t \le \frac{k_1}{r_t}\underset{\sim}{\mathbf{I}} \tag{4.93}$$

$$\frac{\|\psi_t\|^2}{r_t} \le k_2, \qquad \forall t \in [0,\infty), \tag{4.94}$$

$B_2 \cdot r = 0$, or $r > 0$ and all zeros of $\det C(z)$ have negative real parts and there are constants $\delta \geq 0$, $k_3 > 0$ such that

$$S_t \triangleq \int_0^t (\xi_s^\tau - \delta \psi_s^\tau \tilde{\theta}_s) \tilde{\theta}_s^\tau \psi_s \, ds + k_3 > 0, \qquad \forall t \in [0, \infty), \tag{4.95}$$

where ξ_s and $\tilde{\theta}_s$ are defined by (3.127) and (3.128) respectively.

C_2. There are random variables $\alpha > 0$, $\beta > 0$ and $T > 0$ such that

$$\int_{\nu(t)}^{\nu(t+\alpha)} \frac{1}{r_s} \psi_s \psi_s^\tau \, ds \geq \beta \mathbf{I}. \tag{4.96}$$

Then $\theta_t \xrightarrow[t\to\infty]{} \theta$ a.s.

Proof. We retain the notation introduced in (3.127), (3.128), and (3.131) and note that (3.132) and (3.133) remain true. Hence

$$C(S)\xi_t = \tilde{\theta}_t^\tau \psi_t, \tag{4.97}$$

$$d\theta_t = \mathbf{R}_t \psi_t (\xi_t^\tau \, dt + \mathbf{F}_t d\mathbf{w}_t), \qquad d\tilde{\theta}_t = -\mathbf{R}_t \psi_t (\xi_t^\tau \, dt + d\mathbf{w}_t^\tau \mathbf{F}_t^\tau). \tag{4.98}$$

Notice that (3.137)–(3.140) still hold.

Let $\Phi_{t,0}$ be the fundamental matrix solving the matrix ordinary differential equation

$$\frac{d}{dt} \Phi_{t,0} = -\mathbf{R}_t \psi_t \psi_t^\tau \Phi_{t,0}, \qquad \Phi_{0,0} = \mathbf{I}. \tag{4.99}$$

By (4.90) and (4.93), we have

$$\|\Phi_{t,0}\| \leq 1 + \left\| \int_0^t \mathbf{R}_s \psi_s \psi_s^\tau \Phi_{s,0} \, ds \right\| \leq 1 + k_1 \int_0^t \beta_s \|\Phi_{s,0}\| \, ds$$

and by (4.94)

$$\|\Phi_{t,0}\| \leq 1 + k_1 k_2 \int_0^t \|\Phi_{s,0}\| \, ds \leq 1 + k_0 k_2 \int_0^t \left(1 + k_1 k_2 \int_0^t \|\Phi_{\lambda,0}\| \, d\lambda \right) ds$$

$$= 1 + k_1 k_2 t + (k_1 k_2)^2 \int_0^t \int_\lambda^t \|\Phi_{\lambda,0}\| \, ds \, d\lambda$$

$$\leq 1 + k_1 k_2 t + (k_1 k_2)^2 \int_0^t (t - \lambda) \left(1 + k_1 k_2 \int_0^\lambda \|\Phi_{s,0}\| \, ds \right) d\lambda$$

$$= 1 + k_1 k_2 t + \frac{(k_1 k_2)^2 t^2}{2} + (k_1 k_2)^3 \int_0^t (t - \lambda) \int_0^\lambda \|\Phi_{s,0}\| \, ds \, d\lambda$$

$$\leq \cdots \leq e^{k_1 k_2 t}. \tag{4.100}$$

From $\Phi_{t,0}\Phi_{0,t} = I$ and (4.99), it is easy to see that

$$\frac{d}{dt}\Phi_{0,t} = \Phi_{0,t}R_t\psi_t\psi_t^\tau, \tag{4.101}$$

from which one can conclude the analogous inequality to (4.100)

$$\|\Phi_{0,t}\| \le e^{k_1k_2t}. \tag{4.102}$$

Then by (4.92) we have

$$\begin{aligned}
\theta_t &= \Phi_{t,0}\theta_0 + \int_0^t \Phi_{t,s}R_s\psi_s\, dy_s^\tau \\
&= \Phi_{t,0}\theta_0 + \int_0^t \Phi_{t,s}R_s\psi_s\big[A_1y_s\, ds + \cdots + A_pS^{p-1}y_s\, ds \\
&\quad + B_1u_s\, ds + \cdots + B_qS^{q-1}u_s\, ds + d\varepsilon_s\big]^\tau.
\end{aligned} \tag{4.103}$$

Now from the equality

$$S^k\int_0^t F_\lambda\, dw_\lambda = \int_0^t \frac{(t-\lambda)^k}{k!}F_\lambda\, dw_\lambda$$

and (3.114), it follows that

$$\begin{aligned}
&E\left\|\int_0^t \Phi_{0,s}R_s\psi_s\left(S^k\int_0^s F_\lambda\, dw_\lambda\right)^\tau ds\right\|^2 \\
&\le k_1e^{2k_1k_2t}E\left[\int_0^t \frac{\|\psi_s\|}{r_s}\left\|\int_0^s \frac{(s-\lambda)^k}{k!}F_\lambda\, dw_\lambda\right\| ds\right]^2 \\
&\le k_1e^{2k_1k_2t}E\int_0^t \frac{\|\psi_s\|^2}{r_s^2}\, ds \int_0^t\int_0^s \frac{(s-\lambda)^{2k}}{(k!)^2}\operatorname{tr} F_\lambda F_\lambda^\tau\, d\lambda\, ds < \infty
\end{aligned} \tag{4.104}$$

and

$$\begin{aligned}
E\left\|\int_0^t \Phi_{0,s}R_s\psi_s(F_s\, dw_s)^\tau\right\|^2 &\le k_1e^{2k_1k_2t}E\int_0^t \frac{\|\psi_s\|^2}{r_s^2}\operatorname{tr} F_sF_s^\tau\, ds \\
&\le k_0'k_1e^{2k_1k_2t}E\int_0^t \frac{\|\psi_s\|^2}{r_s^{2-\varepsilon}}\, ds < \infty.
\end{aligned} \tag{4.105}$$

Hence

$$E\left\|\int_0^t \underset{\sim}{\Phi}_{0,s}\underset{\sim}{\mathbf{R}}_s\psi_s\,d\varepsilon_s^\tau\right\|^2 < \infty \tag{4.106}$$

and

$$\begin{aligned}E\|\underset{\sim}{\theta}_t\|^2 &\le E\|\underset{\sim}{\Phi}_{t,0}\|^2 E\left\|\underset{\sim}{\theta}_0 + \int_0^t \underset{\sim}{\Phi}_{0,s}\underset{\sim}{\mathbf{R}}_s\psi_s(d\mathbf{y}_s^\tau - d\varepsilon_s^\tau) + \int_0^t \underset{\sim}{\Phi}_{0,s}\underset{\sim}{\mathbf{R}}_s\psi_s\,d\varepsilon_s^\tau\right\|^2\\
&\le e^{2k_1k_2t}\left(3E\|\underset{\sim}{\theta}_0\|^2 + 3e^{2k_1k_2t}k_0'\text{const}\,E\int_0^t \frac{\|\psi_s\|^2}{r_s}\,ds\right.\\
&\qquad\left. + 3E\left\|\int_0^t \underset{\sim}{\Phi}_{0,s}\underset{\sim}{\mathbf{R}}_s\psi_s\,d\varepsilon_s^\tau\right\|^2\right) < \infty,\end{aligned} \tag{4.107}$$

where we have invoked (3.142) and (4.94).

By Ito's formula we know that

$$\begin{aligned}\frac{\operatorname{tr}\underset{\sim}{\tilde{\theta}}_t^\tau \underset{\sim}{\mathbf{R}}_t^{-1}\underset{\sim}{\tilde{\theta}}_t}{r_t} &= \frac{\operatorname{tr}\underset{\sim}{\tilde{\theta}}_s^\tau \underset{\sim}{\mathbf{R}}_s^{-1}\underset{\sim}{\tilde{\theta}}_s}{r_s} - 2\int_s^t \frac{1}{r_\lambda}\xi_\lambda^\tau\underset{\sim}{\tilde{\theta}}_\lambda^\tau\psi_\lambda\,d\lambda - 2\int_s^t \frac{1}{r_\lambda}\psi_\lambda^\tau\underset{\sim}{\tilde{\theta}}_\lambda\underset{\sim}{\mathbf{F}}_\lambda\,d\mathbf{w}_\lambda\\
&\quad + \int_s^t \frac{\operatorname{tr}\underset{\sim}{\mathbf{F}}_\lambda\underset{\sim}{\mathbf{F}}_\lambda^\tau\psi_\lambda^\tau\underset{\sim}{\mathbf{R}}_\lambda\psi_\lambda}{r_\lambda}\,d\lambda + \int_s^t \operatorname{tr}\underset{\sim}{\tilde{\theta}}_\lambda^\tau\frac{d}{d\lambda}\left(\frac{\underset{\sim}{\mathbf{R}}_\lambda^{-1}}{r_\lambda}\right)\underset{\sim}{\tilde{\theta}}_\lambda\,d\lambda\\
&\le \frac{\operatorname{tr}\underset{\sim}{\tilde{\theta}}_s^\tau \underset{\sim}{\mathbf{R}}_s^{-1}\underset{\sim}{\tilde{\theta}}_s}{r_s} - 2\int_s^t \frac{1}{r_\lambda}(\xi_\lambda^\tau - \delta\psi_\lambda^\tau\underset{\sim}{\tilde{\theta}}_\lambda)\underset{\sim}{\tilde{\theta}}_\lambda^\tau\psi_\lambda\,d\lambda\\
&\quad -2\int_s^t \frac{1}{r_\lambda}\psi_\lambda^\tau\underset{\sim}{\tilde{\theta}}_\lambda\underset{\sim}{\mathbf{F}}_\lambda\,d\mathbf{w}_\lambda + \int_s^t \frac{1}{r_\lambda}\operatorname{tr}\underset{\sim}{\mathbf{F}}_\lambda\underset{\sim}{\mathbf{F}}_\lambda^\tau\psi_\lambda^\tau\underset{\sim}{\mathbf{R}}_\lambda\psi_\lambda\,d\lambda\\
&\quad -2\delta\int_s^t \frac{1}{r_\lambda}\|\underset{\sim}{\tilde{\theta}}_\lambda^\tau\psi_\lambda\|^2\,d\lambda.\end{aligned} \tag{4.108}$$

It is clear that (3.145) and (3.146) are still valid. Furthermore,

$$\begin{aligned}E\int_0^t \frac{1}{r_\lambda^2}\|\psi_\lambda\|^2\|\underset{\sim}{\tilde{\theta}}_\lambda\|^2\|\underset{\sim}{\mathbf{F}}_\lambda\|^2\,d\lambda &\le k_0'k_2E\int_0^t \frac{\|\underset{\sim}{\tilde{\theta}}_\lambda\|^2}{r_\lambda^{1-\varepsilon}}\,d\lambda\\
&\le k_0'k_2E\int_0^t \|\underset{\sim}{\tilde{\theta}}_\lambda\|^2\,d\lambda < \infty,\end{aligned} \tag{4.109}$$

$$E\int_s^t \frac{1}{r_\lambda}\operatorname{tr}\underset{\sim}{\mathbf{F}}_\lambda\underset{\sim}{\mathbf{F}}_\lambda^\tau\psi_\lambda^\tau\underset{\sim}{\mathbf{R}}_\lambda\psi_\lambda\,d\lambda \le k_0'k_1E\int_s^t \frac{1}{r_\lambda^{2-\varepsilon}}\|\psi_\lambda\|^2\,d\lambda \le \frac{k_0'k_1}{1-\varepsilon}. \tag{4.110}$$

Hence, on the right-hand side of (4.108), all terms have finite expectation.

Denote

$$m_t = \frac{\operatorname{tr} \tilde{\underset{\sim}{\theta}}_t^\tau \underset{\sim}{\mathbf{R}}_t^{-1} \tilde{\underset{\sim}{\theta}}_t}{r_t} + 2\frac{S_t}{r_t} + 2\int_0^t S_s \frac{\|\psi_s\|^2}{r_s^2}\, ds + \frac{k_1 k_0'}{1-\varepsilon}\frac{1}{r_t^{1-\varepsilon}} + 2\delta \int_0^t \frac{\|\tilde{\underset{\sim}{\theta}}_s^\tau \psi_s\|^2}{r_s}\, ds, \tag{4.111}$$

which has finite expectation according to the preceding analysis and because $E(S_t/r_t) < \infty$, which holds by virtue of (3.145), (3.146), and (4.95).

Taking the conditional expectation with respect to $\mathscr{F}_s$, we obtain

$$\begin{aligned}
E(m_t/\mathscr{F}_s) \le{}& \frac{\operatorname{tr} \tilde{\underset{\sim}{\theta}}_s^\tau \underset{\sim}{\mathbf{R}}_s^{-1} \tilde{\underset{\sim}{\theta}}_s}{r_s} - 2E\left(\int_s^t \frac{dS_\lambda}{r_\lambda} \Big/ \mathscr{F}_s \right) + k_1 k_0' E\left(\int_s^t \frac{\|\psi_\lambda\|^2}{r_\lambda^{2-\varepsilon}}\, d\lambda \Big/ \mathscr{F}_s \right) \\
& - 2\delta E\left(\int_s^t \frac{1}{r_\lambda} \|\tilde{\underset{\sim}{\theta}}_\lambda^\tau \psi_\lambda\|^2\, d\lambda \Big/ \mathscr{F}_s \right) + 2E\left(\frac{S_t}{r_t} \Big/ \mathscr{F}_s \right) \\
& + 2E\left(\int_0^t S_\lambda \frac{\|\psi_\lambda\|^2}{r_\lambda^2}\, d\lambda \Big/ \mathscr{F}_s \right) + \frac{k_1 k_0'}{1-\varepsilon} E\left(\frac{1}{r_t^{1-\varepsilon}} \Big/ \mathscr{F}_s \right) \\
& + 2\delta \int_0^s \frac{\|\tilde{\underset{\sim}{\theta}}_\lambda^\tau \psi_\lambda\|^2}{r_\lambda}\, d\lambda + 2\delta E\left(\int_s^t \frac{\|\tilde{\underset{\sim}{\theta}}_\lambda^\tau \psi_\lambda\|^2}{r_\lambda}\, d\lambda \Big/ \mathscr{F}_s \right) \\
={}& \frac{\operatorname{tr} \tilde{\underset{\sim}{\theta}}_s^\tau \underset{\sim}{\mathbf{R}}_s^{-1} \tilde{\underset{\sim}{\theta}}_s}{r_s} - 2E\left(\frac{S_t}{r_t} - \frac{S_s}{r_s} + \int_s^t S_\lambda \frac{\|\psi_\lambda\|^2}{r_\lambda^2}\, d\lambda \Big/ \mathscr{F}_s \right) \\
& + \frac{k_0' k_1}{1-\varepsilon} E\left(\frac{1}{r_s^{1-\varepsilon}} - \frac{1}{r_t^{1-\varepsilon}} \Big/ \mathscr{F}_s \right) + 2E\left(\frac{S_t}{r_t} \Big/ \mathscr{F}_s \right) \\
& + 2E\left(\int_0^t S_\lambda \frac{\|\psi_\lambda\|^2}{r_\lambda^2}\, d\lambda \Big/ \mathscr{F}_s \right) + \frac{k_0' k_1}{1-\varepsilon} E\left(\frac{1}{r_t^{1-\varepsilon}} \Big/ \mathscr{F}_s \right) \\
& + 2\delta \int_0^s \frac{\|\tilde{\underset{\sim}{\theta}}_\lambda^\tau \psi_\lambda\|^2}{r_\lambda}\, d\lambda = m_s,
\end{aligned}$$

which shows that $(m_t, \mathscr{F}_t)$ is a nonnegative supermartingale. Hence it converges to a finite limit and consequently, by (4.111) and $\underset{\sim}{\mathbf{R}}_t^{-1} \le (r_t/k_0)\underset{\sim}{\mathbf{I}}$ there exists an a.s. finite random variable $\gamma < \infty$ such that

$$\|\tilde{\underset{\sim}{\theta}}_t\| \le \gamma, \qquad \forall t \in [0, \infty). \tag{4.112}$$

By the same reasoning, we also have

$$\int_0^\infty \frac{\|\tilde{\underset{\sim}{\theta}}_s^\tau \psi_s\|^2}{r_s}\, ds < \infty, \text{ a.s.} \tag{4.113}$$

Further, using integration by parts, we see that

$$\frac{1}{r_t}\int_0^t \|\underset{\sim}{\tilde{\theta}}_s^T \psi_s\|^2\, ds = \frac{1}{r_t}\int_0^s r_s\, d\int_0^s \frac{\|\underset{\sim}{\tilde{\theta}}_\lambda^T \psi_\lambda\|^2}{r_\lambda}\, d\lambda$$

$$= \int_0^t \frac{\|\underset{\sim}{\tilde{\theta}}_\lambda^T \psi_\lambda\|^2}{r_\lambda}\, d\lambda - \frac{1}{r_t}\int_0^t\int_0^s \frac{\|\underset{\sim}{\tilde{\theta}}_\lambda^T \psi_\lambda\|^2}{r_\lambda}\, d\lambda\, \|\psi_s\|^2\, ds$$

$$= \int_0^t \frac{\|\underset{\sim}{\tilde{\theta}}_\lambda^T \psi_\lambda\|^2}{r_\lambda}\, d\lambda - \int_0^\infty \frac{\|\underset{\sim}{\tilde{\theta}}_\lambda^T \psi_\lambda\|^2}{r_\lambda}\, d\lambda\, \frac{1}{r_t}\int_0^t \|\psi_s\|^2\, ds$$

$$+ \frac{1}{r_t}\int_0^t\int_s^\infty \frac{\|\underset{\sim}{\tilde{\theta}}_\lambda^T \psi_\lambda\|^2}{r_\lambda}\, d\lambda\, \|\psi_s\|^2\, ds$$

$$= \int_0^t \frac{\|\underset{\sim}{\tilde{\theta}}_\lambda \psi_\lambda\|^2}{r_\lambda}\, d\lambda - \int_0^\infty \frac{\|\underset{\sim}{\tilde{\theta}}_\lambda^T \psi_\lambda\|^2}{r_\lambda}\, d\lambda + \frac{1}{r_t}\int_0^\infty \frac{\|\underset{\sim}{\tilde{\theta}}_\lambda^T \psi_\lambda\|^2}{r\lambda}\, d\lambda$$

$$+ \frac{1}{r_t}\int_0^T\left(\int_s^\infty \frac{\|\underset{\sim}{\tilde{\theta}}_\lambda^T \psi_\lambda\|^2}{r_\lambda}\, d\lambda\right)\|\psi_s\|^2\, ds$$

$$+ \frac{1}{r_t}\int_T^t\left(\int_s^\infty \frac{\|\underset{\sim}{\tilde{\theta}}_\lambda^T \psi_\lambda\|^2}{r_\lambda}\, d\lambda\right)\|\psi_s\|^2\, ds. \tag{4.114}$$

From this it is easy to conclude that

$$\frac{1}{r_t}\int_0^t \|\underset{\sim}{\tilde{\theta}}_s^T \psi_s\|^2\, ds \underset{t\to\infty}{\to} 0 \text{ a.s.} \tag{4.115}$$

since $r_t \underset{t\to\infty}{\to} \infty$ and by (4.113) $\int_s^\infty(\|\underset{\sim}{\tilde{\theta}}_\lambda^T \psi_\lambda\|^2/r_\lambda)\, d\lambda$ can be made arbitrarily small for all $s \geq T$ for T sufficiently large.

We now show that

$$S_t/r_t \underset{t\to\infty}{\to} 0 \text{ a.s.} \tag{4.116}$$

First note representation (3.133) of ξ_t, then from (4.115) it is clear that for (4.116) to hold it is sufficient to prove that

$$\frac{1}{r_t}\int_0^t \psi_s^T \underset{\sim}{\tilde{\theta}}_s \int_0^t \underset{\sim}{G}\underset{\sim}{F} e^{\mathbf{F}(s-\lambda)} \underset{\sim}{G}^T \underset{\sim}{\tilde{\theta}}_\lambda^T \psi_\lambda\, d\lambda\, ds \underset{t\to\infty}{\to} 0. \tag{4.117}$$

But using (2.213),

$$\|e^{\mathbf{F}t}\| \leq k_4 e^{-ct},$$

we have

$$\frac{1}{r_t}\left\| \int_0^t \psi_s^\tau \tilde{\theta}_s \int_0^s \mathbf{G}\mathbf{F} e^{\mathbf{F}(s-\lambda)} \mathbf{G}^\tau \tilde{\theta}_\lambda^\tau \psi_\lambda \, d\lambda \, ds \right\|$$

$$\leq \frac{\|\mathbf{F}\| k_4}{r_t} \int_0^t \|\tilde{\theta}_s^\tau \psi_s\| \int_0^s e^{-c(s-\lambda)} \|\tilde{\theta}_\lambda^\tau \psi_\lambda\| \, d\lambda \, ds. \tag{4.118}$$

This can be shown to converge to zero as $t \to \infty$ by the same argument as used in (4.114) if we can prove

$$\int_0^\infty \frac{1}{r_s} \|\tilde{\theta}_s^\tau \psi_s\| \int_0^s e^{-c(s-\lambda)} \|\tilde{\theta}_\lambda^\tau \psi_\lambda\| \, d\lambda \, ds < \infty. \tag{4.119}$$

By (4.113), we know that

$$\int_0^\infty \frac{1}{r_s} \|\tilde{\theta}_s^\tau \psi_s\| \int_0^s e^{-c(s-\lambda)} \|\tilde{\theta}_\lambda^\tau \psi_\lambda\| \, d\lambda \, ds$$

$$\leq \frac{1}{2} \int_0^\infty \int_0^s \frac{\|\tilde{\theta}_s^\tau \psi_s\|^2}{r_s} e^{-c(s-\lambda)} \, d\lambda \, ds + \frac{1}{2} \int_0^\infty \frac{1}{r_\lambda} \int_0^s e^{-c(s-\lambda)} \|\tilde{\theta}_\lambda^\tau \psi_\lambda\|^2 \, d\lambda \, ds$$

$$\leq \frac{1}{2c} \int_0^\infty \frac{\|\tilde{\theta}_s^\tau \psi_s\|^2}{r_s} \, ds + \frac{1}{2} \int_0^\infty d\lambda \int_\lambda^\infty \frac{1}{r_\lambda} \|\tilde{\theta}_\lambda^\tau \psi_\lambda\|^2 e^{-c(s-\lambda)} \, ds$$

$$= \frac{1}{c} \int_0^\infty \frac{\|\tilde{\theta}_s^\tau \psi_s\|^2}{r_s} \, ds < \infty,$$

which proves (4.119) and hence (4.116).

By the convergence of m_t it follows from (4.111) that $\operatorname{tr} \tilde{\theta}_t^\tau \mathbf{R}_t^{-1} \tilde{\theta}_t / r_t$ tends to a finite limit. That is,

$$\frac{\operatorname{tr} \tilde{\theta}_t^\tau \mathbf{R}_t^{-1} \tilde{\theta}_t}{r_t} \underset{t \to \infty}{\to} v < \infty, \text{ a.s.} \tag{4.120}$$

Let us write

$$\tilde{\theta}(t) = \tilde{\theta}_{\nu(t)}, \qquad \tilde{\theta}_n(t) = \tilde{\theta}(t+n) \tag{4.121}$$

$$\mathbf{J}_t = \int_0^t \mathbf{R}_s \psi_s \, d\mathbf{w}_s^\tau \mathbf{F}_s^\tau \tag{4.122}$$

$$\Pi_t = \int_0^t \mathbf{R}_s \psi_s \int_0^s \psi_\lambda^\tau \tilde{\theta}_\lambda \mathbf{G} e^{\mathbf{F}^\tau(t-\lambda)} \mathbf{F}^\tau \mathbf{G}^\tau \, d\lambda \, ds. \tag{4.123}$$

Then from (3.133) and (4.98), it follows that

$$\tilde{\theta}_t = \tilde{\theta}_0 - \int_0^t R_s \psi_s \psi_s^\tau \tilde{\theta}_s \, ds - \Pi_t - J_t. \tag{4.124}$$

The martingale $(J_t, \mathscr{F}_t)$ tends to a finite limit J as $t \to \infty$, since, by (4.93) and (4.89)

$$E \operatorname{tr} J_t^\tau J_t = \int_0^t \psi_s^\tau R_s^2 \psi_s \operatorname{tr} F_s F_s^\tau \, ds \le k_0' k_1 \int_0^t \frac{\|\psi_s\|^2}{r_s^{2-\varepsilon}} \, ds \le \frac{k_0' k_1}{1-\varepsilon}.$$

Since $\beta_t \neq 0$, $\mu(t)$ is strictly increasing and

$$\mu(t) = \int_0^t \frac{\|\psi_s\|^2}{r_s} \, ds = \int_0^t \frac{dr_s}{r_s} = \ln r_s |_0^t = \ln r_t \underset{t\to\infty}{\longrightarrow} \infty.$$

Consequently, the inverse function $\nu(t)$ of $\mu(t)$ is continuous and $\nu(t) \underset{t\to\infty}{\longrightarrow} \infty$. Hence

$$J_{\nu(t)} \underset{t\to\infty}{\longrightarrow} J \text{ a.s.} \tag{4.125}$$

Notice J_t is continuous, so by (4.125), we find that $J_{\nu(t)}$ is bounded and uniformly continuous on $[0, \infty)$.

Further, by (2.213), we have

$$\begin{aligned}
&\|\Pi_{\nu(t+\Delta)} - \Pi_{\nu(t)}\| \\
&\le \int_{\nu(t)}^{\nu(t+\Delta)} \|R_s \psi_s\| \left| \int_0^s \|\tilde{\theta}_\lambda^\tau \psi_\lambda\| \, \|e^{F^\tau(s-\lambda)}\| \, \|F\| \, d\lambda \, ds \right. \\
&\le \|F\| k_1 k_4 \int_{\nu(t)}^{\nu(t+\Delta)} \frac{\|\psi_s\|}{r_s} \int_0^s \|\tilde{\theta}_\lambda^\tau \psi_\lambda\| e^{-c(s-\lambda)} \, d\lambda \, ds \\
&\le \|F\| k_1 k_4 \int_{\nu(t)}^{\nu(t+\Delta)} \beta_s \frac{1}{r_s^{1/2}} \int_0^s \|\tilde{\theta}_\lambda^\tau \psi_\lambda\| e^{-c(s-\lambda)} \, d\lambda \, ds \\
&\le \|F\| k_1 k_4 \int_{\nu(t)}^{\nu(t+\Delta)} \beta_s \left[\frac{1}{r_s} \int_0^s \|\tilde{\theta}_\lambda^\tau \psi_\lambda\|^2 \, d\lambda \int_0^s e^{-2c(s-\lambda)} \, d\lambda \right]^{1/2} ds \\
&\le \frac{\|F\| k_1 k_4}{\sqrt{2c}} \int_{\nu(t)}^{\nu(t+\Delta)} \beta_s \left[\frac{1}{r_s} \int_0^s \|\tilde{\theta}_\lambda^\tau \psi_\lambda\|^2 \, d\lambda \right]^{1/2} ds.
\end{aligned} \tag{4.126}$$

For any $\varepsilon > 0$, by (4.115) we can select T such that

$$\frac{1}{r_s}\int_0^s \|\tilde{\theta}_\lambda^\tau \psi_\lambda\|^2\,ds < \varepsilon^2, \qquad \forall s \geq \nu(T).$$

Hence

$$\int_{\nu(t)}^{\nu(t+\Delta)} \beta_s \left[\frac{1}{r_s}\int_0^s \|\tilde{\theta}_\lambda^\tau \psi_\lambda\|^2\,d\lambda\right]^{1/2} ds \leq \varepsilon\Delta \quad \text{for } t \geq T. \tag{4.127}$$

Combining (4.126) and (4.127) leads to

$$\left\|\Pi_{\nu(t+\Delta)} - \Pi_{\nu(t)}\right\| \underset{t\to\infty}{\longrightarrow} 0, \tag{4.128}$$

which demonstrates the uniform continuity of $\Pi_{\nu(t)}$ in t because $\Pi_{\nu(t)}$ is a continuous function in t.

Then we have

$$\begin{aligned}
\left\|\int_{\nu(t)}^{\nu(t+\Delta)} \mathbf{R}_s \psi_s \psi_s^\tau \tilde{\theta}_s\,ds\right\| &\leq k_1 \int_{\nu(t)}^{\nu(t+\Delta)} \frac{1}{r_s}\|\psi_s\|\ \|\tilde{\theta}_s^\tau \psi_s\|\,ds \\
&\leq k_1\left(\int_{\nu(t)}^{\nu(t+\Delta)} \beta_s\,ds \int_{\nu(t)}^{\nu(t+\Delta)} \frac{\|\tilde{\theta}_s^\tau \psi_s\|^2}{r_s}\,ds\right)^{1/2} \\
&= k_1\sqrt{\Delta}\left(\int_{\nu(t)}^{\nu(t+\Delta)} \frac{\|\tilde{\theta}_s^\tau \psi_s\|^2}{r_s}\,ds\right)^{1/2} \underset{t\to\infty}{\longrightarrow} 0,
\end{aligned} \tag{4.129}$$

and this yields the uniform continuity of

$$\int_0^{\nu(t)} \mathbf{R}_s \psi_s \psi_s^\tau \tilde{\theta}_s\,ds$$

in t. Thus we have proved that

$$\tilde{\theta}_{\nu(t)} = \tilde{\theta}_0 - \int_0^{\nu(t)} \mathbf{R}_s \psi_s \psi_s^\tau \tilde{\theta}_s\,ds - \Pi_{\nu(t)} - \mathbf{J}_{\nu(t)}$$

is uniformly continuous in t. According to our definition

$$\tilde{\theta}_n(t+\Delta) - \tilde{\theta}_n(t) = \tilde{\theta}(t+n+\Delta) - \tilde{\theta}(t+n) = \tilde{\theta}_{\nu(t+n+\Delta)} - \tilde{\theta}_{\nu(t+n)}.$$

Hence the uniform continuity of $\tilde{\theta}_{\nu(t)}$ implies the equicontinuity of the family $\{\tilde{\theta}_n(t)\}$. In addition, for fixed ω $\{\tilde{\theta}_n(t)\}$ is uniformly bounded by (4.112). From Theorem 2.3, we see that for fixed ω there is a subsequence $\tilde{\theta}_{n_k}(t)$, with $n_k \underset{k\to\infty}{\to} \infty$, and a continuous matrix function $\theta(t)$ which is the uniform limit of $\theta_{n_k}(t)$ over any finite interval.

Then for any $\Delta > 0$ it follows from (4.129) that

$$\begin{aligned}\theta(t+\Delta) - \theta(t) &= \lim_{k\to\infty}\left(\tilde{\theta}_{n_k}(t+\Delta) - \tilde{\theta}_{n_k}(t)\right)\\ &= \lim_{k\to\infty}\left(\tilde{\theta}_{\nu(t+n_k+\Delta)} - \tilde{\theta}_{\nu(t+n_k)}\right)\\ &= -\lim_{k\to\infty}\left(\int_{\nu(t+n_k)}^{\nu(t+n_k+\Delta)} \mathbf{R}_s\psi_s\psi_s^\tau\tilde{\theta}_s\,ds + \Pi_{\nu(t+n_k+\Delta)}\right.\\ &\qquad \left. -\Pi_{\nu(t+n_k)} + \mathbf{J}_{\nu(t+n_k+\Delta)} - \mathbf{J}_{\nu(t+n_k)}\right)\end{aligned}$$

which goes to zero by (4.125), (4.128), and (4.129). Hence $\theta(t)$ is a constant matrix. Denote it by θ^0.

From (4.129), we know that

$$\int_{\nu(t+n_k)}^{\nu(t+n_k+\alpha)} \frac{\psi_s\psi_s^\tau}{r_s}\tilde{\theta}_s\,ds \underset{k\to\infty}{\to} \mathbf{0}.$$

Further, we have

$$\begin{aligned}&\left\| \lim_{k\to\infty}\int_{\nu(t+n_k)}^{\nu(t+n_k+\alpha)} \frac{1}{r_s}\psi_s\psi_s^\tau(\tilde{\theta}_s - \theta^0)\,ds\right\| \le \lim_{k\to\infty}\int_{\nu(t+n_k)}^{\nu(t+n_k+\alpha)} \beta_s\|\tilde{\theta}_s - \theta^0\|\,ds\\ &= \lim_{k\to\infty}\int_{\nu(t+n_k)}^{\nu(t+n_k+\alpha)} \|\tilde{\theta}_s - \theta^0\|\,d\mu(s) = \lim_{k\to\infty}\int_{t+n_k}^{t+n_k+\alpha} \|\tilde{\theta}_{\nu(\lambda)} - \theta^0\|\,d\lambda\\ &\le \lim_{k\to\infty}\max_{0\le s\le t+\alpha}\left\|\tilde{\theta}_{\nu(n_k+s)} - \theta^0\right\|\alpha = \alpha\lim_{k\to\infty}\max_{0\le s\le t+\alpha}\left\|\tilde{\theta}_{n_k}(s) - \theta^0\right\| = 0.\end{aligned}$$

Consequently,

$$\begin{aligned}\lim_{k\to\infty}&\left\|\theta^{0\tau}\int_{\nu(t+n_k)}^{\nu(t+n_k+\alpha)} \frac{1}{r_s}\psi_s\psi_s^\tau\,ds\,\theta^0\right\| \le \lim_{k\to\infty}\|\theta^0\|\left\|\int_{\nu(t+n_k)}^{\nu(t+n_k+\alpha)} \frac{\psi_s\psi_s^\tau}{r_s}\tilde{\theta}_s\,ds\right\|\\ &+ \lim_{k\to\infty}\|\theta^0\|\left\|\int_{\nu(t+n_k)}^{\nu(t+n_k+\alpha)} \frac{1}{r_s}\psi_s\psi_s^\tau(\tilde{\theta}_s - \theta^0)\,ds\right\| = 0.\end{aligned}$$

On the other hand, by condition (4.96),

$$\lim_{k\to\infty} \underset{\sim}{\theta}^{0\tau} \int_{\nu(t+n_k)}^{\nu(t+n_k+\alpha)} \frac{1}{r_s} \psi_s \psi_s^{\tau}\, ds\, \underset{\sim}{\theta}^0 \ge \beta \underset{\sim}{\theta}^{0\tau} \underset{\sim}{\theta}^0,$$

which is possible only when $\underset{\sim}{\theta}^0 = 0$ because $\beta > 0$.

Thus, we have shown that

$$\tilde{\underset{\sim}{\theta}}_{\nu(t+n_k)} \underset{k\to\infty}{\to} \underset{\sim}{0} \text{ a.s.}$$

But $\nu(t+n_k) \underset{k\to\infty}{\to} \infty$, then from (4.120) it follows that

$$\lim_{t\to\infty} \frac{\operatorname{tr} \tilde{\underset{\sim}{\theta}}_t^{\tau} \underset{\sim}{R}_t^{-1} \tilde{\underset{\sim}{\theta}}_t}{r_t} = \lim_{k\to\infty} \frac{\operatorname{tr} \tilde{\underset{\sim}{\theta}}_{\nu(t+n_k)}^{\tau} \underset{\sim}{R}_{\nu(t+n_k)}^{-1} \tilde{\underset{\sim}{\theta}}_{\nu(t+n_k)}}{r_{\nu(t+n_k)}} = 0.$$

Further, from (4.93), $\underset{\sim}{R}_t^{-1} \ge (r_t/k_0)\underset{\sim}{I}$, and this yields

$$0 = \lim_{t\to\infty} \frac{\operatorname{tr} \tilde{\underset{\sim}{\theta}}_t^{\tau} \underset{\sim}{R}_t^{-1} \tilde{\underset{\sim}{\theta}}_t}{r_t} \ge \overline{\lim_{t\to\infty}} \frac{1}{k_0} \operatorname{tr} \tilde{\underset{\sim}{\theta}}_t^{\tau} \tilde{\underset{\sim}{\theta}}_t \ge 0,$$

and

$$\tilde{\underset{\sim}{\theta}}_t \underset{t\to\infty}{\to} \underset{\sim}{0} \text{ a.s.}$$ □

Remark. If $\underset{\sim}{R}_t = (1/r_t)\underset{\sim}{I}$, then the conditions imposed on $\underset{\sim}{R}_t$ are satisfied automatically and the algorithm is called the quasi-least-squares algorithm.

As in the discrete-time case, we shall now give sufficient conditions for (4.96).

As before, denote by $\lambda_{\max}^t$ and $\lambda_{\min}^t$, the maximum and minimum eigenvalues of the matrix

$$\underset{\sim}{P}_t = \left(\int_0^t \psi_s \psi_s^{\tau}\, ds + \frac{1}{mp + lq + mr} \underset{\sim}{I} \right)^{-1}.$$

Theorem 4.7. *If there is an* a.s. *finite random variable* $c > 0$ *such that*

$$\lambda_{\max}^t / \lambda_{\min}^t < c \text{ a.s.} \qquad \forall t, \tag{4.130}$$

then (4.96) *holds; further, if there is a positive matrix* $\underset{\sim}{P}$ *such that*

$$\frac{1}{t} \underset{\sim}{P}_t^{-1} \underset{t\to\infty}{\to} \underset{\sim}{P} > 0, \text{ a.s.} \tag{4.131}$$

then (4.130) *is satisfied.*

Proof. First we note that

$$\underset{\sim}{\mathbf{P}}_t^{-1} \geq \frac{1}{\lambda^t_{\max}}\underset{\sim}{\mathbf{I}} \geq \frac{1}{c\lambda^t_{\min}}\underset{\sim}{\mathbf{I}} = \frac{\|\underset{\sim}{\mathbf{P}}_t^{-1}\|}{c}\underset{\sim}{\mathbf{I}} \geq \frac{r_t}{c(mp+lq+mr)}\underset{\sim}{\mathbf{I}},$$

then

$$\int_{\nu(t)}^{\nu(t+\alpha)} \frac{\psi_s\psi_s^\tau}{r_s}\,ds = \int_{\nu(t)}^{\nu(t+\alpha)} \frac{d\underset{\sim}{\mathbf{P}}_s^{-1}}{r_s} = \left[\frac{\underset{\sim}{\mathbf{P}}_s^{-1}}{r_s}\right]\Bigg|_{\nu(t)}^{\nu(t+\alpha)} + \int_{\nu(t)}^{\nu(t+\alpha)} \frac{\|\psi_s\|^2\underset{\sim}{\mathbf{P}}_s^{-1}}{r_s^2}\,ds$$

$$\geq \frac{1}{c(mp+lq+mr)}\int_{\nu(t)}^{\nu(t+\alpha)}\frac{\|\psi_s\|^2}{r_s}\,ds\,\underset{\sim}{\mathbf{I}} - \underset{\sim}{\mathbf{I}}$$

$$= \frac{\alpha - c(mp+lq+mr)}{c(mp+lq+mr)}\underset{\sim}{\mathbf{I}}.$$

Hence (4.96) holds if $\alpha > 1 + c(mp+lq+mr)$. The fact that (4.130) is the consequence of (4.131) is the trivial part of the theorem. □

The following theorem corresponds to Theorem 4.5 for the discrete-time case

Theorem 4.8. *Suppose that* $\underset{\sim}{\mathbf{F}}_t \equiv \underset{\sim}{\mathbf{D}}$, $r_t/t \underset{t\to\infty}{\to} 1$ *and the conditions of Theorem* 4.6 *are fulfilled. Then for any* t,

$$\lim_{n\to\infty}\frac{1}{t}\sum_{i=0}^{n-1}\left(\mathbf{y}_{t_{n,i+1}} - y_{t_{n,i}} - \int_{t_{n,i}}^{t_{n,i+1}} \underset{\sim}{\theta}_s^\tau\psi_s\,ds\right)$$

$$\left(\mathbf{y}_{t_{n,i+1}} - \mathbf{y}_{t_{n,i}} - \int_{t_{n,i}}^{t_{n,i+1}} \underset{\sim}{\theta}_s\psi_s\,ds\right)^\tau = \underset{\sim}{\mathbf{D}}\underset{\sim}{\mathbf{D}}^\tau,$$

where

$$t_{n,i} = \frac{it}{n}.$$

Proof. It is easy to see that

$$\frac{1}{t}\sum_{i=0}^{n-1}\left(\int_{t_{n,i}}^{t_{n,i+1}}\xi_s\,ds\right)\left(\int_{t_{n,i}}^{t_{n,i+1}}\xi_s\,ds\right)^\tau \underset{n\to\infty}{\to} 0 \tag{4.132}$$

by the continuity of ξ_t. Hence, from (3.128),

$$\frac{1}{t}\sum_{i=0}^{n-1}\left(\mathbf{y}_{t_{n,i+1}} - \mathbf{y}_{t_{n,i}} - \int_{t_{n,i}}^{t_{n,i+1}} \underset{\sim}{\theta}_s^{\tau}\psi_s\,ds\right)\left(\mathbf{y}_{t_{n,i+1}} - \mathbf{y}_{t_{n,i}} - \int_{t_{n,i}}^{t_{n,i+1}} \underset{\sim}{\theta}_s^{\tau}\psi_s\,ds\right)^{\tau}$$

$$= \frac{1}{t}\sum_{i=0}^{n-1}\left(\int_{t_{n,i}}^{t_{n,i+1}} \xi_s\,ds\right)\left(\int_{t_{n,i}}^{t_{n,i+1}} \xi_s\,ds\right)^{\tau}$$

$$+ \frac{1}{t}\sum_{i=0}^{n-1}\left[\left(\int_{t_{n,i}}^{t_{n,i+1}} \xi_s\,ds\right)\left(\int_{t_{n,i}}^{t_{n,i+1}} \underset{\sim}{\mathbf{D}}\,d\mathbf{w}_s\right)^{\tau} + \left(\int_{t_{n,i}}^{t_{n,i+1}} \underset{\sim}{\mathbf{D}}\,d\mathbf{w}_s\right)\left(\int_{t_{n,i}}^{t_{n,i+1}} \xi_s\,ds\right)^{\tau}\right]$$

$$+ \frac{1}{t}\underset{\sim}{\mathbf{D}}\sum_{i=0}^{n-1}(\mathbf{w}_{t_{n,i+1}} - \mathbf{w}_{t_{n,i}})(\mathbf{w}_{t_{n,i+1}} - \mathbf{w}_{t_{n,i}})^{\tau}\underset{\sim}{\mathbf{D}}^{\tau},$$

which by (1.16) and (4.132) and the Schwarz inequality leads to the assertion of the theorem. □

4.4. ADAPTIVE CONTROL

We shall continue to consider the system described by (3.52), (3.53), and (3.62) and use the notation of (3.4) and (3.63)–(3.65). Up to now we have only discussed how to estimate an unknown parameter $\underset{\sim}{\theta}$ based on input–output data. In the final two sections of this chapter, in addition to parameter estimation, the input will be given by a control law designed for some specific purpose. To be precise, in this section we want a general quadratic index to take a minimum value, while in the next section the output is intended to track a given signal.

Let $\{\mathbf{y}_n^*\}$ be a given bounded deterministic reference sequence. We wish to estimate the unknown parameter $\underset{\sim}{\theta}$ and at the same time to give a control law that makes the performance index

$$J(u) = \overline{\lim_{N\to\infty}}\, E\sum_{i=0}^{N-1}\left[(\mathbf{y}_i - \mathbf{y}_i^*)^{\tau}\underset{\sim}{\mathbf{Q}}_1(\mathbf{y}_i - \mathbf{y}_i^*) + \mathbf{u}_i^{\tau}\underset{\sim}{\mathbf{Q}}_2\mathbf{u}_i\right] \tag{4.133}$$

as small as possible when $\underset{\sim}{\mathbf{Q}}_1 \geq 0$, $\underset{\sim}{\mathbf{Q}}_2 \geq 0$. Since the control actions (i.e., the inputs) will depend upon estimates of the unknown parameter $\underset{\sim}{\theta}$, this is known as a (parameter) adaptive control law.

The class of admissible controls $\mathscr{U}$ consists of those input sequences $\{\mathbf{u}_k\}$ such that:

1. $\mathbf{u}_k$ is $\mathscr{F}_k^{\mathbf{y}}$-measurable, where $\mathscr{F}_k^{\mathbf{y}} = \sigma\{\mathbf{y}_0, \ldots, \mathbf{y}_k\}$ denotes the σ-algebra generated by $(\mathbf{y}_0, \ldots, \mathbf{y}_k)$.
2. $E\|\mathbf{u}_k\|^2 < \infty\ \forall k \geq 0$.

In order to obtain the desired adaptive control law, we first clarify the optimal control under which (4.133) reaches its minimum when $\underset{\sim}{\theta}$ is known.

Denote

$$\underset{\sim}{\mathbf{A}} = \begin{bmatrix} \underset{\sim}{\mathbf{A}}_1 & \underset{\sim}{\mathbf{I}} & \underset{\sim}{\mathbf{0}} & \cdots & \underset{\sim}{\mathbf{0}} \\ & \underset{\sim}{\mathbf{0}} & \ddots & \ddots & \vdots \\ \vdots & & & & \underset{\sim}{\mathbf{0}} \\ & \vdots & & \ddots & \underset{\sim}{\mathbf{I}} \\ \underset{\sim}{\mathbf{A}}_{p'} & \underset{\sim}{\mathbf{0}} & \cdots & & \underset{\sim}{\mathbf{0}} \end{bmatrix}, \quad \underset{\sim}{\mathbf{B}} = \begin{bmatrix} \underset{\sim}{\mathbf{B}}_1 \\ \vdots \\ \underset{\sim}{\mathbf{B}}_{p'} \end{bmatrix},$$

$$\underset{\sim}{\mathbf{C}} = \begin{bmatrix} \underset{\sim}{\mathbf{I}} \\ \underset{\sim}{\mathbf{C}}_1 \\ \vdots \\ \underset{\sim}{\mathbf{C}}_{p'-1} \end{bmatrix}, \quad \underset{\sim}{\mathbf{H}} = \underbrace{[\underset{\sim}{\mathbf{I}}\ \underset{\sim}{\mathbf{0}} \cdots \underset{\sim}{\mathbf{0}}]}_{mp'} \}m \tag{4.134}$$

where $p' = p \vee q \vee (r+1)$, and $\underset{\sim}{\mathbf{A}}_i = \underset{\sim}{\mathbf{0}}$, $\underset{\sim}{\mathbf{B}}_j = \underset{\sim}{\mathbf{0}}$, $\underset{\sim}{\mathbf{C}}_l = \underset{\sim}{\mathbf{0}}$ for $i > p$, $j > q$, $l > r$.

Let the mp'-dimensional vector $\mathbf{x}_k$ be defined by

$$\mathbf{x}_{k+1} = \underset{\sim}{\mathbf{A}}\mathbf{x}_k + \underset{\sim}{\mathbf{B}}\mathbf{u}_k + \underset{\sim}{\mathbf{C}}\underset{\sim}{\mathbf{F}}_{k+1}\mathbf{w}_{k+1}, \qquad \mathbf{x}_0^{\tau} = [\mathbf{y}_0^{\tau}\mathbf{0} \cdots \mathbf{0}]. \tag{4.135}$$

It is straightforward to verify that

$$\mathbf{y}_k = \underset{\sim}{\mathbf{H}}\mathbf{x}_k, \qquad \forall k \geq 0 \tag{4.136}$$

Lemma 4.5. *For the system* (4.135) *and* (4.136) *with any random vectors* $\{\mathbf{u}_k\}$, *if* $\hat{\mathbf{x}}_k$ *is calculated by*

$$\hat{\mathbf{x}}_{k+1} = \underset{\sim}{\mathbf{A}}\hat{\mathbf{x}}_k + \underset{\sim}{\mathbf{B}}\mathbf{u}_k + \underset{\sim}{\mathbf{C}}\underset{\sim}{\mathbf{F}}_{k+1}\underset{\sim}{\mathbf{F}}_{k+1}^{\tau}\left(\underset{\sim}{\mathbf{F}}_{k+1}\underset{\sim}{\mathbf{F}}_{k+1}^{\tau}\right)^{+}\left(\mathbf{y}_{k+1} - \underset{\sim}{\mathbf{H}}\underset{\sim}{\mathbf{A}}\hat{\mathbf{x}}_k - \underset{\sim}{\mathbf{H}}\underset{\sim}{\mathbf{B}}\mathbf{u}_k\right) \tag{4.137}$$

$$\hat{\mathbf{x}}_0^{\tau} = [\mathbf{y}_0^{\tau}\underset{\sim}{\mathbf{0}} \cdots \underset{\sim}{\mathbf{0}}],$$

then $\hat{\mathbf{x}}_k \equiv \mathbf{x}_k$, *in particular, if* $\{\mathbf{u}_k\} \in \mathscr{U}$ *then* $\mathbf{u}_k \in \mathscr{F}_k^{\mathbf{y}}$ *and then*

$$\hat{\mathbf{x}}_k = E(\mathbf{x}_k/\mathscr{F}_k^{\mathbf{y}}) \qquad \forall k \geq 0. \tag{4.138}$$

Proof. We first show that $\hat{\mathbf{x}}_k$ defined by (4.137) is identically equal to $\mathbf{x}_k$. We have $\hat{\mathbf{x}}_0 = \mathbf{x}_0$. Assume $\hat{\mathbf{x}}_k = \mathbf{x}_k$. Then from (4.137), (4.135), and

(4.136) we find that

$$\hat{\mathbf{x}}_{k+1} = \underset{\sim}{\mathbf{A}}\mathbf{x}_k + \underset{\sim}{\mathbf{B}}\mathbf{u}_k + \underset{\sim}{\mathbf{C}}\underset{\sim}{\mathbf{F}}_{k+1}\underset{\sim}{\mathbf{F}}^{\tau}_{k+1}(\underset{\sim}{\mathbf{F}}_{k+1}\underset{\sim}{\mathbf{F}}^{\tau}_{k+1})^{+}$$

$$\times (\underset{\sim}{\mathbf{H}}\underset{\sim}{\mathbf{A}}\mathbf{x}_k + \underset{\sim}{\mathbf{H}}\underset{\sim}{\mathbf{B}}\mathbf{u}_k + \underset{\sim}{\mathbf{H}}\underset{\sim}{\mathbf{C}}\underset{\sim}{\mathbf{F}}_{k+1}\mathbf{w}_{k+1} - \underset{\sim}{\mathbf{H}}\underset{\sim}{\mathbf{A}}\hat{\mathbf{x}}_k - \underset{\sim}{\mathbf{H}}\underset{\sim}{\mathbf{B}}\mathbf{u}_k)$$

$$= \underset{\sim}{\mathbf{A}}\mathbf{x}_k + \underset{\sim}{\mathbf{B}}\mathbf{u}_k + \underset{\sim}{\mathbf{C}}\underset{\sim}{\mathbf{F}}_{k+1}\underset{\sim}{\mathbf{F}}^{\tau}_{k+1}(\underset{\sim}{\mathbf{F}}_{k+1}\underset{\sim}{\mathbf{F}}^{\tau}_{k+1})^{+}\underset{\sim}{\mathbf{F}}_{k+1}\mathbf{w}_{k+1}$$

$$= \underset{\sim}{\mathbf{A}}\mathbf{x}_k + \mathbf{B}\mathbf{u}_k + \underset{\sim}{\mathbf{C}}\underset{\sim}{\mathbf{F}}_{k+1}\mathbf{w}_{k+1} = \mathbf{x}_{k+1}.$$

Hence $\hat{\mathbf{x}}_k \equiv \mathbf{x}_k$ for all $k \geq 0$.

From (4.137), it is obvious that $\hat{\mathbf{x}}_{k+1}$ is $\mathscr{F}^{\mathbf{y}}_{k+1}$-measurable if $\mathbf{u}_k$ is $\mathscr{F}^{\mathbf{y}}_k$-measurable. Hence by (1.48)

$$E\left(\mathbf{x}_{k+1} - E\left(\frac{\mathbf{x}_{k+1}}{\mathscr{F}^{\mathbf{y}}_{k+1}}\right)\right)\left(\mathbf{x}_{k+1} - E\left(\frac{\mathbf{x}_{k+1}}{\mathscr{F}^{\mathbf{y}}_{k+1}}\right)\right)^{\tau}$$

$$\leq E(\mathbf{x}_{k+1} - \hat{\mathbf{x}}_{k+1})(\mathbf{x}_{k+1} - \hat{\mathbf{x}}_{k+1})^{\tau} = \underset{\sim}{\mathbf{0}}.$$

This proves that when $\{\mathbf{u}_k\} \in \mathscr{U}$ $E(\mathbf{x}_{k+1}/\mathscr{F}^{\mathbf{y}}_{k+1}) = \hat{\mathbf{x}}_{k+1}$ for all $k \geq 0$, where the latter quantity is defined by (4.137). □

From now on in this section, we assume that

$$\underset{\sim}{\mathbf{F}}_k \equiv \underset{\sim}{\mathbf{F}},$$

which is a constant matrix independent of ω.

Let us introduce the following difference equations:

$$\underset{\sim}{\mathbf{S}}_k(N) = \underset{\sim}{\mathbf{A}}^{\tau}\underset{\sim}{\mathbf{S}}_{k+1}(N)\underset{\sim}{\mathbf{A}} - \underset{\sim}{\mathbf{A}}^{\tau}\underset{\sim}{\mathbf{S}}_{k+1}(N)\underset{\sim}{\mathbf{B}}\underset{\sim}{\mathbf{M}}^{+}_k\underset{\sim}{\mathbf{B}}^{\tau}\underset{\sim}{\mathbf{S}}_{k+1}(N)\underset{\sim}{\mathbf{A}} + \underset{\sim}{\mathbf{H}}^{\tau}\underset{\sim}{\mathbf{Q}}_1\underset{\sim}{\mathbf{H}},\ \underset{\sim}{\mathbf{S}}_N(N) = \underset{\sim}{\mathbf{0}}, \quad (4.139)$$

$$\underset{\sim}{\mathbf{M}}_k = \underset{\sim}{\mathbf{Q}}_2 + \underset{\sim}{\mathbf{B}}^{\tau}\underset{\sim}{\mathbf{S}}_{k+1}(N)\underset{\sim}{\mathbf{B}}, \quad (4.140)$$

$$\mathbf{b}_k(N) = (\underset{\sim}{\mathbf{A}}^{\tau} - \underset{\sim}{\mathbf{A}}^{\tau}\underset{\sim}{\mathbf{S}}_{k+1}(N)\underset{\sim}{\mathbf{B}}\underset{\sim}{\mathbf{M}}^{+}_k\underset{\sim}{\mathbf{B}}^{\tau})\mathbf{b}_{k+1}(N) - \underset{\sim}{\mathbf{H}}^{\tau}\underset{\sim}{\mathbf{Q}}_1\mathbf{y}^{*}_k, \quad \mathbf{b}_N(N) = \mathbf{0}, \quad (4.141)$$

$$d_k(N) = d_{k+1}(N) - \mathbf{b}^{\tau}_{k+1}(N)\underset{\sim}{\mathbf{B}}\underset{\sim}{\mathbf{M}}^{+}_k\underset{\sim}{\mathbf{B}}^{\tau}\mathbf{b}_{k+1}(N) + \mathbf{y}^{*\tau}_k\underset{\sim}{\mathbf{Q}}_1\mathbf{y}^{*}_k, \quad d_N(N) = 0. \quad (4.142)$$

First we note that

$$(\underset{\sim}{\mathbf{I}} - \underset{\sim}{\mathbf{M}}^{+}_k\underset{\sim}{\mathbf{M}}_k)\underset{\sim}{\mathbf{B}}^{\tau}\underset{\sim}{\mathbf{S}}_{k+1}(N)\underset{\sim}{\mathbf{B}}(\underset{\sim}{\mathbf{I}} - \underset{\sim}{\mathbf{M}}^{+}_k\underset{\sim}{\mathbf{M}}_k)$$

$$\leq (\underset{\sim}{\mathbf{I}} - \underset{\sim}{\mathbf{M}}^{+}_k\underset{\sim}{\mathbf{M}}_k)\underset{\sim}{\mathbf{M}}_k(\underset{\sim}{\mathbf{I}} - \underset{\sim}{\mathbf{M}}^{+}_k\underset{\sim}{\mathbf{M}}_k) = \underset{\sim}{\mathbf{0}}, \quad (4.143)$$

Hence

$$\mathbf{\underset{\sim}{S}}_{k+1}(N)\mathbf{\underset{\sim}{B}}\mathbf{\underset{\sim}{M}}_k^+\mathbf{\underset{\sim}{M}}_k = \mathbf{\underset{\sim}{S}}_{k+1}(N)\mathbf{\underset{\sim}{B}}. \tag{4.144}$$

By using (4.144), it may be verified that

$$E\left[\sum_{i=0}^{N-1}(\mathbf{y}_i - \mathbf{y}_i^*)^\tau\mathbf{Q}_1(\mathbf{y}_i - \mathbf{y}_i^*) + \mathbf{u}_i^\tau\mathbf{Q}_2\mathbf{u}_i\right]$$

$$= \left\{\sum_{i=0}^{N-1}\left[(\mathbf{y}_i - \mathbf{y}_i^*)^\tau\mathbf{Q}_1(\mathbf{y}_i - \mathbf{y}_i^*) + \mathbf{u}_i^\tau\mathbf{Q}_2\mathbf{u}_i\right]\right.$$

$$+ \mathbf{x}_0^\tau\mathbf{\underset{\sim}{S}}_0(N)\mathbf{x}_0 + 2\mathbf{x}_0^\tau\mathbf{b}_0(N) + d_0(N)$$

$$+ \sum_{i=0}^{N-1}\left[\mathbf{x}_{i+1}^\tau\mathbf{\underset{\sim}{S}}_{i+1}(N)\mathbf{x}_{i+1} - \mathbf{x}_i^\tau\mathbf{\underset{\sim}{S}}_i(N)\mathbf{x}_i + 2(\mathbf{x}_{i+1}^\tau\mathbf{b}_{i+1}(N)\right.$$

$$\left.\left. - \mathbf{x}_i^\tau\mathbf{b}_i(N)) + d_{i+1}(N) - d_i(N)\right]\right\}$$

$$= d_0(N) + 2(E\mathbf{x}_0)^\tau\mathbf{b}_0(N) + E\mathbf{x}_0^\tau\mathbf{\underset{\sim}{S}}_0(N)\mathbf{x}_0 + \sum_{i=0}^{N-1}\operatorname{tr}\mathbf{\underset{\sim}{S}}_i(N)\mathbf{\underset{\sim}{C}}\mathbf{\underset{\sim}{F}}\mathbf{\underset{\sim}{F}}^\tau\mathbf{\underset{\sim}{C}}^\tau$$

$$+ E\left\{\sum_{i=0}^{N-1}\mathbf{x}_i^\tau\mathbf{\underset{\sim}{H}}^\tau\mathbf{\underset{\sim}{Q}}_1\mathbf{\underset{\sim}{H}}\mathbf{x}_i + \mathbf{y}_i^{*\tau}\mathbf{Q}_1\mathbf{y}_i^* - 2\mathbf{y}_i^\tau\mathbf{Q}_1\mathbf{y}_i^*\right.$$

$$+ \mathbf{u}_i^\tau\mathbf{Q}_2\mathbf{u}_i + \mathbf{x}_i^\tau\mathbf{\underset{\sim}{A}}^\tau\mathbf{\underset{\sim}{S}}_{i+1}(N)\mathbf{\underset{\sim}{A}}\mathbf{x}_i + 2\mathbf{x}_i^\tau\mathbf{\underset{\sim}{A}}^\tau\mathbf{\underset{\sim}{S}}_{i+1}(N)\mathbf{\underset{\sim}{B}}\mathbf{u}_i + \mathbf{u}_i^\tau\mathbf{\underset{\sim}{B}}^\tau\mathbf{\underset{\sim}{S}}_{i+1}(N)\mathbf{\underset{\sim}{B}}\mathbf{u}_i$$

$$- \mathbf{x}_i^\tau\mathbf{\underset{\sim}{A}}^\tau\mathbf{\underset{\sim}{S}}_{i+1}(N)\mathbf{\underset{\sim}{A}}\mathbf{x}_i + \mathbf{x}_i^\tau\mathbf{\underset{\sim}{A}}^\tau\mathbf{\underset{\sim}{S}}_{i+1}(N)\mathbf{\underset{\sim}{B}}\mathbf{\underset{\sim}{M}}_i^+\mathbf{\underset{\sim}{B}}^\tau\mathbf{\underset{\sim}{S}}_{i+1}(N)\mathbf{\underset{\sim}{A}}\mathbf{x}_i$$

$$- \mathbf{x}_i^\tau\mathbf{\underset{\sim}{H}}^\tau\mathbf{\underset{\sim}{Q}}_1\mathbf{\underset{\sim}{H}}\mathbf{x}_i + 2(\mathbf{x}_i^\tau\mathbf{\underset{\sim}{A}}^\tau + \mathbf{u}_i^\tau\mathbf{\underset{\sim}{B}}^\tau)\mathbf{b}_{i+1}(N) - 2\mathbf{x}_i^\tau\mathbf{\underset{\sim}{A}}^\tau\mathbf{b}_{i+1}(N)$$

$$+ 2\mathbf{x}_i^\tau\mathbf{\underset{\sim}{A}}^\tau\mathbf{\underset{\sim}{S}}_{i+1}(N)\mathbf{\underset{\sim}{B}}\mathbf{\underset{\sim}{M}}_i^+\mathbf{\underset{\sim}{B}}^\tau\mathbf{b}_{i+1}(N) + 2\mathbf{x}_i^\tau\mathbf{\underset{\sim}{H}}^\tau\mathbf{\underset{\sim}{Q}}_1\mathbf{y}_i^*$$

$$\left. + \mathbf{b}_{i+1}^\tau(N)\mathbf{\underset{\sim}{B}}\mathbf{\underset{\sim}{M}}_i^+\mathbf{\underset{\sim}{B}}^\tau\mathbf{b}_{i+1}(N) - \mathbf{y}_i^{*\tau}\mathbf{Q}_1\mathbf{y}_i^*\right\}$$

$$= d_0(N) + 2(E\mathbf{x}_0)^\tau \mathbf{b}_0(N) + E\mathbf{x}_0^\tau \underset{\sim}{\mathbf{S}}_0(N)\mathbf{x}_0 + \sum_{i=0}^{N-1} \operatorname{tr} \underset{\sim}{\mathbf{S}}_i(N)\underset{\sim}{\mathbf{C}}\underset{\sim}{\mathbf{F}}\underset{\sim}{\mathbf{F}}^\tau \underset{\sim}{\mathbf{C}}^\tau$$

$$+E \sum_{i=0}^{N-1} \left[\mathbf{u}_i + \underset{\sim}{\mathbf{M}}_i^+ \underset{\sim}{\mathbf{B}}^\tau(\underset{\sim}{\mathbf{S}}_{i+1}(N)\underset{\sim}{\mathbf{A}}\mathbf{x}_i + \mathbf{b}_{i+1}(N))\right]^\tau$$

$$\times \underset{\sim}{\mathbf{M}}_i\left[\mathbf{u}_i + \underset{\sim}{\mathbf{M}}_i^+ \underset{\sim}{\mathbf{B}}^\tau(\underset{\sim}{\mathbf{S}}_{i+1}(N)\underset{\sim}{\mathbf{A}}\mathbf{x}_i + \mathbf{b}_{i+1}(N)\right]. \tag{4.145}$$

If N is fixed, then

$$\mathbf{u}_i = -\underset{\sim}{\mathbf{M}}_i^+ \underset{\sim}{\mathbf{B}}^\tau(\underset{\sim}{\mathbf{S}}_{i+1}(N)\underset{\sim}{\mathbf{A}}\mathbf{x}_i + \mathbf{b}_{i+1}(N)) \tag{4.146}$$

makes (4.145) reach its minimum.

We now need to introduce the following basic system theoretic conditions on the matrices $\underset{\sim}{\mathbf{A}}$, $\underset{\sim}{\mathbf{B}}$ and $\underset{\sim}{\mathbf{D}}$ (see, e.g., [75, 76]):

A_3. $(\underset{\sim}{\mathbf{A}}, \underset{\sim}{\mathbf{B}})$ is controllable, that is, the matrix $[\underset{\sim}{\mathbf{B}} \quad \underset{\sim}{\mathbf{A}}\underset{\sim}{\mathbf{B}}\, \underset{\sim}{\mathbf{A}}^2\underset{\sim}{\mathbf{B}} \cdots \underset{\sim}{\mathbf{A}}^{n-1}\underset{\sim}{\mathbf{B}}]$ with $n = p'm$ is of full rank, and $(\underset{\sim}{\mathbf{A}}, \underset{\sim}{\mathbf{D}})$ is observable. That is, the matrix

$$\begin{bmatrix} \underset{\sim}{\mathbf{D}} \\ \underset{\sim}{\mathbf{D}}\underset{\sim}{\mathbf{A}} \\ \vdots \\ \underset{\sim}{\mathbf{D}}\underset{\sim}{\mathbf{A}}^{n-1} \end{bmatrix}$$

is of full rank. Here $\underset{\sim}{\mathbf{D}}$ is any matrix such that $\underset{\sim}{\mathbf{D}}\underset{\sim}{\mathbf{D}}^\tau = \underset{\sim}{\mathbf{H}}^\tau \underset{\sim}{\mathbf{Q}}_1 \underset{\sim}{\mathbf{H}}$.

B_3. All zeros of $\det(\underset{\sim}{\mathbf{I}} - \underset{\sim}{\mathbf{A}}_1\lambda - \cdots - \underset{\sim}{\mathbf{A}}_p\lambda^p)$ are outside the closed unit disk and $p = p'$ (i.e., $p \geq q$, $p \geq r + 1$).

We shall use the following fact which can be found in many references on linear system theory (see, e.g., [75]).

Under Condition A_3, if $\underset{\sim}{\mathbf{Q}}_2 > 0$ $\underset{\sim}{\mathbf{S}}_k(N)$ tends to $\underset{\sim}{\mathbf{S}}$ in the sense that

$$\underset{\sim}{\mathbf{S}}_k(N) \underset{N-k\to\infty}{\longrightarrow} \underset{\sim}{\mathbf{S}}, \tag{4.147}$$

where $\underset{\sim}{\mathbf{S}}$ is the unique positive definite solution of the algebraic Riccati equation

$$\underset{\sim}{\mathbf{S}} = \underset{\sim}{\mathbf{A}}^\tau \underset{\sim}{\mathbf{S}}\underset{\sim}{\mathbf{A}} - \underset{\sim}{\mathbf{A}}^\tau \underset{\sim}{\mathbf{S}}\underset{\sim}{\mathbf{B}}\left(\underset{\sim}{\mathbf{Q}}_2 + \underset{\sim}{\mathbf{B}}^\tau \underset{\sim}{\mathbf{S}}\underset{\sim}{\mathbf{B}}\right)^{-1} \underset{\sim}{\mathbf{B}}^\tau \underset{\sim}{\mathbf{S}}\underset{\sim}{\mathbf{A}} + \underset{\sim}{\mathbf{H}}^\tau \underset{\sim}{\mathbf{Q}}_1 \underset{\sim}{\mathbf{H}} \tag{4.148}$$

and all eigenvalues of

$$\mathbf{A} - \mathbf{B}\left(\mathbf{Q}_2 + \mathbf{B}^\tau \mathbf{S}\mathbf{B}\right)^{-1} \mathbf{B}^\tau \mathbf{S}\mathbf{A}$$

lie inside the open unit disk.

Consequently, similar to (2.82), there exist $k_0 > 0$ and $\rho \in [0,1)$ such that

$$\left\| \left[\mathbf{A} - \mathbf{B}\left(\mathbf{Q}_2 + \mathbf{B}^\tau \mathbf{S}\mathbf{B}\right)^{-1} \mathbf{B}^\tau \mathbf{S}\mathbf{A} \right]^k \right\| \le k_0 \rho^k, \qquad \forall k = 0,1,\ldots \quad (4.149)$$

Lemma 4.6. *If all eigenvalues of a matrix $\mathbf{\Phi}$ lie inside the open unit disk, then*

$$\mathbf{M} \triangleq \sum_{i=0}^{\infty} \mathbf{\Phi}^{i\tau} \mathbf{\Phi}^i < \infty. \quad (4.150)$$

Further, if there is a constant $\varepsilon \in (0,1]$ such that for random matrices $\mathbf{\Phi}_1, \ldots, \mathbf{\Phi}_s$

$$\mathbf{I} - \mathbf{\Phi}^\tau \mathbf{M} \mathbf{\Gamma}_k - \mathbf{\Gamma}_k^\tau \mathbf{M} \mathbf{\Phi} - \mathbf{\Gamma}_k^\tau \mathbf{M} \mathbf{\Gamma}_k \ge \varepsilon \mathbf{M}, \qquad \forall \omega, \quad \forall k = 1, \ldots, s \quad (4.151)$$

with

$$\mathbf{\Gamma}_k = \mathbf{\Phi}_k - \mathbf{\Phi},$$

then

$$\left\| \prod_{i=1}^{s} \mathbf{\Phi}_i \right\| \le \sqrt{\|\mathbf{M}\|}\, \lambda^s, \quad (4.152)$$

where $0 \le \lambda = \sqrt{1-\varepsilon} < 1$.

Proof. By the condition on the eigenvalues of $\mathbf{\Phi}$, we have

$$\|\mathbf{\Phi}^k\| \le k_1 \mu^k, \quad k_1 > 0, \quad \mu \in [0,1), \qquad \forall k$$

which yields (4.150).

Now consider

$$\mathbf{z}_{k+1} = \Phi_k \mathbf{z}_k, \qquad k = 1, \ldots, s.$$

Noticing that

$$\mathbf{M} = \Phi^\tau \mathbf{M} \Phi + \mathbf{I} > \mathbf{I}, \tag{4.153}$$

by (4.151), we have

$$\begin{aligned}
\mathbf{z}_{k+1}^\tau \mathbf{M} \mathbf{z}_{k+1} &= \mathbf{z}_k^\tau \Phi_k^\tau \mathbf{M} \Phi_k \mathbf{z}_k = \mathbf{z}_k^\tau (\Phi^\tau + \Gamma_k^\tau) \mathbf{M} (\Phi + \Gamma_k) \mathbf{z}_k \\
&= \mathbf{z}_k^\tau (\Phi^\tau \mathbf{M} \Phi + \Phi^\tau \mathbf{M} \Gamma_k + \Gamma_k^\tau \mathbf{M} \Phi + \Gamma_k^\tau \mathbf{M} \Gamma_k) \mathbf{z}_k \\
&= \mathbf{z}_k^\tau (\mathbf{M} - \mathbf{I} + \Phi^\tau \mathbf{M} \Gamma_k + \Gamma_k^\tau \mathbf{M} \Phi + \Gamma_k \mathbf{M} \Gamma_k) \mathbf{z}_k \\
&\le (1 - \varepsilon) \mathbf{z}_k^\tau \mathbf{M} \mathbf{z}_k, \qquad \forall k = 1, \ldots, s.
\end{aligned}$$

Hence it follows that

$$\mathbf{z}_{s+1}^\tau \mathbf{M} \mathbf{z}_{s+1} \le (1 - \varepsilon)^s \mathbf{z}_1^\tau \mathbf{M} \mathbf{z}_1$$

and

$$\mathbf{z}_1^\tau (\Phi_s \cdots \Phi_1)^\tau \mathbf{M} (\Phi_s \cdots \Phi_1) \mathbf{z}_1 \le (1 - \varepsilon)^s \mathbf{z}_1^\tau \mathbf{M} \mathbf{z}_1.$$

Since $\mathbf{z}_1$ was arbitrary, we conclude from here that

$$(\Phi_s \cdots \Phi_1)^\tau \mathbf{M} (\Phi_s \cdots \Phi_1) \le (1 - \varepsilon)^s \mathbf{M}$$

then by (4.153) the estimate (4.152) follows. □

Denote

$$\Phi = \mathbf{A} - \mathbf{B}\left(\mathbf{Q}_2 + \mathbf{B}^\tau \mathbf{S} \mathbf{B}\right)^{-1} \mathbf{B}^\tau \mathbf{S} \mathbf{A} \tag{4.154}$$

$$\Phi_k = \mathbf{A} - \mathbf{B}\left(\mathbf{Q}_2 + \mathbf{B}^\tau \mathbf{S}_k(N) \mathbf{B}\right)^{-1} \mathbf{B}^\tau \mathbf{S}_k(N) \mathbf{A}, \tag{4.155}$$

$$\mathbf{b}_k^0 = -\mathbf{H}^\tau \mathbf{Q}_1 \mathbf{y}_k^* - \sum_{j=1}^{\infty} \Phi^{j\tau} \mathbf{H}^\tau \mathbf{Q}_1 \mathbf{y}_{k+j}^*, \tag{4.156}$$

$$\sup_{1 \le k \le \infty} \|\mathbf{H}^\tau \mathbf{Q}_1 \mathbf{y}_k^*\| \triangleq k_3 < \infty. \tag{4.157}$$

Clearly,

$$\|\mathbf{b}_k^0\| \le k_3 + k_0 k_3 \frac{\rho}{1-\rho} \triangleq k_4, \tag{4.158}$$

and from (4.141)

$$\mathbf{b}_k(N) = \Phi_{k+1}^\tau \mathbf{b}_{k+1}(N) - \mathbf{H}^\tau \mathbf{Q}_1 \mathbf{y}_k^*$$

$$= -\mathbf{H}^\tau \mathbf{Q}_1 \mathbf{y}_k^* - \sum_{j=1}^{N-k-1} \Phi_{k+1}^\tau \cdots \Phi_{k+j}^\tau \mathbf{H}^\tau \mathbf{Q}_1 \mathbf{y}_{k+j}^*.$$

Lemma 4.7. *If* $\mathbf{Q}_2 > 0$ *and Condition* A_3 *is satisfied, then*

$$\left\|\mathbf{b}_k^0 - \mathbf{b}_k(N)\right\| \underset{N\to\infty}{\to} 0, \qquad \forall k, \tag{4.159}$$

$$\lim_{N\to\infty} \frac{1}{N} \sum_{i=0}^{N-1} \operatorname{tr} \mathbf{S}_i(N) \mathbf{C}\mathbf{F}\mathbf{F}^\tau \mathbf{C}^\tau = \operatorname{tr} \mathbf{S}\mathbf{C}\mathbf{F}\mathbf{F}^\tau \mathbf{C}^\tau \tag{4.160}$$

$$\overline{\lim_{N\to\infty}} \frac{1}{N}\left(d_0(N) + 2E\mathbf{x}_0^\tau \mathbf{b}_0(N) + E\mathbf{x}_0^\tau \mathbf{S}_0(N)\mathbf{x}_0\right)$$

$$= \overline{\lim_{N\to\infty}} \frac{1}{N} \sum_{i=1}^{N} \left(\mathbf{y}_i^{*\tau} \mathbf{Q}_1 \mathbf{y}_i^* - \mathbf{b}_i^{0\tau} \mathbf{B}(\mathbf{Q}_2 + \mathbf{B}^\tau \mathbf{S}\mathbf{B})^{-1} \mathbf{B}\mathbf{b}_i^0\right). \tag{4.161}$$

Proof. By Lemma 4.6 and (4.157) it is easy to see that

$$\left\|\mathbf{b}_k^0 - \mathbf{b}_k(N)\right\| \le \left\| \sum_{j=1}^{k'} \left(\Phi_{k+1}^\tau \cdots \Phi_{k+j}^\tau - \Phi^{j\tau}\right) \mathbf{H}^\tau \mathbf{Q}_1 \mathbf{y}_{k+j}^* \right\|$$

$$+ \left\| \sum_{j=k'+1}^{N-k'-1} \Phi_{k+1}^\tau \cdots \Phi_{k+j}^\tau \mathbf{H}^\tau \mathbf{Q}_1 \mathbf{y}_{k+j}^* \right\| + \left\| \sum_{j=k'+1}^{\infty} \Phi^{j\tau} \mathbf{H}^\tau \mathbf{Q}_1 \mathbf{y}_{k+j}^* \right\| \underset{\substack{N\to\infty \\ k'\to\infty}}{\to} 0.$$

Given $\varepsilon > 0$ by (4.147) there exists k' so that

$$\|\underset{\sim}{\mathbf{S}}_k(N) - \underset{\sim}{\mathbf{S}}\| < \varepsilon, \qquad \forall k \le T - k'.$$

Hence (4.160) is trivial.

From (4.142), we have

$$d_k(N) = \sum_{i=k}^{N-1} \mathbf{y}_i^{*\tau}\underset{\sim}{\mathbf{Q}}_1\mathbf{y}_i^* - \sum_{i=k+1}^{N-1} \mathbf{b}_i^\tau(N)\underset{\sim}{\mathbf{B}}\left(\underset{\sim}{\mathbf{Q}}_2 + \underset{\sim}{\mathbf{B}}^\tau\underset{\sim}{\mathbf{S}}_{i+1}(N)\underset{\sim}{\mathbf{B}}\right)^{-1}\underset{\sim}{\mathbf{B}}^\tau\mathbf{b}_i(N).$$

Hence (4.161) immediately follows from (4.147) and (4.159). □

Lemma 4.8. (1) *If Conditions* A_3 *and* B_3 *are satisfied,* $\underset{\sim}{\mathbf{Q}}_2 > 0$, $\{\mathbf{u}_i\} \in \mathscr{U}$ *and* $\{E\|\mathbf{u}_i\|^2\}$ *is bounded then* $E\|\hat{\mathbf{x}}_i\|^2 (\equiv E\|\mathbf{x}_i\|^2)$ *is also bounded and*

$$J(\mathbf{u}) = \operatorname{tr}\underset{\sim}{\mathbf{S}}\underset{\sim}{\mathbf{C}}\underset{\sim}{\mathbf{F}}\underset{\sim}{\mathbf{F}}^\tau\underset{\sim}{\mathbf{C}}^\tau + \overline{\lim_{N\to\infty}}\left\{\frac{1}{N}\sum_{i=1}^{N}\left(\mathbf{y}_i^{*\tau}\underset{\sim}{\mathbf{Q}}_1\mathbf{y}_i^* - \mathbf{b}_i^{0\tau}\underset{\sim}{\mathbf{B}}\left(\underset{\sim}{\mathbf{Q}}_2 + \underset{\sim}{\mathbf{B}}^\tau\underset{\sim}{\mathbf{S}}\underset{\sim}{\mathbf{B}}\right)^{-1}\underset{\sim}{\mathbf{B}}\mathbf{b}_i^0\right)\right.$$

$$+ E\frac{1}{N}\sum_{i=n}^{N-1}\left[\mathbf{u}_i + \left(\underset{\sim}{\mathbf{Q}}_2 + \underset{\sim}{\mathbf{B}}^\tau\underset{\sim}{\mathbf{S}}\underset{\sim}{\mathbf{B}}\right)^{-1}\underset{\sim}{\mathbf{B}}^\tau\left(\underset{\sim}{\mathbf{S}}\underset{\sim}{\mathbf{A}}\mathbf{x}_i + \mathbf{b}_i^0\right)\right]^\tau\left(\underset{\sim}{\mathbf{Q}}_2 + \underset{\sim}{\mathbf{B}}^\tau\underset{\sim}{\mathbf{S}}\underset{\sim}{\mathbf{B}}\right)$$

$$\left.\times\left[\mathbf{u}_i + \left(\underset{\sim}{\mathbf{Q}}_2 + \underset{\sim}{\mathbf{B}}^\tau\underset{\sim}{\mathbf{S}}\underset{\sim}{\mathbf{B}}\right)^{-1}\underset{\sim}{\mathbf{B}}^\tau\left(\underset{\sim}{\mathbf{S}}\underset{\sim}{\mathbf{A}}\mathbf{x}_i + \mathbf{b}_i^0\right)\right]\right\}, \qquad \forall n \ge 0 \tag{4.162}$$

(2) *If Condition* A_3 *is satisfied and* $\underset{\sim}{\mathbf{Q}}_2 > \underset{\sim}{\mathbf{0}}$, *then for any* n

$$\mathbf{u}_i^*(n) = \begin{cases} \mathbf{0}, & i \le n \\ -\left(\underset{\sim}{\mathbf{Q}}_2 + \underset{\sim}{\mathbf{B}}^\tau\underset{\sim}{\mathbf{S}}\underset{\sim}{\mathbf{B}}\right)^{-1}\underset{\sim}{\mathbf{B}}^\tau\left(\underset{\sim}{\mathbf{S}}\underset{\sim}{\mathbf{A}}\mathbf{x}_i + \mathbf{b}_i^0\right), & i > n \end{cases} \tag{4.163}$$

is an optimal control, and

$$J(\mathbf{u}^*) = \operatorname{tr}\underset{\sim}{\mathbf{S}}\underset{\sim}{\mathbf{C}}\underset{\sim}{\mathbf{F}}\underset{\sim}{\mathbf{F}}^\tau\underset{\sim}{\mathbf{C}}^\tau$$

$$+ \overline{\lim_{N\to\infty}}\frac{1}{N}\sum_{i=1}^{N}\left(\mathbf{y}_i^{*\tau}\underset{\sim}{\mathbf{Q}}_1\mathbf{y}_i^* - \mathbf{b}_i^{0\tau}\underset{\sim}{\mathbf{B}}\left(\underset{\sim}{\mathbf{Q}}_2 + \underset{\sim}{\mathbf{B}}^\tau\underset{\sim}{\mathbf{S}}\underset{\sim}{\mathbf{B}}\right)^{-1}\underset{\sim}{\mathbf{B}}^\tau\mathbf{b}_i^0\right) \tag{4.164}$$

Proof. (1) By Condition B_3, it is obvious that $\{E\|\mathbf{x}_i\|^2\}$ is bounded. Hence

$$\lim_{N\to\infty}\frac{1}{N}\Bigg|E\sum_{i=0}^{N-1}\left[\mathbf{u}_i+\underset{\sim}{\mathbf{M}}_i^{-1}\underset{\sim}{\mathbf{B}}^{\tau}(\underset{\sim}{\mathbf{S}}_{i+1}(N)\underset{\sim}{\mathbf{A}}\mathbf{x}_i+\mathbf{b}_{i+1}(N))\right]^{\tau}$$

$$\times\underset{\sim}{\mathbf{M}}_i\left[\mathbf{u}_i+\underset{\sim}{\mathbf{M}}_i^{-1}\underset{\sim}{\mathbf{B}}^{\tau}(\underset{\sim}{\mathbf{S}}_{i+1}(N)\underset{\sim}{\mathbf{A}}\mathbf{x}_i+\mathbf{b}_{i+1}(N)\right]$$

$$-E\sum_{i=n}^{N-1}\left[\mathbf{u}_i+\left(\underset{\sim}{\mathbf{Q}}_2+\underset{\sim}{\mathbf{B}}^{\tau}\underset{\sim}{\mathbf{S}}\underset{\sim}{\mathbf{B}}\right)^{-1}\underset{\sim}{\mathbf{B}}^{\tau}(\underset{\sim}{\mathbf{S}}\underset{\sim}{\mathbf{A}}\mathbf{x}_i+\mathbf{b}_i^0)\right]^{\tau}\left(\underset{\sim}{\mathbf{Q}}_2+\underset{\sim}{\mathbf{B}}^{\tau}\underset{\sim}{\mathbf{S}}\underset{\sim}{\mathbf{B}}\right)$$

$$\times\left[\mathbf{u}_i+\left(\underset{\sim}{\mathbf{Q}}_2+\underset{\sim}{\mathbf{B}}^{\tau}\underset{\sim}{\mathbf{S}}\underset{\sim}{\mathbf{B}}\right)^{-1}\underset{\sim}{\mathbf{B}}^{\tau}(\underset{\sim}{\mathbf{S}}\underset{\sim}{\mathbf{A}}\mathbf{x}_i+\mathbf{b}_i^0)\right]\Bigg|$$

$$\le\lim_{N\to\infty}\frac{1}{N}E\sum_{i=0}^{N-1}\|\underset{\sim}{\mathbf{S}}_{i+1}(N)-\underset{\sim}{\mathbf{S}}\|\,\|\underset{\sim}{\mathbf{B}}\|^2$$

$$\times\left\|\mathbf{u}_i+\underset{\sim}{\mathbf{M}}_i^{-1}\underset{\sim}{\mathbf{B}}^{\tau}(\underset{\sim}{\mathbf{S}}_{i+1}(N)\underset{\sim}{\mathbf{A}}\mathbf{x}_i+\mathbf{b}_{i+1}(N))\right\|^2$$

$$+\lim_{N\to\infty}\frac{1}{N}E\sum_{i=0}^{N-1}2\|\mathbf{u}_i\|\,\|\underset{\sim}{\mathbf{Q}}_2+\underset{\sim}{\mathbf{B}}^{\tau}\underset{\sim}{\mathbf{S}}\underset{\sim}{\mathbf{B}}\|$$

$$\times\left|\underset{\sim}{\mathbf{M}}_i^{-1}\underset{\sim}{\mathbf{B}}^{\tau}(\underset{\sim}{\mathbf{S}}_{i+1}(N)\underset{\sim}{\mathbf{A}}\mathbf{x}_i+\mathbf{b}_{i+1}(N))\right.$$

$$\left.-\left(\underset{\sim}{\mathbf{Q}}_2+\underset{\sim}{\mathbf{B}}^{\tau}\underset{\sim}{\mathbf{S}}\underset{\sim}{\mathbf{B}}\right)^{-1}\underset{\sim}{\mathbf{B}}^{\tau}(\underset{\sim}{\mathbf{S}}\underset{\sim}{\mathbf{A}}\mathbf{x}_i+\mathbf{b}_i^0)\right|$$

$$+\lim_{N\to\infty}\frac{1}{N}E\sum_{i=0}^{N-1}\left\|\left(\underset{\sim}{\mathbf{Q}}_2+\underset{\sim}{\mathbf{B}}^{\tau}\underset{\sim}{\mathbf{S}}\underset{\sim}{\mathbf{B}}\right)\right\|$$

$$\times\left\|\left(\underset{\sim}{\mathbf{Q}}_2+\underset{\sim}{\mathbf{B}}^{\tau}\underset{\sim}{\mathbf{S}}\underset{\sim}{\mathbf{B}}\right)^{-1}\underset{\sim}{\mathbf{B}}^{\tau}(\underset{\sim}{\mathbf{S}}\underset{\sim}{\mathbf{A}}\mathbf{x}_i+\mathbf{b}_i^0)\right.$$

$$\left.-\underset{\sim}{\mathbf{M}}_i^{-1}\underset{\sim}{\mathbf{B}}^{\tau}(\underset{\sim}{\mathbf{S}}_{i+1}(N)\underset{\sim}{\mathbf{A}}\mathbf{x}_i+\mathbf{b}_{i+1}(N))\right\|$$

$$\times\left\|\left(\underset{\sim}{\mathbf{Q}}_2+\underset{\sim}{\mathbf{B}}^{\tau}\underset{\sim}{\mathbf{S}}\underset{\sim}{\mathbf{B}}\right)^{-1}\underset{\sim}{\mathbf{B}}^{\tau}(\underset{\sim}{\mathbf{S}}\underset{\sim}{\mathbf{A}}\mathbf{x}_i+\mathbf{b}_i^0)\right.$$

$$\left.+\underset{\sim}{\mathbf{M}}_i^{-1}\underset{\sim}{\mathbf{B}}^{\tau}(\underset{\sim}{\mathbf{S}}_{i+1}(N)\underset{\sim}{\mathbf{A}}\mathbf{x}_i+\mathbf{b}_{i+1}(N))\right\|=0,$$

where the limit is zero because of (4.147), the boundedness of $\{E\|\mathbf{u}_i\|^2\}$, $\{\mathbf{M}_i^{-1}\}$, $\{\mathbf{S}_{i+1}(N)\}$, and $\{\mathbf{b}_{i+1}(N)\}$ and Lemma 4.7.
(2) When $\{\mathbf{u}_i^*(n)\}$ given by (4.163) is applied then, for $k > n$, we obtain the stable difference equation

$$\mathbf{x}_{k+1} = \mathbf{\Phi}\mathbf{x}_k - \mathbf{B}\left(\mathbf{Q}_2 + \mathbf{B}^\tau\mathbf{S}\mathbf{B}\right)^{-1}\mathbf{B}^\tau\mathbf{b}_k^0 + \mathbf{C}\mathbf{F}\mathbf{w}_{k+1}.$$

Hence $\{E\|\mathbf{x}_k\|^2\}$ and $\{E\|\mathbf{u}_k^*\|^2\}$ are uniformly bounded, and the rest of the conclusions of the lemma follow from (4.162). □

Lemma 4.9. *Let* $\mathbf{Q}_2 > \mathbf{0}$ *and let* $\{\mathbf{A}_n\}$, $\{\mathbf{B}_n\}$, $\{\mathbf{C}_n\}$ *be consistent estimates for* $\mathbf{A}$, $\mathbf{B}$, $\mathbf{C}$, *respectively. If Conditions* A_3 *and* B_3 *hold, then the equation*

$$\mathbf{S}_n = \mathbf{A}_n^\tau\mathbf{S}_n\mathbf{A}_n - \mathbf{A}_n^\tau\mathbf{S}_n\mathbf{B}_n\left(\mathbf{Q}_2 + \mathbf{B}_n^\tau\mathbf{S}_n\mathbf{B}_n\right)^{-1}\mathbf{B}_n^\tau\mathbf{S}_n\mathbf{A}_n + \mathbf{H}^\tau\mathbf{Q}_1\mathbf{H} \tag{4.165}$$

has a unique positive definite solution $\mathbf{S}_n$ *and*

$$\mathbf{S}_n \underset{n\to\infty}{\to} \mathbf{S},$$

where $\mathbf{S}$ *is the solution of* (4.148).

Proof. It is clear that $(\mathbf{A}_n, \mathbf{B}_n)$, $(\mathbf{A}_n, \mathbf{D})$ are controllable and observable, respectively, for sufficiently large n. Hence there exists a unique $\mathbf{S}_n$ satisfying (4.165).

First we show that $\{\|\mathbf{S}_n\|\}$ is bounded in n. By Lemma 4.6 and Condition B_3 there exists k_5, $\lambda \in [0, 1)$ and n' such that

$$\|\mathbf{A}_n^s\| \le k_5\lambda^s, \qquad \forall s, \quad \forall n \ge n'.$$

From (4.165), we have

$$\mathbf{S}_n \le \mathbf{A}_n^\tau\mathbf{S}_n\mathbf{A}_n + \mathbf{H}^\tau\mathbf{Q}_1\mathbf{H} \le \cdots \le \sum_{i=0}^{\infty}\mathbf{A}_n^{i\tau}\mathbf{H}^\tau\mathbf{Q}_1\mathbf{H}\mathbf{A}_n^i.$$

Hence

$$\|\mathbf{S}_n\| \le k_5^2\|\mathbf{H}^\tau\mathbf{Q}_1\mathbf{H}\|\sum_{i=0}^{\infty}\lambda^{2i} = k_5^2\|\mathbf{H}^\tau\mathbf{Q}_1\mathbf{H}\|\frac{1}{1-\lambda^2}. \qquad \forall n$$

For simplicity, we denote the right-hand side of (4.148) and (4.165) by $P(\mathbf{A}, \mathbf{B}, \mathbf{S})$ and $P(\mathbf{A}_n, \mathbf{B}_n, \mathbf{S}_n)$, respectively, and set

$$\boldsymbol{\varepsilon}_n = \mathbf{P}(\mathbf{A}, \mathbf{B}, \mathbf{S}_n) - \mathbf{P}(\mathbf{A}_n, \mathbf{B}_n, \mathbf{S}_n).$$

By the consistency of $\{\mathbf{A}_n\}$ and $\{\mathbf{B}_n\}$ and the boundedness of $\{\mathbf{S}_n\}$ it is easy to prove that

$$\mathbf{A}^T\mathbf{S}_n\mathbf{A} - \mathbf{A}_n^T\mathbf{S}_n\mathbf{A}_n = (\mathbf{A}^T - \mathbf{A}_n^T)\mathbf{S}_n\mathbf{A} + \mathbf{A}_n^T\mathbf{S}_n(\mathbf{A} - \mathbf{A}_n) \underset{n\to\infty}{\to} \mathbf{0},$$

$$\left(\mathbf{Q}_2 + \mathbf{B}^T\mathbf{S}_n\mathbf{B}\right)^{-1} - \left(\mathbf{Q}_2 + \mathbf{B}_n^T\mathbf{S}_n\mathbf{B}_n\right)^{-1}$$

$$= \left(\mathbf{Q}_2 + \mathbf{B}^T\mathbf{S}_n\mathbf{B}\right)^{-1}(\mathbf{B}_n^T\mathbf{S}_n\mathbf{B}_n - \mathbf{B}^T\mathbf{S}_n\mathbf{B})\left(\mathbf{Q}_2 + \mathbf{B}_n^T\mathbf{S}_n\mathbf{B}_n\right)^{-1} \underset{n\to\infty}{\to} \mathbf{0}.$$

From here it follows that

$$\mathbf{A}^T\mathbf{S}_n\mathbf{B}\left(\mathbf{Q}_2 + \mathbf{B}^T\mathbf{S}_n\mathbf{B}\right)^{-1}\mathbf{B}^T\mathbf{S}_n\mathbf{A} - \mathbf{A}_n^T\mathbf{S}_n\mathbf{B}_n\left(\mathbf{Q}_2 + \mathbf{B}_n^T\mathbf{S}_n\mathbf{B}_n\right)^{-1}\mathbf{B}_n^T\mathbf{S}_n\mathbf{A}_n \to \mathbf{0}$$

and hence

$$\boldsymbol{\varepsilon}_n \underset{n\to\infty}{\to} \mathbf{0},$$

$$\|\mathbf{S}_n - \mathbf{P}(\mathbf{A}, \mathbf{B}, \mathbf{S}_n)\| \le \|\boldsymbol{\varepsilon}_n\| \underset{n\to\infty}{\to} 0. \tag{4.166}$$

By the boundedness of $\|\mathbf{S}_n\|$, there is a convergent subsequence $\mathbf{S}_{n_k}$ of $\mathbf{S}_n$:

$$\mathbf{S}_{n_k} \underset{k\to\infty}{\to} \mathbf{S}',$$

and from (4.166)

$$\mathbf{S}' = \mathbf{P}(\mathbf{A}, \mathbf{B}, \mathbf{S}').$$

But the solution of (4.146) is unique. Hence any convergent subsequence of $\mathbf{S}_n$ converges to the same limit $\mathbf{S}$. This means that

$$\mathbf{S}_n \to \mathbf{S}.$$

□

Lemma 4.10. *If* $\underset{\sim}{\mathbf{Q}}_2 > 0$, $\{\underset{\sim}{\mathbf{A}}_n\}$, $\{\underset{\sim}{\mathbf{B}}_n\}$ *are consistent estimates of* $\underset{\sim}{\mathbf{A}}$ *and* $\underset{\sim}{\mathbf{B}}$, *and Conditions* A_3 *and* B_3 *hold, then*

$$\|\mathbf{b}_k^0 - \mathbf{b}_k\| \underset{k\to\infty}{\to} 0$$

where

$$\mathbf{b}_k \triangleq -\underset{\sim}{\mathbf{H}}^\tau \underset{\sim}{\mathbf{Q}}_1 \mathbf{y}_k^* - \sum_{j=1}^{\infty} \left[\underset{\sim}{\mathbf{A}}_k - \underset{\sim}{\mathbf{B}}_k\left(\underset{\sim}{\mathbf{Q}}_2 + \underset{\sim}{\mathbf{B}}_k^\tau \underset{\sim}{\mathbf{S}}_k \underset{\sim}{\mathbf{B}}_k\right)^{-1} \underset{\sim}{\mathbf{B}}_k^\tau \underset{\sim}{\mathbf{S}}_k \underset{\sim}{\mathbf{A}}_k\right]^{j\tau} \underset{\sim}{\mathbf{H}}^\tau \underset{\sim}{\mathbf{Q}}_1 \mathbf{y}_{k+j}^*.$$

Proof. By Lemma 4.9, we know that

$$\underset{\sim}{\mathbf{A}}_k - \underset{\sim}{\mathbf{B}}_k\left(\underset{\sim}{\mathbf{Q}}_2 + \underset{\sim}{\mathbf{B}}_k^\tau \underset{\sim}{\mathbf{S}}_k \underset{\sim}{\mathbf{B}}_k\right)^{-1} \underset{\sim}{\mathbf{B}}_k^\tau \underset{\sim}{\mathbf{S}}_k \underset{\sim}{\mathbf{A}}_k \underset{k\to\infty}{\to} \underset{\sim}{\mathbf{\Phi}},$$

where $\underset{\sim}{\mathbf{\Phi}}$ is given by (4.154). By Lemma 4.6,

$$\left\|\left(\underset{\sim}{\mathbf{A}}_k - \underset{\sim}{\mathbf{B}}_k\left(\underset{\sim}{\mathbf{Q}}_2 + \underset{\sim}{\mathbf{B}}_k^\tau \underset{\sim}{\mathbf{S}}_k \underset{\sim}{\mathbf{B}}_k\right)^{-1} \underset{\sim}{\mathbf{B}}_k^\tau \underset{\sim}{\mathbf{S}}_k \underset{\sim}{\mathbf{A}}_k\right)^s\right\| \le k_6 \lambda^s, \qquad \forall s, \quad \forall k \ge k',$$

where $\lambda \in [0,1)$, and k_6 is a constant and k' is some integer. Hence we have

$$\begin{aligned}
&\|\mathbf{b}_k - \mathbf{b}_k^0\| \\
&\le \left\| \sum_{j=1}^{k'-1} \left\{ \underset{\sim}{\mathbf{\Phi}}^{j\tau} - \left[\underset{\sim}{\mathbf{A}}_k - \underset{\sim}{\mathbf{B}}_k\left(\underset{\sim}{\mathbf{Q}}_2 + \underset{\sim}{\mathbf{B}}_k^\tau \underset{\sim}{\mathbf{S}}_k \underset{\sim}{\mathbf{B}}_k\right)^{-1} \underset{\sim}{\mathbf{B}}_k^\tau \underset{\sim}{\mathbf{S}}_k \underset{\sim}{\mathbf{A}}_k\right]^{j\tau} \right\} \underset{\sim}{\mathbf{H}}^\tau \underset{\sim}{\mathbf{Q}}_1 \mathbf{y}_{k+j}^* \right\| \\
&\quad + \sum_{j=k'}^{\infty} k_1 \mu^j \|\underset{\sim}{\mathbf{H}}^\tau \underset{\sim}{\mathbf{Q}}_1 \mathbf{y}_{k+j}^*\| + \sum_{j=k'}^{\infty} k_6 \lambda^j \|\underset{\sim}{\mathbf{H}}^\tau \underset{\sim}{\mathbf{Q}}_1 \mathbf{y}_{k+j}^*\| \underset{\substack{k\to\infty \\ k'\to\infty}}{\to} 0. \qquad \square
\end{aligned}$$

Let the consistent estimates $\{\underset{\sim}{\mathbf{A}}_n\}$, $\{\underset{\sim}{\mathbf{B}}_n\}$, and $\{\underset{\sim}{\mathbf{C}}_n\}$ be $\mathscr{F}_n^{\mathbf{y}}$-adapted for each $n \ge 0$.

For any $n \ge 0$ and $\varepsilon > 0$, define

$$\tau(\varepsilon, n) = \begin{cases} \inf[k \ge n : \|\underset{\sim}{\mathbf{A}}_k - \underset{\sim}{\mathbf{A}}\| \vee \|\underset{\sim}{\mathbf{B}}_k - \underset{\sim}{\mathbf{B}}\| \vee \|\underset{\sim}{\mathbf{C}}_k - \underset{\sim}{\mathbf{C}}\| \\ \qquad \vee \|\mathbf{b}_k - \mathbf{b}_k^0\| \vee \|\underset{\sim}{\mathbf{S}}_k - \underset{\sim}{\mathbf{S}}\| > \varepsilon] \\ \infty \quad \text{if } \|\underset{\sim}{\mathbf{A}}_k - \underset{\sim}{\mathbf{A}}\| \vee \|\underset{\sim}{\mathbf{B}}_k - \underset{\sim}{\mathbf{B}}\| \vee \|\underset{\sim}{\mathbf{C}}_k - \underset{\sim}{\mathbf{C}}\| \\ \qquad \vee \|\mathbf{b}_k - \mathbf{b}_k^0\| \vee \|\underset{\sim}{\mathbf{S}}_k - \underset{\sim}{\mathbf{S}}\| \le \varepsilon, \qquad \forall k \ge n, \end{cases} \tag{4.167}$$

It is clear that

$$[\tau(\varepsilon, n) \le n + s]$$

is $\mathscr{F}^{\mathbf{y}}_{n+s}$-measurable for any $s \ge 0$.

From now on, we assume the upper bound c for

$$\|\underset{\sim}{\mathbf{A}}\| \vee \|\underset{\sim}{\mathbf{B}}\| \vee \|\underset{\sim}{\mathbf{C}}\| \vee \|\underset{\sim}{\mathbf{S}}\| \vee \|\mathbf{b}^0_k\| < c$$

is known. Define

$$\tau_c(n) = \begin{cases} \inf[k \ge n : \|\underset{\sim}{\mathbf{A}}_k\| \vee \|\underset{\sim}{\mathbf{B}}_k\| \vee \|\underset{\sim}{\mathbf{C}}_k\| \vee \|\underset{\sim}{\mathbf{S}}_k\| \vee \|\mathbf{b}_k\| > c] \\ \infty \quad \text{if } \|\underset{\sim}{\mathbf{A}}_k\| \vee \|\underset{\sim}{\mathbf{B}}_k\| \vee \|\underset{\sim}{\mathbf{C}}_k\| \vee \|\underset{\sim}{\mathbf{S}}_k\| \vee \|\mathbf{b}_k\| \le c, \quad \forall k \ge n, \end{cases} \tag{4.168}$$

$$\underset{\sim}{\mathbf{A}}_k(c) = \underset{\sim}{\mathbf{A}}_k I_{[n<k<\tau_c(n)]}, \qquad \underset{\sim}{\mathbf{B}}_k(c) = \underset{\sim}{\mathbf{B}}_k I_{[\tau<k<\tau_c(n)]},$$

$$\underset{\sim}{\mathbf{C}}_k(c) = \underset{\sim}{\mathbf{C}}_k I_{[n<k<\tau_c(n)]}, \qquad \underset{\sim}{\mathbf{S}}_k(c) = \underset{\sim}{\mathbf{S}}_k I_{[n<k<\tau_c(n)]},$$

$$\mathbf{b}_k(c) = \mathbf{b}_k I_{[n<k<\tau_c(n)]}.$$

If $\tau_c(n)$ is replaced by $\tau(\varepsilon, n)$, then they are denoted by $\underset{\sim}{\mathbf{A}}_k(\tau)$, $\underset{\sim}{\mathbf{B}}_k(\tau)$, $\underset{\sim}{\mathbf{C}}_k(\tau)$, $\underset{\sim}{\mathbf{S}}_k(\tau)$, and $\mathbf{b}_k(\tau)$, respectively.

Define

$$\mathbf{u}^\varepsilon_k(n) = \begin{cases} -\left(\underset{\sim}{\mathbf{Q}}_2 + \underset{\sim}{\mathbf{B}}^\tau_k(\tau)\underset{\sim}{\mathbf{S}}_k(\tau)\underset{\sim}{\mathbf{B}}_k(\tau)\right)^{-1} \underset{\sim}{\mathbf{B}}^\tau_k(\tau)(\underset{\sim}{\mathbf{S}}_k(\tau)\underset{\sim}{\mathbf{A}}_k(\tau)\hat{\mathbf{x}}^\varepsilon_k + \mathbf{b}_k(\tau)), \\ \hfill k > n \\ 0, \hfill k \le n, \end{cases} \tag{4.169}$$

$$\mathbf{u}^a_k(n) = \begin{cases} -\left(\underset{\sim}{\mathbf{Q}}_2 + \underset{\sim}{\mathbf{B}}^\tau_k(c)\underset{\sim}{\mathbf{S}}_k(c)\underset{\sim}{\mathbf{B}}_k(c)\right)^{-1} \underset{\sim}{\mathbf{B}}^\tau_k(c)(\underset{\sim}{\mathbf{S}}_k(c)\underset{\sim}{\mathbf{A}}_k(c)\hat{\mathbf{x}}^a_k + \mathbf{b}_k(c)), \\ \hfill k > n \\ 0, \hfill k \le n \end{cases} \tag{4.170}$$

where $\hat{\mathbf{x}}_k^\varepsilon$ and $\hat{\mathbf{x}}_k^a$ are defined by

$$\hat{\mathbf{x}}_{k+1}^\varepsilon = \underset{\sim}{\mathbf{A}}_k(\tau)\hat{\mathbf{x}}_k^\varepsilon + \underset{\sim}{\mathbf{B}}_k(\tau)\mathbf{u}_k^\varepsilon(n)$$

$$+\underset{\sim}{\mathbf{C}}_k(\tau)(\mathbf{y}_{k+1} - \underset{\sim}{\mathbf{H}}\underset{\sim}{\mathbf{A}}_k(\tau)\hat{\mathbf{x}}_k^\varepsilon - \underset{\sim}{\mathbf{H}}\underset{\sim}{\mathbf{B}}_k(\tau)\mathbf{u}_k^\varepsilon(n)) \quad (4.171)$$

$$\hat{\mathbf{x}}_{k+1}^a = \underset{\sim}{\mathbf{A}}_k(c)\hat{\mathbf{x}}_k^a + \underset{\sim}{\mathbf{B}}_k(c)\mathbf{u}_k^a(n)$$

$$+\underset{\sim}{\mathbf{C}}_k(c)(\mathbf{y}_{k+1} - \underset{\sim}{\mathbf{H}}\underset{\sim}{\mathbf{A}}_k(c)\hat{\mathbf{x}}_k^a - \underset{\sim}{\mathbf{H}}\underset{\sim}{\mathbf{B}}_k(c)\mathbf{u}_k^a(n)), \quad (4.172)$$

respectively, and

$$\hat{\mathbf{x}}_n^\varepsilon = \hat{\mathbf{x}}_n^a = [\mathbf{y}_n^\tau, \mathbf{0} \cdots \mathbf{0}].$$

It is clear that the control $\mathbf{u}_k^\varepsilon(n)$ can be computed at each instant k whenever c has been specified.

Next we need Condition C_3:

C_3. All zeros of $\det(\underset{\sim}{\mathbf{I}} + \underset{\sim}{\mathbf{C}}_1\lambda + \cdots + \underset{\sim}{\mathbf{C}}_r\lambda^r)$ are outside the closed unit disk and $r = p - 1$.

Remark. If $r < p - 1$, then the last elements of $\hat{\mathbf{x}}_k^a$ cannot be corrected by new measurements.

Denote

$$\hat{\mathbf{x}}_k^{\varepsilon\tau} = [\hat{\mathbf{x}}_k^{\varepsilon 1\tau} \cdots \hat{\mathbf{x}}_k^{\varepsilon p\tau}]$$

$$\mathbf{z}_k = [\mathbf{x}_k^{2\tau} \cdots \mathbf{x}_k^{p\tau}], \qquad \hat{\mathbf{z}}_k^\varepsilon = [\hat{\mathbf{x}}_k^{\varepsilon 2\tau} \cdots \hat{\mathbf{x}}_k^{\varepsilon p\tau}]$$

Lemma 4.11. *Suppose that* $\underset{\sim}{\mathbf{Q}}_2 > \underset{\sim}{\mathbf{0}}$, *the* $\{\mathscr{F}_k^y\}$*-adapted processes* $\{\underset{\sim}{\mathbf{A}}_k\}$, $\{\underset{\sim}{\mathbf{B}}_k\}$, *and* $\{\underset{\sim}{\mathbf{C}}_k\}$ *are consistent and Conditions* A_3–C_3 *hold. Then* $E\|\hat{\mathbf{x}}_k^\varepsilon\|^2$ *and* $E\|\mathbf{u}_k^\varepsilon(n)\|^2$ *are uniformly bounded for* $k \geq 0$ *and* $\varepsilon \in [0, \varepsilon_1)$ *for some small* ε_1.

Proof. From (4.17), we obtain the equation for $\hat{\mathbf{x}}_k^{\varepsilon 1}$

$$\hat{\mathbf{x}}_{k+1}^{\varepsilon 1} = \underset{\sim}{\mathbf{A}}_{1k}(\tau)\hat{\mathbf{x}}_k^{\varepsilon 1} + \hat{\mathbf{x}}_k^{\varepsilon 2} + \underset{\sim}{\mathbf{B}}_{1k}(\tau)\mathbf{u}_k^\varepsilon(n)$$

$$+(\underset{\sim}{\mathbf{H}}\underset{\sim}{\mathbf{A}}\mathbf{x}_k + \underset{\sim}{\mathbf{H}}\underset{\sim}{\mathbf{B}}\mathbf{u}_k^\varepsilon(n) + \underset{\sim}{\mathbf{F}}\mathbf{w}_{k+1} - \underset{\sim}{\mathbf{H}}\underset{\sim}{\mathbf{A}}_k(\tau)\hat{\mathbf{x}}_k^\varepsilon - \underset{\sim}{\mathbf{H}}\underset{\sim}{\mathbf{B}}_k(\tau)\mathbf{u}_k^\varepsilon(n))$$

$$= \underset{\sim}{\mathbf{A}}_1\mathbf{x}_k^1 + \mathbf{x}_k^2 + \mathbf{B}_1\mathbf{u}_k^\varepsilon(n) + \underset{\sim}{\mathbf{F}}\mathbf{w}_{k+1} = \mathbf{x}_{k+1}^1,$$

where the $\mathbf{x}_k^i$ are the components of $\mathbf{x}_k$:

$$\mathbf{x}_k = [\mathbf{x}_k^{1\tau} \cdots \mathbf{x}_k^{p\tau}]^\tau.$$

Hence we have

$$\begin{aligned}
&\underset{\sim}{C}_k(\tau)(\mathbf{y}_{k+1} - \underset{\sim}{H}\underset{\sim}{A}_k(\tau)\hat{\mathbf{x}}_k^\varepsilon - \underset{\sim}{H}\underset{\sim}{B}_k(\tau)\mathbf{u}_k^\varepsilon(n)) \\
&\quad = \underset{\sim}{C}_k(\tau)\underset{\sim}{H}\underset{\sim}{A}(\mathbf{x}_k - \hat{\mathbf{x}}_k^\varepsilon) + \underset{\sim}{C}_k(\tau)\underset{\sim}{H}(\underset{\sim}{A} - \underset{\sim}{A}_k(\tau))\hat{\mathbf{x}}_k^\varepsilon \\
&\qquad + \underset{\sim}{C}_k(\tau)\underset{\sim}{H}(\underset{\sim}{B} - \underset{\sim}{B}_k(\tau))\mathbf{u}_k^\varepsilon(n) + \underset{\sim}{C}_k(\tau)\underset{\sim}{F}\mathbf{w}_{k+1} \\
&\quad = \underset{\sim}{C}_k^0(\tau)(\mathbf{z}_k - \hat{\mathbf{z}}_k^\varepsilon) + \underset{\sim}{C}_k(\tau)\underset{\sim}{H}(\underset{\sim}{A} - \underset{\sim}{A}_k(\tau))\hat{\mathbf{x}}_k^\varepsilon \\
&\qquad + \underset{\sim}{C}_k(\tau)\underset{\sim}{H}(\underset{\sim}{B} - \underset{\sim}{B}_k(\tau))\mathbf{u}_k^\varepsilon(n) + \underset{\sim}{C}_k(\tau)\underset{\sim}{F}\mathbf{w}_{k+1},
\end{aligned} \tag{4.173}$$

where

$$\underset{\sim}{C}_k^0(\tau) \triangleq \begin{bmatrix} \underset{\sim}{I} & \underset{\sim}{0} & \cdots & \underset{\sim}{0} \\ \underset{\sim}{C}_{1k}(\tau) & \vdots & & \vdots \\ \vdots & & & \\ \underset{\sim}{C}_{p-1k}(\tau) & \underset{\sim}{0} & \cdots & \underset{\sim}{0} \end{bmatrix}.$$

By (4.173), we can rewrite (4.171) as

$$\begin{aligned}
\hat{\mathbf{x}}_{k+1}^\varepsilon = \Big[&\underset{\sim}{A}_k(\tau) - \underset{\sim}{B}_k(\tau)\Big(\underset{\sim}{Q}_2 + \underset{\sim}{B}_k^\tau(\tau)\underset{\sim}{S}_k(\tau)\underset{\sim}{B}_k(\tau)\Big)^{-1}\underset{\sim}{B}_k^\tau(\tau)\underset{\sim}{A}_k(\tau) \\
&+ \underset{\sim}{C}_k(\tau)\underset{\sim}{H}(\underset{\sim}{A} - \underset{\sim}{A}_k(\tau)) - \underset{\sim}{C}_k(\tau)\underset{\sim}{H}(\underset{\sim}{B} - \underset{\sim}{B}_k(\tau)) \\
&\times\Big(\underset{\sim}{Q}_2 + \underset{\sim}{B}_k^\tau(\tau)\underset{\sim}{S}_k(\tau)\underset{\sim}{B}_k(\tau)\Big)^{-1}\underset{\sim}{B}_k^\tau(\tau)\underset{\sim}{S}_k(\tau)\underset{\sim}{A}_k(\tau)\Big)\Big]\hat{\mathbf{x}}_k^\varepsilon \\
&+ \underset{\sim}{C}_k^0(\tau)(\mathbf{z}_k - \hat{\mathbf{z}}_k^\varepsilon) - \big[\underset{\sim}{B}_k(\tau) + \underset{\sim}{C}_k(\tau)\underset{\sim}{H}(\underset{\sim}{B} - \underset{\sim}{B}_k(\tau))\big] \\
&\times\Big(\underset{\sim}{Q}_2 + \underset{\sim}{B}_k^\tau(\tau)\underset{\sim}{S}_k(\tau)\underset{\sim}{B}_k(\tau)\Big)^{-1}\cdot\underset{\sim}{B}_k^\tau(\tau)\mathbf{b}_k(\tau) + \underset{\sim}{C}_k(\tau)\underset{\sim}{F}\mathbf{w}_{k+1},
\end{aligned}$$

and we have

$$\begin{aligned}
\mathbf{x}_{k+1} - \hat{\mathbf{x}}_{k+1}^\varepsilon = {}&\underset{\sim}{A}(\mathbf{x}_k - \hat{\mathbf{x}}_k^\varepsilon) + (\underset{\sim}{A} - \underset{\sim}{A}_k(\tau))\hat{\mathbf{x}}_k^\varepsilon + (\underset{\sim}{B} - \underset{\sim}{B}_k(\tau))\mathbf{u}_k^\varepsilon(n) \\
&+ (\underset{\sim}{C} - \underset{\sim}{C}_k(\tau))\underset{\sim}{F}\mathbf{w}_{k+1} - \underset{\sim}{C}_k^0(\tau)(\mathbf{z}_k - \hat{\mathbf{z}}_k^\varepsilon) \\
&- \underset{\sim}{C}_k(\tau)\underset{\sim}{H}(\underset{\sim}{A} - \underset{\sim}{A}_k(\tau))\hat{\mathbf{x}}_k^\varepsilon - \underset{\sim}{C}_k(\tau)\underset{\sim}{H}(\underset{\sim}{B} - \underset{\sim}{B}_k(\tau))\mathbf{u}_k^\varepsilon(n).
\end{aligned}$$

From this we obtain

$$\begin{aligned}\mathbf{z}_{k+1} - \hat{\mathbf{z}}^{\varepsilon}_{k+1} = {} & \mathbf{G}_k(\tau)(\mathbf{z}_k - \hat{\mathbf{z}}^{\varepsilon}_k) + \mathbf{A}'_k \hat{\mathbf{x}}^{\varepsilon}_k \\ & + \big(\mathbf{B}' - \mathbf{B}'_k(\tau)\big)\mathbf{u}^{\varepsilon}_k(n) + \big(\mathbf{C}' - \mathbf{C}'_k(\tau)\big)\mathbf{F}\mathbf{w}_{k+1} \\ & - \mathbf{C}'_k(\tau)\mathbf{H}(\mathbf{A} - \mathbf{A}_k(\tau))\hat{\mathbf{x}}^{\varepsilon}_k - \mathbf{C}'_k(\tau)\mathbf{H}(\mathbf{B} - \mathbf{B}_k(\tau))\mathbf{u}^{\varepsilon}_k(n)\end{aligned}$$

and

$$\begin{bmatrix} \hat{\mathbf{x}}^{\varepsilon}_{k+1} \\ \mathbf{z}_{k+1} - \hat{\mathbf{z}}^{\varepsilon}_{k+1} \end{bmatrix} = \mathbf{U}_k \begin{bmatrix} \hat{\mathbf{x}}^{\varepsilon}_k \\ \mathbf{z}_k - \hat{\mathbf{z}}^{\varepsilon}_k \end{bmatrix} + \begin{bmatrix} \mathbf{N}^1_k \\ \mathbf{N}^2_k \end{bmatrix} \mathbf{b}_k(\tau) + \begin{bmatrix} \mathbf{C}_k(\tau) \\ \mathbf{C}' - \mathbf{C}'_k(\tau) \end{bmatrix} \mathbf{F}\mathbf{w}_{k+1}, \tag{4.174}$$

where

$$\mathbf{G}_k(\tau) = \begin{bmatrix} -\mathbf{C}_{1,k}(\tau) & \mathbf{I} & \mathbf{0} & & \mathbf{0} \\ \cdot & \mathbf{0} & & \ddots & \vdots \\ \cdot & \cdot & & & \mathbf{0} \\ \cdot & \cdot & & & \mathbf{I} \\ -\mathbf{C}_{p-1,k}(\tau) & \mathbf{0} & & \cdots & \mathbf{0} \end{bmatrix},$$

$$\mathbf{A}'_k = \begin{bmatrix} \mathbf{A}_2 - \mathbf{A}_{2,k}(\tau) & \mathbf{0} \ldots \mathbf{0} \\ \vdots & \\ \mathbf{A}_p - \mathbf{A}_{p,k}(\tau) & \mathbf{0} \ldots \mathbf{0} \end{bmatrix}, \quad \mathbf{B}' = \begin{bmatrix} \mathbf{B}_2 \\ \vdots \\ \mathbf{B}_p \end{bmatrix}$$

$$\mathbf{B}'_k(\tau) = \begin{bmatrix} \mathbf{B}_{2,k}(\tau) \\ \vdots \\ \mathbf{B}_{p,k}(\tau) \end{bmatrix}, \quad \mathbf{C}' = \begin{bmatrix} \mathbf{C}_1 \\ \vdots \\ \mathbf{C}_{p-1} \end{bmatrix}, \quad \mathbf{C}'_k(\tau) = \begin{bmatrix} \mathbf{C}_{1,k}(\tau) \\ \vdots \\ \mathbf{C}_{p-1,k}(\tau) \end{bmatrix},$$

$$\mathbf{U}_k = \left[\begin{array}{l|c} \mathbf{A}_k(\tau) - \mathbf{B}_k(\tau)\big(\mathbf{Q}_2 + \mathbf{B}^{\tau}_k(\tau)\mathbf{S}_k(\tau)\mathbf{B}_k(\tau)\big)^{-1}\mathbf{B}^{\tau}_k(\tau)\mathbf{A}_k(\tau) & \\ + \mathbf{C}_k(\tau)\mathbf{H}(\mathbf{A} - \mathbf{A}_k(\tau)) & \mathbf{C}^0_k(\tau) \\ - \mathbf{C}_k(\tau)\mathbf{H}(\mathbf{B} - \mathbf{B}_k(\tau))\big(\mathbf{Q}_2 + \mathbf{B}^{\tau}_k(\tau)\mathbf{S}_k(\tau)\mathbf{B}_k(\tau)\big)^{-1}\mathbf{B}^{\tau}_k(\tau)\mathbf{S}_k(\tau)\mathbf{A}_k(\tau) & \\ \hline \mathbf{A}'_k - \big(\mathbf{B}' - \mathbf{B}'_k(\tau)\big)\big(\mathbf{Q}_2 + \mathbf{B}^{\tau}_k(\tau)\mathbf{S}_k(\tau)\mathbf{B}_k(\tau)\big)^{-1}\mathbf{B}^{\tau}_k(\tau)\mathbf{S}_k(\tau)\mathbf{A}_k(\tau) & \\ - \mathbf{C}'_k(\tau)\mathbf{H}(\mathbf{A} - \mathbf{A}_k(\tau)) + \mathbf{C}'_k(\tau)\mathbf{H}(\mathbf{B} - \mathbf{B}_k(\tau)) & \mathbf{G}_k(\tau) \\ \times\big(\mathbf{Q}_2 + \mathbf{B}^{\tau}_k(\tau)\mathbf{S}_k(\tau)\mathbf{B}_k(\tau)\big)^{-1}\mathbf{B}^{\tau}_k(\tau)\mathbf{S}_k(\tau)\mathbf{A}_k(\tau) & \end{array}\right]$$

$$\mathbf{N}^1_k = -\big[\mathbf{B}_k(\tau) + \mathbf{C}_k(\tau)\mathbf{H}(\mathbf{B} - \mathbf{B}_k(\tau))\big]\big(\mathbf{Q}_2 + \mathbf{B}^{\tau}_k(\tau)\mathbf{S}_k(\tau)\mathbf{B}_k(\tau)\big)^{-1}\mathbf{B}^{\tau}_k(\tau),$$

and

$$\underset{\sim}{\mathbf{N}}_k^2 = \left[\underset{\sim}{\mathbf{C}}_k'(\tau)\underset{\sim}{\mathbf{H}}(\underset{\sim}{\mathbf{B}} - \mathbf{B}_k(\tau)) - (\underset{\sim}{\mathbf{B}}' - \underset{\sim}{\mathbf{B}}_k'(\tau))\right]$$

$$\times\left(\underset{\sim}{\mathbf{Q}}_2 + \underset{\sim}{\mathbf{B}}_k^{\tau}(\tau)\underset{\sim}{\mathbf{S}}_k(\tau)\underset{\sim}{\mathbf{B}}_k(\tau)\right)^{-1}\underset{\sim}{\mathbf{B}}_k^{\tau}(\tau).$$

Set

$$\underset{\sim}{\mathbf{U}} = \left[\begin{array}{c|c} \underset{\sim}{\mathbf{A}} - \underset{\sim}{\mathbf{B}}\left(\underset{\sim}{\mathbf{Q}}_2 + \underset{\sim}{\mathbf{B}}^{\tau}\underset{\sim}{\mathbf{S}}\underset{\sim}{\mathbf{B}}\right)^{-1}\underset{\sim}{\mathbf{B}}^{\tau}\underset{\sim}{\mathbf{A}} & \underset{\sim}{\mathbf{C}}\underset{\sim}{\mathbf{0}} \\ \hline \underset{\sim}{\mathbf{0}} & \underset{\sim}{\mathbf{G}} \end{array}\right],$$

$$G = \begin{bmatrix} -\underset{\sim}{\mathbf{C}}_1 & \underset{\sim}{\mathbf{I}} & \underset{\sim}{\mathbf{0}} & \mathbf{0} \\ \cdot & \underset{\sim}{\mathbf{0}} & \ddots & \vdots \\ \cdot & \vdots & \ddots & \underset{\sim}{\mathbf{0}} \\ \cdot & & & \underset{\sim}{\mathbf{I}} \\ -\underset{\sim}{\mathbf{C}}_{p-1} & \underset{\sim}{\mathbf{0}} & \cdots & \underset{\sim}{\mathbf{0}} \end{bmatrix}.$$

Under Conditions A_3 and C_3, all eigenvalues of $\underset{\sim}{\mathbf{U}}$ are inside the closed unit disk. We note that

$$\|\underset{\sim}{\mathbf{U}}_k - \underset{\sim}{\mathbf{U}}\|$$

can be made to be arbitrarily small on $[n < k < \tau(\varepsilon, n)]$ if ε is small enough. Let $\underset{\sim}{\mathbf{M}} \triangleq \sum_{i=0}^{\infty}\underset{\sim}{\mathbf{U}}^{i\tau}\underset{\sim}{\mathbf{U}}^{i}$. Then by Lemma 4.6, we have

$$\|\underset{\sim}{\mathbf{U}}_{k+s}\underset{\sim}{\mathbf{U}}_{k+s-1}\cdots\underset{\sim}{\mathbf{U}}_{k+1}\| I_{[n\le k\le\tau-1]} \le \sqrt{\|\underset{\sim}{\mathbf{M}}\|}\,\lambda^s, \qquad \forall s, \tag{4.175}$$

where henceforth τ denotes $\tau(\varepsilon, n)$, and $\|\underset{\sim}{\mathbf{M}}\|$ and λ are constants independent of ω with $\lambda \in [0,1]$. The stopping time τ $(\equiv \tau(\varepsilon, n))$ should not be confused with the matrix transpose sign.

From (4.174), we have

$$\begin{bmatrix} \hat{\mathbf{x}}_{k+1}^{\varepsilon} \\ \mathbf{z}_{k+1} - \hat{\mathbf{z}}_{k+1}^{\varepsilon} \end{bmatrix} I_{[n<k<\tau]} = \prod_{i=n+1}^{k}\underset{\sim}{\mathbf{U}}_i \begin{bmatrix} \hat{\mathbf{x}}_{n+1}^{\varepsilon} \\ z_{n+1} - \hat{\mathbf{z}}_{n+1}^{\varepsilon} \end{bmatrix} I_{[n<k<\tau]}$$

$$+ \sum_{i=n+1}^{k}\prod_{j=i+1}^{k}\underset{\sim}{\mathbf{U}}_j I_{[n<k<\tau]}$$

$$\times\left(\begin{bmatrix} \underset{\sim}{\mathbf{N}}_i^1 \\ \underset{\sim}{\mathbf{N}}_i^2 \end{bmatrix}\mathbf{b}_i(\tau) + \begin{bmatrix} \underset{\sim}{\mathbf{C}}_i(\tau) \\ \underset{\sim}{\mathbf{C}}' - \underset{\sim}{\mathbf{C}}_k'(\tau) \end{bmatrix}\underset{\sim}{\mathbf{F}}\mathbf{w}_{i+1}\right), \quad \forall k \ge n, \tag{4.176}$$

where

$$\prod_{j=i+1}^{k} \underset{\sim}{\mathbf{U}}_j = \begin{cases} \underset{\sim}{\mathbf{U}}_k \cdots \underset{\sim}{\mathbf{U}}_{i+1}, & k > i \\ \underset{\sim}{\mathbf{I}}, & k = i \end{cases}.$$

By using (4.175), it is easy to see that

$$E\left\| \prod_{i=n+1}^{k} \begin{bmatrix} \hat{\mathbf{x}}_n^{\varepsilon} \\ \mathbf{z}_n - \hat{\mathbf{z}}_n^{\varepsilon} \end{bmatrix} I_{[n<k<\tau]} \right\|^2$$

is bounded with respect to k, since $\mathbf{u}_k^{\varepsilon}(n) = 0$, $k \leq n$. Further, in the event $[n < k < \tau]$ for any i, $n + 1 \leq i \leq k$,

$$\left\| \begin{bmatrix} \underset{\sim}{\mathbf{C}}_i(\tau) \\ \underset{\sim}{\mathbf{C}}' - \underset{\sim}{\mathbf{C}}_k'(\tau) \end{bmatrix} \right\| \leq \|\underset{\sim}{\mathbf{C}}\| + \varepsilon.$$

Hence we have

$$E\left\| \sum_{i=n+1}^{k} \prod_{j=i+1}^{k} \underset{\sim}{\mathbf{U}}_j I_{[n<k<\tau]} \begin{bmatrix} \underset{\sim}{\mathbf{C}}_i(\tau) \\ \underset{\sim}{\mathbf{C}}' - \underset{\sim}{\mathbf{C}}_k'(\tau) \end{bmatrix} \underset{\sim}{\mathbf{F}} \mathbf{w}_{i+1} \right\|^2$$

$$\leq E\left(\sum_{i=n+1}^{k} \|\underset{\sim}{\mathbf{M}}\| \lambda^{k-i} (\|\underset{\sim}{\mathbf{C}}\| + \varepsilon) \|\underset{\sim}{\mathbf{F}}\| \, \|\mathbf{w}_{i+1}\| \right)^2$$

$$\leq \|\underset{\sim}{\mathbf{M}}\| (\|\underset{\sim}{\mathbf{C}}\| + \varepsilon)^2 \|\underset{\sim}{\mathbf{F}}\|^2 \sum_{i=n+1}^{k} \lambda^{k-i} \sum_{i=n+1}^{k} E\lambda^{k-i} \|\mathbf{w}_{i+1}\|^2$$

$$\leq m \|\underset{\sim}{\mathbf{M}}\| (\|\underset{\sim}{\mathbf{C}}\| + \varepsilon)^2 \|\underset{\sim}{\mathbf{F}}\|^2 \frac{1}{(1-\lambda)^2},$$

where m is the dimension of $\mathbf{w}_i$.

Similarly, we can prove that

$$E\left\| \sum_{i=n+1}^{k} \prod_{j=i+1}^{k} \underset{\sim}{\mathbf{U}}_j I_{[n<k<\tau]} \underset{\sim}{\mathbf{N}}_i \mathbf{b}_i(\tau) \right\|^2$$

is bounded with respect to k. Hence by (4.176)

$$E\left\|\begin{bmatrix}\hat{\mathbf{x}}^{\varepsilon}_{k+1}\\ \mathbf{z}_{k+1}-\hat{\mathbf{z}}^{\varepsilon}_{k+1}\end{bmatrix}I_{[n<k<\tau]}\right\|^2$$

is bounded with respect to k. This fact leads us to the boundedness of $E\|\hat{\mathbf{x}}^{\varepsilon}_k\|^2$, and hence $E\|\mathbf{u}^{\varepsilon}_k(n)\|^2$ since

$$\hat{\mathbf{x}}^{\varepsilon}_{k+1}=\mathbf{0}\quad\text{on } [k=n]\cup[k\geq\tau].$$ □

Lemma 4.12. *Under the conditions of Lemma* 4.11

$$E\|\hat{\mathbf{x}}^{\varepsilon}_k-\mathbf{x}_k\|^2 \underset{\substack{n\to\infty\\ \varepsilon\to 0}}{\longrightarrow} 0$$

uniformly in $k\geq n$. *That is, given any* $\varepsilon_1>0$, *there exist* $n(\varepsilon_1)$ *and* $\varepsilon(\varepsilon_1)>0$ *such that if* $n>n(\varepsilon_1)$ *and if* $\varepsilon<\varepsilon(\varepsilon_1)$, *then* $E\|\hat{\mathbf{x}}^{\varepsilon}_k-\mathbf{x}_k\|^2<\varepsilon_1$ *for all* $k>n(\varepsilon_1)$.

Proof. In Lemma 4.11, we have shown $\hat{\mathbf{x}}^{\varepsilon 1}_{k+1}=\mathbf{x}^1_{k+1}\ \forall k$. Therefore, we only need to prove

$$E\|\hat{\mathbf{z}}^{\varepsilon}_k-\mathbf{z}_k\|^2 \underset{\substack{n\to\infty\\ \varepsilon\to 0}}{\longrightarrow} 0$$

uniformly in $k\geq n$.

From (4.174), we have

$$\begin{aligned}\mathbf{z}_{k+1}-\hat{\mathbf{z}}^{\varepsilon}_{k+1}&=\underset{\sim}{\mathbf{G}}^{k-n}(\mathbf{z}_{n+1}-\hat{\mathbf{z}}^{\varepsilon}_{n+1})+\sum_{i=n+1}^{k}\underset{\sim}{\mathbf{G}}^{k-i}(\underset{\sim}{\mathbf{G}}_i(\tau)-\underset{\sim}{\mathbf{G}})(\mathbf{z}_i-\hat{\mathbf{z}}^{\varepsilon}_i)\\&+\sum_{i=n+1}^{k}\underset{\sim}{\mathbf{G}}^{k-i}\underset{\sim}{\mathbf{E}}_i\hat{\mathbf{x}}^{\varepsilon}_i+\sum_{i=n+1}^{k}\underset{\sim}{\mathbf{G}}^{k-i}\underset{\sim}{\mathbf{N}}^2_i\mathbf{b}_k(\tau)\\&+\sum_{i=n+1}^{k}\underset{\sim}{\mathbf{G}}^{k-i}(\underset{\sim}{\mathbf{C}}'-\underset{\sim}{\mathbf{C}}'_i(\tau))\underset{\sim}{\mathbf{F}}\mathbf{w}_{i+1},\end{aligned}\tag{4.177}$$

where $\underset{\sim}{\mathbf{E}}_i$ denotes the lower left block of matrix $\underset{\sim}{\mathbf{U}}_i$.

By Lemmas 4.11 and 4.8, $\{E\|\mathbf{x}_k\|^2\}$ is bounded. Consequently, there are constants $\varepsilon_1 > 0$, k_7, and k_8 such that

$$E\|\hat{\mathbf{x}}_k^\varepsilon\|^2 \vee E\|\mathbf{x}_k\|^2 \vee E\|\mathbf{u}_k^\tau(n)\|^2 \leq k_7,$$

$$\|\underset{\sim}{\mathbf{S}}_k(\tau) - \underset{\sim}{\mathbf{S}}\| \vee \|\underset{\sim}{\mathbf{A}}_k(\tau) - \underset{\sim}{\mathbf{A}}\| \vee \|\underset{\sim}{\mathbf{B}}_k(\tau) - \underset{\sim}{\mathbf{B}}\| \vee \|\underset{\sim}{\mathbf{C}}_k(\tau) - \underset{\sim}{\mathbf{C}}\| \leq k_8,$$

$$\forall k = 1, 2, \ldots, \quad \forall \varepsilon \in [0, \varepsilon_1].$$

Since $\underset{\sim}{\mathbf{C}}_i(\tau)$ is $\mathscr{F}_i^{\mathbf{y}}$- and hence $\mathscr{F}_i$-measurable,

$$E\big(\underset{\sim}{\mathbf{C}}' - \underset{\sim}{\mathbf{C}}_i'(\tau)\big)\underset{\sim}{\mathbf{F}}\mathbf{w}_{i+1}\mathbf{w}_{j+1}^\tau\underset{\sim}{\mathbf{F}}^\tau\big(\underset{\sim}{\mathbf{C}}' - \underset{\sim}{\mathbf{C}}_j'(\tau)\big)^\tau$$

$$= E\left[\big(\underset{\sim}{\mathbf{C}}' - \underset{\sim}{\mathbf{C}}_i'(\tau)\big)\underset{\sim}{\mathbf{F}}\mathbf{w}_{i+1}E\left(\frac{\mathbf{w}_{j+1}^\tau\underset{\sim}{\mathbf{F}}^\tau}{\mathscr{F}_j}\right)\big(\underset{\sim}{\mathbf{C}}' - \underset{\sim}{\mathbf{C}}_j'(\tau)\big)^\tau\right] = 0 \quad \text{for } j > i.$$

By using the estimate,

$$\|\underset{\sim}{\mathbf{G}}_k\| \leq k_9\nu^k, \qquad k_9 > 0, \quad \nu \in [0, 1)$$

it is immediate that

$$E\left\|\sum_{i=n+1}^{k} \underset{\sim}{\mathbf{G}}^{k-i}\big(\underset{\sim}{\mathbf{C}}' - \underset{\sim}{\mathbf{C}}_i'(\tau)\big)\underset{\sim}{\mathbf{F}}\mathbf{w}_{i+1}\right\|^2$$

$$= \left\|\sum_{i=n+1}^{k} \underset{\sim}{\mathbf{G}}^{k-i}\big(\underset{\sim}{\mathbf{C}}' - \underset{\sim}{\mathbf{C}}_i'(\tau)\big)\underset{\sim}{\mathbf{F}}\underset{\sim}{\mathbf{F}}^\tau\big(\underset{\sim}{\mathbf{C}}' - \underset{\sim}{\mathbf{C}}_i'(\tau)\big)^\tau\underset{\sim}{\mathbf{G}}^{\tau k-i}\right\|$$

$$\leq k_9^2\|\underset{\sim}{\mathbf{F}}\|^2 \sum_{i=n+1}^{k} \nu^{2(k-i)}\|\underset{\sim}{\mathbf{C}} - \underset{\sim}{\mathbf{C}}_i(\tau)\|^2\big(I_{[n<i<\tau]} + I_{[\tau=n]\cup[i\geq\tau]}\big)$$

$$\leq k_9^2\|\underset{\sim}{\mathbf{F}}\|^2\left(\frac{\varepsilon^2}{1-\nu^2} + \frac{k_8^2}{1-\nu^2}P[\tau < \infty]\right) \underset{\substack{n\to\infty \\ \varepsilon\to 0}}{\longrightarrow} 0,$$

$$E\left\|\sum_{i=n+1}^{k} \underset{\sim}{\mathbf{G}}^{k-i}\big(\underset{\sim}{\mathbf{G}}_i(\tau) - \underset{\sim}{\mathbf{G}}\big)\mathbf{z}_i\right\|^2$$

$$\leq k_9^2 \sum_{i=n+1}^{k} \nu^{k-i} \sum_{i=n+1}^{k} \nu^{k-i}E\|\underset{\sim}{\mathbf{G}}_i(\tau) - \underset{\sim}{\mathbf{G}}\|^2\|\mathbf{z}_i\|^2$$

$$\leq \frac{k_9^2}{1-\nu}\frac{k_7k_8^2}{1-\nu}P[\tau < \infty] + \frac{\varepsilon^2 k_7}{1-\nu}\Bigg) \underset{\substack{n\to\infty \\ \varepsilon\to 0}}{\longrightarrow} 0.$$

By a similar argument, we can show that the remaining terms in (4.177) also go to zero in mean square as $n \to \infty$ and $\varepsilon \to 0$. □

Theorem 4.9. *Suppose that* $\underset{\sim}{\mathbf{Q}}_2 > 0$, $\{\underset{\sim}{\mathbf{A}}_k, \mathscr{F}_k^{\mathbf{y}}\}$, $\{\underset{\sim}{\mathbf{B}}_k, \mathscr{F}_k^{\mathbf{y}}\}$, *and* $\{\underset{\sim}{\mathbf{C}}_k, \mathscr{F}_k^{\mathbf{y}}\}$ *are strongly consistent estimates for* $\underset{\sim}{\mathbf{A}}$, $\underset{\sim}{\mathbf{B}}$, *and* $\underset{\sim}{\mathbf{C}}$, *respectively. (Sufficient for this is that the conditions of any one of the Theorems 3.5, 4.1, 4.2, and 4.3 are satisfied). Assume that upper bounds on* $\|\underset{\sim}{\mathbf{A}}\|$, $\|\underset{\sim}{\mathbf{B}}\|$, $\|\underset{\sim}{\mathbf{C}}\|$ *are known and that Conditions* A_3–C_3 *are satisfied. Then for any* $\varepsilon > 0$, $\delta > 0$, *there exists the admissible control* $\{\mathbf{u}_k(\delta, \varepsilon)\} \in \mathscr{U}$ *and an adaptive control* $\{\mathbf{u}_k^a(n)\} \in \mathscr{U}$ *such that*

$$J(\mathbf{u}(\delta, \varepsilon)) - J(\mathbf{u}^*) < \delta \tag{4.178}$$

and

$$P\left[\mathbf{u}_k^a(n) = \mathbf{u}_k(\delta, \varepsilon), \forall k\right] > 1 - \varepsilon \tag{4.179}$$

where $J(\mathbf{u}^*)$ *is the minimum value of the performance index given by* (4.164).

Proof. Define

$$\mathbf{u}_k(\delta, \varepsilon) = \begin{cases} \mathbf{u}_k^{\varepsilon}(n), & k < \tau \\ -\left(\underset{\sim}{\mathbf{Q}}_2 + \underset{\sim}{\mathbf{B}}^{\tau}\underset{\sim}{\mathbf{S}}\underset{\sim}{\mathbf{B}}\right)^{-1}\underset{\sim}{\mathbf{B}}^{\tau}\left(\underset{\sim}{\mathbf{S}}\underset{\sim}{\mathbf{A}}\hat{\mathbf{x}}_k + \mathbf{b}_k^0\right), & k \geq \tau, \end{cases} \tag{4.180}$$

where $\hat{\mathbf{x}}_k$ is the solution of (4.137) with $\{\mathbf{u}_k\}$ given by (2.180). Since τ is a stopping time and $\underset{\sim}{\mathbf{A}}_k$, $\underset{\sim}{\mathbf{B}}_k$, and $\underset{\sim}{\mathbf{C}}_k$ are $\mathscr{F}_k^{\mathbf{y}}$-measurable, it is not difficult to see that $\mathbf{u}_k(\delta, \varepsilon)$ is $\mathscr{F}_k^{\mathbf{y}}$-measurable.

Now we show that $E\|\mathbf{u}_k(\delta, \varepsilon)\|^2$ is bounded. Clearly, on $[k < \tau]$ $\mathbf{x}_k$ remains invariant whenever $\mathbf{u}_k(\delta, \varepsilon)$ or $\mathbf{u}_k^{\varepsilon}(n)$ is applied. Hence $\mathbf{u}_k^{\varepsilon}(n)$ itself will be unchanged on $[k < \tau]$ for both cases. Consequently, by Lemma 4.11, we only need to prove that

$$E\left\|\left(\underset{\sim}{\mathbf{Q}}_2 + \underset{\sim}{\mathbf{B}}^{\tau}\underset{\sim}{\mathbf{S}}\underset{\sim}{\mathbf{B}}\right)^{-1}\underset{\sim}{\mathbf{B}}^{\tau}\left(\underset{\sim}{\mathbf{S}}\underset{\sim}{\mathbf{A}}\hat{\mathbf{x}}_k + \mathbf{b}_k^0\right)I_{[\tau \leq k]}\right\|^2$$

or, equivalently,

$$E\left\|\mathbf{x}_k I_{[\tau \leq k]}\right\|^2$$

is bounded with respect to k, since $\{\mathbf{b}_k^0\}$ is bounded and $\hat{\mathbf{x}}_k \equiv \mathbf{x}_k$ by Lemma 4.5.

From (4.135) and (4.154), we have

$$\begin{aligned}
\mathbf{x}_k I_{[\tau \le k]} &= \underset{\sim}{\mathbf{A}}\mathbf{x}_{k-1} I_{[\tau \le k-1]} - \underset{\sim}{\mathbf{B}}\left(\underset{\sim}{\mathbf{Q}}_2 + \underset{\sim}{\mathbf{B}}^\tau \underset{\sim}{\mathbf{S}}\underset{\sim}{\mathbf{B}}\right)^{-1} \underset{\sim}{\mathbf{B}}^\tau\left(\underset{\sim}{\mathbf{S}}\underset{\sim}{\mathbf{A}}\mathbf{x}_{k-1} + \mathbf{b}_{k-1}^0\right) I_{[\tau \le k-1]} \\
&\quad + \left(\underset{\sim}{\mathbf{A}}\mathbf{x}_{k-1} + \underset{\sim}{\mathbf{B}}\mathbf{u}_{k-1}^\varepsilon(n)\right) I_{[\tau = k]} + \underset{\sim}{\mathbf{C}}\underset{\sim}{\mathbf{F}}\mathbf{w}_k I_{[\tau \le k]} \\
&= \underset{\sim}{\mathbf{\Phi}}^k \mathbf{x}_0 I_{[\tau \le 0]} - \sum_{i=0}^{k-1} \underset{\sim}{\mathbf{\Phi}}^{k-1-i} \underset{\sim}{\mathbf{B}}\left(\underset{\sim}{\mathbf{Q}}_2 + \underset{\sim}{\mathbf{B}}^\tau \underset{\sim}{\mathbf{S}}\underset{\sim}{\mathbf{B}}\right)^{-1} \underset{\sim}{\mathbf{B}}^\tau \mathbf{b}_i^0 I_{[\tau \le i]} \\
&\quad + \sum_{i=0}^{k-1} \underset{\sim}{\mathbf{\Phi}}^{k-1-i}\left(\underset{\sim}{\mathbf{A}}\mathbf{x}_i + \underset{\sim}{\mathbf{B}}\mathbf{u}_i^\varepsilon(n)\right) I_{[i = \tau - 1]} \\
&\quad + \sum_{i=0}^{k-1} \underset{\sim}{\mathbf{\Phi}}^{k-1-i} \underset{\sim}{\mathbf{C}}\underset{\sim}{\mathbf{F}}\mathbf{w}_{i+1} I_{[\tau \le i+1]}.
\end{aligned}$$

By (3.52), (4.149), and the boundedness of $\{\mathbf{b}_i^0\}$, it is clear that

$$E\left\|\underset{\sim}{\mathbf{\Phi}}^k I_{[\tau \le 0]}\right\|^2, \qquad E\left\|\sum_{i=0}^{k-1} \underset{\sim}{\mathbf{\Phi}}^{k-1-i} \underset{\sim}{\mathbf{B}}\left(\underset{\sim}{\mathbf{Q}}_2 + \underset{\sim}{\mathbf{B}}^\tau \underset{\sim}{\mathbf{S}}\underset{\sim}{\mathbf{B}}\right)^{-1} \underset{\sim}{\mathbf{B}}^\tau \mathbf{b}_i^0 I_{[\tau \le i]}\right\|^2$$

and

$$E\left\|\sum_{i=0}^{k-1} \underset{\sim}{\mathbf{\Phi}}^{k-1-i} \underset{\sim}{\mathbf{C}}\underset{\sim}{\mathbf{F}}\mathbf{w}_i I_{[\tau \le i+1]}\right\|^2$$

are bounded with respect to k.

Again by Lemma 4.11, $\{E\|\mathbf{u}_i^\varepsilon(n) I_{[i \le \tau - 1]}\|^2\}$ is bounded and hence $\{E\|\mathbf{x}_i I_{[i \le \tau - 1]}\|^2\}$ is also bounded. Thus by (4.149),

$$E\left\|\sum_{i=0}^{k-1} \underset{\sim}{\mathbf{\Phi}}^{k-1-i}\left(\underset{\sim}{\mathbf{A}}\mathbf{x}_i + \underset{\sim}{\mathbf{B}}\mathbf{u}_i^\varepsilon(n)\right) I_{[i = \tau - 1]}\right\|^2$$

is bounded with respect to k, and we have proved that $\{\mathbf{u}_k(\delta, \varepsilon)\} \in \mathscr{U}$, $E\|\mathbf{u}_k(\delta, \varepsilon)\|^2$ is uniformly bounded and hence (4.162) holds with $\mathbf{u}_i$ equal to $\mathbf{u}_i(\delta, \varepsilon)$.

Now we have

$$\begin{aligned}
J(\mathbf{u}(\delta, \varepsilon)) &= J(\mathbf{u}^*) + \overline{\lim_{N \to \infty}} \frac{1}{N} \\
&\quad \times E \sum_{i=n}^{N-1} \left\{\left[\mathbf{u}_i^\varepsilon(n) + \left(\underset{\sim}{\mathbf{Q}}_2 + \underset{\sim}{\mathbf{B}}^\tau \underset{\sim}{\mathbf{S}}\underset{\sim}{\mathbf{B}}\right)^{-1} \underset{\sim}{\mathbf{B}}^\tau\left(\underset{\sim}{\mathbf{S}}\underset{\sim}{\mathbf{A}}\mathbf{x}_i + \mathbf{b}_i^0\right)\right]^\tau \right. \\
&\quad \left. \times \left(\underset{\sim}{\mathbf{Q}}_2 + \underset{\sim}{\mathbf{B}}^\tau \underset{\sim}{\mathbf{S}}\underset{\sim}{\mathbf{B}}\right)\left[\mathbf{u}_i^\varepsilon(n) + \left(\underset{\sim}{\mathbf{Q}}_2 + \underset{\sim}{\mathbf{B}}^\tau \underset{\sim}{\mathbf{S}}\underset{\sim}{\mathbf{B}}\right)^{-1} \underset{\sim}{\mathbf{B}}^\tau\left(\underset{\sim}{\mathbf{S}}\underset{\sim}{\mathbf{A}}\mathbf{x}_i + \mathbf{b}_i^0\right)\right] I_{[i < \tau]}\right\}
\end{aligned}$$

$$
\begin{aligned}
&\le J(\mathbf{u}^*) + \varlimsup_{N\to\infty} \frac{2\|\underset{\sim}{\mathbf{Q}}_2 + \underset{\sim}{\mathbf{B}}^\tau \underset{\sim}{\mathbf{S}}\underset{\sim}{\mathbf{B}}\|}{N} \\
&\quad \times E\sum_{i=n}^{N-1} \Bigg\{ \bigg\| \Big[\big(\underset{\sim}{\mathbf{Q}}_2 + \underset{\sim}{\mathbf{B}}^\tau \underset{\sim}{\mathbf{S}}\underset{\sim}{\mathbf{B}}\big)^{-1} \underset{\sim}{\mathbf{B}}^\tau \underset{\sim}{\mathbf{S}}\underset{\sim}{\mathbf{A}}\mathbf{x}_i - \big(\underset{\sim}{\mathbf{Q}}_2 + \underset{\sim}{\mathbf{B}}_i^\tau(\tau)\underset{\sim}{\mathbf{S}}_i(\tau)\underset{\sim}{\mathbf{B}}_i(\tau)\big)^{-1} \\
&\quad \times \underset{\sim}{\mathbf{B}}_i^\tau(\tau)\underset{\sim}{\mathbf{S}}_i(\tau)\underset{\sim}{\mathbf{A}}_i(\tau)\hat{\mathbf{x}}_i^\varepsilon \Big] I_{[i<\tau]} \bigg\|^2 \\
&\quad + \bigg\| \Big[\big(\underset{\sim}{\mathbf{Q}}_2 + \underset{\sim}{\mathbf{B}}^\tau \underset{\sim}{\mathbf{S}}\underset{\sim}{\mathbf{B}}\big)^{-1} \underset{\sim}{\mathbf{B}}^\tau \mathbf{b}_i^0 \\
&\quad - \big(\underset{\sim}{\mathbf{Q}}_2 + \underset{\sim}{\mathbf{B}}_i^\tau(\tau)\underset{\sim}{\mathbf{S}}_i(\tau)\underset{\sim}{\mathbf{B}}_i(\tau)\big)^{-1} \underset{\sim}{\mathbf{B}}_i^\tau(\tau)\mathbf{b}_i(\tau) \Big] I_{[i<\tau]} \bigg\|^2 \Bigg\} \\
&\triangleq J(\mathbf{u}^*) + 2(\zeta_1 + \zeta_2)\|\underset{\sim}{\mathbf{Q}}_2 + \underset{\sim}{\mathbf{B}}^\tau \underset{\sim}{\mathbf{S}}\underset{\sim}{\mathbf{B}}\|,
\end{aligned}
\tag{4.181}
$$

where

$$
\begin{aligned}
\zeta_1 &= \varlimsup_{N\to\infty} \frac{1}{N} E \sum_{i=n}^{N-1} \bigg\| \Big[\big(\underset{\sim}{\mathbf{Q}}_2 + \underset{\sim}{\mathbf{B}}^\tau \underset{\sim}{\mathbf{S}}\underset{\sim}{\mathbf{B}}\big)^{-1} \underset{\sim}{\mathbf{B}}^\tau \underset{\sim}{\mathbf{S}}\underset{\sim}{\mathbf{A}}\mathbf{x}_i \\
&\quad - \big(\underset{\sim}{\mathbf{Q}}_2 + \underset{\sim}{\mathbf{B}}_i^\tau(\tau)\underset{\sim}{\mathbf{S}}_i(\tau)\underset{\sim}{\mathbf{B}}_i(\tau)\big)^{-1} \underset{\sim}{\mathbf{B}}_i^\tau(\tau)\underset{\sim}{\mathbf{S}}_i(\tau)\underset{\sim}{\mathbf{A}}_i(\tau)\hat{\mathbf{x}}_i^\varepsilon \Big] I_{[i<\tau]} \bigg\|^2 \\
&\le \varlimsup_{N\to\infty} \frac{3}{N} \sum_{i=n}^{N-1} \Bigg\{ \Big\| \big(\underset{\sim}{\mathbf{Q}}_2 + \underset{\sim}{\mathbf{B}}^\tau \underset{\sim}{\mathbf{S}}\underset{\sim}{\mathbf{B}}\big)^{-1} \underset{\sim}{\mathbf{B}}^\tau \underset{\sim}{\mathbf{S}}\underset{\sim}{\mathbf{A}}(\mathbf{x}_i - \hat{\mathbf{x}}_i^\varepsilon) I_{[i<\tau]} \Big\|^2 \\
&\quad + \bigg\| \Big[\big(\underset{\sim}{\mathbf{Q}}_2 + \underset{\sim}{\mathbf{B}}^\tau \underset{\sim}{\mathbf{S}}\underset{\sim}{\mathbf{B}}\big)^{-1} \underset{\sim}{\mathbf{B}}^\tau \underset{\sim}{\mathbf{S}}\underset{\sim}{\mathbf{A}} \\
&\quad - \big(\underset{\sim}{\mathbf{Q}}_2 + \underset{\sim}{\mathbf{B}}_i^\tau(\tau)\underset{\sim}{\mathbf{S}}_i(\tau)\underset{\sim}{\mathbf{B}}_i(\tau)\big)^{-1} \underset{\sim}{\mathbf{B}}^\tau \underset{\sim}{\mathbf{S}}\underset{\sim}{\mathbf{A}} \Big] \hat{\mathbf{x}}_i^\varepsilon I_{[i<\tau]} \bigg\|^2 \\
&\quad + \Big\| \big(\underset{\sim}{\mathbf{Q}}_2 + \underset{\sim}{\mathbf{B}}_i^\tau(\tau)\underset{\sim}{\mathbf{S}}_i(\tau)\underset{\sim}{\mathbf{B}}_i(\tau)\big)^{-1} \\
&\quad \times \big(\underset{\sim}{\mathbf{B}}^\tau \underset{\sim}{\mathbf{S}}\underset{\sim}{\mathbf{A}} - \underset{\sim}{\mathbf{B}}_i^\tau(\tau)\underset{\sim}{\mathbf{S}}_i(\tau)\underset{\sim}{\mathbf{A}}_i(\tau)\big)\hat{\mathbf{x}}_i^\varepsilon I_{[i<\tau]} \Big\|^2 \Bigg\} \underset{\substack{n\to\infty \\ \varepsilon\to 0}}{\longrightarrow} 0,
\end{aligned}
\tag{4.182}
$$

where the first term goes to zero because, on $[i < \tau]$, $\mathbf{x}_i$ and $\hat{\mathbf{x}}_i^\varepsilon$ are the same as in Lemma 4.12, which, consequently, we may invoke. The second term is estimated by

$$\overline{\lim_{N\to\infty}} \frac{3}{N} \sum_{i=n}^{N-1} \Big\| \left(\underset{\sim}{\mathbf{Q}}_2 + \underset{\sim}{\mathbf{B}}_i^\tau(\tau)\underset{\sim}{\mathbf{S}}_i(\tau)\underset{\sim}{\mathbf{B}}_i(\tau)\right)^{-1} \left(\underset{\sim}{\mathbf{B}}^\tau\underset{\sim}{\mathbf{S}}\underset{\sim}{\mathbf{B}} - \underset{\sim}{\mathbf{B}}_i^\tau(\tau)\underset{\sim}{\mathbf{S}}_i(\tau)\underset{\sim}{\mathbf{B}}_i(\tau)\right)$$

$$\times \left(\underset{\sim}{\mathbf{Q}}_2 + \underset{\sim}{\mathbf{B}}^\tau\underset{\sim}{\mathbf{S}}\underset{\sim}{\mathbf{B}}\right)^{-1} \underset{\sim}{\mathbf{B}}^\tau\underset{\sim}{\mathbf{S}}\underset{\sim}{\mathbf{A}}\hat{\mathbf{x}}_i^\varepsilon I_{[i<\tau]} \Big\|^2$$

$$\leq \overline{\lim_{N\to\infty}} \frac{3}{N} \sum_{i=n}^{N-1} \|\underset{\sim}{\mathbf{Q}}_2^{-1}\|^4 9\varepsilon^2 c^4 \|\underset{\sim}{\mathbf{B}}^\tau\underset{\sim}{\mathbf{S}}\underset{\sim}{\mathbf{A}}\|^2 E\|\hat{\mathbf{x}}_i^\varepsilon I_{[i<\tau]}\|^2 \underset{\varepsilon\to 0}{\to} 0,$$

and the third term is estimated in a similar way. Finally,

$$\zeta_2 = \overline{\lim_{N\to\infty}} \frac{1}{N} E \sum_{i=n}^{N-1} \Big\| \Big[\left(\underset{\sim}{\mathbf{Q}}_2 + \underset{\sim}{\mathbf{B}}^\tau\underset{\sim}{\mathbf{S}}\underset{\sim}{\mathbf{B}}\right)^{-1} \underset{\sim}{\mathbf{B}}\mathbf{b}_i^0$$

$$- \left(\underset{\sim}{\mathbf{Q}}_2 + \underset{\sim}{\mathbf{B}}_i^\tau(\tau)\underset{\sim}{\mathbf{S}}_i(\tau)\underset{\sim}{\mathbf{B}}_i(\tau)\right)^{-1} \underset{\sim}{\mathbf{B}}_i(\tau)\mathbf{b}_i(\tau) \Big] I_{[i<\tau]} \Big\|^2$$

$$\leq \overline{\lim_{N\to\infty}} \frac{2}{N} E \sum_{i=n}^{N-1} \Big[\Big\| \left(\underset{\sim}{\mathbf{Q}}_2 + \underset{\sim}{\mathbf{B}}^\tau\underset{\sim}{\mathbf{S}}\underset{\sim}{\mathbf{B}}\right)^{-1} \underset{\sim}{\mathbf{B}}\left(\mathbf{b}_i^0 - \mathbf{b}_i(\tau)\right) I_{[i<\tau]} \Big\|^2$$

$$+ \Big\| \Big[\left(\underset{\sim}{\mathbf{Q}}_2 + \underset{\sim}{\mathbf{B}}^\tau\underset{\sim}{\mathbf{S}}\underset{\sim}{\mathbf{B}}\right)^{-1} \underset{\sim}{\mathbf{B}}^\tau$$

$$- \left(\underset{\sim}{\mathbf{Q}}_2 + \underset{\sim}{\mathbf{B}}_i^\tau(\tau)\underset{\sim}{\mathbf{S}}_i(\tau)\underset{\sim}{\mathbf{B}}_i(\tau)\right)^{-1} \underset{\sim}{\mathbf{B}}_i^\tau(\tau) \Big] \mathbf{b}_i(\tau) I_{[i<\tau]} \Big\|^2 \underset{\varepsilon\to 0}{\to} 0, \qquad (4.183)$$

Formulas (4.181)–(4.183) prove (4.178).

From the strong consistency hypothesis, it follows that

$$P[\tau_c(n) < \infty] \leq P[\tau(\varepsilon, n) < \infty] \underset{n\to\infty}{\to} 0,$$

but on the set $[\omega : \tau(\varepsilon, n) = \infty]$,

$$\mathbf{u}_k^a(n) = \mathbf{u}_k(\delta, \varepsilon), \qquad \forall k \geq 0,$$

where $\mathbf{u}_k^a(n)$ is given by (4.170). This proves (4.179). □

Remark 1. If instead of $J(\mathbf{u})$, we consider

$$J_1(\mathbf{u}) = \overline{\lim_{N\to\infty}} \frac{1}{N} E\Big\{(\mathbf{y}_N - \mathbf{y}_N^*)^T \underset{\sim}{\mathbf{Q}}_0 (\mathbf{y}_N - \mathbf{y}_N^*) + \sum_{i=0}^{N-1}\Big[(\mathbf{y}_i - \mathbf{y}_i^*)^T \underset{\sim}{\mathbf{Q}}_1 (\mathbf{y}_i - \mathbf{y}_i^*) + \mathbf{u}_i^T \underset{\sim}{\mathbf{Q}}_2 \mathbf{u}_i\Big]\Big\},$$

where $\underset{\sim}{\mathbf{Q}}_0 \geq 0$, then

$$J(\mathbf{u}) \leq J_1(\mathbf{u}).$$

But under the control given by (4.180),

$$\frac{1}{N} E\|\mathbf{y}_N - \mathbf{y}_N^*\|^2 \underset{N\to\infty}{\to} 0.$$

Therefore, Theorem 4.9 holds true for $J_1(\mathbf{u})$.

Remark 2. Similar results can be extended to the continuous-time case.

4.5. ADAPTIVE TRACKING

In this section, we shall consider the adaptive tracking problem for the system (3.52), (4.53), and (3.62); we shall retain the notation adopted in (3.4) and (3.63)–(3.65). Let $\{\mathbf{y}_i^*\}$ be a deterministic bounded reference sequence. In an adaptive tracking problem, we wish the output of the system to follow the reference signal as closely as possible but, in contrast to the previous problem, we do not penalize the use of the control input.

Theorem 4.10. *Suppose that* (1) $r = 0$ *or* $r \geq 1$ *and the transfer function* $\underset{\sim}{\mathbf{C}}^{-1}(z) - \frac{1}{2}\underset{\sim}{\mathbf{I}}$ *is strictly positive real and all zeros of* $\det \underset{\sim}{\mathbf{C}}(\lambda)$ *lie outside of the closed unit disk,* (2) $l \leq m$, $\underset{\sim}{\mathbf{B}}_1$ *is of full rank, and all zeros of* $\underset{\sim}{\mathbf{B}}_1^+ \underset{\sim}{\mathbf{B}}(\lambda)$ *are outside the closed unit disk,* (3) *An* $\mathscr{F}_n$*-measurable control* $\mathbf{u}_n$ *can be selected at each n such that*

$$\underset{\sim}{\boldsymbol{\theta}}_n^T \boldsymbol{\phi}_n = \mathbf{y}_{n+1}^*, \tag{4.184}$$

where $\underset{\sim}{\boldsymbol{\theta}}_n$ *is given by MLS algorithm* (4.54)–(4.59),

(4)
$$\lim_{n\to\infty} \frac{1}{n} \sum_{i=1}^{n} \underset{\sim}{\mathbf{F}}_i \mathbf{w}_i \mathbf{w}_i^T \underset{\sim}{\mathbf{F}}_i^T = \underset{\sim}{\mathbf{R}}. \tag{4.185}$$

Then

$$r_n \to \infty, \qquad \overline{\lim_{n\to\infty}} \frac{1}{n} \sum_{i=1}^{n} \|\mathbf{u}_i\|^2 < \infty, \qquad \overline{\lim_{n\to\infty}} \frac{1}{n} \sum_{i=1}^{n} \|\mathbf{y}_i\|^2 < \infty, \tag{4.186}$$

and

$$\lim_{n\to\infty} \frac{1}{n} \sum_{i=1}^{n} (\mathbf{y}_i - \mathbf{y}_i^*)(\mathbf{y}_i - \mathbf{y}_i^*)^\tau = R. \tag{4.187}$$

If, in addition, (4.12) *or* (4.82) *or* (4.82′) *is satisfied, then* $\underset{\sim}{\theta}_n \xrightarrow[n\to\infty]{} \underset{\sim}{\theta}$ a.s.

Proof. The fact that $r_n \to \infty$ follows from (4.185) and Lemma 3.3.

Since $\underset{\sim}{\mathbf{B}}_1^+ \underset{\sim}{\mathbf{B}}(\lambda)$ is asymptotically stable, (4.185) and (4.53) directly yield the existence of constants c_1 and c_2, which may depend on ω, such that

$$\frac{1}{n} \sum_{i=1}^{n} \|\mathbf{u}_i\|^2 \le \frac{c_1}{n} \sum_{i=1}^{n} \|\mathbf{y}_{i+1}\|^2 + c_2, \qquad \forall n. \tag{4.188}$$

Similarly, by the asymptotic stability of $\underset{\sim}{\mathbf{C}}(z)$, (4.8), and (3.70), we can find constants, possibly depending on ω, such that

$$\frac{1}{n} \sum_{i=1}^{n} \|\mathbf{y}_i - \underset{\sim}{\theta}_i^\tau \underset{\sim}{\Phi}_{i-1}\|^2 \le \frac{c_3}{n} \sum_{i=1}^{n} \|\tilde{\underset{\sim}{\theta}}_i^\tau \underset{\sim}{\Phi}_{i-1}\|^2 + c_4. \tag{4.189}$$

Now by (4.184) and (4.61), we have

$$(\mathbf{y}_{n+1} - \mathbf{y}_{n+1}^*)(\mathbf{y}_{n+1} - \mathbf{y}_{n+1}^*)^\tau = \frac{1}{(1 - a_n \phi_n^\tau \underset{\sim}{\mathbf{R}}_{n+1} \phi_n)^2} \times (\mathbf{y}_{n+1} - \underset{\sim}{\theta}_{n+1}^\tau \phi_n)(\mathbf{y}_{n+1} - \underset{\sim}{\theta}_{n+1}^\tau \phi_n)^\tau. \tag{4.190}$$

By using the boundedness of $\{\mathbf{y}_n^*\}$, and (4.67), (4.189), and (4.190), we conclude that there are constants c_5 and c_6 such that

$$\frac{1}{n} \sum_{i=1}^{n} \|\mathbf{y}_i\|^2 \le \frac{c_5}{n} \sum_{i=1}^{n} \|\tilde{\underset{\sim}{\theta}}_i^\tau \phi_{i-1}\|^2 + c_6. \tag{4.191}$$

Combining (4.189)–(4.191), we have

$$\frac{r_{n+1}}{n} \le \frac{c_7}{n} \sum_{i=1}^{n} \|\tilde{\underset{\sim}{\theta}}_i^\tau \phi_{i-1}\|^2 + c_8 = c_7 \frac{r_{n+1}}{n} \sum_{i=1}^{n} \|\tilde{\underset{\sim}{\theta}}_i^\tau \phi_{i-1}\|^2 + c_8,$$

and by (4.77) and Lemma 3.2, we have

$$\frac{r_{n+1}}{n} \le c_8 \left(1 - c_7 \frac{1}{r_{n+1}} \sum_{i=1}^{n} \|\tilde{\underset{\sim}{\theta}}_i^\tau \phi_{i-1}\|^2\right)^{-1} \underset{n\to\infty}{\longrightarrow} c_8.$$

Hence

$$\overline{\lim_{n\to\infty}} \frac{r_{n+1}}{n} < \infty \text{ a.s.,} \tag{4.192}$$

and from (4.84)

$$\frac{1}{n} \sum_{i=1}^{n} \|\xi_i\|^2 \underset{n\to\infty}{\longrightarrow} 0 \text{ a.s.} \tag{4.193}$$

By (4.185), (4.193), and (3.70), we have

$$\overline{\lim_{n\to\infty}} \frac{n}{r_{n+1}} \le \overline{\lim_{n\to\infty}} \frac{n}{\sum_{i=1}^{n} \|\mathbf{y}_i - \underset{\sim}{\theta}_i^\tau \phi_{i-1}\|^2}$$

$$= \overline{\lim_{n\to\infty}} \frac{n}{\sum_{i=1}^{n} \|\xi_i + \underset{\sim}{\mathbf{F}}_i \mathbf{w}_i\|^2} = \frac{1}{\operatorname{tr} \underset{\sim}{\mathbf{R}}} < \infty, \tag{4.194}$$

and by (4.185)

$$\frac{1}{n} \|\underset{\sim}{\mathbf{F}}_i \mathbf{w}_i\|^2 = \frac{1}{n} \sum_{i=1}^{n} \|\underset{\sim}{\mathbf{F}}_i \mathbf{w}_i\|^2 - \frac{n-1}{n} \frac{1}{n-1} \sum_{i=1}^{n-1} \|\underset{\sim}{\mathbf{F}}_i \mathbf{w}_i\|^2 \to 0.$$

Similarly, from (4.196), we have

$$\frac{1}{n} \|\xi_n\|^2 \underset{n\to\infty}{\longrightarrow} 0 \text{ a.s.} \tag{4.195}$$

Hence from (3.70), we see that

$$\frac{1}{n} \|\mathbf{y}_n - \underset{\sim}{\theta}_n^\tau \phi_{n-1}\|^2 \underset{n\to\infty}{\longrightarrow} 0 \text{ a.s.} \tag{4.196}$$

Hence, by (4.67) and the boundedness of $\{\mathbf{y}_i^*\}$, it follows from (4.190) that

$$\frac{\|\mathbf{y}_n\|^2}{n} \underset{n\to\infty}{\to} 0 \text{ a.s.} \tag{4.197}$$

and so

$$\frac{\|\mathbf{u}_n\|^2}{n} \underset{n\to\infty}{\to} 0 \text{ a.s.} \tag{4.198}$$

From (4.196)–(4.198), we obtain

$$\frac{\|\phi_n\|^2}{n} \underset{n\to\infty}{\to} 0 \text{ a.s.} \tag{4.199}$$

Setting

$$b_n = a_n \phi_n^\tau \underset{\sim}{\mathbf{R}}_{n+1} \phi_n$$

and using (4.79), (4.194), and (4.199), we have

$$|b_n| \leq \|\underset{\sim}{\mathbf{R}}_{n+1}\| \, \|\phi_n\|^2 \leq \frac{k_0(mp + lq + mr)\|\underset{\sim}{\phi}_n\|^2 n}{r_n n} \underset{n\to\infty}{\to} 0. \tag{4.200}$$

From (4.190) and (3.70), we have

$$\begin{aligned}
\frac{1}{n}\sum_{i=1}^{n}(\mathbf{y}_i - \mathbf{y}_i^*)(\mathbf{y}_i - \mathbf{y}_i^*)^\tau &= \frac{1}{n}\sum_{i=1}^{n}\frac{(\xi_i + \underset{\sim}{\mathbf{F}}_i\mathbf{w}_i)(\xi_i + \underset{\sim}{\mathbf{F}}_i\mathbf{w}_i)^\tau}{1 - b_{i-1}} \\
&= \frac{1}{n}\sum_{i=1}^{n}\left(\frac{\xi_i}{1 - b_{i-1}} + \frac{b_{i-1}}{1 - b_{i-1}}\underset{\sim}{\mathbf{F}}_i\mathbf{w}_i + \underset{\sim}{\mathbf{F}}_i\mathbf{w}_i\right) \\
&\quad \times \left(\frac{\xi_i}{1 - b_{i-1}} + \frac{b_{i-1}}{1 - b_{i-1}}\underset{\sim}{\mathbf{F}}_i\mathbf{w}_i + \underset{\sim}{\mathbf{F}}_i\mathbf{w}_i\right)^\tau.
\end{aligned} \tag{4.201}$$

Next, by (4.185), (4.195), and (4.200), it is easy to see that

$$\frac{1}{n}\sum_{i=1}^{n}\frac{\|\xi_i\|^2}{(1 - b_{k-1})^2} \underset{n\to\infty}{\to} 0, \qquad \frac{1}{n}\sum_{i=1}^{n}\left(\frac{b_{i-1}}{1 - b_{i-1}}\right)^2 \|\mathbf{w}_i\|^2 \underset{n\to\infty}{\to} 0.$$

Thus (4.187) immediately follows from (4.201), and (4.186) follows from (4.187), the boundedness of $\{\mathbf{y}_i^*\}$ and the asymptotic stability of $\underset{\sim}{\mathbf{B}}_1^+ \underset{\sim}{\mathbf{B}}(\lambda)$. Finally, strong consistency follows from Theorems 4.3 and 4.4. □

CHAPTER 5

Recursive Estimation and Control for Discrete-Time Systems

5.1. RECURSIVE (KALMAN) FILTERING, INTERPOLATION, AND PREDICTION FOR CONDITIONALLY GAUSSIAN SYSTEMS

We first introduce the concept of conditional Gaussian random vector.

DEFINITION 5.1. A random vector $\mathbf{x}$ with $E\|\mathbf{x}\|^2 < \infty$ is called a conditionally Gaussian given the random vector $\mathbf{y}$, if the conditional characteristic function of $\mathbf{x}$ given $\mathbf{y}$ is expressed by

$$E\left(\frac{e^{i\lambda^T\mathbf{x}}}{\mathbf{y}}\right) = e^{i\lambda^T E(\mathbf{x}/\mathbf{y})}e^{-\lambda^T \underset{\sim}{\mathbf{R}}_{\mathbf{x}}^{\mathbf{y}}\lambda/2}, \tag{5.1}$$

where

$$\underset{\sim}{\mathbf{R}}_{\mathbf{x}}^{\mathbf{y}} \triangleq E\left[\left(\mathbf{x} - E\left(\frac{\mathbf{x}}{\mathbf{y}}\right)\right)\left(\mathbf{x} - E\left(\frac{\mathbf{x}}{\mathbf{y}}\right)\right)^T \Big/ \mathbf{y}\right].$$

This definition means that, given $\mathbf{y}$, the random vector $\mathbf{x}$ is normally distributed with expectation $E(\mathbf{x}/\mathbf{y})$ and covariance matrix $\underset{\sim}{\mathbf{R}}_{\mathbf{x}}^{\mathbf{y}}$.

In Chapter 1, we mentioned that for Gaussian vectors independence is equivalent to the uncorrelated property and that Gaussian random variables remain Gaussian after a linear transformation. These properties are also true for conditional Gaussian vectors.

Lemma 5.1. *Let $f_{\mathbf{x}}^{\mathbf{z}}(\lambda)$ and $f_{\mathbf{y}}^{\mathbf{z}}(\mu)$, respectively, denote the conditional characteristic functions of* $\mathbf{x}$ *and* $\mathbf{y}$. *If $f_{\mathbf{x}}^{\mathbf{z}}(\lambda)$ and $f_{\mathbf{y}}^{\mathbf{z}}(\mu)$ are conditionally*

Gaussian, then so is their product $f_{\mathbf{x}}^{\mathbf{z}}(\lambda)f_{\mathbf{y}}^{\mathbf{z}}(\mu)$. *If* $\mathbf{w} \triangleq [\mathbf{x}^T\mathbf{y}^T]^T$ *is conditionally Gaussian given* $\mathbf{z}$, *then*

1. $\mathbf{x}$ *and* $\mathbf{y}$ *are conditionally independent given* $\mathbf{z}$ *if and only if*

$$\underset{\sim}{\mathbf{R}}_{\mathbf{xy}}^{\mathbf{z}} \triangleq E\left[\left(\mathbf{x} - E\left(\frac{\mathbf{x}}{\mathbf{z}}\right)\right)\left(\mathbf{y} - E\left(\frac{\mathbf{y}}{\mathbf{z}}\right)\right)^T \Big/ \mathbf{z}\right] = 0,$$

2. $\underset{\sim}{\mathbf{A}}(\mathbf{z}) + \mathbf{w} + \mathbf{b}(\mathbf{z})$ is conditionally Gaussian if the matrix $\underset{\sim}{\mathbf{A}}(\mathbf{z})$ and vector $\mathbf{b}(\mathbf{z})$ are measurable with respect to $\mathbf{z}$ and $E\|\underset{\sim}{\mathbf{A}}(\mathbf{z})\mathbf{w} + \mathbf{b}(\mathbf{z})\|^2 < \infty$.

Proof. (1) Let $f_{\mathbf{x}}^{\mathbf{z}}(\lambda)$ and $f_{\mathbf{y}}^{\mathbf{z}}(\mu)$ be conditionally Gaussian. Then

$$f_{\mathbf{x}}^{\mathbf{z}}(\lambda)f_{\mathbf{y}}^{\mathbf{z}}(\mu) = \exp\left\{i\begin{bmatrix}\lambda\\ \mu\end{bmatrix}^T\begin{bmatrix}E(\mathbf{x}/\mathbf{z})\\ E(\mathbf{y}/\mathbf{z})\end{bmatrix} - \frac{1}{2}[\lambda^T\mu^T]\begin{bmatrix}\underset{\sim}{\mathbf{R}}_{\mathbf{x}}^{\mathbf{z}} & \underset{\sim}{\mathbf{0}}\\ \underset{\sim}{\mathbf{0}} & \underset{\sim}{\mathbf{R}}_{\mathbf{y}}^{\mathbf{z}}\end{bmatrix}\begin{bmatrix}\lambda\\ \mu\end{bmatrix}\right\}, \tag{5.2}$$

which is obviously a conditional Gaussian characteristic function.
(2) If, given $\mathbf{z}$, $\mathbf{w}$ is conditionally Gaussian and $\mathbf{x}$ and $\mathbf{y}$ are uncorrelated, then its conditional characteristic function $f_{\mathbf{w}}^{\mathbf{z}}(\gamma)$ is equal to $f_{\mathbf{x}}^{\mathbf{z}}(\lambda)f_{\mathbf{y}}^{\mathbf{z}}(\mu)$ given by (5.2) with $\gamma = [\lambda^T\mu^T]^T$. By Theorem 1.8, $\mathbf{x}$ and $\mathbf{y}$ are conditionally independent. By the same theorem and (5.2), the converse implication is trivial.

The remaining part of the lemma is obvious. □

Lemma 5.2. *Suppose that given* $\mathbf{z}$ *the vector* $[\mathbf{x}^T\mathbf{y}^T]^T$ *is conditionally Gaussian. Then*

1. $$E^{\mathbf{z},\mathbf{y}}\mathbf{x} \triangleq E(\mathbf{x}/\mathbf{z},\mathbf{y}) = E^{\mathbf{z}}\mathbf{x} + \underset{\sim}{\mathbf{R}}_{\mathbf{xy}}^{\mathbf{z}}(\underset{\sim}{\mathbf{R}}_{\mathbf{y}}^{\mathbf{z}})^{+}(\mathbf{y} - E^{\mathbf{z}}\mathbf{y}) \tag{5.3}$$

$$\underset{\sim}{\mathbf{R}}_{\mathbf{x}}^{\mathbf{z},\mathbf{y}} \triangleq E^{\mathbf{z},\mathbf{y}}(\mathbf{x} - E^{\mathbf{z},\mathbf{y}}\mathbf{x})(\mathbf{x} - E^{\mathbf{z},\mathbf{y}}\mathbf{x})^T$$

$$= \underset{\sim}{\mathbf{R}}_{\mathbf{x}}^{\mathbf{z}} - \underset{\sim}{\mathbf{R}}_{\mathbf{x},\mathbf{y}}^{\mathbf{z}}\left(\underset{\sim}{\mathbf{R}}_{\mathbf{y}}^{\mathbf{z}}\right)^{+}\underset{\sim}{\mathbf{R}}_{\mathbf{y},\mathbf{x}}^{\mathbf{z}}, \tag{5.4}$$

where

$$\underset{\sim}{\mathbf{R}}_{\mathbf{x},\mathbf{y}}^{\mathbf{z}} = E^{\mathbf{z}}(\mathbf{x} - E^{\mathbf{z}}\mathbf{x})(\mathbf{y} - E^{\mathbf{z}}\mathbf{y})^T,$$

$E^{\mathbf{zy}}$ *denotes the conditional expectation given* $\mathbf{z}$ *and* $\mathbf{y}$.

2. *Given* $\mathbf{z}, \mathbf{x} - E^{\mathbf{z},\mathbf{y}}\mathbf{x}$ *is independent of* $\mathbf{y}$ *and is conditionally Gaussian.*
3. *Given* $(\mathbf{z},\mathbf{y})$, $\mathbf{x}$ *is conditionally Gaussian with expectation* $E^{\mathbf{z},\mathbf{y}}\mathbf{x}$ *and covariance* $\underset{\sim}{\mathbf{R}}_{\mathbf{x}}^{\mathbf{z},\mathbf{y}}$.

Proof. $\mathbf{R}^{\mathbf{z}}_{\mathbf{y}}$ is measurable with respect to $\mathbf{z}$. It can be shown (see [66]) that the eigenvalues $\lambda_1 \cdots \lambda_m$ ($\mathbf{y}$ is assumed m-dimensional) and the orthonormal eigenvectors $\mathbf{v}_1, \ldots, \mathbf{v}_m$ of $\mathbf{R}^{\mathbf{z}}_{\mathbf{y}}$ can be selected so as to be measurable with respect to $\mathbf{z}$.

Set

$$\mathbf{V} = [\mathbf{v}_1, \ldots, \mathbf{v}_m], \qquad \mathbf{\Lambda} = \begin{bmatrix} \lambda_1 & & \mathbf{0} \\ & \ddots & \\ \mathbf{0} & & \lambda_m \end{bmatrix},$$

then $(\mathbf{R}^{\mathbf{z}}_{\mathbf{y}})^{+} = \mathbf{V}\mathbf{\Lambda}^{+}\mathbf{V}^{T}$ remains $\mathbf{z}$-measurable.

Consequently,

$$E^{\mathbf{z}}\left\{\left[\mathbf{I} - (\mathbf{R}^{\mathbf{z}}_{\mathbf{y}})^{+}\mathbf{R}^{\mathbf{z}}_{\mathbf{y}}\right][\mathbf{y} - E^{\mathbf{z}}\mathbf{y}]\right\}\left\{\left[\mathbf{I} - (\mathbf{R}^{\mathbf{z}}_{\mathbf{y}})^{+}\mathbf{R}^{\mathbf{z}}_{\mathbf{y}}\right][\mathbf{y} - E^{\mathbf{z}}\mathbf{y}]\right\}^{T}$$

$$= \left[\mathbf{I} - (\mathbf{R}^{\mathbf{z}}_{\mathbf{y}})^{+}\mathbf{R}^{\mathbf{z}}_{\mathbf{y}}\right]E^{\mathbf{z}}(\mathbf{y} - E^{\mathbf{z}}\mathbf{y})(\mathbf{y} - E^{\mathbf{z}}\mathbf{y})^{T}\left[\mathbf{I} - (\mathbf{R}^{\mathbf{z}}_{\mathbf{y}})^{+}\mathbf{R}^{\mathbf{z}}_{\mathbf{y}}\right] = \mathbf{0},$$

and evaluating the conditional expectation leads to

$$(\mathbf{y} - E^{\mathbf{z}}\mathbf{y})^{T} = (\mathbf{y} - E^{\mathbf{z}}\mathbf{y})^{T}(\mathbf{R}^{\mathbf{z}}_{\mathbf{y}})^{+}\mathbf{R}^{\mathbf{z}}_{\mathbf{y}} \tag{5.5}$$

and

$$\mathbf{R}^{\mathbf{z}}_{\mathbf{x},\mathbf{y}}(\mathbf{R}^{\mathbf{z}}_{\mathbf{y}})^{+}\mathbf{R}^{\mathbf{z}}_{\mathbf{y}} = \mathbf{R}^{\mathbf{z}}_{\mathbf{x},\mathbf{y}}. \tag{5.6}$$

By using (5.5), we obtain

$$E^{\mathbf{z}}\left\{\left[\mathbf{x} - E^{\mathbf{z}}\mathbf{x} - \mathbf{R}^{\mathbf{z}}_{\mathbf{x},\mathbf{y}}(\mathbf{R}^{\mathbf{z}}_{\mathbf{y}})^{+}(\mathbf{y} - E^{\mathbf{z}}\mathbf{y})\right](\mathbf{y} - E^{\mathbf{z}}\mathbf{y})^{T}\right\} = \mathbf{0}$$

and

$$E^{\mathbf{z}}\left\{\left[\mathbf{x} - E^{\mathbf{z}}\mathbf{x} - \mathbf{R}^{\mathbf{z}}_{\mathbf{x},\mathbf{y}}(\mathbf{R}^{\mathbf{z}}_{\mathbf{y}})^{+}(\mathbf{y} - E^{\mathbf{z}}\mathbf{y})\right]\mathbf{y}^{T}\right\} = \mathbf{0}.$$

Therefore, by Lemma 5.1, given an observation on $\mathbf{z}$, the random variables $\mathbf{y}$ and $\mathbf{x} - E^{\mathbf{z}}\mathbf{x} - \mathbf{R}^{\mathbf{z}}_{\mathbf{x},\mathbf{y}}(\mathbf{R}^{\mathbf{z}}_{\mathbf{y}})^{+}(\mathbf{y} - E^{\mathbf{z}}\mathbf{y})$ are independent, and hence by Theorem 1.8

$$E^{\mathbf{zy}}\left[\mathbf{x} - E^{\mathbf{z}}\mathbf{x} - \mathbf{R}^{\mathbf{z}}_{\mathbf{x},\mathbf{y}}(\mathbf{R}^{\mathbf{z}}_{\mathbf{y}})^{+}(\mathbf{y} - E^{\mathbf{z}}\mathbf{y})\right]$$

$$= E^{\mathbf{z}}\left[\mathbf{x} - E^{\mathbf{z}}\mathbf{x} - \mathbf{R}^{\mathbf{z}}_{\mathbf{x},\mathbf{y}}(\mathbf{R}^{\mathbf{z}}_{\mathbf{y}})^{+}(\mathbf{y} - E^{\mathbf{z}}\mathbf{y})\right] = \mathbf{0}$$

which yields (5.3) and (5.4). At the same time Conclusion 2 has also been established.

By using Conclusion 2 of the lemma, we have

$$\underset{\sim}{\mathbf{R}}_{\mathbf{x}}^{\mathbf{z},\mathbf{y}} = E^{\mathbf{z},\mathbf{y}}(\mathbf{x} - E^{\mathbf{z},\mathbf{y}}\mathbf{x})(\mathbf{x} - E^{\mathbf{z},\mathbf{y}}\mathbf{x})^{\tau} = E^{\mathbf{z}}(\mathbf{x} - E^{\mathbf{z},\mathbf{y}}\mathbf{x})(\mathbf{x} - E^{\mathbf{z},\mathbf{y}}\mathbf{x})^{\tau}$$

and

$$\begin{aligned} E^{\mathbf{z},\mathbf{y}}\exp(i\boldsymbol{\lambda}^{\tau}\mathbf{x}) &= \exp(i\boldsymbol{\lambda}^{\tau}E^{\mathbf{z},\mathbf{y}}\mathbf{x})E^{\mathbf{z},\mathbf{y}}\exp[i\boldsymbol{\lambda}^{\tau}(\mathbf{x} - E^{\mathbf{z},\mathbf{y}}\mathbf{x})] \\ &= \exp(i\boldsymbol{\lambda}^{\tau}E^{\mathbf{z},\mathbf{y}}\mathbf{x})E^{\mathbf{z}}\exp\{i\boldsymbol{\lambda}^{\tau}(\mathbf{x} - E^{\mathbf{z},\mathbf{y}}\mathbf{x})\} \\ &= \exp(i\boldsymbol{\lambda}^{\tau}E^{\mathbf{z},\mathbf{y}}\mathbf{x})\exp(-\tfrac{1}{2}\boldsymbol{\lambda}^{\tau}\underset{\sim}{\mathbf{R}}_{\mathbf{x}}^{\mathbf{z},\mathbf{y}}\boldsymbol{\lambda}) \end{aligned}$$

which proves Conclusion 3. □

Corollary. *If in Lemma 5.2, we take* $\mathbf{z}$ *equal to a constant vector then it leads to the following special case. Let* $[\mathbf{x}^{\tau}, \mathbf{y}^{\tau}]$ *be a Gaussian vector. Then*

1. $E^{\mathbf{y}}\mathbf{x} = E\mathbf{x} + \underset{\sim}{\mathbf{R}}_{\mathbf{x},\mathbf{y}}\underset{\sim}{\mathbf{R}}_{\mathbf{y}}^{+}(\mathbf{y} - E\mathbf{y})$,

$$E(\mathbf{x} - E^{\mathbf{y}}\mathbf{x})(\mathbf{x} - E^{\mathbf{y}}\mathbf{x})^{\tau} = \underset{\sim}{\mathbf{R}}_{\mathbf{x}} - \underset{\sim}{\mathbf{R}}_{\mathbf{x},\mathbf{y}}\underset{\sim}{\mathbf{R}}_{\mathbf{y}}^{+}\underset{\sim}{\mathbf{R}}_{\mathbf{y},\mathbf{x}} = \underset{\sim}{\mathbf{R}}_{\mathbf{x}}^{\mathbf{y}}, \quad (5.7)$$

2. $\mathbf{x} - E^{\mathbf{y}}\mathbf{x}$ *is independent of* $\mathbf{y}$, *and* $\mathbf{x} - E^{\mathbf{y}}\mathbf{x}$ *is normally distributed,*
3. *Given* $\mathbf{y}$, $\mathbf{x}$ *is conditionally Gaussian with expectation* $E^{\mathbf{y}}\mathbf{x}$ *and covariance matrix* $\underset{\sim}{\mathbf{R}}_{\mathbf{x}}^{\mathbf{y}}$.

Now we consider systems which are linear in the state but possibly nonlinear with respect to measurements. Namely, the system is of the form

$$\mathbf{x}_{k+1} = \underset{\sim}{\boldsymbol{\Phi}}_{k+1}\mathbf{x}_k + \underset{\sim}{\mathbf{B}}_{k+1}\mathbf{u}_k + \underset{\sim}{\mathbf{D}}_{k+1}\boldsymbol{\zeta}_{k+1}, \quad (5.8)$$

$$\mathbf{y}_k = \underset{\sim}{\mathbf{C}}_k\mathbf{x}_{k-1} + \underset{\sim}{\mathbf{H}}_k\mathbf{v}_{k-1} + \underset{\sim}{\mathbf{F}}_k\boldsymbol{\zeta}_k, \quad (5.9)$$

where $\mathbf{x}_k$, $\mathbf{y}_k$, $\mathbf{v}_k$, $\mathbf{u}_k$, and $\boldsymbol{\zeta}_k$ are respectively n-dimensional state, m-dimensional measurement, s-dimensional measurement control, r-dimensional state control, and l-dimensional noise, and $\underset{\sim}{\boldsymbol{\Phi}}_{k+1}$, $\underset{\sim}{\mathbf{D}}_{k+1}$, and $\underset{\sim}{\mathbf{B}}_{k+1}\mathbf{u}_k$, may depend nonlinearly on

$$\mathbf{y}^k \triangleq [\mathbf{y}_0^{\tau} \cdots \mathbf{y}_k^{\tau}]^{\tau}$$

and $\underset{\sim}{\mathbf{C}}_k$, $\underset{\sim}{\mathbf{F}}_k$, and $\underset{\sim}{\mathbf{H}}_k\mathbf{v}_{k-1}$ may depend nonlinearly on $\mathbf{y}^{k-1}$.

We assume the following conditions hold:

A_1. $\underset{\sim}{\Phi}_{k+1}$, $\underset{\sim}{D}_{k+1}$, $\underset{\sim}{B}_{k+1}$ and $\mathbf{u}_k$ are $\mathbf{y}^k$-measurable, and for any k, $\underset{\sim}{\Phi}_{k+1}$ and $\underset{\sim}{B}_{k+1}$ are bounded by a constant (the bounds may depend on k). Further,

$$E\|\mathbf{u}_k\|^2 < \infty, E\|\underset{\sim}{D}_{k+1}\|^2 < \infty.$$

B_1. $\underset{\sim}{C}_k$, $\underset{\sim}{F}_k$, $\underset{\sim}{H}_k$ and $\mathbf{v}_{k-1}$ are $\mathbf{y}^{k-1}$-measurable, and for all k, $\underset{\sim}{C}_k$ is bounded by a constant. Further,

$$E\|\underset{\sim}{H}_k\mathbf{v}_{k-1}\|^2 < \infty, \qquad E\|\underset{\sim}{F}_k\|^2 < \infty,$$

C_1. $E(\|\mathbf{x}_0\|^2 + \|\mathbf{y}_0\|^2) < \infty$ and, given $\mathbf{y}_0$, $\mathbf{x}_0$ is conditionally Gaussian with $E^{\mathbf{y}_0}\mathbf{x}_0$ and $\underset{\sim}{R}^{\mathbf{y}_0}_{\mathbf{x}_0}$ as parameters.

D_1. $\{\zeta_k\}$ are mutually independent and independent of $[\mathbf{x}_0^\tau, \mathbf{y}_0^\tau]^\tau$ and $\zeta_k \in N(0, I)$.

Lemma 5.3. *Assume Conditions* A_1–D_1 *hold. Then*

1. *Given* $\mathbf{y}^k$, *both* $\mathbf{x}_k$ *and* $[\mathbf{x}_{k+1}^\tau \mathbf{y}_{k+1}^\tau]^\tau$ *are conditionally Gaussian,* $k = 0, 1 \dots$. *(In this case the process* $(\mathbf{x}_k, \mathbf{y}_k)$ *is called conditionally Gaussian.)*
2. *Given* $(\mathbf{y}_k, \mathbf{x}_j)$, *with* $k \ge j$, $\mathbf{x}_k$ *and* $[\mathbf{x}_{k+1}^\tau \mathbf{y}_{k+1}^\tau]^\tau$ *are also conditionally Gaussian.*

Proof. (1) Condition C_1 and Lemma 5.1 guarantee the conclusion is true for $k = 0$. Now let it hold for k. We compute the conditional characteristic function as follows:

$$E^{\mathbf{y}^k}\exp\{i(\boldsymbol{\lambda}^\tau\mathbf{x}_{k+1} + \boldsymbol{\mu}^\tau\mathbf{y}_{k+1})\} = E^{\mathbf{y}^k}\{\exp i(\boldsymbol{\lambda}^\tau\underset{\sim}{\Phi}_{k+1} + \boldsymbol{\mu}^\tau\underset{\sim}{C}_{k+1})\mathbf{x}_k$$

$$\cdot\exp i(\boldsymbol{\lambda}^\tau\underset{\sim}{B}_{k+1}\mathbf{u}_k + \boldsymbol{\mu}^\tau\underset{\sim}{H}_{k+1}\mathbf{v}_k)\exp i(\boldsymbol{\lambda}^\tau\underset{\sim}{D}_{k+1} + \boldsymbol{\mu}^\tau\underset{\sim}{F}_{k+1})\zeta_{k+1}\}$$

$$= E^{\mathbf{y}^k}\Big\{f(\mathbf{x}_k, \mathbf{y}^k)E^{\mathbf{y}^k, \mathbf{x}_k}\exp i(\boldsymbol{\lambda}^\tau\underset{\sim}{D}_{k+1} + \boldsymbol{\mu}^\tau\underset{\sim}{F}_{k+1})\zeta_{k+1}\Big\}$$

$$= E^{\mathbf{y}^k}\{f(\mathbf{x}_k, \mathbf{y}^k)\exp - \tfrac{1}{2}(\boldsymbol{\lambda}^\tau\underset{\sim}{D}_{k+1} + \boldsymbol{\mu}^\tau\underset{\sim}{F}_{k+1})(\boldsymbol{\lambda}^\tau\underset{\sim}{D}_{k+1} + \boldsymbol{\mu}^\tau\underset{\sim}{F}_{k+1})^\tau\}. \tag{5.10}$$

The last equality follows from Theorem 1.9 since ζ_{k+1} is normal $N(0, I)$ and is independent of $(\mathbf{y}^k, \mathbf{x}_k)$, and $f(\mathbf{x}_k, \mathbf{y}^k)$ denotes the $(\mathbf{y}^k, \mathbf{x}_k)$-measurable function:

$$f(\mathbf{x}_k, \mathbf{y}^k) \triangleq \exp i(\boldsymbol{\lambda}^\tau \underset{\sim}{\boldsymbol{\Phi}}_{k+1}, +\boldsymbol{\mu}^\tau \underset{\sim}{\mathbf{C}}_{k+1})\mathbf{x}_k \cdot \exp i(\boldsymbol{\lambda}^\tau \underset{\sim}{\mathbf{B}}_{k+1}\mathbf{u}_k + \boldsymbol{\mu}^\tau \underset{\sim}{\mathbf{H}}_{k+1}\mathbf{v}_k).$$

Noticing that $\underset{\sim}{\boldsymbol{\Phi}}_{k+1}, \underset{\sim}{\mathbf{C}}_{k+1}, \underset{\sim}{\mathbf{B}}_{k+1}\mathbf{u}_k, \underset{\sim}{\mathbf{H}}_{k+1}\mathbf{V}_k$ are all $\mathbf{y}^k$-measurable, we find that $E^{\mathbf{y}^k} f(\mathbf{x}_k, \mathbf{y}^k)$ is a conditionally Gaussian characteristic function since $\mathbf{x}_k$ is conditionally Gaussian given $\mathbf{y}^k$ by the inductive assumption. Hence the right-hand side of (5.10) is a product of the conditionally Gaussian characteristic functions, so given $\mathbf{y}^k$, $[\mathbf{x}_{k+1}^\tau, \mathbf{y}_{k+1}^\tau]^\tau$ is conditionally Gaussian. Then, by Conclusion 3 of Lemma 5.2 $\mathbf{x}_{k+1}$ is conditionally Gaussian given $\mathbf{y}^{k+1}$, this completes the cycle to the next inductive step.

(2) Since

$$E^{\mathbf{y}^j, \mathbf{x}_j}\mathbf{x}_j = \mathbf{x}_j, \qquad E^{\mathbf{y}^j, \mathbf{x}_j}\mathbf{y}_j = \mathbf{y}_j,$$

$$E^{\mathbf{y}^j, \mathbf{x}_j}\exp i(\boldsymbol{\lambda}^\tau \mathbf{x}_j + \boldsymbol{\mu}^\tau \mathbf{y}_j) = \exp i(\boldsymbol{\lambda}^\tau \mathbf{x}_j + \boldsymbol{\mu}^\tau \mathbf{y}_j),$$

$[\mathbf{x}_j^\tau \mathbf{y}_j^\tau]^\tau$ is conditionally Gaussian given $(\mathbf{y}^j, \mathbf{x}_j)$. Hence by induction, in a manner similar to the preceding argument, we know that $E^{\mathbf{y}^k, \mathbf{x}_j}\exp i(\boldsymbol{\lambda}^\tau \mathbf{x}_{k+1} + \boldsymbol{\mu}^\tau \mathbf{y}_{k+1})$ is a conditionally Gaussian characteristic function provided that $E^{\mathbf{y}^k, \mathbf{x}_j}\exp i\boldsymbol{\lambda}^\tau \mathbf{x}_k$ is also. Consequently, $E^{\mathbf{y}^{k+1}, \mathbf{x}_j} e^{i\boldsymbol{\lambda}^\tau \mathbf{x}_{k+1}}$ is also conditionally Gaussian. □

We shall adopt the following usual notation:

$$\hat{\mathbf{x}}_k \triangleq E^{\mathbf{y}^k}\mathbf{x}_k, \qquad \hat{\mathbf{x}}'_k \triangleq E^{\mathbf{y}^{k-1}}\mathbf{x}_k,$$

$$\underset{\sim}{\mathbf{P}}_k \triangleq E^{\mathbf{y}^k}(\mathbf{x}_k - \hat{\mathbf{x}}_k)(\mathbf{x}_k - \hat{\mathbf{x}}_k)^\tau,$$

$$\underset{\sim}{\mathbf{P}}'_k \triangleq E^{\mathbf{y}^{k-1}}(\mathbf{x}_k - \hat{\mathbf{x}}'_k)(\mathbf{x}_k - \hat{\mathbf{x}}'_k)^\tau.$$

Theorem 5.1. *Assume Conditions* A_1–D_1 *hold. Then for the system* (5.8) *and* (5.9), *the minimum covariance unbiased estimate can be calculated* *recursively via the Kalman filter equations*:

$$\hat{\mathbf{x}}_{k+1} = \boldsymbol{\Phi}_{k+1}\hat{\mathbf{x}}_k + \mathbf{B}_{k+1}\mathbf{u}_k + \mathbf{K}_{k+1}(\mathbf{y}_{k+1} - \mathbf{C}_{k+1}\hat{\mathbf{x}}_k - \mathbf{H}_{k+1}\mathbf{v}_k) \quad (5.11)$$

$$\mathbf{K}_{k+1} = \left(\boldsymbol{\Phi}_{k+1}\mathbf{P}_k\mathbf{C}_{k+1}^{\tau} + \mathbf{D}_{k+1}\mathbf{F}_{k+1}^{\tau}\right)\left(\mathbf{C}_{k+1}\mathbf{P}_k\mathbf{C}_{k+1}^{\tau} + \mathbf{F}_{k+1}\mathbf{F}_{k+1}^{\tau}\right)^{+}$$

$$(5.12)$$

$$\mathbf{P}_{k+1} = \boldsymbol{\Phi}_{k+1}\mathbf{P}_k\boldsymbol{\Phi}_{k+1}^{\tau} - \left(\boldsymbol{\Phi}_{k+1}\mathbf{P}_k\mathbf{C}_{k+1}^{\tau} + \mathbf{D}_{k+1}\mathbf{F}_{k+1}^{\tau}\right)$$

$$\cdot \left(\mathbf{C}_{k+1}\mathbf{P}_k\mathbf{C}_{k+1}^{\tau} + \mathbf{F}_{k+1}\mathbf{F}_{k+1}^{\tau}\right)^{+}$$

$$\cdot \left(\boldsymbol{\Phi}_{k+1}\mathbf{P}_k\mathbf{C}_{k+1}^{\tau} + \mathbf{D}_{k+1}\mathbf{F}_{k+1}^{\tau}\right)^{\tau} + \mathbf{D}_{k+1}\mathbf{D}_{k+1}^{\tau} \quad (5.13)$$

$$\hat{\mathbf{x}}_0 = E^{\mathbf{y}_0}\mathbf{x}_0, \qquad \mathbf{P}_0 = \mathbf{R}_{\mathbf{x}_0}^{\mathbf{y}_0} \quad (5.14)$$

Proof. We notice that $\mathbf{y}^k$ is independent of $\boldsymbol{\zeta}_{k+1}$ because it depends on $\mathbf{x}_0$, $\mathbf{y}_0$, and $\boldsymbol{\zeta}^k$ only. Then

$$\hat{\mathbf{y}}'_{k+1} = E^{\mathbf{y}^k}\mathbf{y}_{k+1} = \mathbf{C}_{k+1}\hat{\mathbf{x}}_k + \mathbf{H}_{k+1}\mathbf{v}_k + \mathbf{F}_{k+1}E^{\mathbf{y}^k}\boldsymbol{\zeta}_{k+1} = \mathbf{C}_{k+1}\hat{\mathbf{x}}_k + \mathbf{H}_{k+1}\mathbf{v}_k$$

$$(5.15)$$

and

$$\hat{\mathbf{x}}'_{k+1} = \boldsymbol{\Phi}_{k+1}\hat{\mathbf{x}}_k + \mathbf{B}_{k+1}\mathbf{u}_k. \quad (5.16)$$

We make $\mathbf{y}^k$, $\mathbf{x}_{k+1}$, and $\mathbf{y}_{k+1}$ correspond to $\mathbf{z}$, $\mathbf{x}$ and $\mathbf{y}$, respectively, in Lemma 5.2; this is applicable to the present case because of Lemma 5.3. Hence from (5.3) we have

$$\hat{\mathbf{x}}_{k+1} = \boldsymbol{\Phi}_{k+1}\hat{\mathbf{x}}_k + \mathbf{B}_{k+1}\mathbf{u}_k + \mathbf{R}_{\mathbf{x}_{k+1}\mathbf{y}_{k+1}}^{\mathbf{y}^k}\left(\mathbf{R}_{\mathbf{y}_{k+1}}^{\mathbf{y}^k}\right)^{+}[\mathbf{y}_{k+1} - \mathbf{C}_{k+1}\hat{\mathbf{x}}_k - \mathbf{H}_{k+1}\mathbf{v}_k].$$

Comparing it with (5.11), we find $\mathbf{K}_{k+1}$ must be

$$\mathbf{R}_{\mathbf{x}_{k+1}\mathbf{y}_{k+1}}^{\mathbf{y}^k}\left(\mathbf{R}_{\mathbf{y}_{k+1}}^{\mathbf{y}^k}\right)^{+}.$$

By the independence of $\boldsymbol{\zeta}_{k+1}$ and $(\mathbf{y}^k, \mathbf{x}_k)$, it follows that

$$E^{\mathbf{y}^k}\boldsymbol{\Phi}_{k+1}(\mathbf{x}_k - \hat{\mathbf{x}}_k)\boldsymbol{\zeta}_{k+1}^{\tau}\mathbf{F}_{k+1}^{\tau} = E^{\mathbf{y}^k}\boldsymbol{\Phi}_{k+1}(\mathbf{x}_k - \hat{\mathbf{x}}_k)\left(E^{\mathbf{y}^k\mathbf{x}_k}\boldsymbol{\zeta}_{k+1}^{\tau}\right)\mathbf{F}_{k+1}^{\tau}$$

$$= E^{\mathbf{y}^k}\boldsymbol{\Phi}_{k+1}(\mathbf{x}_k - \hat{\mathbf{x}}_k)E\boldsymbol{\zeta}_{k+1}^{\tau}\mathbf{F}_{k+1}^{\tau} = \mathbf{0};$$

then by (5.15) and (5.16),

$$\mathbf{R}^{\mathbf{y}^k}_{\underset{\sim}{\mathbf{x}}_{k+1}\mathbf{y}_{k+1}} = E^{\mathbf{y}^k}\left[\underset{\sim}{\boldsymbol{\Phi}}_{k+1}(\mathbf{x}_k - \hat{\mathbf{x}}_k) + \underset{\sim}{\mathbf{D}}_{k+1}\underset{\sim}{\boldsymbol{\zeta}}_{k+1}\right]$$

$$\times \left[\underset{\sim}{\mathbf{C}}_{k+1}(\mathbf{x}_k - \hat{\mathbf{x}}_k) + \underset{\sim}{\mathbf{F}}_{k+1}\underset{\sim}{\boldsymbol{\zeta}}_{k+1}\right]^{\tau}$$

$$= \underset{\sim}{\boldsymbol{\Phi}}_{k+1}\underset{\sim}{\mathbf{P}}_k\underset{\sim}{\mathbf{C}}^{\tau}_{k+1} + \underset{\sim}{\mathbf{D}}_{k+1}\underset{\sim}{\mathbf{F}}^{\tau}_{k+1} \qquad (5.17)$$

and similarly,

$$\mathbf{R}^{\mathbf{y}^k}_{\mathbf{y}_{k+1}} = \underset{\sim}{\mathbf{C}}_{k+1}\underset{\sim}{\mathbf{P}}_k\underset{\sim}{\mathbf{C}}^{\tau}_{k+1} + \underset{\sim}{\mathbf{F}}_{k+1}\underset{\sim}{\mathbf{F}}^{\tau}_{k+1}. \qquad (5.18)$$

Combining (5.17) and (5.18) gives the expression for $\underset{\sim}{\mathbf{K}}_{k+1}$.
Finally, by (5.16), it is easy to see that

$$\underset{\sim}{\mathbf{P}}'_{k+1} = \underset{\sim}{\boldsymbol{\Phi}}_{k+1}\underset{\sim}{\mathbf{P}}_k\underset{\sim}{\boldsymbol{\Phi}}^{\tau}_{k+1} + \underset{\sim}{\mathbf{D}}_{k+1}\underset{\sim}{\mathbf{D}}^{\tau}_{k+1}. \qquad (5.19)$$

Substituting (5.17)–(5.19) into (5.4) leads to (5.13). □

We now consider a system slightly different from (5.8) and (5.9):

$$\mathbf{x}_{k+1} = \underset{\sim}{\boldsymbol{\Phi}}_{k+1}\mathbf{x}_k + \underset{\sim}{\mathbf{B}}_{k+1}\mathbf{u}_k + \underset{\sim}{\mathbf{D}}_{k+1}\underset{\sim}{\boldsymbol{\zeta}}_{k+1}, \qquad (5.20)$$

$$\mathbf{y}_k = \underset{\sim}{\mathbf{C}}_k\mathbf{x}_k + \underset{\sim}{\mathbf{H}}_k\mathbf{v}_{k-1} + \underset{\sim}{\mathbf{F}}_k\underset{\sim}{\boldsymbol{\zeta}}_k, \qquad (5.21)$$

the difference consisting in the first term on the right of (5.21).

Theorem 5.2. *Under Conditions* $\mathbf{A}_1$–$\mathbf{D}_1$ *for the system* (5.20) *and* (5.21), *the Kalman filter is given by*

$$\hat{\mathbf{x}}_{k+1} = \hat{\mathbf{x}}'_{k+1} + \underset{\sim}{\mathbf{K}}_{k+1}\left(\mathbf{y}_{k+1} - \underset{\sim}{\mathbf{C}}_{k+1}\hat{\mathbf{x}}'_{k+1} - \underset{\sim}{\mathbf{H}}_{k+1}\mathbf{v}_k\right), \qquad (5.22)$$

$$\hat{\mathbf{x}}'_{k+1} = \underset{\sim}{\boldsymbol{\Phi}}_{k+1}\hat{\mathbf{x}}_k + \underset{\sim}{\mathbf{B}}_{k+1}\mathbf{u}_k, \qquad (5.23)$$

$$\underset{\sim}{\mathbf{K}}_{k+1} = \left(\underset{\sim}{\mathbf{P}}'_{k+1}\underset{\sim}{\mathbf{C}}^{\tau}_{k+1} + \underset{\sim}{\mathbf{D}}_{k+1}\underset{\sim}{\mathbf{F}}^{\tau}_{k+1}\right)\left(\underset{\sim}{\mathbf{C}}_{k+1}\underset{\sim}{\mathbf{P}}'_{k+1}\underset{\sim}{\mathbf{C}}^{\tau}_{k+1} + \underset{\sim}{\mathbf{F}}_{k+1}\underset{\sim}{\mathbf{F}}^{\tau}_{k+1}\right.$$

$$\left. + \underset{\sim}{\mathbf{C}}_{k+1}\underset{\sim}{\mathbf{D}}_{k+1}\underset{\sim}{\mathbf{F}}^{\tau}_{k+1} + \underset{\sim}{\mathbf{F}}_{k+1}\underset{\sim}{\mathbf{D}}^{\tau}_{k+1}\underset{\sim}{\mathbf{C}}^{\tau}_{k+1}\right)^{+}, \qquad (5.24)$$

$$\underset{\sim}{\mathbf{P}}_{k+1} = \underset{\sim}{\mathbf{P}}'_{k+1} - \underset{\sim}{\mathbf{K}}_{k+1}\left(\underset{\sim}{\mathbf{C}}_{k+1}\underset{\sim}{\mathbf{P}}'_{k+1} + \underset{\sim}{\mathbf{F}}_{k+1}\underset{\sim}{\mathbf{D}}^{\tau}_{k+1}\right), \qquad (5.25)$$

$$\underset{\sim}{\mathbf{P}}'_{k+1} = \underset{\sim}{\boldsymbol{\Phi}}_{k+1}\underset{\sim}{\mathbf{P}}_k\underset{\sim}{\boldsymbol{\Phi}}^{\tau}_{k+1} + \underset{\sim}{\mathbf{D}}_{k+1}\underset{\sim}{\mathbf{D}}^{\tau}_{k+1}, \qquad (5.26)$$

$$\hat{\mathbf{x}}'_0 = E\mathbf{x}_0, \qquad \underset{\sim}{\mathbf{P}}'_0 = \underset{\sim}{\mathbf{R}}_{\mathbf{x}_0}, \qquad \underset{\sim}{\mathbf{D}}_0 = \underset{\sim}{\mathbf{0}} \qquad (5.27)$$

Proof. It is easy to rewrite $\mathbf{y}_k$ (given by (5.21)) in the form of (5.9):

$$\mathbf{y}_k = \underset{\sim}{\mathbf{C}}_k \underset{\sim}{\mathbf{\Phi}}_k \mathbf{x}_{k-1} + (\underset{\sim}{\mathbf{C}}_k \underset{\sim}{\mathbf{B}}_k \mathbf{u}_{k-1} + \underset{\sim}{\mathbf{H}}_k \mathbf{v}_{k-1}) + (\underset{\sim}{\mathbf{C}}_k \underset{\sim}{\mathbf{D}}_k + \underset{\sim}{\mathbf{F}}_k)\zeta_k.$$

Then Theorem 5.2 directly follows from Theorem 5.1. □

Remark 1. The gain matrix $\underset{\sim}{\mathbf{K}}_k$ and covariance matrices $\underset{\sim}{\mathbf{P}}_k$ and $\underset{\sim}{\mathbf{P}}'_k$ in Theorems 5.1 and 5.2 depend on $\mathbf{y}^k$. Hence, generally speaking, they are control-dependent. However for the special case where $\underset{\sim}{\mathbf{\Phi}}_{k+1}, \underset{\sim}{\mathbf{D}}_{k+1}, \underset{\sim}{\mathbf{C}}_k, \underset{\sim}{\mathbf{F}}_k$ are deterministic and $\mathbf{x}_0$ is independent of $\mathbf{y}_0$ then all $\underset{\sim}{\mathbf{K}}_k$, $\underset{\sim}{\mathbf{P}}_k$, and $\underset{\sim}{\mathbf{P}}'_k$ are deterministic and control-independent.

Remark. Let $k \geq j \geq 0$. Denote

$$\hat{\mathbf{x}}_k^j = E^{\mathbf{y}^k, \mathbf{x}_j}\mathbf{x}_k, \qquad \underset{\sim}{\mathbf{P}}_k^j = E^{\mathbf{y}^k, \mathbf{x}_j}(\mathbf{x}_k - \hat{\mathbf{x}}_k^j)(\mathbf{x}_k - \hat{\mathbf{x}}_k^j)^\tau.$$

Theorems 5.1 and 5.2 have given recursive formulas for $\hat{\mathbf{x}}_k^j$ for the systems (5.8), (5.9) and (5.20), (5.21), respectively, if in (5.11)–(5.13) and (5.22)–(5.26) we replace $\hat{\mathbf{x}}_k$ by $\hat{\mathbf{x}}_k^j$, $\underset{\sim}{\mathbf{P}}_k$ by $\underset{\sim}{\mathbf{P}}_k^j$ and set $\hat{\mathbf{x}}_j^j = \mathbf{x}_j$, $\underset{\sim}{\mathbf{P}}_j^j = 0$. This is because in Lemma 5.3 we have proved that if $(\mathbf{y}^k, \mathbf{x}_j)$, $\mathbf{x}_k$, and $[\mathbf{x}_{k+1}^\tau \mathbf{y}_{k+1}^\tau]^\tau$ are conditionally Gaussian then the argument used in Theorems 5.1 and 5.2 can be applied.

We now give an example of the application of the Kalman filter to parameter estimation.

Example. Let the random vector $\mathbf{x}$ be such that $\mathbf{x} \in N(E\mathbf{x}, P_0)$, let it be independent of $\mathbf{y}_0$, and let

$$\mathbf{y}_k = \underset{\sim}{\mathbf{C}}_k \mathbf{x} + \underset{\sim}{\mathbf{H}}_k \mathbf{v}_{k-1} + \underset{\sim}{\mathbf{F}}_k \zeta_k \tag{5.28}$$

with Conditions B_1–D_1 satisfied and $\underset{\sim}{\mathbf{F}}_k \underset{\sim}{\mathbf{F}}_k^\tau > \underset{\sim}{\mathbf{0}}$ a.s. Then

$$\begin{aligned} E^{\mathbf{y}^k}\mathbf{x} \triangleq \hat{\mathbf{x}}_k = {} & \underset{\sim}{\mathbf{S}}_k \underset{\sim}{\mathbf{P}}_0 \left[\underset{\sim}{\mathbf{P}}_0^+ E\mathbf{x} + \sum_{i=1}^{k} \underset{\sim}{\mathbf{C}}_i^\tau (\underset{\sim}{\mathbf{F}}_i \underset{\sim}{\mathbf{F}}_i^\tau)^{-1} (\mathbf{y}_i - \underset{\sim}{\mathbf{H}}_i \mathbf{v}_{i-1}) \right] \\ & + \left[\underset{\sim}{\mathbf{I}} - \underset{\sim}{\mathbf{S}}_k \underset{\sim}{\mathbf{P}}_0 \sum_{i=1}^{k} \underset{\sim}{\mathbf{C}}_i^\tau (\underset{\sim}{\mathbf{F}}_i \underset{\sim}{\mathbf{F}}_i^\tau)^{-1} \underset{\sim}{\mathbf{C}}_i \right] [\underset{\sim}{\mathbf{I}} - \underset{\sim}{\mathbf{P}}_0 \underset{\sim}{\mathbf{P}}_0^+] E\mathbf{x}, \end{aligned} \tag{5.29}$$

$$\underset{\sim}{\mathbf{P}}_k \triangleq E(\mathbf{x} - \hat{\mathbf{x}}_k)(\mathbf{x} - \hat{\mathbf{x}}_k)^\tau = \underset{\sim}{\mathbf{S}}_k \underset{\sim}{\mathbf{P}}_0, \tag{5.30}$$

where

$$\underset{\sim}{\mathbf{S}}_k = \left[\underset{\sim}{\mathbf{I}} + \underset{\sim}{\mathbf{P}}_0 \sum_{i=1}^{k} \underset{\sim}{\mathbf{C}}_i^\tau (\underset{\sim}{\mathbf{F}}_i \underset{\sim}{\mathbf{F}}_i^\tau)^{-1} \underset{\sim}{\mathbf{C}}_i \underset{\sim}{\mathbf{P}}_0 \underset{\sim}{\mathbf{P}}_0^+ \right]^{-1}.$$

Proof. We apply Theorem 5.1 with $\mathbf{x}_k \equiv \mathbf{x}$, $\boldsymbol{\Phi}_{k+1} \equiv \mathbf{I}$, $\mathbf{B}_k = \mathbf{0}$, $\mathbf{D}_k \equiv \mathbf{0}$. From (5.13), we have

$$\mathbf{P}_{k+1} = \mathbf{P}_k - \mathbf{P}_k\mathbf{C}_{k+1}^T(\mathbf{C}_{k+1}\mathbf{P}_k\mathbf{C}_{k+1}^T + \mathbf{F}_{k+1}\mathbf{F}_{k+1}^T)^{-1}\mathbf{C}_{k+1}\mathbf{P}_k. \quad (5.31)$$

We first prove (5.29) and (5.30) for the $\mathbf{P}_0 > \mathbf{0}$ case. By the Matrix Inversion Lemma (3.18) and (5.31), it follows that

$$\mathbf{P}_{k+1}^{-1} = \mathbf{P}_k^{-1} + \mathbf{C}_{k+1}^T(\mathbf{F}_{k+1}\mathbf{F}_{k+1}^T)^{-1}\mathbf{C}_{k+1}. \quad (5.32)$$

Clearly, (5.29) and (5.30) hold for $k = 0$; let them be true for k. Then by (5.32) and the induction assumption it is immediate that (5.30) is valid for $k + 1$.

By (5.11), (5.12), and (5.31), we obtain

$$\begin{aligned}\hat{\mathbf{x}}_{k+1} &= \left[\mathbf{I} - \mathbf{P}_k\mathbf{C}_{k+1}^T(\mathbf{C}_{k+1}\mathbf{P}_k\mathbf{C}_{k+1}^T + \mathbf{F}_{k+1}\mathbf{F}_{k+1}^T)^{-1}\mathbf{C}_{k+1}\right]\hat{\mathbf{x}}_k \\ &\quad + \mathbf{P}_k\mathbf{C}_{k+1}^T(\mathbf{C}_{k+1}\mathbf{P}_k\mathbf{C}_{k+1}^T + \mathbf{F}_{k+1}\mathbf{F}_{k+1}^T)^{-1}(\mathbf{y}_{k+1} - \mathbf{H}_{k+1}\mathbf{v}_k) \\ &= \mathbf{P}_{k+1}\mathbf{P}_k^{-1}\hat{\mathbf{x}}_k + \mathbf{P}_k\mathbf{C}_{k+1}^T(\mathbf{C}_{k+1}\mathbf{P}_k\mathbf{C}_{k+1}^T + \mathbf{F}_{k+1}\mathbf{F}_{k+1}^T)^{-1}(\mathbf{y}_{k+1} - \mathbf{H}_{k+1}\mathbf{v}_k)\end{aligned}$$

and by induction

$$\begin{aligned}\hat{\mathbf{x}}_{k+1} &= \mathbf{P}_{k+1}\mathbf{P}_0^{-1}\mathbf{S}_k^{-1}\mathbf{S}_k\left[E\mathbf{x} + \mathbf{P}_0\sum_{i=1}^{k}\mathbf{C}_i^T(\mathbf{F}_i\mathbf{F}_i^T)^{-1}(\mathbf{y}_i - \mathbf{H}_i\mathbf{v}_{i-1})\right] \\ &\quad + \mathbf{P}_k\mathbf{C}_{k+1}^T(\mathbf{C}_{k+1}\mathbf{P}_k\mathbf{C}_{k+1}^T + \mathbf{F}_{k+1}\mathbf{F}_{k+1}^T)^{-1}(\mathbf{y}_{k+1} - \mathbf{H}_{k+1}\mathbf{v}_k),\end{aligned}$$

in which, by using (5.32), the second term becomes

$$\begin{aligned}&\mathbf{P}_{k+1}\left[\mathbf{P}_k^{-1} + \mathbf{C}_{k+1}^T(\mathbf{F}_{k+1}\mathbf{F}_{k+1}^T)^{-1}\mathbf{C}_{k+1}\right] \\ &\qquad \times \mathbf{P}_k\mathbf{C}_{k+1}^T(\mathbf{C}_{k+1}\mathbf{P}_k\mathbf{C}_{k+1}^T + \mathbf{F}_{k+1}\mathbf{F}_{k+1}^T)^{-1}(\mathbf{y}_{k+1} - \mathbf{H}_{k+1}\mathbf{v}_k) \\ &\quad = \mathbf{P}_{k+1}\left[\mathbf{C}_{k+1}^T(\mathbf{C}_{k+1}\mathbf{P}_k\mathbf{C}_{k+1}^T + \mathbf{F}_{k+1}\mathbf{F}_{k+1}^T)^{-1}(\mathbf{y}_{k+1} - \mathbf{H}_{k+1}\mathbf{v}_k)\right] \\ &\quad + \mathbf{P}_{k+1}\mathbf{C}_{k+1}^T(\mathbf{F}_{k+1}\mathbf{F}_{k+1}^T)^{-1}(\mathbf{y}_{k+1} - \mathbf{H}_{k+1}\mathbf{v}_k) \\ &\quad - \mathbf{P}_{k+1}\mathbf{C}_{k+1}^T(\mathbf{C}_{k+1}\mathbf{P}_k\mathbf{C}_{k+1}^T + \mathbf{F}_{k+1}\mathbf{F}_{k+1}^T)^{-1}(\mathbf{y}_{k+1} - \mathbf{H}_{k+1}\mathbf{v}_k) \\ &\quad = \mathbf{P}_{k+1}\mathbf{C}_{k+1}^T(\mathbf{F}_{k+1}\mathbf{F}_{k+1}^T)^{-1}(\mathbf{y}_{k+1} - \mathbf{H}_{k+1}\mathbf{v}_k),\end{aligned}$$

which proves that (5.29) is valid for $k + 1$.

Now consider the general case when $\mathbf{P}_0$ is possibly degenerate. Let

$$\mathbf{V} = \begin{bmatrix} \mathbf{V}_1 \\ \mathbf{V}_2 \end{bmatrix}$$

be an orthogonal matrix which makes

$$\mathbf{V}\mathbf{P}_0\mathbf{V}^\tau = \begin{bmatrix} \mathbf{\Lambda} & \mathbf{0} \\ \mathbf{0} & \mathbf{0} \end{bmatrix}$$

with $\mathbf{\Lambda} = \mathbf{V}_1\mathbf{P}_0\mathbf{V}_1^\tau$ nondegenerate.

It is easy to see that

$$\mathbf{V}^\tau \begin{bmatrix} \mathbf{\Lambda} & \mathbf{0} \\ \mathbf{0} & \mathbf{0} \end{bmatrix} \mathbf{V} = \mathbf{V}_1^\tau \mathbf{\Lambda} \mathbf{V}_1 = \mathbf{P}_0, \tag{5.33}$$

$$\mathbf{P}_0^+ = \mathbf{V}_1^\tau \mathbf{\Lambda}^{-1} \mathbf{V}_1, \qquad \mathbf{P}_0 \mathbf{P}_0^+ = \mathbf{V}_1^\tau \mathbf{V}_1, \tag{5.34}$$

and

$$\mathbf{V}_2 \mathbf{x} = \mathbf{V}_2 E\mathbf{x},$$

since $\mathbf{V}_2\mathbf{P}_0\mathbf{V}_2^\tau = \mathbf{0}$. Denote $\mathbf{z} = \mathbf{V}_1\mathbf{x}$. Then

$$\mathbf{R}_\mathbf{z} = \mathbf{V}_1\mathbf{P}_0\mathbf{V}_1^\tau = \mathbf{\Lambda},$$

and (5.29) and (5.30) are verified for $\mathbf{z}$. Since

$$\begin{bmatrix} \mathbf{z} \\ \mathbf{V}_2 E\mathbf{x} \end{bmatrix} = \begin{bmatrix} \mathbf{V}_1\mathbf{x} \\ \mathbf{V}_2\mathbf{x} \end{bmatrix} = \mathbf{V}\mathbf{x},$$

we have

$$\hat{\mathbf{x}}_k = \mathbf{V}^\tau \begin{bmatrix} \hat{\mathbf{z}}_k \\ \mathbf{V}_2 E\mathbf{x} \end{bmatrix} = \mathbf{V}_1^\tau \hat{\mathbf{z}}_k + \mathbf{V}_2^\tau \mathbf{V}_2 E\mathbf{x}, \tag{5.35}$$

$$\mathbf{P}_k = \mathbf{V}^\tau \begin{bmatrix} \mathbf{P}_k^\mathbf{z} & \mathbf{0} \\ \mathbf{0} & \mathbf{0} \end{bmatrix} \mathbf{V}, \tag{5.36}$$

where $\hat{\mathbf{z}}_k$ and $\mathbf{P}_k^\mathbf{z}$ denote the minimum variance unbiased estimate of $\mathbf{z}$ based on $\mathbf{y}^k$ and its estimation error covariance matrix, respectively.

By (5.33) and (5.34), it follows that

$$\mathbf{V}_1^T\left[\mathbf{I} + \mathbf{\Lambda}\mathbf{V}_1 \sum_{i=1}^{k} \mathbf{C}_i^T(\mathbf{F}_i\mathbf{F}_i^T)^{-1}\mathbf{C}_i\mathbf{V}_i^T\right]^{-1} \mathbf{V}_1$$

$$= \mathbf{V}^T\begin{bmatrix} \mathbf{I} + \mathbf{\Lambda}\mathbf{V}_1 \sum_{i=1}^{k} \mathbf{C}_i^T(\mathbf{F}_i\mathbf{F}_i^T)^{-1}\mathbf{C}_i\mathbf{V}_1^T & \mathbf{0} \\ \mathbf{0} & \mathbf{I} \end{bmatrix}^{-1} \begin{bmatrix} \mathbf{I} & \mathbf{0} \\ \mathbf{0} & \mathbf{0} \end{bmatrix}\mathbf{V}$$

$$= \left\{\mathbf{I} + \mathbf{V}^T\begin{bmatrix} \mathbf{\Lambda}\mathbf{V}_1 \sum_{i=1}^{k} \mathbf{C}_i^T(\mathbf{F}_i\mathbf{F}_i^T)^{-1}\mathbf{C}_i\mathbf{V}_1^T & \mathbf{0} \\ \mathbf{0} & \mathbf{0} \end{bmatrix}\mathbf{V}\right\}^{-1} \mathbf{V}^T\begin{bmatrix} \mathbf{I} & \mathbf{0} \\ \mathbf{0} & \mathbf{0} \end{bmatrix}\mathbf{V}$$

$$= \left[\mathbf{I} + \mathbf{V}_1^T\mathbf{\Lambda}\mathbf{V}_1 \sum_{i=1}^{k} \mathbf{C}_i^T(\mathbf{F}_i\mathbf{F}_i^T)^{-1}\mathbf{C}_i\mathbf{V}_1^T\right]^{-1} \mathbf{V}_1^T\mathbf{V}_1 = \mathbf{S}_k\mathbf{P}_0\mathbf{P}_0^+. \tag{5.37}$$

Noticing that

$$\mathbf{y}_k = \mathbf{C}_k\mathbf{V}_1^T\mathbf{z} + \mathbf{C}_k\mathbf{V}_2^T\mathbf{V}_2E\mathbf{x} + \mathbf{H}_k\mathbf{V}_{k-1} + \mathbf{F}_k\boldsymbol{\zeta}_k$$

and then by applying (5.30) to $\mathbf{z}$ and using (5.36) and (5.37), we conclude that

$$\mathbf{P}_k = \mathbf{V}_1^T\mathbf{P}_k^z\mathbf{V}_1 = \mathbf{V}_1^T\left[\mathbf{I} + \mathbf{\Lambda}\mathbf{V}_1 \sum_{i=1}^{k} \mathbf{C}_i^T(\mathbf{F}_i\mathbf{F}_i^T)^{-1}\mathbf{C}_i\mathbf{V}_1^T\right]^{-1} \mathbf{\Lambda}\mathbf{V}_1$$

$$= \mathbf{V}_1^T\left[\mathbf{I} + \mathbf{\Lambda}\mathbf{V}_1 \sum_{i=1}^{k} \mathbf{C}_i^T(\mathbf{F}_i\mathbf{F}_i^T)^{-1}\mathbf{C}_i\mathbf{V}_1^T\right]^{-1} \mathbf{V}_1\mathbf{V}_1^T\mathbf{\Lambda}\mathbf{V}_1 = \mathbf{S}_k\mathbf{P}_0.$$

Finally, from (5.29) applied to $\mathbf{z}$ and (5.35) and (5.37) it follows that

$$\hat{\mathbf{x}}_k = \mathbf{V}_1^T\left[\mathbf{I} + \mathbf{\Lambda}\mathbf{V}_1 \sum_{i=1}^{k} \mathbf{C}_i^T(\mathbf{F}_i\mathbf{F}_i^T)^{-1}\mathbf{C}_i\mathbf{V}_1^T\right]^{-1}$$

$$\times\left[\mathbf{V}_1E\mathbf{x} + \mathbf{\Lambda}\mathbf{V}_1 \sum_{i=1}^{k} \mathbf{C}_i^T(\mathbf{F}_i\mathbf{F}_i^T)^{-1}(\mathbf{y}_i - \mathbf{C}_i\mathbf{V}_2^T\mathbf{V}_2E\mathbf{x} - \mathbf{H}_i\mathbf{v}_{i-1})\right]\mathbf{V}_2^T\mathbf{V}_2E\mathbf{x}$$

$$= \mathbf{S}_k\mathbf{P}_0\mathbf{P}_0^+E\mathbf{x} + \mathbf{S}_k\mathbf{P}_0 \sum_{i=1}^{k} \mathbf{C}_i^T(\mathbf{F}_i\mathbf{F}_i^T)^{-1}(\mathbf{y}_i - \mathbf{H}_i\mathbf{V}_{i-1})$$

$$-\mathbf{S}_k\mathbf{P}_0 \sum_{i=1}^{k} \mathbf{C}_i^T(\mathbf{F}_i\mathbf{F}_i^T)^{-1}\mathbf{C}_i\mathbf{V}_2^T\mathbf{V}_2E\mathbf{x} + \mathbf{V}_2^T\mathbf{V}_2E\mathbf{x},$$

which together with

$$\underset{\sim}{\mathbf{V}}_2^{\tau}\underset{\sim}{\mathbf{V}}_2 = [\underset{\sim}{\mathbf{I}} - \underset{\sim}{\mathbf{P}}_0\underset{\sim}{\mathbf{P}}_0^{+}]$$

yields (5.29). □

In the following theorem, the reader should notice that the random variable ζ_i is not assumed to be Gaussian.

Theorem 5.3. *Let* $\underset{\sim}{\mathbf{\Phi}}_{k+1}$, $\underset{\sim}{\mathbf{D}}_{k+1}$, $\underset{\sim}{\mathbf{C}}_k$, $\underset{\sim}{\mathbf{F}}_k$, $\underset{\sim}{\mathbf{B}}_{k+1}\mathbf{u}_k$, *and* $\underset{\sim}{\mathbf{H}}_k\mathbf{v}_{k-1}$ *be deterministic and let* $E\zeta_i = 0$, $E\zeta_i[\mathbf{x}_0^{\tau}\mathbf{y}_0^{\tau}] = 0$, $E\zeta_i\zeta_j^{\tau} = \underset{\sim}{\mathbf{I}}\delta_{ij}$ *for all i, j. Then* (5.11)–(5.14) [(5.22)–(5.27)] *are the recursive formulas of the* LUMVE *of* $\mathbf{x}_k$ *based on* $\mathbf{y}^k$ *for system* (5.8), (5.9) [(5.20), (5.21)].

Proof. Under the assumption that $\{\xi_i\}$ is Gaussian, the process $(\mathbf{x}_k, \mathbf{y}_k)$ is Gaussian and $\hat{\mathbf{x}}_k$ given by (5.11)–(5.14) [or (5.22)–(5.27)] coincides with LUMVE of $\mathbf{x}_k$ based on $\mathbf{y}^k$. (See Section 1.11). But the LUMVE does not depend on the distribution of $\{\xi_i\}$ and thus the theorem is true. □

Now we proceed to describe the interpolation and prediction formulas. Denote

$$\hat{\mathbf{x}}_k^j = E^{\mathbf{y}^k,\mathbf{x}_j}\mathbf{x}_k, \qquad \underset{\sim}{\mathbf{P}}_k^j = E^{\mathbf{y}^k,\mathbf{x}_j}(\mathbf{x}_k - \hat{\mathbf{x}}_k^j)(\mathbf{x}_k - \hat{\mathbf{x}}_k^j)^{\tau},$$

$$\hat{\mathbf{x}}_j(k) = E^{\mathbf{y}^k}\mathbf{x}_j, \qquad \underset{\sim}{\mathbf{P}}_j(k) = E^{\mathbf{y}^k}(\mathbf{x}_j - \hat{\mathbf{x}}_j(k))(\mathbf{x}_j - \hat{\mathbf{x}}_j(k))^{\tau}.$$

Clearly,

$$\hat{\mathbf{x}}_k(k) = \hat{\mathbf{x}}_k, \qquad \underset{\sim}{\mathbf{P}}_k(k) = \underset{\sim}{\mathbf{P}}_k.$$

Lemma 5.4. *Under Conditions* A_1–D_1 *for the system* (5.8), (5.9),

$$\hat{\mathbf{x}}_k^j = \underset{\sim}{\mathbf{\Psi}}_{k,j}^j\mathbf{x}_j + \sum_{i=j+1}^{k}\underset{\sim}{\mathbf{\Psi}}_{k,i}^j[\underset{\sim}{\mathbf{B}}_i\mathbf{u}_{i-1} + \underset{\sim}{\mathbf{K}}_i^j(\mathbf{y}_i - \underset{\sim}{\mathbf{H}}_i\mathbf{v}_{i-1})] \tag{5.38}$$

$$\underset{\sim}{\mathbf{P}}_{k+1}^j = \underset{\sim}{\mathbf{\Phi}}_{k+1}\underset{\sim}{\mathbf{P}}_k^j\underset{\sim}{\mathbf{\Phi}}_{k+1}^{\tau} - \underset{\sim}{\mathbf{K}}_{k+1}^j(\underset{\sim}{\mathbf{\Phi}}_{k+1}\underset{\sim}{\mathbf{P}}_k^j\underset{\sim}{\mathbf{C}}_{k+1}^{\tau} + \underset{\sim}{\mathbf{D}}_{k+1}\underset{\sim}{\mathbf{F}}_{k+1}^{\tau})^{\tau} + \underset{\sim}{\mathbf{D}}_{k+1}\underset{\sim}{\mathbf{D}}_{k+1}^{\tau} \tag{5.39}$$

$$\underset{\sim}{\mathbf{P}}_j^j = \underset{\sim}{\mathbf{0}},$$

where

$$\underset{\sim}{\mathbf{K}}_{k+1}^{j} = \left(\underset{\sim}{\Phi}_{k+1}\underset{\sim}{\mathbf{P}}_{k}^{j}\underset{\sim}{\mathbf{C}}_{k+1}^{\tau} + \underset{\sim}{\mathbf{D}}_{k+1}\underset{\sim}{\mathbf{F}}_{k+1}^{\tau}\right)\left(\underset{\sim}{\mathbf{C}}_{k+1}\underset{\sim}{\mathbf{P}}_{k}^{j}\underset{\sim}{\mathbf{C}}_{k+1}^{\tau} + \underset{\sim}{\mathbf{F}}_{k+1}\underset{\sim}{\mathbf{F}}_{k+1}^{\tau}\right)^{+}$$

$$\underset{\sim}{\Psi}_{k,i}^{j} = \left(\underset{\sim}{\Phi}_{k} - \underset{\sim}{\mathbf{K}}_{k}^{j}\underset{\sim}{\mathbf{C}}_{k}\right)\left(\underset{\sim}{\Phi}_{k-1} - \underset{\sim}{\mathbf{K}}_{k-1}^{j}\underset{\sim}{\mathbf{C}}_{k-1}\right)\cdots\left(\underset{\sim}{\Phi}_{i+1} - \underset{\sim}{\mathbf{K}}_{i+1}^{j}\underset{\sim}{\mathbf{C}}_{i+1}\right), \qquad i \geq j.$$

Proof. From Remark 2, (5.39) follows.

When $k = j$, (5.38) becomes $\hat{\mathbf{x}}_j^j = \mathbf{x}_j$, which is obviously true. Let us verify it for $k = l$. By Remark 2, we have

$$\begin{aligned}
\hat{\mathbf{x}}_{l+1}^{j} &= \underset{\sim}{\Phi}_{l+1}\hat{\mathbf{x}}_{l}^{j} + \underset{\sim}{\mathbf{B}}_{l+1}\mathbf{u}_{l} + \underset{\sim}{\mathbf{K}}_{l+1}^{j}\left(\mathbf{y}_{l+1} - \underset{\sim}{\mathbf{C}}_{l+1}\hat{\mathbf{x}}_{l}^{j} - \underset{\sim}{\mathbf{H}}_{l+1}\mathbf{v}_{l}\right)\\
&= \left(\underset{\sim}{\Phi}_{l+1} - \underset{\sim}{\mathbf{K}}_{l+1}^{j}\underset{\sim}{\mathbf{C}}_{l+1}\right)\hat{\mathbf{x}}_{l}^{j} + \underset{\sim}{\mathbf{K}}_{l+1}^{j}\left(\mathbf{y}_{l+1} - \underset{\sim}{\mathbf{H}}_{l+1}\mathbf{v}_{l}\right) + \underset{\sim}{\mathbf{B}}_{l+1}\mathbf{u}_{l}\\
&= \left(\underset{\sim}{\Phi}_{l+1} - \underset{\sim}{\mathbf{K}}_{l+1}^{j}\underset{\sim}{\mathbf{C}}_{l+1}\right)\left\{\underset{\sim}{\Psi}_{l,j}^{j}\mathbf{x}_{j} + \sum_{i=j+1}^{l}\underset{\sim}{\Psi}_{l,i}^{j}\left[\underset{\sim}{\mathbf{B}}_{i}\mathbf{u}_{i-1} + \underset{\sim}{\mathbf{K}}_{i}^{j}\left(\mathbf{y}_{i} - \underset{\sim}{\mathbf{H}}_{i}\mathbf{v}_{i-1}\right)\right]\right\}\\
&\quad + \underset{\sim}{\mathbf{B}}_{l+1}\mathbf{u}_{l} + \underset{\sim}{\mathbf{K}}_{l+1}^{j}\left(\mathbf{y}_{l+1} - \underset{\sim}{\mathbf{H}}_{l+1}\mathbf{v}_{l}\right)\\
&= \underset{\sim}{\Psi}_{l+1,j}^{j}\mathbf{x}_{j} + \sum_{i=j+1}^{l+1}\underset{\sim}{\Psi}_{l+1,i}^{j}\left[\underset{\sim}{\mathbf{B}}_{i}\mathbf{u}_{i-1} + \underset{\sim}{\mathbf{K}}_{i}^{j}\left(\mathbf{y}_{i} - \underset{\sim}{\mathbf{H}}_{i}\mathbf{v}_{i-1}\right)\right],
\end{aligned}$$

which shows (5.38) holds for $l + 1$. □

Lemma 5.5. *Under Conditions* A_1–D_1, $[\mathbf{x}_j^{\tau}\mathbf{x}_k^{\tau}]^{\tau}$ *is conditionally Gaussian given* $\mathbf{y}^l$ *for* $l \geq j$, $l \geq k$.

Proof. For $l = k = j$, the lemma follows from Lemma 5.3. Hence we only need to prove it for the two cases $l \geq k > j$ and $l > k = j$.

$$\begin{aligned}
&E^{\mathbf{y}^j}\exp i\left(\lambda^{\tau}\mathbf{x}_j + \mu^{\tau}\mathbf{y}_{j+1}\right)\\
&\qquad = E^{\mathbf{y}^j}\exp i\left[\lambda^{\tau}\mathbf{x}_j + \mu^{\tau}\left(\underset{\sim}{\mathbf{C}}_{j+1}\mathbf{x}_j + \underset{\sim}{\mathbf{H}}_{j+1}\mathbf{v}_j + \underset{\sim}{\mathbf{F}}_{j+1}\zeta_{j+1}\right)\right]\\
&\qquad = \exp\left(i\mu^{\tau}\underset{\sim}{\mathbf{H}}_{j+1}\mathbf{v}_j\right)E^{\mathbf{y}^j}\exp i\left(\lambda^{\tau} + \mu^{\tau}\underset{\sim}{\mathbf{C}}_{j+1}\right)\mathbf{x}_j E^{\mathbf{y}^j\mathbf{x}_j}\exp i\mu^{\tau}\underset{\sim}{\mathbf{F}}_{j+1}\zeta_{j+1}.
\end{aligned}$$

Since $\mathbf{y}^j$ and $\mathbf{x}_j$ are independent of ζ_{j+1}, Theorem 1.9 is applicable. Then by Lemma 5.3 it follows from the preceding calculation that given $\mathbf{y}^j$,

$[\mathbf{x}_j^\tau \mathbf{y}_{j+1}^\tau]^\tau$ is conditionally Gaussian. Then by 3 of Lemma 5.2, $\mathbf{x}_j$ is conditionally Gaussian given $\mathbf{y}^{j+1}$.

Assume $\mathbf{x}_j$ is conditionally Gaussian given $\mathbf{y}^t$ with $t > j$; we are going to prove that the $\mathbf{x}_j$ is conditionally Gaussian on $\mathbf{y}^{t+1}$. For this, by Lemma 5.2, it is sufficient to prove that $[\mathbf{x}_j^\tau \mathbf{y}_{t+1}^\tau]^\tau$ is conditionally Gaussian given $\mathbf{y}^t$. We have

$$\begin{aligned} E^{\mathbf{y}^t} \exp i\left(\boldsymbol{\lambda}^\tau \mathbf{x}_j + \boldsymbol{\mu}^\tau \mathbf{y}_{t+1}\right) &= E^{\mathbf{y}^t}\left[\exp i\boldsymbol{\lambda}^\tau \mathbf{x}_j E^{\mathbf{y}^t, \mathbf{x}_j} \exp i\boldsymbol{\mu}^\tau \mathbf{y}_{t+1}\right] \\ &= E^{\mathbf{y}^t}\left\{\exp i\boldsymbol{\lambda}^\tau \mathbf{x}_j \exp\left[i\boldsymbol{\mu}^\tau\left(\mathbf{C}_{t+1}\hat{\mathbf{x}}_t^j + \mathbf{H}_{t+1}\mathbf{v}_t\right)\right.\right. \\ &\quad \left.\left. - \tfrac{1}{2}\boldsymbol{\mu}^\tau\left(\mathbf{C}_{t+1}\mathbf{P}_t^j\mathbf{C}_{t+1}^\tau + \mathbf{F}_{t+1}\mathbf{F}_{t+1}^\tau\right)\mu\right]\right\} \end{aligned} \tag{5.40}$$

since, by Lemma 5.3, $\mathbf{y}_{t+1}$ is conditionally Gaussian given $(\mathbf{y}^t, \mathbf{x}_j)$ and since

$$E^{\mathbf{y}^t, \mathbf{x}_j}\mathbf{y}_{t+1} = \mathbf{C}_{t+1}\hat{\mathbf{x}}_t^j + \mathbf{H}_{t+1}\mathbf{v}_t.$$

Putting (5.38) into (5.40) and using Lemma 5.1, we see that (5.40) yields a conditionally Gaussian characteristic function. Thus we have shown that given $\mathbf{y}^l, \mathbf{x}_j$ is conditionally Gaussian for $l > j$, and the lemma has been verified for the $l > k = j$ case since $[\mathbf{x}_j^\tau \mathbf{x}_j^\tau]^\tau = [\mathbf{I} \quad \mathbf{I}]^\tau \mathbf{x}_j$ and Conclusion 2 of Lemma 5.1 holds.

Now assume $l \geq k > j$. By Conclusion 2 of Lemma 5.3 and (5.38), we know that

$$\begin{aligned} &E^{\mathbf{y}^l} \exp i\left(\boldsymbol{\lambda}_1^\tau \mathbf{x}_j + \boldsymbol{\lambda}_2^\tau \mathbf{x}_k\right) \\ &= E^{\mathbf{y}^l} \exp\left(i\boldsymbol{\lambda}_1^\tau \mathbf{x}_j\right) E^{\mathbf{y}^k, \mathbf{x}_j} \exp\left(i\boldsymbol{\lambda}_2^\tau \mathbf{x}_k\right) \\ &= E^{\mathbf{y}^l} \exp\left(i\boldsymbol{\lambda}_1^\tau \mathbf{x}_j\right) \exp\left(i\boldsymbol{\lambda}_2^\tau \hat{\mathbf{x}}_k^j - \tfrac{1}{2}\boldsymbol{\lambda}_2^\tau \mathbf{P}_k^j \boldsymbol{\lambda}_2\right) \\ &= \exp\left\{ i\boldsymbol{\lambda}_2^\tau \sum_{i=j+1}^{k} \boldsymbol{\Psi}_{k,i}^j\left[\mathbf{B}_i \mathbf{u}_{i-1} + \mathbf{K}_i^j(\mathbf{y}_i - \mathbf{H}_i \mathbf{v}_{i-1})\right] - \tfrac{1}{2}\boldsymbol{\lambda}_2^\tau \mathbf{P}_k^j \boldsymbol{\lambda}_2 \right\} \\ &\quad \cdot E^{\mathbf{y}^l} \exp i\left(\boldsymbol{\lambda}_1^\tau + \boldsymbol{\lambda}_2^\tau \boldsymbol{\Psi}_{k,j}^j\right)\mathbf{x}_j, \end{aligned}$$

which, by Lemma 5.1, is a conditionally Gaussian characteristic function since $\mathbf{x}_j$ is conditionally Gaussian given $\mathbf{y}^l$. □

Corollary. *Applying Conclusion* 3 *of Lemma* 5.2 *to Lemma* 5.5 *yields that given* $(\mathbf{y}^l, \mathbf{x}_k)$, $\mathbf{x}_j$ *is conditionally Gaussian for* $l \geq j$, $l \geq k$.

The minimum variance unbiased estimate of $\mathbf{x}_j$ based on $\mathbf{y}^k$ is called an interpolation of $\mathbf{x}_j$ if $j \leq k$ and a prediction of $\mathbf{x}_j$ if $j \geq k$.

For $k \geq j$, if j is fixed and k is changing, then the equation for $\hat{\mathbf{x}}_j(k)$ $(\triangleq E^{\mathbf{y}^k}\mathbf{x}_j)$ is called forward. It is called backward if k is fixed and j is changing.

For state space systems, the prediction and interpolation (or smoothing) problems have a huge literature; in order to access this work we shall only refer the reader to the current text [77]. A key feature of the following exposition is that, as in Ref. 2, we carry out the derivations under the conditionally Gaussian hypothesis; roughly, this states that given observations on the output process $\mathbf{y}$, a linear system whose parameters depend nonlinearly on the history of $\mathbf{y}$ can be treated like a conventional Markovian state space system with Gaussian input, state, and output processes.

Theorem 5.4. *Under Conditions* A_1–D_1 *for system* (5.8), (5.9), *the forward interpolation equations are given by*

$$\hat{\mathbf{x}}_j(k+1) = \hat{\mathbf{x}}_j(k) + \tilde{\mathbf{P}}_j(k)\tilde{\mathbf{\Psi}}_{k,j}^{j\tau}\tilde{\mathbf{C}}_{k+1}^{\tau}\left(\tilde{\mathbf{C}}_{k+1}\tilde{\mathbf{P}}_k\tilde{\mathbf{C}}_{k+1}^{\tau} + \tilde{\mathbf{F}}_{k+1}^{\tau}\tilde{\mathbf{F}}_{k+1}\right)^{+}$$

$$\times\left(\mathbf{y}_{k+1} - \tilde{\mathbf{C}}_{k+1}\hat{\mathbf{x}}_k - \tilde{\mathbf{H}}_{k+1}\mathbf{v}_k\right) \tag{5.41}$$

$$\tilde{\mathbf{P}}_j(k+1) = \tilde{\mathbf{P}}_j(k) - \tilde{\mathbf{P}}_j(k)\tilde{\mathbf{\Psi}}_{k,j}^{j\tau}\tilde{\mathbf{C}}_{k+1}^{\tau}\left(\tilde{\mathbf{C}}_{k+1}\tilde{\mathbf{P}}_k\tilde{\mathbf{C}}_{k+1}^{\tau} + \tilde{\mathbf{F}}_{k+1}\tilde{\mathbf{F}}_{k+1}^{\tau}\right)^{+}$$

$$\times\,\tilde{\mathbf{C}}_{k+1}\tilde{\mathbf{\Psi}}_{k,j}^{j}\tilde{\mathbf{P}}_j(k) \tag{5.42}$$

$$\hat{\mathbf{x}}_j(j) = \hat{\mathbf{x}}_j, \qquad \tilde{\mathbf{P}}_j(j) = \tilde{\mathbf{P}}_j.$$

Proof. From the proof of Lemma 5.5 [see (5.40)], we have seen that given $\mathbf{y}^k$, $[\mathbf{x}_j^{\tau}\mathbf{y}_{k+1}^{\tau}]^{\tau}$ is conditionally Gaussian. Hence by (5.3)

$$\hat{\mathbf{x}}_j(k+1) = \hat{\mathbf{x}}_j(k) + \tilde{\mathbf{R}}_{\mathbf{x}_j,\mathbf{y}_{k+1}}^{\mathbf{y}^k}\left(\tilde{\mathbf{R}}_{\mathbf{y}_{k+1}}^{\mathbf{y}^k}\right)^{+}\left(\mathbf{y}_{k+1} - \hat{\mathbf{y}}_{k+1}'\right), \tag{5.43}$$

where $\hat{\mathbf{y}}_{k+1}'$ and $\tilde{\mathbf{R}}_{\mathbf{y}_{k+1}}^{\mathbf{y}^k}$ are given by (5.15) and (5.18).

Now we give a formula for $\tilde{\mathbf{R}}_{\mathbf{x}_j,\mathbf{y}_{k+1}}^{\mathbf{y}^k}$.

Taking $E^{\mathbf{y}^k}(\cdot)$ of both sides of (5.38) leads to

$$\hat{\mathbf{x}}_k = \tilde{\mathbf{\Psi}}_{k,j}^{j}\hat{\mathbf{x}}_j(k) + \sum_{i=j+1}^{k}\tilde{\mathbf{\Psi}}_{k,i}^{j}\left[\tilde{\mathbf{B}}_i\mathbf{u}_{i-1} + \tilde{\mathbf{K}}_i^{j}(\mathbf{y}_i - \tilde{\mathbf{H}}_i\mathbf{v}_{i-1})\right]$$

and hence

$$E^{\mathbf{y}^k,\mathbf{x}_j}(\mathbf{x}_k - \hat{\mathbf{x}}_k) = \hat{\mathbf{x}}_k^j - \hat{\mathbf{x}}_k = \underset{\sim}{\mathbf{\Psi}}_{k,j}^{i}[\mathbf{x}_j - \hat{\mathbf{x}}_j(k)]. \tag{5.44}$$

Noticing that $E^{\mathbf{y}^k,\mathbf{x}_j}\zeta_{k+1}^{\tau}\underset{\sim}{\mathbf{F}}_{k+1}^{\tau} = 0$, we have

$$\begin{aligned}\underset{\sim}{\mathbf{R}}_{\mathbf{x}_j,\mathbf{y}_{k+1}}^{\mathbf{y}^k} &= E^{\mathbf{y}^k}E^{\mathbf{y}^k,\mathbf{x}_j}[\mathbf{x}_j - \hat{\mathbf{x}}_j(k)][\underset{\sim}{\mathbf{C}}_{k+1}(\mathbf{x}_k - \hat{\mathbf{x}}_k) + \underset{\sim}{\mathbf{F}}_{k+1}\zeta_{k+1}]^{\tau}\\ &= E^{\mathbf{y}^k}[\mathbf{x}_j - \hat{\mathbf{x}}_j(k)]E^{\mathbf{y}^k}(\mathbf{x}_k - \hat{\mathbf{x}}_k)\underset{\sim}{\mathbf{C}}_{k+1}^{\tau}\\ &= E^{\mathbf{y}^k}[\mathbf{x}_j - \hat{\mathbf{x}}_j(k)][\mathbf{x}_j - \hat{\mathbf{x}}_j(k)]^{\tau}\underset{\sim}{\mathbf{\Psi}}_{k,j}^{j\tau}\underset{\sim}{\mathbf{C}}_{k+1}^{\tau} = \underset{\sim}{\mathbf{P}}_j(k)\underset{\sim}{\mathbf{\Psi}}_{k,j}^{j\tau}\underset{\sim}{\mathbf{C}}_{k+1}^{\tau}.\end{aligned}$$

Substituting this, (5.15), and (5.18) into (5.43) gives (5.41) and from (5.4), (5.42) follows immediately.

□

Theorem 5.5. *Under Conditions* A_1–D_1 *for the system* (5.8), (5.9), *the forward interpolation equation based on both* $\mathbf{y}^k$ *and* $\mathbf{x}_k$ *can be represented as*

$$\hat{\mathbf{x}}_j^k(k) \triangleq E^{\mathbf{y}^k,\mathbf{x}_k}\mathbf{x}_j = \hat{\mathbf{x}}_j(k) + \underset{\sim}{\mathbf{P}}_j(k)\underset{\sim}{\mathbf{\Psi}}_{k,j}^{j\tau}\underset{\sim}{\mathbf{P}}_k^{+}(\mathbf{x}_k - \hat{\mathbf{x}}_k) \tag{5.45}$$

$$\begin{aligned}\underset{\sim}{\mathbf{P}}_j^k(k) &\triangleq E^{\mathbf{y}^k\mathbf{x}_k}\left(\mathbf{x}_j - \hat{\mathbf{x}}_j^k(k)\right)\left(\mathbf{x}_j - \hat{\mathbf{x}}_j^k(k)\right)^{\tau}\\ &= \underset{\sim}{\mathbf{P}}_j(k) - \underset{\sim}{\mathbf{P}}_j(k)\underset{\sim}{\mathbf{\Psi}}_{k,j}^{j\tau}\underset{\sim}{\mathbf{P}}_k^{+}\underset{\sim}{\mathbf{\Psi}}_{k,j}^{j}\underset{\sim}{\mathbf{P}}_j(k).\end{aligned} \tag{5.46}$$

Remark. Of course, unless $\mathbf{x}_k$ is available the recursive equations (5.45) and (5.46) are not implementable. The same is true of (5.48) and (5.49). The implementable interpolation filter is given in Theorems 5.7 and 5.8.

Proof. By the corollary of Lemma 5.5, given $(\mathbf{y}^k, \mathbf{x}_k)$, $\mathbf{x}_j$ is conditionally Gaussian, hence by (5.3)

$$E^{\mathbf{y}^k,\mathbf{x}_k}\mathbf{x}_j = \hat{\mathbf{x}}_j(k) + \underset{\sim}{\mathbf{R}}_{\mathbf{x}_j,\mathbf{x}_k}^{\mathbf{y}^k}\left(\underset{\sim}{\mathbf{R}}_{\mathbf{x}_k}^{\mathbf{y}^k}\right)^{+}(\mathbf{x}_k - \hat{\mathbf{x}}_k). \tag{5.47}$$

Similarly to (5.44), we can prove that

$$\underset{\sim}{\mathbf{R}}_{\mathbf{x}_j,\mathbf{x}_k}^{\mathbf{y}^k} = \underset{\sim}{\mathbf{P}}_j(k)\underset{\sim}{\mathbf{\Psi}}_{k,j}^{j\tau}.$$

Since $\underset{\sim}{\mathbf{R}}_{\mathbf{x}_k}^{\mathbf{y}^k} = \underset{\sim}{\mathbf{P}}_k$, (5.45) follows from (5.47) immediately, and (5.46) follows from (5.4).

□

For the backward interpolation, we first give a lemma.

Lemma 5.6. *For the system* (5.8), (5.9) *under Conditions* A_1–D_1, *given* $(\mathbf{y}^k, \mathbf{x}_k)$, $k \geq j$, $[\mathbf{x}_j^\tau, \mathbf{y}_j^\tau]^\tau$ *is conditionally independent of* $[\mathbf{x}_{k+1}^\tau, \mathbf{y}_{k+1}^\tau]^\tau, \ldots, [\mathbf{x}_l^\tau, \mathbf{y}_l^\tau]^\tau$, $k < l$.

Remark. This lemma constitutes a generalization of the familiar fact that the joint-state output process $(\mathbf{x}_k, \mathbf{y}_k)$ of a system of the form (5.8), (5.9), with deterministic coefficient matrices is a Markov process.

Proof. Denote $\mathbf{z}_p = [\mathbf{x}_p^\tau \mathbf{y}_p^\tau]^\tau$ and by $\boldsymbol{\gamma}_p$ the $n+m$-dimensional deterministic vector. Then

$$E^{\mathbf{y}^k \mathbf{x}_k} \exp\left(i\boldsymbol{\gamma}_j^\tau \mathbf{z}_j + i \sum_{p=k+1}^{l} \boldsymbol{\gamma}_p^\tau \mathbf{z}_p \right)$$

$$= E^{\mathbf{y}^k \mathbf{x}_k} \left\{ \exp\left(i\boldsymbol{\gamma}_j^\tau \mathbf{z}_j \right) E^{\mathbf{y}^k \mathbf{x}_k \mathbf{z}_j} \exp\left(i \sum_{p=k+1}^{l} \boldsymbol{\gamma}_p^\tau \mathbf{z}_p \right) \right\}.$$

Obviously, for fixed $(\mathbf{y}^k, \mathbf{x}_k, \mathbf{z}_j)$ the random variability of $\mathbf{z}_{k+1}, \ldots, \mathbf{z}_l$ is determined by $\zeta_{k+1}, \ldots, \zeta_l$ only. Hence if in $\mathbf{z}_{k+1}, \ldots, \mathbf{z}_l$, $(\mathbf{y}^k, \mathbf{x}_k, \mathbf{z}_j)$ is a fixed deterministic vector, then they are independent of $(\mathbf{y}^k, \mathbf{x}_k, \mathbf{z}_j)$. Applying Theorem 1.9, we know that

$$E^{\mathbf{y}^k, \mathbf{x}_k, \mathbf{z}_j} \exp\left(i \sum_{p=k+1}^{l} \boldsymbol{\gamma}_p^\tau \mathbf{z}_p \right)$$

is $(\mathbf{y}^k, \mathbf{x}_k)$-measurable, since for fixed $(\mathbf{y}^k, \mathbf{x}_k)$, $\mathbf{z}_{k+1}, \ldots, \mathbf{z}_l$ are independent of $\mathbf{z}_j$.

Consequently,

$$E^{\mathbf{y}^k, \mathbf{x}_k, \mathbf{z}_j} \exp\left(i \sum_{p=k+1}^{l} \boldsymbol{\gamma}_p^\tau \mathbf{z}_p \right) = E^{\mathbf{y}^k, \mathbf{x}_k} \exp\left(i \sum_{p=k+1}^{l} \boldsymbol{\gamma}_p^\tau \mathbf{z}_p \right),$$

and

$$E^{\mathbf{y}^k, \mathbf{x}_k} \exp\left(i\boldsymbol{\gamma}_j^\tau \mathbf{z}_j + i \sum_{p=k+1}^{l} \boldsymbol{\gamma}_p^\tau \mathbf{z}_p \right)$$

$$= E^{\mathbf{y}^k, \mathbf{x}_k} \exp\left(i\boldsymbol{\gamma}_j^\tau z_j \right) E^{\mathbf{y}^k, \mathbf{x}_k} \exp\left(i \sum_{p=k+1}^{l} \boldsymbol{\gamma}_p^\tau \mathbf{z}_p \right),$$

from which the conclusion of the lemma follows by using Theorem 1.8. □

Theorem 5.6. *Under Conditions* A_1–D_1 *for the system* (5.8), (5.9), *the following backward interpolation equations hold*:

$$\hat{\mathbf{x}}_j^k(k) = \hat{\mathbf{x}}_j(j+1) + \underset{\sim}{\mathbf{P}}_j(j+1)\underset{\sim}{\mathbf{\Psi}}_{j+1,j}^{j\tau}\underset{\sim}{\mathbf{P}}_{j+1}^{+}\left(\hat{\mathbf{x}}_{j+1}^k(k) - \hat{\mathbf{x}}_{j+1}\right), \qquad j < k, \tag{5.48}$$

$$\underset{\sim}{\mathbf{P}}_j^k(k) = \underset{\sim}{\mathbf{P}}_j^{j+1}(j+1) + \underset{\sim}{\mathbf{P}}_j(j+1)\underset{\sim}{\mathbf{\Psi}}_{j+1,j}^{j\tau}\underset{\sim}{\mathbf{P}}_{j+1}^{+}\underset{\sim}{\mathbf{P}}_{j+1}^{k}(k)\underset{\sim}{\mathbf{P}}_{j+1}^{+}\underset{\sim}{\mathbf{\Psi}}_{j+1,j}^{j}\underset{\sim}{\mathbf{P}}_j(j+1) \tag{5.49}$$

with initial values $\hat{\mathbf{x}}_k^k(k) = \mathbf{x}_k$, $\underset{\sim}{\mathbf{P}}_k^k(k) = 0$.

Proof. Using Lemma 5.6 and Theorem 1.9, we have

$$E^{\mathbf{y}^k, \mathbf{x}_{j+1}, \mathbf{x}_{j+2} \ldots \mathbf{x}_k}\mathbf{x}_j = E^{\mathbf{y}^{j+1}, \mathbf{x}_{j+1}}\mathbf{x}_j, \qquad j < k$$

and hence

$$E^{\mathbf{y}^k, \mathbf{x}_k}\mathbf{x}_j = E^{\mathbf{y}^k, \mathbf{x}_k}E^{\mathbf{y}^{j+1}, \mathbf{x}_{j+1}}\mathbf{x}_j.$$

Using (5.45), it follows that

$$\begin{aligned}\hat{\mathbf{x}}_j^k(k) &= E^{\mathbf{y}^k, \mathbf{x}_k}\left[\hat{\mathbf{x}}_j(j+1) + \underset{\sim}{\mathbf{P}}_j(j+1)\underset{\sim}{\mathbf{\Psi}}_{j+1,j}^{j\tau}\underset{\sim}{\mathbf{P}}_{j+1}^{+}(\mathbf{x}_{j+1} - \hat{\mathbf{x}}_{j+1})\right]\\ &= \hat{\mathbf{x}}_j(j+1) + \underset{\sim}{\mathbf{P}}_j(j+1)\underset{\sim}{\mathbf{\Psi}}_{j+1,j}^{j\tau}\underset{\sim}{\mathbf{P}}_{j+1}^{+}\left(\hat{\mathbf{x}}_{j+1}^k(k) - \hat{\mathbf{x}}_{j+1}\right).\end{aligned}$$

Now we proceed to prove (5.49). We have

$$\begin{aligned}\underset{\sim}{\mathbf{P}}_j^k(k) &= E^{\mathbf{y}^k, \mathbf{x}_k}\left(\mathbf{x}_j - E^{\mathbf{y}^k, \mathbf{x}_k}\mathbf{x}_j\right)\left(\mathbf{x}_j - E^{\mathbf{y}^k, \mathbf{x}_k}\mathbf{x}_j\right)^\tau\\ &= E^{\mathbf{y}^k, \mathbf{x}_k}\Big\{\Big[E^{\mathbf{y}^k, \mathbf{x}_k, \mathbf{x}_{j+1}}\left(\mathbf{x}_j\mathbf{x}_j^\tau\right) - E^{\mathbf{y}^k, \mathbf{x}_k, \mathbf{x}_{j+1}}\mathbf{x}_j E^{\mathbf{y}^k, \mathbf{x}_k}\mathbf{x}_j^\tau\\ &\quad - E^{\mathbf{y}^k, \mathbf{x}_k}\mathbf{x}_j E^{\mathbf{y}^k, \mathbf{x}_k, \mathbf{x}_{j+1}}\mathbf{x}_j^\tau\\ &\quad + E^{\mathbf{y}^k, \mathbf{x}_k, \mathbf{x}_{j+1}}\mathbf{x}_j E^{\mathbf{y}^k, \mathbf{x}_k, \mathbf{x}_{j+1}}\mathbf{x}_j^\tau\Big] + E^{\mathbf{y}^k, \mathbf{x}_k}\mathbf{x}_j E^{\mathbf{y}^k, \mathbf{x}_k}\mathbf{x}_j^\tau - E^{\mathbf{y}^k, \mathbf{x}_k, \mathbf{x}_{j+1}}\mathbf{x}_j E^{\mathbf{y}^k, \mathbf{x}_k, \mathbf{x}_{j+1}}\mathbf{x}_j^\tau\Big\}\\ &= E^{\mathbf{y}^k, \mathbf{x}_k}\Big\{E^{\mathbf{y}^k, \mathbf{x}_k, \mathbf{x}_{j+1}}\left(\mathbf{x}_j - E^{\mathbf{y}^k, \mathbf{x}_k, \mathbf{x}_{j+1}}\mathbf{x}_j\right)\left(\mathbf{x}_j - E^{\mathbf{y}^k, \mathbf{x}_k, \mathbf{x}_{j+1}}\mathbf{x}_j\right)^\tau\\ &\quad + \left(E^{\mathbf{y}^k, \mathbf{x}_k, \mathbf{x}_{j+1}}\mathbf{x}_j - E^{\mathbf{y}^k, \mathbf{x}_k}\mathbf{x}_j\right)\left(E^{\mathbf{y}^k, \mathbf{x}_k, \mathbf{x}_{j+1}}\mathbf{x}_j - E^{\mathbf{y}^k, \mathbf{x}_k}\mathbf{x}_j\right)^\tau\Big\}.\end{aligned} \tag{5.50}$$

Between the braces, the first term equals

$$E^{\mathbf{y}^{j+1},\mathbf{x}_{j+1}}\left(\mathbf{x}_j - E^{\mathbf{y}^{j+1},\mathbf{x}_{j+1}}\mathbf{x}_j\right)\left(\mathbf{x}_j - E^{\mathbf{y}^{j+1},\mathbf{x}_{j+1}}\mathbf{x}_j\right)^{\tau} = \underset{\sim}{\mathbf{P}}_j^{j+1}(j+1) \tag{5.51}$$

by Lemma 5.6. From (5.13), we see that $\underset{\sim}{\mathbf{P}}_{k+1}$ is $\mathbf{y}^k$-measurable and from Lemma 5.4 $\underset{\sim}{\mathbf{P}}_{k+1}^j, \underset{\sim}{\mathbf{K}}_{k+1}^j$ are $\mathbf{y}^k$-measurable. $\underset{\sim}{\mathbf{\Psi}}_{k,i}^j$ is $\mathbf{y}^{k-1}$-measurable, and by (5.42), $\underset{\sim}{\mathbf{P}}_j(k+1)$ is $\mathbf{y}^k$-measurable. Hence, by (5.46), we know that $\underset{\sim}{\mathbf{P}}_j^k(k)$ is $\mathbf{y}^{k-1}$-measurable, and $\underset{\sim}{\mathbf{P}}_j^{j+1}(j+1)$ is $(\mathbf{y}^k, \mathbf{x}_k)$-measurable. Applying Lemma 5.6 to the second term in the braces in (5.50), we see that it equals

$$\left(\hat{\mathbf{x}}_j^{j+1}(j+1) - \hat{\mathbf{x}}_j^k(k)\right)\left(\hat{\mathbf{x}}_j^{j+1}(j+1) - \hat{\mathbf{x}}_j^k(k)\right)^{\tau}.$$

But by (5.48), it follows that

$$\hat{\mathbf{x}}_j^{j+1}(j+1) - \hat{\mathbf{x}}_j^k(k) = \underset{\sim}{\mathbf{P}}_j(j+1)\underset{\sim}{\mathbf{\Psi}}_{j+1,j}^{j\tau}\underset{\sim}{\mathbf{P}}_{j+1}^{+}\left(\mathbf{x}_{j+1} - \hat{\mathbf{x}}_{j+1}^k(k)\right).$$

Hence

$$E^{\mathbf{y}^k,\mathbf{x}_k}\left[\hat{\mathbf{x}}_j^{j+1}(j+1) - \hat{\mathbf{x}}_j^k(k)\right]\left[\hat{\mathbf{x}}_j^{j+1}(j+1) - \hat{\mathbf{x}}_j^k(k)\right]^{\tau}$$

$$= \underset{\sim}{\mathbf{P}}_j(j+1)\underset{\sim}{\mathbf{\Psi}}_{j+1,j}^{j\tau}\underset{\sim}{\mathbf{P}}_{j+1}^{+}\underset{\sim}{\mathbf{P}}_{j+1}^{k}(k)\underset{\sim}{\mathbf{P}}_{j+1}^{+}\underset{\sim}{\mathbf{\Psi}}_{j+1,j}^{j}\underset{\sim}{\mathbf{P}}_j(j+1). \tag{5.52}$$

Putting (5.51) and (5.52) into (5.50) gives (5.49). □

Theorem 5.7. *Under Conditions* A_1–D_1 *on the system* (5.8) *and* (5.9), *the backward interpolation equations can be represented as*

$$\hat{\mathbf{x}}_j(k) = \hat{\mathbf{x}}_j(j+1) + \underset{\sim}{\mathbf{P}}_j(j+1)$$

$$\underset{\sim}{\mathbf{\Psi}}_{j+1,j}^{j\tau}\underset{\sim}{\mathbf{P}}_{j+1}^{+}\left[\hat{\mathbf{x}}_{j+1}(k) - \hat{\mathbf{x}}_{j+1}\right], \qquad j < k \tag{5.53}$$

$$\underset{\sim}{\mathbf{P}}_j(k) = \underset{\sim}{\mathbf{P}}_j^{j+1}(j+1) + \underset{\sim}{\mathbf{P}}_j(j+1)\underset{\sim}{\mathbf{\Psi}}_{j+1,j}^{j\tau}\underset{\sim}{\mathbf{P}}_{j+1}^{+}\underset{\sim}{\mathbf{P}}_{j+1}(k)$$

$$\underset{\sim}{\mathbf{P}}_{j+1}^{+}\underset{\sim}{\mathbf{\Psi}}_{j+1,j}^{j}\underset{\sim}{\mathbf{P}}_j(j+1), \qquad j < k. \tag{5.54}$$

Proof. Taking conditional expectations of both sides of (5.48), we obtain (5.53).

In a manner similar to the derivation of (5.50) (by removing $\mathbf{x}_k$ therein), we can show that

$$\begin{aligned}\underset{\sim}{\mathbf{P}}_j(k) &= E^{\mathbf{y}^k}\Big\{E^{\mathbf{y}^k,\mathbf{x}_{j+1}}\Big(\mathbf{x}_j - E^{\mathbf{y}^k,\mathbf{x}_{j+1}}\mathbf{x}_j\Big)\Big(\mathbf{x}_j - E^{\mathbf{y}^k,\mathbf{x}_{j+1}}\mathbf{x}_j\Big)^{\tau} \\ &\quad + \Big(E^{\mathbf{y}^k,\mathbf{x}_{j+1}}\mathbf{x}_j - E^{\mathbf{y}^k}\mathbf{x}_j\Big)\Big(E^{\mathbf{y}^k,\mathbf{x}_{j+1}}\mathbf{x}_j - E^{\mathbf{y}^k}\mathbf{x}_j\Big)^{\tau}\Big\} \\ &= E^{\mathbf{y}^k}\Big\{E^{\mathbf{y}^{j+1},\mathbf{x}_{j+1}}\big(\mathbf{x}_j - \hat{\mathbf{x}}_j^{j+1}(j+1)\big)\big(\mathbf{x}_j - \hat{\mathbf{x}}_j^{j+1}(j+1)\big)^{\tau} \\ &\quad + \big(\hat{\mathbf{x}}_j^{j+1}(j+1) - \hat{\mathbf{x}}_j(k)\big)\big(\hat{\mathbf{x}}_j^{j+1}(j+1) - \hat{\mathbf{x}}_j(k)\big)^{\tau}\Big\}.\end{aligned}$$

By using (5.48) and (5.53), we can deduce that

$$\begin{aligned}\underset{\sim}{\mathbf{P}}_j(k) &= E^{\mathbf{y}^k}\underset{\sim}{\mathbf{P}}_j^{j+1}(j+1) \\ &\quad + E^{\mathbf{y}^k}\underset{\sim}{\mathbf{P}}_j(j+1)\underset{\sim}{\mathbf{\Psi}}_{j+1,j}^{j\tau}\underset{\sim}{\mathbf{P}}_{j+1}^{+}\underset{\sim}{\mathbf{P}}_j(k)\underset{\sim}{\mathbf{P}}_{j+1}^{+}\underset{\sim}{\mathbf{\Psi}}_{j+1,j}^{j}\underset{\sim}{\mathbf{P}}_j(j+1),\end{aligned}$$

in which the terms under the conditional expectation are $\mathbf{y}^k$-measurable. Hence (5.54) follows from this immediately. □

For prediction, we consider only a special case of system (5.8), (5.9):

$$\mathbf{x}_{k+1} = \underset{\sim}{\mathbf{\Phi}}_{k+1}\mathbf{x}_k + \underset{\sim}{\mathbf{B}}_{k+1}\mathbf{y}_k + \underset{\sim}{\mathbf{D}}_{k+1}\boldsymbol{\zeta}_{k+1}, \tag{5.55}$$

$$\mathbf{y}_{k+1} = \underset{\sim}{\mathbf{C}}_{k+1}\mathbf{x}_k + \underset{\sim}{\mathbf{H}}_{k+1}\mathbf{y}_k + \underset{\sim}{\mathbf{F}}_{k+1}\boldsymbol{\zeta}_{k+1}, \tag{5.56}$$

where $\{\boldsymbol{\zeta}_k\}$ and $[\mathbf{x}_0^{\tau}\mathbf{y}_0^{\tau}]$ are the same as before but $\underset{\sim}{\mathbf{\Phi}}_{k+1}$, $\underset{\sim}{\mathbf{B}}_k$, $\underset{\sim}{\mathbf{D}}_k$, $\underset{\sim}{\mathbf{C}}_k$, $\underset{\sim}{\mathbf{H}}_k$, and $\underset{\sim}{\mathbf{F}}_k$ are all deterministic.

Theorem 5.8. *For (5.55) and (5.56), the forward prediction equations are given by*

$$\hat{\mathbf{x}}_{k+1}(j) = \underset{\sim}{\mathbf{\Phi}}_{k+1}\hat{\mathbf{x}}_k(j) + \underset{\sim}{\mathbf{B}}_{k+1}\hat{\mathbf{y}}_k(j), \qquad k \geq j, \tag{5.57}$$

$$\hat{\mathbf{y}}_{k+1}(j) = \underset{\sim}{\mathbf{C}}_{k+1}\hat{\mathbf{x}}_k(j) + \underset{\sim}{\mathbf{H}}_{k+1}\hat{\mathbf{y}}_k(j), \qquad \hat{\mathbf{x}}_j(j) = \hat{\mathbf{x}}_j, \qquad \hat{\mathbf{y}}_j(j) = \mathbf{y}_j, \tag{5.58}$$

where $\hat{\mathbf{x}}_k(j) = E^{\mathbf{y}^j}\mathbf{x}_k$, $\hat{\mathbf{y}}_k(j) = E^{\mathbf{y}^j}\mathbf{y}_k$.

Proof. By the independence of $\boldsymbol{\zeta}_{k+1}$ and $\mathbf{y}^j$, taking $E^{\mathbf{y}^j}(\cdot)$ of (5.55) and (5.56) leads to the desired result. □

By showing that $[\mathbf{x}_k^{\tau}\mathbf{y}_k^{\tau}\mathbf{y}_{j+1}^{\tau}]^{\tau}$ is conditionally Gaussian given $\mathbf{y}^j$, it can be verified that the backward prediction equations of (5.55) and (5.56) are as follows:

$$\hat{\mathbf{z}}_k(j+1) = \hat{\mathbf{z}}_k(j) + \underset{\sim}{\mathbf{A}}_{k,j+1}$$

$$\times \begin{bmatrix} (\underset{\sim}{\mathbf{\Phi}}_{j+1}\underset{\sim}{\mathbf{P}}_j\underset{\sim}{\mathbf{C}}_{j+1}^\tau + \underset{\sim}{\mathbf{D}}_{j+1}\underset{\sim}{\mathbf{F}}_{j+1}^\tau)(\underset{\sim}{\mathbf{C}}_{j+1}\underset{\sim}{\mathbf{P}}_j\underset{\sim}{\mathbf{C}}_{j+1}^\tau + \underset{\sim}{\mathbf{F}}_{j+1}\underset{\sim}{\mathbf{F}}_{j+1}^\tau)^+ \\ \underset{\sim}{\mathbf{I}} \end{bmatrix}$$

$$\cdot (\mathbf{y}_{j+1} - \underset{\sim}{\mathbf{C}}_{j+1}\hat{\mathbf{x}}_j - \underset{\sim}{\mathbf{H}}_{j+1}\mathbf{y}_j), \qquad k \geq j, \tag{5.59}$$

where

$$\hat{\mathbf{z}}_k(j) = \begin{bmatrix} E^{\mathbf{y}^j}\mathbf{x}_k \\ E^{\mathbf{y}^j}\mathbf{y}_k \end{bmatrix}, \qquad \underset{\sim}{\mathbf{A}}_{k+1,k} = \begin{bmatrix} \underset{\sim}{\mathbf{\Phi}}_{k+1} & \underset{\sim}{\mathbf{B}}_{k+1} \\ \underset{\sim}{\mathbf{C}}_{k+1} & \underset{\sim}{\mathbf{H}}_{k+1} \end{bmatrix},$$

$$\underset{\sim}{\mathbf{A}}_{k+1,j} = \underset{\sim}{\mathbf{A}}_{k+1,k}\underset{\sim}{\mathbf{A}}_{k,j}, \underset{\sim}{\mathbf{A}}_{jj} = \underset{\sim}{\mathbf{I}}.$$

Corollary 1. *Until now the interpolation and prediction equations have been obtained for system* (5.8), (5.9). *But the system* (5.20), (5.21) *is the special case of* (5.8), (5.9) *where*

$$\mathbf{y}_k = \underset{\sim}{\mathbf{C}}_k\underset{\sim}{\mathbf{\Phi}}_k\mathbf{x}_{k-1} + \underset{\sim}{\mathbf{C}}_k\underset{\sim}{\mathbf{B}}_k\mathbf{u}_{k-1} + \underset{\sim}{\mathbf{H}}_k\mathbf{v}_{k-1} + (\underset{\sim}{\mathbf{C}}_k\underset{\sim}{\mathbf{D}}_k + \underset{\sim}{\mathbf{F}}_k)\zeta_k,$$

so it is easy to write down the interpolation and prediction equations for (5.20) *and* (5.21).

Corollary 2. *Let* $\underset{\sim}{\mathbf{\Phi}}_{k+1}, \underset{\sim}{\mathbf{D}}_{k+1}, \underset{\sim}{\mathbf{C}}_k, \underset{\sim}{\mathbf{F}}_k, \underset{\sim}{\mathbf{B}}_{k+1}\mathbf{u}_k,$ *and* $\underset{\sim}{\mathbf{H}}_k\mathbf{v}_{k-1}$ *be deterministic and* $E\zeta_i = 0$, $E\zeta_i[\mathbf{x}_0^\tau\mathbf{y}_0^\tau] = \underset{\sim}{\mathbf{0}}$, $E\zeta_i\zeta_j^\tau = \underset{\sim}{\mathbf{I}}\delta_{ij}$ *for all i, j, but assume* $\{\zeta_j\}$ *is not necessarily Gaussian. Then Theorems* 5.4–5.8 *are valid as recursive formula of linear unbiased minimum variance interpolation and prediction, since the LUMVE expressions do not depend on the distribution of* $\{\zeta_i\}$. *Thus to derive it we can artificially assume that the* $\{\zeta_i\}$ *process is normal.*

5.2. STOCHASTIC OBSERVABILITY AND ESTIMATION WITHOUT KNOWLEDGE OF INITIAL VALUES (EWKIV)

Now we consider system (5.20), (5.21) with $\underset{\sim}{\mathbf{\Phi}}_{k+1}, \underset{\sim}{\mathbf{C}}_k, \underset{\sim}{\mathbf{F}}_k, \underset{\sim}{\mathbf{B}}_{k+1}\mathbf{u}_k$ and $\underset{\sim}{\mathbf{H}}_k\mathbf{v}_{k-1}$ being deterministic and $E\zeta_i = 0$, $E\zeta_i\zeta_i^\tau = \underset{\sim}{\mathbf{I}}\delta_{ij}$ for all i, j, with $E\zeta_i[\mathbf{x}_0^\tau\mathbf{y}_0^\tau] = \underset{\sim}{\mathbf{0}}$, $E\mathbf{x}_0\mathbf{y}_0^\tau = \underset{\sim}{\mathbf{0}}$, but we make no Gaussian assumption on $\{\zeta_i\}$. Such a system is still referred to as (5.20), (5.21) in what follows.

For simplicity, assume $\mathbf{B}_k \equiv \underset{\sim}{\mathbf{0}}$, $\underset{\sim}{\mathbf{H}}_k \equiv \underset{\sim}{\mathbf{0}}$. From Section 5.1, we have seen that all recursive estimation equations have to be given initial conditions. In the present case, this means that the $E\mathbf{x}_0$ and $\underset{\sim}{\mathbf{R}} \triangleq E(\mathbf{x}_0 - E\mathbf{x}_0)(\mathbf{x}_0 - E\mathbf{x}_0)^\tau$ have to be known. In order to emphasize the dependence of the LUMVE of $\mathbf{x}_j$ based on $\mathbf{y}^k$ on the initial values $E\mathbf{x}_0, \mathbf{R}$, we write $\hat{\mathbf{x}}_j(k)$, $\mathbf{P}_j(k)$ as $\hat{\mathbf{x}}_j(k, E\mathbf{x}_0, \mathbf{R})$ and $\mathbf{P}_j(k, E\mathbf{x}_0, \mathbf{R})$, respectively.

In practice, $E\mathbf{x}_0$ and $\mathbf{R}$ usually cannot be given and a physically meaningful approach is to set $E\mathbf{x}_0 = \mathbf{0}$, $\mathbf{R} = \alpha\mathbf{I}$, $\alpha > 0$ and then to let α tend to infinity. We now rigorously consider this an intuitive treatment. In the next section, its optimality will be demonstrated.

After substituting for $E\mathbf{x}_0$ by $\mathbf{0}$ and $\mathbf{R}$ by $\alpha\mathbf{I}$, we denote $\hat{\mathbf{x}}_j(k)$ and $\mathbf{P}_j(k)$ by $\bar{\mathbf{x}}_j(k, \alpha)$ and $\bar{\mathbf{P}}_j(k, \alpha)$, respectively. $\bar{\mathbf{x}}_j(k, \alpha)$ is still a linear estimate of $\mathbf{x}_j$ based on $\mathbf{y}^k$, but it is not necessarily unbiased, and of course the minimality of the estimation error covariance is lost; furthermore, $\bar{\mathbf{P}}_j(k, \alpha)$ is no longer its estimation error covariance matrix.

We now introduce a technical definition giving a condition that characterizes those cases in which the preceding approximate procedure outlined yields the optimal estimation procedure in the limit as $\alpha \to \infty$. It is termed stochastic observability (detectability) because it reduces to the familiar observability (detectability) condition when the system in question is deterministic.

DEFINITION 5.2. If for some k, $\bar{\mathbf{P}}_0(k, \alpha)$, $(\bar{\mathbf{P}}_k(k, \alpha))$ tends to a finite limit $\bar{\mathbf{P}}_0(k)$, $(\bar{\mathbf{P}}_k)$ as $\alpha \to \infty$, then the system (5.20), (5.21) is called stochastically observable (detectable). □

Let us write

$$\mathbf{\Phi}_{k,i} = \mathbf{\Phi}_k \mathbf{\Phi}_{k-1,i}, \quad \mathbf{\Phi}_{i,i} = \mathbf{I}, \qquad i \le k$$

$$\mathbf{H}_k = \begin{bmatrix} \mathbf{0} & \mathbf{0} & \mathbf{0} & \cdots & \mathbf{0} \\ & \mathbf{C}_1\mathbf{D}_1 & \mathbf{0} & & \vdots \\ \vdots & \mathbf{C}_2\mathbf{\Phi}_{2,1}\mathbf{D}_1 & \mathbf{C}_2\mathbf{D}_2 & & \vdots \\ \vdots & \vdots & \vdots & \ddots & \mathbf{0} \\ \mathbf{0} & \mathbf{C}_k\mathbf{\Phi}_{k,1}\mathbf{D}_1 & \mathbf{C}_k\mathbf{\Phi}_{k,2}\mathbf{D}_2 & \cdots & \mathbf{C}_k\mathbf{D}_k \end{bmatrix} + \begin{bmatrix} \mathbf{F}_0 & & \mathbf{0} \\ & \ddots & \\ \mathbf{0} & & \mathbf{F}_k \end{bmatrix}$$

$$\mathbf{G}_k^T = [\mathbf{C}_0^T \vdots \mathbf{\Phi}_{1,0}^T\mathbf{C}_1^T \vdots \cdots \vdots \mathbf{\Phi}_{k,0}^T\mathbf{C}_k^T], \qquad \mathbf{L}_k = [\underbrace{\mathbf{0}}_{l} \vdots \mathbf{\Phi}_{k,1}\mathbf{D}_1 \vdots \mathbf{\Phi}_{k,2}\mathbf{D}_2 \vdots \cdots \vdots \mathbf{D}_k].$$

then for $j \le k$

$$\mathbf{y}^k = \underset{\sim}{\mathbf{G}}_k \mathbf{x}_0 + \underset{\sim}{\mathbf{H}}_k \boldsymbol{\zeta}^k \tag{5.60}$$

$$\mathbf{x}_j = \underset{\sim}{\mathbf{\Phi}}_{j,0} \mathbf{x}_0 + \Big[\underset{\sim}{\mathbf{L}}_j \vdots \underbrace{\underset{\sim}{\mathbf{0}}}_{(k-j)l}\Big] \boldsymbol{\zeta}^k \tag{5.61}$$

and for $j > k$

$$\mathbf{y}^k = \underset{\sim}{\mathbf{G}}_k \mathbf{x}_0 + \Big[\underset{\sim}{\mathbf{H}}_k \vdots \underbrace{\underset{\sim}{\mathbf{0}}}_{(j-k)l}\Big] \boldsymbol{\zeta}^j, \tag{5.62}$$

$$\mathbf{x}_j = \underset{\sim}{\mathbf{\Phi}}_{j,0} \mathbf{x}_0 + \underset{\sim}{\mathbf{L}}_j \boldsymbol{\zeta}^j. \tag{5.63}$$

We only need to discuss the $j \le k$ case. For the other case, the only thing we should do is to replace $\underset{\sim}{\mathbf{H}}_k$ and $[\underset{\sim}{\mathbf{L}}_j \vdots \underset{\sim}{\mathbf{0}}]$ by $[\underset{\sim}{\mathbf{H}}_k \vdots \underset{\sim}{\mathbf{0}}]$ and $\underset{\sim}{\mathbf{L}}_j$, respectively.

Notice that $\mathbf{x}_0$ is uncorrelated with $\{\boldsymbol{\zeta}_k\}$. Thus by (1.45) and (1.46), we have

$$\bar{\mathbf{x}}_j(k,\alpha) = \big(\alpha \underset{\sim}{\mathbf{\Phi}}_{j,0} \underset{\sim}{\mathbf{G}}_k^\tau + [\underset{\sim}{\mathbf{L}}_j \vdots \underset{\sim}{\mathbf{0}}] \underset{\sim}{\mathbf{H}}_k^\tau\big)\big(\alpha \underset{\sim}{\mathbf{G}}_k \underset{\sim}{\mathbf{G}}_k^\tau + \underset{\sim}{\mathbf{H}}_k \underset{\sim}{\mathbf{H}}_k^\tau\big)^+ \mathbf{y}^k \tag{5.64}$$

$$\begin{aligned}\bar{\underset{\sim}{\mathbf{P}}}_j(k,\alpha) = {} & \alpha \underset{\sim}{\mathbf{\Phi}}_{j,0} \underset{\sim}{\mathbf{\Phi}}_{j,0}^\tau + \underset{\sim}{\mathbf{L}}_j \underset{\sim}{\mathbf{L}}_j^\tau - \big(\alpha \underset{\sim}{\boldsymbol{\phi}}_{j,0} \underset{\sim}{\mathbf{G}}_k^\tau + [\underset{\sim}{\mathbf{L}}_j \vdots \underset{\sim}{\mathbf{0}}] \underset{\sim}{\mathbf{H}}_k^\tau\big)\big(\alpha \underset{\sim}{\mathbf{G}}_k \underset{\sim}{\mathbf{G}}_k^\tau + \underset{\sim}{\mathbf{H}}_k \underset{\sim}{\mathbf{H}}_k^\tau\big)^+ \\ & \cdot \big(\alpha \underset{\sim}{\mathbf{\Phi}}_{j,0} \underset{\sim}{\mathbf{G}}_k^\tau + [\underset{\sim}{\mathbf{L}}_j \vdots \underset{\sim}{\mathbf{0}}] \underset{\sim}{\mathbf{H}}_k^\tau\big)^\tau \end{aligned} \tag{5.65}$$

Denote by $L(\underset{\sim}{\mathbf{A}})$, the linear subspace spanned by columns of matrix $\underset{\sim}{\mathbf{A}}$.

Lemma 5.7. *Let $\underset{\sim}{\mathbf{G}}, \underset{\sim}{\mathbf{H}}$ be two matrices with the same number of columns. Let the columns of $\underset{\sim}{\mathbf{V}}_1$ compose the orthonormal basis of $\mathscr{L}(\underset{\sim}{\mathbf{G}})$ and the columns of $\underset{\sim}{\mathbf{V}}_2$ compose an orthonormal basis of $\mathscr{L}(\underset{\sim}{\mathbf{H}} - \underset{\sim}{\mathbf{V}}_1 \underset{\sim}{\mathbf{V}}_1^\tau \underset{\sim}{\mathbf{H}})$. Finally, let $\underset{\sim}{\mathbf{V}}_3$ be such that*

$$\underset{\sim}{\mathbf{V}} = \Big[\underbrace{\underset{\sim}{\mathbf{V}}_1}_{\beta} \;\; \underbrace{\underset{\sim}{\mathbf{V}}_2}_{\gamma} \;\; \underbrace{\underset{\sim}{\mathbf{V}}_3}_{\delta}\Big] \tag{5.66}$$

is an orthogonal matrix. Then

1.
$$\underset{\sim}{\mathbf{G}}^\tau \underset{\sim}{\mathbf{V}} = \Big[\underbrace{\underset{\sim}{\mathbf{G}}^\tau \underset{\sim}{\mathbf{V}}_1}_{\beta} \vdots \underbrace{\underset{\sim}{\mathbf{0}}}_{\gamma} \vdots \underbrace{\underset{\sim}{\mathbf{0}}}_{\delta}\Big], \tag{5.67}$$

$$\underset{\sim}{\mathbf{D}} \triangleq \underset{\sim}{\mathbf{V}}_1^\tau \underset{\sim}{\mathbf{G}} \underset{\sim}{\mathbf{G}}^\tau \underset{\sim}{\mathbf{V}}_1 > \underset{\sim}{\mathbf{0}} \quad \text{if } \beta > 0$$

$$\underset{\sim}{\mathbf{C}} \triangleq \underset{\sim}{\mathbf{V}}_2^\tau \underset{\sim}{\mathbf{H}} \underset{\sim}{\mathbf{H}}^\tau \underset{\sim}{\mathbf{V}}_2 > \underset{\sim}{\mathbf{0}} \quad \text{if } \gamma > 0$$

2. $(\alpha \mathbf{G}\mathbf{G}^{\tau} + \mathbf{H}\mathbf{H}^{\tau})^{+}$

$$= \frac{1}{\alpha}\mathbf{V}\begin{bmatrix} \mathbf{M}_{\alpha}^{-1} & -\mathbf{M}_{\alpha}^{-1}\mathbf{B}\mathbf{C}^{-1} & \mathbf{0} \\ -\mathbf{C}^{-1}\mathbf{B}^{\tau}\mathbf{M}_{\alpha}^{-1} & \alpha\mathbf{C}^{-1} + \mathbf{C}^{-1}\mathbf{B}^{\tau}\mathbf{M}_{\alpha}^{-1}\mathbf{B}\mathbf{C}^{-1} & \mathbf{0} \\ \mathbf{0} & \mathbf{0} & \mathbf{0} \end{bmatrix}\mathbf{V}^{\tau}, \tag{5.68}$$

where

$$\mathbf{A} \triangleq \mathbf{V}_1^{\tau}\mathbf{H}\mathbf{H}^{\tau}\mathbf{V}_2, \qquad \mathbf{B} \triangleq \mathbf{V}_1^{\tau}\mathbf{H}\mathbf{H}^{\tau}\mathbf{V}_2,$$

$$\mathbf{M}_{\alpha} \triangleq \mathbf{D} + \frac{1}{\alpha}(\mathbf{A} - \mathbf{B}\mathbf{C}^{-1}\mathbf{B}^{\tau}).$$

Proof. (1) It is clear that there are matrices $\mathbf{S}$ and $\mathbf{T}$ such that

$$(\mathbf{I} - \mathbf{V}_1\mathbf{V}_1^{\tau})\mathbf{H}\mathbf{S} = \mathbf{V}_2, \qquad \mathbf{G} = \mathbf{V}_1\mathbf{T}, \tag{5.69}$$

since the columns of $\mathbf{V}_1$ and $\mathbf{V}_2$ are bases of the corresponding subspaces. Then

$$\mathbf{G}^{\tau}\mathbf{V}_2 = \mathbf{T}^{\tau}\mathbf{V}_1^{\tau}(\mathbf{I} - \mathbf{V}_1\mathbf{V}_1^{\tau})\mathbf{H}\mathbf{S} = \mathbf{T}^{\tau}(\mathbf{V}_1^{\tau} - \mathbf{V}_1^{\tau})\mathbf{H}\mathbf{S} = \mathbf{0},$$

$$\mathbf{G}^{\tau}\mathbf{V}_3 = \mathbf{T}^{\tau}\mathbf{V}_1^{\tau}\mathbf{V}_3 = \mathbf{0},$$

and (5.67) follows.

If $\beta > 0$, then rank of $\mathbf{G}$ is β, but $\mathbf{V}_1$ is of β columns and hence $\mathbf{T}$ has β rows and

$$\mathbf{D} = \mathbf{V}_1^{\tau}\mathbf{V}_1\mathbf{T}\mathbf{T}^{\tau}\mathbf{V}_1^{\tau}\mathbf{V}_1 = \mathbf{T}\mathbf{T}^{\tau} > \mathbf{0}.$$

Further, there is a matrix $\mathbf{U}$ of rank γ with γ rows such that

$$(\mathbf{I} - \mathbf{V}_1\mathbf{V}_1^{\tau})\mathbf{H} = \mathbf{V}_2\mathbf{U}.$$

Therefore,

$$\mathbf{H} = \mathbf{V}_1\mathbf{V}_1^{\tau}\mathbf{H} + \mathbf{V}_2\mathbf{U}, \qquad \mathbf{V}_2^{\tau}\mathbf{H} = \mathbf{U} \tag{5.70}$$

and

$$\underset{\sim}{\mathbf{C}} = \underset{\sim}{\mathbf{V}}_2^T\underset{\sim}{\mathbf{H}}\underset{\sim}{\mathbf{H}}^T\underset{\sim}{\mathbf{V}}_2 = \underset{\sim}{\mathbf{U}}\underset{\sim}{\mathbf{U}}^T > \underset{\sim}{\mathbf{0}}.$$

(2) By (5.67) and (5.70),

$$\underset{\sim}{\mathbf{V}}^T\underset{\sim}{\mathbf{G}}\underset{\sim}{\mathbf{G}}^T\underset{\sim}{\mathbf{V}} = \begin{bmatrix} \underset{\sim}{\mathbf{D}} & \underset{\sim}{\mathbf{0}} & \underset{\sim}{\mathbf{0}} \\ \underset{\sim}{\mathbf{0}} & \underset{\sim}{\mathbf{0}} & \underset{\sim}{\mathbf{0}} \\ \underbrace{\underset{\sim}{\mathbf{0}}}_{\beta} & \underbrace{\underset{\sim}{\mathbf{0}}}_{\gamma} & \underbrace{\underset{\sim}{\mathbf{0}}}_{\delta} \end{bmatrix},$$

$$\underset{\sim}{\mathbf{V}}^T\underset{\sim}{\mathbf{H}}\underset{\sim}{\mathbf{H}}^T\underset{\sim}{\mathbf{V}} = \begin{bmatrix} \underset{\sim}{\mathbf{V}}_1^T\underset{\sim}{\mathbf{H}}\underset{\sim}{\mathbf{H}}^T\underset{\sim}{\mathbf{V}}_1 & \underset{\sim}{\mathbf{V}}_1^T\underset{\sim}{\mathbf{H}}\underset{\sim}{\mathbf{H}}^T\underset{\sim}{\mathbf{V}}_2 & \underset{\sim}{\mathbf{0}} \\ \underset{\sim}{\mathbf{V}}_2^T\underset{\sim}{\mathbf{H}}\underset{\sim}{\mathbf{H}}^T\underset{\sim}{\mathbf{V}}_1 & \underset{\sim}{\mathbf{V}}_2^T\underset{\sim}{\mathbf{H}}\underset{\sim}{\mathbf{H}}^T\underset{\sim}{\mathbf{V}}_2 & \underset{\sim}{\mathbf{0}} \\ \underset{\sim}{\mathbf{0}} & \underset{\sim}{\mathbf{0}} & \underset{\sim}{\mathbf{0}} \end{bmatrix},$$

hence

$$(\alpha\underset{\sim}{\mathbf{G}}\underset{\sim}{\mathbf{G}}^T + \underset{\sim}{\mathbf{H}}\underset{\sim}{\mathbf{H}}^T)^+ = \frac{1}{\alpha}\underset{\sim}{\mathbf{V}}\left[\underset{\sim}{\mathbf{V}}^T\underset{\sim}{\mathbf{G}}\underset{\sim}{\mathbf{G}}^T\underset{\sim}{\mathbf{V}} + \frac{1}{\alpha}\underset{\sim}{\mathbf{V}}^T\underset{\sim}{\mathbf{H}}\underset{\sim}{\mathbf{H}}^T\underset{\sim}{\mathbf{V}}\right]^+\underset{\sim}{\mathbf{V}}^T$$

$$= \frac{1}{\alpha}\underset{\sim}{\mathbf{V}}\begin{bmatrix} \underset{\sim}{\mathbf{D}} + \frac{1}{\alpha}\underset{\sim}{\mathbf{A}} & \frac{1}{\alpha}\underset{\sim}{\mathbf{B}} & \underset{\sim}{\mathbf{0}} \\ \frac{1}{\alpha}\underset{\sim}{\mathbf{B}}^T & \frac{1}{\alpha}\underset{\sim}{\mathbf{C}} & \underset{\sim}{\mathbf{0}} \\ \underset{\sim}{\mathbf{0}} & \underset{\sim}{\mathbf{0}} & \underset{\sim}{\mathbf{0}} \end{bmatrix}^+\underset{\sim}{\mathbf{V}}^T$$

$$= \frac{1}{\alpha}\underset{\sim}{\mathbf{V}}\begin{bmatrix} \begin{bmatrix} \underset{\sim}{\mathbf{D}} + \frac{1}{\alpha}\underset{\sim}{\mathbf{A}} & \frac{1}{\alpha}\underset{\sim}{\mathbf{B}} \\ \frac{1}{\alpha}\underset{\sim}{\mathbf{B}}^T & \frac{1}{\alpha}\underset{\sim}{\mathbf{C}} \end{bmatrix}^{-1} & \begin{matrix} \underset{\sim}{\mathbf{0}} \\ \underset{\sim}{\mathbf{0}} \end{matrix} \\ \begin{matrix} \underset{\sim}{\mathbf{0}} & \underset{\sim}{\mathbf{0}} \end{matrix} & \underset{\sim}{\mathbf{0}} \end{bmatrix}\underset{\sim}{\mathbf{V}}^T,$$

from which (5.68) is directly verified. □

Lemma 5.8. *As* $\alpha \to \infty$,

$$\underset{\sim}{\mathbf{S}}(\alpha) \triangleq \alpha\underset{\sim}{\mathbf{G}}^T(\underset{\sim}{\mathbf{G}}\underset{\sim}{\mathbf{G}}^T)^+\underset{\sim}{\mathbf{G}} - \alpha^2\underset{\sim}{\mathbf{G}}^T(\alpha\underset{\sim}{\mathbf{G}}\underset{\sim}{\mathbf{G}}^T + \underset{\sim}{\mathbf{H}}\underset{\sim}{\mathbf{H}}^T)^+\underset{\sim}{\mathbf{G}}$$

$$\to \underset{\sim}{\mathbf{S}} \triangleq \underset{\sim}{\mathbf{G}}^+\underset{\sim}{\mathbf{H}}(\underset{\sim}{\mathbf{I}} - \underset{\sim}{\mathbf{H}}^T\underset{\sim}{\mathbf{W}}\underset{\sim}{\mathbf{H}})\underset{\sim}{\mathbf{H}}^T\underset{\sim}{\mathbf{G}}^{T+}, \tag{5.71}$$

where

$$\mathbf{W} = [(\mathbf{I} - \mathbf{G}\mathbf{G}^+)\mathbf{H}\mathbf{H}^T(\mathbf{I} - \mathbf{G}\mathbf{G}^+)]^+. \tag{5.72}$$

Proof. By (5.67) and (5.68), we have

$$\mathbf{S}(\alpha) = \alpha\mathbf{G}^T\mathbf{V}\left[(\mathbf{V}^T\mathbf{G}\mathbf{G}^T\mathbf{V})^+ - \alpha(\alpha\mathbf{V}^T\mathbf{G}\mathbf{G}^T\mathbf{V} + \mathbf{V}^T\mathbf{H}\mathbf{H}^T\mathbf{V})^+\right]\mathbf{V}^T\mathbf{G}$$

$$= \alpha\left[\mathbf{G}^T\mathbf{V}_1 \quad \mathbf{0} \quad \mathbf{0}\right]\left\{\begin{bmatrix} \mathbf{D}^{-1} & \mathbf{0} & \mathbf{0} \\ \mathbf{0} & \mathbf{0} & \mathbf{0} \\ \mathbf{0} & \mathbf{0} & \mathbf{0} \end{bmatrix}\right.$$

$$\left. - \begin{bmatrix} \mathbf{M}_\alpha^{-1} & -\mathbf{M}_\alpha^{-1}\mathbf{B}\mathbf{C}^{-1} & \mathbf{0} \\ -\mathbf{C}^{-1}\mathbf{B}^T\mathbf{M}_\alpha^{-1} & \alpha\mathbf{C}^{-1} + \mathbf{C}^{-1}\mathbf{B}^T\mathbf{M}_\alpha^{-1}\mathbf{B}\mathbf{C}^{-1} & \mathbf{0} \\ \mathbf{0} & \mathbf{0} & \mathbf{0} \end{bmatrix}\right\}\begin{bmatrix} \mathbf{V}_1^T\mathbf{G} \\ \mathbf{0} \\ \mathbf{0} \end{bmatrix}$$

$$= \alpha\mathbf{G}^T\mathbf{V}_1(\mathbf{D}^{-1} - \mathbf{M}_\alpha^{-1})\mathbf{V}_1^T\mathbf{G}. \tag{5.73}$$

It is easy to verify that

$$\alpha(\mathbf{D}^{-1} - \mathbf{M}_\alpha^{-1}) = \mathbf{D}^{-1}(\mathbf{A} - \mathbf{B}\mathbf{C}^{-1}\mathbf{B}^T)\left[\mathbf{D} + \frac{1}{\alpha}(\mathbf{A} - \mathbf{B}\mathbf{C}^{-1}\mathbf{B}^T)\right]^{-1}.$$

Therefore, by (5.73),

$$\mathbf{S}(\alpha) = \mathbf{G}^T\mathbf{V}_1\mathbf{D}^{-1}(\mathbf{A} - \mathbf{B}\mathbf{C}^{-1}\mathbf{B}^T)\left[\mathbf{D} + \frac{1}{\alpha}(\mathbf{A} - \mathbf{B}\mathbf{C}^{-1}\mathbf{B}^T)\right]^{-1}\mathbf{V}_1^T\mathbf{G}$$

$$\underset{\alpha\to\infty}{\to} \mathbf{G}^T\mathbf{V}_1\mathbf{D}^{-1}(\mathbf{A} - \mathbf{B}\mathbf{C}^{-1}\mathbf{B}^T)\mathbf{D}^{-1}\mathbf{V}_1^T\mathbf{G}$$

$$= \mathbf{G}^T\mathbf{V}_1\mathbf{D}^{-1}\mathbf{V}_1^T\mathbf{H}(\mathbf{I} - \mathbf{H}^T\mathbf{V}_2\mathbf{C}^{-1}\mathbf{V}_2^T\mathbf{H})\mathbf{H}^T\mathbf{V}_1\mathbf{D}^{-1}\mathbf{V}_1^T\mathbf{G}. \tag{5.74}$$

Noticing that

$$\mathbf{G} = \mathbf{V}\begin{bmatrix} \mathbf{V}_1^T\mathbf{G} \\ \mathbf{0} \\ \mathbf{0} \end{bmatrix} = \mathbf{V}_1\mathbf{V}_1^T\mathbf{G}, \tag{5.75}$$

by (1.44), we have

$$\mathbf{G}^{+} = \mathbf{G}^{T}\mathbf{V}_1(\mathbf{V}_1^{T}\mathbf{G}\mathbf{G}^{T}\mathbf{V}_1)^{-1}\mathbf{V}_1^{T} = \mathbf{G}^{T}\mathbf{V}_1\mathbf{D}^{-1}\mathbf{V}_1^{T}. \tag{5.76}$$

Hence from (5.74),

$$\mathbf{S}(\alpha) \underset{\alpha\to\infty}{\to} \mathbf{S} \triangleq \mathbf{G}^{+}\mathbf{H}(\mathbf{I} - \mathbf{H}^{T}\mathbf{V}_2\mathbf{C}^{-1}\mathbf{V}_2^{T}\mathbf{H})\mathbf{H}^{T}\mathbf{G}^{T+}. \tag{5.77}$$

From (5.75) and (5.76), it follows that

$$\mathbf{G}\mathbf{G}^{+} = \mathbf{V}_1\mathbf{V}_1^{T},$$

Consequently,

$$\mathbf{I} - \mathbf{G}\mathbf{G}^{+} = \mathbf{V}_2\mathbf{V}_2^{T} + \mathbf{V}_3\mathbf{V}_3^{T},$$

and

$$\mathbf{W} = [(\mathbf{V}_2\mathbf{V}_2^{T} + \mathbf{V}_3\mathbf{V}_3^{T})\mathbf{H}\mathbf{H}^{T}(\mathbf{V}_2\mathbf{V}_2^{T} + \mathbf{V}_3\mathbf{V}_3^{T})]^{+}.$$

But (5.70) implies $\mathbf{V}_3^{T}\mathbf{H} = \mathbf{0}$. Hence by (1.36), the preceding expression gives

$$\mathbf{W} = [\mathbf{V}_2(\mathbf{V}_2^{T}\mathbf{H}\mathbf{H}^{T}\mathbf{V}_2\mathbf{V}_2^{T})]^{+} = \mathbf{V}_2(\mathbf{V}_2^{T}\mathbf{H}\mathbf{H}^{T}\mathbf{V}_2)^{-1}\mathbf{V}_2^{T}, \tag{5.78}$$

which, together with (5.77), gives (5.71). □

Theorem 5.9. *As* $\alpha \to \infty$ $\overline{\mathbf{P}}_j(k,\alpha)$ *given by* (5.65) *tends to a finite limit* $\overline{\mathbf{P}}_j(k)$ *if and only if*

$$\mathbf{\Phi}_{j,0}\mathbf{G}_k^{+}\mathbf{G}_k = \mathbf{\Phi}_{j,0}, \tag{5.79}$$

and if (5.79) *holds then*

$$\overline{\mathbf{P}}_j(k,\alpha) \underset{\alpha\to\infty}{\to} \overline{\mathbf{P}}_j(k) \triangleq \{[\mathbf{L}_j\ \ \mathbf{0}] - \mathbf{\Phi}_{j,0}\mathbf{G}_k^{+}\mathbf{H}_k\}$$
$$\cdot\{\mathbf{I} - \mathbf{H}_k^{T}\mathbf{W}_k\mathbf{H}_k\}\{[\mathbf{L}_j\ \ \mathbf{0}] - \mathbf{\Phi}_{j,0}\mathbf{G}_k^{+}\mathbf{H}_k\}^{T}, \tag{5.80}$$

where

$$\mathbf{W}_k = [(\mathbf{I} - \mathbf{G}_k\mathbf{G}_k^{+})\mathbf{H}_k\mathbf{H}_k^{T}(\mathbf{I} - \mathbf{G}_k\mathbf{G}_k^{+})]^{+}. \tag{5.81}$$

Proof. We put $\mathbf{H}_k$ and $\mathbf{G}_k$ into correspondence with $\mathbf{H}$ and $\mathbf{G}$, respectively, in Lemmas 5.7 and 5.8, rewrite $\mathbf{S}(\alpha), \mathbf{W}, \mathbf{S}$ in (5.71) and (5.72) as $\mathbf{S}_k(\alpha), \mathbf{W}_k, \mathbf{S}_k$, and retain the other notation therein. In particular, we shall not subscript the matrix $\mathbf{V}$ or its blocks $\mathbf{V}_1, \mathbf{V}_2, \mathbf{V}_3$ by k.

Assume (5.79) holds. By (1.48), we have

$$\mathbf{G}_k^+ = \mathbf{G}_k^\tau(\mathbf{G}_k\mathbf{G}_k^\tau)^+. \tag{5.82}$$

Hence

$$\mathbf{\Phi}_{j,0}\mathbf{\Phi}_{j,0}^\tau = \mathbf{\Phi}_{j,0}\mathbf{G}_k^+\mathbf{G}_k\mathbf{\Phi}_{j,0}^\tau = \mathbf{\Phi}_{j,0}\mathbf{G}_k^\tau(\mathbf{G}_k\mathbf{G}_k^\tau)^+\mathbf{G}_k\mathbf{\Phi}_{j,0}^\tau,$$

which makes (5.65) into

$$\begin{aligned}\overline{\mathbf{P}}_j(k,\alpha) = {} & \mathbf{\Phi}_{j,0}\mathbf{S}_k(\alpha)\mathbf{\Phi}_{j,0}^\tau + \mathbf{L}_j\mathbf{L}_j^\tau \\ & -\alpha\mathbf{\Phi}_{j,0}\mathbf{G}_k^\tau(\alpha\mathbf{G}_k\mathbf{G}_k^\tau + \mathbf{H}_k\mathbf{H}_k^\tau)^+\mathbf{H}_k[\mathbf{L}_j \;\; \mathbf{0}]^\tau \\ & -\alpha[\mathbf{L}_j \;\; \mathbf{0}]\mathbf{H}_k^\tau(\alpha\mathbf{G}_k\mathbf{G}_k^\tau + \mathbf{H}_k\mathbf{H}_k^\tau)^+\mathbf{G}_k\mathbf{\Phi}_{j,0}^\tau \\ & -[\mathbf{L}_j \;\; \mathbf{0}]\mathbf{H}_k^\tau(\alpha\mathbf{G}_k\mathbf{G}_k^\tau + \mathbf{H}_k\mathbf{H}_k^\tau)^+\mathbf{H}_k[\mathbf{L}_j \;\; \mathbf{0}]^\tau. \end{aligned} \tag{5.83}$$

Then by (5.67), (5.68), (5.76), and (5.78)

$$\begin{aligned} & \alpha\mathbf{\Phi}_{j,0}\mathbf{G}_k^\tau(\alpha\mathbf{G}_k\mathbf{G}_k^\tau + \mathbf{H}_k\mathbf{H}_k^\tau)^+\mathbf{H}_k[\mathbf{L}_j \;\; \mathbf{0}]^\tau \\ & \underset{\alpha\to\infty}{\rightarrow} \mathbf{\Phi}_{j,0}[\mathbf{G}_k^\tau\mathbf{V}_1\mathbf{D}^{-1} \vdots -\mathbf{G}_k^\tau\mathbf{V}_1\mathbf{D}^{-1}\mathbf{B}\mathbf{C}^{-1} \vdots \mathbf{0}]\mathbf{V}^\tau\mathbf{H}_k[\mathbf{L}_j \;\; \mathbf{0}]^\tau \\ & = \mathbf{\Phi}_{j,0}(\mathbf{G}_k^\tau\mathbf{V}_1\mathbf{D}^{-1}\mathbf{V}_1^\tau\mathbf{H}_k - \mathbf{G}_k^\tau\mathbf{V}_1\mathbf{D}^{-1}\mathbf{B}\mathbf{C}^{-1}\mathbf{V}_2^\tau\mathbf{H}_k)[\mathbf{L}_j \;\; \mathbf{0}]^\tau \\ & = \mathbf{\Phi}_{j,0}\mathbf{G}_k^+\mathbf{H}_k(\mathbf{I} - \mathbf{H}_k^\tau\mathbf{V}_2\mathbf{C}^{-1}\mathbf{V}_2^\tau\mathbf{H}_k)[\mathbf{L}_j \;\; \mathbf{0}]^\tau \\ & = \mathbf{\Phi}_{j,0}\mathbf{G}_k^+\mathbf{H}_k(\mathbf{I} - \mathbf{H}_k^\tau\mathbf{W}_k\mathbf{H}_k)[\mathbf{L}_j \;\; \mathbf{0}]^\tau \end{aligned} \tag{5.84}$$

Further, by (5.67), (5.68), and $\mathbf{M}_\alpha^{-1} \underset{\alpha \to \infty}{\to} \mathbf{D}^{-1}$, we see that

$$[\mathbf{L}_j \ \mathbf{0}]\mathbf{H}_k^\tau(\alpha\mathbf{G}_k\mathbf{G}_k^\tau + \mathbf{H}_k\mathbf{H}_k^\tau)^+ \mathbf{H}_k[\mathbf{L}_j \ \mathbf{0}]^\tau$$

$$\underset{\alpha \to \infty}{\to} [\mathbf{L}_j \ \mathbf{0}]\mathbf{H}_k^\tau\mathbf{V}\begin{bmatrix} \mathbf{0} & \mathbf{0} & \mathbf{0} \\ \mathbf{0} & \mathbf{C}^{-1} & \mathbf{0} \\ \mathbf{0} & \mathbf{0} & \mathbf{0} \end{bmatrix}\mathbf{V}^\tau\mathbf{H}_k[\mathbf{L}_j \ \mathbf{0}]^\tau$$

$$= [\mathbf{L}_j \ \mathbf{0}]\mathbf{H}_k^\tau\mathbf{W}_k\mathbf{H}_k[\mathbf{L}_j \ \mathbf{0}]^\tau \qquad (5.85)$$

by use of (5.81).

By using (5.71), (5.84), and (5.85), the desired relationship (5.80) follows from (5.83).

Now we prove the necessity part of the theorem. Let $\overline{\mathbf{P}}_j(k, \alpha)$ have a finite limit as $\alpha \to \infty$. Denote the right-hand side of (5.83) by $\tilde{\mathbf{P}}_j(k, \alpha)$ for the moment. According to the preceding discussion, $\tilde{\mathbf{P}}_j(k, \alpha)$, as $\alpha \to \infty$, goes to a limit expressed by (5.80).

From (5.65), (5.82), and (5.83), we have

$$\overline{\mathbf{P}}_j(k, \alpha) = \tilde{\mathbf{P}}_j(k, \alpha) + \alpha\mathbf{\Phi}_{j,0}\mathbf{\Phi}_{j,0}^\tau - \alpha\mathbf{\Phi}_{j,0}\mathbf{G}_k^+\mathbf{G}_k\mathbf{\Phi}_{j,0}^\tau$$

$$= \tilde{\mathbf{P}}_j(k, \alpha) + \alpha\mathbf{\Phi}_{j,0}(\mathbf{I} - \mathbf{G}_k^+\mathbf{G}_k)\mathbf{\Phi}_{j,0}^\tau$$

and hence

$$\mathbf{\Phi}_{j,0}(\mathbf{I} - \mathbf{G}_k^+\mathbf{G}_k)\mathbf{\Phi}_{j,0}^\tau = \mathbf{0}, \qquad (5.86)$$

because both $\overline{\mathbf{P}}_j(k, \alpha)$ and $\tilde{\mathbf{P}}_j(k, \alpha)$ converge as $\alpha \to \infty$.

Notice that

$$(\mathbf{I} - \mathbf{G}_k^+\mathbf{G}_k)(\mathbf{I} - \mathbf{G}_k^+\mathbf{G}_k)^\tau = \mathbf{I} - \mathbf{G}_k^+\mathbf{G}_k.$$

Hence (5.86) yields

$$\mathbf{\Phi}_{j,0}(\mathbf{I} - \mathbf{G}_k^+\mathbf{G}_k) = \mathbf{0}. \qquad \square$$

Theorem 5.10. *System* (5.20), (5.21) *is stochastically observable if and only if the rank of $\mathbf{G}_k$ is n starting from some k and is stochastically detectable* *if and only if for some k*

$$\Phi_{k,0} G_k^+ G_k = \Phi_{k,0}.$$

Stochastic observability implies stochastic detectability. Conversely, if the system is stochastically detectable (i.e., *if* $\bar{P}_k(k,\alpha) \underset{\alpha\to\infty}{\to} \bar{P}_k(k) \triangleq \bar{P}_k$, *and if for some* k, $\Phi_{k,0}$ *is of full rank*), *then the system is stochastically observable.*

Proof. According to Theorem 5.9, a necessary and sufficient condition for a system to be stochastic observable is the existence of some k so that $G_k^+ G_k = I$. If the rank of G_k is n, then $G_k^+ = (G_k^T G_k)^{-1} G_k^T$ and $G_k^+ G_k = I$. Conversely, if $G_k^+ G_k = I$, then the rank of G_k is obviously n. The rest of the assertions follow easily from Theorem 5.9. □

Corollary. *For deterministic systems, that is, those for which* $D_k \equiv 0$, $F_k \equiv 0$, *stochastic observability (detectability) coincides with complete observability (detectability).*

Proof. We only have to show that for a deterministic system there is a matrix $T_j(k)$ such that $T_j(k) y^k = x_j$ for any x_0 if and only if

$$\Phi_{j,0} G_k^+ G_k = \Phi_{j,0}.$$

From (5.60), we see for that in the deterministic case

$$y^k = G_k x_0.$$

Hence if (5.79) holds, then we can take

$$T_j(k) = \Phi_{j,0} G_k^+.$$

Conversely, if for any x_0,

$$T_j(k) G_k x_0 = \Phi_{j,0} x_0,$$

then $T_j(k) G_k = \Phi_{j,0}$, and

$$\Phi_{j,0} G_k^+ G_k = T_j(k) G_k G_k^+ G_k = T_j(k) G_k = \Phi_{j,0}.$$ □

In the next theorem, we show that as $\alpha \to \infty$ $\bar{x}_j(k,\alpha)$ given by (5.64) tends to a limit $\bar{x}_j(k)$ a.s. and in the mean square sense. We call $\bar{x}_j(k)$ an EWKIV of x_j based on y^k.

Theorem 5.11. *For the system* (5.20), (5.21),

$$\bar{\mathbf{x}}_j(k,\alpha) \underset{\alpha\to\infty}{\to} \mathbf{T}_j^0(k)\mathbf{y}^k, \tag{5.87}$$

where

$$\mathbf{T}_j^0(k) = \mathbf{\Phi}_{j,0}\mathbf{G}_k^+ + \left(\left[\mathbf{L}_j \;\; \mathbf{0}\right] - \mathbf{\Phi}_{j,0}\mathbf{G}_k^+\mathbf{H}_k\right)\mathbf{H}_k^\tau\mathbf{W}_k, \tag{5.88}$$

and $\mathbf{W}_k$ *is given by* (5.81), *and for any* $E\mathbf{x}_0$, $\bar{\mathbf{x}}_j(k) \triangleq \mathbf{T}_j^0(k)\mathbf{y}^k$ *is unbiased for* $\mathbf{x}_j$ *if and only if* (5.79) *holds. When the condition is satisfied, then*

$$E\left(\mathbf{x}_j - \bar{\mathbf{x}}_j(k)\right)\left(\mathbf{x}_j - \bar{\mathbf{x}}_j(k)\right)^\tau = \bar{\mathbf{P}}_j(k),$$

where $\bar{\mathbf{P}}_j(k)$ *is defined by* (5.80).

Proof. From (5.64) and (5.68), we have

$$\bar{\mathbf{x}}_j(k,\alpha) = \left(\alpha\mathbf{\Phi}_{j,0}\mathbf{G}_k^\tau + \left[\mathbf{L}_j \;\; \mathbf{0}\right]\mathbf{H}_k^\tau\right)\frac{1}{\alpha}\mathbf{V}$$

$$\cdot\begin{bmatrix} \mathbf{M}_\alpha^{-1} & -\mathbf{M}_\alpha^{-1}\mathbf{B}\mathbf{C}^{-1} & \mathbf{0} \\ -\mathbf{C}^{-1}\mathbf{B}^\tau\mathbf{M}_\alpha^{-1} & \alpha\mathbf{C}^{-1} + \mathbf{C}^{-1}\mathbf{B}^\tau\mathbf{M}_\alpha^{-1}\mathbf{B}\mathbf{C}^{-1} & \mathbf{0} \\ \mathbf{0} & \mathbf{0} & \mathbf{0} \end{bmatrix}\mathbf{V}^\tau\mathbf{y}^k.$$

Noting (5.67) and observing that $M_\alpha^{-1} \underset{\alpha\to\infty}{\to} D^{-1}$, it is clear that

$$\bar{\mathbf{x}}_j(k,\alpha) \underset{\alpha\to\infty}{\to} \mathbf{\Phi}_{j,0}\mathbf{G}_k^\tau\mathbf{V}_1\left[\mathbf{D}^{-1} \vdots -\mathbf{D}^{-1}\mathbf{B}\mathbf{C}^{-1} \vdots \mathbf{0}\right]\mathbf{V}^\tau\mathbf{y}^k$$

$$+\left[\mathbf{L}_j \;\; \mathbf{0}\right]\mathbf{H}_k^\tau\mathbf{V}\begin{bmatrix} \mathbf{0} & \mathbf{0} & \mathbf{0} \\ \mathbf{0} & \mathbf{C}^{-1} & \mathbf{0} \\ \mathbf{0} & \mathbf{0} & \mathbf{0} \end{bmatrix}\mathbf{V}^\tau\mathbf{y}^k$$

$$= \mathbf{\Phi}_{j,0}\mathbf{G}_k^\tau\mathbf{V}_1\left(\mathbf{D}^{-1}\mathbf{V}_1^\tau - \mathbf{D}^{-1}\mathbf{B}\mathbf{C}^{-1}\mathbf{V}_2^\tau\right)\mathbf{y}^k + \left[\mathbf{L}_j \;\; \mathbf{0}\right]\mathbf{H}_k^\tau\mathbf{V}_2\mathbf{C}^{-1}\mathbf{V}_2^\tau\mathbf{y}^k$$

$$= \left[\mathbf{\Phi}_{j,0}\mathbf{G}_k^\tau\mathbf{V}_1\mathbf{D}^{-1}\mathbf{V}_1^\tau + \left(\left[\mathbf{L}_j \;\; \mathbf{0}\right] - \mathbf{\Phi}_{j,0}\mathbf{G}_k^\tau\mathbf{V}_1\mathbf{D}^{-1}\mathbf{V}_1^\tau\mathbf{H}_k\right)\mathbf{H}_k^\tau\mathbf{V}_2\mathbf{C}^{-1}\mathbf{V}_2^\tau\right]\mathbf{y}^k \tag{5.89}$$

which is exactly the right-hand side of (5.87) by using (5.76) and (5.78). Substituting (5.60) into (5.89) yields

$$\bar{\mathbf{x}}_j(k) = \underset{\sim}{\Phi}_{j,0}\underset{\sim}{\mathbf{G}}_k^+\underset{\sim}{\mathbf{G}}_k\mathbf{x}_0 + \left[\underset{\sim}{\Phi}_{j,0}\underset{\sim}{\mathbf{G}}_k^+ + \left(\left[\underset{\sim}{\mathbf{L}}_j \;\; \underset{\sim}{\mathbf{0}}\right] - \underset{\sim}{\Phi}_{j,0}\underset{\sim}{\mathbf{G}}_k^+\underset{\sim}{\mathbf{H}}_k\right)\underset{\sim}{\mathbf{H}}_k^\tau\underset{\sim}{\mathbf{W}}_k\right]\underset{\sim}{\mathbf{H}}_k\zeta^k \tag{5.90}$$

which implies

$$E\bar{\mathbf{x}}_j(k) = \Phi_{j,0}\underset{\sim}{\mathbf{G}}_k^+\underset{\sim}{\mathbf{G}}_k E\mathbf{x}_0.$$

By noticing that $E\mathbf{x}_j = \underset{\sim}{\Phi}_{j,0}E\mathbf{x}_0$, it follows that for any $E\mathbf{x}_0$, $\bar{\mathbf{x}}_j(k)$ is unbiased for $\mathbf{x}_j$ if and only if (5.79) holds, and if the condition is satisfied, then from (5.90) we have

$$\begin{aligned}\bar{\mathbf{x}}_j(k) &= \underset{\sim}{\Phi}_{j,0}\mathbf{x}_0 + \left[\underset{\sim}{\mathbf{L}}_j \;\; \underset{\sim}{\mathbf{0}}\right]\underset{\sim}{\mathbf{H}}_k^\tau\underset{\sim}{\mathbf{W}}_k\underset{\sim}{\mathbf{H}}_k\zeta^k + \underset{\sim}{\Phi}_{j,0}\underset{\sim}{\mathbf{G}}_k^+\underset{\sim}{\mathbf{H}}_k(\underset{\sim}{\mathbf{I}} - \underset{\sim}{\mathbf{H}}_k^\tau\underset{\sim}{\mathbf{W}}_k\underset{\sim}{\mathbf{H}}_k)\zeta^k \\ &= \underset{\sim}{\Phi}_{j,0}\mathbf{x}_0 + \left[\underset{\sim}{\mathbf{L}}_j \;\; \underset{\sim}{\mathbf{0}}\right]\zeta^k + \left(\underset{\sim}{\Phi}_{j,0}\underset{\sim}{\mathbf{G}}_k^+\mathbf{H}_k - \left[\underset{\sim}{\mathbf{L}}_j \;\; \underset{\sim}{\mathbf{0}}\right]\right)(\underset{\sim}{\mathbf{I}} - \underset{\sim}{\mathbf{H}}_k^\tau\underset{\sim}{\mathbf{W}}_k\underset{\sim}{\mathbf{H}}_k)\zeta^k,\end{aligned}$$

Further, by (5.61),

$$\bar{\mathbf{x}}_j(k) - \mathbf{x}_j = \left(\underset{\sim}{\Phi}_{j,0}\underset{\sim}{\mathbf{G}}_k^+\underset{\sim}{\mathbf{H}}_k - \left[\underset{\sim}{\mathbf{L}}_j \;\; \underset{\sim}{\mathbf{0}}\right]\right)(\underset{\sim}{\mathbf{I}} - \underset{\sim}{\mathbf{H}}_k^\tau\underset{\sim}{\mathbf{W}}_k\underset{\sim}{\mathbf{H}}_k)\zeta^k,$$

which gives the last assertion of the theorem because by (5.78)

$$(\underset{\sim}{\mathbf{I}} - \underset{\sim}{\mathbf{H}}_k^\tau\underset{\sim}{\mathbf{W}}_k\underset{\sim}{\mathbf{H}}_k)(\underset{\sim}{\mathbf{I}} - \underset{\sim}{\mathbf{H}}_k^\tau\underset{\sim}{\mathbf{W}}_k\underset{\sim}{\mathbf{H}}_k)^\tau = \underset{\sim}{\mathbf{I}} - \underset{\sim}{\mathbf{H}}_k^\tau\underset{\sim}{\mathbf{W}}_k\underset{\sim}{\mathbf{H}}_k.$$ □

5.3. GAUSS–MARKOV ESTIMATION

We continue to consider the system (5.20), (5.21) with deterministic coefficients and $\underset{\sim}{\mathbf{B}}_k \equiv \underset{\sim}{\mathbf{0}}$, $\underset{\sim}{\mathbf{H}}_k \equiv \underset{\sim}{\mathbf{0}}$. In Section 5.2 we have seen that the unbiasedness of EWKIV is closely related to stochastic observability and detectability. We now show the optimality of estimates of this kind.

Denote by $\mathscr{L}_j(k)$ the totality of $\underset{\sim}{\mathbf{T}}_j(k)\mathbf{y}^k$, where $\underset{\sim}{\mathbf{T}}_j(k)$ is an $n \times (k+1)m$ matrix and for any $E\mathbf{x}_0$, $\underset{\sim}{\mathbf{T}}_j(k)\mathbf{y}^k$ is an unbiased estimate for $\mathbf{x}_j$.

When $\mathscr{L}_j(k)$ is nonempty, let

$$\begin{aligned}&E\left[\bar{\underset{\sim}{\mathbf{T}}}_j(k)\mathbf{y}^k - \mathbf{x}_j\right]\left[\bar{\underset{\sim}{\mathbf{T}}}_j(k)\mathbf{y}^k - \mathbf{x}_j\right]^\tau \\ &\quad = \min_{\underset{\sim}{\mathbf{T}}_j(k)\mathbf{y}^k \in \mathscr{L}_j(k)} E\left[\underset{\sim}{\mathbf{T}}_j(k)\mathbf{y}^k - \mathbf{x}_k\right]\left[\underset{\sim}{\mathbf{T}}_j(k)\mathbf{y}^k - \mathbf{x}_k\right]^\tau.\end{aligned} \tag{5.91}$$

$\bar{\mathbf{T}}_j(k)$ is known as a Gauss–Markov estimate of $\mathbf{x}_j$ based on $\mathbf{y}^k$. It is a linear unbiased minimum variance estimate and uses no initial statistical characteristics of $\mathbf{x}_0$, by which it differs from LUMVE defined in Section 1.11.

Theorem 5.12. *$\mathscr{L}_j(k)$ is nonempty if and only if* (5.79) *holds. When this condition is satisfied, the* EWKIV $\bar{\mathbf{x}}_j(k)$ *is just the Gauss–Markov estimate, that is,*

$$\bar{\mathbf{x}}_j(k) = \mathbf{T}_j^0(k)\mathbf{y}^k = \bar{\mathbf{T}}_j(k)\mathbf{y}^k \tag{5.92}$$

and

$$\|\mathbf{T}_j^0(k)\| \le \|\bar{\mathbf{T}}_j(k)\|, \tag{5.93}$$

where $\mathbf{T}_j^0(k)$ is given by (5.88).

Proof. Now

$$\mathbf{T}_j(k)\mathbf{y}^k = \mathbf{T}_j(k)\mathbf{G}_k\mathbf{x}_0 + \mathbf{T}_j(k)\mathbf{H}_k\zeta^k,$$

which should be unbiased for $\mathbf{x}_j$ for any $E\mathbf{x}_0$ if $\mathscr{L}_j(k)$ is nonempty. In this case,

$$\mathbf{T}_j(k)\mathbf{G}_k = \mathbf{\Phi}_{j,0} \tag{5.94}$$

or

$$\mathbf{T}_j(k)\mathbf{G}_k\mathbf{G}_k^+\mathbf{G}_k = \mathbf{\Phi}_{j,0}.$$

These two formulas lead to (5.79). Conversely, if (5.79) is valid, then we can take $\mathbf{T}_j(k) = \mathbf{\Phi}_{j,0}\mathbf{G}_k^+$, which means $\mathscr{L}_j(k)$ is nonempty.

Now assume $\mathscr{L}_j(k)$ nonempty; as proved, $\mathbf{T}_j(k)$ must satisfy the algebraic equation (5.94), for which the general solution is

$$\mathbf{T}_j(k) = \mathbf{\Phi}_{j,0}\mathbf{G}_k^+ + \mathbf{Z}(\mathbf{I} - \mathbf{G}_k\mathbf{G}_k^+) \tag{5.95}$$

with $\mathbf{Z}$ arbitrary $n \times m(k+1)$ matrix. This is because if $\mathbf{T}$ is a solution of (5.94), then

$$\begin{aligned}\mathbf{T} &= \mathbf{\Phi}_{j,0}\mathbf{G}_k^+ + \mathbf{T} - \mathbf{\Phi}_{j,0}\mathbf{G}_k^+ \\ &= \mathbf{\Phi}_{j,0}\mathbf{G}_k^+ + \mathbf{T} - \mathbf{T}\mathbf{G}_k\mathbf{G}_k^+ \\ &= \mathbf{\Phi}_{j,0}\mathbf{G}_k^+ + \mathbf{T}(\mathbf{I} - \mathbf{G}_k\mathbf{G}_k^+),\end{aligned}$$

which is in the form of (5.95), and, conversely, by (5.79), it is clear that (5.95) satisfies (5.94).

We shall use the notation agreed on in Theorem 5.9. Using the equality

$$\mathbf{I} - \mathbf{G}_k \mathbf{G}_k^+ = \mathbf{V}_2 \mathbf{V}_2^\tau + \mathbf{V}_3 \mathbf{V}_3^\tau$$

proved in Lemma 5.8, we have from (5.95)

$$\mathbf{T}_j(k) = \mathbf{\Phi}_{j,0} \mathbf{G}_k^+ + \mathbf{Z} \mathbf{V}_2 \mathbf{V}_2^\tau + \mathbf{Z} \mathbf{V}_3 \mathbf{V}_3^\tau. \tag{5.96}$$

Notice that

$$\mathbf{V}_3^\tau \mathbf{G}_k = \mathbf{0}, \qquad \mathbf{V}_2^\tau \mathbf{G}_k = \mathbf{0}, \qquad \mathbf{V}_3^\tau \mathbf{H}_k = \mathbf{0}, \tag{5.97}$$

then

$$\mathbf{T}_j(k)\mathbf{y}^k = \left(\mathbf{\Phi}_{j,0} \mathbf{G}_k^+ + \mathbf{Z} \mathbf{V}_2 \mathbf{V}_2^\tau \right) \mathbf{y}^k$$

$$= \mathbf{\Phi}_{j,0} \mathbf{G}_k^+ \mathbf{G}_k \mathbf{x}_0 + \left(\mathbf{\Phi}_{j,0} \mathbf{G}_k^+ \mathbf{H}_k + \mathbf{Z} \mathbf{V}_2 \mathbf{V}_2^\tau \mathbf{H}_k \right) \zeta^k$$

$$= \mathbf{\Phi}_{j,0} \mathbf{x}_0 + \left(\mathbf{\Phi}_{j,0} \mathbf{G}_k^+ \mathbf{H}_k + \mathbf{Z} \mathbf{V}_2 \mathbf{V}_2^\tau \mathbf{H}_k \right) \zeta^k.$$

Therefore,

$$E\left(\mathbf{T}_j(k)\mathbf{y}^k - \mathbf{x}_j \right) \left(\mathbf{T}_j(k)\mathbf{y}^k - \mathbf{x}_j \right)^\tau$$

$$= \left(\mathbf{\Phi}_{j,0} \mathbf{G}_k^+ \mathbf{H}_k + \mathbf{Z} \mathbf{V}_2 \mathbf{V}_2^\tau \mathbf{H}_k - \begin{bmatrix} \mathbf{L}_j & \mathbf{0} \end{bmatrix} \right) \left(\mathbf{\Phi}_{j,0} \mathbf{G}_k^+ \mathbf{H}_k + \mathbf{Z} \mathbf{V}_2 \mathbf{V}_2^\tau \mathbf{H}_k - \begin{bmatrix} \mathbf{L}_j & \mathbf{0} \end{bmatrix} \right)^\tau$$

$$= \left(\begin{bmatrix} \mathbf{L}_j & \mathbf{0} \end{bmatrix} - \mathbf{\Phi}_{j,0} \mathbf{G}_k^+ \mathbf{H}_k \right) \left[\mathbf{I} - \mathbf{H}_k^\tau \mathbf{V}_2 \mathbf{C}^{-1} \mathbf{V}_2^\tau \mathbf{H}_k \right] \left(\begin{bmatrix} \mathbf{L}_j & \mathbf{0} \end{bmatrix} - \mathbf{\Phi}_{j,0} \mathbf{G}_k^+ \mathbf{H}_k \right)^\tau$$

$$+ \left[\mathbf{Z} \mathbf{V}_2 + \mathbf{\Phi}_{j,0} \mathbf{G}_k^+ \mathbf{H}_k \mathbf{H}_k^\tau \mathbf{V}_2 \mathbf{C}^{-1} - \begin{bmatrix} \mathbf{L}_j & \mathbf{0} \end{bmatrix} \mathbf{H}_k^\tau \mathbf{V}_2 \mathbf{C}^{-1} \right] \mathbf{C}$$

$$\cdot \left[\mathbf{Z} \mathbf{V}_2 + \mathbf{\Phi}_{j,0} \mathbf{G}_k^+ \mathbf{H}_k \mathbf{H}_k^\tau \mathbf{V}_2 \mathbf{C}^{-1} - \begin{bmatrix} \mathbf{L}_j & \mathbf{0} \end{bmatrix} \mathbf{H}_k^\tau \mathbf{V}_2 \mathbf{C}^{-1} \right]^\tau,$$

which reaches its minimum when

$$\mathbf{Z} \mathbf{V}_2 = \begin{bmatrix} \mathbf{L}_j & \mathbf{0} \end{bmatrix} \mathbf{H}_k^\tau \mathbf{V}_2 \mathbf{C}^{-1} - \mathbf{\Phi}_{j,0} \mathbf{G}_k^+ \mathbf{H}_k \mathbf{H}_k^\tau \mathbf{V}_2 \mathbf{C}^{-1}.$$

Hence by (5.78), (5.96) yields

$$\overline{\mathbf{T}}_j(k) = \boldsymbol{\Phi}_{j,0}\mathbf{G}_k^+ + \left(\left[\mathbf{L}_j \ \ \mathbf{0}\right] - \boldsymbol{\Phi}_{j,0}\mathbf{G}_k^+\mathbf{H}_k\right)\mathbf{H}_k^\tau \mathbf{V}_2\mathbf{C}^{-1}\mathbf{V}_2^\tau + \mathbf{Z}\mathbf{V}_3\mathbf{V}_3^\tau$$

$$= \boldsymbol{\Phi}_{j,0}\mathbf{G}_k^+ + \left(\left[\mathbf{L}_j \ \ \mathbf{0}\right] - \boldsymbol{\Phi}_{j,0}\mathbf{G}_k^+\mathbf{H}_k\right)\mathbf{H}_k^\tau \mathbf{W}_k + \mathbf{Z}\mathbf{V}_3\mathbf{V}_3^\tau. \qquad (5.98)$$

But $\mathbf{G}_k^+\mathbf{V}_3 = \mathbf{0}$, $\mathbf{W}_k\mathbf{V}_3 = \mathbf{0}$ by (5.76) and (5.78). Therefore, from the last expression for $\overline{\mathbf{T}}_j(k)$, we find

$$\overline{\mathbf{T}}_j(k)\overline{\mathbf{T}}_j^\tau(k) = \mathbf{T}_j^0(k)\mathbf{T}_j^{0\tau}(k) + \mathbf{Z}\mathbf{V}_3\mathbf{V}_3^\tau\mathbf{Z}^\tau$$

and hence

$$\|\mathbf{T}_j^0(k)\| \le \|\overline{\mathbf{T}}_j(k)\|.$$

Finally, by (5.97) we see that $\mathbf{V}_3^\tau\mathbf{y}^k = 0$, and hence from (5.88) and (5.98), we conclude (5.92) holds. □

To summarize, the following facts proved in Theorems 5.9–5.12 are equivalent:

1. $\overline{\mathbf{P}}_j(k, \alpha)$ tends to a finite limit $\overline{\mathbf{P}}_j(k)$ as $\alpha \to \infty$.
2. For any $E\mathbf{x}_0$, EWKIV $\overline{\mathbf{x}}_j(k)$ is unbiased for $\mathbf{x}_j$.
3. $\mathscr{L}_j(k)$ is nonempty.
4. $\boldsymbol{\Phi}_{j,0}\mathbf{G}_k^+\mathbf{G}_k = \boldsymbol{\Phi}_{j,0}$.

Any of these conditions can be taken as the definition of stochastic observability if $j = 0$ and of stochastic detectability if $j = k$.

Remark. In mathematical statistics, the classical linear model is

$$\mathbf{y} = \mathbf{G}\mathbf{x} + \mathbf{H}\boldsymbol{\zeta},$$

with $E\boldsymbol{\zeta} = \mathbf{0}$, $E\boldsymbol{\zeta}\boldsymbol{\zeta}^\tau = \mathbf{I}$, $\mathbf{H}\mathbf{H}^\tau > \mathbf{0}$. The Gauss–Markov estimate for $\mathbf{x}$ is given by

$$\hat{\mathbf{x}} = \left[\mathbf{G}^\tau(\mathbf{H}\mathbf{H}^\tau)^{-1}\mathbf{G}\right]^{-1}\mathbf{G}^\tau(\mathbf{H}\mathbf{H}^\tau)^{-1}\mathbf{y}$$

when $\mathbf{G}^+\mathbf{G} = \mathbf{I}$.

Here we have extended this classical result in several directions. First, we consider a more general model: Instead of $\mathbf{x}$, the vector $\mathbf{z}$ is estimated by $\mathbf{y}$,

where

$$\mathbf{z} = \underset{\sim}{\Phi}\mathbf{x} + \underset{\sim}{\mathbf{L}}\zeta,$$

$$\mathbf{y} = \underset{\sim}{\mathbf{G}}\mathbf{x} + \underset{\sim}{\mathbf{H}}\zeta. \tag{5.99}$$

Second, we do not require $\underset{\sim}{\mathbf{H}}\underset{\sim}{\mathbf{H}}^{\tau} > \underset{\sim}{\mathbf{0}}$. Theorems 5.11 and 5.12 tell us that for any $\mathbf{x}$ there exists a matrix $\underset{\sim}{\mathbf{T}}$ such that $\underset{\sim}{\mathbf{T}}\mathbf{y}$ is an unbiased estimator for $\mathbf{z}$ if and only if

$$\underset{\sim}{\Phi}\underset{\sim}{\mathbf{G}}^{+}\underset{\sim}{\mathbf{G}} = \underset{\sim}{\Phi}. \tag{5.100}$$

Furthermore, if this condition is fulfilled, then the Gauss–Markov estimate for $\mathbf{z}$ is

$$\hat{\mathbf{z}} = \underset{\sim}{\mathbf{T}}\mathbf{y}, \tag{5.101}$$

where

$$\underset{\sim}{\mathbf{T}} = [\underset{\sim}{\Phi}\underset{\sim}{\mathbf{G}}^{+} + (\underset{\sim}{\mathbf{L}} - \underset{\sim}{\Phi}\underset{\sim}{\mathbf{G}}^{+}\underset{\sim}{\mathbf{H}})\underset{\sim}{\mathbf{H}}^{\tau}\underset{\sim}{\mathbf{W}}]\mathbf{y}, \tag{5.102}$$

$$\underset{\sim}{\mathbf{W}} = [(\underset{\sim}{\mathbf{I}} - \underset{\sim}{\mathbf{G}}\underset{\sim}{\mathbf{G}}^{+})\underset{\sim}{\mathbf{H}}\underset{\sim}{\mathbf{H}}^{\tau}(\underset{\sim}{\mathbf{I}} - \underset{\sim}{\mathbf{G}}\underset{\sim}{\mathbf{G}}^{+})]^{+} \tag{5.103}$$

and the estimation error covariance matrix is

$$(\underset{\sim}{\mathbf{L}} - \underset{\sim}{\Phi}\underset{\sim}{\mathbf{G}}^{+}\underset{\sim}{\mathbf{H}})(\underset{\sim}{\mathbf{I}} - \underset{\sim}{\mathbf{H}}^{\tau}\underset{\sim}{\mathbf{W}}\underset{\sim}{\mathbf{H}})(\underset{\sim}{\mathbf{L}} - \underset{\sim}{\Phi}\underset{\sim}{\mathbf{G}}^{+}\underset{\sim}{\mathbf{H}})^{\tau}. \tag{5.104}$$

5.4. RECURSIVE FORMULAS FOR GAUSS–MARKOV ESTIMATES

Lemma 5.9. *As* $\alpha \to \infty$, $\overline{\underset{\sim}{\mathbf{P}}}_0(k, \alpha)$ *and* $\overline{\underset{\sim}{\mathbf{P}}}_k(k, \alpha)$ *are nondecreasing.*

Proof. From (5.65), we have

$$\overline{\underset{\sim}{\mathbf{P}}}_0(k, \alpha) = \alpha\underset{\sim}{\mathbf{I}} - \alpha^2\underset{\sim}{\mathbf{G}}_k^{\tau}(\alpha\underset{\sim}{\mathbf{G}}_k\underset{\sim}{\mathbf{G}}_k^{\tau} + \underset{\sim}{\mathbf{H}}_k\underset{\sim}{\mathbf{H}}_k^{\tau})^{+}\underset{\sim}{\mathbf{G}}_k, \tag{5.105}$$

since

$$(\alpha\underset{\sim}{\mathbf{G}}_k\underset{\sim}{\mathbf{G}}_k^{\tau} + \underset{\sim}{\mathbf{H}}_k\underset{\sim}{\mathbf{H}}_k^{\tau})(\alpha\underset{\sim}{\mathbf{G}}_k\underset{\sim}{\mathbf{G}}_k^{\tau} + \underset{\sim}{\mathbf{H}}_k\underset{\sim}{\mathbf{H}}_k^{\tau})^{+}\underset{\sim}{\mathbf{G}}_k = \underset{\sim}{\mathbf{G}}_k, \tag{5.106}$$

which follows from the identity

$$\left[(\alpha \mathbf{G}_k \mathbf{G}_k^\tau + \mathbf{H}_k \mathbf{H}_k^\tau)(\alpha \mathbf{G}_k \mathbf{G}_k^\tau + \mathbf{H}_k \mathbf{H}_k^\tau)^+ - \mathbf{I}\right](\alpha \mathbf{G}_k \mathbf{G}_k^\tau + \mathbf{H}_k \mathbf{H}_k^\tau)$$
$$\cdot\left[(\alpha \mathbf{G}_k \mathbf{G}_k^\tau + \mathbf{H}_k \mathbf{H}_k^\tau)(\alpha \mathbf{G}_k \mathbf{G}_k^\tau + \mathbf{H}_k \mathbf{H}_k^\tau)^+ - \mathbf{I}\right] = \mathbf{0},$$

it can be directly verified that for any $\mathbf{x}$ and $\mathbf{u}$

$$\alpha(\mathbf{x} + \mathbf{G}_k^\tau \mathbf{u})^\tau (\mathbf{x} + \mathbf{G}_k^\tau \mathbf{u}) + \mathbf{u}^\tau \mathbf{H}_k \mathbf{H}_k^\tau \mathbf{u}$$
$$= \mathbf{x}^\tau \overline{\mathbf{P}}_0(k, \alpha)\mathbf{x} + (\mathbf{u} + \mathbf{L}_0(\alpha)\mathbf{x})^\tau (\alpha \mathbf{G}_k \mathbf{G}_k^\tau + \mathbf{H}_k \mathbf{H}_k^\tau)(\mathbf{u} + \mathbf{L}_0(\alpha)\mathbf{x}) \tag{5.107}$$

where

$$\mathbf{L}_0(\alpha) = \alpha(\alpha \mathbf{G}_k \mathbf{G}_k^\tau + \mathbf{H}_k \mathbf{H}_k^\tau)^+ \mathbf{G}_k.$$

Let $\alpha_1 \leq \alpha_2$. Then, by (5.107),

$$\mathbf{x}^\tau \overline{\mathbf{P}}_0(k, \alpha_1)\mathbf{x} \leq \mathbf{x}^\tau \overline{\mathbf{P}}_0(k, \alpha_1)\mathbf{x} + (-\mathbf{L}_0(\alpha_2)\mathbf{x} + \mathbf{L}_0(\alpha_1)\mathbf{x})^\tau$$
$$\cdot(\alpha_1 \mathbf{G}_k \mathbf{G}_k^\tau + \mathbf{H}_k \mathbf{H}_k^\tau)(-\mathbf{L}_0(\alpha_2)\mathbf{x} + \mathbf{L}_0(\alpha_1)\mathbf{x})$$
$$= \alpha_1(\mathbf{x} - \mathbf{G}_k^\tau \mathbf{L}_0(\alpha_2)\mathbf{x})^\tau(\mathbf{x} - \mathbf{G}_k^\tau \mathbf{L}_0(\alpha_2)\mathbf{x}) + \mathbf{x}^\tau \mathbf{L}_0^\tau(\alpha_2)\mathbf{H}_k \mathbf{H}_k^\tau \mathbf{L}_0(\alpha_2)\mathbf{x}$$
$$\leq \alpha_2(\mathbf{x} - \mathbf{G}_k^\tau \mathbf{L}_0(\alpha_2)\mathbf{x})^\tau(\mathbf{x} - \mathbf{G}_k^\tau \mathbf{L}_0(\alpha_2)\mathbf{x}) + \mathbf{x}^\tau \mathbf{L}_0^\tau(\alpha_2)\mathbf{H}_k \mathbf{H}_k^\tau \mathbf{L}_0(\alpha_2)\mathbf{x}$$
$$= \mathbf{x}^\tau \overline{\mathbf{P}}_0(k, \alpha_2)\mathbf{x}.$$

From here, $\overline{\mathbf{P}}_0(k, \alpha_2)$ is nondecreasing in α because of the arbitrarity nature of $\mathbf{x}$.

Now we show that

$$\overline{\mathbf{P}}_k(k, \alpha) = \mathbf{P}_k^0 + \boldsymbol{\Psi}_{k,0}^0 \overline{\mathbf{P}}_0(k, \alpha) \boldsymbol{\Psi}_{k,0}^{0\tau} \tag{5.108}$$

where $\boldsymbol{\Psi}_{k,0}^0$ is defined in Lemma 5.4. Once (5.108) has been established, $\overline{\mathbf{P}}_k(k, \alpha)$ nondecreasing as $\alpha \to \infty$ will follow from the same property of $\overline{\mathbf{P}}_0(k, \alpha)$.

Without loss of generality, we may assume $\{\zeta_i\}$ are normally distributed. Notice that

$$E^{\mathbf{y}^k}(\mathbf{x}_k - \hat{\mathbf{x}}_k^j)(\hat{\mathbf{x}}_k^j - \hat{\mathbf{x}}_k)^\tau = E^{\mathbf{y}^k}\left[E^{\mathbf{y}^k, \mathbf{x}_j}(\mathbf{x}_k - \hat{\mathbf{x}}_k^j)\right](\hat{\mathbf{x}}_k^j - \hat{\mathbf{x}}_k)^\tau = \mathbf{0},$$

and so

$$\mathbf{P}_k = E^{\mathbf{y}^k}(\mathbf{x}_k - \hat{\mathbf{x}}_k)(\mathbf{x}_k - \hat{\mathbf{x}}_k)^{\tau}$$

$$= E^{\mathbf{y}^k}(\mathbf{x}_k - \hat{\mathbf{x}}_k^j)(\mathbf{x}_k - \hat{\mathbf{x}}_k^j)^{\tau} + E^{\mathbf{y}^k}(\hat{\mathbf{x}}_k^j - \hat{\mathbf{x}}_k)(\hat{\mathbf{x}}_k^j - \hat{\mathbf{x}}_k)^{\tau}.$$

From this and (5.44), (5.108) follows. □

Lemma 5.10. *Let $\mathbf{A}_k$ be $n \times n$ symmetric and nonnegative. If $\mathbf{A}_k \le \mathbf{A}_{k+1}$, for all k, and $\mathbf{A}_k \underset{k\to\infty}{\to} \mathbf{A} < \infty$, then $\mathbf{A}_k^+ \underset{k\to\infty}{\to} \mathbf{A}^+$.*

Proof. Let $\mathbf{U}$ be an orthogonal matrix so that

$$\mathbf{U}\mathbf{A}\mathbf{U}^{\tau} = \begin{bmatrix} \mathbf{\Lambda} & \mathbf{0} \\ \mathbf{0} & \mathbf{0} \end{bmatrix}, \qquad \mathbf{U} = \begin{bmatrix} \mathbf{U}_1 \\ \mathbf{U}_2 \end{bmatrix}, \qquad \mathbf{\Lambda} = \mathbf{U}_1\mathbf{A}\mathbf{U}_1^{\tau} > \mathbf{0}.$$

Since $\mathbf{A} \ge \mathbf{A}_k$ and $\mathbf{U}_2\mathbf{A}\mathbf{U}_2^{\tau} = \mathbf{0}$, then

$$\mathbf{U}\mathbf{A}_k\mathbf{U}^{\tau} = \begin{bmatrix} \mathbf{U}_1\mathbf{A}_k\mathbf{U}_1^{\tau} & \mathbf{0} \\ \mathbf{0} & \mathbf{0} \end{bmatrix},$$

and $\mathbf{U}_1\mathbf{A}_k\mathbf{U}_1^{\tau} \to \mathbf{U}_1\mathbf{A}\mathbf{U}_1^{\tau}$. Hence for sufficiently large k, $\mathbf{U}_1\mathbf{A}_k\mathbf{U}_1^{\tau}$ is of full rank, and

$$\mathbf{A}_k^+ = \mathbf{U}^{\tau}\begin{bmatrix} (\mathbf{U}_1\mathbf{A}_k\mathbf{U}_1^{\tau})^{-1} & \mathbf{0} \\ \mathbf{0} & \mathbf{0} \end{bmatrix}\mathbf{U} \to \mathbf{U}^{\tau}\begin{bmatrix} \mathbf{\Lambda}^{-1} & \mathbf{0} \\ \mathbf{0} & \mathbf{0} \end{bmatrix}\mathbf{U} = \mathbf{A}^+. \qquad \square$$

Theorem 5.13. *If for some k, $\mathscr{L}_k(k)$ is nonempty, then the Gauss–Markov estimate can be calculated recursively via*

$$\bar{\mathbf{x}}_{k+1} = \mathbf{\Phi}_{k+1}\bar{\mathbf{x}}_k + \mathbf{K}_{k+1}(\mathbf{y}_{k+1} - \mathbf{C}_{k+1}\mathbf{\Phi}_{k+1}\bar{\mathbf{x}}_k), \tag{5.109}$$

$$\begin{aligned}\mathbf{K}_{k+1} = {} & [(\mathbf{\Phi}_{k+1}\bar{\mathbf{P}}_k\mathbf{\Phi}_{k+1}^{\tau} + \mathbf{D}_{k+1}\mathbf{D}_{k+1}^{\tau})\mathbf{C}_{k+1}^{\tau} + \mathbf{D}_{k+1}\mathbf{F}_{k+1}^{\tau}] \\ & \times [\mathbf{C}_{k+1}(\mathbf{\Phi}_{k+1}\bar{\mathbf{P}}_k\mathbf{\Phi}_{k+1}^{\tau} + \mathbf{D}_{k+1}\mathbf{D}_{k+1}^{\tau})\mathbf{C}_{k+1}^{\tau} \\ & + \mathbf{F}_{k+1}\mathbf{F}_{k+1}^{\tau} + \mathbf{C}_{k+1}\mathbf{D}_{k+1}\mathbf{F}_{k+1}^{\tau} + \mathbf{F}_{k+1}\mathbf{D}_{k+1}^{\tau}\mathbf{C}_{k+1}^{\tau}]^{+},\end{aligned} \tag{5.110}$$

$$\begin{aligned}\bar{\mathbf{P}}_{k+1} = {} & \mathbf{\Phi}_{k+1}\bar{\mathbf{P}}_k\mathbf{\Phi}_{k+1}^{\tau} + \mathbf{D}_{k+1}\mathbf{D}_{k+1}^{\tau} \\ & - \mathbf{K}_{k+1}[\mathbf{C}_{k+1}(\mathbf{\Phi}_{k+1}\bar{\mathbf{P}}_k\mathbf{\Phi}_{k+1}^{\tau} + \mathbf{D}_{k+1}\mathbf{D}_{k+1}^{\tau}) + \mathbf{F}_{k+1}\mathbf{D}_{k+1}^{\tau}]\end{aligned} \tag{5.111}$$

for $k \geq k_0 \triangleq \min\{k\colon \mathscr{L}_k(k) \neq \phi\}$, *where, for simplicity, we write*

$$\bar{\mathbf{x}}_k \triangleq \bar{\mathbf{x}}_k(k), \qquad \bar{\mathbf{P}}_k \triangleq \bar{\mathbf{P}}_k(k).$$

Proof. According to the definitions of $\bar{\mathbf{x}}_k(k,\alpha)$ and $\bar{\mathbf{P}}_k(k,\alpha)$, their recursive formulas are the same as (5.22)–(5.26) with the $\mathbf{B}_k$ and $\mathbf{H}_k$ sequences set to $\mathbf{0}$. By Theorem 5.12 and Lemmas 5.9 and 5.10, letting $\alpha \to \infty$ in the recursive formulas for $\bar{\mathbf{x}}_k(k,\alpha)$ and $\bar{\mathbf{P}}_k(k,\alpha)$ leads to (5.109)–(5.111). □

Remark. If $\mathscr{L}_j(k)$ is nonempty, then by the same argument used in Theorem 5.13, it is easy to show that from the instant $k_0 \triangleq \min\{k\colon \mathscr{L}_j(k) \neq \phi\}$ onwards the Gauss–Markov interpolation and prediction of $\mathbf{x}_k$ are generated by recursive formulas as in the conditionally Gaussian case with $\mathbf{B}_k \equiv \mathbf{0}$, $\mathbf{H}_k \equiv \mathbf{0}$.

5.5. CONTROL UNDER QUADRATIC INDEX

We consider system (5.20), (5.21) with the following conditions:

1. $\mathbf{\Phi}_{k+1}, \mathbf{D}_{k+1}, \mathbf{B}_{k+1}$ are deterministic, the control $\mathbf{u}_k$ is $\mathbf{y}^k$-measurable and $E\|\mathbf{u}_k\|^2 < \infty$.
2. $\mathbf{C}_k, \mathbf{F}_k$ are deterministic, $\mathbf{H}_k\mathbf{v}_{k-1}$ is $\mathbf{y}^{k-1}$-measurable with $E\|\mathbf{H}_k\mathbf{v}_{k-1}\|^2 < \infty$.
3. $E(\|\mathbf{x}_0\|^2 + \|\mathbf{y}_0\|^2) < \infty$ and, given $\mathbf{y}_0$, $\mathbf{x}_0$ is conditionally Gaussian with the conditional mean and covariance $E^{\mathbf{y}_0}\mathbf{x}_0$ and $\mathbf{R}_{\mathbf{x}_0}^{\mathbf{y}_0}$, respectively.
4. $\{\zeta_k\}$ are mutually independent, identically distributed with $\zeta_k \in N(\mathbf{0}, \mathbf{I})$ and the ζ_k process is independent of $[\mathbf{x}_0^\tau, \mathbf{y}_0^\tau]^\tau$.

We hope to find an admissible control (i.e., $\mathbf{y}^k$-measurable with $E\|\mathbf{u}_k\|^2 < \infty$) to minimize $EJ(\mathbf{u})$, where

$$J(\mathbf{u}) = \mathbf{x}_N^\tau \mathbf{Q}_0 \mathbf{x}_N + \sum_{k=0}^{N-1} \left(\mathbf{x}_k^\tau \mathbf{Q}_1(k)\mathbf{x}_k + \mathbf{u}_k^\tau \mathbf{Q}_2(k)\mathbf{u}_k\right) \qquad (5.112)$$

with $\mathbf{Q}_1 \geq \mathbf{0}$, $\mathbf{Q}_1(k) \geq \mathbf{0}$, $\mathbf{Q}_2(k) \geq \mathbf{0}$.

Theorem 5.14. *Under Conditions* 1–4 *on the system* (5.20), (5.21) *with the performance index* (5.112), *the optimal control* $\mathbf{u}_k^0$ *is*

$$\mathbf{u}_k^0 = -\underset{\sim}{\mathbf{L}}_k \hat{\mathbf{x}}_k + (\underset{\sim}{\mathbf{I}} - \underset{\sim}{\mathbf{M}}_k^+ \underset{\sim}{\mathbf{M}}_k)\mathbf{u}_k, \qquad k = 0, \ldots, N-1 \quad (5.113)$$

and the minimal value of the performance index is

$$EJ(\mathbf{u}^0) = E\hat{\mathbf{x}}_0^\tau \underset{\sim}{\mathbf{S}}_0 E\hat{\mathbf{x}}_0 + \operatorname{tr} \underset{\sim}{\mathbf{S}}_0 \underset{\sim}{\mathbf{R}}_{\mathbf{x}_0}^{\mathbf{y}_0} + \sum_{k=0}^{N-1} \operatorname{tr}(\underset{\sim}{\mathbf{S}}_{k+1} \underset{\sim}{\mathbf{D}}_{k+1} \underset{\sim}{\mathbf{D}}_{k+1}^\tau + \underset{\sim}{\mathbf{L}}_k \underset{\sim}{\mathbf{P}}_k \underset{\sim}{\mathbf{L}}_k^\tau \underset{\sim}{\mathbf{M}}_k),$$

$$(5.114)$$

where $\mathbf{u}_k$ *is an arbitrary admissible control, and the* $n \times n$ *matrix* $\underset{\sim}{\mathbf{S}}_k$ *is defined recursively via*

$$\underset{\sim}{\mathbf{S}}_k = (\underset{\sim}{\mathbf{\Phi}}_{k+1} - \underset{\sim}{\mathbf{B}}_{k+1}\underset{\sim}{\mathbf{L}}_k)^\tau \underset{\sim}{\mathbf{S}}_{k+1} (\underset{\sim}{\mathbf{\Phi}}_{k+1} - \underset{\sim}{\mathbf{B}}_{k+1}\underset{\sim}{\mathbf{L}}_k)$$

$$+ \underset{\sim}{\mathbf{L}}_k^\tau \mathbf{Q}_2(k) \underset{\sim}{\mathbf{L}}_k + \underset{\sim}{\mathbf{Q}}_1(k), \qquad \underset{\sim}{\mathbf{S}}_N = \underset{\sim}{\mathbf{Q}}_0,$$

$$\underset{\sim}{\mathbf{L}}_k = \underset{\sim}{\mathbf{M}}_k^+ \underset{\sim}{\mathbf{B}}_{k+1}^\tau \underset{\sim}{\mathbf{S}}_{k+1} \underset{\sim}{\mathbf{\Phi}}_{k+1}, \qquad \underset{\sim}{\mathbf{M}}_k = \mathbf{Q}_2(k) + \underset{\sim}{\mathbf{B}}_{k+1}^\tau \underset{\sim}{\mathbf{S}}_{k+1} \underset{\sim}{\mathbf{B}}_{k+1}.$$

Proof. Similar to (5.106), it is easy to see that

$$\underset{\sim}{\mathbf{S}}_{k+1} \underset{\sim}{\mathbf{B}}_{k+1} \underset{\sim}{\mathbf{M}}_k^+ \underset{\sim}{\mathbf{M}}_k = \underset{\sim}{\mathbf{S}}_{k+1} \underset{\sim}{\mathbf{B}}_{k+1}.$$

Hence

$$\underset{\sim}{\mathbf{L}}_k^\tau \underset{\sim}{\mathbf{M}}_k = \underset{\sim}{\mathbf{\Phi}}_{k+1}^\tau \underset{\sim}{\mathbf{S}}_{k+1} \underset{\sim}{\mathbf{B}}_{k+1},$$

$$\underset{\sim}{\mathbf{L}}_k^\tau \underset{\sim}{\mathbf{M}}_k \underset{\sim}{\mathbf{L}}_k = \underset{\sim}{\mathbf{\Phi}}_{k+1}^\tau \underset{\sim}{\mathbf{S}}_{k+1} \underset{\sim}{\mathbf{B}}_{k+1} \underset{\sim}{\mathbf{L}}_k = \underset{\sim}{\mathbf{L}}_k^\tau \underset{\sim}{\mathbf{B}}_{k+1}^\tau \underset{\sim}{\mathbf{S}}_{k+1} \underset{\sim}{\mathbf{\Phi}}_{k+1}.$$

Then, using the completion of the square technique (see, e.g. [59]), $J(\mathbf{u})$ can be rewritten as

$$J(\mathbf{u}) = \mathbf{x}_0^\tau \mathbf{S}_0 \mathbf{x}_0 + \sum_{k=0}^{N-1} \Big\{ (\mathbf{x}_{k+1}^\tau \mathbf{S}_{k+1} \mathbf{x}_{k+1} - \mathbf{x}_k^\tau \mathbf{S}_k \mathbf{x}_k)$$

$$+ \mathbf{x}_k^\tau \mathbf{Q}_1(k) \mathbf{x}_k + \mathbf{u}_k^\tau \mathbf{Q}_2(k) \mathbf{u}_k \Big\}$$

$$= \mathbf{x}_0^\tau \mathbf{S}_0 \mathbf{x}_0 + \sum_{k=0}^{N-1} \Big\{ (\boldsymbol{\Phi}_{k+1} \mathbf{x}_k + \mathbf{B}_{k+1} \mathbf{u}_k + \mathbf{D}_{k+1} \boldsymbol{\zeta}_{k+1})^\tau$$

$$\times \mathbf{S}_{k+1} (\boldsymbol{\Phi}_{k+1} \mathbf{x}_k + \mathbf{B}_{k+1} \mathbf{u}_k + \mathbf{D}_{k+1} \boldsymbol{\zeta}_{k+1})$$

$$- \mathbf{x}_k^\tau \Big[(\boldsymbol{\Phi}_{k+1} - \mathbf{B}_{k+1} \mathbf{L}_k)^\tau \mathbf{S}_{k+1} (\boldsymbol{\Phi}_{k+1} - \mathbf{B}_{k+1} \mathbf{L}_k)$$

$$+ \mathbf{L}_k^\tau \mathbf{Q}_2(k) \mathbf{L}_k + \mathbf{Q}_1(k) \Big] \mathbf{x}_k + \mathbf{x}_k^\tau \mathbf{Q}_1(k) \mathbf{x}_k + \mathbf{u}_k^\tau \mathbf{Q}_2(k) \mathbf{u}_k \Big\}$$

$$= \mathbf{x}_0^\tau \mathbf{S}_0 \mathbf{x}_0 + \sum_{k=0}^{N-1} \{ \boldsymbol{\zeta}_{k+1}^\tau \mathbf{D}_{k+1}^\tau \mathbf{S}_{k+1} (\boldsymbol{\Phi}_{k+1} \mathbf{x}_k + \mathbf{B}_{k+1} \mathbf{u}_k)$$

$$+ (\boldsymbol{\Phi}_{k+1} \mathbf{x}_k + \mathbf{B}_{k+1} \mathbf{u}_k)^\tau \mathbf{S}_{k+1} \mathbf{D}_{k+1} \boldsymbol{\zeta}_{k+1} + \boldsymbol{\zeta}_{k+1}^\tau \mathbf{D}_{k+1}^\tau \mathbf{S}_{k+1} \mathbf{D}_{k+1} \boldsymbol{\zeta}_{k+1} \}$$

$$+ \sum_{k=0}^{N-1} \{ \mathbf{x}_k^\tau \boldsymbol{\Phi}_{k+1}^\tau \mathbf{S}_{k+1} \boldsymbol{\Phi}_{k+1} \mathbf{x}_k + \mathbf{x}_k^\tau \mathbf{L}_k^\tau \mathbf{M}_k \mathbf{u}_k$$

$$+ \mathbf{u}_k^\tau \mathbf{M}_k \mathbf{L}_k \mathbf{x}_k + \mathbf{u}_k^\tau \mathbf{B}_{k+1}^\tau \mathbf{S}_{k+1} \mathbf{B}_{k+1} \mathbf{u}_k - \mathbf{x}_k^\tau \boldsymbol{\Phi}_{k+1}^\tau \mathbf{S}_{k+1} \boldsymbol{\Phi}_{k+1} \mathbf{x}_k$$

$$+ \mathbf{x}_k^\tau \mathbf{L}_k^\tau \mathbf{M}_k \mathbf{L}_k \mathbf{x}_k + \mathbf{x}_k^\tau \mathbf{L}_k^\tau \mathbf{M}_k \mathbf{L}_k \mathbf{x}_k - \mathbf{x}_k^\tau \mathbf{L}_k^\tau \mathbf{B}_k^\tau \mathbf{S}_{k+1} \mathbf{B}_{k+1} \mathbf{L}_k \mathbf{x}_k$$

$$- \mathbf{x}_k^\tau \mathbf{L}_k^\tau \mathbf{Q}_2(k) \mathbf{L}_k \mathbf{x}_k + \mathbf{u}_k^\tau \mathbf{Q}_2(k) \mathbf{u}_k \Big\}$$

$$= \mathbf{x}_0^\tau \mathbf{S}_0 \mathbf{x}_0 + \sum_{k=0}^{N-1} \Big\{ \boldsymbol{\zeta}_{k+1}^\tau \mathbf{D}_{k+1}^\tau \mathbf{S}_{k+1} (\boldsymbol{\Phi}_{k+1} \mathbf{x}_k + \mathbf{B}_{k+1} \mathbf{u}_k)$$

$$+ (\boldsymbol{\Phi}_{k+1} \mathbf{x}_k + \mathbf{B}_{k+1} \mathbf{u}_k)^\tau \mathbf{S}_{k+1} \mathbf{D}_{k+1} \boldsymbol{\zeta}_{k+1} + \boldsymbol{\zeta}_{k+1}^\tau \mathbf{D}_{k+1}^\tau \mathbf{S}_{k+1} \mathbf{D}_{k+1} \boldsymbol{\zeta}_{k+1}$$

$$+ \sum_{k=0}^{N-1} (\mathbf{u}_k + \mathbf{L}_k \mathbf{x}_k)^\tau \mathbf{M}_k (\mathbf{u}_k + \mathbf{L}_k \mathbf{x}_k).$$

Therefore, by Theorem 5.2,

$$EJ(\mathbf{u}) = E\mathbf{x}_0^\tau \mathbf{S}_0 \mathbf{x}_0 + \sum_{k=0}^{N-1} \operatorname{tr} \mathbf{S}_{k+1}\mathbf{D}_{k+1}\mathbf{D}_{k+1}^\tau$$

$$+ E \sum_{k=0}^{N-1} E^{\mathbf{y}^k}(\mathbf{u}_k + \mathbf{L}_k \mathbf{x}_k)^\tau \mathbf{M}_k (\mathbf{u}_k + \mathbf{L}_k \mathbf{x}_k)$$

$$= E\hat{\mathbf{x}}_0^\tau \mathbf{S}_0 E\hat{\mathbf{x}}_0 + \operatorname{tr} \mathbf{S}_0 \mathbf{R}_{\mathbf{x}_0}^{\mathbf{y}_0} + \sum_{k=0}^{N-1} \operatorname{tr}(\mathbf{S}_{k+1}\mathbf{D}_{k+1}\mathbf{D}_{k+1}^\tau + \mathbf{L}_k \mathbf{P}_k \mathbf{L}_k^\tau \mathbf{M}_k)$$

$$+ E \sum_{k=0}^{N-1} (\mathbf{u}_k + \mathbf{L}_k \hat{\mathbf{x}}_k)^\tau \mathbf{M}_k (\mathbf{u}_k + \mathbf{L}_k \hat{\mathbf{x}}_k).$$

Hence the theorem follows. □

Remark 1. We can relax the conditions on $\{\boldsymbol{\zeta}_k\}$; in fact we can omit the assumption on the distribution of $\boldsymbol{\zeta}_k$, merely requiring

$$E\boldsymbol{\zeta}_k = \mathbf{0}, \quad E\boldsymbol{\zeta}_k\boldsymbol{\zeta}_j^\tau = \delta_{kj}\mathbf{I}, \quad E\boldsymbol{\zeta}_k[\mathbf{x}_0^\tau \mathbf{y}_0^\tau] = \mathbf{0}, \qquad \forall k, j,$$

but for this we have to restrict $\mathbf{u}_k$ to be a linear function of past measurements $\mathbf{u}_k = \mathbf{T}_k \mathbf{y}^k + \mathbf{a}_k$, where $\mathbf{T}_k$ and $\mathbf{a}_k$ are deterministic. In this case, Theorem 5.14 remains valid if by $\hat{\mathbf{x}}_k$ we mean the LUMVE of $\mathbf{x}_k$ based on $\mathbf{y}^k$.

Remark 2. The covariance matrix $\mathbf{P}_k$ gives the filtering accuracy and is independent of the control law, as can be seen from Theorem 5.2.

Remark 3. If $\mathbf{D}_k \equiv \mathbf{0}$, $\mathbf{F}_k \equiv \mathbf{0}$ and $\mathbf{x}_0$ is given the system becomes deterministic. In this case, for the performance index (5.112), the optimal control is the same as that given by (5.113) with $\hat{\mathbf{x}}_k$ replaced by $\mathbf{x}_k$.

CHAPTER 6

Linear Unbiased Minimum Variance Estimates for Continuous-Time Systems

6.1. INTRODUCTION

Let $(\Omega, \mathscr{F}, P)$ be the underlying probability space, $\{\mathbf{w}_t\}$ be an l-dimensional Wiener process or a generalized Wiener process and let $\mathbf{x}_t, \mathbf{y}_t, \mathbf{u}_t$ be the n-, m-, and r-dimensional state, measurement, and control vectors, respectively, for $t \in [0, T]$, $T < \infty$.

Consider the dynamic system

$$d\mathbf{x}_t = \underset{\sim}{\mathbf{A}}_t \mathbf{x}_t\, dt + \underset{\sim}{\mathbf{B}}_t \mathbf{u}_t\, dt + \underset{\sim}{\mathbf{D}}_t\, d\mathbf{w}_t \tag{6.1}$$

$$d\mathbf{y}_t = \underset{\sim}{\mathbf{C}}_t \mathbf{x}_t\, dt + \underset{\sim}{\mathbf{F}}_t\, d\mathbf{w}_t, \qquad \underset{\sim}{\mathbf{F}}_t \underset{\sim}{\mathbf{F}}_t^T > \underset{\sim}{\mathbf{0}}, \tag{6.2}$$

where the coefficients are real measurable matrix-valued functions of compatible dimension.

Assume that

$$\int_0^T \|\underset{\sim}{\mathbf{A}}_s\|\, ds < \infty, \qquad \int_0^T \left\| \underset{\sim}{\mathbf{C}}_s^T (\underset{\sim}{\mathbf{F}}_s \underset{\sim}{\mathbf{F}}_s^T)^{-1} \underset{\sim}{\mathbf{C}}_s \right\| ds < \infty,$$

$$\int_0^T \left(\|\underset{\sim}{\mathbf{C}}_s\|^2 + \|\underset{\sim}{\mathbf{B}}_s\|^2 + \|\underset{\sim}{\mathbf{D}}_s\|^2 + \|\underset{\sim}{\mathbf{F}}_s\|^2 \right) ds < \infty, \tag{6.3}$$

and that

$$\hat{\mathbf{x}}_0 \triangleq E\mathbf{x}_0 + \underset{\sim}{\mathbf{R}}_{\mathbf{x}_0\mathbf{y}_0} \underset{\sim}{\mathbf{R}}_{\mathbf{y}_0}^{+} (\mathbf{y}_0 - E\mathbf{y}_0) \tag{6.4}$$

$$\underset{\sim}{\mathbf{R}}_0 \triangleq \underset{\sim}{\mathbf{R}}_{\mathbf{x}_0} - \underset{\sim}{\mathbf{R}}_{\mathbf{x}_0\mathbf{y}_0} \underset{\sim}{\mathbf{R}}_{\mathbf{y}_0}^{+} \underset{\sim}{\mathbf{R}}_{\mathbf{y}_0\mathbf{x}_0} \tag{6.5}$$

$$E\mathbf{w}_t [\mathbf{x}_0^T \mathbf{y}_0^T] = \underset{\sim}{\mathbf{0}}, \qquad \forall t \in [0, T] \tag{6.6}$$

are given.

We are going to discuss the LUMVE of $\mathbf{x}_t$ based on $(\mathbf{y}_s, 0 \le s \le t)$. Denote by Π_t the totality of linear estimates of the form:

$$\Pi_t = \left\{ \mathbf{c}_t + \underset{\sim}{\mathbf{G}}_0(t)\mathbf{y}_0 + \underset{\sim}{\mathbf{G}}_1(t)\int_0^t \underset{\sim}{\mathbf{G}}_2(s)\, d\mathbf{y}_s : \int_0^t \|\underset{\sim}{\mathbf{G}}_2(s)\underset{\sim}{\mathbf{C}}_s\|\, ds < \infty, \right.$$

$$\left. \int_0^t \|\underset{\sim}{\mathbf{G}}_2(s)\underset{\sim}{\mathbf{F}}_s\|^2\, ds < \infty \right\} \tag{6.7}$$

If

$$E(\mathbf{x}_t - \hat{\mathbf{x}}_t)(\mathbf{x}_t - \hat{\mathbf{x}}_t)^\tau = \min_{\mathbf{x} \in \Pi_t} E(\mathbf{x}_t - \mathbf{x})(\mathbf{x}_t - \mathbf{x})^\tau,$$

then $\hat{\mathbf{x}}_t$ is called LUMVE of $\mathbf{x}_t$ based on $\{\mathbf{y}_s, 0 \le s \le t\}$.

When $\underset{\sim}{\mathbf{B}}_t \equiv \underset{\sim}{\mathbf{0}}$, the stochastic differential equations satisfied by $\hat{\mathbf{x}}_t$ are called the Kalman–Bucy filter.

When $\mathbf{u}_t$ linearly depends on $\{\mathbf{y}_s, 0 \le s \le t\}$ and $\{\mathbf{w}_t\}$ is a Wiener process, then $[\mathbf{x}_t^\tau, \mathbf{y}_t^\tau]^\tau$ is a normal random process, hence the LUMVE of $\mathbf{x}_s$ based on $\{\mathbf{y}_\lambda, 0 \le \lambda \le t\}$ coincides with the minimum variance estimate $E(\mathbf{x}_s/\mathbf{y}_\lambda, 0 \le \lambda \le t)$. Therefore, for the generalized Wiener process case (see Chapter 1), without loss of generality, we can first assume $\{\mathbf{w}_t\}$ is a Wiener process and find $E(\mathbf{x}_s/\mathbf{y}_\lambda, 0 \le \lambda \le t)$, which is just the LUMVE of $\mathbf{x}_s$ based on $(\mathbf{y}_\lambda, 0 \le \lambda \le t)$ for $\{\mathbf{w}_t\}$ being a generalized Wiener process. We shall frequently adopt this treatment.

In filtering and control problems, the following Riccati equation is of great importance:

$$\dot{\underset{\sim}{\mathbf{P}}}_t = \underset{\sim}{\mathbf{P}}_t\underset{\sim}{\mathbf{A}}_t^\tau + \underset{\sim}{\mathbf{A}}_t\underset{\sim}{\mathbf{P}}_t + \underset{\sim}{\mathbf{D}}_t\underset{\sim}{\mathbf{D}}_t^\tau - (\underset{\sim}{\mathbf{D}}_t\underset{\sim}{\mathbf{F}}_t^\tau + \underset{\sim}{\mathbf{P}}_t\underset{\sim}{\mathbf{C}}_t^\tau)(\underset{\sim}{\mathbf{F}}_t\underset{\sim}{\mathbf{F}}_t^\tau)^{-1}(\underset{\sim}{\mathbf{D}}_t\underset{\sim}{\mathbf{F}}_t^\tau + \underset{\sim}{\mathbf{P}}_t\underset{\sim}{\mathbf{C}}_t^\tau)^\tau,$$

$$\underset{\sim}{\mathbf{P}}_0 = \underset{\sim}{\mathbf{R}}_0. \tag{6.8}$$

Lemma 6.1. *Under condition* (6.3) *with* $\|\underset{\sim}{\mathbf{B}}_t\|^2$ *deleted*, (6.8) *has a unique solution* $\underset{\sim}{\mathbf{P}}_t$ *on* $[0, T]$ *satisfying the initial value* $\underset{\sim}{\mathbf{P}}_0 = \underset{\sim}{\mathbf{R}}_0 \ge \underset{\sim}{\mathbf{0}}$ *and* $\underset{\sim}{\mathbf{P}}_t$ *is symmetric, nonnegative definite.*

Proof. Denote

$$\bar{\underset{\sim}{\mathbf{A}}}_t = \underset{\sim}{\mathbf{A}}_t - \underset{\sim}{\mathbf{D}}_t\underset{\sim}{\mathbf{F}}_t^\tau(\underset{\sim}{\mathbf{F}}_t\underset{\sim}{\mathbf{F}}_t^\tau)^{-1}\underset{\sim}{\mathbf{C}}_t, \tag{6.9}$$

$$\underset{\sim}{\mathbf{Q}}_t = \underset{\sim}{\mathbf{D}}_t\underset{\sim}{\mathbf{D}}_t^\tau - \underset{\sim}{\mathbf{D}}_t\underset{\sim}{\mathbf{F}}_t^\tau(\underset{\sim}{\mathbf{F}}_t\underset{\sim}{\mathbf{F}}_t^\tau)^{-1}\underset{\sim}{\mathbf{F}}_t\underset{\sim}{\mathbf{D}}_t^\tau = \underset{\sim}{\mathbf{D}}_t\left(\underset{\sim}{\mathbf{I}} - \underset{\sim}{\mathbf{F}}_t^\tau(\underset{\sim}{\mathbf{F}}_t\underset{\sim}{\mathbf{F}}_t^\tau)^{-1}\underset{\sim}{\mathbf{F}}_t\right)\underset{\sim}{\mathbf{D}}_t^\tau. \tag{6.10}$$

The eigenvalues of $\mathbf{F}_t^T(\mathbf{F}_t\mathbf{F}_t^T)^{-1}\mathbf{F}_t$ are either 0 or 1, consequently

$$\mathbf{I} - \mathbf{F}_t^T(\mathbf{F}_t\mathbf{F}_t^T)^{-1}\mathbf{F}_t \geq \mathbf{0}, \qquad \mathbf{Q}_t \geq \mathbf{0}, \tag{6.11}$$

$$\mathbf{D}_t\mathbf{F}_t^T(\mathbf{F}_t\mathbf{F}_t^T)^{-1}\mathbf{F}_t\mathbf{D}_t^T \leq \mathbf{D}_t\mathbf{D}_t^T, \tag{6.12}$$

and hence

$$\int_0^T \|\overline{\mathbf{A}}_t\|\, dt \leq 2\int_0^T \|\mathbf{A}_t\|\, dt$$

$$+2\left[\int_0^T \left\|\mathbf{D}_t\mathbf{F}_t^T(\mathbf{F}_t\mathbf{F}_t^T)^{-1/2}\right\|^2 dt \int_0^T \left\|(\mathbf{F}_t\mathbf{F}_t^T)^{-1/2}\mathbf{C}_t\right\|^2 dt\right]^{1/2} < \infty.$$

Equation (6.8) can be rewritten as

$$\dot{\mathbf{P}}_t = \mathbf{P}_t\overline{\mathbf{A}}_t^T + \overline{\mathbf{A}}_t\mathbf{P}_t + \mathbf{Q}_t - \mathbf{P}_t\mathbf{C}_t^T(\mathbf{F}_t\mathbf{F}_t^T)^{-1}\mathbf{C}_t\mathbf{P}_t. \tag{6.13}$$

We first prove the uniqueness of its solution. Let $\mathbf{P}_{1t}$ and $\mathbf{P}_{2t}$ be solutions of (6.13) such that

$$\mathbf{P}_{10} = \mathbf{P}_{20} = \mathbf{R}_0.$$

Clearly, they are continuous, and hence

$$\int_0^T \left\|\overline{\mathbf{A}}_t - \mathbf{P}_{1t}\mathbf{C}_t^T(\mathbf{F}_t\mathbf{F}_t^T)^{-1}\mathbf{C}_t\right\| dt < \infty.$$

Let

$$\dot{\mathbf{G}}_t = \left(\overline{\mathbf{A}}_t - \mathbf{P}_{1t}\mathbf{C}_t^T(\mathbf{F}_t\mathbf{F}_t^T)^{-1}\mathbf{C}_t\right)\mathbf{G}_t, \qquad \mathbf{G}_0 = \mathbf{I} \tag{6.14}$$

$$\dot{\mathbf{H}}_t = \mathbf{H}_t\left(\overline{\mathbf{A}}_t - \mathbf{P}_{1t}\mathbf{C}_t^T(\mathbf{F}_t\mathbf{F}_t^T)^{-1}\mathbf{C}_t\right)^T, \qquad \mathbf{H}_0 = \mathbf{I} \tag{6.15}$$

$$\mathbf{T}_t = \mathbf{P}_{1t} - \mathbf{P}_{2t}, \qquad \mathbf{M}_t = \mathbf{G}_t^{-1}\mathbf{T}_t\mathbf{H}_t^{-1}, \qquad \mathbf{M}_0 = \mathbf{0}.$$

We have

$$\dot{\underset{\sim}{M}}_t = -\underset{\sim}{G}_t^{-1}\big(\overline{\underset{\sim}{A}}_t - \underset{\sim}{P}_{1t}\underset{\sim}{C}_t^\tau(\underset{\sim}{F}_t\underset{\sim}{F}_t^\tau)^{-1}\underset{\sim}{C}_t\big)\underset{\sim}{G}_t\underset{\sim}{G}_t^{-1}\underset{\sim}{T}_t\underset{\sim}{H}_t^{-1}$$

$$+\underset{\sim}{G}_t^{-1}\Big[\underset{\sim}{T}_t\overline{\underset{\sim}{A}}_t^\tau + \overline{\underset{\sim}{A}}_t\underset{\sim}{T}_t - \underset{\sim}{P}_{1t}\underset{\sim}{C}_t^\tau(\underset{\sim}{F}_t\underset{\sim}{F}_t^\tau)^{-1}\underset{\sim}{C}_t\underset{\sim}{P}_{1t}$$

$$+\underset{\sim}{P}_{2t}\underset{\sim}{C}_t^\tau(\underset{\sim}{F}_t\underset{\sim}{F}_t^\tau)^{-1}\underset{\sim}{C}_t\underset{\sim}{P}_{2t}\Big]\underset{\sim}{H}_t^{-1}$$

$$-\underset{\sim}{G}_t^{-1}\underset{\sim}{T}_t\underset{\sim}{H}_t^{-1}\underset{\sim}{H}_t\big(\overline{\underset{\sim}{A}}_t - \underset{\sim}{P}_{1t}\underset{\sim}{C}_t^\tau(\underset{\sim}{F}_t\underset{\sim}{F}_t^\tau)^{-1}\underset{\sim}{C}_t\big)\underset{\sim}{H}_t^{-1}$$

$$= \underset{\sim}{G}_t^{-1}\Big[\underset{\sim}{P}_{1t}\underset{\sim}{C}_t^\tau(\underset{\sim}{F}_t\underset{\sim}{F}_t^\tau)^{-1}\underset{\sim}{C}_t(\underset{\sim}{P}_{1t} - \underset{\sim}{P}_{2t}) - \underset{\sim}{P}_{1t}\underset{\sim}{C}_t^\tau(\underset{\sim}{F}_t\underset{\sim}{F}_t^\tau)^{-1}\underset{\sim}{C}_t\underset{\sim}{P}_{1t}$$

$$+\underset{\sim}{P}_{2t}\underset{\sim}{C}_t^\tau(\underset{\sim}{F}_t\underset{\sim}{F}_t^\tau)^{-1}\underset{\sim}{C}_t\underset{\sim}{P}_{2t} + \underset{\sim}{G}_t^{-1}(\underset{\sim}{P}_{1t} - \underset{\sim}{P}_{2t})\underset{\sim}{C}_t^\tau(\underset{\sim}{F}_t\underset{\sim}{F}_t^\tau)^{-1}\underset{\sim}{C}_t\underset{\sim}{P}_{1t}\Big]\underset{\sim}{H}_t^{-1}$$

$$= \underset{\sim}{G}_t^{-1}\Big[(\underset{\sim}{P}_{1t} - \underset{\sim}{P}_{2t})\underset{\sim}{C}_t^\tau(\underset{\sim}{F}_t\underset{\sim}{F}_t^\tau)^{-1}\underset{\sim}{C}_t(\underset{\sim}{P}_{1t} - \underset{\sim}{P}_{2t})\Big]\underset{\sim}{H}_t^{-1}$$

$$= \underset{\sim}{M}_t\underset{\sim}{H}_t\underset{\sim}{C}_t^\tau(\underset{\sim}{F}_t\underset{\sim}{F}_t^\tau)^{-1}\underset{\sim}{C}_t\underset{\sim}{G}_t\underset{\sim}{M}_t, \qquad \underset{\sim}{M}_0 = \underset{\sim}{0}. \tag{6.16}$$

Notice that

$$\underset{\sim}{N}_t \triangleq \underset{\sim}{M}_t\underset{\sim}{H}_t\underset{\sim}{C}_t^\tau(\underset{\sim}{F}_t\underset{\sim}{F}_t^\tau)^{-1}\underset{\sim}{C}_t\underset{\sim}{G}_t$$

is an absolutely integrable matrix since $\underset{\sim}{M}_t$ is continuous and condition (6.3) is in force; so, from the uniqueness of solutions for linear equations for which the coefficient matrices are absolutely integrable, it follows that

$$\dot{\underset{\sim}{M}}_t = \underset{\sim}{N}_t\underset{\sim}{M}_t, \qquad \underset{\sim}{M}_0 = \underset{\sim}{0}$$

implies

$$\underset{\sim}{M}_t \equiv \underset{\sim}{0} \quad \text{and} \quad \underset{\sim}{P}_{1t} \equiv \underset{\sim}{P}_{2t}.$$

Now we show that if (6.13) has a solution $\underset{\sim}{P}_t$, then it is necessarily a symmetric, nonnegative definite matrix. Let

$$\dot{\underset{\sim}{z}}_t = -\big(\overline{\underset{\sim}{A}}_t^\tau - \underset{\sim}{C}_t^\tau(\underset{\sim}{F}_t\underset{\sim}{F}_t^\tau)^{-1}\underset{\sim}{C}_t\underset{\sim}{P}_t\big)\underset{\sim}{z}_t, \qquad \underset{\sim}{z}_s = \underset{\sim}{z},$$

where $\mathbf{z}$ is any deterministic vector. Then

$$\frac{d\mathbf{z}_t^T\mathbf{P}_t\mathbf{z}_t}{dt} = -\mathbf{z}_t^T\left(\overline{\mathbf{A}}_t - \mathbf{P}_t\mathbf{C}_t^T(\mathbf{F}_t\mathbf{F}_t^T)^{-1}\mathbf{C}_t\right)\mathbf{P}_t\mathbf{z}_t$$

$$-\mathbf{z}_t^T\mathbf{P}_t\left(\overline{\mathbf{A}}_t^T - \mathbf{C}_t^T(\mathbf{F}_t\mathbf{F}_t^T)^{-1}\mathbf{C}_t\mathbf{P}_t\right)\mathbf{z}_t$$

$$+\mathbf{z}_t^T\left(\mathbf{P}_t\overline{\mathbf{A}}_t^T + \overline{\mathbf{A}}_t\mathbf{P}_t + \mathbf{Q}_t - \mathbf{P}_t\mathbf{C}_t^T(\mathbf{F}_t\mathbf{F}_t^T)^{-1}\mathbf{C}_t\mathbf{P}_t\right)\mathbf{z}_t$$

$$= \mathbf{z}_t^T\left(\mathbf{Q}_t + \mathbf{P}_t\mathbf{C}_t^T(\mathbf{F}_t\mathbf{F}_t^T)^{-1}\mathbf{C}_t\mathbf{P}_t\right)\mathbf{z}_t$$

and

$$\mathbf{z}^T\mathbf{P}_s\mathbf{z} = \mathbf{z}_0^T\mathbf{R}_0\mathbf{z}_0 + \int_0^s \mathbf{z}_\lambda^T\left[\mathbf{Q}_\lambda + \mathbf{P}_\lambda\mathbf{C}_\lambda^T(\mathbf{F}_\lambda\mathbf{F}_\lambda^T)^{-1}\mathbf{C}_\lambda\mathbf{P}_\lambda\right]\mathbf{z}_\lambda\,d\lambda. \quad (6.17)$$

By the arbitrary nature of $\mathbf{z}$ it follows that

$$\mathbf{P}_s \geq \mathbf{0}, \qquad \forall s \in [0, T].$$

Now we proceed to prove that (6.13) has a solution. Let

$$\frac{d}{dt}\mathbf{\Phi}_{t,s} = \overline{\mathbf{A}}_t\mathbf{\Phi}_{t,s}, \qquad \mathbf{\Phi}_{s,s} = \mathbf{I} \quad (6.18)$$

Because of the continuity of $\mathbf{\Phi}_{t,s}$ on $[0,T]\times[0,T]$, there is a constant $k_0 > 0$ such that

$$\|\mathbf{\Phi}_{t,s}\| \leq k_0, \qquad \forall t, s \in [0, T]. \quad (6.19)$$

Equation (6.13) is equivalent to

$$\mathbf{P}_t = \mathbf{\Phi}_{t,0}\mathbf{R}_0\mathbf{\Phi}_{t,0}^T + \int_0^t \mathbf{\Phi}_{t,s}\left(\mathbf{Q}_s - \mathbf{P}_s\mathbf{C}_s^T(\mathbf{F}_s\mathbf{F}_s^T)^{-1}\mathbf{C}_s\mathbf{P}_s\right)\mathbf{\Phi}_{t,s}^T\,ds. \quad (6.20)$$

If (6.20) has a solution on $[0, a)$, $a \in (0, T]$, then it has an upper bound which is independent of a, since

$$0 \leq \|\mathbf{P}_t\| \leq \|\mathbf{\Phi}_{t,0}\mathbf{R}_0\mathbf{\Phi}_{t,0}^T\| + \left\|\int_0^t \mathbf{\Phi}_{t,s}\mathbf{Q}_s\mathbf{\Phi}_{t,s}^T\,ds\right\|$$

$$\leq k_0^2\|\mathbf{R}_0\| + k_0^2\int_0^T\|\mathbf{Q}_s\|\,ds \triangleq k_1. \quad (6.21)$$

Without loss of generality assume $\|\mathbf{R}_0\| \leq k_1$.

Let

$$\underset{\sim}{L}_t = \begin{bmatrix} -\bar{\underset{\sim}{A}}_t^{\tau} & \underset{\sim}{C}_t^{\tau}(\underset{\sim}{F}_t\underset{\sim}{F}_t^{\tau})^{-1}\underset{\sim}{C}_t \\ \underset{\sim}{Q}_t & \bar{\underset{\sim}{A}}_t \end{bmatrix}, \tag{6.22}$$

$$\dot{\underset{\sim}{H}}_{t,s} = \underset{\sim}{L}_t\underset{\sim}{H}_{t,s}, \qquad \underset{\sim}{H}_{s,s} = \underset{\sim}{I}, \tag{6.23}$$

and write $\underset{\sim}{H}_{t,s}$ in four blocks of dimension $n \times n$:

$$\underset{\sim}{H}_{t,s} = \begin{bmatrix} \underset{\sim}{H}_{t,s}^{(1)} & \underset{\sim}{H}_{t,s}^{(2)} \\ \underset{\sim}{H}_{t,s}^{(3)} & \underset{\sim}{H}_{t,s}^{(4)} \end{bmatrix}. \tag{6.24}$$

Since it is continuous on $[0, T] \times [0, T]$ and $\underset{\sim}{H}_{s,s} = \underset{\sim}{I}$, for all $s \in [0, T]$, for any given $\delta > 0$ there is an $\varepsilon > 0$ such that

$$(1-\delta)\underset{\sim}{I} \le \underset{\sim}{H}_{t,s}^{(1)}\underset{\sim}{H}_{t,s}^{(1)\tau} \le (1+\delta)\underset{\sim}{I}, \quad \|\underset{\sim}{H}_{t,s}^{(2)}\| < \delta, \qquad \forall s, t \in [0, T]$$

if $|t - s| < 2\varepsilon$. We take δ so that

$$1 - \delta - 2k_1\delta\sqrt{1+\delta} > 0$$

and denote

$$\underset{\sim}{U}_t = \underset{\sim}{H}_{t,0}^{(1)} + \underset{\sim}{H}_{t,0}^{(2)}\underset{\sim}{R}_0 \tag{6.25}$$

$$\underset{\sim}{V}_t = \underset{\sim}{H}_{t,0}^{(3)} + \underset{\sim}{H}_{t,0}^{(4)}\underset{\sim}{R}_0. \tag{6.26}$$

Then for $t \in [0, 2\varepsilon]$

$$\begin{aligned} \underset{\sim}{U}_t\underset{\sim}{U}_t^{\tau} &\ge \underset{\sim}{H}_{t,0}^{(1)}\underset{\sim}{H}_{t,0}^{(1)\tau} + \underset{\sim}{H}_{t,0}^{(1)}\underset{\sim}{R}_0\underset{\sim}{H}_{t,0}^{(2)\tau} + \underset{\sim}{H}_{t,0}^{(2)}\underset{\sim}{R}_0\underset{\sim}{H}_{t,0}^{(1)\tau} \\ &\ge \left[(1-\delta) - 2\sqrt{1+\delta}\,k_1\delta\right]\underset{\sim}{I} > 0, \end{aligned}$$

which means $\underset{\sim}{U}_t$ is invertible on $t \in [0, 2\varepsilon]$.

Define

$$\underset{\sim}{P}_t \triangleq \underset{\sim}{V}_t\underset{\sim}{U}_t^{-1}. \tag{6.27}$$

Clearly, $\underset{\sim}{\mathbf{P}}_0 = \underset{\sim}{\mathbf{V}}_0\underset{\sim}{\mathbf{U}}_0^{-1} = \underset{\sim}{\mathbf{R}}_0$, and by (6.22)–(6.24) we have

$$
\begin{aligned}
\dot{\underset{\sim}{\mathbf{P}}}_t &= \dot{\underset{\sim}{\mathbf{V}}}_t\underset{\sim}{\mathbf{U}}_t^{-1} - \underset{\sim}{\mathbf{V}}_t\underset{\sim}{\mathbf{U}}_t^{-1}\dot{\underset{\sim}{\mathbf{U}}}_t\underset{\sim}{\mathbf{U}}_t^{-1} \\
&= \left[\underset{\sim}{\mathbf{Q}}_t\underset{\sim}{\mathbf{H}}_{t,0}^{(1)} + \overline{\underset{\sim}{\mathbf{A}}}_t\underset{\sim}{\mathbf{H}}_{t,0}^{(3)} + \left(\underset{\sim}{\mathbf{Q}}_t\underset{\sim}{\mathbf{H}}_{t,0}^{(2)} + \overline{\underset{\sim}{\mathbf{A}}}_t\underset{\sim}{\mathbf{H}}_{t,0}^{(4)}\right)\underset{\sim}{\mathbf{R}}_0\right]\underset{\sim}{\mathbf{U}}_t^{-1} \\
&\quad - \underset{\sim}{\mathbf{V}}_t\underset{\sim}{\mathbf{U}}_t^{-1}\left[-\overline{\underset{\sim}{\mathbf{A}}}_t^{\tau}\underset{\sim}{\mathbf{H}}_{t,0}^{(1)} + \underset{\sim}{\mathbf{C}}_t^{\tau}(\underset{\sim}{\mathbf{F}}_t\underset{\sim}{\mathbf{F}}_t^{\tau})^{-1}\underset{\sim}{\mathbf{C}}_t\underset{\sim}{\mathbf{H}}_{t,0}^{(3)}\right. \\
&\quad \left.\times\left(-\overline{\underset{\sim}{\mathbf{A}}}_t^{\tau}\underset{\sim}{\mathbf{H}}_{t,0}^{(2)} + \underset{\sim}{\mathbf{C}}_t^{\tau}(\underset{\sim}{\mathbf{F}}_t\underset{\sim}{\mathbf{F}}_t^{\tau})^{-1}\underset{\sim}{\mathbf{C}}_t\underset{\sim}{\mathbf{H}}_{t,0}^{(4)}\right)\underset{\sim}{\mathbf{R}}_0\right]\underset{\sim}{\mathbf{U}}_t^{-1} \\
&= \overline{\underset{\sim}{\mathbf{A}}}_t\underset{\sim}{\mathbf{P}}_t + \underset{\sim}{\mathbf{P}}_t\overline{\underset{\sim}{\mathbf{A}}}_t^{\tau} + \underset{\sim}{\mathbf{Q}}_t - \underset{\sim}{\mathbf{P}}_t\underset{\sim}{\mathbf{C}}_t^{\tau}(\underset{\sim}{\mathbf{F}}_t\underset{\sim}{\mathbf{F}}_t^{\tau})^{-1}\underset{\sim}{\mathbf{C}}_t\underset{\sim}{\mathbf{P}}_t,
\end{aligned}
\tag{6.28}
$$

which means that $\underset{\sim}{\mathbf{P}}_t$ given by (6.27) on $[0, 2\varepsilon]$ satisfies (6.13) and the initial condition. Furthermore, from (6.28) we see $\underset{\sim}{\mathbf{P}}_t$ satisfies (6.13) on $[0, a]$, $\forall a \in [0, T]$ whenever $\det \underset{\sim}{\mathbf{U}}_t \neq 0$ on $[0, a]$.

Suppose

$$
\det \underset{\sim}{\mathbf{U}}_t \begin{cases} \neq 0, & t \in [0, t_1) \\ 0, & t = t_1, \quad 2\varepsilon < t_1 \leq T, \end{cases}
$$

and on $[t_1 - \varepsilon, t_1 + \varepsilon]$ consider

$$
\underset{\sim}{\mathbf{U}}_t^{(1)} = \underset{\sim}{\mathbf{H}}_{t,t_1-\varepsilon}^{(1)} + \underset{\sim}{\mathbf{H}}_{t,t_1-\varepsilon}^{(2)}\underset{\sim}{\mathbf{P}}_{t_1-\varepsilon}, \tag{6.29}
$$

$$
\underset{\sim}{\mathbf{V}}_t^{(1)} = \underset{\sim}{\mathbf{H}}_{t,t_1-\varepsilon}^{(3)} + \underset{\sim}{\mathbf{H}}_{t,t_1-\varepsilon}^{(4)}\underset{\sim}{\mathbf{P}}_{t_1-\varepsilon}. \tag{6.30}
$$

Since $\|\underset{\sim}{\mathbf{P}}_t\| \leq k_1$ for $t \in [0, t_1)$ by (6.21), we have, as in the proof for $\underset{\sim}{\mathbf{U}}_t\underset{\sim}{\mathbf{U}}_t^{\tau}$,

$$
\begin{aligned}
\underset{\sim}{\mathbf{U}}_t^{(1)}\underset{\sim}{\mathbf{U}}_t^{(1)\tau} &\geq \underset{\sim}{\mathbf{H}}_{t,t_1-\varepsilon}^{(1)}\underset{\sim}{\mathbf{H}}_{t,t_1-\varepsilon}^{(1)\tau} + \underset{\sim}{\mathbf{H}}_{t,t_1-\varepsilon}^{(1)}\underset{\sim}{\mathbf{P}}_{t_1-\varepsilon}\underset{\sim}{\mathbf{H}}_{t,t_1-\varepsilon}^{(2)\tau} + \underset{\sim}{\mathbf{H}}_{t,t_1-\varepsilon}^{(2)}\underset{\sim}{\mathbf{P}}_{t_1-\varepsilon}\underset{\sim}{\mathbf{H}}_{t,t_1-\varepsilon}^{(1)\tau} \\
&\geq \left[(1-\delta) - 2\sqrt{1+\delta}\,k_1\delta\right]\underset{\sim}{\mathbf{I}} > \underset{\sim}{\mathbf{0}},
\end{aligned}
$$

and $\det \underset{\sim}{\mathbf{U}}_t^{(1)} \neq 0$ for $t \in [t_1 - \varepsilon, t_1 + \varepsilon]$.

Hence $\underset{\sim}{\mathbf{V}}_t^{(1)}(\underset{\sim}{\mathbf{U}}_t^{(1)})^{-1}$ satisfies (6.13) on $[t_1 - \varepsilon, t_1 + \varepsilon]$ and by uniqueness of the solution it coincides with $\underset{\sim}{\mathbf{V}}_t\underset{\sim}{\mathbf{U}}_t^{-1}$ on $[t_1 - \varepsilon, t_1)$. By such a technique, we can extend the solution step by step and the length of extension is not less than ε for each step. This means that (6.13) has a unique solution on $[0, T]$. □

6.2. THE KALMAN–BUCY FILTER

Consider (6.1) and (6.2) with $\underset{\sim}{\mathbf{B}}_t \equiv \mathbf{0}$.

Theorem 6.1. *Under conditions of Lemma* 6.1, *the* LUMVE $\hat{\mathbf{x}}_t$ *of* $\mathbf{x}_t$ *based on* $\{\mathbf{y}_\lambda,\ c \le \lambda \le t\}$ *satisfies the stochastic differential equation*:

$$d\hat{\mathbf{x}}_t = \underset{\sim}{\mathbf{A}}_t\hat{\mathbf{x}}_t\,dt + (\underset{\sim}{\mathbf{D}}_t\underset{\sim}{\mathbf{F}}_t^\tau + \underset{\sim}{\mathbf{P}}_t\underset{\sim}{\mathbf{C}}_t^\tau)(\underset{\sim}{\mathbf{F}}_t\underset{\sim}{\mathbf{F}}_t^\tau)^{-1}(d\mathbf{y}_t - \underset{\sim}{\mathbf{C}}_t\hat{\mathbf{x}}_t\,dt) \tag{6.31}$$

with initial value given by (6.4), *and its estimation error covariance matrix* $\underset{\sim}{\mathbf{P}}_t$ *satisfies the Riccati equation* (6.8) *with initial value* (6.5).

Proof. Let

$$\frac{d}{dt}\underset{\sim}{\mathbf{\Phi}}_{t,s} = \underset{\sim}{\mathbf{A}}_t\underset{\sim}{\mathbf{\Phi}}_{t,s}, \qquad \underset{\sim}{\mathbf{\Phi}}_{s,s} = \underset{\sim}{\mathbf{I}}. \tag{6.32}$$

From (1.41), we know that

$$\mathbf{x}_t = \underset{\sim}{\mathbf{\Phi}}_{t,0}\left(\mathbf{x}_0 + \int_0^t \underset{\sim}{\mathbf{\Phi}}_{0,s}\underset{\sim}{\mathbf{D}}_s\,d\mathbf{w}_s\right).$$

Let us put

$$\underset{\sim}{\mathbf{H}}_t = \underset{\sim}{\mathbf{\Phi}}_{t,0} - \int_0^t \underset{\sim}{\mathbf{G}}_1(t)\underset{\sim}{\mathbf{G}}_2(\lambda)\underset{\sim}{\mathbf{C}}_\lambda\underset{\sim}{\mathbf{\Phi}}_{\lambda,0}\,d\lambda \tag{6.33}$$

and let $\underset{\sim}{\mathbf{G}}_2(t)$ satisfy the conditions in (6.7). Then

$$\begin{aligned} x_t^G \triangleq{}& \mathbf{c}_t + \underset{\sim}{\mathbf{G}}_0(t)\mathbf{y}_0 + \underset{\sim}{\mathbf{G}}_1(t)\int_0^t \underset{\sim}{\mathbf{G}}_2(s)\,d\mathbf{y}_s \\ ={}& \mathbf{c}_t + \underset{\sim}{\mathbf{G}}_0(t)\mathbf{y}_0 + \underset{\sim}{\mathbf{G}}_1(t)\int_0^t \underset{\sim}{\mathbf{G}}_2(s)\underset{\sim}{\mathbf{F}}_s\,d\mathbf{w}_s \\ &+ \underset{\sim}{\mathbf{G}}_1(t)\int_0^t \underset{\sim}{\mathbf{G}}_2(s)\underset{\sim}{\mathbf{C}}_s\underset{\sim}{\mathbf{\Phi}}_{s,0}\left(\mathbf{x}_0 + \int_0^s \underset{\sim}{\mathbf{\Phi}}_{0,\lambda}\underset{\sim}{\mathbf{D}}_\lambda\,d\mathbf{w}_\lambda\right)ds \\ ={}& \mathbf{c}_t + \underset{\sim}{\mathbf{G}}_0(t)\mathbf{y}_0 + \underset{\sim}{\mathbf{G}}_1(t)\int_0^t \underset{\sim}{\mathbf{G}}_2(s)\underset{\sim}{\mathbf{F}}_s\,d\mathbf{w}_s + \underset{\sim}{\mathbf{G}}_1(t)\int_0^t \underset{\sim}{\mathbf{G}}_2(s)\underset{\sim}{\mathbf{C}}_s\underset{\sim}{\mathbf{\Phi}}_{s,0}\,ds\,\mathbf{x}_0 \\ &+ \underset{\sim}{\mathbf{G}}_1(t)\int_0^t\int_\lambda^t \underset{\sim}{\mathbf{G}}_2(s)\underset{\sim}{\mathbf{C}}_s\underset{\sim}{\mathbf{\Phi}}_{s,0}\,ds\,\underset{\sim}{\mathbf{\Phi}}_{0,\lambda}\underset{\sim}{\mathbf{D}}_\lambda\,d\mathbf{w}_\lambda, \end{aligned} \tag{6.34}$$

and

$$\begin{aligned} \mathbf{x}_t - \mathbf{x}_t^G ={}& \underset{\sim}{\mathbf{H}}_t\left(\mathbf{x}_0 + \int_0^t \underset{\sim}{\mathbf{\Phi}}_{0,s}\underset{\sim}{\mathbf{D}}_s\,d\mathbf{w}_s\right) \\ &+ \underset{\sim}{\mathbf{G}}_1(t)\int_0^t\int_0^\lambda \underset{\sim}{\mathbf{G}}_2(s)\underset{\sim}{\mathbf{C}}_s\underset{\sim}{\mathbf{\Phi}}_{s,0}\,ds\,\underset{\sim}{\mathbf{\Phi}}_{0,\lambda}\underset{\sim}{\mathbf{D}}_\lambda\,d\mathbf{w}_\lambda \\ &- \mathbf{c}_t - \underset{\sim}{\mathbf{G}}_0(t)\mathbf{y}_0 - \underset{\sim}{\mathbf{G}}_1(t)\int_0^t \underset{\sim}{\mathbf{G}}_2(s)\underset{\sim}{\mathbf{F}}_s\,d\mathbf{w}_s. \end{aligned} \tag{6.35}$$

Let

$$\underset{\sim}{L}_s^\tau = \left(\underset{\sim}{G}_1(t)\int_0^s \underset{\sim}{G}_2(\lambda)\underset{\sim}{C}_\lambda\underset{\sim}{\Phi}_{\lambda,0}\,d\lambda + \underset{\sim}{H}_t\right)\underset{\sim}{\Phi}_{0,s}. \tag{6.36}$$

Taking its derivative, we have

$$\frac{d}{ds}\underset{\sim}{L}_s^\tau = -\underset{\sim}{L}_s^\tau\underset{\sim}{A}_s + \underset{\sim}{G}_1(t)\underset{\sim}{G}_2(s)\underset{\sim}{C}_s, \qquad \underset{\sim}{L}_t^\tau = \underset{\sim}{I}, \quad \underset{\sim}{L}_0^\tau = \underset{\sim}{H}_t. \tag{6.37}$$

From (6.35) and (6.36), we have

$$\mathbf{x}_t - \mathbf{x}_t^G = \underset{\sim}{H}_t\mathbf{x}_0 - \mathbf{c}_t - \underset{\sim}{G}_0(t)\mathbf{y}_0 + \int_0^t \underset{\sim}{L}_s^\tau\underset{\sim}{D}_s\,d\mathbf{w}_s - \int_0^t \underset{\sim}{G}_1(t)\underset{\sim}{G}_2(s)\underset{\sim}{F}_s\,d\mathbf{w}_s$$

and

$$\begin{aligned}
&E(\mathbf{x}_t - \mathbf{x}_t^G)(\mathbf{x}_t - \mathbf{x}_t^G)^\tau \\
&\quad = E(\underset{\sim}{H}_t\mathbf{x}_0 - \mathbf{c}_t - \underset{\sim}{G}_0(t)\mathbf{y}_0)(\underset{\sim}{H}_t\mathbf{x}_0 - \mathbf{c}_t - \underset{\sim}{G}_0(t)\mathbf{y}_0)^\tau \\
&\qquad + \int_0^t [\underset{\sim}{L}_s^\tau\underset{\sim}{D}_s - \underset{\sim}{G}_1(t)\underset{\sim}{G}_2(s)\underset{\sim}{F}_s][\underset{\sim}{L}_s^\tau\underset{\sim}{D}_s - \underset{\sim}{G}_1(t)\underset{\sim}{G}_2(s)\underset{\sim}{F}_s]^\tau\,ds,
\end{aligned} \tag{6.38}$$

because $[\mathbf{x}_0^\tau\mathbf{y}_0^\tau]^\tau$ and $\{\mathbf{w}_t\}$ are uncorrelated.

From (6.37) and (6.8), it follows that

$$\begin{aligned}
\underset{\sim}{P}_t &= \underset{\sim}{L}_t^\tau\underset{\sim}{P}_t\underset{\sim}{L}_t = \underset{\sim}{L}_0^\tau\underset{\sim}{R}_0\underset{\sim}{L}_0 + \int_0^t d(\underset{\sim}{L}_s^\tau\underset{\sim}{P}_s\underset{\sim}{L}_s) \\
&= \underset{\sim}{H}_t\underset{\sim}{R}_0\underset{\sim}{H}_t^\tau + \int_0^t \Big[(\underset{\sim}{G}_1(t)\underset{\sim}{G}_2(s)\underset{\sim}{C}_s - \underset{\sim}{L}_s^\tau\underset{\sim}{A}_s)\underset{\sim}{P}_s\underset{\sim}{L}_s \\
&\qquad + \underset{\sim}{L}_s^\tau\underset{\sim}{P}_s(\underset{\sim}{C}_s^\tau\underset{\sim}{G}_2^\tau(s)\underset{\sim}{G}_1^\tau(t) - \underset{\sim}{A}_s^\tau\underset{\sim}{L}_s) + \underset{\sim}{L}_s^\tau\big(\underset{\sim}{P}_s\underset{\sim}{A}_s^\tau + \underset{\sim}{A}_s\underset{\sim}{P}_s + \underset{\sim}{D}_s\underset{\sim}{D}_s^\tau \\
&\qquad - (\underset{\sim}{D}_s\underset{\sim}{F}_s^\tau + \underset{\sim}{P}_s\underset{\sim}{C}_s^\tau)(\underset{\sim}{F}_s\underset{\sim}{F}_s^\tau)^{-1}(\underset{\sim}{D}_s\underset{\sim}{F}_s^\tau + \underset{\sim}{P}_s\underset{\sim}{C}_s^\tau)^\tau\underset{\sim}{L}_s\Big]\,ds \\
&= \underset{\sim}{L}_0^\tau\underset{\sim}{R}_0\underset{\sim}{L}_0 - \int_0^t \Big[\underset{\sim}{G}_1(t)\underset{\sim}{G}_2(s) - \underset{\sim}{L}_s^\tau(\underset{\sim}{D}_s\underset{\sim}{F}_s^\tau + \underset{\sim}{P}_s\underset{\sim}{C}_s^\tau)(\underset{\sim}{F}_s\underset{\sim}{F}_s^\tau)^{-1}\Big]\underset{\sim}{F}_s\underset{\sim}{F}_s^\tau \\
&\qquad \cdot\Big[\underset{\sim}{G}_1(t)\underset{\sim}{G}_2(s) - \underset{\sim}{L}_s^\tau(\underset{\sim}{D}_s\underset{\sim}{F}_s^\tau + \underset{\sim}{P}_s\underset{\sim}{C}_s^\tau)(\underset{\sim}{F}_s\underset{\sim}{F}_s^\tau)^{-1}\Big]^\tau\,ds \\
&\qquad + \int_0^t [\underset{\sim}{L}_s^\tau\underset{\sim}{D}_s - \underset{\sim}{G}_1(t)\underset{\sim}{G}_2(s)\underset{\sim}{F}_s][\underset{\sim}{L}_s^\tau\underset{\sim}{D}_s - \underset{\sim}{G}_1(t)\underset{\sim}{G}_2(s)\underset{\sim}{F}_s]^\tau\,ds.
\end{aligned} \tag{6.39}$$

By (6.4) and (1.53), it is easy to see that

$$E(\mathbf{x}_0 - \hat{\mathbf{x}}_0)\mathbf{y}_0^\tau = \underset{\sim}{\mathbf{0}}, \qquad E(\mathbf{x}_0 - \hat{\mathbf{x}}_0)\hat{\mathbf{x}}_0^\tau = \underset{\sim}{\mathbf{0}}.$$

Hence

$$\begin{aligned}
&E(\underset{\sim}{\mathbf{H}}_t\mathbf{x}_0 - \mathbf{c}_t - \underset{\sim}{\mathbf{G}}_0(t)\mathbf{y}_0)(\underset{\sim}{\mathbf{H}}_t\mathbf{x}_0 - \mathbf{c}_t - \underset{\sim}{\mathbf{G}}_0(t)\mathbf{y}_0)^\tau \\
&\quad = E[\underset{\sim}{\mathbf{H}}_t(\mathbf{x}_0 - \hat{\mathbf{x}}_0) + \underset{\sim}{\mathbf{H}}_t\hat{\mathbf{x}}_0 - \mathbf{c}_t - \underset{\sim}{\mathbf{G}}_0(t)\mathbf{y}_0] \\
&\qquad \times [\underset{\sim}{\mathbf{H}}_t(\mathbf{x}_0 - \hat{\mathbf{x}}_0) + \underset{\sim}{\mathbf{H}}_t\hat{\mathbf{x}}_0 - \mathbf{c}_t - \underset{\sim}{\mathbf{G}}_0(t)\mathbf{y}_0]^\tau \\
&\quad = \underset{\sim}{\mathbf{H}}_t\underset{\sim}{\mathbf{R}}_0\underset{\sim}{\mathbf{H}}_t^\tau + E(\underset{\sim}{\mathbf{H}}_t\hat{\mathbf{x}}_0 - \mathbf{c}_t - \underset{\sim}{\mathbf{G}}_0(t)\mathbf{y}_0)(\underset{\sim}{\mathbf{H}}_t\hat{\mathbf{x}}_0 - \mathbf{c}_t - \underset{\sim}{\mathbf{G}}_0(t)\mathbf{y}_0)^\tau.
\end{aligned} \tag{6.40}$$

Combining (6.38)–(6.40), we obtain

$$\begin{aligned}
E(\mathbf{x}_t - \mathbf{x}_t^G)(\mathbf{x}_t - \mathbf{x}_t^G)^\tau &= E(\underset{\sim}{\mathbf{H}}_t\hat{\mathbf{x}}_0 - \mathbf{c}_t - \underset{\sim}{\mathbf{G}}_0(t)\mathbf{y}_0)(\underset{\sim}{\mathbf{H}}_t\hat{\mathbf{x}}_0 - \mathbf{c}_t - \underset{\sim}{\mathbf{G}}_0(t)\mathbf{y}_0)^\tau \\
&\quad + \underset{\sim}{\mathbf{P}}_t + \int_0^t [\underset{\sim}{\mathbf{G}}_1(t)\underset{\sim}{\mathbf{G}}_2(s) - \underset{\sim}{\mathbf{L}}_s^\tau(\underset{\sim}{\mathbf{D}}_s\underset{\sim}{\mathbf{F}}_s^\tau + \underset{\sim}{\mathbf{P}}_s\underset{\sim}{\mathbf{C}}_s^\tau)(\underset{\sim}{\mathbf{F}}_s\underset{\sim}{\mathbf{F}}_s^\tau)^{-1}]\underset{\sim}{\mathbf{F}}_s\underset{\sim}{\mathbf{F}}_s^\tau \\
&\quad \cdot [\underset{\sim}{\mathbf{G}}_1(t)\underset{\sim}{\mathbf{G}}_2(s) - \underset{\sim}{\mathbf{L}}_2^\tau(\underset{\sim}{\mathbf{D}}_s\underset{\sim}{\mathbf{F}}_s^\tau + \underset{\sim}{\mathbf{P}}_s\underset{\sim}{\mathbf{C}}_s^\tau)(\underset{\sim}{\mathbf{F}}_s\underset{\sim}{\mathbf{F}}_s^\tau)^{-1}]^\tau\, ds.
\end{aligned} \tag{6.41}$$

Denote

$$\underset{\sim}{\mathbf{K}}_t = (\underset{\sim}{\mathbf{D}}_t\underset{\sim}{\mathbf{F}}_t^\tau + \underset{\sim}{\mathbf{P}}_t\underset{\sim}{\mathbf{C}}_t^\tau)(\underset{\sim}{\mathbf{F}}_t\underset{\sim}{\mathbf{F}}_t^\tau)^{-1}. \tag{6.42}$$

Notice $\underset{\sim}{\mathbf{P}}_t$ is independent of $\mathbf{c}_t$, $\underset{\sim}{\mathbf{G}}_0(t)$, $\underset{\sim}{\mathbf{G}}_1(t)$ and $\underset{\sim}{\mathbf{G}}_2(t)$. Hence in order for (6.41) to reach its minimum the necessary and sufficient conditions are

$$\underset{\sim}{\mathbf{H}}_t\hat{\mathbf{x}}_0 - \mathbf{c}_t - \underset{\sim}{\mathbf{G}}_0(t)\mathbf{y}_0 = \mathbf{0}, \tag{6.43}$$

$$\underset{\sim}{\mathbf{G}}_1(t)\underset{\sim}{\mathbf{G}}_2(s) = \underset{\sim}{\mathbf{L}}_s^\tau\underset{\sim}{\mathbf{K}}_s, \tag{6.44}$$

and if these conditions are fulfilled, then $\underset{\sim}{\mathbf{P}}_t$ is the estimation error covariance matrix.

Further, from (6.44) and (6.37), we find that $G_1(t)G_2(s)$ is optimal if and only if L_s^τ satisfies the following equation

$$\frac{d}{ds}L_s^\tau = L_s^\tau[-A_s + K_s C_s], \qquad L_t^\tau = I. \tag{6.45}$$

Let $\Psi_{t,s}$ be the fundamental solution matrix to the equation

$$\frac{d}{dt}\Psi_{t,s} = (A_t - K_t C_t)\Psi_{t,s}, \qquad \Psi_{s,s} = I, \tag{6.46}$$

which is completely determined by A_t, D_t, F_t, C_t, and P_t given by (6.8). Clearly, for fixed t, $\Psi_{t,s}$, as a function of s, satisfies (6.45), and by the uniqueness of the solution, we have

$$L_s^\tau = \Psi_{t,s}. \tag{6.47}$$

Thus from (6.44), we conclude that the optimal gain matrix $G_1^0(t)G_2^0(s)$ is expressed by

$$G_1^0(t)G_2^0(s) = \Psi_{t,s}K_s \tag{6.48}$$

or

$$G_1^0(t) = \Psi_{t,0}, \qquad G_2^0(s) = \Psi_{0,s}(D_s F_s^\tau + P_s C_s^\tau)(F_s F_s^\tau)^{-1}.$$

Because $\|\Psi_{0,s}\|$ and $\|P_s\|$ are uniformly bounded on $[0, T]$, by (6.12) and $\int_0^T \|C_s^\tau(F_s F_s^\tau)^{-1}C_s\|\, ds < \infty$, it is easy to verify that

$$\int_0^T \|G_2^0(s)F_s\|^2\, ds < \infty, \qquad \int_0^T \|G_2^0(s)C_s\|\, ds < \infty.$$

Hence $G_2^0(s)$ actually meets the requirements in (6.7).

Having obtained L_s from (6.36), we know that

$$H_t = \Psi_{t,0}.$$

Then to satisfy (6.43) we should take

$$c_t = c_t^0 \triangleq \Psi_{t,0}\left(E\mathbf{x}_0 - R_{\mathbf{x}_0\mathbf{y}_0}R_{\mathbf{y}_0}^{+}E\mathbf{y}_0\right),$$

$$G_0(t) = G_0^0(t) \triangleq \Psi_{t,0}R_{\mathbf{x}_0\mathbf{y}_0}R_{\mathbf{y}_0}^{+}.$$

Substituting these optimal gains into the formula for $\mathbf{x}_t^G$, we finally find that the LUMVE is

$$\hat{\mathbf{x}}_t = \underset{\sim}{\boldsymbol{\Psi}}_{t,0}\Big(E\mathbf{x}_0 - \underset{\sim}{\mathbf{R}}_{\mathbf{x}_0\mathbf{y}_0}\underset{\sim}{\mathbf{R}}_{\mathbf{y}_0}^{+}E\mathbf{y}_0 + \underset{\sim}{\mathbf{R}}_{\mathbf{x}_0\mathbf{y}_0}\underset{\sim}{\mathbf{R}}_{\mathbf{y}_0}^{+}\mathbf{y}_0\Big)$$

$$+\underset{\sim}{\boldsymbol{\Psi}}_{t,0}\int_0^t \underset{\sim}{\boldsymbol{\Psi}}_{0,s}(\underset{\sim}{\mathbf{D}}_s\underset{\sim}{\mathbf{F}}_s^T + \underset{\sim}{\mathbf{P}}_s\underset{\sim}{\mathbf{C}}_s^T)(\underset{\sim}{\mathbf{F}}_s\underset{\sim}{\mathbf{F}}_s^T)^{-1}\,d\mathbf{y}_s$$

$$= \underset{\sim}{\boldsymbol{\Psi}}_{t,0}\hat{\mathbf{x}}_0 + \underset{\sim}{\boldsymbol{\Psi}}_{t,0}\int_0^t \underset{\sim}{\boldsymbol{\Psi}}_{0,s}(\underset{\sim}{\mathbf{D}}_s\underset{\sim}{\mathbf{F}}_s^T + \underset{\sim}{\mathbf{P}}_s\underset{\sim}{\mathbf{C}}_s^T)(\underset{\sim}{\mathbf{F}}_s\underset{\sim}{\mathbf{F}}_s^T)^{-1}\,d\mathbf{y}_s$$

for which the stochastic differential is

$$d\hat{\mathbf{x}}_t = (\underset{\sim}{\mathbf{A}}_t - \underset{\sim}{\mathbf{K}}_t\underset{\sim}{\mathbf{C}}_t)\hat{\mathbf{x}}_t\,dt + \underset{\sim}{\mathbf{K}}_t\,d\mathbf{y}_t$$

with $\hat{\mathbf{x}}_0$ given by (6.4). □

Denote

$$\mathscr{F}_t^{\mathbf{y}} = \sigma\{\mathbf{y}_\lambda, 0 \le \lambda \le t\}, \qquad \mathscr{F}_t^{\mathbf{x}_s,\mathbf{y}} = \sigma\{\mathbf{x}_s, \mathbf{y}_\lambda, 0 \le \lambda \le t\}. \tag{6.49}$$

Remark 1. If $\{\mathbf{w}_t\}$ is a Wiener process, then the estimate $\hat{\mathbf{x}}_t$ given by (6.31) is the minimum variance estimate. That is,

$$\hat{\mathbf{x}}_t = E\left(\frac{\mathbf{x}_t}{\mathscr{F}_t^{\mathbf{y}}}\right),$$

while in case $\{\mathbf{w}_t\}$ is a generalized Wiener process, then $\hat{\mathbf{x}}_t$ is a LUMVE of $\mathbf{x}_t$ based on $\{\mathbf{y}_\lambda, 0 \le \lambda \le t\}$. This fact will remain true for the subsequent theorems on interpolation and prediction.

Remark 2. When $\{\mathbf{w}_t\}$ is a Wiener process, then $(\overline{\mathbf{w}}_t, \mathscr{F}_t^{\mathbf{y}})$ is a Wiener process and

$$\mathscr{F}_t^{\overline{\mathbf{w}}} = \mathscr{F}_t^{\mathbf{y}},$$

where

$$\overline{\mathbf{w}}_t = \int_0^t (\underset{\sim}{\mathbf{F}}_s\underset{\sim}{\mathbf{F}}_s^T)^{-1}(d\mathbf{y}_s - \underset{\sim}{\mathbf{C}}_s\hat{\mathbf{x}}_s\,ds).$$

This means that the information contained in $\{\overline{\mathbf{w}}_\lambda, 0 \le \lambda \le t\}$ equals that in $\{\mathbf{y}_\lambda, 0 \le \lambda \le t\}$. In this case, $\{\overline{\mathbf{w}}_t\}$ is called the innovations process of the process $\mathbf{y}$.

6.3. INTERPOLATION AND PREDICTION

Denote by $\hat{\mathbf{x}}_s(t)$ and $\underset{\sim}{\mathbf{P}}_s(t)$, respectively, the LUMVE of $\mathbf{x}_s$ based on $\{\mathbf{y}_\lambda, 0 \le \lambda \le t\}$ and its estimation error covariance matrix when $s \le t$. For s fixed, the equation satisfied by $\hat{\mathbf{x}}_s(t)$ is called forward; conversely, if t is fixed, it is called backward.

Assume $\underset{\sim}{\mathbf{P}}_t^s$ satisfies the Riccati equation

$$\frac{d}{dt}\underset{\sim}{\mathbf{P}}_t^s = \underset{\sim}{\mathbf{A}}_t\underset{\sim}{\mathbf{P}}_t^s + \underset{\sim}{\mathbf{P}}_t^s\underset{\sim}{\mathbf{A}}_t^T + \underset{\sim}{\mathbf{D}}_t\underset{\sim}{\mathbf{D}}_t^T - (\underset{\sim}{\mathbf{D}}_t\underset{\sim}{\mathbf{F}}_t^T + \underset{\sim}{\mathbf{P}}_t^s\underset{\sim}{\mathbf{C}}_t^T)(\underset{\sim}{\mathbf{F}}_t\underset{\sim}{\mathbf{F}}_t^T)^{-1}(\underset{\sim}{\mathbf{D}}_t\underset{\sim}{\mathbf{F}}_t^T + \underset{\sim}{\mathbf{P}}_t^s\underset{\sim}{\mathbf{C}}_t^T)^T,$$

$$\underset{\sim}{\mathbf{P}}_s^s = \underset{\sim}{\mathbf{0}} \tag{6.50}$$

which has a unique positive definite solution by Lemma 6.1. Let $\underset{\sim}{\mathbf{G}}_1(t)$ and $\underset{\sim}{\mathbf{G}}_2(t)$ meet the conditions required in (6.7) and denote

$$\underset{\sim}{\mathbf{K}}_t^s = (\underset{\sim}{\mathbf{D}}_t\underset{\sim}{\mathbf{F}}_t^T + \underset{\sim}{\mathbf{P}}_t^s\underset{\sim}{\mathbf{C}}_t^T)(\underset{\sim}{\mathbf{F}}_t\underset{\sim}{\mathbf{F}}_t^T)^{-1}, \qquad t \ge s, \tag{6.51}$$

$$\frac{d}{dt}\underset{\sim}{\mathbf{\Psi}}_{t,\lambda}^s = (\underset{\sim}{\mathbf{A}}_t - \underset{\sim}{\mathbf{K}}_t^s\underset{\sim}{\mathbf{C}}_t)\underset{\sim}{\mathbf{\Psi}}_{t,\lambda}^s, \qquad \underset{\sim}{\mathbf{\Psi}}_{\lambda,\lambda}^s = \underset{\sim}{\mathbf{I}}, \tag{6.52}$$

$$\underset{\sim}{\mathbf{H}}_t^s = \underset{\sim}{\mathbf{\Phi}}_{t,s} - \underset{\sim}{\mathbf{G}}_1(t)\int_s^t \underset{\sim}{\mathbf{G}}_2(\lambda)\underset{\sim}{\mathbf{C}}_\lambda\underset{\sim}{\mathbf{\Phi}}_{\lambda,s}\,d\lambda, \tag{6.53}$$

$$\underset{\sim}{\mathbf{L}}_\mu^{sT} = \left(\underset{\sim}{\mathbf{G}}_1(t)\int_s^\mu \underset{\sim}{\mathbf{G}}_2(\lambda)\underset{\sim}{\mathbf{C}}_\lambda\underset{\sim}{\mathbf{\Phi}}_{\lambda,s}\,d\lambda + \underset{\sim}{\mathbf{H}}_t^s\right)\underset{\sim}{\mathbf{\Phi}}_{s,\mu}, \tag{6.54}$$

where $\underset{\sim}{\mathbf{\Phi}}_{t,s}$ is defined by (6.32).

Clearly,

$$\frac{d}{d\mu}\underset{\sim}{\mathbf{L}}_\mu^{sT} = -\underset{\sim}{\mathbf{L}}_\mu^{sT}\underset{\sim}{\mathbf{A}}_\mu + \underset{\sim}{\mathbf{G}}_1(t)\underset{\sim}{\mathbf{G}}_2(\mu)\underset{\sim}{\mathbf{C}}_\mu \tag{6.55}$$

$$\underset{\sim}{\mathbf{L}}_t^{sT} = \underset{\sim}{\mathbf{I}}, \qquad \underset{\sim}{\mathbf{L}}_s^{sT} = \underset{\sim}{\mathbf{H}}_t^s. \tag{6.56}$$

We first seek the LUMVE $\hat{\mathbf{x}}_t^s$ of $\mathbf{x}_t$ based on $\mathbf{x}_s$ and $(\mathbf{y}_\lambda, 0 \le \lambda \le t)$. As already mentioned, if $\{\mathbf{w}_t\}$ is a Wiener process, then

$$\hat{\mathbf{x}}_t^s = E\left(\frac{\mathbf{x}_t}{\mathscr{F}_t^{\mathbf{x}_s,\mathbf{y}}}\right). \tag{6.57}$$

Lemma 6.2. *Under the conditions of Lemma* 6.1,

$$\hat{\mathbf{x}}_t^s = \underset{\sim}{\mathbf{\Psi}}_{t,s}^s\mathbf{x}_s + \underset{\sim}{\mathbf{\Psi}}_{t,s}^s\int_s^t \underset{\sim}{\mathbf{\Psi}}_{s\lambda}^s\underset{\sim}{\mathbf{K}}_\lambda^s\,d\mathbf{y}_\lambda, \tag{6.58}$$

and its estimation error covariance matrix is $\underset{\sim}{\mathbf{P}}_t^s$ *given by* (6.50).

Proof. Denote again by $\mathbf{x}_t^G$ the linear estimate for $\mathbf{x}_t$ given by

$$\mathbf{x}_t^G = \mathbf{c}_t + \underset{\sim}{\mathbf{G}}_0(t)\mathbf{x}_s + \underset{\sim}{\mathbf{G}}_1(t)\int_0^t \underset{\sim}{\mathbf{G}}_2(\lambda)\, d\mathbf{y}_\lambda. \tag{6.59}$$

We hope to define $\mathbf{c}_t^0$, $\underset{\sim}{\mathbf{G}}_0^0(t)$, $\underset{\sim}{\mathbf{G}}_1^0(t)$, and $\underset{\sim}{\mathbf{G}}_2^0(t)$ so that

$$E\left(\mathbf{x}_t - \mathbf{c}_t^0 - \underset{\sim}{\mathbf{G}}_0^0(t)\mathbf{x}_s - \underset{\sim}{\mathbf{G}}_1^0(t)\int_0^t \underset{\sim}{\mathbf{G}}_2^0(\lambda)\, d\mathbf{y}_\lambda\right)$$

$$\times\left(\mathbf{x}_t - \mathbf{c}_t^0 - \underset{\sim}{\mathbf{G}}_0^0(t)\mathbf{x}_s - \underset{\sim}{\mathbf{G}}_1^0(t)\int_0^t \underset{\sim}{\mathbf{G}}_2^0(\lambda)\, d\mathbf{y}_\lambda\right)^\tau$$

$$= \min_{\mathbf{c}_t,\, \underset{\sim}{\mathbf{G}}_0(t),\, \underset{\sim}{\mathbf{G}}_1(t),\, \underset{\sim}{\mathbf{G}}_2(t)} E\left(\mathbf{x}_t - \mathbf{x}_t^G\right)\left(\mathbf{x}_t - \mathbf{x}_t^G\right)^\tau.$$

Since

$$\mathbf{x}_t^G = \mathbf{c}_t + \underset{\sim}{\mathbf{G}}_0(t)\mathbf{x}_s + \underset{\sim}{\mathbf{G}}_1(t)\int_0^s \underset{\sim}{\mathbf{G}}_2(\lambda)\, d\mathbf{y}_\lambda + \underset{\sim}{\mathbf{G}}_1(t)\int_s^t \underset{\sim}{\mathbf{G}}_2(\lambda)\mathbf{C}_\lambda \underset{\sim}{\mathbf{\Phi}}_{\lambda,s}\, d\lambda\, \mathbf{x}_s$$

$$+\underset{\sim}{\mathbf{G}}_1(t)\int_s^t\int_\mu^t \underset{\sim}{\mathbf{G}}_2(\lambda)\underset{\sim}{\mathbf{C}}_\lambda \underset{\sim}{\mathbf{\Phi}}_{\lambda,s}\, d\lambda\, \underset{\sim}{\mathbf{\Phi}}_{s,\mu} \underset{\sim}{\mathbf{D}}_\mu\, d\mathbf{w}_\mu + \underset{\sim}{\mathbf{G}}_1(t)\int_s^t \underset{\sim}{\mathbf{G}}_2(\lambda)\underset{\sim}{\mathbf{F}}_\lambda\, d\mathbf{w}_\lambda.$$

As in (6.35), we obtain

$$\mathbf{x}_t - \mathbf{x}_t^G = \left(\underset{\sim}{\mathbf{H}}_t^s - \underset{\sim}{\mathbf{G}}_0(t)\right)\mathbf{x}_s - \mathbf{c}_t + \int_s^t \left(\underset{\sim}{\mathbf{L}}_\lambda^{s\tau}\underset{\sim}{\mathbf{D}}_\lambda - \underset{\sim}{\mathbf{G}}_1(t)\underset{\sim}{\mathbf{G}}_2(\lambda)\underset{\sim}{\mathbf{F}}_\lambda\right) d\mathbf{w}_\lambda$$

$$-\underset{\sim}{\mathbf{G}}_1(t)\int_0^s \underset{\sim}{\mathbf{G}}_2(\lambda)\, d\mathbf{y}_\lambda.$$

Obviously, $\mathbf{x}_s$ and $\mathbf{y}_\lambda$, $0 \le \lambda \le s$, are uncorrelated with $\{\mathbf{w}_\lambda,\ s \le \lambda \le t\}$ because they depend linearly on $\{\mathbf{w}_\lambda,\ 0 \le \lambda \le s\}$ only. Thus we have

$$E\left(\mathbf{x}_t - \mathbf{x}_t^G\right)\left(\mathbf{x}_t - \mathbf{x}_t^G\right)^\tau$$

$$= E\left[\left(\underset{\sim}{\mathbf{H}}_t^s - \underset{\sim}{\mathbf{G}}_0(t)\right)\mathbf{x}_s - \mathbf{c}_t - \underset{\sim}{\mathbf{G}}_1(t)\int_0^s \underset{\sim}{\mathbf{G}}_2(\lambda)\, d\mathbf{y}_\lambda\right]$$

$$\times\left[\left(\mathbf{H}_t^s - \underset{\sim}{\mathbf{G}}_0(t)\right)\mathbf{x}_s - \mathbf{c}_t - \underset{\sim}{\mathbf{G}}_1(t)\int_0^s \underset{\sim}{\mathbf{G}}_2(\lambda)\, d\mathbf{y}_\lambda\right]^\tau$$

$$+\int_s^t \left(\underset{\sim}{\mathbf{L}}_\lambda^{s\tau}\underset{\sim}{\mathbf{D}}_\lambda - \underset{\sim}{\mathbf{G}}_1(t)\underset{\sim}{\mathbf{G}}_2(\lambda)\underset{\sim}{\mathbf{F}}_\lambda\right)\left(\underset{\sim}{\mathbf{L}}_\lambda^{s\tau}\underset{\sim}{\mathbf{D}}_\lambda - \underset{\sim}{\mathbf{G}}_1(t)\underset{\sim}{\mathbf{G}}_2(\lambda)\underset{\sim}{\mathbf{F}}_\lambda\right)^\tau d\lambda. \tag{6.60}$$

Using the same procedure as with (6.39), we obtain

$$\underset{\sim}{\mathbf{P}}_t^s = \underset{\sim}{\mathbf{L}}_t^{sT}\underset{\sim}{\mathbf{P}}_t^s\underset{\sim}{\mathbf{L}}_t^s = \int_s^t d(\underset{\sim}{\mathbf{L}}_\lambda^{sT}\underset{\sim}{\mathbf{P}}_\lambda^s\underset{\sim}{\mathbf{L}}_\lambda^s)$$

$$= \int_s^t [\underset{\sim}{\mathbf{L}}_\lambda^{sT}\underset{\sim}{\mathbf{D}}_\lambda - \underset{\sim}{\mathbf{G}}_1(t)\underset{\sim}{\mathbf{G}}_2(\lambda)\underset{\sim}{\mathbf{F}}_\lambda][\underset{\sim}{\mathbf{L}}_\lambda^{sT}\underset{\sim}{\mathbf{D}}_\lambda - \underset{\sim}{\mathbf{G}}_1(t)\underset{\sim}{\mathbf{G}}_2(\lambda)\underset{\sim}{\mathbf{F}}_\lambda]\, d\lambda$$

$$- \int_s^t (\underset{\sim}{\mathbf{G}}_1(t)\underset{\sim}{\mathbf{G}}_2(\lambda) - \underset{\sim}{\mathbf{L}}_\lambda^{sT}\underset{\sim}{\mathbf{K}}_\lambda^s)\underset{\sim}{\mathbf{F}}_\lambda\underset{\sim}{\mathbf{F}}_\lambda^T(\underset{\sim}{\mathbf{G}}_1(t)\underset{\sim}{\mathbf{G}}_2(\lambda) - \underset{\sim}{\mathbf{L}}_\lambda^{sT}\underset{\sim}{\mathbf{K}}_\lambda^s)^T\, d\lambda.$$

Hence

$$E(\mathbf{x}_t - \mathbf{x}_t^G)(\mathbf{x}_t - \mathbf{x}_t^G)^T =$$

$$\underset{\sim}{\mathbf{P}}_t^s + E\left[(\underset{\sim}{\mathbf{H}}_t^s - \underset{\sim}{\mathbf{G}}_0(t))\mathbf{x}_s - \mathbf{c}_t - \underset{\sim}{\mathbf{G}}_1(t)\int_0^s \underset{\sim}{\mathbf{G}}_2(\lambda)\, d\mathbf{y}_\lambda\right]$$

$$\cdot\left[(\underset{\sim}{\mathbf{H}}_t^s - \underset{\sim}{\mathbf{G}}_0(t))\mathbf{x}_s - \mathbf{c}_t - \underset{\sim}{\mathbf{G}}_1(t)\int_0^s \underset{\sim}{\mathbf{G}}_2(\lambda)\, d\mathbf{y}_\lambda\right]^T$$

$$+ \int_s^t [\underset{\sim}{\mathbf{G}}_1(t)\underset{\sim}{\mathbf{G}}_2(\lambda) - \underset{\sim}{\mathbf{L}}_\lambda^{sT}\underset{\sim}{\mathbf{K}}_\lambda^s]\underset{\sim}{\mathbf{F}}_\lambda\underset{\sim}{\mathbf{F}}_\lambda^T[\underset{\sim}{\mathbf{G}}_1(t)\underset{\sim}{\mathbf{G}}_2(\lambda) - \underset{\sim}{\mathbf{L}}_\lambda^{sT}\underset{\sim}{\mathbf{K}}_\lambda^s]^T\, d\lambda. \tag{6.61}$$

From here, it is clear that (6.61) reaches its minimum if and only if

$$\underset{\sim}{\mathbf{G}}_1(t)\underset{\sim}{\mathbf{G}}_2(\lambda) - \underset{\sim}{\mathbf{L}}_\lambda^{sT}\underset{\sim}{\mathbf{K}}_\lambda^s = \mathbf{0}, \qquad \forall\lambda \in [s, t] \tag{6.62}$$

$$(\underset{\sim}{\mathbf{H}}_t^s - \underset{\sim}{\mathbf{G}}_0(t))\mathbf{x}_s - \mathbf{c}_t - \underset{\sim}{\mathbf{G}}_1(t)\int_0^s \underset{\sim}{\mathbf{G}}_2(\lambda)\, d\mathbf{y}_\lambda = \mathbf{0}, \tag{6.63}$$

and if these conditions hold, then the estimation error covariance matrix is $\underset{\sim}{\mathbf{P}}_t^s$.

From (6.55) and (6.62), it is clear that $\underset{\sim}{\mathbf{G}}_1(t)\underset{\sim}{\mathbf{G}}_2(\lambda)$, $\lambda \in [s, t]$ is optimal if and only if $\underset{\sim}{\mathbf{L}}_\lambda^{sT}$ satisfies equation:

$$\frac{d}{d\mu}\underset{\sim}{\mathbf{L}}_\mu^{sT} = -\underset{\sim}{\mathbf{L}}_\mu^{sT}\underset{\sim}{\mathbf{A}}_\mu + \underset{\sim}{\mathbf{L}}_\mu^{sT}\underset{\sim}{\mathbf{K}}_\mu^s\underset{\sim}{\mathbf{C}}_\mu, \qquad \underset{\sim}{\mathbf{L}}_t^{sT} = \underset{\sim}{\mathbf{I}}.$$

But by (6.52), we have

$$\underset{\sim}{\mathbf{L}}_\mu^{sT} = \underset{\sim}{\mathbf{\Psi}}_{t,\mu}^s.$$

Hence for $\lambda \in [s, t]$, the optimal gain is

$$\underset{\sim}{G}_1^0(t)\underset{\sim}{G}_2^0(\lambda) = \underset{\sim}{\Psi}_{t,\lambda}^s \underset{\sim}{K}_\lambda^s. \tag{6.64}$$

Further, (6.63) will be fulfilled if we take

$$\mathbf{c}_t^0 \equiv \mathbf{0}, \qquad \underset{\sim}{G}_2^0(\lambda) = \mathbf{0}, \qquad \lambda \in [0, s], \qquad \underset{\sim}{G}_0^0(t) = \underset{\sim}{H}_t^s.$$

Noticing that

$$\underset{\sim}{H}_t^s = \underset{\sim}{L}_s^{s\tau} = \underset{\sim}{\Psi}_{t,s}^s$$

together with (6.64) implies (6.58). Thus we have

$$\underset{\sim}{G}_1^0(t) = \underset{\sim}{\Psi}_{t,0}^s, \qquad \underset{\sim}{G}_2^0(t) = \begin{cases} \underset{\sim}{0}, & 0 \le t \le s \\ \underset{\sim}{\Psi}_{0,t}^s \underset{\sim}{K}_t^s, & s \le \lambda \le T, \end{cases}$$

and just as in Theorem 6.1, it is easy to see

$$\int_0^T \|\underset{\sim}{G}_2^0(t)\underset{\sim}{C}_t\|\, dt < \infty, \qquad \int_0^T \|\underset{\sim}{G}_2^0(t)\underset{\sim}{F}_t\|^2\, dt < \infty. \qquad \square$$

By this lemma we obtain the following forward interpolation equation.

Theorem 6.2. *Under conditions of Lemma* 6.1, *for fixed* $s \in [0, t]$, $\hat{\mathbf{x}}_s(t)$, *and its estimation error covariance matrix* $\underset{\sim}{P}_s(t)$ *satisfy*

$$\hat{\mathbf{x}}_s(t) = \hat{\mathbf{x}}_s + \int_s^t \underset{\sim}{P}_s(\lambda)\underset{\sim}{\Psi}_{\lambda,s}^{s\tau}\underset{\sim}{C}_\lambda^\tau(\underset{\sim}{F}_\lambda\underset{\sim}{F}_\lambda^\tau)^{-1}(d\mathbf{y}_\lambda - \underset{\sim}{C}_\lambda\hat{\mathbf{x}}_\lambda\, d\lambda) \tag{6.65}$$

$$\underset{\sim}{P}_s(t) = \left[\underset{\sim}{I} + \underset{\sim}{P}_s\int_s^t \underset{\sim}{\Psi}_{\lambda,s}^{s\tau}\underset{\sim}{C}_\lambda^\tau(\underset{\sim}{F}_\lambda\underset{\sim}{F}_\lambda^\tau)^{-1}\underset{\sim}{C}_\lambda\underset{\sim}{\Psi}_{\lambda,s}^s\, d\lambda\right]^{-1}\underset{\sim}{P}_s, \qquad t \ge s. \tag{6.66}$$

Proof. Since we are only concerned with linear estimation, without loss of generality we can assume $\{\mathbf{w}_t\}$ is a Wiener process. This assumption will simplify the proof. Of course, we could also proceed without this assumption by the method used earlier.

By (6.57) and by taking $E(\cdot/\mathscr{F}_t^y)$ in (6.58), we have

$$\hat{\mathbf{x}}_t = \underset{\sim}{\Psi}_{t,s}^s\hat{\mathbf{x}}_s(t) + \underset{\sim}{\Psi}_{t,s}^s\int_s^t \underset{\sim}{\Psi}_{s,\lambda}^s\underset{\sim}{K}_\lambda^s\, d\mathbf{y}_\lambda, \qquad t \ge s, \tag{6.67}$$

and by (6.58) and (6.67),

$$\hat{\mathbf{x}}_t^s - \hat{\mathbf{x}}_t = \underset{\sim}{\boldsymbol{\Psi}}_{t,s}^s(\mathbf{x}_s - \hat{\mathbf{x}}_s(t)). \qquad (6.68)$$

Applying Ito's formula to (6.65) and by (6.31), (6.42), and (6.51), this leads to

$$\begin{aligned} d\hat{\mathbf{x}}_s(t) &= d\left[\underset{\sim}{\boldsymbol{\Psi}}_{s,t}^s \hat{\mathbf{x}}_t - \int_s^t \underset{\sim}{\boldsymbol{\Psi}}_{s,\lambda}^s \underset{\sim}{\mathbf{K}}_\lambda^s \, d\mathbf{y}_\lambda\right] \\ &= -\underset{\sim}{\boldsymbol{\Psi}}_{s,t}^s(\underset{\sim}{\mathbf{A}}_t - \underset{\sim}{\mathbf{K}}_t^s \underset{\sim}{\mathbf{C}}_t)\hat{\mathbf{x}}_t \, dt - \underset{\sim}{\boldsymbol{\Psi}}_{s,t}^s \underset{\sim}{\mathbf{K}}_t^s \, d\mathbf{y}_t \\ &\quad + \underset{\sim}{\boldsymbol{\Psi}}_{s,t}^s[\underset{\sim}{\mathbf{A}}_t \hat{\mathbf{x}}_t \, dt + \underset{\sim}{\mathbf{K}}_t(d\mathbf{y}_t - \underset{\sim}{\mathbf{C}}_t \hat{\mathbf{x}}_t) \, dt] \\ &= \underset{\sim}{\boldsymbol{\Psi}}_{s,t}^s[(\underset{\sim}{\mathbf{K}}_t - \underset{\sim}{\mathbf{K}}_t^s)(d\mathbf{y}_t - \underset{\sim}{\mathbf{C}}_t \hat{\mathbf{x}}_t) \, dt] \\ &= \underset{\sim}{\boldsymbol{\Psi}}_{s,t}^s(\underset{\sim}{\mathbf{P}}_t - \underset{\sim}{\mathbf{P}}_t^s)\underset{\sim}{\mathbf{C}}_t^\tau(\underset{\sim}{\mathbf{F}}_t \underset{\sim}{\mathbf{F}}_t^\tau)^{-1}(d\mathbf{y}_t - \underset{\sim}{\mathbf{C}}_t \hat{\mathbf{x}}_t \, dt) \end{aligned}$$

or

$$\hat{\mathbf{x}}_s(t) = \hat{\mathbf{x}}_s + \int_s^t \underset{\sim}{\boldsymbol{\Psi}}_{s,\lambda}^s(\underset{\sim}{\mathbf{P}}_\lambda - \underset{\sim}{\mathbf{P}}_\lambda^s)\mathbf{C}_\lambda^\tau(\underset{\sim}{\mathbf{F}}_\lambda \underset{\sim}{\mathbf{F}}_\lambda^\tau)^{-1}(d\mathbf{y}_\lambda - \underset{\sim}{\mathbf{C}}_\lambda \hat{\mathbf{x}}_\lambda \, d\lambda). \qquad (6.69)$$

Noticing that

$$E\left[\frac{(\mathbf{x}_t - \hat{\mathbf{x}}_t^s)(\hat{\mathbf{x}}_t^s - \hat{\mathbf{x}}_t)^\tau}{\mathscr{F}_t^{\mathbf{x}_s\mathbf{y}}}\right] = E\left[\frac{(\mathbf{x}_t - \hat{\mathbf{x}}_t^s)}{\mathscr{F}_t^{\mathbf{x}_s\mathbf{y}}}\right][\hat{\mathbf{x}}_t^s - \hat{\mathbf{x}}_t]^\tau = \underset{\sim}{\mathbf{0}},$$

it follows from (6.69) and (6.68) that

$$\begin{aligned} \underset{\sim}{\mathbf{P}}_t &= E(\mathbf{x}_t - \hat{\mathbf{x}}_t^s + \hat{\mathbf{x}}_t^s - \hat{\mathbf{x}}_t)(\mathbf{x}_t - \hat{\mathbf{x}}_t^s + \hat{\mathbf{x}}_t^s - \hat{\mathbf{x}}_t)^\tau \\ &= EE\left[\frac{(\mathbf{x}_t - \hat{\mathbf{x}}_t^s + \hat{\mathbf{x}}_t^s - \hat{\mathbf{x}}_t)(\mathbf{x}_t - \hat{\mathbf{x}}_t^s + \hat{\mathbf{x}}_t^s - \hat{\mathbf{x}}_t)^\tau}{\mathscr{F}_t^{\mathbf{x}_s\mathbf{y}}}\right] \\ &= E(\mathbf{x}_t - \hat{\mathbf{x}}_t^s)(\mathbf{x}_t - \hat{\mathbf{x}}_t^s)^\tau + E(\hat{\mathbf{x}}_t^s - \hat{\mathbf{x}}_t)(\hat{\mathbf{x}}_t^s - \hat{\mathbf{x}}_t)^\tau \\ &= \underset{\sim}{\mathbf{P}}_t^s + \underset{\sim}{\boldsymbol{\Psi}}_{t,s}^s E(\mathbf{x}_s - \hat{\mathbf{x}}_s(t))(\mathbf{x}_s - \hat{\mathbf{x}}_s(t))^\tau \underset{\sim}{\boldsymbol{\Psi}}_{t,s}^{s\tau}. \end{aligned}$$

That is,

$$\underset{\sim}{\mathbf{P}}_t = \underset{\sim}{\mathbf{P}}_t^s + \underset{\sim}{\mathbf{\Psi}}_{t,s}^s \underset{\sim}{\mathbf{P}}_s(t) \underset{\sim}{\mathbf{\Psi}}_{t,s}^{sT}, \tag{6.70}$$

which with (6.69) yields (6.65).

From (6.70), it follows that

$$\underset{\sim}{\mathbf{P}}_s(t) = \underset{\sim}{\mathbf{\Psi}}_{s,t}^s (\underset{\sim}{\mathbf{P}}_t - \underset{\sim}{\mathbf{P}}_t^s) \underset{\sim}{\mathbf{\Psi}}_{s,t}^{sT}, \tag{6.71}$$

and by taking derivatives and observing that

$$\underset{\sim}{\mathbf{K}}_t - \underset{\sim}{\mathbf{K}}_t^s = (\underset{\sim}{\mathbf{P}}_t - \underset{\sim}{\mathbf{P}}_t^s) \underset{\sim}{\mathbf{C}}_t^T (\underset{\sim}{\mathbf{F}}_t \underset{\sim}{\mathbf{F}}_t^T)^{-1},$$

we obtain

$$\begin{aligned}
\frac{d}{dt} \underset{\sim}{\mathbf{P}}_s(t) &= \underset{\sim}{\mathbf{\Psi}}_{s,t}^s \big[-(\underset{\sim}{\mathbf{A}}_t - \underset{\sim}{\mathbf{K}}_t^s \underset{\sim}{\mathbf{C}}_t)(\underset{\sim}{\mathbf{P}}_t - \underset{\sim}{\mathbf{P}}_t^s) - (\underset{\sim}{\mathbf{P}}_t - \underset{\sim}{\mathbf{P}}_t^s)(\underset{\sim}{\mathbf{A}}_t^T - \underset{\sim}{\mathbf{C}}_t^T \underset{\sim}{\mathbf{K}}_t^{sT}) \\
&\quad + \underset{\sim}{\mathbf{A}}_t \underset{\sim}{\mathbf{P}}_t + \underset{\sim}{\mathbf{P}}_t \underset{\sim}{\mathbf{A}}_t^T - \underset{\sim}{\mathbf{K}}_t (\underset{\sim}{\mathbf{F}}_t \underset{\sim}{\mathbf{D}}_t^T + \underset{\sim}{\mathbf{C}}_t \underset{\sim}{\mathbf{P}}_t) - \underset{\sim}{\mathbf{A}}_t \underset{\sim}{\mathbf{P}}_t^s - \underset{\sim}{\mathbf{P}}_t^s \underset{\sim}{\mathbf{A}}_t^T \\
&\quad + \underset{\sim}{\mathbf{K}}_t^s (\underset{\sim}{\mathbf{F}}_t \underset{\sim}{\mathbf{D}}_t^T + \underset{\sim}{\mathbf{C}}_t \underset{\sim}{\mathbf{P}}_t^s) \big] \underset{\sim}{\mathbf{\Psi}}_{s,t}^{sT} \\
&= \underset{\sim}{\mathbf{\Psi}}_{s,t}^s \big[\underset{\sim}{\mathbf{K}}_t^s \underset{\sim}{\mathbf{C}}_t (\underset{\sim}{\mathbf{P}}_t - \underset{\sim}{\mathbf{P}}_t^s) + (\underset{\sim}{\mathbf{P}}_t - \underset{\sim}{\mathbf{P}}_t^s) \underset{\sim}{\mathbf{C}}_t^T \underset{\sim}{\mathbf{K}}_t^{sT} \\
&\quad - (\underset{\sim}{\mathbf{K}}_t - \underset{\sim}{\mathbf{K}}_t^s)(\underset{\sim}{\mathbf{F}}_t \underset{\sim}{\mathbf{D}}_t^T + \underset{\sim}{\mathbf{C}}_t \underset{\sim}{\mathbf{P}}_t) - \underset{\sim}{\mathbf{K}}_t^s \underset{\sim}{\mathbf{C}}_t (\underset{\sim}{\mathbf{P}}_t - \underset{\sim}{\mathbf{P}}_t^s) \big] \underset{\sim}{\mathbf{\Psi}}_{s,t}^{sT} \\
&= \underset{\sim}{\mathbf{\Psi}}_{s,t}^s \big[(\underset{\sim}{\mathbf{P}}_t - \underset{\sim}{\mathbf{P}}_t^s) \underset{\sim}{\mathbf{C}}_t^T \underset{\sim}{\mathbf{K}}_t^{sT} - (\underset{\sim}{\mathbf{P}}_t - \underset{\sim}{\mathbf{P}}_t^s) \underset{\sim}{\mathbf{C}}_t^T \underset{\sim}{\mathbf{K}}_t^T \big] \underset{\sim}{\mathbf{\Psi}}_{s,t}^{sT} \\
&= -\underset{\sim}{\mathbf{\Psi}}_{s,t}^s (\underset{\sim}{\mathbf{P}}_t - \underset{\sim}{\mathbf{P}}_t^s) \underset{\sim}{\mathbf{C}}_t^T (\underset{\sim}{\mathbf{F}}_t \underset{\sim}{\mathbf{F}}_t^T)^{-1} \underset{\sim}{\mathbf{C}}_t (\underset{\sim}{\mathbf{P}}_t - \underset{\sim}{\mathbf{P}}_t^s) \underset{\sim}{\mathbf{\Psi}}_{s,t}^{sT} \\
&= -\underset{\sim}{\mathbf{P}}_s(t) \underset{\sim}{\mathbf{\Psi}}_{t,s}^{sT} \underset{\sim}{\mathbf{C}}_t^T (\underset{\sim}{\mathbf{F}}_t \underset{\sim}{\mathbf{F}}_t^T)^{-1} \underset{\sim}{\mathbf{C}}_t \underset{\sim}{\mathbf{\Psi}}_{t,s}^s \underset{\sim}{\mathbf{P}}_s(t), \qquad t \geq s,
\end{aligned} \tag{6.72}$$

which is the same as (6.66). □

Now we proceed to deal with backward interpolation equation.

Lemma 6.3. *Assume that the conditions of Lemma* 6.1 *hold and that* $\det \underset{\sim}{\mathbf{P}}_t \neq 0$ $\forall t \in [0, T]$. *Let* $\underset{\sim}{\mathbf{V}}_{t,s}$ *satisfy*

$$\frac{d\underset{\sim}{\mathbf{V}}_{t,s}}{dt} = \big[\underset{\sim}{\mathbf{A}}_t + \underset{\sim}{\mathbf{D}}_t (\underset{\sim}{\mathbf{D}}_t^T - \underset{\sim}{\mathbf{F}}_t^T \underset{\sim}{\mathbf{K}}_t^T) \underset{\sim}{\mathbf{P}}_t^{-1}\big] \underset{\sim}{\mathbf{V}}_{t,s}, \qquad \underset{\sim}{\mathbf{V}}_{s,s} = \underset{\sim}{\mathbf{I}}. \tag{6.73}$$

Then

$$\mathbf{P}_s(t)\boldsymbol{\Psi}_{t,s}^{s,\tau} = \mathbf{V}_{s,t}\mathbf{P}_t, \qquad 0 \le s \le t. \tag{6.74}$$

Proof. Denote

$$\mathbf{U}_{s,t} \triangleq \mathbf{P}_s(t)\boldsymbol{\Psi}_{t,s}^{s,\tau}\mathbf{P}_t^{-1}.$$

Clearly,

$$\mathbf{U}_{s,s} = \mathbf{P}_s(s)\boldsymbol{\Psi}_{s,s}^{s,\tau}\mathbf{P}_s^{-1} = \mathbf{I},$$

since $\mathbf{P}_s(s) = \mathbf{P}_s$.

By (6.52), (6.70), and (6.72), we have

$$\begin{aligned}\frac{d}{dt}\mathbf{P}_s(t)\boldsymbol{\Psi}_{t,s}^{s,\tau} &= -\mathbf{P}_s(t)\boldsymbol{\Psi}_{t,s}^{s,\tau}\mathbf{C}_t^\tau(\mathbf{F}_t\mathbf{F}_t^\tau)^{-1}\mathbf{C}_t\boldsymbol{\Psi}_{t,s}^{s}\mathbf{P}_s(t)\boldsymbol{\Psi}_{t,s}^{s,\tau}\\ &\quad + \mathbf{P}_s(t)\boldsymbol{\Psi}_{t,s}^{s,\tau}(\mathbf{A}_t^\tau - \mathbf{C}_t\mathbf{K}_t^{s\tau})\\ &= \mathbf{P}_s(t)\boldsymbol{\Psi}_{t,s}^{s,\tau}\left[\mathbf{A}_t^\tau - \mathbf{C}_t^\tau\mathbf{K}_t^{s,\tau} - \mathbf{C}_t^\tau(\mathbf{F}_t\mathbf{F}_t^\tau)^{-1}\mathbf{C}_t(\mathbf{P}_t - \mathbf{P}_t^s)\right]\\ &= \mathbf{P}_s(t)\boldsymbol{\Psi}_{t,s}^{s,\tau}(\mathbf{A}_t^\tau - \mathbf{C}_t^\tau\mathbf{K}_t^\tau).\end{aligned}$$

Hence

$$\begin{aligned}\frac{d}{dt}\mathbf{U}_{s,t} &= \mathbf{P}_s(t)\boldsymbol{\Psi}_{t,s}^{s,\tau}(\mathbf{A}_t^\tau - \mathbf{C}_t^\tau\mathbf{K}_t^\tau)\mathbf{P}_t^{-1}\\ &\quad - \mathbf{U}_{s,t}\left[\mathbf{P}_t\mathbf{A}_t^\tau + \mathbf{A}_t\mathbf{P}_t + \mathbf{D}_t\mathbf{D}_t^\tau - (\mathbf{D}_t\mathbf{F}_t^\tau + \mathbf{P}_t\mathbf{C}_t^\tau)\mathbf{K}_t^\tau\right]\mathbf{P}_t^{-1}\\ &= -\mathbf{U}_{s,t}\left[\mathbf{A}_t + \mathbf{D}_t(\mathbf{D}_t^\tau - \mathbf{F}_t^\tau\mathbf{K}_t^\tau)\mathbf{P}_t^{-1}\right].\end{aligned}$$

From this we know that $\mathbf{U}_{s,t}$ is a fundamental solution matrix and that

$$\frac{d}{dt}\mathbf{U}_{t,s} = \left[\mathbf{A}_t + \mathbf{D}_t(\mathbf{D}_t^\tau - \mathbf{F}_t^\tau\mathbf{K}_t^\tau)\mathbf{P}_t^{-1}\right]\mathbf{U}_{t,s}, \qquad \mathbf{U}_{s,s} = \mathbf{I}.$$

By comparing it with (6.73) and by the uniqueness of the solution, we conclude that $\mathbf{U}_{t,s} = \mathbf{V}_{t,s}$. Further,

$$\mathbf{U}_{t,s} = \mathbf{U}_{s,t}^{-1} = \left[\mathbf{P}_s(t)\boldsymbol{\Psi}_{t,s}^{s,\tau}\mathbf{P}_t^{-1}\right]^{-1},$$

which yields the assertion of the lemma. □

Theorem 6.3. *Assume that the conditions of Lemma* 6.1 *hold. For fixed* t, *the backward interpolation equations are as follows*:

$$d_s\hat{\mathbf{x}}_s(t) = \underset{\sim}{\mathbf{A}}_s\hat{\mathbf{x}}_s(t)\,ds + \underset{\sim}{\mathbf{D}}_s(\underset{\sim}{\mathbf{D}}_s^{\tau} - \underset{\sim}{\mathbf{F}}_s^{\tau}\underset{\sim}{\mathbf{K}}_s^{\tau})\underset{\sim}{\mathbf{P}}_s^{-1}(\hat{\mathbf{x}}_s(t) - \hat{\mathbf{x}}_s)\,ds$$

$$+\underset{\sim}{\mathbf{D}}_s\underset{\sim}{\mathbf{F}}_s^{\tau}(\underset{\sim}{\mathbf{F}}_s\underset{\sim}{\mathbf{F}}_s^{\tau})^{-1}(d\mathbf{y}_s - \underset{\sim}{\mathbf{C}}_s\hat{\mathbf{x}}_s\,ds), \tag{6.75}$$

$$\frac{d}{ds}\underset{\sim}{\mathbf{P}}_s(t) = [\underset{\sim}{\mathbf{A}}_s + \underset{\sim}{\mathbf{D}}_s(\underset{\sim}{\mathbf{D}}_s^{\tau} - \underset{\sim}{\mathbf{F}}_s^{\tau}\underset{\sim}{\mathbf{K}}_s^{\tau})\underset{\sim}{\mathbf{P}}_s^{-1}]\underset{\sim}{\mathbf{P}}_s(t) + \underset{\sim}{\mathbf{P}}_s(t)$$

$$\times[\underset{\sim}{\mathbf{A}}_s + \underset{\sim}{\mathbf{D}}_s(\underset{\sim}{\mathbf{D}}_s^{\tau} - \underset{\sim}{\mathbf{F}}_s^{\tau}\underset{\sim}{\mathbf{K}}_s^{\tau})\underset{\sim}{\mathbf{P}}_s^{-1}]^{\tau}$$

$$-\underset{\sim}{\mathbf{D}}_s\left[\underset{\sim}{\mathbf{I}} - \underset{\sim}{\mathbf{F}}_s^{\tau}(\underset{\sim}{\mathbf{F}}_s\underset{\sim}{\mathbf{F}}_s^{\tau})^{-1}\underset{\sim}{\mathbf{F}}_s\right]\underset{\sim}{\mathbf{D}}_s^{\tau}, \qquad s \le t, \quad \underset{\sim}{\mathbf{P}}_t(t) = \underset{\sim}{\mathbf{P}}_t. \tag{6.76}$$

Proof. By (6.65) and Lemma 6.3, we know that

$$\hat{\mathbf{x}}_s(t) = \hat{\mathbf{x}}_s + \underset{\sim}{\mathbf{V}}_{s,0}\int_s^t \underset{\sim}{\mathbf{V}}_{0,\lambda}\underset{\sim}{\mathbf{P}}_\lambda\underset{\sim}{\mathbf{C}}_\lambda^{\tau}(\underset{\sim}{\mathbf{F}}_\lambda\underset{\sim}{\mathbf{F}}_\lambda^{\tau})^{-1}(d\mathbf{y}_\lambda - \underset{\sim}{\mathbf{C}}_\lambda\hat{\mathbf{x}}_\lambda\,d\lambda), \tag{6.77}$$

and by Theorem 6.1 and (6.73),

$$d_s\hat{\mathbf{x}}_s(t) = \underset{\sim}{\mathbf{A}}_s\hat{\mathbf{x}}_s\,ds + \underset{\sim}{\mathbf{K}}_s(d\mathbf{y}_s - \underset{\sim}{\mathbf{C}}_s\hat{\mathbf{x}}_s\,ds)$$

$$+[\underset{\sim}{\mathbf{A}}_s + \underset{\sim}{\mathbf{D}}_s(\underset{\sim}{\mathbf{D}}_s^{\tau} - \underset{\sim}{\mathbf{F}}_s^{\tau}\underset{\sim}{\mathbf{K}}_s^{\tau})\underset{\sim}{\mathbf{P}}_s^{-1}]\underset{\sim}{\mathbf{V}}_{s,0}$$

$$\times\int_s^t \underset{\sim}{\mathbf{V}}_{0,\lambda}\underset{\sim}{\mathbf{P}}_\lambda\underset{\sim}{\mathbf{C}}_\lambda^{\tau}(\underset{\sim}{\mathbf{F}}_\lambda\underset{\sim}{\mathbf{F}}_\lambda^{\tau})^{-1}(d\mathbf{y}_\lambda - \underset{\sim}{\mathbf{C}}_\lambda\hat{\mathbf{x}}_\lambda\,d\lambda)\,ds$$

$$-\underset{\sim}{\mathbf{P}}_s\underset{\sim}{\mathbf{C}}_s^{\tau}(\underset{\sim}{\mathbf{F}}_s\underset{\sim}{\mathbf{F}}_s^{\tau})^{-1}(d\mathbf{y}_s - \underset{\sim}{\mathbf{C}}_s\hat{\mathbf{x}}_s\,ds)$$

$$= \underset{\sim}{\mathbf{A}}_s\hat{\mathbf{x}}_s\,ds + \underset{\sim}{\mathbf{D}}_s\underset{\sim}{\mathbf{F}}_s^{\tau}(\underset{\sim}{\mathbf{F}}_s\underset{\sim}{\mathbf{F}}_s^{\tau})^{-1}(d\mathbf{y}_s - \underset{\sim}{\mathbf{C}}_s\hat{\mathbf{x}}_s\,ds)$$

$$+[\underset{\sim}{\mathbf{A}}_s + \underset{\sim}{\mathbf{D}}_s(\underset{\sim}{\mathbf{D}}_s^{\tau} - \underset{\sim}{\mathbf{F}}_s^{\tau}\underset{\sim}{\mathbf{K}}_s^{\tau})\underset{\sim}{\mathbf{P}}_s^{-1}](\hat{\mathbf{x}}_s(t) - \hat{\mathbf{x}}_s)\,ds$$

$$= \underset{\sim}{\mathbf{A}}_s\hat{\mathbf{x}}_s(t)\,ds + \underset{\sim}{\mathbf{D}}_s(\underset{\sim}{\mathbf{D}}_s^{\tau} - \underset{\sim}{\mathbf{F}}_s^{\tau}\underset{\sim}{\mathbf{K}}_s^{\tau})\underset{\sim}{\mathbf{P}}_s^{-1}(\hat{\mathbf{x}}_s(t) - \hat{\mathbf{x}}_s)\,ds$$

$$+\underset{\sim}{\mathbf{D}}_s\underset{\sim}{\mathbf{F}}_s^{\tau}(\underset{\sim}{\mathbf{F}}_s\underset{\sim}{\mathbf{F}}_s^{\tau})^{-1}(d\mathbf{y}_s - \underset{\sim}{\mathbf{C}}_s\hat{\mathbf{x}}_s\,ds).$$

This proves (6.75). For (6.76), by (6.72) and (6.74) we have

$$\underset{\sim}{\mathbf{P}}_s(t) = \underset{\sim}{\mathbf{P}}_s - \int_s^t \underset{\sim}{\mathbf{V}}_{s,\lambda}\underset{\sim}{\mathbf{P}}_\lambda\underset{\sim}{\mathbf{C}}_\lambda^\tau(\underset{\sim}{\mathbf{F}}_\lambda\underset{\sim}{\mathbf{F}}_\lambda^\tau)^{-1}\underset{\sim}{\mathbf{C}}_\lambda\underset{\sim}{\mathbf{P}}_\lambda\underset{\sim}{\mathbf{V}}_{s,\lambda}^\tau\,d\lambda.$$

Taking its derivative and using (6.73) leads to

$$\begin{aligned}\frac{d\underset{\sim}{\mathbf{P}}_s(t)}{ds} &= \underset{\sim}{\mathbf{P}}_s\underset{\sim}{\mathbf{A}}_s^\tau + \underset{\sim}{\mathbf{A}}_s\underset{\sim}{\mathbf{P}}_s + \underset{\sim}{\mathbf{D}}_s\underset{\sim}{\mathbf{D}}_s^\tau - \underset{\sim}{\mathbf{K}}_s(\underset{\sim}{\mathbf{D}}_s\underset{\sim}{\mathbf{F}}_s^\tau + \underset{\sim}{\mathbf{P}}_s\underset{\sim}{\mathbf{C}}_s^\tau)^\tau \\ &\quad + \left[\underset{\sim}{\mathbf{A}}_s + \underset{\sim}{\mathbf{D}}_s(\underset{\sim}{\mathbf{D}}_s^\tau - \underset{\sim}{\mathbf{F}}_s^\tau\underset{\sim}{\mathbf{K}}_s^\tau)\underset{\sim}{\mathbf{P}}_s^{-1}\right](\underset{\sim}{\mathbf{P}}_s(t) - \underset{\sim}{\mathbf{P}}_s) \\ &\quad + (\underset{\sim}{\mathbf{P}}_s(t) - \underset{\sim}{\mathbf{P}}_s)\left[\underset{\sim}{\mathbf{A}}_s^\tau + \underset{\sim}{\mathbf{P}}_s^{-1}(\underset{\sim}{\mathbf{D}}_s - \underset{\sim}{\mathbf{K}}_s\underset{\sim}{\mathbf{F}}_s)\underset{\sim}{\mathbf{D}}_s^\tau\right] + \underset{\sim}{\mathbf{P}}_s\underset{\sim}{\mathbf{C}}_s^\tau(\underset{\sim}{\mathbf{F}}_s\underset{\sim}{\mathbf{F}}_s^\tau)^{-1}\underset{\sim}{\mathbf{C}}_s\underset{\sim}{\mathbf{P}}_s \\ &= \left[\underset{\sim}{\mathbf{A}}_s + \underset{\sim}{\mathbf{D}}_s(\underset{\sim}{\mathbf{D}}_s^\tau - \underset{\sim}{\mathbf{F}}_s^\tau\underset{\sim}{\mathbf{K}}_s^\tau)\underset{\sim}{\mathbf{P}}_s^{-1}\right]\underset{\sim}{\mathbf{P}}_s(t) \\ &\quad + \underset{\sim}{\mathbf{P}}_s(t)\left[\underset{\sim}{\mathbf{A}}_s + \underset{\sim}{\mathbf{D}}_s(\underset{\sim}{\mathbf{D}}_s^\tau - \underset{\sim}{\mathbf{F}}_s^\tau\underset{\sim}{\mathbf{K}}_s^\tau)\underset{\sim}{\mathbf{P}}_s^{-1}\right]^\tau \\ &\quad - \underset{\sim}{\mathbf{D}}_s\left[\underset{\sim}{\mathbf{I}} - \underset{\sim}{\mathbf{F}}_s^\tau(\underset{\sim}{\mathbf{F}}_s\underset{\sim}{\mathbf{F}}_s^\tau)^{-1}\underset{\sim}{\mathbf{F}}_s\right]\underset{\sim}{\mathbf{D}}_s^\tau.\end{aligned}$$

□

Remark. The solution to (6.75) and (6.76) exists and is unique since these equations linearly depend on $\hat{\mathbf{x}}_s(t)$ and $\underset{\sim}{\mathbf{P}}_s(t)$.

Now we shall consider the prediction problem. Let $t \geq s$, and denote by $\hat{\mathbf{x}}_t(s)$ the LUMVE of $\hat{\mathbf{x}}_t$, based on $\{\mathbf{y}_\lambda,\ 0 \leq \lambda \leq s\}$, and by $\underset{\sim}{\mathbf{P}}_t(s)$ its estimation error covariance matrix.

Theorem 6.4. *Under the conditions of Lemma* 6.1

$$\hat{\mathbf{x}}_t(s) = \underset{\sim}{\mathbf{\Phi}}_{t,s}\hat{\mathbf{x}}_s, \tag{6.78}$$

$$\underset{\sim}{\mathbf{P}}_t(s) = \underset{\sim}{\mathbf{\Phi}}_{t,s}\underset{\sim}{\mathbf{P}}_s\underset{\sim}{\mathbf{\Phi}}_{t,s}^\tau + \int_s^t \underset{\sim}{\mathbf{\Phi}}_{t,\lambda}\underset{\sim}{\mathbf{D}}_\lambda\underset{\sim}{\mathbf{D}}_\lambda^\tau\underset{\sim}{\mathbf{\Phi}}_{t,\lambda}^\tau\,d\lambda, \qquad t \geq s. \tag{6.79}$$

Proof. Without loss of generality, assume $\{\mathbf{w}_t\}$ is a Wiener process. Then

$$\hat{\mathbf{x}}_t(s) = E\left(\frac{\mathbf{x}_t}{\mathscr{F}_s^{\mathbf{y}}}\right).$$

Since

$$\mathbf{x}_t = \underset{\sim}{\mathbf{\Phi}}_{t,s}\mathbf{x}_s + \int_s^t \underset{\sim}{\mathbf{\Phi}}_{t,\lambda}\underset{\sim}{\mathbf{D}}_\lambda\,d\mathbf{w}_\lambda, \qquad E\left(\int_s^t \frac{\underset{\sim}{\mathbf{\Phi}}_{t,\lambda}\underset{\sim}{\mathbf{D}}_\lambda\,d\underset{\sim}{\mathbf{w}}_\lambda}{\mathscr{F}_s^{\mathbf{y}}}\right) = \mathbf{0},$$

it is immediate that

$$\hat{\mathbf{x}}_t(s) = \underset{\sim}{\Phi}_{t,s}\hat{\mathbf{x}}_s$$

and from here (6.78) and (6.79) follow at once. □

6.4. FILTERING AND CONTROL WITH LINEAR FEEDBACK

We now return to the system (6.1), (6.2) with the control term included. The set of admissible controls $\mathscr{U}$ is now

$$\mathscr{U} = \left\{\mathbf{u}_t : \mathbf{u}_t = \underset{\sim}{U}_1(t)\left(\mathbf{c} + \underset{\sim}{C}\mathbf{y}_0 + \int_0^t \underset{\sim}{U}_2(s)\, d\mathbf{y}_s\right), \quad \int_0^T \| \underset{\sim}{U}_2(t)\underset{\sim}{C}_t \| \, dt < \infty, \right.$$

$$\left. \times \int_0^T \| \underset{\sim}{U}_2(t)\underset{\sim}{F}_t \|^2 \, dt < \infty, \quad \int_0^T \| \underset{\sim}{B}_t\underset{\sim}{U}_1(t) \| \, dt < \infty \right\}, \tag{6.80}$$

where $\underset{\sim}{U}_1(t)$ and $\underset{\sim}{U}_2(t)$ are $r \times n$ and $n \times m$ matrices, respectively, of measurable functions, $\mathbf{c}$ is an n-dimensional vector, and $\underset{\sim}{C}$ is an $n \times m$ matrix.

Let us put

$$\mathbf{z}_t = \mathbf{c} + \underset{\sim}{C}\mathbf{y}_0 + \int_0^t \underset{\sim}{U}_2(s)\, d\mathbf{y}_s.$$

Then (6.1) and (6.2) can be rewritten as

$$\left.\begin{aligned} d\mathbf{x}_t &= \underset{\sim}{A}_t\mathbf{x}_t\, dt + \underset{\sim}{B}_t\underset{\sim}{U}_1(t)\mathbf{z}_t\, dt + \underset{\sim}{D}_t\, d\mathbf{w}_t, \\ d\mathbf{z}_t &= \underset{\sim}{U}_2(t)(\underset{\sim}{C}_t\mathbf{x}_t\, dt + \underset{\sim}{F}_t\, d\mathbf{w}_t), \\ d\mathbf{y}_t &= \underset{\sim}{C}_t\mathbf{x}_t\, dt + \underset{\sim}{F}_t\, d\mathbf{w}_t, \end{aligned}\right\} \tag{6.81}$$

which is a system of linear stochastic differential equations with the appropriate integrability conditions satisfied (see [2]) and hence has a solution. This means that for $\{\mathbf{u}_t\} \in \mathscr{U}$ the system (6.1) and (6.2) is solvable.

Let us write

$$\mathbf{x}_t^0 = \underset{\sim}{\Phi}_{t,0}\mathbf{x}_0 + \int_0^t \underset{\sim}{\Phi}_{t,\lambda}\underset{\sim}{D}_\lambda\, d\mathbf{w}_\lambda \tag{6.82}$$

$$\mathbf{y}_t^0 = \mathbf{y}_0 + \int_0^t \underset{\sim}{C}_\lambda\mathbf{x}_\lambda^0\, d\lambda + \int_0^t \underset{\sim}{F}_\lambda\, d\mathbf{w}_\lambda \tag{6.83}$$

$$\mathbf{v}_t = \int_0^t \underset{\sim}{\Phi}_{t,\lambda}\underset{\sim}{B}_\lambda\mathbf{u}_\lambda\, d\lambda, \tag{6.84}$$

then

$$\mathbf{x}_t = \mathbf{x}_t^0 + \mathbf{v}_t \tag{6.85}$$

$$d\mathbf{y}_t = \underset{\sim}{\mathbf{C}}_t(\mathbf{x}_t^0 + \mathbf{v}_t)\,dt + \underset{\sim}{\mathbf{F}}_t\,d\mathbf{w}_t = d\mathbf{y}_t^0 + \underset{\sim}{\mathbf{C}}_t\mathbf{v}_t\,dt. \tag{6.86}$$

From (6.81), (6.84), and (6.86), we have

$$d\mathbf{v}_t = \underset{\sim}{\mathbf{A}}_t\mathbf{v}_t\,dt + \underset{\sim}{\mathbf{B}}_t\mathbf{u}_t\,dt = \underset{\sim}{\mathbf{A}}_t\mathbf{v}_t\,dt + \underset{\sim}{\mathbf{B}}_t\underset{\sim}{\mathbf{U}}_1(t)\mathbf{z}_t\,dt, \tag{6.87}$$

$$d\mathbf{z}_t = \underset{\sim}{\mathbf{U}}_2(t)\,d\mathbf{y}_t^0 + \underset{\sim}{\mathbf{U}}_2(t)\underset{\sim}{\mathbf{C}}_t\mathbf{v}_t\,dt. \tag{6.88}$$

Set

$$\underset{\sim}{\mathbf{A}}_t^0 = \begin{bmatrix} \underset{\sim}{\mathbf{A}}_t & \underset{\sim}{\mathbf{B}}_t\underset{\sim}{\mathbf{U}}_1(t) \\ \underset{\sim}{\mathbf{0}} & \underset{\sim}{\mathbf{U}}_2(t)\underset{\sim}{\mathbf{C}}_t \end{bmatrix},$$

which clearly is absolutely integrable on $[0, T]$.

Let $\underset{\sim}{\mathbf{L}}_{t,s}$ be the fundamental solution matrix given by

$$\frac{d}{dt}\underset{\sim}{\mathbf{L}}_{t,s} = \underset{\sim}{\mathbf{A}}_t^0\underset{\sim}{\mathbf{L}}_{t,s}, \qquad \underset{\sim}{\mathbf{L}}_{s,s} = \underset{\sim}{\mathbf{I}}.$$

Noticing that

$$\mathbf{v}_0 = \mathbf{0}, \qquad \mathbf{z}_0 = \mathbf{c} + \underset{\sim}{\mathbf{C}}\mathbf{y}_0,$$

we have

$$\begin{bmatrix} \mathbf{v}_t \\ \mathbf{z}_t \end{bmatrix} = \underset{\sim}{\mathbf{L}}_{t,0}\begin{bmatrix} \mathbf{0} \\ \mathbf{c} + \underset{\sim}{\mathbf{C}}\mathbf{y}^0 \end{bmatrix} + \int_0^t \underset{\sim}{\mathbf{L}}_{t,\lambda}\begin{bmatrix} \underset{\sim}{\mathbf{0}} \\ \underset{\sim}{\mathbf{U}}_2(\lambda) \end{bmatrix} d\mathbf{y}_\lambda^0.$$

Hence $\mathbf{v}_t$ is $\mathscr{F}_t^{\mathbf{y}^0}$-measurable and from (6.86) it is known that

$$\mathscr{F}_t^{\mathbf{y}} \subset \mathscr{F}_t^{\mathbf{y}^0}. \tag{6.89}$$

On the other hand, $\mathbf{u}_t$ is $\mathscr{F}_t^{\mathbf{y}}$-measurable, therefore from (6.84) we see that $\mathbf{v}_t$ is $\mathscr{F}_t^{\mathbf{y}}$-measurable. Furthermore, by (5.86),

$$\mathbf{y}_t^0 = \mathbf{y}_t - \int_0^t \underset{\sim}{\mathbf{C}}_\lambda\mathbf{v}_\lambda\,d\lambda.$$

Hence $\mathbf{y}_t^0$ is $\mathscr{F}_t^{\mathbf{y}}$-measurable (i.e., $\mathscr{F}_t^{\mathbf{y}^0} \subset \mathscr{F}_t^{\mathbf{y}}$), which together with (6.89) implies

$$\mathscr{F}_t^{\mathbf{y}} = \mathscr{F}_t^{\mathbf{y}^0}. \tag{6.90}$$

Theorem 6.5. *Under the conditions* (6.3), *if* $\{\mathbf{u}_t\} \in \mathscr{U}$, *then for the system* (6.1) *and* (6.2) *the linear unbiased minimum variance filter estimate* $\hat{\mathbf{x}}_t$ *and its filtering error covariance matrix* $\underset{\sim}{\mathbf{P}}_t$ *are given by* (6.8) *and*

$$d\hat{\mathbf{x}}_t = \underset{\sim}{\mathbf{A}}_t\hat{\mathbf{x}}_t\,dt + \underset{\sim}{\mathbf{B}}_t\mathbf{u}_t\,dt + (\underset{\sim}{\mathbf{D}}_t\underset{\sim}{\mathbf{F}}_t^{\tau} + \underset{\sim}{\mathbf{P}}_t\underset{\sim}{\mathbf{C}}_t^{\tau})(\underset{\sim}{\mathbf{F}}_t\underset{\sim}{\mathbf{F}}_t^{\tau})^{-1}(d\mathbf{y}_t - \underset{\sim}{\mathbf{C}}_t\hat{\mathbf{x}}_t\,dt) \tag{6.91}$$

with initial values (6.4) *and* (6.5).

Proof. Without loss of generality, assume $\{\mathbf{w}_t\}$ is a Wiener process. From (6.85) and (6.90), we have

$$E\left(\frac{\mathbf{x}_t}{\mathscr{F}_t^{\mathbf{y}}}\right) = E\left(\frac{\mathbf{x}_t^0}{\mathscr{F}_t^{\mathbf{y}}}\right) + \int_0^t \underset{\sim}{\mathbf{\Phi}}_{t,\lambda}\mathbf{B}_\lambda\mathbf{u}_\lambda\,d\lambda$$

$$= E\left(\frac{\mathbf{x}_t^0}{\mathscr{F}_t^{\mathbf{y}^0}}\right) + \int_0^t \underset{\sim}{\mathbf{\Phi}}_{t,\lambda}\mathbf{B}_\lambda\mathbf{u}_\lambda\,d\lambda,$$

and

$$\hat{\mathbf{x}}_t = \hat{\mathbf{x}}_t^0 + \int_0^t \underset{\sim}{\mathbf{\Phi}}_{t,\lambda}\mathbf{B}_\lambda\mathbf{u}_\lambda\,d\lambda, \tag{6.92}$$

where

$$\hat{\mathbf{x}}_t^0 \triangleq E\left(\frac{\mathbf{x}_t^0}{\mathscr{F}_t^{\mathbf{y}^0}}\right).$$

Since (6.82) and (6.83) are independent of the control, Theorem 6.1 shows that $\hat{\mathbf{x}}_t^0$ satisfies (6.31). Hence from (6.92), we have

$$d\hat{\mathbf{x}}_t = \underset{\sim}{\mathbf{A}}_t\hat{\mathbf{x}}_t^0\,dt + \underset{\sim}{\mathbf{K}}_t(d\mathbf{y}_t^0 - \underset{\sim}{\mathbf{C}}_t\hat{\mathbf{x}}_t^0\,dt) + \underset{\sim}{\mathbf{A}}_t\int_0^t \underset{\sim}{\mathbf{\Phi}}_{t,\lambda}\mathbf{B}_\lambda\mathbf{u}_\lambda\,d\lambda + \underset{\sim}{\mathbf{B}}_t\mathbf{u}_t\,dt$$

$$= \underset{\sim}{\mathbf{A}}_t\hat{\mathbf{x}}_t\,dt + \underset{\sim}{\mathbf{B}}_t\mathbf{u}_t\,dt + \underset{\sim}{\mathbf{K}}_t(d\mathbf{y}_t^0 - \underset{\sim}{\mathbf{C}}_t\hat{\mathbf{x}}_t^0\,dt), \tag{6.93}$$

and by (6.85), (6.86), and (6.92)

$$
\begin{aligned}
d\mathbf{y}_t^0 - \underset{\sim}{\mathbf{C}}_t\hat{\mathbf{x}}_t^0\,dt &= d\mathbf{y}_t - \underset{\sim}{\mathbf{C}}_t\int_0^t \underset{\sim}{\mathbf{\Phi}}_{t,\lambda}\underset{\sim}{\mathbf{B}}_\lambda\mathbf{u}_\lambda\,d\lambda - \underset{\sim}{\mathbf{C}}_t\hat{\mathbf{x}}_t^0\,dt \\
&= d\mathbf{y}_t - \underset{\sim}{\mathbf{C}}_t\int_0^t \underset{\sim}{\mathbf{\Phi}}_{t,\lambda}\underset{\sim}{\mathbf{B}}_\lambda\mathbf{u}_\lambda\,d\lambda - \underset{\sim}{\mathbf{C}}_t\hat{\mathbf{x}}_t\,dt + \underset{\sim}{\mathbf{C}}_t\int_0^t \underset{\sim}{\mathbf{\Phi}}_{t,\lambda}\underset{\sim}{\mathbf{B}}_\lambda\mathbf{u}_\lambda\,d\lambda \\
&= d\mathbf{y}_t - \underset{\sim}{\mathbf{C}}_t\hat{\mathbf{x}}_t\,dt. \qquad (6.94)
\end{aligned}
$$

Substituting (6.94) into (6.93) leads to (6.91), and from (6.84), (6.85), and (6.92), it follows that

$$\mathbf{x}_t - \hat{\mathbf{x}}_t = \mathbf{x}_t^0 - \hat{\mathbf{x}}_t^0.$$

Hence

$$E(\mathbf{x}_t - \hat{\mathbf{x}}_t)(\mathbf{x}_t - \hat{\mathbf{x}}_t)^\tau = E\left(\mathbf{x}_t^0 - \hat{\mathbf{x}}_t^0\right)\left(\mathbf{x}_t^0 - \hat{\mathbf{x}}_t^0\right)^\tau = \underset{\sim}{\mathbf{P}}_t,$$

which is given by (6.8). □

Remark 1. Equation (6.8) is independent of the control. This fact means that for $\{\mathbf{u}_t\} \in \mathscr{U}$, the filtering accuracy is not effected by the control and in this case it is said that the separation principle holds.

Remark 2. If $\{\mathbf{w}_t\}$ is a Wiener process, then starting from the general nonlinear filtering equations, it can be proved that Theorem 6.5 remains valid for the more general set of admissible controls, which includes nonlinear feedback controls (see [2] and [61]).

We shall now solve a stochastic control problem with quadratic cost index. For the system (6.1) and (6.2), we assume $\{\mathbf{w}_t\}$ is a Wiener process and $\{\mathbf{u}_t\} \in \mathscr{U}$ given by (6.80). We hope to find the control $\{\mathbf{u}_t^*\} \in \mathscr{U}$, which minimizes $EJ(\mathbf{u})$, where

$$EJ(\mathbf{u}) = E\left\{\mathbf{x}_T^\tau\underset{\sim}{\mathbf{Q}}_0\mathbf{x}_T + \int_0^T\left(\mathbf{x}_t^\tau\underset{\sim}{\mathbf{Q}}_1(t)\mathbf{x}_t + \mathbf{u}_t^\tau\underset{\sim}{\mathbf{Q}}_2(t)\mathbf{u}_t\right)dt\right\} \qquad (6.95)$$

with $\underset{\sim}{\mathbf{Q}}_0 \geq \underset{\sim}{\mathbf{0}}$, $\underset{\sim}{\mathbf{Q}}_1(t) \geq \underset{\sim}{\mathbf{0}}$, and $\underset{\sim}{\mathbf{Q}}_2(t) > \underset{\sim}{\mathbf{0}}$. That is,

$$EJ(\mathbf{u}^*) = \inf_{u\in\mathscr{U}} EJ(\mathbf{u}). \qquad (6.96)$$

This problem is usually called the linear quadratic Gaussian (LQG) problem.

Theorem 6.6. *In addition to condition* (6.3), *it is assumed that*

$$\int_0^T \left(\left\| \mathbf{Q}_1(t) \right\| + \left\| \mathbf{B}_t \mathbf{Q}_2^{-1}(t) \mathbf{B}_t^T \right\| \right) dt < \infty.$$

Then the Riccati equation

$$-\dot{\mathbf{S}}_t = \mathbf{A}_t^T \mathbf{S}_t + \mathbf{S}_t \mathbf{A}_t + \mathbf{Q}_1(t) - \mathbf{S}_t \mathbf{B}_t \mathbf{Q}_2^{-1}(t) \mathbf{B}_t^T \mathbf{S}_t, \qquad \mathbf{S}_T = \mathbf{Q}_0 \quad (6.97)$$

is solvable, and for the LQG problem the optimal control is given by

$$\mathbf{u}_t^* = \mathbf{L}_t \hat{\mathbf{x}}_t \quad (6.98)$$

with

$$\mathbf{L}_t = -\mathbf{Q}_2^{-1}(t) \mathbf{B}_t^T \mathbf{S}_t, \quad (6.99)$$

and the minimum value of the performance index is

$$EJ(\mathbf{u}^*) = E\mathbf{x}_0^T \mathbf{S}_0 \mathbf{x}_0 + \operatorname{tr} \mathbf{S}_0 \mathbf{R}_{\mathbf{x}_0} + \int_0^T \operatorname{tr} \left[\mathbf{D}_t \mathbf{D}_t^T \mathbf{S}_t + \mathbf{L}_t^T \mathbf{Q}_2(t) \mathbf{L}_t \mathbf{P}_t \right] dt, \quad (6.100)$$

where $\hat{\mathbf{x}}_t$ *and* $\mathbf{P}_t$ *are given by Theorem* 6.5 *with* $\mathbf{u}_t$ *in* (6.91) *replaced by* $\mathbf{L}_t \hat{\mathbf{x}}_t$.

Proof. With the substitution of variables $t = T - s$ and $dt = -ds$, (6.97) becomes

$$\frac{d}{ds} \mathbf{S}_{T-s} = \mathbf{A}_{T-s}^T \mathbf{S}_{T-s} + \mathbf{S}_{T-s} \mathbf{A}_{T-s} + \mathbf{Q}_1(T-s) - \mathbf{S}_{T-s} \mathbf{B}_{T-s} \mathbf{Q}_2^{-1}(T-s) \mathbf{B}_{T-s}^T \mathbf{S}_{T-s} \quad (6.101)$$

or

$$\frac{d}{ds} \mathbf{S}_s' = \mathbf{A}_s'^T \mathbf{S}_s' + \mathbf{S}_s' \mathbf{A}_s' + \mathbf{Q}_1'(s) - \mathbf{S}_s' \mathbf{B}_s' \mathbf{Q}_2'^{-1}(s) \mathbf{B}_s'^T \mathbf{S}_s', \qquad \mathbf{S}_0' = \mathbf{Q}_0,$$

where

$$\mathbf{S}_s' = \mathbf{S}_{T-s}, \qquad \mathbf{A}_s' = \mathbf{A}_{T-s}, \qquad \mathbf{Q}_1'(s) = \mathbf{Q}_1(T-s)$$

$$\mathbf{B}_s' = \mathbf{B}_{T-s}, \qquad \mathbf{Q}_2'(s) = \mathbf{Q}_2(T-s).$$

By Lemma 6.1, the preceding equation and hence (6.101) each have a unique nonnegative solution.

Again, shall we obtain the optimal control by the completion of the square technique [59].

Applying Ito's formula to $\mathbf{x}_t^T \mathbf{S}_t \mathbf{x}_t$ leads to

$$\mathbf{x}_T^T \mathbf{Q}_0 \mathbf{x}_T - \mathbf{x}_0^T \mathbf{S}_0 \mathbf{x}_0 = \int_0^T d\mathbf{x}_t^T \mathbf{S}_t \mathbf{x}_t = \int_0^T \Big[2\mathbf{x}_t^T \mathbf{S}_t (\mathbf{A}_t \mathbf{x}_t + \mathbf{B}_t \mathbf{u}_t)$$

$$+ \operatorname{tr} \mathbf{S}_t \mathbf{D}_t \mathbf{D}_t^T - \mathbf{x}_t^T \Big(\mathbf{A}_t^T \mathbf{S}_t + \mathbf{S}_t \mathbf{A}_t + \mathbf{Q}_1(t) - \mathbf{S}_t \mathbf{B}_t \mathbf{Q}_2^{-1}(t) \mathbf{B}_t^T \mathbf{S}_t \Big) \mathbf{x}_t \Big] \, dt$$

$$+ 2 \int_0^T \mathbf{x}_t^T \mathbf{S}_t \mathbf{D}_t \, d\mathbf{w}_t .$$

Hence

$$J(\mathbf{u}) = \mathbf{x}_0^T \mathbf{S}_0 \mathbf{x}_0 + \int_0^T d\mathbf{x}_t^T \mathbf{S}_t \mathbf{x}_t + \int_0^T \Big(\mathbf{x}_t^T \mathbf{Q}_1(t) \mathbf{x}_t + \mathbf{u}_t^T \mathbf{Q}_2(t) \mathbf{u}_t \Big) \, dt$$

$$= \mathbf{x}_0^T \mathbf{S}_0 \mathbf{x}_0 + \int_0^T \Big(2\mathbf{x}_t^T \mathbf{S}_t \mathbf{B}_t \mathbf{u}_t + \operatorname{tr} \mathbf{S}_t \mathbf{D}_t \mathbf{D}_t^T + \mathbf{x}_t^T \mathbf{S}_t \mathbf{B}_t \mathbf{Q}_2^{-1}(t) \mathbf{B}_t^T \mathbf{S}_t \mathbf{x}_t$$

$$+ \mathbf{u}_t^T \mathbf{Q}_2(t) \mathbf{u}_t \Big) \, dt + 2 \int_0^T \mathbf{x}_t^T \mathbf{S}_t \mathbf{D}_t \, d\mathbf{w}_t$$

$$= \mathbf{x}_0^T \mathbf{S}_0 \mathbf{x}_0 + \int_0^T \operatorname{tr} \mathbf{S}_t \mathbf{D}_t \mathbf{D}_t^T \, dt + \int_0^T (\mathbf{u}_t - \mathbf{L}_t \mathbf{x}_t)^T \mathbf{Q}_2(t) (\mathbf{u}_t - \mathbf{L}_t \mathbf{x}_t) \, dt$$

$$+ 2 \int_0^T \mathbf{x}_t^T \mathbf{S}_t \mathbf{D}_t \, d\mathbf{w}_t . \tag{6.102}$$

Now we show that

$$E \int_0^T \| \mathbf{x}_t^T \mathbf{S}_t \mathbf{D}_t \|^2 \, dt < \infty . \tag{6.103}$$

For this, it is sufficient to prove that

$$E \int_0^T \| \mathbf{x}_t \|^2 \| \mathbf{D}_t \|^2 \, dt < \infty , \tag{6.104}$$

since $\underset{\sim}{\mathbf{S}}_t$ is continuous on $[0, T]$ and so $\|\underset{\sim}{\mathbf{S}}_t\|$ is bounded on $[0, T]$. But for (6.104), it is enough to show that $E\|\mathbf{x}_t\|^2$ is bounded on $[0, T]$ because $\int_0^T \|\underset{\sim}{\mathbf{D}}_t\|^2 \, dt < \infty$.

Set

$$\bar{\underset{\sim}{\mathbf{A}}}_t = \begin{bmatrix} \underset{\sim}{\mathbf{A}}_t & \underset{\sim}{\mathbf{B}}_t \underset{\sim}{\mathbf{U}}_1(t) \\ \underset{\sim}{\mathbf{U}}_2(t) \underset{\sim}{\mathbf{C}}_t & \underset{\sim}{\mathbf{0}} \end{bmatrix}$$

and denote by $\underset{\sim}{\mathbf{M}}_{t,s}$ the fundamental solution matrix

$$\frac{d}{dt} \underset{\sim}{\mathbf{M}}_{t,s} = \bar{\underset{\sim}{\mathbf{A}}}_t \underset{\sim}{\mathbf{M}}_{t,s}, \qquad \underset{\sim}{\mathbf{M}}_{s,s} = \underset{\sim}{\mathbf{I}}.$$

By the continuity of $\underset{\sim}{\mathbf{M}}_{t,s}$ there is a constant $k_0 > 0$, so that

$$\|\underset{\sim}{\mathbf{M}}_{t,s}\| \le k_0, \qquad \forall t,\, s \in [0, T].$$

From (6.81), we have

$$\begin{bmatrix} \mathbf{x}_t \\ \mathbf{z}_t \end{bmatrix} = \underset{\sim}{\mathbf{M}}_{t,0} \begin{bmatrix} \mathbf{x}_0 \\ \mathbf{0} \end{bmatrix} + \int_0^t \underset{\sim}{\mathbf{M}}_{t,\lambda} \begin{bmatrix} \underset{\sim}{\mathbf{D}}_\lambda \\ \underset{\sim}{\mathbf{F}}_\lambda \end{bmatrix} d\mathbf{w}_\lambda.$$

Therefore,

$$E\left(\|\mathbf{x}_t\|^2 + \|\mathbf{z}_t\|^2\right) \le 2k_0^2 E\|\mathbf{x}_0\|^2 + 2k_0^2 \int_0^T \operatorname{tr}(\underset{\sim}{\mathbf{D}}_t \underset{\sim}{\mathbf{D}}_t^\tau + \underset{\sim}{\mathbf{F}}_t \underset{\sim}{\mathbf{F}}_t^\tau)\, dt \triangleq k_1, \tag{6.105}$$

and thus (6.103) has been established. By (1.24), we obtain

$$E\int_0^T \mathbf{x}_t^\tau \underset{\sim}{\mathbf{S}}_t \underset{\sim}{\mathbf{D}}_t \, d\mathbf{w}_t = 0. \tag{6.106}$$

Notice that $\mathbf{u}_t$ is $\mathscr{F}_t^{\mathbf{y}}$-measurable, then

$$E(\mathbf{u}_t - \hat{\mathbf{x}}_t)^\tau \mathbf{Q}_2(t)(\hat{\mathbf{x}}_t - \mathbf{x}_t) = E\left\{(\mathbf{u}_t - \hat{\mathbf{x}}_t)^\tau \underset{\sim}{\mathbf{Q}}_2(t) E\left[\frac{\hat{\mathbf{x}}_t - \mathbf{x}_t}{\mathscr{F}_t^{\mathbf{y}}}\right]\right\} = 0.$$

By Theorem 6.5, we see that

$$E(\mathbf{u}_t - \underset{\sim}{\mathbf{L}}_t\mathbf{x}_t)^T\underset{\sim}{\mathbf{Q}}_2(t)(\mathbf{u}_t - \underset{\sim}{\mathbf{L}}_t\mathbf{x}_t)$$

$$= E(\mathbf{u}_t - \underset{\sim}{\mathbf{L}}_t\hat{\mathbf{x}}_t)^T\underset{\sim}{\mathbf{Q}}_2(t)(\mathbf{u}_t - \underset{\sim}{\mathbf{L}}_t\hat{\mathbf{x}}_t)$$

$$+ E(\hat{\mathbf{x}}_t - \mathbf{x}_t)^T\underset{\sim}{\mathbf{L}}_t^T\underset{\sim}{\mathbf{Q}}_2(t)\underset{\sim}{\mathbf{L}}_t(\hat{\mathbf{x}}_t - \mathbf{x}_t)$$

$$= E(\mathbf{u}_t - \underset{\sim}{\mathbf{L}}_t\hat{\mathbf{x}}_t)^T\underset{\sim}{\mathbf{Q}}_2(t)(\mathbf{u}_t - \underset{\sim}{\mathbf{L}}_t\hat{\mathbf{x}}_t)$$

$$+ \operatorname{tr} E\underset{\sim}{\mathbf{L}}_t^T\underset{\sim}{\mathbf{Q}}_2(t)\underset{\sim}{\mathbf{L}}_t(\hat{\mathbf{x}}_t - \mathbf{x}_t)(\hat{\mathbf{x}}_t - \mathbf{x}_t)^T$$

$$= E(\mathbf{u}_t - \underset{\sim}{\mathbf{L}}_t\hat{\mathbf{x}}_t)^T\underset{\sim}{\mathbf{Q}}_2(t)(\mathbf{u}_t - \underset{\sim}{\mathbf{L}}_t\hat{\mathbf{x}}_t) + \operatorname{tr} \underset{\sim}{\mathbf{L}}_t^T\underset{\sim}{\mathbf{Q}}_2(t)\underset{\sim}{\mathbf{L}}_t\underset{\sim}{\mathbf{P}}_t \tag{6.107}$$

Similarly, we obtain

$$E\mathbf{x}_0^T\underset{\sim}{\mathbf{S}}_0\mathbf{x}_0 = E\mathbf{x}_0^T\underset{\sim}{\mathbf{S}}_0E\mathbf{x}_0 + \operatorname{tr} \underset{\sim}{\mathbf{S}}_0\underset{\sim}{\mathbf{R}}_{\mathbf{x}_0}. \tag{6.108}$$

Taking the expectation of (6.102) and using (6.106)–(6.108) leads to

$$EJ(\mathbf{u}) = E\mathbf{x}_0^T\underset{\sim}{\mathbf{S}}_0E\mathbf{x}_0 + \operatorname{tr} \underset{\sim}{\mathbf{S}}_0\underset{\sim}{\mathbf{R}}_{\mathbf{x}_0} + \int_0^T\left(\operatorname{tr} \underset{\sim}{\mathbf{S}}_t\underset{\sim}{\mathbf{D}}_t\underset{\sim}{\mathbf{D}}_t^T\,dt + \operatorname{tr} \underset{\sim}{\mathbf{L}}_t^T\underset{\sim}{\mathbf{Q}}_2(t)\underset{\sim}{\mathbf{L}}_t\underset{\sim}{\mathbf{P}}_t\right)dt$$

$$+ \int_0^T(\mathbf{u}_t - \underset{\sim}{\mathbf{L}}_t\hat{\mathbf{x}}_t)^T\underset{\sim}{\mathbf{Q}}_2(t)(\mathbf{u}_t - \underset{\sim}{\mathbf{L}}_t\hat{\mathbf{x}}_t)\,dt, \tag{6.109}$$

in which the only term depending on the control is the last one. Consequently, $EJ(\mathbf{u})$ reaches its infimum if and only if

$$\mathbf{u}_t^* = \underset{\sim}{\mathbf{L}}_t\hat{\mathbf{x}}_t, \tag{6.110}$$

and under this control, (6.100) is valid.

Finally, we have to prove $\{\mathbf{u}_t^*\} \in U$. From (6.31) and (6.46), it is known that

$$\hat{\mathbf{x}}_t = \underset{\sim}{\boldsymbol{\Psi}}_{t,0}\hat{\mathbf{x}}_0 + \underset{\sim}{\boldsymbol{\Psi}}_{t,0}\int_0^t\underset{\sim}{\boldsymbol{\Psi}}_{0,\lambda}\underset{\sim}{\mathbf{K}}_\lambda\,d\mathbf{y}_\lambda.$$

Hence

$$\mathbf{u}_t^* = -\mathbf{Q}_2^{-1}(t)\mathbf{B}_t^T\mathbf{S}_t\left(\boldsymbol{\Psi}_{t,0}\hat{\mathbf{x}}_0 + \boldsymbol{\Psi}_{t,0}\int_0^t \boldsymbol{\Psi}_{0,\lambda}(\mathbf{D}_\lambda\mathbf{F}_\lambda^T + \mathbf{P}_\lambda\mathbf{C}_\lambda^T)(\mathbf{F}_\lambda\mathbf{F}_\lambda^T)^{-1}\,d\mathbf{y}_\lambda.\right. \tag{6.111}$$

By (6.111) and (6.4), with $\{\mathbf{u}_t\}$ belonging to $\mathscr{U}$ given by (6.80), we find that

$$\mathbf{c} = E\mathbf{x}_0 - \mathbf{R}_{\mathbf{x}_0\mathbf{y}_0}\mathbf{R}_{\mathbf{y}_0}^+ E\mathbf{y}_0, \qquad \mathbf{C} = \mathbf{R}_{\mathbf{x}_0\mathbf{y}_0}\mathbf{R}_{\mathbf{y}_0}^+$$

$$\mathbf{U}_1(t) = -\mathbf{Q}_2^{-1}(t)\mathbf{B}_t^T\mathbf{S}_t\boldsymbol{\Psi}_{t,0}, \qquad \mathbf{U}_2(t) = \boldsymbol{\Psi}_{0,t}(\mathbf{D}_t\mathbf{F}_t^T + \mathbf{P}_t\mathbf{C}_t^T)(\mathbf{F}_t\mathbf{F}_t^T)^{-1}.$$

Now the matrix $\mathbf{S}_t\boldsymbol{\Psi}_{t,0}$ is continuous on $[0, T]$ and hence is bounded on this interval. By the conditions of the theorem, it follows that

$$\int_0^T \left\|\mathbf{B}_t\mathbf{Q}_2^{-1}(t)\mathbf{B}_t^T\right\| dt < \infty,$$

and this implies that $\int_0^T \|\mathbf{B}_t\mathbf{U}_1(t)\|\,dt < \infty$. Since in Theorem 6.1 it is proved that

$$\int_0^T \left(\|\mathbf{U}_2(t)\mathbf{C}_t\| + \|\mathbf{U}_2(t)\mathbf{F}_t\|^2\right) dt < \infty,$$

we conclude that $\{\mathbf{u}_t^*\} \in \mathscr{U}$. □

Remark. Suppose $\mathbf{D}_t \equiv \mathbf{0}$, $\mathbf{F}_t \equiv \mathbf{0}$, and $\mathbf{x}_0$ and $\mathbf{y}_0$ are both deterministic, then, from (6.102),

$$J(\mathbf{u}) = \mathbf{x}_0^T\mathbf{S}_0\mathbf{x}_0 + \int_0^T (\mathbf{u}_t - \mathbf{L}_t\mathbf{x}_t)^T\mathbf{Q}_2(t)(\mathbf{u}_t - \mathbf{L}_t\mathbf{x}_t)\,dt.$$

Hence for deterministic systems with a quadratic performance index the optimal control is given by

$$\mathbf{u}_t^* = \mathbf{L}_t\mathbf{x}_t. \tag{6.112}$$

Theorem 6.6 shows that for the LQG problem, the optimal stochastic control has the same feedback gain as for the deterministic case; the only difference is that the state is replaced by its filtering value. This is sometimes called the certainty equivalence principle.

6.5. STOCHASTIC DIFFERENTIAL GAMES

The problem solved in the last section is a one-side control problem in the sense that there was only one controller. In this section, we shall give a result on a two-side control problem, that is to say, a problem in which there are two controllers.

Let the dynamic system and the measurement equations be given by

$$d\mathbf{x}_t = \underset{\sim}{\mathbf{A}}_t \mathbf{x}_t\, dt + \underset{\sim}{\mathbf{B}}_{pt} \mathbf{u}_{pt}\, dt + \underset{\sim}{\mathbf{B}}_{et} \mathbf{u}_{et}\, dt + \underset{\sim}{\mathbf{D}}_t\, d\mathbf{w}_t \tag{6.113}$$

$$d\mathbf{y}_t = \underset{\sim}{\mathbf{C}}_t \mathbf{x}_t\, dt + \underset{\sim}{\mathbf{F}}_t\, d\mathbf{w}_t, \qquad 0 \le t \le T < \infty, \tag{6.114}$$

where $\mathbf{w}_t$ is a Wiener process independent of $[\mathbf{x}_0^\tau \mathbf{y}_0^\tau]^\tau$ and $\underset{\sim}{\mathbf{A}}_t, \underset{\sim}{\mathbf{B}}_t \triangleq [\underset{\sim}{\mathbf{B}}_{pt}, \underset{\sim}{\mathbf{B}}_{et}]$ $\underset{\sim}{\mathbf{C}}_t, \underset{\sim}{\mathbf{D}}_t, \underset{\sim}{\mathbf{F}}_t$ satisfy (6.3).

Equation (6.113) has two controls $\mathbf{u}_{pt}$ and $\mathbf{u}_{et}$ with r_p and r_e dimensions, respectively. We denote $\mathscr{U}$ defined by (6.80) by $\mathscr{U}_p$ if $\underset{\sim}{\mathbf{U}}_1(t)$ is a $r_p \times n$ matrix and by $\mathscr{U}_e$ if $\underset{\sim}{\mathbf{U}}_1(t)$ is a $r_e \times n$ matrix.

Denote

$$J(\mathbf{u}_p, \mathbf{u}_e) = \mathbf{x}_T^\tau \underset{\sim}{\mathbf{Q}}_0 \mathbf{x}_T + \int_0^T \left(\mathbf{x}_t^\tau \underset{\sim}{\mathbf{Q}}_{1t} \mathbf{x}_t + \mathbf{u}_{pt}^\tau \underset{\sim}{\mathbf{Q}}_{2t} \mathbf{u}_{pt} - \mathbf{u}_{et}^\tau \underset{\sim}{\mathbf{Q}}_{3t} \mathbf{u}_{et} \right) dt, \tag{6.115}$$

where $\underset{\sim}{\mathbf{Q}}_0 \ge \underset{\sim}{\mathbf{0}}$, $\underset{\sim}{\mathbf{Q}}_{1t} \ge \underset{\sim}{\mathbf{0}}$, $\underset{\sim}{\mathbf{Q}}_{2t} > \underset{\sim}{\mathbf{0}}$, $\underset{\sim}{\mathbf{Q}}_{3t} > \underset{\sim}{\mathbf{0}}$.

The performance index is

$$\sup_{\mathbf{u}_e \in \mathscr{U}_e} \inf_{\mathbf{u}_p \in \mathscr{U}_p} EJ(\mathbf{u}_p, \mathbf{u}_e), \tag{6.116}$$

in which the purposes of $\mathbf{u}_e$ and $\mathbf{u}_p$ are opposed: $\mathbf{u}_p$ intends to make EJ as small as possible while $\mathbf{u}_e$ intends to make it as large as possible. If

$$\sup_{\mathbf{u}_e \in \mathscr{U}_e} \inf_{\mathbf{u}_p \in \mathscr{U}_p} EJ(\mathbf{u}_p, \mathbf{u}_e) = \inf_{\mathbf{u}_p \in \mathscr{U}_p} \sup_{\mathbf{u}_e \in \mathscr{U}_e} EJ(\mathbf{u}_p, \mathbf{u}_e) \triangleq V, \tag{6.117}$$

then the game is said to have a value, and the quantity V is called the value of the game. If, in addition, there is a control $\mathbf{u}^* = [\mathbf{u}_p^{*\tau} \mathbf{u}_e^{*\tau}]^\tau$,

$$\mathbf{u}_p^* \in \mathscr{U}_p, \qquad \mathbf{u}_e^* \in \mathscr{U}_e$$

so that

$$EJ\big(\mathbf{u}_p^*, \mathbf{u}_e\big) \le EJ\big(\mathbf{u}_p^*, \mathbf{u}_e^*\big) = V \le EJ\big(\mathbf{u}_p, \mathbf{u}_e^*\big), \qquad \forall \mathbf{u}_p \in \mathscr{U}_p, \quad \mathbf{u}_e \in \mathscr{U}_e \tag{6.118}$$

then $\mathbf{u}^*$ is called the optimal strategy or the saddle point.

We need the following auxiliary equation:

$$-\dot{\mathbf{S}}_t = \mathbf{A}_t^T\mathbf{S}_t + \mathbf{S}_t\mathbf{A}_t + \mathbf{Q}_{1t} - \mathbf{S}_t\mathbf{B}_{pt}\mathbf{Q}_{2t}^{-1}\mathbf{B}_{pt}^T\mathbf{S}_t + \mathbf{S}_t\mathbf{B}_{et}\mathbf{Q}_{3t}^{-1}\mathbf{B}_{et}^T\mathbf{S}_t, \qquad \mathbf{S}_T = \mathbf{Q}_0 \tag{6.119}$$

Theorem 6.7. *Assume that for the system* (6.113), (6.114) $\|\mathbf{C}_t^T(\mathbf{F}_t\mathbf{F}_t^T)^{-1}\mathbf{C}_t\|$, $\|\mathbf{B}_{pt}\mathbf{Q}_{2t}^{-1}\mathbf{B}_{pt}^T\|$, $\|\mathbf{B}_{et}\mathbf{Q}_{3t}^{-1}\mathbf{B}_{et}^T\|$, $\|\mathbf{A}_t\|$, $\|\mathbf{C}_t\|^2$, $\|\mathbf{D}_t\|^2$, $\|\mathbf{F}_t\|^2$, $\|\mathbf{B}_{pt}\|^2$ *and* $\|\mathbf{B}_{et}\|^2$ *are integrable on* $[0, T]$ *and that* (6.119) *is solvable. Then for the index* (6.116), *the game has a value V and a saddle point* $\mathbf{u}^* = [\mathbf{u}_p^{*T}\mathbf{u}_e^{*T}]^T$ *such that*

$$V = E\mathbf{x}_0^T\mathbf{S}_0\mathbf{x}_0 + \operatorname{tr}\mathbf{S}_0\mathbf{R}_{\mathbf{x}_0} + \int_0^t \operatorname{tr}\Big[\mathbf{D}_t\mathbf{D}_t^T\mathbf{S}_t + \mathbf{S}_t\mathbf{B}_{pt}\mathbf{Q}_{2t}^{-1}\mathbf{B}_{pt}^T\mathbf{S}_t\mathbf{P}_t - \mathbf{S}_t\mathbf{B}_{et}\mathbf{Q}_{3t}^{-1}\mathbf{B}_{et}^T\mathbf{S}_t\mathbf{P}_t\Big]\,dt \tag{6.120}$$

$$\mathbf{u}_{pt}^* = -\mathbf{Q}_{2t}^{-1}\mathbf{B}_{pt}^T\mathbf{S}_t\hat{\mathbf{x}}_t, \qquad \mathbf{u}_{et}^* = \mathbf{Q}_{3t}^{-1}\mathbf{B}_{et}^T\mathbf{S}_t\hat{\mathbf{x}}_t. \tag{6.121}$$

Proof. Set

$$\mathbf{u}_t = \begin{bmatrix} \mathbf{u}_{pt} \\ \mathbf{u}_{et} \end{bmatrix}, \qquad \mathbf{B}_t = \big[\mathbf{B}_{pt}\mathbf{B}_{et}\big], \qquad \mathbf{Q}_t = \begin{bmatrix} \mathbf{Q}_{2t} & \mathbf{0} \\ \mathbf{0} & -\mathbf{Q}_{3t} \end{bmatrix}.$$

Then (6.113), (6.115), and (6.119) can be rewritten, respectively, as

$$d\mathbf{x}_t = \mathbf{A}_t\mathbf{x}_t\,dt + \mathbf{B}_t\mathbf{u}_t\,dt + \mathbf{D}_t\,d\mathbf{w}_t, \tag{6.122}$$

$$J(\mathbf{u}_p, \mathbf{u}_e) = \mathbf{x}_T^T\mathbf{Q}_0\mathbf{x}_T + \int_0^T\Big(\mathbf{x}_t^T\mathbf{Q}_{1t}\mathbf{x}_t + \mathbf{u}_t^T\mathbf{Q}_t\mathbf{u}_t\Big)\,dt, \tag{6.123}$$

$$-\dot{\mathbf{S}}_t = \mathbf{A}_t^T\mathbf{S}_t + \mathbf{S}_t\mathbf{A}_t + \mathbf{Q}_{1t} - \mathbf{S}_t\mathbf{B}_t\mathbf{Q}_t^{-1}\mathbf{B}_t^T\mathbf{S}_t, \tag{6.124}$$

which are in the same form as (6.1), (6.95), and (6.119). Hence by analogy with (6.102)–(6.109), we obtain

$$EJ(\mathbf{u}_p, \mathbf{u}_e) = E\mathbf{x}_0^T \mathbf{S}_0 E\mathbf{x}_0 + \operatorname{tr} \mathbf{S}_0 \mathbf{R}_{\mathbf{x}_0}$$

$$+ \int_0^T \operatorname{tr}\left[\mathbf{S}_t \mathbf{D}_t \mathbf{D}_t^T + \mathbf{S}_t \mathbf{B}_t \mathbf{Q}_t^{-1} \mathbf{B}_t^T \mathbf{S}_t \mathbf{P}_t\right] dt$$

$$+ \int_0^T \left(\mathbf{u}_t + \mathbf{Q}_t^{-1} \mathbf{B}_t^T \mathbf{S}_t \hat{\mathbf{x}}_t\right)^T \mathbf{Q}_t \left(\mathbf{u}_t + \mathbf{Q}_t^{-1} \mathbf{B}_t^T \mathbf{S}_t \hat{\mathbf{x}}_t\right) dt$$

$$= E\mathbf{x}_0^T \mathbf{S}_0 E\mathbf{x}_0 + \operatorname{tr} \mathbf{S}_0 \mathbf{R}_{\mathbf{x}_0} + \int_0^T \operatorname{tr}\left[\mathbf{S}_t \mathbf{D}_t \mathbf{D}_t^T + \mathbf{S}_t \mathbf{B}_{pt} \mathbf{Q}_{2t}^{-1} \mathbf{B}_{pt}^T \mathbf{S}_t \mathbf{P}_t\right.$$

$$\left. - \mathbf{S}_t \mathbf{B}_{et} \mathbf{Q}_{3t}^{-1} \mathbf{B}_{et}^T \mathbf{S}_t \mathbf{P}_t\right] dt$$

$$+ \int_0^T \left[\left(\mathbf{u}_t + \mathbf{Q}_{2t}^{-1} \mathbf{B}_{pt}^T \mathbf{S}_t \hat{\mathbf{x}}_t\right)^T \mathbf{Q}_{2t} \left(\mathbf{u}_p + \mathbf{Q}_{2t}^{-1} \mathbf{B}_{pt}^T \mathbf{S}_t \hat{\mathbf{x}}_t\right)\right.$$

$$\left. - \left(\mathbf{u}_{et} - \mathbf{Q}_{3t}^{-1} \mathbf{B}_{et}^T \mathbf{S}_t \hat{\mathbf{x}}_t\right)^T \mathbf{Q}_{3t} \left(\mathbf{u}_{et} - \mathbf{Q}_{3t}^{-1} \mathbf{B}_{et}^T \mathbf{S}_t \hat{\mathbf{x}}_t\right)\right] dt,$$

from which it is immediate that (6.117) and (6.118) hold and the game value is given by (6.120) if the control is given by (6.121). The fact that $\{\mathbf{u}_{pt}^*\} \in \mathscr{U}_p$ $\{\mathbf{u}_{et}^*\} \in \mathscr{U}_e$ is proved in a similar way as in Theorem 6.6. □

CHAPTER 7

Singular Problems

7.1. INTRODUCTION

In Chapter 6, when we sought the equations satisfied by $\hat{\mathbf{x}}_t$ and $\mathbf{P}_t$, the matrix $\mathbf{F}_t\mathbf{F}_t^\tau$ was supposed to be positive definite since once $\mathbf{F}_t\mathbf{F}_t^\tau$ is degenerate, $\hat{\mathbf{x}}_t$ and $\mathbf{P}_t$ may be discontinuous and so they obviously cannot satisfy a differential equation. This point can be seen from the following simple example.

Let $n = m = 1$, $x_t \equiv \theta$. Then the state equation is

$$dx_t = 0.$$

Let the measurement equation be given by

$$dy_t = \theta\, dt + F_t\, dw_t,$$

$$F_t = \begin{cases} 1, & t \in [0,1), \\ 0, & 1 \le t \le T, \end{cases}$$

with $\hat{x}_0 = 0$ and $R_0 = 1$.

According to Theorem 5.1, $\hat{x}_t$ on $[0,1)$ satisfies

$$d\hat{x}_t = \hat{x}_t\, dt + P_t(dy_t - \hat{x}_t\, dt)$$

and P_t satisfies

$$\dot{P}_t = -P_t^2, \qquad P_0 = 1.$$

Then

$$P_t = \frac{1}{1+t}, \qquad t \in [0,1).$$

But for $t \geq 1$, $dy_t = \theta\, dt$. Hence

$$\theta = \frac{1}{t-1}(y_t - y_1), \qquad t > 1$$

and $P_t = 0$ for $t > 1$. Consequently, P_t is discontinuous at $t = 1$, and P_t cannot be a solution of a differential equation on $[0, T]$. The discontinuity of P_t implies the discontinuity of $\hat{x}_t$ since x_t is continuous, and so $\hat{x}_t$ itself cannot satisfy a stochastic differential equation. This example shows that for filtering equations, the condition $\underset{\sim}{\mathbf{F}}_t \underset{\sim}{\mathbf{F}}_t^{\tau} > \underset{\sim}{\mathbf{0}}$ is natural.

However, in practice, $\underset{\sim}{\mathbf{F}}_t \underset{\sim}{\mathbf{F}}_t^{\tau}$ may be degenerate. The filtering and interpolation problems for this case of practical importance will be considered in this chapter. Further, in the LQG problem, $\underset{\sim}{\mathbf{Q}}_2(t)$ is assumed positive definite. When $\underset{\sim}{\mathbf{Q}}_2(t)$ is possibly degenerate, it is called the singular control problem, which will also be considered in this chapter.

We shall need a fact from measure theory, and it is given here without proof.

Lemma 7.1. *Denote by Π the class of subsets in Ω with the property that $A_1 \cap A_2 \in \Pi$ if $A_1 \in \Pi$, $A_2 \in \Pi$ and denote by $\mathscr{F}(\Pi)$ the minimal σ-algebra containing Π. Let $\mathscr{L}$ be a family of functions defined on Ω such that:*

1. *All constants $\in \mathscr{L}$.*
2. *$af_1 + bf_2 \in \mathscr{L}$ if $f_1 \in \mathscr{L}$, $f_2 \in \mathscr{L}$ and a, b are constants.*
3. *$f \in \mathscr{L}$ if $f_n \in \mathscr{L}$ and $0 \leq f_n \uparrow f$.*

Then $\mathscr{L}$ includes all $\mathscr{F}(\Pi)$-measurable functions if $I_A \in \mathscr{L}$ for all $A \in \Pi$.

This lemma can be proved by the standard method (termed the Π-system and λ-system) used in measure theory (see, e.g., [4]).

Let $\boldsymbol{\zeta}_t$ $0 \leq t \leq T$ be a stochastic process, and let $\mathscr{F}_t^{\boldsymbol{\zeta}}$ be the σ-algebra generated by $\{\boldsymbol{\zeta}_s, 0 \leq s \leq t\}$.

Lemma 7.2. *Let $\boldsymbol{\xi}$ be a random variable measurable with respect to $\mathscr{F}_t^{\boldsymbol{\zeta}}$, $E\boldsymbol{\xi}^2 < \infty$, where $\boldsymbol{\zeta}_t = [\zeta_t^1 \cdots \zeta_t^m]^{\tau}$. Then there are linear combinations*

$$\boldsymbol{\xi}_n \triangleq \sum_{k=1}^{n} a_k(n) \prod_{j=1}^{v(k,n)} \prod_{s,i=1}^{m} \sin \lambda_{sj}(n,k) \zeta_{t_j(n,k)}^{s} \cos \mu_{ij}(n,k) \zeta_{t_j(n,k)}^{i} \tag{7.1}$$

of the random variables

$$\prod_{j=1}^{v} \prod_{s,i=1}^{m} \sin \lambda_{sj} \zeta_{t_j}^{s} \cos \mu_{ij} \zeta_{t_j}^{i}, \qquad 0 \leq t_j \leq t, \tag{7.2}$$

such that the L^2 approximation

$$E|\xi - \xi_n I_{[|\xi_n| < 2|\xi|]}|^2 \underset{n\to\infty}{\to} 0 \tag{7.3}$$

holds.

Proof. Denote by $\mathscr{L}$ the closed set of random variables of type (7.2) obtained by passing to the limit. If each f_i, $i \geq 1$, is a limit of linear combinations of random variables of type (7.2) and if $f_i \to f$, then f can also be expressed as a limit of the aforementioned linear combinations by the method of selecting the diagonal sequence. Then, by induction, it is known that any $f \in \mathscr{L}$ is a limit of such combinations.

Clearly, $\mathscr{L}$ meets the conditions required in Lemma 7.1. Let Γ_i^s be an interval on the real line, and let Π be composed of sets of the form

$$\prod_{i=1}^{k} \left[\zeta_{t_i}^1 \in \Gamma_i^1\right] \cdots \left[\zeta_{t_i}^m \in \Gamma_i^m\right], \qquad 0 \leq t_i \leq t.$$

Define

$$f = \prod_{i=1}^{k} I_{[\zeta_{t_i}^1 \in \Gamma_i^1]} \cap \cdots \cap I_{[\zeta_{t_i}^m \in \Gamma_i^m]}.$$

Then f can be approximated pointwise by linear combinations of random variables of type (7.1). Obviously $I_{[\zeta_{t_i}^1 \in \Gamma_i^1]}$ can be approximated pointwise by trigonometrical series of the form

$$\sum_{i=1}^{n} \left(a_i \sin \lambda_i \zeta_{t_i}^1 + b_i \cos \mu_i \zeta_{t_i}^1\right).$$

Therefore, $f \in \mathscr{L}$ and by Lemma 7.1, $\mathscr{L}$ contains all $\mathscr{F}(\Pi)$-measurable random variables. But clearly $\mathscr{F}(\Pi) = \mathscr{F}_t^\zeta$. Hence for an $\mathscr{F}_t^\zeta$-measurable random variable ξ, there are random variables ξ_n expressed by (7.2) such that

$$\lim_{n\to\infty} \xi_n = \xi,$$

and so

$$\lim_{n\to\infty} \xi_n I_{[|\xi_n| < 2|\xi|]} = \xi,$$

and by the dominated convergence theorem

$$E|\xi_n I_{[|\xi_n| < 2|\xi|]} - \xi|^2 \underset{n\to\infty}{\to} 0. \qquad \square$$

Lemma 7.3. *Let ζ_t and η_t be m-dimensional stochastic processes and let*

$$\zeta_t^\varepsilon \triangleq \zeta_t + \sqrt{\varepsilon}\,\eta_t. \tag{7.4}$$

Then

$$E\left(\xi - E\left(\frac{\xi}{\mathscr{F}_t^{\zeta^\varepsilon}}\right)\right)^2 \underset{\varepsilon\to 0}{\to} 0 \tag{7.5}$$

for any $\mathscr{F}_t^{\zeta}$-measurable integrable random variable ξ.

Proof. First assume ξ is bounded by a constant k_0: $|\xi| < k_0$. Let ξ_n be given by (7.2). Then by Lemma 7.2, it is easy to see that

$$\left(\xi - \xi_n I_{[|\xi_n| < 2k_0]}\right)^2 \underset{n\to\infty}{\to} 0, \tag{7.6}$$

$$\begin{aligned} E\left(\xi - E\left(\frac{\xi}{\mathscr{F}_t^{\zeta^\varepsilon}}\right)\right)^2 &\le 3E\left(\xi - \xi_n I_{[|\xi_n| < 2k_0]}\right)^2 \\ &\quad + 3E\left[\xi_n I_{[|\xi_n| < 2k_0]} - E\left(\frac{\xi_n I_{[|\xi_n| < 2k_0]}}{\mathscr{F}_t^{\zeta^\varepsilon}}\right)\right]^2 \\ &\quad + 3E\left[E\left(\frac{\xi_n I_{[|\xi_n| < 2k_0]}}{\mathscr{F}_t^{\zeta^\varepsilon}}\right) - E\left(\xi/\mathscr{F}_t^{\zeta^\varepsilon}\right)\right]^2, \end{aligned} \tag{7.7}$$

for which the first term by (7.6) can be made less than $\delta/3$ if $n \ge N$. Then the last term will be also less than $\delta/3$ since it is dominated by

$$3E\left\{E\left[\frac{\left(\xi_n I_{[|\xi_n| < 2k_0]} - \xi\right)^2}{\mathscr{F}_t^{\zeta^\varepsilon}}\right]\right\} < \frac{\delta}{3}.$$

For fixed n, denote

$$\xi_n^\varepsilon = \sum_{l=1}^{n} a_l(n) \prod_{j=1}^{v(k,n)} \prod_{s,i=1}^{m} \sin\lambda_{sj}(n,k)\zeta_{t_j(n,k)}^{\varepsilon s} \cos\mu_{ij}(n,k)\zeta_{t_j(n,k)}^{\varepsilon s}$$

where

$$\zeta^{\varepsilon s}_{t_j(n,k)} = \zeta^{s}_{t_j(n,k)} + \sqrt{\varepsilon}\,\eta^{s}_{t_j(n,k)}.$$

Clearly ξ^ε_n is uniformly bounded in ε for fixed n and $\xi^\varepsilon_n \underset{\varepsilon\to 0}{\to} \xi_n$. Thus

$$3E\left[\xi_n I_{[|\xi_n| < 2k_0]} - E\left(\frac{\xi_n I_{[|\xi_n| < 2k_0]}}{\mathscr{F}_t^{\zeta^\varepsilon}}\right)\right]^2$$

$$\leq 3E\left[\xi_n I_{[|\xi_n| < 2k_0]} - \xi^\varepsilon_n I_{[|\xi^\varepsilon_n| < 2k_0]}\right]^2 \underset{\varepsilon\to 0}{\to} 0.$$

Hence the right-hand side of (7.7) goes to zero as $\varepsilon \to 0$ and then $n \to \infty$.

Now we make no boundedness assumption on ξ. Then we have

$$E\left(\xi - E\left(\frac{\xi}{\mathscr{F}_t^{\zeta^\varepsilon}}\right)\right)^2 \leq 3E(\xi - \xi I_{[|\xi| < c]})^2$$

$$+3E\left[\xi I_{[|\xi| < c]} - E\left(\frac{\xi I_{[|\xi| < c]}}{\mathscr{F}_t^{\zeta^\varepsilon}}\right)\right]^2$$

$$+3E\left[E\left(\frac{\xi I_{[|\xi| < c]}}{\mathscr{F}_t^{\zeta^\varepsilon}}\right) - E\left(\frac{\xi}{\mathscr{F}_t^{\zeta^\varepsilon}}\right)\right]^2 \underset{\varepsilon\to 0}{\to} 0, \quad (7.8)$$

since we have already shown that the lemma is valid for $\xi I_{[|\xi| < c]}$. □

7.2. SUBOPTIMAL FILTERING

Consider on $[0, T]$, the system of stochastic differential equations:

$$d\mathbf{x}_t = \underset{\sim}{\mathbf{A}}_t \mathbf{x}_t\,dt + \underset{\sim}{\mathbf{D}}_t\,d\mathbf{w}_t, \quad (7.9)$$

$$d\mathbf{y}_t = \underset{\sim}{\mathbf{C}}_t \mathbf{x}_t\,dt + \underset{\sim}{\mathbf{F}}_t\,d\mathbf{w}_t, \quad (7.10)$$

where $\{\mathbf{w}_t\}$ is a generalized Wiener process, and $\|\underset{\sim}{\mathbf{A}}_t\|$, $\|\underset{\sim}{\mathbf{C}}_t\|^2$, $\|\underset{\sim}{\mathbf{D}}_t\|^2$, and $\|\underset{\sim}{\mathbf{F}}_t\|^2$ are integrable on $[0, T]$.

Notice we do not require $\underset{\sim}{\mathbf{F}}_t \underset{\sim}{\mathbf{F}}_t^\tau$ to be positive definite.

Assume $\{\mathbf{w}_t\}$ is uncorrelated with $[\mathbf{x}_0^\tau, \mathbf{y}_0^\tau]^\tau$ and $\{\tilde{\mathbf{w}}_t\}$ is an m-dimensional generalized Wiener process uncorrelated with $\{\mathbf{w}_t\}$ and $[\mathbf{x}_0^\tau \mathbf{y}_0^\tau]^\tau$.

Denote

$$\mathbf{y}_t^\varepsilon = \mathbf{y}_t + \sqrt{\varepsilon}\,\tilde{\mathbf{w}}_t, \tag{7.11}$$

$$\overline{\mathbf{w}}_t = \left[\mathbf{w}_t^T, \tilde{\mathbf{w}}_t^T\right]^T, \qquad \underset{\sim}{\mathbf{F}}_t' = \left[\underset{\sim}{\mathbf{F}}_t\, \sqrt{\varepsilon}\,\underset{\sim}{\mathbf{I}}\right], \qquad \underset{\sim}{\mathbf{D}}_t' = \left[\underset{\sim}{\mathbf{D}}_t\, \underset{\sim}{\mathbf{0}}\right].$$

Clearly, $\{\overline{\mathbf{w}}_t\}$ is an $l + m$-dimensional generalized Wiener process and (7.9) and (7.11) can be rewritten as

$$d\mathbf{x}_t = \underset{\sim}{\mathbf{A}}_t \mathbf{x}_t\,dt + \underset{\sim}{\mathbf{D}}_t'\,d\overline{\mathbf{w}}_t, \tag{7.12}$$

$$d\mathbf{y}_t^\varepsilon = \underset{\sim}{\mathbf{C}}_t \mathbf{x}_t\,dt + \underset{\sim}{\mathbf{F}}_t'\,d\overline{\mathbf{w}}_t, \tag{7.13}$$

with $\underset{\sim}{\mathbf{F}}_t'\underset{\sim}{\mathbf{F}}_t'^T = \underset{\sim}{\mathbf{F}}_t\underset{\sim}{\mathbf{F}}_t^T + \varepsilon\underset{\sim}{\mathbf{I}} > 0$.

Hence Theorem 5.1 is applicable to the system (7.12), (7.13). Denote by $\hat{\mathbf{x}}_t^\varepsilon$ and $\underset{\sim}{\mathbf{P}}_t^\varepsilon$ the LUMVE of $\mathbf{x}_t$ based on $\{\mathbf{y}_\lambda^\varepsilon, 0 \le \lambda \le t\}$ and its estimation error covariance matrix. Then

$$d\hat{\mathbf{x}}_t^\varepsilon = \underset{\sim}{\mathbf{A}}_t\hat{\mathbf{x}}_t^\varepsilon\,dt + \left(\underset{\sim}{\mathbf{D}}_t\underset{\sim}{\mathbf{F}}_t^T + \underset{\sim}{\mathbf{P}}_t^\varepsilon\underset{\sim}{\mathbf{C}}_t^T\right)\left(\underset{\sim}{\mathbf{F}}_t\underset{\sim}{\mathbf{F}}_t^T + \varepsilon\underset{\sim}{\mathbf{I}}\right)^{-1}\left(d\mathbf{y}_t^\varepsilon - \underset{\sim}{\mathbf{C}}_t\hat{\mathbf{x}}_t^\varepsilon\,dt\right) \tag{7.14}$$

$$\begin{aligned}\dot{\underset{\sim}{\mathbf{P}}}_t^\varepsilon &= \underset{\sim}{\mathbf{P}}_t^\varepsilon\underset{\sim}{\mathbf{A}}_t^T + \underset{\sim}{\mathbf{A}}_t\underset{\sim}{\mathbf{P}}_t^{\varepsilon T} + \underset{\sim}{\mathbf{D}}_t\underset{\sim}{\mathbf{D}}_t^T\\ &\quad - \left(\underset{\sim}{\mathbf{D}}_t\underset{\sim}{\mathbf{F}}_t^T + \underset{\sim}{\mathbf{P}}_t^\varepsilon\underset{\sim}{\mathbf{C}}_t^T\right)\left(\underset{\sim}{\mathbf{F}}_t\underset{\sim}{\mathbf{F}}_t^T + \varepsilon\underset{\sim}{\mathbf{I}}\right)^{-1}\left(\underset{\sim}{\mathbf{D}}_t\underset{\sim}{\mathbf{F}}_t^T + \underset{\sim}{\mathbf{P}}_t^\varepsilon\underset{\sim}{\mathbf{C}}_t^T\right)^T\end{aligned} \tag{7.15}$$

with the initial conditions $\hat{\mathbf{x}}_0^\varepsilon = \hat{\mathbf{x}}_0$, $\underset{\sim}{\mathbf{P}}_0^\varepsilon = \underset{\sim}{\mathbf{R}}_0$, where $\hat{\mathbf{x}}_0$ and $\underset{\sim}{\mathbf{R}}_0$ are defined by (6.4) and (6.5).

For a given ε, Equation (7.15) is computable but (7.14) is not, because the process $\mathbf{y}_t^\varepsilon$ cannot be obtained as real observations. Now define the process $\tilde{\mathbf{x}}_t^\varepsilon$ which is computable:

$$d\tilde{\mathbf{x}}_t^\varepsilon = \underset{\sim}{\mathbf{A}}_t\tilde{\mathbf{x}}_t^\varepsilon\,dt + \left(\underset{\sim}{\mathbf{D}}_t\underset{\sim}{\mathbf{F}}_t^T + \underset{\sim}{\mathbf{P}}_t^\varepsilon\underset{\sim}{\mathbf{C}}_t^T\right)\left(\underset{\sim}{\mathbf{F}}_t\underset{\sim}{\mathbf{F}}_t^T + \varepsilon\underset{\sim}{\mathbf{I}}\right)^{-1}\left(d\mathbf{y}_t - \underset{\sim}{\mathbf{C}}_t\tilde{\mathbf{x}}_t^\varepsilon\,dt\right) \tag{7.16}$$

and write

$$\tilde{\underset{\sim}{\mathbf{P}}}_t^\varepsilon = E\left(\mathbf{x}_t - \tilde{\mathbf{x}}_t^\varepsilon\right)\left(\mathbf{x}_t - \tilde{\mathbf{x}}_t^\varepsilon\right)^T. \tag{7.17}$$

In what follows, l.i.m. denotes the limit in mean square sense, that is to say,

$$\underset{n\to\infty}{\text{l.i.m.}}\,\zeta_n = \zeta \quad \text{means} \quad \lim_{n\to\infty} E\|\zeta_n - \zeta\|^2 = 0.$$

Theorem 7.1. *Provided* $\|\underset{\sim}{\mathbf{A}}_t$, $\|\underset{\sim}{\mathbf{C}}_t\|^2$, $\|\underset{\sim}{\mathbf{D}}_t\|^2$, *and* $\|\underset{\sim}{\mathbf{F}}_t\|^2$ *are integrable on* $[0, T]$, *then for the system* (7.9), (7.10), $\tilde{\mathbf{x}}_t^\varepsilon$ *and* $\tilde{\underset{\sim}{\mathbf{P}}}_t^\varepsilon$ *defined by the suboptimal filtering equations* (7.16) *and* (7.17) *have the following properties*:

$$E\tilde{\mathbf{x}}_t^\varepsilon = E\mathbf{x}_t, \qquad \underset{\sim}{\mathbf{P}}_t \leq \tilde{\underset{\sim}{\mathbf{P}}}_t^\varepsilon \leq \underset{\sim}{\mathbf{P}}_t^\varepsilon, \tag{7.18}$$

$$\lim_{\varepsilon \to 0} \tilde{\underset{\sim}{\mathbf{P}}}_t^\varepsilon = \lim_{\varepsilon \to 0} \underset{\sim}{\mathbf{P}}_t^\varepsilon = \underset{\sim}{\mathbf{P}}_t, \tag{7.19}$$

$$\underset{\varepsilon \to 0}{\text{l.i.m.}}\, \tilde{\mathbf{x}}_t^\varepsilon = \hat{\mathbf{x}}_t. \tag{7.20}$$

Proof. Without loss of generality, we can assume that $\{\mathbf{w}_t\}$ and $\{\tilde{\mathbf{w}}_t\}$ are Wiener processes and $[\mathbf{x}_0^\tau \;\; \mathbf{y}_0^\tau]^\tau$ is normally distributed.

Since $\hat{\mathbf{x}}_t = E(\mathbf{x}_t/\mathscr{F}_t^{\mathbf{y}})$ is $\mathscr{F}_t^{\mathbf{y}}$-measurable and $E\|\hat{\mathbf{x}}_t\|^2 \leq E\|\mathbf{x}_t\|^2 < \infty$, by Lemma 7.3, we have

$$\underset{\varepsilon \to 0}{\text{l.i.m.}}\, E\left(\frac{\hat{\mathbf{x}}_t}{\mathscr{F}_t^{\mathbf{y}^\varepsilon}}\right) = \hat{\mathbf{x}}_t. \tag{7.21}$$

By the definition of $\mathbf{y}_t^\varepsilon$, it is easy to see that

$$\mathscr{F}_t^{\mathbf{y},\mathbf{y}^\varepsilon} = \mathscr{F}_t^{\mathbf{y},\tilde{\mathbf{w}}}.$$

Notice that $[\mathbf{x}_t^\tau \mathbf{y}_t^\tau]^\tau$ is independent of $\{\tilde{\mathbf{w}}_t\}$, by Theorem 1.8 with $\boldsymbol{\xi}$, $\boldsymbol{\zeta}$, and $\boldsymbol{\eta}$ corresponding to $\mathbf{x}_t$, $\mathscr{F}_t^{\mathbf{y}}$, and $\mathscr{F}_t^{\tilde{\mathbf{w}}}$, respectively. It follows that given $\mathscr{F}_t^{\mathbf{y}}$, $\mathbf{x}_t$ and $\mathscr{F}_t^{\tilde{\mathbf{w}}}$ are conditionally independent, so

$$E\left(\frac{\mathbf{x}_t}{\mathscr{F}_t^{\mathbf{y},\tilde{\mathbf{w}}}}\right) = E\left(\frac{\mathbf{x}_t}{\mathscr{F}_t^{\mathbf{y}}}\right), \tag{7.22}$$

and hence

$$E\left(\frac{\mathbf{x}_t}{\mathscr{F}_t^{\mathbf{y},\mathbf{y}^\varepsilon}}\right) = E\left(\frac{\mathbf{x}_t}{\mathscr{F}_t^{\mathbf{y},\tilde{\mathbf{w}}}}\right) = E\left(\frac{\mathbf{x}_t}{\mathscr{F}_t^{\mathbf{y}}}\right) = \hat{\mathbf{x}}_t. \tag{7.23}$$

By taking the conditional expectation $E(\cdot/\mathscr{F}_t^{\mathbf{y}^\varepsilon})$, (7.23) yields

$$\hat{\mathbf{x}}_t^\varepsilon = E\left(\frac{\mathbf{x}_t}{\mathscr{F}_t^{\mathbf{y}^\varepsilon}}\right) = E\left(\frac{\hat{\mathbf{x}}_t}{\mathscr{F}_t^{\mathbf{y}^\varepsilon}}\right), \tag{7.24}$$

and so by (7.21) and (7.24)

$$\underset{\varepsilon \to 0}{\text{l.i.m.}}\, \hat{\mathbf{x}}_t^\varepsilon = \hat{\mathbf{x}}_t. \tag{7.25}$$

Hence

$$\underset{\sim}{P}_t^\varepsilon = E(\hat{\mathbf{x}}_t^\varepsilon - \hat{\mathbf{x}}_t)(\hat{\mathbf{x}}_t^\varepsilon - \hat{\mathbf{x}}_t)^\tau + \underset{\sim}{P}_t + E(\hat{\mathbf{x}}_t^\varepsilon - \hat{\mathbf{x}}_t)(\hat{\mathbf{x}}_t - \mathbf{x}_t)^\tau$$

$$+ E(\hat{\mathbf{x}}_t - \mathbf{x}_t)(\hat{\mathbf{x}}_t^\varepsilon - \hat{\mathbf{x}}_t)^\tau .$$

From this, (7.25), and the Schwarz inequality, it follows that

$$\lim_{\varepsilon \to 0} \underset{\sim}{P}_t^\varepsilon = \underset{\sim}{P}_t. \tag{7.26}$$

Introduce

$$d\boldsymbol{\delta}_t^\varepsilon = \underset{\sim}{A}_t \boldsymbol{\delta}_t^\varepsilon \, dt + (\underset{\sim}{D}_t \underset{\sim}{F}_t^\tau + \underset{\sim}{P}_t^\varepsilon \underset{\sim}{C}_t^\tau)(\underset{\sim}{F}_t \underset{\sim}{F}_t^\tau + \varepsilon \underset{\sim}{I})^{-1}(\sqrt{\varepsilon}\, d\tilde{\mathbf{w}}_t - \underset{\sim}{C}_t \boldsymbol{\delta}_t^\varepsilon \, dt),$$

$$\boldsymbol{\delta}_0^\varepsilon = \mathbf{0}. \tag{7.27}$$

From (7.14), (7.16), and (7.27), we have

$$\hat{\mathbf{x}}_t^\varepsilon = \tilde{\mathbf{x}}_t^\varepsilon + \boldsymbol{\delta}_t^\varepsilon. \tag{7.28}$$

On the other hand, $E\boldsymbol{\delta}_t^\varepsilon \equiv \mathbf{0}$ and $E\hat{\mathbf{x}}_t^\varepsilon = E\mathbf{x}_t$. Hence $\tilde{\mathbf{x}}_t^\varepsilon$ is an unbiased estimate for $\mathbf{x}_t$:

$$E\tilde{\mathbf{x}}_t^\varepsilon = E\mathbf{x}_t. \tag{7.29}$$

Because $\tilde{\mathbf{x}}_t^\varepsilon$ depends on $\{\mathbf{y}_s,\ s \le t\}$ only, it is independent of $\{\tilde{\mathbf{w}}_t\}$. This means that $\{\tilde{\mathbf{x}}_t^\varepsilon\}$ and $\{\boldsymbol{\delta}_t^\varepsilon\}$ are mutually independent. Furthermore, $\{\boldsymbol{\delta}_t^\varepsilon\}$ and $\{\mathbf{x}_t\}$ are also independent, so that by (7.28), it follows that

$$\underset{\sim}{P}_t^\varepsilon = E(\mathbf{x}_t - \hat{\mathbf{x}}_t^\varepsilon)(\mathbf{x}_t - \hat{\mathbf{x}}_t^\varepsilon)^\tau = E(\mathbf{x}_t - \tilde{\mathbf{x}}_t^\varepsilon - \boldsymbol{\delta}_t^\varepsilon)(\mathbf{x}_t - \tilde{\mathbf{x}}_t^\varepsilon - \boldsymbol{\delta}_t^\varepsilon)^\tau$$

$$= \tilde{\underset{\sim}{P}}_t^\varepsilon + E\boldsymbol{\delta}_t^\varepsilon \boldsymbol{\delta}_t^{\varepsilon\tau} \ge \tilde{\underset{\sim}{P}}_t^\varepsilon, \tag{7.30}$$

which, together with (7.26), (7.29), and (7.30) proves (7.18) and (7.19) since $\tilde{\underset{\sim}{P}}_t^\varepsilon \ge \underset{\sim}{P}_t$ is trivial. In addition, (7.26) and (7.30) imply that

$$E\boldsymbol{\delta}_t^\varepsilon \boldsymbol{\delta}_t^{\varepsilon\tau} \underset{\varepsilon \to 0}{\to} \underset{\sim}{0}. \tag{7.31}$$

Now observe that

$$E(\mathbf{x}_t - \tilde{\mathbf{x}}_t^\varepsilon)(\tilde{\mathbf{x}}_t^\varepsilon - \hat{\mathbf{x}}_t)^\tau = E\left[E\left(\frac{\mathbf{x}_t - \tilde{\mathbf{x}}_t^\varepsilon}{\mathscr{F}_t^{\mathbf{y}}}\right)\right](\tilde{\mathbf{x}}_t^\varepsilon - \hat{\mathbf{x}}^\tau)^\tau$$

$$= E\left\{E\left[(\mathbf{x}_t - \hat{\mathbf{x}}_t) + \frac{(\hat{\mathbf{x}}_t - \tilde{\mathbf{x}}_t^\varepsilon)}{\mathscr{F}_t^{\mathbf{y}}}\right](\tilde{\mathbf{x}}_t^\varepsilon - \hat{\mathbf{x}}_t)^\tau\right\}$$

$$= -E(\tilde{\mathbf{x}}_t^\varepsilon - \hat{\mathbf{x}}_t)(\tilde{\mathbf{x}}_t^\varepsilon - \hat{\mathbf{x}}_t)^\tau. \tag{7.32}$$

Then

$$\underset{\sim}{\mathbf{P}}_t = E(\mathbf{x}_t - \tilde{\mathbf{x}}_t^\varepsilon + \tilde{\mathbf{x}}_t^\varepsilon - \hat{\mathbf{x}}_t)(\mathbf{x}_t - \tilde{\mathbf{x}}_t^\varepsilon + \tilde{\mathbf{x}}_t^\varepsilon - \hat{\mathbf{x}}_t)^\tau$$

$$= \underset{\sim}{\tilde{\mathbf{P}}}_t^\varepsilon - E(\tilde{\mathbf{x}}_t^\varepsilon - \hat{\mathbf{x}}_t)(\tilde{\mathbf{x}}_t^\varepsilon - \hat{\mathbf{x}}_t)^\tau,$$

and so

$$E(\tilde{\mathbf{x}}_t^\varepsilon - \hat{\mathbf{x}}_t)(\tilde{\mathbf{x}}_t^\varepsilon - \hat{\mathbf{x}}_t)^\tau = \underset{\sim}{\mathbf{P}}_t - \underset{\sim}{\tilde{\mathbf{P}}}_t^\varepsilon \underset{\varepsilon\to 0}{\to} \underset{\sim}{\mathbf{0}}. \qquad \square$$

Denote

$$\underset{\sim}{\mathbf{K}}_t^\varepsilon = (\underset{\sim}{\mathbf{D}}_t \underset{\sim}{\mathbf{F}}_t^\tau + \underset{\sim}{\mathbf{P}}_t^\varepsilon \underset{\sim}{\mathbf{C}}_t^\tau)(\underset{\sim}{\mathbf{F}}_t \underset{\sim}{\mathbf{F}}_t^\tau + \varepsilon \underset{\sim}{\mathbf{I}})^{-1} \tag{7.33}$$

$$\frac{d\underset{\sim}{\mathbf{\Psi}}_{t,s}^\varepsilon}{dt} = (\underset{\sim}{\mathbf{A}}_t - \underset{\sim}{\mathbf{K}}_t^\varepsilon \underset{\sim}{\mathbf{C}}_t)\underset{\sim}{\mathbf{\Psi}}_{t,s}^\varepsilon, \qquad \underset{\sim}{\mathbf{\Psi}}_{s,s}^\varepsilon = \underset{\sim}{\mathbf{I}}. \tag{7.34}$$

Then with

$$\tilde{\mathbf{x}}_t^\varepsilon = \underset{\sim}{\mathbf{\Psi}}_{t,0}^\varepsilon \hat{\mathbf{x}}_0 + \int_0^t \underset{\sim}{\mathbf{\Psi}}_{t,s}^\varepsilon \underset{\sim}{\mathbf{K}}_s^\varepsilon \, d\mathbf{y}_s, \tag{7.35}$$

$\tilde{\mathbf{x}}_t^\varepsilon$ is a linear estimate depending on past measurements only. By Theorem 7.1, in the limit as $\varepsilon \to 0$, this estimate is unbiased and equals the value of the optimal estimate of $\mathbf{x}_t$. Hence $\tilde{\mathbf{x}}_t^\varepsilon$ is a suboptimal estimate for fixed $\varepsilon > 0$, which is asymptotically optimal.

7.3. SUBOPTIMAL INTERPOLATION AND PREDICTION

In this section, we shall discuss the interpolation and prediction problems with possibly degenerate $\underset{\sim}{\mathbf{F}}_t \underset{\sim}{\mathbf{F}}_t^\tau$. We shall use the notation introduced in

Chapter 6 and, in addition, denote

$$\hat{\mathbf{x}}_t^{\varepsilon s} = E\left(\frac{\mathbf{x}_t}{\mathscr{F}^{\mathbf{y}_t^\varepsilon}, \mathbf{x}_s}\right), \qquad \hat{\mathbf{x}}_s^\varepsilon(t) = E\left(\frac{\mathbf{x}_s}{\mathscr{F}_t^{\mathbf{y}^\varepsilon}}\right), \tag{7.36}$$

$$\mathbf{P}_s^\varepsilon(t) = E(\mathbf{x}_s - \hat{\mathbf{x}}_s^\varepsilon(t))(\mathbf{x}_s - \hat{\mathbf{x}}_s^\varepsilon(t))^\tau,$$

$$\frac{d}{dt}\underset{\sim}{\mathbf{P}}_t^{\varepsilon s} = \underset{\sim}{\mathbf{A}}_t\underset{\sim}{\mathbf{P}}_t^{\varepsilon s} + \underset{\sim}{\mathbf{P}}_t^{\varepsilon s}\underset{\sim}{\mathbf{A}}_t^\tau + \underset{\sim}{\mathbf{D}}_t\underset{\sim}{\mathbf{D}}_t^\tau - (\underset{\sim}{\mathbf{D}}_t\underset{\sim}{\mathbf{F}}_t^\tau + \underset{\sim}{\mathbf{P}}_t^{\varepsilon s}\underset{\sim}{\mathbf{C}}_t^\tau)(\underset{\sim}{\mathbf{F}}_t\underset{\sim}{\mathbf{F}}_t^\tau + \varepsilon\underset{\sim}{\mathbf{I}})^{-1}$$

$$\cdot(\underset{\sim}{\mathbf{D}}_t\underset{\sim}{\mathbf{F}}_t^\tau + \underset{\sim}{\mathbf{P}}_t^\varepsilon\underset{\sim}{\mathbf{C}}_t^\tau)^\tau, \qquad \underset{\sim}{\mathbf{P}}_s^{\varepsilon s} = \underset{\sim}{\mathbf{0}}, \tag{7.37}$$

$$\underset{\sim}{\mathbf{K}}_t^{\varepsilon s} = (\underset{\sim}{\mathbf{D}}_t\underset{\sim}{\mathbf{F}}_t^\tau + \underset{\sim}{\mathbf{P}}_t^{\varepsilon s}\underset{\sim}{\mathbf{C}}_t^\tau)(\underset{\sim}{\mathbf{F}}_t\underset{\sim}{\mathbf{F}}_t^\tau + \varepsilon\mathbf{I})^{-1}, \tag{7.38}$$

$$\frac{d}{dt}\underset{\sim}{\mathbf{\Psi}}_{t,\lambda}^{\varepsilon s} = (\underset{\sim}{\mathbf{A}}_t - \underset{\sim}{\mathbf{K}}_t^{\varepsilon s}\underset{\sim}{\mathbf{C}}_t)\underset{\sim}{\mathbf{\Psi}}_{t,\lambda}^{\varepsilon s}, \qquad \underset{\sim}{\mathbf{\Psi}}_{\lambda,\lambda}^{\varepsilon s} = \underset{\sim}{\mathbf{I}}. \tag{7.39}$$

By Theorem 6.2, for $t \geq s$, $\hat{\mathbf{x}}_s^\varepsilon(t)$ and $\underset{\sim}{\mathbf{P}}_s^\varepsilon$ are defined by

$$\hat{\mathbf{x}}_s^\varepsilon(t) = \hat{\mathbf{x}}_s^\varepsilon + \int_s^t \underset{\sim}{\mathbf{P}}_s^\varepsilon(\lambda)(\underset{\sim}{\mathbf{\Psi}}_{\lambda,s}^{\varepsilon s})^\tau \mathbf{C}_\lambda(\underset{\sim}{\mathbf{F}}_\lambda\underset{\sim}{\mathbf{F}}_\lambda^\tau + \varepsilon\mathbf{I})^{-1}(d\mathbf{y}_\lambda^\varepsilon - \underset{\sim}{\mathbf{C}}_\lambda\hat{\mathbf{x}}_\lambda^\varepsilon\, d\lambda), \tag{7.40}$$

$$\underset{\sim}{\mathbf{P}}_s^\varepsilon(t) = \left[\underset{\sim}{\mathbf{I}} + \underset{\sim}{\mathbf{P}}_s^\varepsilon \int_s^t (\underset{\sim}{\mathbf{\Psi}}_{\lambda,s}^{\varepsilon s})^\tau \underset{\sim}{\mathbf{C}}_\lambda^\tau(\underset{\sim}{\mathbf{F}}_\lambda\underset{\sim}{\mathbf{F}}_\lambda^\tau + \varepsilon\underset{\sim}{\mathbf{I}})^{-1}\underset{\sim}{\mathbf{C}}_\lambda\underset{\sim}{\mathbf{\Psi}}_{\lambda,s}^{\varepsilon s}\, d\lambda\right]^{-1}\underset{\sim}{\mathbf{P}}_s^\varepsilon, \qquad t \geq s. \tag{7.41}$$

Define

$$\hat{\mathbf{x}}_s^\varepsilon(t) = \tilde{\mathbf{x}}_s^\varepsilon + \int_s^t \underset{\sim}{\mathbf{P}}_s^\varepsilon(\lambda)(\underset{\sim}{\mathbf{\Psi}}_{\lambda,s}^{\varepsilon s})^\tau \underset{\sim}{\mathbf{C}}_\lambda(\underset{\sim}{\mathbf{F}}_\lambda\underset{\sim}{\mathbf{F}}_\lambda^\tau + \varepsilon\underset{\sim}{\mathbf{I}})^{-1}(d\mathbf{y}_\lambda - \underset{\sim}{\mathbf{C}}_\lambda\tilde{\mathbf{x}}_\lambda^\varepsilon\, d\lambda) \tag{7.42}$$

$$\tilde{\underset{\sim}{\mathbf{P}}}_s^\varepsilon(t) = E(\mathbf{x}_s - \tilde{\mathbf{x}}^s(t))(\mathbf{x}_s - \tilde{\mathbf{x}}_s^\varepsilon(t))^\tau. \tag{7.43}$$

Since $\tilde{\mathbf{x}}_t^\varepsilon$ depends on past measurements only, so does $\tilde{\mathbf{x}}_s^\varepsilon(t)$. Now we establish the asymptotic optimality of the suboptimal estimate $\tilde{\mathbf{x}}_s^\varepsilon(t)$.

Theorem 7.2. *Under the conditions of Theorem* 7.1, *for* $s \le t$,

$$E\tilde{\mathbf{x}}_s^\varepsilon(t) = E\mathbf{x}_s, \qquad \underset{\sim}{\mathbf{P}}_s(t) \le \tilde{\underset{\sim}{\mathbf{P}}}_s^\varepsilon(t) \le \underset{\sim}{\mathbf{P}}_s^\varepsilon(t), \tag{7.44}$$

$$\lim_{\varepsilon\to 0} \tilde{\underset{\sim}{\mathbf{P}}}_s^\varepsilon(t) = \lim_{\varepsilon\to 0} \underset{\sim}{\mathbf{P}}_s^\varepsilon(t) = \underset{\sim}{\mathbf{P}}_s(t), \tag{7.45}$$

$$\underset{\varepsilon\to 0}{\text{l.i.m.}}\, \hat{\mathbf{x}}_s^\varepsilon(t) = \hat{\mathbf{x}}_s(t), \qquad \underset{\varepsilon\to 0}{\text{l.i.m.}}\, \hat{\mathbf{x}}_t^{\varepsilon s} = \hat{\mathbf{x}}_t^s, \tag{7.46}$$

$$\underset{\sim}{\mathbf{P}}_t^{\varepsilon s} = E(\mathbf{x}_t - \hat{\mathbf{x}}_t^{\varepsilon s})(\mathbf{x}_t - \hat{\mathbf{x}}_t^{\varepsilon s})^\tau, \tag{7.47}$$

and as s varies, $\tilde{\mathbf{x}}_s^\varepsilon(t)$ *satisfies the following equation*

$$d_s\tilde{\mathbf{x}}_s^\varepsilon(t) = \underset{\sim}{\mathbf{A}}_s\tilde{\mathbf{x}}_s^\varepsilon(t)\,ds + \underset{\sim}{\mathbf{D}}_s(\underset{\sim}{\mathbf{D}}_s^\tau - \underset{\sim}{\mathbf{F}}_s\underset{\sim}{\mathbf{K}}_s^{\varepsilon\tau})\underset{\sim}{\mathbf{P}}_s^{\varepsilon-1}(\tilde{\mathbf{x}}_s^\varepsilon(t) - \tilde{\mathbf{x}}_s^\varepsilon)\,ds$$

$$+\underset{\sim}{\mathbf{D}}_s\underset{\sim}{\mathbf{F}}_s^\tau(\underset{\sim}{\mathbf{F}}_s\underset{\sim}{\mathbf{F}}_s^\tau + \varepsilon\underset{\sim}{\mathbf{I}})^{-1}(d\mathbf{y}_s - \underset{\sim}{\mathbf{C}}_s\tilde{\mathbf{x}}_s^\varepsilon\,ds), \tag{7.48}$$

where and hereafter $\underset{\sim}{\mathbf{P}}_s^{\varepsilon-1}$ *denotes* $(\underset{\sim}{\mathbf{P}}_s^\varepsilon)^{-1}$.

Proof. Without loss of generality, assume $\{\mathbf{w}_t\}$ is a Wiener process and $[\mathbf{x}_0^\tau \;\; \mathbf{y}_0^\tau]^\tau$ is normally distributed. The proof is similar to that of Theorem 7.1. Equations (7.21)–(7.26) remain valid if $\mathbf{x}_t$, $\hat{\mathbf{x}}_t$, $\hat{\mathbf{x}}_t^\varepsilon$, and $\tilde{\mathbf{x}}_t^\varepsilon$ are replaced by $\mathbf{x}_s$, $\hat{\mathbf{x}}_s(t)$, $\hat{\mathbf{x}}_s^\varepsilon(t)$, and $\tilde{\mathbf{x}}_s^\varepsilon(t)$, respectively.

Denote

$$\boldsymbol{\delta}_s^\varepsilon(t) = \delta_s^\varepsilon + \int_s^t \underset{\sim}{\mathbf{P}}_s^\varepsilon(\lambda)\underset{\sim}{\boldsymbol{\Psi}}_{\lambda,s}^{\varepsilon s\tau}\underset{\sim}{\mathbf{C}}_\lambda(\underset{\sim}{\mathbf{F}}_\lambda\underset{\sim}{\mathbf{F}}_\lambda^\tau + \varepsilon\underset{\sim}{\mathbf{I}})^{-1}\left(\sqrt{\varepsilon}\,d\tilde{\mathbf{w}}_\lambda - \underset{\sim}{\mathbf{C}}_\lambda\boldsymbol{\delta}_\lambda^\varepsilon\,d\lambda\right). \tag{7.49}$$

From (7.28), (7.40), and (7.42), we know

$$\hat{\mathbf{x}}_s^\varepsilon(t) = \tilde{\mathbf{x}}_s^\varepsilon(t) + \boldsymbol{\delta}_s^\varepsilon(t). \tag{7.50}$$

Since $E\boldsymbol{\delta}_s^\varepsilon \equiv \mathbf{0}$ and $E\boldsymbol{\delta}_s^\varepsilon(t) \equiv \mathbf{0}$ for $s \le t$, $\tilde{\mathbf{x}}_s^\varepsilon(t)$ is unbiased. The proof of Theorem 7.1 is applicable to the present case, for this one only requires the replacement mentioned and replacing $\boldsymbol{\delta}_t^\varepsilon$ with $\boldsymbol{\delta}_s^\varepsilon(t)$. Thus (7.44), (7.45), and the first relationship of (7.46) have been established, while the second relationship of (7.46) is proved analogously and (7.47) is obvious.

According to Theorem 6.3, $\hat{\mathbf{x}}_s^\varepsilon(t)$ satisfies the backward interpolation equation

$$d_s\hat{\mathbf{x}}_s^\varepsilon(t) = \underset{\sim}{\mathbf{A}}_s\hat{\mathbf{x}}_s^\varepsilon(t)\,ds + \underset{\sim}{\mathbf{D}}_s(\underset{\sim}{\mathbf{D}}_s^\tau - \underset{\sim}{\mathbf{F}}_s^\tau\underset{\sim}{\mathbf{K}}_s^{\varepsilon\tau})\underset{\sim}{\mathbf{P}}_s^{\varepsilon-1}(\hat{\mathbf{x}}_s^\varepsilon(t) - \hat{\mathbf{x}}_s^\varepsilon)\,ds$$

$$+\underset{\sim}{\mathbf{D}}_s\underset{\sim}{\mathbf{F}}_s^\tau(\underset{\sim}{\mathbf{F}}_s\underset{\sim}{\mathbf{F}}_s^\tau + \varepsilon\underset{\sim}{\mathbf{I}})^{-1}(d\mathbf{y}_s^\varepsilon - \underset{\sim}{\mathbf{C}}_s\hat{\mathbf{x}}_s^\varepsilon\,ds), \tag{7.51}$$

and $\tilde{\mathbf{x}}_s^\varepsilon(t)$ is obtained from the expression for $\hat{\mathbf{x}}_s^\varepsilon(t)$ with $\mathbf{y}_s^\varepsilon$ replaced by $\mathbf{y}_s$. Putting $\mathbf{y}_s$ instead of $\mathbf{y}_s^\varepsilon$ in (7.51) leads to (7.48). □

By use of Theorem 7.1, it is easy to write down the suboptimal value of the estimate given by

$$\tilde{\mathbf{x}}_t^\varepsilon(s) \triangleq \underset{\sim}{\mathbf{\Phi}}_{t,s}\tilde{\mathbf{x}}_s^\varepsilon.$$

Clearly,

$$\underset{\varepsilon\to 0}{\text{l.i.m.}}\,\tilde{\mathbf{x}}_t^\varepsilon(s) = \hat{\mathbf{x}}_t(s),$$

and the estimation error covariance matrix is

$$\underset{\sim}{\tilde{\mathbf{P}}}_t^\varepsilon(s) = \underset{\sim}{\mathbf{\Phi}}_{t,s}\underset{\sim}{\mathbf{P}}_s^\varepsilon\underset{\sim}{\mathbf{\Phi}}_{t,s}^\tau + \int_0^t \underset{\sim}{\mathbf{\Phi}}_{t,\lambda}\underset{\sim}{\mathbf{D}}_\lambda\underset{\sim}{\mathbf{D}}_\lambda^\tau\underset{\sim}{\mathbf{\Phi}}_{t,\lambda}^\tau\,d\lambda.$$

7.4. UNIFIED CONTROL APPLICABLE TO POSSIBLY SINGULAR CASES

We continue to consider the system (6.1), (6.2) and performance index (6.95), but now we allow $\underset{\sim}{\mathbf{Q}}_2(t)$ to be degenerate. In this case, it is called a singular control problem. In addition, our consideration also includes the case of the possible degeneracy of $\underset{\sim}{\mathbf{F}}_t\underset{\sim}{\mathbf{F}}_t^\tau$.

Let the set $\mathscr{U}$ of admissible controls be composed of all $\mathbf{u}$ satisfying the conditions:

1. $\mathbf{u}_t$ is a measurable process and for any t is $\mathscr{F}_t^{\mathbf{y}}$-measurable. (7.52)
2. $E\int_0^T\|\mathbf{u}_t\|^2\,dt < \infty$. (7.53)
3. Under $\mathbf{u}_t$ equations (6.1), (6.2) are solvable and $\underset{\sim}{\mathbf{P}}_t$ is the same as with $\mathbf{u}_t$ removed. (7.54)

In Theorem 6.6, where the $\underset{\sim}{\mathbf{F}}_t\underset{\sim}{\mathbf{F}}_t^\tau > 0$ case is considered, the set of admissible controls is defined by (6.80). By Theorem 6.5, for $\underset{\sim}{\mathbf{F}}_t\underset{\sim}{\mathbf{F}}_t^\tau > \underset{\sim}{\mathbf{0}}$, any $\{\mathbf{u}_t\} \in \mathscr{U}$ given by (6.80) must satisfy (7.52)–(7.54).

Denote the system (6.1), (6.2) with $\mathbf{u}_t \equiv \mathbf{0}$ by

$$d\boldsymbol{\xi}_t = \underset{\sim}{\mathbf{A}}_t \boldsymbol{\xi}_t\, dt + \underset{\sim}{\mathbf{D}}_t\, d\mathbf{w}_t, \qquad \boldsymbol{\xi}_0 = \mathbf{x}_0, \tag{7.55}$$

$$d\boldsymbol{\eta}_t = \underset{\sim}{\mathbf{C}}_t \boldsymbol{\xi}_t\, dt + \underset{\sim}{\mathbf{F}}_t\, d\mathbf{w}_t, \qquad \boldsymbol{\eta}_0 = \mathbf{y}_0, \tag{7.56}$$

and denote by $\hat{\boldsymbol{\xi}}_t$ the LUMVE of $\boldsymbol{\xi}_t$ based on $\{\boldsymbol{\eta}_s,\ s \le t\}$ and its estimation error covariance matrix by $\underset{\sim}{\mathbf{P}}_t$. Notice that $\hat{\boldsymbol{\xi}}_t$ and $\underset{\sim}{\mathbf{P}}_t$ may be discontinuous.

We continue to use the notation given by (7.15)–(7.17), but with $\mathbf{x}_t, \mathbf{y}_t$ replaced by $\boldsymbol{\xi}_t, \boldsymbol{\eta}_t$, respectively. For example, (7.16) defines $\tilde{\boldsymbol{\xi}}_t^{\varepsilon}$ via

$$d\tilde{\boldsymbol{\xi}}_t^{\varepsilon} = \underset{\sim}{\mathbf{A}}_t \tilde{\boldsymbol{\xi}}_t^{\varepsilon}\, dt + \left(\underset{\sim}{\mathbf{D}}_t \underset{\sim}{\mathbf{F}}_t^{\tau} + \underset{\sim}{\mathbf{P}}_t^{\varepsilon} \underset{\sim}{\mathbf{C}}_t^{\tau}\right)\left(\underset{\sim}{\mathbf{F}}_t \underset{\sim}{\mathbf{F}}_t^{\tau} + \varepsilon \underset{\sim}{\mathbf{I}}\right)^{-1}\left(d\boldsymbol{\eta}_t - \underset{\sim}{\mathbf{C}}_t \tilde{\boldsymbol{\xi}}_t^{\varepsilon}\, dt\right), \qquad \tilde{\boldsymbol{\xi}}_0 = \hat{\mathbf{x}}_0. \tag{7.57}$$

Lemma 7.4. *Assume* $\|\underset{\sim}{\mathbf{A}}_t\|, \|\underset{\sim}{\mathbf{B}}_t\|^2, \|\underset{\sim}{\mathbf{C}}_t\|^2, \|\underset{\sim}{\mathbf{F}}_t\|^2, \|\underset{\sim}{\mathbf{D}}_t\|^2$ *are integrable on* $[0, T]$, *and* $\underset{\sim}{\mathbf{M}}_t$ *is a* $r \times n$ *deterministic matrix of measurable functions with* $\int_0^T \|\underset{\sim}{\mathbf{M}}_t\|^2\, dt < \infty$. *Then*

$$\mathbf{u}_t \triangleq \underset{\sim}{\mathbf{M}}_t \boldsymbol{\zeta}_t^{\varepsilon} \tag{7.58}$$

belongs to $\mathscr{U}$, *where*

$$\boldsymbol{\zeta}_t^{\varepsilon} = \tilde{\boldsymbol{\xi}}_t^{\varepsilon} + \int_0^t \underset{\sim}{\boldsymbol{\Phi}}_{t,s} \underset{\sim}{\mathbf{B}}_s \underset{\sim}{\mathbf{M}}_s \boldsymbol{\zeta}_s^{\varepsilon}\, ds. \tag{7.59}$$

Proof. To start with, we show that (7.59) indeed defines a process $\boldsymbol{\zeta}_t^{\varepsilon}$. Take the stochastic differential of (7.59) and substitute (7.57). This yields

$$\begin{aligned} d\boldsymbol{\zeta}_t^{\varepsilon} &= \underset{\sim}{\mathbf{A}}_t \tilde{\boldsymbol{\xi}}_t^{\varepsilon}\, dt + \underset{\sim}{\mathbf{K}}_t^{\varepsilon}\left(d\boldsymbol{\eta}_t - \underset{\sim}{\mathbf{C}}_t \tilde{\boldsymbol{\xi}}_t^{\varepsilon}\, dt\right) + \underset{\sim}{\mathbf{A}}_t \int_0^t \underset{\sim}{\boldsymbol{\Phi}}_{t,s} \underset{\sim}{\mathbf{B}}_s \underset{\sim}{\mathbf{M}}_s \boldsymbol{\zeta}_s^{\varepsilon}\, ds + \underset{\sim}{\mathbf{B}}_t \underset{\sim}{\mathbf{M}}_t \boldsymbol{\zeta}_t^{\varepsilon}\, dt \\ &= \left(\underset{\sim}{\mathbf{A}}_t + \underset{\sim}{\mathbf{B}}_t \underset{\sim}{\mathbf{M}}_t\right)\boldsymbol{\zeta}_t^{\varepsilon}\, dt + \underset{\sim}{\mathbf{K}}_t^{\varepsilon}\left(d\boldsymbol{\eta}_t - \underset{\sim}{\mathbf{C}}_t \tilde{\boldsymbol{\xi}}_t^{\varepsilon}\, dt\right), \end{aligned} \tag{7.60}$$

where $\underset{\sim}{\mathbf{K}}_t^{\varepsilon}$ is given by (7.33).

$\boldsymbol{\eta}_t$ and $\tilde{\boldsymbol{\xi}}_t^{\varepsilon}$ are well-defined processes and (7.60) is a linear stochastic differential equation with respect to $\boldsymbol{\zeta}_t^{\varepsilon}$. Hence it can be defined.

As $\{\mathbf{u}_t\}$ is given by (7.58), we have

$$\mathbf{x}_t = \boldsymbol{\xi}_t + \int_0^t \underset{\sim}{\boldsymbol{\Phi}}_{t,s} \underset{\sim}{\mathbf{B}}_s \underset{\sim}{\mathbf{M}}_s \boldsymbol{\zeta}_s^{\varepsilon}\, ds,$$

$$d\mathbf{y}_t = d\boldsymbol{\eta}_t + \underset{\sim}{\mathbf{C}}_t \int_0^t \underset{\sim}{\boldsymbol{\Phi}}_{t,s} \underset{\sim}{\mathbf{B}}_s \underset{\sim}{\mathbf{M}}_s \boldsymbol{\zeta}_s^{\varepsilon}\, ds\, dt. \tag{7.61}$$

Hence, from (7.59),

$$d\mathbf{y}_t - \underset{\sim}{\mathbf{C}}_t\zeta_t^\varepsilon\,dt = d\boldsymbol{\eta}_t + \underset{\sim}{\mathbf{C}}_t\left[\int_0^t \underset{\sim}{\boldsymbol{\Phi}}_{t,s}\underset{\sim}{\mathbf{B}}_s\underset{\sim}{\mathbf{M}}_s\zeta_s^\varepsilon\,ds - \zeta_t^\varepsilon\right]dt$$

$$= d\boldsymbol{\eta}_t - \underset{\sim}{\mathbf{C}}_t\tilde{\xi}_t^\varepsilon\,dt.$$

Substituting this into (7.60) yields

$$d\zeta_t^\varepsilon = (\underset{\sim}{\mathbf{A}}_t + \underset{\sim}{\mathbf{B}}_t\underset{\sim}{\mathbf{M}}_t)\zeta_t^\varepsilon\,dt + \underset{\sim}{\mathbf{K}}_t^\varepsilon(d\mathbf{y}_t - \underset{\sim}{\mathbf{C}}_t\zeta_t^\varepsilon\,dt), \qquad \zeta_0^\varepsilon = \hat{\mathbf{x}}_0. \tag{7.62}$$

Notice that

$$\|\underset{\sim}{\mathbf{C}}_t^\tau(\underset{\sim}{\mathbf{F}}_t\underset{\sim}{\mathbf{F}}_t^\tau + \varepsilon\underset{\sim}{\mathbf{I}})^{-1}\underset{\sim}{\mathbf{C}}_t\| \le \frac{1}{\varepsilon}\|\underset{\sim}{\mathbf{C}}_t\|^2,$$

$$\|\underset{\sim}{\mathbf{D}}_t\underset{\sim}{\mathbf{F}}_t^\tau(\underset{\sim}{\mathbf{F}}_t\underset{\sim}{\mathbf{F}}_t^\tau + \varepsilon\underset{\sim}{\mathbf{I}})^{-1}\underset{\sim}{\mathbf{C}}_t\| \le \|\underset{\sim}{\mathbf{D}}_t\|\,\|\underset{\sim}{\mathbf{C}}_t\|\,\|\underset{\sim}{\mathbf{F}}_t^\tau(\underset{\sim}{\mathbf{F}}_t\underset{\sim}{\mathbf{F}}_t^\tau + \varepsilon\underset{\sim}{\mathbf{I}})^{-1}\|$$

$$= \|\underset{\sim}{\mathbf{D}}_t\|\,\|\underset{\sim}{\mathbf{C}}_t\|\,\|(\underset{\sim}{\mathbf{F}}_t\underset{\sim}{\mathbf{F}}_t^\tau + \varepsilon\underset{\sim}{\mathbf{I}})^{-1}\underset{\sim}{\mathbf{F}}_t\underset{\sim}{\mathbf{F}}_t^\tau(\underset{\sim}{\mathbf{F}}_t\underset{\sim}{\mathbf{F}}_t^\tau + \varepsilon\underset{\sim}{\mathbf{I}})^{-1}\|^{1/2}$$

$$\le \|\underset{\sim}{\mathbf{D}}_t\|\,\|\underset{\sim}{\mathbf{C}}_t\|\,\|(\underset{\sim}{\mathbf{F}}_t\underset{\sim}{\mathbf{F}}_t^\tau + \varepsilon\underset{\sim}{\mathbf{I}})^{-1}\|^{1/2} \le \frac{1}{\sqrt{\varepsilon}}\|\underset{\sim}{\mathbf{D}}_t\|\,\|\underset{\sim}{\mathbf{C}}_t\|,$$

and so by the conditions of the lemma, it is easy to see that $\|\underset{\sim}{\mathbf{K}}_t^\varepsilon\underset{\sim}{\mathbf{C}}_t\|$ is integrable on $[0, T]$. Hence there is a fundamental solution matrix:

$$\frac{d}{dt}\underset{\sim}{\mathbf{H}}_{t,s} = (\underset{\sim}{\mathbf{A}}_t + \underset{\sim}{\mathbf{B}}_t\underset{\sim}{\mathbf{M}}_t - \underset{\sim}{\mathbf{K}}_t^\varepsilon\underset{\sim}{\mathbf{C}}_t)\underset{\sim}{\mathbf{H}}_{t,s}, \qquad \underset{\sim}{\mathbf{H}}_{s,s} = \underset{\sim}{\mathbf{I}} \tag{7.63}$$

and so

$$\zeta_t^\varepsilon = \underset{\sim}{\mathbf{H}}_{t,0}\hat{\mathbf{x}}_0 + \int_0^t \underset{\sim}{\mathbf{H}}_{t,\lambda}\underset{\sim}{\mathbf{K}}_\lambda^\varepsilon\,d\mathbf{y}_\lambda. \tag{7.64}$$

Consequently, $\mathbf{u}_t$ defined by (7.58) is $\mathscr{F}_t^{\mathbf{y}}$-measurable.

Now (6.1), (6.2), (7.62), and (7.58) gives us a system of linear stochastic differential equations with respect to $\mathbf{x}_t$ and ζ_t^ε as follows

$$d\mathbf{x}_t = \underset{\sim}{\mathbf{A}}_t\mathbf{x}_t\,dt + \underset{\sim}{\mathbf{B}}_t\underset{\sim}{\mathbf{M}}_t\zeta_t^\varepsilon\,dt + \underset{\sim}{\mathbf{D}}_t\,d\mathbf{w}_t \tag{7.65}$$

$$d\zeta_t^\varepsilon = \underset{\sim}{\mathbf{K}}_t^\varepsilon\underset{\sim}{\mathbf{C}}_t\mathbf{x}_t\,dt + (\underset{\sim}{\mathbf{A}}_t + \underset{\sim}{\mathbf{B}}_t\underset{\sim}{\mathbf{M}}_t - \underset{\sim}{\mathbf{K}}_t^\varepsilon\underset{\sim}{\mathbf{C}}_t)\zeta_t^\varepsilon\,dt + \underset{\sim}{\mathbf{K}}_t^\varepsilon\underset{\sim}{\mathbf{F}}_t\,d\mathbf{w}_t. \tag{7.66}$$

Let $\underset{\sim}{\mathbf{V}}_{t,s}$ be the fundamental solution matrix:

$$\frac{d}{dt}\underset{\sim}{\mathbf{V}}_{t,s} = \begin{bmatrix} \underset{\sim}{\mathbf{A}}_t & \underset{\sim}{\mathbf{B}}_t\underset{\sim}{\mathbf{M}}_t \\ \underset{\sim}{\mathbf{K}}_t^\varepsilon\underset{\sim}{\mathbf{C}}_t & \underset{\sim}{\mathbf{A}}_t + \underset{\sim}{\mathbf{B}}_t\underset{\sim}{\mathbf{M}}_t - \underset{\sim}{\mathbf{K}}_t^\varepsilon\underset{\sim}{\mathbf{C}}_t \end{bmatrix}\underset{\sim}{\mathbf{V}}_{t,s}, \qquad \underset{\sim}{\mathbf{V}}_{s,s} = \underset{\sim}{\mathbf{I}}, \tag{7.67}$$

then

$$\begin{bmatrix} \mathbf{x}_t \\ \boldsymbol{\zeta}_t^\varepsilon \end{bmatrix} = \underset{\sim}{\mathbf{V}}_{t,0}\begin{bmatrix} \mathbf{x}_0 \\ \hat{\mathbf{x}}_0 \end{bmatrix} + \int_0^t \underset{\sim}{\mathbf{V}}_{t,s}\begin{bmatrix} \underset{\sim}{\mathbf{D}}_s \\ \underset{\sim}{\mathbf{K}}_s^\varepsilon\underset{\sim}{\mathbf{F}}_s \end{bmatrix} d\mathbf{w}_s. \tag{7.68}$$

Since

$$\left(\underset{\sim}{\mathbf{F}}_t^T(\underset{\sim}{\mathbf{F}}_t\underset{\sim}{\mathbf{F}}_t^T + \varepsilon\underset{\sim}{\mathbf{I}})^{-1}\underset{\sim}{\mathbf{F}}_t\right)^2 \le \underset{\sim}{\mathbf{F}}_t^T(\underset{\sim}{\mathbf{F}}_t\underset{\sim}{\mathbf{F}}_t^T + \varepsilon\underset{\sim}{\mathbf{I}})^{-1}\underset{\sim}{\mathbf{F}}_t, \tag{7.69}$$

the eigenvalues of $\underset{\sim}{\mathbf{F}}_t^T(\underset{\sim}{\mathbf{F}}_t\underset{\sim}{\mathbf{F}}_t^T + \varepsilon\underset{\sim}{\mathbf{I}})^{-1}\underset{\sim}{\mathbf{F}}_t$ are less than or equal to 1, and so $\|\underset{\sim}{\mathbf{D}}_t\underset{\sim}{\mathbf{F}}_t^T(\underset{\sim}{\mathbf{F}}_t\underset{\sim}{\mathbf{F}}_t^T + \varepsilon\underset{\sim}{\mathbf{I}})^{-1}\underset{\sim}{\mathbf{F}}_t\|^2$ and hence $\|\underset{\sim}{\mathbf{K}}_t^\varepsilon\underset{\sim}{\mathbf{F}}_t\|^2$ are integrable on $[0, T]$.

Therefore, from (7.66), $E\|\boldsymbol{\zeta}_t^\varepsilon\|^2$ is bounded on $[0, T]$, then $\mathbf{u}_t$ defined by (7.58) satisfies (7.53) since $\|\underset{\sim}{\mathbf{M}}_t\|^2$ is integrable on $[0, T]$.

Both $\boldsymbol{\zeta}_t^\varepsilon$ and $\hat{\boldsymbol{\xi}}_t^\varepsilon$ are $\mathscr{F}_t^{\mathbf{y}}$-measurable, so from (7.61)

$$\mathscr{F}_t^{\boldsymbol{\eta}} \subset \mathscr{F}_t^{\mathbf{y}}. \tag{7.70}$$

On the other hand, $\boldsymbol{\zeta}_t^\varepsilon$ is $\mathscr{F}_t^{\boldsymbol{\eta}}$-measurable by (7.60). Hence from (7.61) $\mathscr{F}_t^{\mathbf{y}} \subset \mathscr{F}_t^{\boldsymbol{\eta}}$, which, together with (7.70) shows that under a control of the type given in (7.58), we have

$$\mathscr{F}_t^{\mathbf{y}} = \mathscr{F}_t^{\boldsymbol{\eta}}. \tag{7.71}$$

Then

$$\begin{aligned} \hat{\mathbf{x}}_t &= E\left(\boldsymbol{\xi}_t + \int_0^t \frac{\underset{\sim}{\boldsymbol{\Phi}}_{t,s}\underset{\sim}{\mathbf{B}}_s\underset{\sim}{\mathbf{M}}_s\boldsymbol{\zeta}_s^\varepsilon\, ds}{\mathscr{F}_t^{\mathbf{y}}}\right) \\ &= E\left(\boldsymbol{\xi}_t - \int_0^t \frac{\underset{\sim}{\boldsymbol{\Phi}}_{t,s}\underset{\sim}{\mathbf{B}}_s\underset{\sim}{\mathbf{M}}_s\boldsymbol{\zeta}_s^\varepsilon\, ds}{\mathscr{F}_t^{\boldsymbol{\eta}}}\right) \\ &= \hat{\boldsymbol{\xi}}_t + \int_0^t \underset{\sim}{\boldsymbol{\Phi}}_{t,s}\underset{\sim}{\mathbf{B}}_s\underset{\sim}{\mathbf{M}}_s\boldsymbol{\zeta}_s^\varepsilon\, ds. \end{aligned}$$

Hence

$$\mathbf{x}_t - \hat{\mathbf{x}}_t = \boldsymbol{\xi}_t - \hat{\boldsymbol{\xi}}_t. \tag{7.72}$$

This means that

$$E(\mathbf{x}_t - \hat{\mathbf{x}}_t)(\mathbf{x}_t - \hat{\mathbf{x}}_t)^\tau = E(\boldsymbol{\xi}_t - \hat{\boldsymbol{\xi}}_t)(\boldsymbol{\xi}_t - \hat{\boldsymbol{\xi}}_t)^\tau = \underset{\sim}{\mathbf{P}}_t. \tag{7.73}$$

That is, for $\{\mathbf{u}_t\} \in \mathscr{U}$, $\underset{\sim}{\mathbf{P}}_t$ is independent of the control. □

DEFINITION 7.1. If $\{\mathbf{u}_t(n)\} \in \mathscr{U}$ and

$$\lim_{n\to\infty} EJ(\mathbf{u}(n)) = \inf_{\mathbf{u}\in\mathscr{U}} EJ(\mathbf{u}),$$

then $\{\mathbf{u}_t(n)\}$ is called the optimal stochastic control sequence and for fixed n, $\{\mathbf{u}_t(n)\}$ is called the suboptimal control.

Lemma 7.5. *Assume $\underset{\sim}{\mathbf{A}}_t$, $\underset{\sim}{\mathbf{B}}_t$, $\underset{\sim}{\mathbf{C}}_t$, $\underset{\sim}{\mathbf{D}}_t$, and $\underset{\sim}{\mathbf{F}}_t$ satisfy the integrability condition required in Lemma 7.4 and assume $\|\underset{\sim}{\mathbf{Q}}_1(t)\|$ is integrable on $[0, T]$, then the Riccati equation*

$$-\dot{\underset{\sim}{\mathbf{S}}}{}_t^\delta = \underset{\sim}{\mathbf{A}}_t^\tau \underset{\sim}{\mathbf{S}}_t^\delta + \underset{\sim}{\mathbf{S}}_t^\delta \underset{\sim}{\mathbf{A}}_t + \underset{\sim}{\mathbf{Q}}_1(t) - \underset{\sim}{\mathbf{S}}_t^\delta \underset{\sim}{\mathbf{B}}_t\left(\underset{\sim}{\mathbf{Q}}_2(t) + \delta \underset{\sim}{\mathbf{I}}\right)^{-1} \underset{\sim}{\mathbf{B}}_t^\tau \underset{\sim}{\mathbf{S}}_t^\delta,$$

$$\underset{\sim}{\mathbf{S}}_T^\delta = \underset{\sim}{\mathbf{Q}}_0, \quad \delta > 0$$

for any $\delta > 0$ has a unique nonnegative definite solution.

The proof is the same as that for (6.97). □

Let us set

$$\underset{\sim}{\mathbf{L}}_t^\delta = -\left(\underset{\sim}{\mathbf{Q}}_2(t) + \delta \underset{\sim}{\mathbf{I}}\right)^{-1} \underset{\sim}{\mathbf{B}}_t^\tau \underset{\sim}{\mathbf{S}}_t^\delta,$$

a quantity which is obviously square integrable.

Replace $\underset{\sim}{\mathbf{M}}_t$ in (7.63) by $\underset{\sim}{\mathbf{L}}_t^\delta$ and correspondingly denote $\underset{\sim}{\mathbf{H}}_{t,s}$ by $\underset{\sim}{\mathbf{H}}_{t,s}(\delta)$. Denote

$$\boldsymbol{\zeta}_t^\varepsilon(\delta) = \tilde{\boldsymbol{\xi}}_t^\varepsilon + \int_0^t \underset{\sim}{\boldsymbol{\Phi}}_{t,s} \underset{\sim}{\mathbf{B}}_s \underset{\sim}{\mathbf{L}}_s^\delta \boldsymbol{\zeta}_s^\varepsilon(\delta)\, ds. \tag{7.74}$$

From (7.60)–(7.64), $\tilde{\boldsymbol{\zeta}}_t^\varepsilon$ can be expressed by

$$\boldsymbol{\zeta}_t^\varepsilon(\delta) = \underset{\sim}{\mathbf{H}}_{t,0}(\delta)\hat{\mathbf{x}}_0 + \int_0^t \underset{\sim}{\mathbf{H}}_{t,\lambda}(\delta) \underset{\sim}{\mathbf{K}}_\lambda^\varepsilon\, d\mathbf{y}_\lambda. \tag{7.75}$$

Theorem 7.3. *Suppose that the conditions of Lemma 7.5 are satisfied. Denote*

$$\mathbf{u}_t(\varepsilon, \delta) \triangleq \underset{\sim}{\mathbf{L}}_t^\delta \boldsymbol{\zeta}_t^\varepsilon(\delta). \tag{7.76}$$

Then there are sequences of real numbers $\varepsilon_n \underset{n\to\infty}{\to} 0$, $\delta_n \underset{n\to\infty}{\to} 0$, $\varepsilon_n > 0$, $\delta_n > 0$ *such that* $\mathbf{u}_t(n) \triangleq \mathbf{u}_t(\varepsilon_n, \delta_n)$ *is a stochastic control sequence such that*

$$EJ(\mathbf{u}(n)) \underset{n\to\infty}{\to} \inf_{u\in\mathscr{U}} EJ(\mathbf{u})$$

Proof. Denote

$$J_\delta(\mathbf{u}) = \mathbf{x}_T^\tau \underset{\sim}{\mathbf{Q}}_0 \mathbf{x}_T + \int_0^T \left[\mathbf{x}_t^\tau \underset{\sim}{\mathbf{Q}}_1(t)\mathbf{x}_t + \mathbf{u}_t^\tau\left(\underset{\sim}{\mathbf{Q}}_2(t) + \delta\underset{\sim}{\mathbf{I}}\right)\mathbf{u}_t\right] dt$$

$$= J(\mathbf{u}) + \delta\int_0^T \|\mathbf{u}_t\|^2\, dt. \tag{7.77}$$

By considering $\int_0^T d(\mathbf{x}_t^\tau \underset{\sim}{\mathbf{S}}_t^\delta \mathbf{x}_t)$, as in (6.102), we obtain

$$J_\delta(\mathbf{u}) = \mathbf{x}_0^\tau \underset{\sim}{\mathbf{S}}_0^\delta \mathbf{x}_0 + \int_0^T \operatorname{tr} \underset{\sim}{\mathbf{S}}_t^\delta \underset{\sim}{\mathbf{D}}_t \underset{\sim}{\mathbf{D}}_t^\tau\, dt$$

$$+ \int_0^T (\mathbf{u}_t - \underset{\sim}{\mathbf{L}}_t^\delta \mathbf{x}_t)^\tau \left(\underset{\sim}{\mathbf{Q}}_2(t) + \delta\underset{\sim}{\mathbf{I}}\right)(\mathbf{u}_t - \underset{\sim}{\mathbf{L}}_t^\delta \mathbf{x}_t)\, dt + 2\int_0^T \mathbf{x}_t^\tau \underset{\sim}{\mathbf{S}}_t^\delta \underset{\sim}{\mathbf{D}}_t\, d\mathbf{w}_t. \tag{7.78}$$

Notice that

$$\mathbf{x}_t = \underset{\sim}{\mathbf{\Phi}}_{t,0}\mathbf{x}_0 + \int_0^t \underset{\sim}{\mathbf{\Phi}}_{t,s}\underset{\sim}{\mathbf{B}}_s \mathbf{u}_s\, ds + \int_0^t \underset{\sim}{\mathbf{\Phi}}_{t,s}\underset{\sim}{\mathbf{D}}_s\, d\mathbf{w}_s$$

and

$$E\left\|\int_0^t \underset{\sim}{\mathbf{\Phi}}_{t,s}\underset{\sim}{\mathbf{B}}_s \mathbf{u}_s\, ds\right\|^2 \le \left[\int_0^t \|\underset{\sim}{\mathbf{\Phi}}_{t,s}\underset{\sim}{\mathbf{B}}_s\|^2\, ds\, E\int_0^t \|\mathbf{u}_s\|^2\, ds\right]^{1/2},$$

so for $\mathbf{u} \in \mathscr{U}$, $E\|\mathbf{x}_t\|^2$ is bounded on $[0, T]$; hence

$$E\int_0^T \|\mathbf{x}_t\|^2 \|\underset{\sim}{\mathbf{S}}_t^\delta \underset{\sim}{\mathbf{D}}_t\|^2\, dt < \infty$$

and

$$E\int_0^T \mathbf{x}_t^\tau \underset{\sim}{\mathbf{S}}_t^\delta \underset{\sim}{\mathbf{D}}_t\, d\mathbf{w}_t = 0. \tag{7.79}$$

Let

$$V_\delta = E\mathbf{x}_0^\tau \mathbf{S}_0^\delta E\mathbf{x}_0 + \operatorname{tr} \mathbf{R}_{\mathbf{x}_0}\mathbf{S}_0^\delta$$

$$+ \int_0^T \operatorname{tr}\left[\mathbf{D}_t\mathbf{D}_t^\tau \mathbf{S}_t^\delta + \mathbf{S}_t^\delta \mathbf{B}_t\left(\mathbf{Q}_2(t) + \delta\mathbf{I}\right)^{-1}\mathbf{B}_t^\tau \mathbf{S}_t^\delta \mathbf{P}_t\right] dt,$$

which is independent of the control.

By (7.79), as in (6.109), we obtain

$$EJ_\delta(\mathbf{u}) = V_\delta + E\int_0^T (\mathbf{u}_t - \mathbf{L}_t^\delta \hat{\mathbf{x}}_t)^\tau \left(\mathbf{Q}_2(t) + \delta\mathbf{I}\right)(\mathbf{u}_t - \mathbf{L}_t^\delta \hat{\mathbf{x}}_t)\, dt, \quad (7.80)$$

which means that

$$\inf_{n \in \mathscr{U}} EJ_\delta(\mathbf{u}) \geq V_\delta. \quad (7.81)$$

Generally speaking, this lower bound is not attainable, since $\mathbf{L}_t^\delta \hat{\mathbf{x}}_t$ does not necessarily belong to $\mathscr{U}$, but we shall now show that V_δ is indeed the precise lower bound.

By Lemma 7.4, we know that $\mathbf{u}(\varepsilon, \delta) \in \mathscr{U}$ for all $\varepsilon > 0$, $\delta > 0$, where $\mathbf{u}(\varepsilon,\delta)$ is given by (7.75).

We denote by $\mathbf{x}_t^\varepsilon(\delta)$ the state $\mathbf{x}_t$ with the control $\mathbf{u}(\varepsilon, \delta)$ applied. Then

$$\mathbf{x}_t^\varepsilon(\delta) = \boldsymbol{\xi}_t + \int_0^t \boldsymbol{\Phi}_{t,\lambda}\mathbf{B}_\lambda \mathbf{L}_\lambda^\delta \boldsymbol{\zeta}_\lambda^\varepsilon(\delta)\, d\lambda, \quad (7.82)$$

and by (7.74),

$$\mathbf{x}_t^\varepsilon(\delta) - \boldsymbol{\zeta}_t^\varepsilon(\delta) = \boldsymbol{\xi}_t - \tilde{\boldsymbol{\xi}}_t^\varepsilon. \quad (7.83)$$

By Theorem 7.1,

$$E(\boldsymbol{\xi}_t - \tilde{\boldsymbol{\xi}}_t^\varepsilon)(\boldsymbol{\xi}_t - \tilde{\boldsymbol{\xi}}_t^\varepsilon)^\tau = \tilde{\mathbf{P}}_t^\varepsilon \underset{\varepsilon \to 0}{\to} \mathbf{P}_t,$$

Hence by the dominated convergence theorem and (7.78), it follows that

$$EJ_\delta(\mathbf{u}(\varepsilon, \delta)) = V_\delta + \int_0^T \operatorname{tr} \mathbf{L}_t^{\delta\tau}\left(\mathbf{Q}_2(t) + \delta\mathbf{I}\right)\mathbf{L}_t^\delta(\tilde{\mathbf{P}}_t^\varepsilon - \mathbf{P}_t)\, dt \underset{\varepsilon \to 0}{\to} V_\delta.$$

Paying attention to (7.81), we have

$$\lim_{\varepsilon \to 0} EJ_\delta(\mathbf{u}(\varepsilon, \delta)) = V_\delta = \inf_{\mathbf{u} \in \mathscr{U}} EJ_\delta(\mathbf{u}). \quad (7.84)$$

For fixed $\mathbf{u} \in \mathscr{U}$ as $\delta \to 0$, $EJ_\delta(\mathbf{u})$ is nonincreasing, hence so is V_δ by (7.84) and so V_δ goes to a nonnegative limit as $\delta \to 0$. From (7.84), we have

$$\lim_{\delta \to 0} \inf_{\mathbf{u} \in \mathscr{U}} EJ_\delta(\mathbf{u}) = \lim_{\delta \to 0} \lim_{\varepsilon \to 0} EJ_\delta(\mathbf{u}(\varepsilon, \delta)). \tag{7.85}$$

For any fixed $\mathbf{u} \in \mathscr{U}$, we have

$$\inf_{u \in \mathscr{U}} EJ_\delta(\mathbf{u}) \le EJ_\delta(\mathbf{u}) = EJ(\mathbf{u}) + \delta E \int_0^T \|\mathbf{u}_t\|^2 \, dt,$$

from which by letting $\delta \to 0$ it follows that

$$\lim_{\delta \to 0} \inf_{u \in \mathscr{U}} EJ_\delta(\mathbf{u}) \le EJ(\mathbf{u})$$

or

$$\lim_{\delta \to 0} \inf_{u \in \mathscr{U}} EJ_\delta(\mathbf{u}) \le \inf_{u \in \mathscr{U}} EJ(\mathbf{u}).$$

Hence from (7.85)

$$\lim_{\delta \to 0} \lim_{\varepsilon \to 0} EJ_\delta(\mathbf{u}(\varepsilon, \delta)) \le \inf_{u \in \mathscr{U}} EJ(\mathbf{u}), \tag{7.86}$$

but

$$EJ_\delta(\mathbf{u}(\varepsilon, \delta)) \ge EJ(\mathbf{u}(\varepsilon, \delta)) \ge \inf_{\mathbf{u} \in \mathscr{U}} EJ(\mathbf{u}).$$

Then

$$\begin{aligned} \lim_{\delta \to 0} \lim_{\varepsilon \to 0} EJ_\delta(\mathbf{u}(\varepsilon, \delta)) &\ge \overline{\lim_{\delta \to 0}}\, \overline{\lim_{\varepsilon \to 0}}\, EJ(\mathbf{u}(\varepsilon, \delta)) \\ &\ge \underline{\lim_{\delta \to 0}}\, \underline{\lim_{\varepsilon \to 0}}\, EJ(\mathbf{u}(\varepsilon, \delta)) \ge \inf_{u \in \mathscr{U}} EJ(\mathbf{u}). \end{aligned} \tag{7.87}$$

From (7.84), (7.86), and (7.87), we conclude that

$$\lim_{\delta \to 0} \lim_{\varepsilon \to 0} EJ(\mathbf{u}(\varepsilon, \delta)) = \inf_{u \in \mathscr{U}} EJ(\mathbf{u}) = \lim_{\delta \to 0} V_\delta.$$

Let $\alpha_n > 0$ be a real decreasing sequence converging to zero. Take $\delta_n > 0$ such that

$$\left| V_\delta - \lim_{\delta \to 0} V_\delta \right| < \frac{\alpha_n}{2}, \qquad \forall \delta \in (0, \delta_n],$$

and then take $\varepsilon_n > 0$ such that

$$|EJ_{\delta_n}(\mathbf{u}(\varepsilon, \delta_n)) - V_{\delta_n}| < \frac{\alpha_n}{2}, \qquad \forall t \in (0, \varepsilon_n].$$

Thus

$$EJ(\mathbf{u}(\varepsilon_n, \delta_n)) \underset{n \to \infty}{\to} \inf_{u \in \mathscr{U}} EJ(\mathbf{u}). \qquad \square$$

7.5. SINGULAR CONTROLS OF SPECIAL CASES

In the last section both $\underset{\sim}{\mathbf{F}}_t \underset{\sim}{\mathbf{F}}_t^{\tau}$ and $\underset{\sim}{\mathbf{Q}}_2(t)$ are possibly degenerate, we now consider the special cases where they are not simultaneously degenerate.

Lemma 7.6. *Let $\underset{\sim}{\mathbf{A}}_t(j)$, $j \in J$, be $n \times n$ matrices such that*

$$\|\underset{\sim}{\mathbf{A}}_t(j)\| \le \|\underset{\sim}{\mathbf{A}}_t\|, \qquad \forall j \in J$$

with $\|\underset{\sim}{\mathbf{A}}_t\|$ integrable on $[0, T)$, and let

$$\frac{d}{dt}\underset{\sim}{\mathbf{\Phi}}_{t,s}(j) = \underset{\sim}{\mathbf{A}}_t(j)\underset{\sim}{\mathbf{\Phi}}_{t,s}(j), \qquad \underset{\sim}{\mathbf{\Phi}}_{s,s}(j) = \underset{\sim}{\mathbf{I}}.$$

Then $\|\underset{\sim}{\mathbf{\Phi}}_{t,0}(j)\|$ and $\|\underset{\sim}{\mathbf{\Phi}}_{0,t}(j)\|$, $j \in J$, are uniformly bounded for $t \in [0, T]$. Further, if $\underset{\sim}{\mathbf{A}}_t(j)$, as a function of j, is uniformly continuous with respect to $t \in [0, T]$, then so are $\underset{\sim}{\mathbf{\Phi}}_{t,0}(j)$ and $\underset{\sim}{\mathbf{\Phi}}_{0,t}(j)$.

Proof. For fixed j, $\|\underset{\sim}{\mathbf{\Phi}}_{0,t}(j)\|$ is bounded on $[0, T]$. Then from

$$\|\underset{\sim}{\mathbf{\Phi}}_{t,0}(j)\| \le 1 + \int_0^t \|\underset{\sim}{\mathbf{A}}_s\| \, \|\underset{\sim}{\mathbf{\Phi}}_{s,0}(j)\| \, ds$$

it is immediate that

$$\|\underset{\sim}{\mathbf{\Phi}}_{t,0}(j)\| \le \exp \int_0^t \|\underset{\sim}{\mathbf{A}}_s\| \, ds \le \exp \int_0^T \|\underset{\sim}{\mathbf{A}}_s\| \, ds \triangleq k.$$

The boundedness of $\|\underset{\sim}{\mathbf{\Phi}}_{0,t}(j)\|$ is proved similarly. The uniform continuity follows from the following inequalities:

$$\|\Phi_{t,0}(j) - \Phi_{t,0}(j')\|$$

$$\leq \int_0^t \|A_s(j) - A_s(j')\| \, \|\Phi_{s,0}(j)\| \, ds$$

$$+ \int_0^t \|A_s(j')\| \, \|\Phi_{s,0}(j) - \Phi_{s,0}(j')\| \, ds$$

$$\leq k \max_{0 \leq s \leq T} \|A_s(j) - A_s(j')\| + \int_0^t \|A_s\| \, \|\Phi_{s,0}(j) - \Phi_{s,0}(j')\| \, ds$$

$$\leq k \max_{0 \leq s \leq T} \|A_s(j) - A_s(j')\| \exp \int_0^T \|A_s\| \, ds. \qquad \square$$

Lemma 7.7. *Assume that* A_t, B_t, C_t, D_t, *and* F_t *satisfy the conditions of Lemma 7.4. Then as* $\varepsilon \to 0$ *the solution of* (7.15) P_t^ε *is nonincreasing, and as* R_0 *increases it is nondecreasing. If* $F_t F_t^\tau > 0$ *and* $C_t^\tau (F_t F_t^\tau)^{-1} C_t$ *is integrable on* $[0, T]$, *then, as* $\varepsilon \to 0$, *the convergence of* P_t^ε *to* P_t *is uniform in* $[0, T)$, *where* P_t *is the solution of* (6.8).

Proof. Let us put

$$F_t' = \left[F_t \quad \sqrt{\varepsilon} I\right], \qquad D_t' = [D_t \quad 0],$$

then (7.15) is rewritten as

$$P_t^\varepsilon = R_0 + \int_0^t \Big[A_s P_s^\varepsilon + P_s^\varepsilon A_s^\tau - (D_s' F_s'^\tau + P_s^\varepsilon C_s^\tau)(F_s' F_s'^\tau)^{-1}$$

$$\times (D_s' F_s'^\tau + P_s^\varepsilon C_s^\tau)^\tau + D_s' D_s'^\tau\Big] \, ds.$$

Consider the auxiliary control problem:

$$dz_s = A_{t-s}^\tau z_s \, ds + C_{t-s}^\tau u_s \, ds, \qquad z_0 = z$$

with performance index

$$J(u) = z_t^\tau R_0 z_t + \int_0^t (z_s^\tau D_{t-s}' + u_s^\tau F_{t-s}')(D_{t-s}'^\tau z_s + F_{t-s}'^\tau u_s) \, ds.$$

By Lemma 6.1, the following equation is solvable

$$-\dot{\underset{\sim}{\mathbf{S}}}_s = \underset{\sim}{\mathbf{A}}_{t-s}\underset{\sim}{\mathbf{S}}_s + \underset{\sim}{\mathbf{S}}_s\underset{\sim}{\mathbf{A}}^{\tau}_{t-s} - \left(\underset{\sim}{\mathbf{D}}'_{t-s}\underset{\sim}{\mathbf{F}}'^{\tau}_{t-s} + \underset{\sim}{\mathbf{S}}_s\underset{\sim}{\mathbf{C}}^{\tau}_{t-s}\right)\left(\underset{\sim}{\mathbf{F}}'_{t-s}\underset{\sim}{\mathbf{F}}'^{\tau}_{t-s}\right)^{-1}$$

$$\cdot\left(\underset{\sim}{\mathbf{D}}'_{t-s}\underset{\sim}{\mathbf{F}}'^{\tau}_{t-s} + \underset{\sim}{\mathbf{S}}_s\underset{\sim}{\mathbf{C}}^{\tau}_{t-s}\right)^{\tau} + \underset{\sim}{\mathbf{D}}'_{t-s}\underset{\sim}{\mathbf{D}}'^{\tau}_{t-s}, \qquad \underset{\sim}{\mathbf{S}}_t = \underset{\sim}{\mathbf{R}}_0.$$

Clearly,

$$\underset{\sim}{\mathbf{S}}_{t-s} = \underset{\sim}{\mathbf{P}}^{\varepsilon}_s.$$

As in (6.102), by considering the integral

$$\int_0^t d\left(\mathbf{z}_s^{\tau}\underset{\sim}{\mathbf{S}}_s\mathbf{z}_s\right),$$

it is not difficult to verify that

$$J(\mathbf{u}) = \mathbf{z}^{\tau}\underset{\sim}{\mathbf{P}}^{\varepsilon}_t\mathbf{z} + \int_0^t\left[\mathbf{u}_s + \left(\underset{\sim}{\mathbf{F}}'_{t-s}\underset{\sim}{\mathbf{F}}'^{\tau}_{t-s}\right)^{-1}(\underset{\sim}{\mathbf{C}}_{t-s}\underset{\sim}{\mathbf{S}}_s + \underset{\sim}{\mathbf{F}}_{t-s}\underset{\sim}{\mathbf{D}}^{\tau}_{t-s})\mathbf{z}_s\right]^{\tau}$$

$$\cdot\left(\underset{\sim}{\mathbf{F}}'_{t-s}\underset{\sim}{\mathbf{F}}'^{\tau}_{t-s}\right)\left[\mathbf{u}_s + \left(\underset{\sim}{\mathbf{F}}'_{t-s}\underset{\sim}{\mathbf{F}}'^{\tau}_{t-s}\right)^{-1}(\underset{\sim}{\mathbf{C}}_{t-s}\underset{\sim}{\mathbf{S}}_s + \underset{\sim}{\mathbf{F}}_{t-s}\underset{\sim}{\mathbf{D}}^{\tau}_{t-s})\mathbf{z}_s\right] ds.$$

Hence the optimal control is

$$\mathbf{u}^*_s = -\left(\underset{\sim}{\mathbf{F}}'_{t-s}\underset{\sim}{\mathbf{F}}'^{\tau}_{t-s}\right)^{-1}(\underset{\sim}{\mathbf{C}}_{t-s}\underset{\sim}{\mathbf{S}}_s + \underset{\sim}{\mathbf{F}}_{t-s}\underset{\sim}{\mathbf{D}}^{\tau}_{t-s})\mathbf{z}_s.$$

and

$$\min_{\mathbf{u}\in\mathscr{U}} J(\mathbf{u}) = J(\mathbf{u}^*) = \mathbf{z}^{\tau}\underset{\sim}{\mathbf{P}}^{\varepsilon}_t\mathbf{z}.$$

Let $0 < \varepsilon_1 < \varepsilon_2$ and denote by J_i the performance index corresponding to the case $\varepsilon = \varepsilon_i$, $\underset{\sim}{\mathbf{F}}'_t = [\underset{\sim}{\mathbf{F}}_t \;\; \sqrt{\varepsilon_i}\,\underset{\sim}{\mathbf{I}}]$, $i = 1, 2$. Let $\mathbf{z}_s(i)$, $\mathbf{u}^*_s(i)$ and $\underset{\sim}{\mathbf{S}}_s(i)$ denote the optimal solution for $\varepsilon = \varepsilon_i$.

Then we have

$$\mathbf{z}^{\tau}\underset{\sim}{\mathbf{P}}^{\varepsilon_1}_t\mathbf{z} = J_1(\mathbf{u}^*(1)) \le J_1(\mathbf{u}^*(2)) = \mathbf{z}_t^{\tau}(2)\underset{\sim}{\mathbf{R}}_0\mathbf{z}_t(2)$$

$$+ \int_0^t\left[\left(\mathbf{z}_s^{\tau}(2)\underset{\sim}{\mathbf{D}}_{t-s} + \mathbf{u}^{*\tau}_s(2)\underset{\sim}{\mathbf{F}}_{t-s}\right)\left(\mathbf{z}_s^{\tau}(2)\underset{\sim}{\mathbf{D}}_{t-s} + \mathbf{u}^{*\tau}_s(2)\underset{\sim}{\mathbf{F}}_{t-s}\right)^{\tau}\right.$$

$$\left. + \varepsilon_1\mathbf{u}^{*\tau}_s(2)\mathbf{u}^*_s(2)\right] ds$$

$$\le J_2(\mathbf{u}^*(2)) = \mathbf{z}^{\tau}\underset{\sim}{\mathbf{P}}^{\varepsilon_2}_t\mathbf{z}.$$

Because of the arbitrary nature of $\mathbf{z}$, we obtain $\mathbf{P}_t^{\varepsilon_1} \le \mathbf{P}_t^{\varepsilon_2}$. By a similar argument, it is easy to show that $\mathbf{P}_t^{\varepsilon}$ is nondecreasing as $\mathbf{R}_0$ increases.

Since $\mathbf{P}_t^{\varepsilon}$ is continuous, there is a constant k_1 so that $\|\mathbf{P}_t^{\varepsilon}\| \le k_1$, for all $t \in [0, T]$ and for all $t \in [0, \varepsilon_1]$. If $\mathbf{F}_t\mathbf{F}_t^{\tau} > \mathbf{0}$, then

$$\|\mathbf{F}_\lambda^{\tau}(\mathbf{F}_\lambda\mathbf{F}_\lambda^{\tau} + \varepsilon\mathbf{I})^{-1}\mathbf{C}_\lambda\| = \|\mathbf{C}_\lambda^{\tau}(\mathbf{F}_\lambda\mathbf{F}_\lambda^{\tau} + \varepsilon\mathbf{I})^{-1}\mathbf{F}_\lambda\mathbf{F}_\lambda^{\tau}(\mathbf{F}_\lambda\mathbf{F}_\lambda^{\tau} + \varepsilon\mathbf{I})^{-1}\mathbf{C}_\lambda\|^{1/2}$$

$$\le \|\mathbf{C}_\lambda^{\tau}(\mathbf{F}_\lambda\mathbf{F}_\lambda^{\tau} + \varepsilon\mathbf{I})^{-1}\mathbf{C}_\lambda\|^{1/2} \le \|\mathbf{C}_\lambda^{\tau}(\mathbf{F}_\lambda\mathbf{F}_\lambda^{\tau})^{-1}\mathbf{C}_\lambda\|^{1/2},$$

$$\mathbf{F}_t^{\tau}(\mathbf{F}_t\mathbf{F}_t^{\tau} + \varepsilon\mathbf{I})^{-1}\mathbf{F}_t \le \mathbf{I}$$

and hence for any $\varepsilon \in [0, \varepsilon_1]$

$$\|\mathbf{P}_t^{\varepsilon} - \mathbf{P}_s^{\varepsilon}\| = \int_s^t \|\mathbf{P}_\lambda^{\varepsilon}\mathbf{A}_\lambda^{\tau} + \mathbf{A}_\lambda\mathbf{P}_\lambda^{\varepsilon} + \mathbf{D}_\lambda\mathbf{D}_\lambda^{\tau} - (\mathbf{D}_\lambda\mathbf{F}_\lambda^{\tau} + \mathbf{P}_\lambda^{\varepsilon}\mathbf{C}_\lambda^{\tau})$$

$$\cdot(\mathbf{F}_\lambda\mathbf{F}_\lambda^{\tau} + \varepsilon\mathbf{I})^{-1}(\mathbf{D}_\lambda\mathbf{F}_\lambda^{\tau} + \mathbf{P}_\lambda^{\varepsilon}\mathbf{C}_\lambda^{\tau})^{\tau}\| \, d\lambda$$

$$\le 2k_1\int_s^t \|\mathbf{A}_\lambda\| \, d\lambda + \int_s^t \|\mathbf{D}_\lambda\|^2 \, d\lambda +$$

$$\times \int_s^t \Big[\|\mathbf{D}_\lambda\|^2 + k_1^2\|\mathbf{C}_\lambda^{\tau}(\mathbf{F}_\lambda\mathbf{F}_\lambda^{\tau})^{-1}\mathbf{C}_\lambda\|$$

$$+2k_1\|\mathbf{D}_\lambda\| \, \|\mathbf{C}_\lambda^{\tau}(\mathbf{F}_\lambda\mathbf{F}_\lambda^{\tau})^{-1}\mathbf{C}_\lambda\|^{1/2}\Big] \, d\lambda$$

$$\le \int_s^t \Big[2k_1\|\mathbf{A}_\lambda\| + 2\|\mathbf{D}_\lambda\|^2 + k_1^2\|\mathbf{C}_\lambda^{\tau}(\mathbf{F}_\lambda\mathbf{F}_\lambda^{\tau})^{-1}\mathbf{C}_\lambda\|\Big] \, d\lambda$$

$$+2k_1\int_s^t \|\mathbf{D}_\lambda\|^2 \, d\lambda \int_s^t \|\mathbf{C}_\lambda^{\tau}(\mathbf{F}_\lambda\mathbf{F}_\lambda^{\tau})^{-1}\mathbf{C}_\lambda\| \, d\lambda,$$

which can be made arbitrarily small independently of ε if $|t - s|$ is small enough. Hence $\mathbf{P}_t^{\varepsilon}$ is equicontinuous on $[0, T]$ with ε as a parameter. By noticing that $0 \le \mathbf{P}_t^{\varepsilon} \le \mathbf{P}_t^{\varepsilon_1}$, for all $\varepsilon \in [0, \varepsilon_1]$, it is obvious that $\mathbf{P}_t^{\varepsilon}$ is uniformly bounded on $t \in [0, T]$, $\varepsilon \in [0, \varepsilon_1]$.

By Theorem 2.3, there is a subsequence $\varepsilon_k \underset{k\to\infty}{\to} 0$ so that $\mathbf{P}_t^{\varepsilon_k} \underset{k\to\infty}{\to} \mathbf{P}_t$ uniformly in $[0, T]$. From here and the fact that $\mathbf{P}_t \le \mathbf{P}_t^{\varepsilon} \le \mathbf{P}_t^{\varepsilon_k}$ for all $\varepsilon \in [0, \varepsilon_k]$, it follows that $\mathbf{P}_t^{\varepsilon} \underset{\varepsilon\to 0}{\to} \mathbf{P}$ uniformly in $t \in [0, T]$. □

Theorem 7.4. *Assume that the conditions of Lemma* 7.5 *hold. For the deterministic system* (i.e., *that for which* $\underset{\sim}{\mathbf{D}}_t \equiv \underset{\sim}{\mathbf{0}}$, $\underset{\sim}{\mathbf{F}}_t \equiv \underset{\sim}{\mathbf{0}}$ *and* $\underset{\sim}{\mathbf{R}}_{\mathbf{x}_0} = \underset{\sim}{\mathbf{0}}$), *the control defined by* (7.76) *becomes*

$$\mathbf{u}_t(\delta) = \underset{\sim}{\mathbf{L}}_t^{\delta}\mathbf{x}_t(\delta), \tag{7.88}$$

and this is the suboptimal control for deterministic systems, that for which

$$\lim_{\delta\to 0} J(\mathbf{u}_t(\delta)) = \inf_{\mathbf{u}\in\mathscr{U}} J(\mathbf{u}).$$

If, in addition, $\underset{\sim}{\mathbf{Q}}_2(t) > 0$ a.e. $t \in [0, T]$ *and* $\underset{\sim}{\mathbf{Q}}_2(t)$ *and* $\underset{\sim}{\mathbf{B}}_t\underset{\sim}{\mathbf{Q}}_2^{-1}(t)\underset{\sim}{\mathbf{B}}_t^{\tau}$ *are integrable on* $[0, T]$, *then the control given by* (7.88) *as* $\delta \to 0$ *tends to the limit* $\mathbf{u}_t$, *that is,*

$$\mathbf{u}_t = -\underset{\sim}{\mathbf{Q}}_2^{-1}(t)\underset{\sim}{\mathbf{B}}_t^{\tau}\underset{\sim}{\mathbf{S}}_t\mathbf{x}_t, \tag{7.89}$$

which is the optimal control. Here $\mathbf{x}_t(\delta)$ *and* $\mathbf{x}_t$ *denote the state resulting from* $\mathbf{u}_t(\delta)$ *and* $\mathbf{u}_t$, *respectively, and* $\underset{\sim}{\mathbf{S}}_t$ *is the solution of* (6.97).

Proof. According to Lemma 6.1, the solution of (7.15) is unique, hence $\underset{\sim}{\mathbf{P}}_t^{\varepsilon} \equiv 0$ for deterministic systems. Notice that $\hat{\mathbf{x}}_0 = \mathbf{x}_0$, so from (7.57) $\tilde{\boldsymbol{\xi}}_t^{\varepsilon} = \boldsymbol{\xi}_t$, and from (7.74) and (7.76), we have

$$\boldsymbol{\zeta}_t^{\varepsilon}(\delta) = \boldsymbol{\xi}_t + \int_0^t \underset{\sim}{\mathbf{\Phi}}_{t,s}\underset{\sim}{\mathbf{B}}_s\mathbf{u}_s\,ds.$$

Thus in this case $\boldsymbol{\zeta}_t^{\varepsilon}(\delta)$ is nothing but the state under the control $\{\mathbf{u}_s\}$, and it does not depend on ε at all. Denote this state process by $\mathbf{x}_t(\delta)$, that is, $\boldsymbol{\zeta}_t^{\varepsilon}(\delta) = \mathbf{x}(\delta)$. Then (7.88) follows from (7.76), and (7.86), (7.87) become

$$\inf_{u\in\mathscr{U}} J(\mathbf{u}) \geq \lim_{\delta\to 0} J_{\delta}(\mathbf{u}(\delta)) \geq \lim_{\delta\to 0} J(\mathbf{u}(\delta)) \geq \inf_{u\in\mathscr{U}} J(\mathbf{u}).$$

Hence (7.88) indeed gives a control converging to the optimal control.

With the additional conditions $\underset{\sim}{\mathbf{Q}}_2(t) > \underset{\sim}{\mathbf{0}}$ and $\int_0^T \|\underset{\sim}{\mathbf{B}}_t\underset{\sim}{\mathbf{Q}}_2^{-1}(t)\underset{\sim}{\mathbf{B}}_t^{\tau}\|\,dt < \infty$ imposed in Theorem 6.6 we have proved that (6.97) is a special case of (6.8). Hence the equation for $\underset{\sim}{\mathbf{S}}_t^{\delta}$ is a special case of (7.15). By Lemma 7.7, as $\delta \to 0$, $\underset{\sim}{\mathbf{S}}_t^{\delta}$ tends nonincreasingly to a finite limit $\underset{\sim}{\mathbf{S}}_t$, and if $\|\underset{\sim}{\mathbf{Q}}_2(t)\|$ and

$\|\underset{\sim}{\mathbf{B}}_t\underset{\sim}{\mathbf{Q}}_2^{-1}(t)\underset{\sim}{\mathbf{B}}_t^\tau\|$ are integrable on $[0, T]$, $\underset{\sim}{\mathbf{S}}_t$ is the solution of (6.97). Write

$$\underset{\sim}{\mathbf{A}}_s(\delta) = \underset{\sim}{\mathbf{A}}_s + \underset{\sim}{\mathbf{B}}_s\underset{\sim}{\mathbf{L}}_s^\delta = \underset{\sim}{\mathbf{A}}_s - \underset{\sim}{\mathbf{B}}_s\left[\underset{\sim}{\mathbf{Q}}_2(s) + \delta\underset{\sim}{\mathbf{I}}\right]^{-1}\underset{\sim}{\mathbf{B}}_s^\tau\underset{\sim}{\mathbf{S}}_s^\delta,$$

$$\frac{d}{dt}\underset{\sim}{\mathbf{\Phi}}_{t,s}(\delta) = \underset{\sim}{\mathbf{A}}_t(\delta)\underset{\sim}{\mathbf{\Phi}}_{t,s}(\delta), \qquad \underset{\sim}{\mathbf{\Phi}}_{s,s}(\delta) = \underset{\sim}{\mathbf{I}}.$$

As $\delta \to 0$, $\underset{\sim}{\mathbf{S}}_t^\delta$ is nonincreasing, hence

$$\|\underset{\sim}{\mathbf{A}}_t(\delta)\| \le \|\underset{\sim}{\mathbf{A}}_t\| + \|\underset{\sim}{\mathbf{B}}_t\underset{\sim}{\mathbf{Q}}_2^{-1}(t)\underset{\sim}{\mathbf{B}}_t^\tau\|\,\|\underset{\sim}{\mathbf{S}}_t^{\delta_1}\|, \qquad \forall\delta \in [0, \delta_1], \quad \delta_1 > 0.$$

By Lemma 7.6, $\|\underset{\sim}{\mathbf{\Phi}}_{t,s}(\delta)\|$ is uniformly bounded for $\delta \in [0, \delta_1]$ and $t, s \in [0, T]$. Hence by the dominated convergence theorem,

$$\mathbf{x}_t(\delta) - \mathbf{x}_t = \int_0^t \underset{\sim}{\mathbf{\Phi}}_{t,s}(\delta)\Big[\underset{\sim}{\mathbf{B}}_s\underset{\sim}{\mathbf{Q}}_2^{-1}(s)\underset{\sim}{\mathbf{B}}_s^\tau\underset{\sim}{\mathbf{S}}_s$$
$$-\underset{\sim}{\mathbf{B}}_s(\underset{\sim}{\mathbf{Q}}_2(s) + \delta\underset{\sim}{\mathbf{I}})^{-1}\underset{\sim}{\mathbf{B}}_s^\tau\underset{\sim}{\mathbf{S}}_s^\delta\Big]\mathbf{x}_s\,ds \underset{\delta\to 0}{\to} 0,$$

and as $\delta \to 0$, the right-hand side of (7.88) converges to the right-hand side of (7.89).

Letting $\delta \to 0$ in (7.78), we know that (7.89) gives the optimal control. □

Theorem 7.5. *Under conditions of Lemma* 7.5, *if* $\underset{\sim}{\mathbf{F}}_t\underset{\sim}{\mathbf{F}}_t^\tau > 0$ a.e. *on* $[0, T]$ *and if* $\underset{\sim}{\mathbf{C}}_t^\tau(\underset{\sim}{\mathbf{F}}_t\underset{\sim}{\mathbf{F}}_t^\tau)^{-1}\underset{\sim}{\mathbf{C}}_t$ *is integrable on* $[0, T]$, *then as* $\varepsilon \to 0$ $\mathbf{u}_t(\varepsilon, \delta)$ *given by* (7.76) *has the limit* $\mathbf{u}_t(\delta)$ *in the mean square sense given by*

$$\mathbf{u}_t(\delta) = \underset{\sim}{\mathbf{L}}_t^\delta\hat{\mathbf{x}}_t(\delta), \tag{7.90}$$

where $\hat{\mathbf{x}}_t(\delta)$ *denotes the state filtering under* $\mathbf{u}_t(\delta)$, *and, further,*

$$\lim_{\delta\to 0} EJ(\mathbf{u}(\delta)) = \inf_{\mathbf{u}\in\mathscr{U}} EJ(\mathbf{u}).$$

If in addition $\underset{\sim}{\mathbf{Q}}_2(t) > \underset{\sim}{\mathbf{0}}$ a.e., $t \in [0, T]$, *and* $\underset{\sim}{\mathbf{B}}_t\underset{\sim}{\mathbf{Q}}_2^{-1}(t)\underset{\sim}{\mathbf{B}}_t^\tau$ *is integrable on* $[0, T]$, *then*

$$\mathbf{u}_t = -\underset{\sim}{\mathbf{Q}}_2^{-1}(t)\underset{\sim}{\mathbf{B}}_t^\tau\underset{\sim}{\mathbf{S}}_t\hat{\mathbf{x}}_t \tag{7.91}$$

is the optimal control of the system (6.1), (6.2) *under performance index*

(6.95). *If, further,* $\mathbf{Q}_2^{-3}(t)$ *is integrable on* $[0, T]$, *then as* $\delta \to 0$, $\mathbf{u}_t(\delta)$ *given by* (7.90) *converges in the mean square sense to* $\mathbf{u}_t$ *given by* (7.91).

Proof. From the proof of Lemma 7.4, it is seen that, as in (7.64), $\zeta_t(\delta)$ can be defined by

$$\zeta_t(\delta) = \xi_t + \int_0^t \Phi_{t,s}\mathbf{B}_s\mathbf{L}_s^\delta \zeta_s(\delta)\, ds. \tag{7.92}$$

Then

$$\zeta_t^\varepsilon(\delta) - \zeta_t(\delta) = \tilde{\xi}_t^\varepsilon - \xi_t - \int_0^t \Phi_{t,s}\mathbf{B}_s\mathbf{L}_s^\delta(\zeta_s^\varepsilon(\delta) - \zeta_s(\delta))\, ds,$$

$$E\|\zeta_t^\varepsilon(\delta) - \zeta_t(\delta)\|^2 \le 2E\|\tilde{\xi}_t^\varepsilon - \xi_t\|^2 + 2\int_0^t \|\Phi_{t,s}\mathbf{B}_s\mathbf{L}_s^\delta\|\, ds$$

$$\times \int_0^t \|\Phi_{t,s}\mathbf{B}_s\mathbf{L}_s^\delta\| E\|\zeta_s^\varepsilon(\delta) - \zeta_s(\delta)\|^2\, ds. \tag{7.93}$$

By the continuity of $\mathbf{S}_t^\delta$ and $\Phi_{t,s}$, there is a constant k_2 so that for any $t \in [0, T]$

$$2\int_0^t \|\Phi_{t,s}\mathbf{B}_s\mathbf{L}_s^\delta\|\, ds \le 2k_2^2 \int_0^T \mathbf{B}_t\Big(\mathbf{Q}_2(t) + \delta\mathbf{I}\Big)^{-1}\mathbf{B}_t^\tau\, dt$$

$$\le \frac{2k_2^2}{\delta}\int_0^T \|\mathbf{B}_t\|^2\, dt \triangleq k_3.$$

Hence from (7.93), we have

$$E\|\zeta_t^\varepsilon(\delta) - \zeta_t(\delta)\|^2 \le 2\Big(\max_{0\le t\le T} E\|\tilde{\xi}_t^\varepsilon - \xi_t\|^2\Big)\exp k_3\int_0^t \|\Phi_{t,s}\mathbf{B}_s\mathbf{L}_s^\delta\|\, ds$$

$$\le 2\Big(\max_{0\le t\le T} E\|\tilde{\xi}_t^\varepsilon - \xi_t\|^2\Big)e^{k_3^2/2}. \tag{7.94}$$

By (7.28) and (7.30),

$$E\|\tilde{\xi}_t^\varepsilon - \xi_t\|^2 = E\|\delta_t^\varepsilon\|^2 = \operatorname{tr}\big(\mathbf{P}_t^\varepsilon - \tilde{\mathbf{P}}_t^\varepsilon\big) \le \operatorname{tr}\big(\mathbf{P}_t^\varepsilon - \mathbf{P}_t\big),$$

which, as $\varepsilon \to 0$, tends to zero uniformly in $t \in [0, T]$ by Lemma 7.7. From this and (7.94), it follows that

$$E\|\zeta_t^\varepsilon(\delta) - \zeta_t(\delta)\|^2 \le 2 \max_{0\le t\le T} \operatorname{tr}(\underset{\sim}{\mathbf{P}}_t^\varepsilon - \underset{\sim}{\mathbf{P}}_t)e^{k_3^2/2} \underset{\varepsilon\to 0}{\to} 0. \tag{7.95}$$

By Lemma 7.4, we know that

$$\mathbf{u}_t(\delta) \triangleq \underset{\sim}{\mathbf{L}}_t^\delta \zeta_t(\delta)$$

is an admissible control, and under it

$$\mathscr{F}_t^{\boldsymbol{\eta}} = \mathscr{F}_t^{\mathbf{y}}.$$

Hence

$$E\left(\frac{\mathbf{x}_t}{\mathscr{F}_t^{\mathbf{y}}}\right) = \boldsymbol{\xi}_t + \int_0^t \underset{\sim}{\boldsymbol{\Phi}}_{t,s} \underset{\sim}{\mathbf{B}}_s \underset{\sim}{\mathbf{L}}_s^\delta \zeta_s(\delta)\, ds. \tag{7.96}$$

Comparing (7.92) with (7.97), we find that

$$\zeta_t(\delta) = E\left(\frac{\mathbf{x}_t}{\mathscr{F}_t^{\mathbf{y}}}\right) = \hat{\mathbf{x}}_t(\delta).$$

Then, from (7.95),

$$E\|\zeta_t^\varepsilon(\delta) - \hat{\mathbf{x}}_t(\delta)\|^2 \underset{\varepsilon\to 0}{\to} 0,$$

and this proves (7.90).

Since

$$\mathbf{u}_t(\delta) = \underset{\sim}{\mathbf{L}}_t^\delta \zeta_t(\delta) = \underset{\sim}{\mathbf{L}}_t^\delta \hat{\mathbf{x}}_t(\delta)$$

is an admissible control, from (7.80),

$$EJ_\delta(\mathbf{u}(\delta)) = \inf_{\mathbf{u}\in\mathscr{U}} EJ_\delta(\mathbf{u}), \tag{7.97}$$

but in Theorem 7.3 it has been shown that

$$\lim_{\delta\to 0} \inf_{\mathbf{u}\in\mathscr{U}} EJ_\delta(\mathbf{u}) \le \inf_{\mathbf{u}\in\mathscr{U}} EJ(\mathbf{u}), \tag{7.98}$$

Then (7.97) and (7.98) yield

$$\lim_{\delta\to 0} EJ_\delta(\mathbf{u}(\delta)) = \lim_{\delta\to 0}\inf_{\mathbf{u}\in\mathscr{U}} EJ_\delta(\mathbf{u}) \le \inf_{\mathbf{u}\in\mathscr{U}} EJ(\mathbf{u}).$$

On the other hand,

$$EJ_\delta(\mathbf{u}(\delta)) \ge EJ(\mathbf{u}(\delta)) \ge \inf_{\mathbf{u}\in\mathscr{U}} EJ(\mathbf{u}).$$

Hence

$$\lim_{\delta\to 0} EJ(\mathbf{u}(\delta)) = \inf_{\mathbf{u}\in\mathscr{U}} EJ(\mathbf{u}).$$

In other words, $\{\mathbf{u}_t(\delta)\}$ is a suboptimal sequence of controls converging to the optimal control as $\delta \to 0$.

Let the additional conditions on $\underset{\sim}{\mathbf{Q}}_2(t)$ hold. Then, similarly to (7.92), we shall define

$$\hat{\mathbf{x}}_t = \boldsymbol{\xi}_t - \int_0^t \underset{\sim}{\boldsymbol{\Phi}}_{t,s}\underset{\sim}{\mathbf{B}}_s\underset{\sim}{\mathbf{Q}}_2^{-1}(s)\underset{\sim}{\mathbf{B}}_s^T\underset{\sim}{\mathbf{S}}_s\hat{\mathbf{x}}_s\,ds, \tag{7.99}$$

and in a similar way it is shown that $\hat{\mathbf{x}}_t = E(\mathbf{x}_t/\mathscr{F}_t^{\mathbf{y}})$. Noticing that $\boldsymbol{\zeta}_t(\delta) = \hat{\mathbf{x}}_t(\delta)$, from (7.92) and (7.99), it follows that

$$\hat{\mathbf{x}}_t(\delta) - \hat{\mathbf{x}}_t = \int_0^t \underset{\sim}{\boldsymbol{\Phi}}_{t,s}\underset{\sim}{\mathbf{B}}_s\left[\underset{\sim}{\mathbf{Q}}_2^{-1}(s)\underset{\sim}{\mathbf{B}}_s^T\underset{\sim}{\mathbf{S}}_s\hat{\mathbf{x}}_s - \left(\underset{\sim}{\mathbf{Q}}_2(s) + \delta\underset{\sim}{\mathbf{I}}\right)^{-1}\underset{\sim}{\mathbf{B}}_s^T\underset{\sim}{\mathbf{S}}_s^\delta\hat{\mathbf{x}}_s(\delta)\right]ds,$$

and

$$\begin{aligned}
E\|\hat{\mathbf{x}}_t(\delta) - \hat{\mathbf{x}}_t\|^2 \le{}& 3E\left\|\int_0^t\left[\underset{\sim}{\mathbf{Q}}_2^{-1}(s)\underset{\sim}{\mathbf{B}}_s^T\underset{\sim}{\mathbf{S}}_s - \left(\underset{\sim}{\mathbf{Q}}_2(s) + \delta\underset{\sim}{\mathbf{I}}\right)^{-1}\underset{\sim}{\mathbf{B}}_s^T\underset{\sim}{\mathbf{S}}_s\right]\hat{\mathbf{x}}_s\,ds\right\|^2\\
&+3E\left\|\int_0^t\left(\underset{\sim}{\mathbf{Q}}_2(s) + \delta\underset{\sim}{\mathbf{I}}\right)^{-1}\underset{\sim}{\mathbf{B}}_s^T(\underset{\sim}{\mathbf{S}}_s - \underset{\sim}{\mathbf{S}}_s^\delta)\hat{\mathbf{x}}_s\,ds\right\|^2\\
&+3E\left\|\int_0^t\left(\underset{\sim}{\mathbf{Q}}_2(s) + \delta\underset{\sim}{\mathbf{I}}\right)^{-1}\underset{\sim}{\mathbf{B}}_s^T\underset{\sim}{\mathbf{S}}_s(\hat{\mathbf{x}}_s - \hat{\mathbf{x}}_s(\delta))\,ds\right\|^2.
\end{aligned} \tag{7.100}$$

Since $\underset{\sim}{\mathbf{P}}_t^\varepsilon$ converges uniformly to $\underset{\sim}{\mathbf{P}}_t$ as $\varepsilon \to 0$, the convergence of $\underset{\sim}{\mathbf{S}}_t^\delta$ to $\underset{\sim}{\mathbf{S}}_t$ as $\delta \to 0$ is also uniform in $t \in [0, T]$. Hence there is a constant $k_4 > 0$ such that

$$\|\underset{\sim}{\mathbf{S}}_t\| \le k_4, \quad E\|\hat{\mathbf{x}}_t\|^2 \le k_4, \quad \|\underset{\sim}{\mathbf{S}}_t^\delta\| \le k_4, \qquad \forall t \in [0, T], \quad \forall\delta \in [0, \delta_1].$$

On the right-hand side of (7.100), the first term can be estimated by

$$3E\left\| \int_0^t \delta\left(\mathbf{Q}_2^2(s) + \delta \mathbf{Q}_2(s)\right)^{-1} \mathbf{B}_s^\tau \mathbf{S}_s \hat{\mathbf{x}}_s \, ds \right\|^2$$

$$\leq 3\delta^2 \int_0^T \left\| \left(\mathbf{Q}_2^2(s) + \delta \mathbf{Q}_2(s)\right)^{-1/2} \left(\mathbf{Q}_2^2(s) + \delta \mathbf{Q}_2(s)\right)^{-1}\right.$$

$$\left. \times \left(\mathbf{Q}_2^2(s) + \delta \mathbf{Q}_2(s)\right)^{-1/2} \right\| ds \int_0^T \|\mathbf{B}_s\|^2 \|\mathbf{S}_s\|^2 E \|\hat{\mathbf{x}}_s\|^2 \, ds$$

$$\leq 3\delta^2 k_4^3 \int_0^T \frac{1}{\delta} \left\| \mathbf{Q}_2^{-1}(s) \left(\mathbf{Q}_2^2(s) + \delta \mathbf{Q}_2(s)\right)^{-1} \right\| ds \int_0^T \|\mathbf{B}_s\|^2 \, ds$$

$$\leq 3\delta k_4^3 \int_0^T \|\mathbf{B}_s\|^2 \, ds \int_0^T \|\mathbf{Q}_2^{-3}(s)\| \, ds,$$

and the second term by

$$3 \max_{0 \leq s \leq T} \|\mathbf{S}_s - \mathbf{S}_s^\delta\|^2 \int_0^T \left\| \left(\mathbf{Q}_2(s) + \delta \mathbf{I}\right)^{-1} \right\|^2 ds \int_0^T \|\mathbf{B}_s\|^2 E \|\hat{\mathbf{x}}_s\|^2 \, ds$$

$$\leq 3k_4^2 \int_0^T \|\mathbf{B}_s\|^2 \, ds \max_{0 \leq s \leq T} \|\mathbf{S}_s - \mathbf{S}_s^\delta\|^2 \int_0^T \|\mathbf{Q}_2^{-2}(s)\| \, ds.$$

Hence these two terms are bounded by a function $f(\delta)$ which is independent of t and $f(\delta) \underset{\delta \to 0}{\to} 0$. Then, from (7.100), we have

$$E\|\hat{\mathbf{x}}_t(\delta) - \hat{\mathbf{x}}_t\|^2$$

$$\leq f(\delta) + 3k_4^2 \int_0^T \left\| \left(\mathbf{Q}_2(s) + \delta \mathbf{I}\right)^{-1} \mathbf{B}_s^\tau \right\|^2 ds \int_0^t E\|\hat{\mathbf{x}}_s(\delta) - \hat{\mathbf{x}}_s\|^2 \, ds$$

$$\leq f(\delta) + 3k_4^2 \int_0^T \left\| \left(\mathbf{Q}_2(s) + \delta \mathbf{I}\right)^{-1} \right\| ds \int_0^T \left\| \left(\mathbf{Q}_2(s) + \delta \mathbf{I}\right)^{-1/2} \mathbf{B}_s^\tau \right\|^2 ds$$

$$\cdot \int_0^t E\|\hat{\mathbf{x}}_s(\delta) - \hat{\mathbf{x}}_s\|^2 \, ds$$

$$\leq f(\delta) + 3k_4^2 \int_0^T \|\mathbf{Q}_2^{-1}(s)\| \, ds \int_0^T \|\mathbf{B}_s \mathbf{Q}_2^{-1}(s) \mathbf{B}_s^\tau\| \, ds \int_0^t E\|\hat{\mathbf{x}}_s(\delta) - \hat{\mathbf{x}}_s\|^2 \, ds$$

$$\leq f(\delta) e^{k_5 t} \underset{\delta \to 0}{\to} 0,$$

where

$$k_5 = 3k_4^2 \int_0^T \|\underset{\sim}{\mathbf{Q}}_2^{-1}(s)\|\, ds \int_0^T \|\underset{\sim}{\mathbf{B}}_s \underset{\sim}{\mathbf{Q}}_2^{-1}(s) \underset{\sim}{\mathbf{B}}_s^T\|\, ds.$$

Hence

$$\underset{\delta \to 0}{\text{l.i.m.}}\, \mathbf{u}_t(\delta) = \mathbf{u}_t.$$

Letting $\delta \to 0$ in (7.80) leads to the optimality of $\mathbf{u}_t$ given by (7.91). Notice that in this procedure of passing to the limit no integrability condition on $\underset{\sim}{\mathbf{Q}}_2^{-3}(t)$ is required. □

Theorem 7.6. *Under the conditions of Lemma* 7.5 *if* $\underset{\sim}{\mathbf{Q}}_2(t) > \underset{\sim}{\mathbf{0}}$ *and* $\underset{\sim}{\mathbf{B}}_t \underset{\sim}{\mathbf{Q}}_2^{-1} \underset{\sim}{\mathbf{B}}_t^T$ *is integrable on* $[0, T]$, *then*

$$\mathbf{u}_t^\varepsilon \triangleq -\underset{\sim}{\mathbf{Q}}_2^{-1}(t) \underset{\sim}{\mathbf{B}}_t^T \underset{\sim}{\mathbf{S}}_t \boldsymbol{\zeta}_t^\varepsilon \in \mathscr{U},$$

and

$$\lim_{\varepsilon \to 0} EJ(\mathbf{u}_t) = \inf_{u \in \mathscr{U}} E(\mathbf{J}),$$

where

$$\boldsymbol{\zeta}_t^\varepsilon = \tilde{\boldsymbol{\xi}}_t^\varepsilon - \int_0^t \underset{\sim}{\boldsymbol{\Phi}}_{t,s} \underset{\sim}{\mathbf{B}}_s \underset{\sim}{\mathbf{Q}}_2^{-1}(s) \underset{\sim}{\mathbf{B}}_s^T \underset{\sim}{\mathbf{S}}_s \boldsymbol{\zeta}_s^\varepsilon\, ds. \tag{7.101}$$

If, in addition, $\underset{\sim}{\mathbf{Q}}_2^{-3}(t)$ *is integrable on* $[0, T]$, *then*

$$\underset{\delta \to 0}{\text{l.i.m.}}\, \mathbf{u}_t(\varepsilon, \delta) = \mathbf{u}_t^\varepsilon. \tag{7.102}$$

Proof. Since

$$\boldsymbol{\zeta}_t^\varepsilon(\delta) - \boldsymbol{\zeta}_t^\varepsilon = \int_0^t \underset{\sim}{\boldsymbol{\Phi}}_{t,s} \underset{\sim}{\mathbf{B}}_s \left[\underset{\sim}{\mathbf{Q}}_2^{-1}(s) \underset{\sim}{\mathbf{B}}_s^T \underset{\sim}{\mathbf{S}}_s \boldsymbol{\zeta}_s^\varepsilon - \left(\underset{\sim}{\mathbf{Q}}_2(s) + \delta \underset{\sim}{\mathbf{I}} \right)^{-1} \underset{\sim}{\mathbf{B}}_s^T \underset{\sim}{\mathbf{S}}_s^\delta \boldsymbol{\zeta}_s^\varepsilon(\delta) \right] ds,$$

by replacing $\hat{\mathbf{x}}_t(\delta)$ and $\hat{\mathbf{x}}_t$ by $\boldsymbol{\zeta}_t^\varepsilon(\delta)$ and $\boldsymbol{\zeta}_t^\varepsilon$, respectively, in (7.100), a similar treatment immediately leads to (7.102).

Letting δ tend to 0 in (7.80) [without use of an integrability condition on $\underset{\sim}{\mathbf{Q}}_2^{-3}(t)$], we have

$$EJ(\mathbf{u}) = \lim_{\delta \to 0} V_\delta + E \int_0^T (\mathbf{u}_t - \underset{\sim}{\mathbf{L}}_t \hat{\mathbf{x}}_t)^T \underset{\sim}{\mathbf{Q}}_2(t) (\mathbf{u}_t - \underset{\sim}{\mathbf{L}}_t \hat{\mathbf{x}}_t)\, dt. \tag{7.103}$$

Subject to the control $\mathbf{u}_t^\varepsilon$ the state $\mathbf{x}_t^\varepsilon$ is given by

$$\mathbf{x}_t^\varepsilon = \boldsymbol{\xi}_t + \int_0^t \underset{\sim}{\boldsymbol{\Phi}}_{t,\lambda} \underset{\sim}{\mathbf{B}}_\lambda \underset{\sim}{\mathbf{L}}_\lambda \boldsymbol{\zeta}_\lambda^\varepsilon \, d\lambda,$$

then from (7.101)

$$\mathbf{x}_t^\varepsilon - \boldsymbol{\zeta}_t^\varepsilon = \boldsymbol{\xi}_t - \tilde{\boldsymbol{\xi}}_t.$$

Hence

$$\lim_{\varepsilon \to 0} EJ(\mathbf{u}^\varepsilon) = \lim_{\varepsilon \to 0} \left[\lim_{\delta \to 0} V_\delta + \int_0^T \operatorname{tr} \underset{\sim}{\mathbf{L}}_t^T \underset{\sim}{\mathbf{Q}}_2(t) \underset{\sim}{\mathbf{L}}_t (\tilde{\underset{\sim}{\mathbf{P}}}_t^\varepsilon - \underset{\sim}{\mathbf{P}}_t) \, dt \right]$$

$$= \lim_{\delta \to 0} V_\delta = \inf_{\mathbf{u} \in \mathscr{U}} EJ(\mathbf{u}). \qquad \square$$

Remark 1. When $\underset{\sim}{\mathbf{F}}_t \underset{\sim}{\mathbf{F}}_t^T > 0$ and $\underset{\sim}{\mathbf{C}}_t^T (\underset{\sim}{\mathbf{F}}_t \underset{\sim}{\mathbf{F}}_t^T)^{-1} \underset{\sim}{\mathbf{C}}_t$ are integrable on $[0, T]$, $\{\mathbf{u}_t(\varepsilon_n, \delta_n)\}$ given by Theorem 7.3 and $\{\mathbf{u}_t(\delta)\}$ given by (7.90) are both the optimal sequence of controls for the system (6.1), (6.2), (6.95). Hence the optimal control sequence is not unique.

Remark 2. By comparing (7.88) with (7.90), we find that for the $\underset{\sim}{\mathbf{Q}}_2(t)$ singular case the optimal control sequence for stochastic systems may be obtained from that for the corresponding deterministic systems if the state is replaced by its filtering value.

7.6. UNIFIED STRATEGIES

Now we return to the stochastic differential game problem described by (6.113)–(6.116).

Assume $\|\underset{\sim}{\mathbf{A}}_t\|$, $\|\underset{\sim}{\mathbf{D}}_t\|^2$, $\|\underset{\sim}{\mathbf{F}}_t\|^2$, $\|\underset{\sim}{\mathbf{C}}_t\|^2$, $\|\underset{\sim}{\mathbf{B}}_{pt}\|^2$, and $\|\underset{\sim}{\mathbf{B}}_{et}\|^2$ are integrable on $[0, T]$. $\underset{\sim}{\mathbf{Q}}_{2t}$ and $\underset{\sim}{\mathbf{Q}}_{3t}$ in (6.115) are now allowed to be degenerate. That is, we only assume that $\underset{\sim}{\mathbf{Q}}_{2t} \geq \underset{\sim}{\mathbf{0}}$ and $\underset{\sim}{\mathbf{Q}}_{3t} \geq \underset{\sim}{\mathbf{0}}$. This is the main point of difference between the current problem and that discussed Section 6.5. The set $\mathscr{U}$ of admissible controls is also different from (6.80); it is now composed of

$$\mathbf{u}_t = \begin{bmatrix} \mathbf{u}_{pt}^T & \mathbf{u}_{et}^T \end{bmatrix}^T,$$

satisfying (7.52)–(7.54).

For $\delta_1 > 0$, $\delta_2 > 0$, introduce an auxiliary equation and performance index as follows:

$$\dot{\underset{\sim}{S}}_t = -\underset{\sim}{A}_t^\tau \underset{\sim}{S}_t - \underset{\sim}{S}_t \underset{\sim}{A}_t - \underset{\sim}{Q}_{1t} + \underset{\sim}{S}_t \underset{\sim}{B}_{pt}\left(\underset{\sim}{Q}_{2t} + \delta_1 \underset{\sim}{I}\right)^{-1} \underset{\sim}{B}_{pt}^\tau \underset{\sim}{S}_t$$

$$- \underset{\sim}{S}_t \underset{\sim}{B}_{et}\left(\underset{\sim}{Q}_{3t} + \delta_2 \underset{\sim}{I}\right)^{-1} \underset{\sim}{B}_{et}^\tau \underset{\sim}{S}_t, \qquad \underset{\sim}{S}_T = \underset{\sim}{Q}_0, \qquad (7.104)$$

$$J_{\delta_1 \delta_2}(\mathbf{u}_p, \mathbf{u}_e) = J(\mathbf{u}_p, \mathbf{u}_e) + \delta_1 \int_0^T \|\mathbf{u}_{pt}\|^2\, dt - \delta_2 \int_0^T \|\mathbf{u}_{et}\|^2\, dt, \quad (7.105)$$

We need the following conditions:

A_1. There exists $\Delta > 0$ such that (7.104) is solvable or any $\delta_1 \in (0, \Delta]$ and $\delta_2 \in (0, \Delta]$.

B_1. $\underset{\sim}{Q}_{2t} > \underset{\sim}{0}$, $\underset{\sim}{Q}_{3t} > \underset{\sim}{0}$ and $\underset{\sim}{B}_{pt}\underset{\sim}{Q}_{2t}^{-i}\underset{\sim}{B}_{pt}^\tau$, $\underset{\sim}{B}_{et}\underset{\sim}{Q}_{3t}^{-i}\underset{\sim}{B}_{et}^\tau$ are integrable on $[0, T]$, $i = 1, 2$.

C_1. The system is degenerate to the deterministic one or $\underset{\sim}{F}_t\underset{\sim}{F}_t^\tau > \underset{\sim}{0}$, but $\underset{\sim}{C}_t^\tau(\underset{\sim}{F}_t\underset{\sim}{F}_t^\tau)^{-1}\underset{\sim}{C}_t$ is integrable on $[0, T]$.

If Condition A_1 holds, set

$$V(\delta_1, \delta_2) = E\mathbf{x}_0^\tau \underset{\sim}{S}_0(\delta_1, \delta_2) E\mathbf{x}_0 + \operatorname{tr} \underset{\sim}{R}_{\mathbf{x}_0}\underset{\sim}{S}_0(\delta_1, \delta_2)$$

$$+ \int_0^T \operatorname{tr}\left[\underset{\sim}{D}_t\underset{\sim}{D}_t^\tau\underset{\sim}{S}_t(\delta_1, \delta_2) + \underset{\sim}{S}_t(\delta_1, \delta_2)\underset{\sim}{B}_{pt}\left(\underset{\sim}{Q}_{2t} + \delta_1\underset{\sim}{I}\right)^{-1}\underset{\sim}{B}_{pt}^\tau\underset{\sim}{S}_t(\delta_1, \delta_2)\underset{\sim}{P}_t \right.$$

$$\left. - \underset{\sim}{S}_t(\delta_1, \delta_2)\underset{\sim}{B}_{et}\left(\underset{\sim}{Q}_{3t} + \delta_2\underset{\sim}{I}\right)^{-1}\underset{\sim}{B}_{et}^\tau\underset{\sim}{S}_t(\delta_1, \delta_2)\underset{\sim}{P}_t \right] dt, \qquad (7.106)$$

then for $\{\mathbf{u}_t\} \in \mathscr{U}$ in a manner similar to Theorem 6.7, it can be shown that

$$EJ_{\delta_1 \delta_2}(\mathbf{u}_p, \mathbf{u}_e) = V(\delta_1, \delta_2) + E\int_0^T \left[\mathbf{u}_{pt} + \left(\underset{\sim}{Q}_{2t} + \delta_1\underset{\sim}{I}\right)^{-1}\underset{\sim}{B}_{pt}^\tau\underset{\sim}{S}_t(\delta_1, \delta_2)\hat{\mathbf{x}}_t\right]^\tau$$

$$\cdot \left(\underset{\sim}{Q}_{2t} + \delta_1\underset{\sim}{I}\right)\left[\mathbf{u}_{pt} + \left(\underset{\sim}{Q}_{2t} + \delta_1\underset{\sim}{I}\right)^{-1}\underset{\sim}{B}_{pt}^\tau\underset{\sim}{S}_t(\delta_1, \delta_2)\hat{\mathbf{x}}_t\right] dt$$

$$- E\int_0^T \left[\mathbf{u}_{et} - \left(\underset{\sim}{Q}_{3t} + \delta_2\underset{\sim}{I}\right)^{-1}\underset{\sim}{B}_{et}\underset{\sim}{S}_t(\delta_1, \delta_2)\hat{\mathbf{x}}_t\right]^\tau \left(\underset{\sim}{Q}_{3t} + \delta_2\underset{\sim}{I}\right)$$

$$\cdot \left[\mathbf{u}_{et} - \left(\underset{\sim}{Q}_{3t} + \delta_2\underset{\sim}{I}\right)^{-1}\underset{\sim}{B}_{et}\underset{\sim}{S}_t(\delta_1, \delta_2)\hat{\mathbf{x}}_t\right] dt. \qquad (7.107)$$

Lemma 7.8. (1) *Provided Condition* A_1 *holds, then*

$$-\infty < V_1 \triangleq \lim_{\delta_2\to 0}\lim_{\delta_1\to 0} V(\delta_1,\delta_2) \le \lim_{\delta_1\to 0}\lim_{\delta_2\to 0} V(\delta_1,\delta_2) \triangleq V_2 \tag{7.108}$$

$$-\infty < S_{1t} \triangleq \lim_{\delta_2\to 0}\lim_{\delta_1\to 0} S_t(\delta_1,\delta_2) \le \lim_{\delta_1\to 0}\lim_{\delta_2\to 0} S_t(\delta_1,\delta_2) \triangleq S_{2t}, \tag{7.109}$$

(2) *If Conditions* A_1 *and* B_1 *hold, then*

$$V_1 = V_2 \triangleq V, \qquad |V| < \infty, \qquad S_{1t} = S_{2t} = S_t(0.0) \tag{7.110}$$

(3) *If Conditions* A_1–C_1 *hold, then* V *is the game value and the optimal strategies are*

$$\mathbf{u}^*_{pt} = -\underset{\sim}{\mathbf{Q}}_{2t}^{-1}\underset{\sim}{\mathbf{B}}^{\tau}_{pt}\underset{\sim}{\mathbf{S}}_t(0,0)\hat{\mathbf{x}}_t, \qquad \mathbf{u}^*_{et} = \underset{\sim}{\mathbf{Q}}_{3t}^{-1}\underset{\sim}{\mathbf{B}}^{\tau}_{et}\underset{\sim}{\mathbf{S}}_t(0,0)\hat{\mathbf{x}}_t, \tag{7.111}$$

Proof. (1) If in (6.113) and (6.114), $\underset{\sim}{\mathbf{D}}_t$ and $\underset{\sim}{\mathbf{F}}_t$ are replaced by

$$\Big[\underset{\sim}{\mathbf{D}}_t \vdots \overbrace{\underset{\sim}{\mathbf{0}}}^{m}\Big] \quad \text{and} \quad \Big[\underset{\sim}{\mathbf{F}}_t \vdots \overbrace{\sqrt{\varepsilon}\;\;\underset{\sim}{\mathbf{I}}}^{m}\Big],$$

respectively, and the dimension of $\mathbf{w}_t$ is extended from l to $l+m$, then any vector of the form $\underset{\sim}{\mathbf{M}}_t\hat{\mathbf{x}}_t$ satisfies (7.53) and (7.54) if (7.52) is fulfilled. In (7.106), instead of $\underset{\sim}{\mathbf{P}}_t$, put $\underset{\sim}{\mathbf{P}}^{\varepsilon}_t$ and denote $V(\delta_1,\delta_2)$ by $V^{\varepsilon}(\delta_1,\delta_2)$, then by Theorem 6.7, the latter is the game value if the performance index is $EJ_{\delta_1\delta_2}(\mathbf{u}_p,\mathbf{u}_e)$.

Obviously, as $\delta_1 \to 0$, $V^{\varepsilon}(\delta_1,\delta_2)$ is nonincreasing, and $V^{\varepsilon}(\delta_1,\delta_2) \underset{\varepsilon\to 0}{\to} V(\delta_1,\delta_2)$ since $\underset{\sim}{\mathbf{P}}^{\varepsilon}_t \underset{\varepsilon\to 0}{\to} \underset{\sim}{\mathbf{P}}_t$, hence as $\delta_1 \to 0$ $V(\delta_1,\delta_2)$ is also nonincreasing. But

$$\inf_{\mathbf{u}_p} EJ_{\delta_1\delta_2}(\mathbf{u}_p,\mathbf{u}_e) \ge \inf_{\mathbf{u}_p} E\left\{\mathbf{x}^{\tau}_T\underset{\sim}{\mathbf{Q}}_0\mathbf{x}_T + \int_0^T\left(\mathbf{x}^{\tau}_t\underset{\sim}{\mathbf{Q}}_{1t}\mathbf{x}_t + \mathbf{u}^{\tau}_{pt}\underset{\sim}{\mathbf{Q}}_{2t}\mathbf{u}_{pt}\right)dt\right\}$$

$$-E\int_0^T \mathbf{u}^{\tau}_{et}\left(\underset{\sim}{\mathbf{Q}}_{3t} + \delta_2\underset{\sim}{\mathbf{I}}\right)\mathbf{u}_{et}\,dt.$$

Hence $V(\delta_1,\delta_2)$ is bounded from below and, consequently, converges to a finite limit as $\delta_1 \to 0$. As $\delta_2 \to 0$, $V^{\varepsilon}(\delta_1,\delta_2)$ is nondecreasing. Hence $V(\delta_1,\delta_2)$ is nondecreasing, and from this (7.108) follows.

By considering the deterministic system on $[t, T]$, it follows that

$$V(\delta_1, \delta_2) = \mathbf{x}_t^\tau \underset{\sim}{\mathbf{S}}_t(\delta_1, \delta_2)\mathbf{x}_t,$$

and, since $\mathbf{x}_t$ is arbitrary, (7.109) is implied in (7.108).

(2) Since (7.104) is assumed to have a solution for $\delta_1 = \delta_2 = 0$ denoted by $\underset{\sim}{\mathbf{S}}_t(0,0)$, it can be solved by iteration for $\delta_1 > 0$ and $\delta_2 > 0$. For simplicity of notation, denote the right-hand side of (7.104) by $\underset{\sim}{\mathbf{F}}(\underset{\sim}{\mathbf{S}}_t, t, \delta_1, \delta_2)$.

Set $\delta_1, \delta_2 \in (0, \Delta]$. Then

$$\underset{\sim}{\mathbf{S}}_t(0,0) = \underset{\sim}{\mathbf{Q}}_0 - \int_t^T \underset{\sim}{\mathbf{F}}(\underset{\sim}{\mathbf{S}}_\lambda(0,0), \lambda, 0, 0)\, d\lambda.$$

Define

$$\underset{\sim}{\mathbf{S}}_t^0(\delta_1, \delta_2) = \underset{\sim}{\mathbf{S}}_t(0,0),$$

$$\underset{\sim}{\mathbf{S}}_t^N(\delta_1, \delta_2) = \underset{\sim}{\mathbf{Q}}_0 - \int_t^T \underset{\sim}{\mathbf{F}}\left(\underset{\sim}{\mathbf{S}}_\lambda^{N-1}(\delta_1, \delta_2), \lambda, \delta_1, \delta_2\right) d\lambda. \qquad (7.112)$$

By continuity there is a constant k so that $\|\underset{\sim}{\mathbf{S}}_t(0,0)\| \leq k/2$. Denote

$$G_t = k^2\left(\|\underset{\sim}{\mathbf{B}}_{pt}\underset{\sim}{\mathbf{Q}}_{2t}^{-2}\underset{\sim}{\mathbf{B}}_{pt}^\tau\| + \|\underset{\sim}{\mathbf{B}}_{et}\underset{\sim}{\mathbf{Q}}_{3t}^{-2}\underset{\sim}{\mathbf{B}}_{et}^\tau\|\right) + \|\underset{\sim}{\mathbf{A}}_t\|$$

$$+ 2k\left(\|\underset{\sim}{\mathbf{B}}_{pt}\underset{\sim}{\mathbf{Q}}_{2t}^{-1}\underset{\sim}{\mathbf{B}}_{pt}^\tau\| + \|\underset{\sim}{\mathbf{B}}_{et}\underset{\sim}{\mathbf{Q}}_{3t}^{-1}\underset{\sim}{\mathbf{B}}_{et}^\tau\|\right)$$

and select $\Delta > 0$ sufficiently small that

$$\Delta \exp \int_0^T G_t\, dt \leq \frac{k}{2}. \qquad (7.113)$$

By Taylor's expansion, we have

$$\|\underset{\sim}{\mathbf{F}}(\underset{\sim}{\mathbf{S}}_\lambda(0,0), \lambda, \delta_1, \delta_2) - \underset{\sim}{\mathbf{F}}(\underset{\sim}{\mathbf{S}}_\lambda(0,0), \lambda, 0, 0)\|$$

$$\leq \|\underset{\sim}{\mathbf{S}}_\lambda(0,0)\underset{\sim}{\mathbf{B}}_{p\lambda}\left(\underset{\sim}{\mathbf{Q}}_{2\lambda} + \delta_1\underset{\sim}{\mathbf{I}}\right)^{-1}\underset{\sim}{\mathbf{B}}_{p\lambda}^\tau\underset{\sim}{\mathbf{S}}_\lambda(0,0) - \underset{\sim}{\mathbf{S}}_\lambda(0,0)\underset{\sim}{\mathbf{B}}_{p\lambda}\underset{\sim}{\mathbf{Q}}_{2\lambda}^{-1}\underset{\sim}{\mathbf{B}}_{p\lambda}^\tau\underset{\sim}{\mathbf{S}}_\lambda(0,0)\|$$

$$+ \|\underset{\sim}{\mathbf{S}}_\lambda(0,0)\underset{\sim}{\mathbf{B}}_{e\lambda}\left(\underset{\sim}{\mathbf{Q}}_{3\lambda} + \delta_2\underset{\sim}{\mathbf{I}}\right)^{-1}\underset{\sim}{\mathbf{B}}_{e\lambda}^\tau\underset{\sim}{\mathbf{S}}_\lambda(0,0) - \underset{\sim}{\mathbf{S}}_\lambda(0,0)\underset{\sim}{\mathbf{B}}_{e\lambda}\underset{\sim}{\mathbf{Q}}_{3\lambda}^{-1}\underset{\sim}{\mathbf{B}}_{e\lambda}^\tau\underset{\sim}{\mathbf{S}}_\lambda(0,0)\|$$

$$\leq \Delta\|\underset{\sim}{\mathbf{S}}_\lambda(0,0)\underset{\sim}{\mathbf{B}}_{p\lambda}\underset{\sim}{\mathbf{Q}}_{2\lambda}^{-2}\underset{\sim}{\mathbf{B}}_{p\lambda}^\tau\underset{\sim}{\mathbf{S}}_\lambda(0,0)\| + \Delta\|\underset{\sim}{\mathbf{S}}_\lambda(0,0)\underset{\sim}{\mathbf{B}}_{e\lambda}\underset{\sim}{\mathbf{Q}}_{3\lambda}^{-2}\underset{\sim}{\mathbf{B}}_{e\lambda}^\tau\underset{\sim}{\mathbf{S}}_\lambda(0,0)\|.$$

Hence

$$\|\underset{\sim}{\mathbf{S}}_t^1(\delta_1,\delta_2) - \underset{\sim}{\mathbf{S}}_t^0(\delta_1,\delta_2)\| \le \Delta \int_t^T G_\lambda \, d\lambda \tag{7.114}$$

and

$$\|\underset{\sim}{\mathbf{S}}_t^1(\delta_1,\delta_2)\| \le \|\underset{\sim}{\mathbf{S}}_t^0(\delta_1,\delta_2)\| + \Delta \exp \int_0^T G_t \, dt \le k. \tag{7.115}$$

Suppose (7.114) and (7.115) are valid for $i \le N$, that is,

$$\|\underset{\sim}{\mathbf{S}}_t^N(\delta_1,\delta_2) - \underset{\sim}{\mathbf{S}}_t^{N-1}(\delta_1,\delta_2)\| \le \Delta \frac{1}{N!}\left(\int_t^T G_\lambda \, d\lambda\right)^N \tag{7.116}$$

$$\|\underset{\sim}{\mathbf{S}}_t^N(\delta_1,\delta_2)\| \le k. \tag{7.117}$$

We now show that (7.116) and (7.117) also hold for $N+1$.

We have

$$\begin{aligned}
&\|\underset{\sim}{\mathbf{F}}\big(\underset{\sim}{\mathbf{S}}_t^N(\delta_1,\delta_2),\lambda,\delta_1,\delta_2\big) - \underset{\sim}{\mathbf{F}}\big(\underset{\sim}{\mathbf{S}}_t^{N-1}(\delta_1,\delta_2),\lambda,\delta_1,\delta_2\big)\| \\
&\le 2\|\underset{\sim}{\mathbf{A}}_t\| \, \|\underset{\sim}{\mathbf{S}}_t^N(\delta_1,\delta_2) - \underset{\sim}{\mathbf{S}}_t^{N-1}(\delta_1,\delta_2)\| \\
&+ \big(\|\underset{\sim}{\mathbf{S}}_t^N(\delta_1,\delta_2)\| + \|\underset{\sim}{\mathbf{S}}_t^{N-1}(\delta_1,\delta_2)\|\big)\|\underset{\sim}{\mathbf{B}}_{p\lambda}\underset{\sim}{\mathbf{Q}}_{2\lambda}^{-1}\underset{\sim}{\mathbf{B}}_{p\lambda}^{\tau}\| \\
&\|\underset{\sim}{\mathbf{S}}_t^N(\delta_1,\delta_2) - \underset{\sim}{\mathbf{S}}_t^{N-1}(\delta_1,\delta_2)\| \\
&+ \big(\|\underset{\sim}{\mathbf{S}}_t^N(\delta_1,\delta_2)\| + \|\underset{\sim}{\mathbf{S}}_t^{N-1}(\delta_1,\delta_2)\|\big)\|\underset{\sim}{\mathbf{B}}_{e\lambda}\underset{\sim}{\mathbf{Q}}_{3\lambda}^{-1}\underset{\sim}{\mathbf{B}}_{e\lambda}^{\tau}\| \\
&\|\underset{\sim}{\mathbf{S}}_t^N(\delta_1,\delta_2) - \underset{\sim}{\mathbf{S}}_t^{N-1}(\delta_1,\delta_2)\| \\
&\le G_t\big(\|\underset{\sim}{\mathbf{S}}_t^N(\delta_1,\delta_2) - \underset{\sim}{\mathbf{S}}_t^{N-1}(\delta_1,\delta_2)\|\big).
\end{aligned}$$

Hence

$$\begin{aligned}
\|\mathbf{S}_t^{N+1}(\delta_1,\delta_2) - \mathbf{S}_t^N(\delta_1,\delta_2)\| &\le \int_t^T G_\lambda \|\underset{\sim}{\mathbf{S}}_\lambda^N(\delta_1,\delta_2) - \underset{\sim}{\mathbf{S}}_\lambda^{N-1}(\delta_1,\delta_2)\| \, d\lambda \\
&\le \frac{\Delta}{N!}\int_t^T G_\lambda \left(\int_\lambda^T G_s \, ds\right)^N d\lambda \\
&= \frac{\Delta}{(N+1)!}\left(\int_t^T G_\lambda \, d\lambda\right)^{N+1},
\end{aligned} \tag{7.118}$$

and then

$$\|\underset{\sim}{\mathbf{S}}_t^{N+1}(\delta_1,\delta_2) - \underset{\sim}{\mathbf{S}}_t^0(\delta_1,\delta_2)\| \le \sum_{i=0}^{N} \|\underset{\sim}{\mathbf{S}}_t^{i+1}(\delta_1,\delta_2) - \underset{\sim}{\mathbf{S}}_t^{i}(\delta_1,\delta_2)\|$$

$$\le \Delta \sum_{i=0}^{N} \frac{1}{(i+1)!}\left(\int_t^T G_\lambda \, d\lambda\right)^{i+1}$$

$$\le \Delta \exp \int_t^T G_\lambda \, d\lambda \le \frac{k}{2}$$

$$\|\underset{\sim}{\mathbf{S}}_t^{N+1}(\delta_1,\delta_2)\| \le \|\underset{\sim}{\mathbf{S}}_t^0(\delta_1,\delta_2)\| + \frac{k}{2} \le k. \tag{7.119}$$

Thus (7.116) and (7.117) are verified for all N.

From (7.118), it is seen that as $N \to \infty$ $\underset{\sim}{\mathbf{S}}_t^N(\delta_1,\delta_2)$ converges to a finite limit $\underset{\sim}{\mathbf{S}}_t(\delta_1,\delta_2)$, which is the solution of (7.104) as can be seen from (7.111) by letting $N \to \infty$.

Letting $N \to \infty$ in (7.119), we obtain

$$\|\underset{\sim}{\mathbf{S}}_t(\delta_1,\delta_2) - \underset{\sim}{\mathbf{S}}_t^0(0,0)\| \le \Delta \exp \int_t^T G_\lambda \, d\lambda,$$

and (7.110) follows from this by letting $\Delta \to 0$.

(3) If the system degenerates to a deterministic one, then $\underset{\sim}{\mathbf{P}}_t \equiv \underset{\sim}{\mathbf{0}}$ and $\hat{\mathbf{x}}_t \equiv \mathbf{x}_t$. If $\underset{\sim}{\mathbf{F}}_t \underset{\sim}{\mathbf{F}}_t^T > \underset{\sim}{\mathbf{0}}$, then we can take $\delta_1 = \delta_2 = 0$ in (7.106). Hence, for these two cases, V is the game value, and the optimal strategies are given by (7.111). This is just the result given by Theorem 6.7, but with a different definition of an admissible control. □

Under the quadratic performance index for the one-side control problem the extremum of the index may not be attained in the set of admissible controls $\mathscr{U}$ but always exists. On the other hand, in the game problems, the game value not only may not be attained in $\mathscr{U}$, but also may not exist at all.

If in $\mathscr{U}$ there is no game value, then we hope to find a subset $\mathscr{U}^\varepsilon$ of $\mathscr{U}$ such that

$$\lim_{\varepsilon \to 0} \inf_{\mathbf{u}_p} \sup_{\substack{\mathbf{u}_e \\ \mathbf{u} \in \mathscr{U}^\varepsilon}} EJ(\mathbf{u}_p, \mathbf{u}_e) = \lim_{\varepsilon \to 0} \sup_{\mathbf{u}_e} \inf_{\substack{\mathbf{u}_p \\ \mathbf{u} \in \mathscr{U}^\varepsilon}} EJ(\mathbf{u}_p, \mathbf{u}_e) \triangleq V \tag{7.120}$$

and a sequence of admissible strategies $\{\mathbf{u}_t(k)\} \in \mathscr{U}$ applicable to both singular and nonsingular cases, such that Conditions A_2, B_2, and C_2 hold.

A_2. As $k \to \infty$, $\mathbf{u}_t(k)$ converges in the mean square sense to the saddle point $\mathbf{u}_t^*$ defined by (7.111) if Conditions B_1 and C_1 hold,

B_2. $EJ(\mathbf{u}_p(k), \mathbf{u}_e(k)) \underset{k\to\infty}{\to} V$.

C_2. (6.118) is satisfied in some limiting sense.

We use the notation $\boldsymbol{\xi}_t$, $\boldsymbol{\eta}_t$, $\tilde{\boldsymbol{\xi}}_t^\varepsilon$, $\underset{\sim}{\mathbf{P}}_t^\varepsilon$, $\underset{\sim}{\tilde{\mathbf{P}}}_t^\varepsilon$, and $\underset{\sim}{\boldsymbol{\Phi}}_{t,s}$ introduced in Section 7.5. Assume

$$\int_0^T \|\underset{\sim}{\mathbf{K}}_{pt}\|^2\,dt < \infty, \qquad \int_0^T \|\underset{\sim}{\mathbf{K}}_{et}\|^2\,dt < \infty, \tag{7.121}$$

$$\boldsymbol{\zeta}_t^\varepsilon = \tilde{\boldsymbol{\xi}}_t^\varepsilon + \int_0^t \underset{\sim}{\boldsymbol{\Phi}}_{t,s}\left[\underset{\sim}{\mathbf{B}}_{ps}\underset{\sim}{\mathbf{K}}_{ps} + \underset{\sim}{\mathbf{B}}_{es}\underset{\sim}{\mathbf{K}}_{es}\right]\boldsymbol{\zeta}_s^\varepsilon\,ds, \tag{7.122}$$

$$\mathbf{u}_{pt}^\varepsilon = \underset{\sim}{\mathbf{K}}_{pt}\boldsymbol{\zeta}_t^\varepsilon, \qquad \mathbf{u}_{et}^\varepsilon = \underset{\sim}{\mathbf{K}}_{et}\boldsymbol{\zeta}_t^\varepsilon, \tag{7.123}$$

and denote by $\mathscr{U}^\varepsilon$ all controls of the form

$$\mathbf{u}_t^\varepsilon \triangleq \left[\mathbf{u}_{pt}^{\varepsilon T} \quad \mathbf{u}_{et}^{\varepsilon T}\right]^T$$

satisfying (7.121)–(7.123).

By Lemma 7.4, $\mathscr{U}^\varepsilon \subset \mathscr{U}$ and $\mathbf{u}_t^\varepsilon$ linearly depends on $(\mathbf{y}_s, 0 \le s \le t)$.

In particular, if

$$\underset{\sim}{\mathbf{K}}_{pt} = -\left(\underset{\sim}{\mathbf{Q}}_{2t} + \delta_1\underset{\sim}{\mathbf{I}}\right)^{-1}\underset{\sim}{\mathbf{B}}_{pt}^T\underset{\sim}{\mathbf{S}}_t(\delta_1, \delta_2), \tag{7.124}$$

then we shall denote $\boldsymbol{\zeta}_t^\varepsilon$ by $\boldsymbol{\zeta}_t^\varepsilon(\delta, \mathbf{u}_t^\varepsilon)$, where δ stands for (δ_1, δ_2), and, correspondingly, denote $\mathbf{u}_{pt}^\varepsilon$ by

$$\mathbf{u}_{pt}^\varepsilon(\delta, \mathbf{u}_e^\varepsilon) = -\left(\underset{\sim}{\mathbf{Q}}_{2t} + \delta_1\underset{\sim}{\mathbf{I}}\right)^{-1}\underset{\sim}{\mathbf{B}}_{pt}^T\underset{\sim}{\mathbf{S}}_t(\delta_1, \delta_2)\boldsymbol{\zeta}_t^\varepsilon(\delta, \mathbf{u}_e^\varepsilon). \tag{7.125}$$

Similarly, if

$$\underset{\sim}{\mathbf{K}}_{et} = \left(\underset{\sim}{\mathbf{Q}}_{3t} + \mu_2\underset{\sim}{\mathbf{I}}\right)^{-1}\underset{\sim}{\mathbf{B}}_{et}^T\underset{\sim}{\mathbf{S}}_t(\mu_1, \mu_2), \tag{7.126}$$

then $\boldsymbol{\zeta}_t^\varepsilon$ and $\mathbf{u}_{et}^\varepsilon$ are denoted by $\boldsymbol{\zeta}_t^\varepsilon(\mathbf{u}_p^\varepsilon, \mu)$ and

$$\mathbf{u}_{et}^\varepsilon(\mathbf{u}_p^\varepsilon, \mu) = \left(\underset{\sim}{\mathbf{Q}}_{3t} + \mu_2\underset{\sim}{\mathbf{I}}\right)^{-1}\underset{\sim}{\mathbf{B}}_{et}^T\underset{\sim}{\mathbf{S}}_t(\mu_1, \mu_2)\boldsymbol{\zeta}_t^\varepsilon(\mathbf{u}_p^\varepsilon, \mu), \tag{7.127}$$

respectively.

If $\underset{\sim}{\mathbf{K}}_{pt}$ and $\underset{\sim}{\mathbf{K}}_{et}$, given by (7.124) and (7.126), respectively, are applied simultaneously, then $\boldsymbol{\zeta}_t^\varepsilon$ is denoted by $\boldsymbol{\zeta}_t^\varepsilon(\delta, \mu)$ and the control by

$$\mathbf{u}_{pt}^\varepsilon(\delta, \mu) = -\left(\underset{\sim}{\mathbf{Q}}_{2t} + \delta_1 \underset{\sim}{\mathbf{I}}\right)^{-1} \underset{\sim}{\mathbf{B}}_{pt}^T \underset{\sim}{\mathbf{S}}_t(\delta_1, \delta_2) \boldsymbol{\zeta}_t^\varepsilon(\delta, \mu), \tag{7.128}$$

$$\mathbf{u}_{et}^\varepsilon(\delta, \mu) = \left(\underset{\sim}{\mathbf{Q}}_{3t} + \mu_2 \underset{\sim}{\mathbf{I}}\right)^{-1} \underset{\sim}{\mathbf{B}}_{et}^T \underset{\sim}{\mathbf{S}}_t(\mu_1, \mu_2) \boldsymbol{\zeta}_t^\varepsilon(\delta, \mu). \tag{7.129}$$

□

Lemma 7.9. *Assume that Condition A_1 holds. If $\underset{\sim}{\mathbf{K}}_{pt}, \underset{\sim}{\mathbf{K}}_{et}$ in (7.123) depend on some parameter λ and $\|\underset{\sim}{\mathbf{K}}_{pt}(\lambda)\| \le f_t$, $\|\underset{\sim}{\mathbf{K}}_{et}(\lambda)\| \le f_t$, where f_t is independent of λ and*

$$\int_0^T f_t^2 \, dt < \infty,$$

then

$$E\int_0^T \|\mathbf{u}_{pt}^\varepsilon\|^2 \, dt \quad \text{and} \quad E\int_0^T \|\mathbf{u}_{et}^\varepsilon\|^2 \, dt$$

are uniformly bounded in λ. If $\underset{\sim}{\mathbf{K}}_{pt}(\lambda)$ and $\underset{\sim}{\mathbf{K}}_{et}(\lambda)$ are uniformly continuous functions of λ in $t \in [0, T]$, then $E\|\mathbf{x}_t\|^2$, $E\|\boldsymbol{\zeta}_t^\varepsilon\|^2$, and $E\|\mathbf{u}_t^\varepsilon\|^2$ are continuous in λ.

Proof. Let

$$\underset{\sim}{\mathbf{B}}_t = \left[\underset{\sim}{\mathbf{B}}_{pt} \underset{\sim}{\mathbf{B}}_{et}\right], \qquad \underset{\sim}{\mathbf{M}}_t(\lambda) = \begin{bmatrix} \underset{\sim}{\mathbf{K}}_{pt}(\lambda) \\ \underset{\sim}{\mathbf{K}}_{et}(\lambda) \end{bmatrix}.$$

By Lemma 7.4, $\mathbf{x}_t$ and $\boldsymbol{\zeta}_t^\varepsilon$ satisfy the system of equations (7.65) and (7.66). By Lemma 7.6, we can conclude from the conditions of the lemma that there is a constant k_1 so that

$$\|\underset{\sim}{\mathbf{V}}_{t,s}(\lambda)\| \le k_1, \qquad \forall \lambda, \quad \forall t, s \in [0, T],$$

where $\underset{\sim}{\mathbf{V}}_{t,s}$ is given by (7.67).

Consequently, there exists a constant $k_2 > 0$ such that

$$E\|\boldsymbol{\zeta}_t^\varepsilon\|^2 \le k_2, \qquad \forall \lambda, \quad \forall t \in [0, T]$$

and

$$E\int_0^T \left(\|\mathbf{u}_{pt}^\varepsilon\|^2 + \|\mathbf{u}_{et}^\varepsilon\|^2\right) dt \le \int_0^T E\left(\|\underset{\sim}{\mathbf{K}}_{pt}\|^2 + \|\underset{\sim}{\mathbf{K}}_{et}\|^2\right) \|\boldsymbol{\zeta}_t^\varepsilon\|^2 \, dt$$

$$\le 2k_2 \int_0^T f_t^2 \, dt < \infty,$$

which is independent of λ. Hence, when $\underset{\sim}{\mathbf{K}}_{pt}(\lambda)$ and $\underset{\sim}{\mathbf{K}}_{et}(\lambda)$ are uniformly continuous in $t \in [0, T]$, by Lemma 7.6 so is $\underset{\sim}{\mathbf{V}}_{t,s}(\lambda)$. Then from (7.65) and (7.66), it is easy to see that $E\|\mathbf{x}_t\|^2$, $E\|\zeta_t^{\varepsilon}\|^2$, and hence $E\|\mathbf{u}_t^{\varepsilon}\|^2$, are continuous in λ. □

On the right-hand side of (7.106) replace the first $\underset{\sim}{\mathbf{P}}_t$ by $\underset{\sim}{\tilde{\mathbf{P}}}_t^{\varepsilon}$, then denote $V(\delta_1, \delta_2)$ by $V_1^{\varepsilon}(\delta_1, \delta_2)$ and replace the second $\underset{\sim}{\mathbf{P}}_t$ by $\underset{\sim}{\tilde{\mathbf{P}}}_t^{\varepsilon}$ with the first $\underset{\sim}{\mathbf{P}}_t$ kept invariant and, correspondingly, denote $V(\delta_1, \delta_2)$ by $V_2^{\varepsilon}(\delta_1, \delta_2)$.

By (7.19), we have

$$\lim_{\varepsilon \to 0} V_1^{\varepsilon}(\delta_1, \delta_2) = \lim_{\varepsilon \to 0} V_2^{\varepsilon}(\delta_1, \delta_2) = V(\delta_1, \delta_2). \tag{7.130}$$

Theorem 7.7. *Suppose that Condition A_1 holds and there exists $\Delta > 0$ such that for sufficiently small $\delta_1 > 0$, $\underset{\sim}{\mathbf{S}}_t(\delta_1, \delta_2)$ is bounded for $t \in [0, T]$, $\delta_2 \in (0, \Delta]$. Further assume that*

$$\lim_{\varepsilon \to 0} \lim_{\delta_1 \to 0} \lim_{\delta_2 \to 0} V_1^{\varepsilon}(\delta_1, \delta_2) = \lim_{\varepsilon \to 0} \lim_{\delta_2 \to 0} \lim_{\delta_1 \to 0} V_2^{\varepsilon}(\delta_1, \delta_2) = V. \tag{7.131}$$

Then (7.120) holds and from (7.128) and (7.129), there is a subsequence

$$\mathbf{u}_t(k) \triangleq \begin{bmatrix} \mathbf{u}_{pt}(k) \\ \mathbf{u}_{et}(k) \end{bmatrix} \triangleq \begin{bmatrix} \mathbf{u}_{pt}^{\varepsilon(k)}(\delta(k), \mu(k)) \\ \mathbf{u}_{et}^{\varepsilon(k)}(\delta(k), \mu(k)) \end{bmatrix},$$

such that A_2 and B_2 are verified and C_2 holds in the following sense:

$$\begin{aligned} \overline{\lim_{\varepsilon \to 0}}\ \overline{\lim_{\delta_1 \to 0}}\ \overline{\lim_{\delta_2 \to 0}}\ EJ\big(\mathbf{u}_p^{\varepsilon}(\delta, \mathbf{u}_e^{\varepsilon}), \mathbf{u}_e^{\varepsilon}\big) &\leq \lim_{k \to \infty} EJ\big(\mathbf{u}_p(k), \mathbf{u}_e(k)\big) \\ &= V \leq \underline{\lim_{\varepsilon \to 0}}\ \underline{\lim_{\mu_2 \to 0}}\ \underline{\lim_{\mu_1 \to 0}}\ EJ\big(\mathbf{u}_p^{\varepsilon}, \mathbf{u}_e^{\varepsilon}(\mathbf{u}_p^{\varepsilon}, \mu)\big) \end{aligned} \tag{7.132}$$

for any $\mathbf{u}_p^{\varepsilon}, \mathbf{u}_e^{\varepsilon}$ if

$$\begin{bmatrix} \mathbf{u}_p^{\varepsilon}(\delta, \mathbf{u}_e^{\varepsilon}) \\ \mathbf{u}_e^{\varepsilon} \end{bmatrix} \in \mathscr{U}^{\varepsilon} \quad \text{and} \quad \begin{bmatrix} \mathbf{u}_p^{\varepsilon} \\ \mathbf{u}_e^{\varepsilon}(\mathbf{u}_p^{\varepsilon}, \mu) \end{bmatrix} \in \mathscr{U}^{\varepsilon}.$$

Proof. Let

$$\big[\mathbf{u}_p^{\varepsilon\tau}, \mathbf{u}_e^{\varepsilon\tau}(\mathbf{u}_p^{\varepsilon}, \mu)\big]^{\tau} \in \mathscr{U}^{\varepsilon}.$$

From (6.113), (7.122), and (7.123), we have

$$\mathbf{x}_t - \zeta_t^\varepsilon(\mathbf{u}_p^\varepsilon, \mu) = \xi_t - \tilde{\xi}_t^\varepsilon.$$

Substituting $\mathbf{u}_p^\varepsilon$ and $\mathbf{u}_e^\varepsilon(\mathbf{u}_p^\varepsilon, \mu)$ into (7.107) leads to

$$\inf_{\mathbf{u}_p^\varepsilon} EJ_{\mu_1\mu_2}\big(\mathbf{u}_p^\varepsilon, \mathbf{u}_e^\varepsilon(\mathbf{u}_p^\varepsilon, \mu)\big) \geq V_2^\varepsilon(\mu_1, \mu_2). \tag{7.133}$$

In particular, if we take

$$\mathbf{u}_p^\varepsilon = \mathbf{u}_p^\varepsilon(\delta, \mathbf{u}_e^\varepsilon),$$

then

$$\mathbf{u}_p^\varepsilon(\delta, \mathbf{u}_e^\varepsilon) = \mathbf{u}_p^\varepsilon(\delta, \mu), \qquad \mathbf{u}_e^\varepsilon(\mathbf{u}_p^\varepsilon, \mu) = \mathbf{u}_e^\varepsilon(\delta, \mu)$$

and hence

$$V_2^\varepsilon(\mu_1, \mu_2) \leq EJ\big(\mathbf{u}_p^\varepsilon(\delta, \mu), \mathbf{u}_e^\varepsilon(\delta, \mu)\big) + \mu_1 E\int_0^T \|\mathbf{u}_{pt}^\varepsilon\|^2\, dt. \tag{7.134}$$

According to the proof of Lemma 7.8, $V(\delta_1, \delta_2)$ is uniformly bounded in $\delta_1 \in (0, \Delta]$ if Δ is small enough. Hence $\underset{\sim}{\mathbf{S}}_t(\delta_1, \delta_2)$ is uniformly bounded in $\delta_1 \in (0, \Delta]$, then by the condition of the theorem, we see that $\underset{\sim}{\mathbf{S}}_t(\delta_1, \delta_2)$ is uniformly bounded for $\delta_1, \delta_2 \in (0, \Delta]$ and $t \in [0, T]$. Consequently, $\mathbf{u}_{pt}^\varepsilon(\delta, \mu)$ given by (7.128) for $\delta_2 \in (0, \Delta]$ and $\mathbf{u}_{et}^\varepsilon(\delta, \mu)$ given by (7.129) for $\mu_1 \in (0, \Delta]$ are uniformly bounded. Hence for fixed δ_1 and μ_2, $\mathbf{u}_{pt}^\varepsilon(\delta, \mu)$ and $\mathbf{u}_{et}^\varepsilon(\delta, \mu)$, with δ_2 and μ_1 as parameters, satisfy the conditions of Lemma 7.9. Then

$$E\int_0^T \|\mathbf{u}_{pt}^\varepsilon\|^2\, dt \quad \text{and} \quad E\int_0^T \|\mathbf{u}_{et}^\varepsilon\|^2\, dt$$

are uniformly bounded in δ_2 and μ_1.

Hence from (7.131) and (7.134), we have

$$\begin{aligned} V &= \lim_{\varepsilon\to 0} \lim_{\mu_2\to 0} \lim_{\mu_1\to 0} V_2^\varepsilon(\mu_1, \mu_2) \\ &\leq \underline{\lim}_{\varepsilon\to 0} \underline{\lim}_{\delta_1, \delta_2, \mu_2\to 0} \underline{\lim}_{\mu_1\to 0} EJ\big(\mathbf{u}_p^\varepsilon(\delta, \mu), \mathbf{u}_e^\varepsilon(\delta, \mu)\big). \end{aligned} \tag{7.135}$$

Similar to (7.133), one can prove that

$$\sup_{\mathbf{u}_e^\varepsilon} EJ_{\delta_1, \delta_2}\big(\mathbf{u}_p^\varepsilon(\delta, \mathbf{u}_e^\varepsilon), \mathbf{u}_e^\varepsilon\big) \leq V_1^\varepsilon(\delta_1, \delta_2),$$

and

$$V = \lim_{\varepsilon \to 0} \lim_{\delta_1 \to 0} \lim_{\delta_2 \to 0} V_1^{\varepsilon}(\delta_1, \delta_2)$$

$$\geq \overline{\lim_{\varepsilon \to 0}} \; \overline{\lim_{\mu_1, \mu_2, \delta_1 \to 0}} \; \overline{\lim_{\delta_2 \to 0}} EJ\big(\mathbf{u}_p^{\varepsilon}(\delta, \mu), \mathbf{u}_e^{\varepsilon}(\delta, \mu)\big). \tag{7.136}$$

If for (7.135) or (7.136) equality holds, then it is trivial to see that from (7.128) and (7.129), a subsequence $\{\mathbf{u}_t(k)\}$ can be selected such that

$$EJ\big(\mathbf{u}_p(k), \mathbf{u}_e(k)\big) \underset{k \to \infty}{\rightarrow} V.$$

If both (7.135) and (7.136) are strict inequalities, then, for any real sequence $\varepsilon(k) \underset{k \to \infty}{\rightarrow} 0$, $\varepsilon(k) > 0$, we must have

$$\overline{\lim_{\mu_1, \mu_2, \delta_1 \to 0}} \; \overline{\lim_{\delta_2 \to 0}} EJ\big(\mathbf{u}_p^{\varepsilon(k)}(\delta, \mu), \mathbf{u}_e^{\varepsilon(k)}(\delta, \mu)\big)$$

$$< V < \underline{\lim_{\delta_1, \delta_2, \mu \to 0}} \; \underline{\lim_{\mu_1 \to 0}} EJ\big(\mathbf{u}_p^{\varepsilon(k)}(\delta, \mu), \mathbf{u}_e^{\varepsilon(k)}(\delta, \mu)\big) \tag{7.137}$$

for k sufficiently large.

For arbitrarily small $\Delta_1 > 0$, let $\delta_1 > \delta_1'$, $\delta_2 > \delta_2'$, and $\delta_1, \delta_2, \delta_1', \delta_2' \in [\Delta_1, \Delta_2]$. In Lemma 7.8, in the proof of Conclusion 2, take $\underset{\sim}{\mathbf{Q}}_{2t} + \delta_1' \underset{\sim}{\mathbf{I}}$ and $\underset{\sim}{\mathbf{Q}}_{3t} + \delta_2' \underset{\sim}{\mathbf{I}}$ instead of $\underset{\sim}{\mathbf{Q}}_{2t}$ and $\underset{\sim}{\mathbf{Q}}_{3t}$, respectively, and let $\underset{\sim}{\mathbf{S}}_t(\delta_1', \delta_2')$ correspond to $\underset{\sim}{\mathbf{S}}_t(0,0)$. Then it is easy to see that $\underset{\sim}{\mathbf{S}}_t(\delta_1, \delta_2)$ is continuous in $\delta_1, \delta_2 \in [\Delta_1, \Delta_2]$.

Subject to the controls (7.128) and (7.129), it follows from the continuity of $\underset{\sim}{\mathbf{S}}_t(\delta_1, \delta_2)$ in δ_1, δ_2 that there is a constant k for which

$$\left\| \big(\underset{\sim}{\mathbf{Q}}_{2t} + \delta_1 \underset{\sim}{\mathbf{I}}\big)^{-1} \underset{\sim}{\mathbf{B}}_{pt}^{\tau} \underset{\sim}{\mathbf{S}}_t(\delta_1, \delta_2) \right\| \leq k \|\underset{\sim}{\mathbf{B}}_{pt}\| \tag{7.138}$$

and

$$\left\| \big(\underset{\sim}{\mathbf{Q}}_{3t} + \mu_2 \underset{\sim}{\mathbf{I}}\big)^{-1} \underset{\sim}{\mathbf{B}}_{et}^{\tau} \underset{\sim}{\mathbf{S}}_t(\delta_1, \delta_2) \right\| \leq k \|\underset{\sim}{\mathbf{B}}_{et}\| \tag{7.139}$$

if δ_1, δ_2, μ_1, and μ_2 all lie in $[\Delta_1, \Delta_2]$. Since the left-hand sides of (7.138) and (7.139) are continuous in δ_1, δ_2, μ_1, and μ_2 belonging to the interval

$[\Delta_1, \Delta_2]$, $EJ(\mathbf{u}_p^\varepsilon(\delta, \mu)\mathbf{u}_e^\varepsilon(\delta, \mu))$ is continuous in these variables over this range by Lemma 7.9. Since Δ_1 is an arbitrary number greater than zero, the continuity of $EJ(\mathbf{u}_p^\varepsilon(\delta, \mu), \mathbf{u}_e^\varepsilon(\delta, \mu))$ actually holds in the region $\delta_1, \delta_2, \mu_1, \mu_2 \in (0, \Delta_2]$.

Let $\Delta_k \underset{k\to\infty}{\to} 0$. Then by (7.137), for any k in the region $\delta_1 + \delta_2 + \mu_1 + \mu_2 < \Delta_k$, there must be $\delta_1(k)$, $\delta_2(k)$, $\mu_1(k)$, and $\mu_2(k)$ such that

$$EJ\left[\mathbf{u}_p^{\varepsilon(k)}(\delta(k), \mu(k)), \mathbf{u}_e^{\varepsilon(k)}(\delta(k), \mu(k))\right] = V.$$

This, together with (7.136), proves S_2.

For fixed $\varepsilon > 0$, let $[\mathbf{u}_p^{\varepsilon\tau} \quad \mathbf{u}_e^{\varepsilon\tau}]^\tau \in \mathscr{U}^\varepsilon$. By (7.133),

$$\underline{\lim}_{\mu_2\to 0} \sup_{\mathbf{u}_e^\varepsilon} \underline{\lim}_{\mu_1\to 0} \inf_{\mathbf{u}_p^\varepsilon} EJ_{\mu_1,\mu_2}(\mathbf{u}_p^\varepsilon, \mathbf{u}_e^\varepsilon) \geq \lim_{\mu_2\to 0} \lim_{\mu_1\to 0} V_2^\varepsilon(\mu_1, \mu_2), \quad (7.140)$$

where the sup should be taken over all $\mathbf{u}_e^\varepsilon$, including those of the form

$$\mathbf{u}_{et}^\varepsilon(\mu_1) \triangleq \underset{\sim}{\mathbf{K}}_{et}(\mu_1)\zeta_t^\varepsilon,$$

where $\|\underset{\sim}{\mathbf{K}}_{et}(\mu_1)\|$ is dominated by a square integrable function independent of μ_1 and $\underset{\sim}{\mathbf{K}}_{et}(\mu_1) \underset{\mu_1\to 0}{\to} \underset{\sim}{\mathbf{K}}_{et}$. Dependence of $\underset{\sim}{\mathbf{K}}_{et}(\mu_1)$ on μ_1 implies that ζ_t^ε depends on μ_1, but for this case it is easy to show by Lemma 7.6 that

$$\lim_{\mu_1\to 0} \mathbf{u}_{et}^\varepsilon(\mu_1) = \underset{\sim}{\mathbf{K}}_{et}\zeta_t^\varepsilon \triangleq \mathbf{u}_{et}^\varepsilon.$$

Hence

$$\lim_{\mu_1\to 0} EJ(\mathbf{u}_p^\varepsilon, \mathbf{u}_e^\varepsilon(\mu_1)) = EJ(\mathbf{u}_p^\varepsilon, \mathbf{u}_e^\varepsilon). \quad (7.141)$$

Notice that

$$\inf_{\mathbf{u}_p^\varepsilon} EJ_{\mu_1\mu_2}(\mathbf{u}_p^\varepsilon, \mathbf{u}_e^\varepsilon(\mu_1)) \leq EJ_{\mu_1\mu_2}(\mathbf{u}_p^\varepsilon, \mathbf{u}_e^\varepsilon(\mu_1))$$

$$\leq EJ(\mathbf{u}_p^\varepsilon, \mathbf{u}_e^\varepsilon(\mu_1)) + \mu_1 E\int_0^T \|\mathbf{u}_{pt}^\varepsilon\|^2\, dt. \quad (7.142)$$

Then by Lemma 7.9 and (7.141), it follows that

$$\overline{\lim}_{\mu_1\to 0} \inf_{\mathbf{u}_p^\varepsilon} EJ_{\mu_1\mu_2}(\mathbf{u}_p^\varepsilon, \mathbf{u}_e^\varepsilon(\mu_1)) \leq \lim_{\mu_1\to 0} EJ(\mathbf{u}_p^\varepsilon, \mathbf{u}_e^\varepsilon(\mu_1)) = EJ(\mathbf{u}_p^\varepsilon, \mathbf{u}_e^\varepsilon).$$

(7.143)

Hence

$$\varlimsup_{\mu_2 \to 0} \sup_{\mathbf{u}_e^\varepsilon} \varlimsup_{\mu_1 \to 0} \inf_{\mathbf{u}_p^\varepsilon} EJ_{\mu_1\mu_2}\big(\mathbf{u}_p^\varepsilon, \mathbf{u}_e^\varepsilon\big) \le \sup_{\mathbf{u}_e^\varepsilon} \inf_{\mathbf{u}_p^\varepsilon} EJ\big(\mathbf{u}_p^\varepsilon, \mathbf{u}_e^\varepsilon\big). \tag{7.144}$$

From (7.140) and (7.144), we have

$$\lim_{\mu_2 \to 0} \lim_{\mu_1 \to 0} V_2^\varepsilon(\mu_1, \mu_2) \le \sup_{\mathbf{u}_e^\varepsilon} \inf_{\mathbf{u}_p^\varepsilon} EJ\big(\mathbf{u}_p^\varepsilon, \mathbf{u}_e^\varepsilon\big). \tag{7.145}$$

By using the condition imposed on $\underset{\sim}{\mathbf{S}}_t(\delta_1, \delta_2)$ in the theorem, we can obtain a series of converse inequalities corresponding to those subsequent to (7.140). To be specific, we obtain the following:

$$\lim_{\delta_1 \to 0} \lim_{\delta_2 \to 0} V_1^\varepsilon(\delta_1, \delta_2) \ge \varlimsup_{\delta_1 \to 0} \inf_{\mathbf{u}_p^\varepsilon} \varlimsup_{\delta_2 \to 0} \sup_{\mathbf{u}_e^\varepsilon} EJ_{\delta_1\delta_2}\big(\mathbf{u}_p^\varepsilon, \mathbf{u}_e^\varepsilon\big), \tag{7.146}$$

$$\sup_{\mathbf{u}_e^\varepsilon} EJ_{\delta_1\delta_2}\big(\mathbf{u}_p^\varepsilon(\delta_2), \mathbf{u}_e^\varepsilon\big) \ge EJ_{\delta_1\delta_2}\big(\mathbf{u}_p^\varepsilon(\delta_2), \mathbf{u}_e^\varepsilon\big)$$

$$\ge EJ\big(\mathbf{u}_p^\varepsilon(\delta_2), \mathbf{u}_e^\varepsilon\big) - \delta_2 E\int_0^T \|\mathbf{u}_{et}^\varepsilon\|^2\, dt,$$

$$\varliminf_{\delta_2 \to 0} \sup_{\mathbf{u}_e^\varepsilon} EJ_{\delta_1\delta_2}\big(\mathbf{u}_p^\varepsilon(\delta_2), \mathbf{u}_e^\varepsilon\big) \ge \lim_{\delta_2 \to 0} EJ\big(\mathbf{u}_p^\varepsilon(\delta_2), \mathbf{u}_e^\varepsilon\big) = EJ\big(\mathbf{u}_p^\varepsilon, \mathbf{u}_e^\varepsilon\big) \tag{7.147}$$

$$\varliminf_{\delta_1 \to 0} \inf_{\mathbf{u}_p^\varepsilon} \varliminf_{\delta_2 \to 0} \sup_{\mathbf{u}_e^\varepsilon} EJ_{\delta_1\delta_2}\big(\mathbf{u}_p^\varepsilon(\delta_2), \mathbf{u}_e^\varepsilon\big) \ge \inf_{\mathbf{u}_p^\varepsilon} \sup_{\mathbf{u}_e^\varepsilon} EJ\big(\mathbf{u}_p^\varepsilon, \mathbf{u}_e^\varepsilon\big). \tag{7.148}$$

From (7.146) and (7.148), it follows that

$$\inf_{\mathbf{u}_p^\varepsilon} \sup_{\mathbf{u}_e^\varepsilon} EJ\big(\mathbf{u}_p^\varepsilon, \mathbf{u}_e^\varepsilon\big) \le \lim_{\delta_1 \to 0} \lim_{\delta_2 \to 0} V_1^\varepsilon(\delta_1, \delta_2).$$

Hence (7.131) implies (7.120).

If in (7.143), we set

$$\mathbf{u}_e^\varepsilon(\mu_1) = \mathbf{u}_e^\varepsilon\big(\mathbf{u}_p^\varepsilon, \mu\big),$$

then by (7.133), we know that

$$\lim_{\mu_1 \to 0} EJ\big(\mathbf{u}_p^\varepsilon, \mathbf{u}_e^\varepsilon(\mathbf{u}_p^\varepsilon, \mu)\big) \geq \overline{\lim_{\mu_1 \to 0}} \inf_{\mathbf{u}_p^\varepsilon} EJ_{\mu_1\mu_2}\big(\mathbf{u}_p^\varepsilon, \mathbf{u}_e^\varepsilon(\mathbf{u}_p^\varepsilon, \mu)\big)$$

$$\geq \overline{\lim_{\mu_1 \to 0}} V_2^\varepsilon(\mu_1, \mu_2),$$

which leads us to the last inequality of (7.132) by passing to the limit $\underline{\lim}_{\varepsilon \to 0} \underline{\lim}_{\mu_2 \to 0}$ and by using (7.131).

$$\mathbf{u}_p^\varepsilon(\delta) = \mathbf{u}_p^\varepsilon(\delta, \mathbf{u}_e^\varepsilon)$$

in (7.147), then by (7.135), it follows that

$$\underline{\lim}_{\delta_2 \to 0} V_1^\varepsilon(\delta_1, \delta_2) \geq \underline{\lim}_{\delta_2 \to 0} \sup_{\mathbf{u}_e^\varepsilon} EJ_{\delta_1\delta_2}\big(\mathbf{u}_p^\varepsilon(\delta, \mathbf{u}_e^\varepsilon), \mathbf{u}_e^\varepsilon\big) \geq \lim_{\delta_2 \to 0} EJ\big(\mathbf{u}_p^\varepsilon(\delta_2), \mathbf{u}_e^\varepsilon\big).$$

By use of (7.131), taking $\underline{\lim}$ as $\varepsilon \to 0$ $\underline{\lim}$ as $\delta_1 \to 0$ on both sides of the preceding formula, we obtain the first inequality of (7.132).

When Condition B_1 holds, then the left-hand sides of (7.138), (7.139) are continuous in $\delta_1, \delta_2, \mu_1, \mu_2 \in [0, \Delta_2]$ and by Lemma 7.9, $\mathbf{u}_p^\varepsilon(\delta, \mu)$ and $\mathbf{u}_e^\varepsilon(\delta, \mu)$ are also continuous in this region.

If, in addition, Condition C_1 holds, then by Theorem 7.5

$$\operatorname*{l.i.m.}_{\varepsilon \to 0} \boldsymbol{\zeta}_t^\varepsilon = \hat{\mathbf{x}}_t.$$

Hence if Conditions B_1 and C_1 hold, $\mathbf{u}_t(k)$ converges in the mean square sense to the saddle point as $k \to \infty$, that is, A_2 holds. □

Remark. If A_1 and C_1 hold, then by Theorem 6.5

$$\operatorname*{l.i.m.}_{\varepsilon \to 0} \boldsymbol{\zeta}_t^\varepsilon = \hat{\mathbf{x}}_t,$$

hence when we select a sequence from (7.123) or (7.125), (7.127)–(7.129) we may first replace $\boldsymbol{\zeta}_t^\varepsilon$ by $\hat{\mathbf{x}}_t$ before the selection.

Denote (7.123) with $\boldsymbol{\zeta}_t^\varepsilon$ replaced by $\hat{\mathbf{x}}_t$ by

$$\mathbf{u}_{pt} = \underset{\sim}{\mathbf{K}}_{pt}\hat{\mathbf{x}}_t, \qquad \mathbf{u}_{et} = \underset{\sim}{\mathbf{K}}_{et}\hat{\mathbf{x}}_t,$$

and denote $\mathscr{U}^\varepsilon$ by $\mathscr{U}^0$ correspondingly. Clearly, $\mathscr{U}^0 \subset \mathscr{U}$ and $\mathscr{U}^0$ is independent of ε. After replacing $\boldsymbol{\zeta}_t^\varepsilon$ by $\hat{\mathbf{x}}_t$ in (7.125), (7.127)–(7.129), we delete the script ε for $\mathbf{u}_{pt}^\varepsilon$ and $\mathbf{u}_{et}^\varepsilon$.

Theorem 7.8. *Suppose that Conditions* A_1 *and* C_1 *hold and there is a* $\Delta > 0$ *such that for sufficiently small* $\delta_1 > 0$, $\underset{\sim}{\mathbf{S}}_t(\delta_1, \delta_2)$ *is bounded for* $t \in [0, T]$, $\delta_2 \in (0, \Delta]$. *Then, if* $V_1 = V_2 = V$, *the game has the value in* $\mathscr{U}^0$ *which is equal to* V, *and from* $\mathbf{u}_{pt}(\delta, \mu)$ *and* $\mathbf{u}_{et}(\delta, \mu)$ *the subsequences* $\mathbf{u}_{pt}(k)$, $\mathbf{u}_{et}(k)$ *can be selected such that* A_2 *and* B_2 *hold and* C_2 *is satisfied in the following sense*:

$$\overline{\lim_{\delta_1 \to 0}}\ \overline{\lim_{\delta_2 \to 0}}\ EJ\big(\mathbf{u}_p(\delta, \mathbf{u}_e), \mathbf{u}_e\big) \le \lim_{k \to \infty} EJ\big(\mathbf{u}_p(k), \mathbf{u}_e(k)\big) = V$$

$$\le \lim_{\mu_2 \to 0} \lim_{\mu_1 \to 0} EJ\big(\mathbf{u}_p, \mathbf{u}_e(\mathbf{u}_p, \mu)\big) \quad \text{for } \mathbf{u} \in \mathscr{U}^0.$$

Proof. The proof can be completed analogously to that of Theorem 7.7 with $\mathscr{U}^\varepsilon$ replaced by $\mathscr{U}^0$. The theorem can also be obtained by directly passing to the limit $\varepsilon \to 0$ in Theorem 7.7. □

Theorem 7.9. *Assume Conditions* A_1 *and* B_1 *hold, then* (7.120) *is satisfied and, as* δ_1, δ_2, μ_1, *and* μ_2 *go to zero, the controls defined by* (7.128) *and* (7.129) *converge a.s. to the limits* $\mathbf{u}_{pt}^\varepsilon(0,0)$, $\mathbf{u}_{et}^\varepsilon(0,0)$ *which satisfy* A_2, B_2, *and fulfill* C_2 *in the following sense*:

$$\lim_{\varepsilon \to 0} EJ\big(\mathbf{u}_p^\varepsilon(0, \mathbf{u}_e^\varepsilon), \mathbf{u}_e^\varepsilon\big) \le \lim_{\varepsilon \to 0} EJ\big(\mathbf{u}_p^\varepsilon(0,0), \mathbf{u}_e^\varepsilon(0,0)\big)$$

$$= V \le \lim_{\varepsilon \to 0} EJ\big(\mathbf{u}_p^\varepsilon, \mathbf{u}_e^\varepsilon(\mathbf{u}_p^\varepsilon, 0)\big).$$

Proof. This theorem is a consequence of Theorem 7.7. Since Condition B_1 holds, $\tilde{\underset{\sim}{\mathbf{P}}}_t^\varepsilon \underset{\varepsilon \to 0}{\to} \underset{\sim}{\mathbf{P}}_t$, and hence the conditions imposed on $\underset{\sim}{\mathbf{S}}_t(\delta_1, \delta_2)$ in Theorem 6.7 and (7.131) are satisfied. □

CHAPTER 8

Gauss–Markov Estimation for Continuous-Time Systems

8.1. ESTIMATION WITHOUT KNOWLEDGE OF INITIAL VALUES

In Chapter 5, for discrete-time systems, we have discussed the linear estimation problem for the case where the initial statistical characteristics of $\mathbf{x}_0$ are unavailable. We have proved that if instead of the unknown $E\mathbf{x}_0$ and $\mathbf{R}_{\mathbf{x}_0}$ we simply put 0 and $\alpha\mathbf{I}$ in the LUMVE $\hat{\mathbf{x}}_j(k)$ (of $\mathbf{x}_j$ based on the measurement up to instant k), then there is a simple necessary and sufficient condition for the resulting value $\bar{\mathbf{x}}_j(k,\alpha)$ to be convergent as $\alpha \to \infty$; further, if this condition is satisfied, the limit is nothing else but the Gauss–Markov estimate.

For the continuous-time system, we cannot put $E\mathbf{x}_0 = 0$, $\mathbf{R}_{\mathbf{x}_0} = \alpha\mathbf{I}$ directly into the LUMVE, since the possible degeneracy of the measurement noise is involved and the LUMVE may not satisfy a differential equation.

We still consider system (7.9), (7.10) with $\mathbf{F}_t\mathbf{F}_t^\tau$ possibly degenerate, $\|\mathbf{A}_t\|$, $\|\mathbf{C}_t\|^2$, $\|\mathbf{D}_t\|^2$ and $\|\mathbf{F}_t\|^2$ are assumed integrable on $[0,T]$, and $[\mathbf{x}_0^\tau \mathbf{y}_0^\tau]^\tau$ is uncorrelated with $\{\mathbf{w}_t\}$.

We retain the notation introduced in Sections 7.2 and 7.3, in particular, $\mathbf{P}_t^{\varepsilon s}$, $\mathbf{K}_t^{\varepsilon s}$, and $\mathbf{\Psi}_{t,\lambda}^{\varepsilon s}$ are defined by (7.36)–(7.39).

Write

$$\mathbf{\Gamma}_t^\varepsilon = \int_0^t \mathbf{\Psi}_{s,0}^{\varepsilon 0\tau}\mathbf{C}_s^\tau\left(\mathbf{F}_s\mathbf{F}_s^\tau + \varepsilon\mathbf{I}\right)^{-1}\mathbf{C}_s\mathbf{\Psi}_{s,0}^{\varepsilon 0}\,ds, \tag{8.1}$$

$$\boldsymbol{\xi}_0 = \mathbf{x} + \sqrt{\alpha}\,\boldsymbol{\xi}_0', \qquad \boldsymbol{\xi}_0' \in \mathcal{N}(\mathbf{0},\mathbf{I}). \tag{8.2}$$

Clearly, $\boldsymbol{\xi}_0 \in \mathcal{N}(\mathbf{x}, \alpha \underset{\sim}{\mathbf{I}})$. Together with (7.9) and (7.10), we consider

$$d\boldsymbol{\xi}_t = \underset{\sim}{\mathbf{A}}_t \boldsymbol{\xi}_t \, dt + \underset{\sim}{\mathbf{D}}_t \, d\mathbf{w}_t, \tag{8.3}$$

$$d\boldsymbol{\eta}_t = \underset{\sim}{\mathbf{C}}_t \boldsymbol{\xi}_t \, dt + \underset{\sim}{\mathbf{F}}_t d\mathbf{w}_t, \qquad \boldsymbol{\eta}_0 = \mathbf{y}_0 \tag{8.4}$$

$$\boldsymbol{\eta}_t^{\varepsilon} = \boldsymbol{\eta}_t + \sqrt{\varepsilon}\, \overline{\mathbf{w}}_t, \tag{8.5}$$

where $\{\mathbf{w}_t\}$ and $\{\overline{\mathbf{w}}_t\}$ are Wiener processes and $\boldsymbol{\xi}_0$, $\boldsymbol{\eta}_0$, $\{\mathbf{w}_t\}$, and $\{\overline{\mathbf{w}}_t\}$ are mutually independent.

By (7.41), the LUMVE $\boldsymbol{\xi}_s^{\varepsilon}(t)$ of $\boldsymbol{\xi}_s$ based on $\{\boldsymbol{\eta}_\lambda^{\varepsilon}, 0 \le \lambda \le t\}$, $s \le t$, has estimation error covariance matrix $\overline{\underset{\sim}{\mathbf{P}}}_s^{\varepsilon}(t, \alpha)$

$$\overline{\underset{\sim}{\mathbf{P}}}_s^{\varepsilon}(t, \alpha) = \left[\underset{\sim}{\mathbf{I}} + \underset{\sim}{\mathbf{P}}_s^{\varepsilon}(\alpha) \int_s^t \underset{\sim}{\boldsymbol{\Psi}}_{\lambda, s}^{\varepsilon s \tau} \underset{\sim}{\mathbf{C}}_\lambda^{\tau} (\underset{\sim}{\mathbf{F}}_\lambda \underset{\sim}{\mathbf{F}}_\lambda^{\tau} + \varepsilon \underset{\sim}{\mathbf{I}})^{-1} \underset{\sim}{\mathbf{C}}_\lambda \underset{\sim}{\boldsymbol{\Psi}}_{\lambda, s}^{\varepsilon s} \, d\lambda \right]^{-1} \underset{\sim}{\mathbf{P}}_s^{\varepsilon}(\alpha). \tag{8.6}$$

For $s = t$, denote

$$\overline{\underset{\sim}{\mathbf{P}}}_t^{\varepsilon}(t, \alpha) = \overline{\underset{\sim}{\mathbf{P}}}_t^{\varepsilon}(\alpha).$$

According to Theorem 7.2 as $\varepsilon \to 0$, $\overline{\underset{\sim}{\mathbf{P}}}_s^{\varepsilon}(t, \alpha)$ tends to a finite limit, which now is denoted by $\overline{\mathbf{P}}_s(t, \alpha)$ for $s < t$ and by $\overline{\underset{\sim}{\mathbf{P}}}_t(\alpha)$ for $s = t$.

For $\{\boldsymbol{\xi}_t\}$ and $\{\boldsymbol{\eta}_t\}$, the supoptimal filtering equation corresponding to (7.16) is

$$d\tilde{\boldsymbol{\xi}}_t^{\varepsilon} = \underset{\sim}{\mathbf{A}}_t \tilde{\boldsymbol{\xi}}_t^{\varepsilon} \, dt + \left(\underset{\sim}{\mathbf{D}}_t \underset{\sim}{\mathbf{F}}_t^{\tau} + \underset{\sim}{\mathbf{P}}_t^{\varepsilon}(\alpha) \underset{\sim}{\mathbf{C}}_t^{\tau} \right) (\underset{\sim}{\mathbf{F}}_t \underset{\sim}{\mathbf{F}}_t^{\tau} + \varepsilon \underset{\sim}{\mathbf{I}})^{-1} \left(d\boldsymbol{\eta}_t - \underset{\sim}{\mathbf{C}}_t \tilde{\boldsymbol{\xi}}_t^{\varepsilon} \, dt \right),$$

$$\tilde{\boldsymbol{\xi}}_0^{\varepsilon} = \mathbf{x}, \tag{8.7}$$

and the filtering error covariance matrix is $\tilde{\underset{\sim}{\mathbf{P}}}_t^{\varepsilon}$. By Theorem 7.1, we have

$$\underset{\varepsilon \to 0}{\text{l.i.m.}}\, \tilde{\boldsymbol{\xi}}_t^{\varepsilon} = \hat{\boldsymbol{\xi}}_t = E\left(\frac{\boldsymbol{\xi}_t}{\mathscr{F}_t^{\eta}} \right), \qquad \tilde{\underset{\sim}{\mathbf{P}}}_t^{\varepsilon} \underset{\varepsilon \to 0}{\to} \underset{\sim}{\mathbf{P}}_t. \tag{8.8}$$

We shall artificially prescribe the initial condition $\tilde{\mathbf{x}}_0^{\varepsilon} = \mathbf{0}$ for $\tilde{\mathbf{x}}_t^{\varepsilon}$ given by (7.16) and $\underset{\sim}{\mathbf{P}}_0^{\varepsilon} = \alpha \underset{\sim}{\mathbf{I}}$ for $\underset{\sim}{\mathbf{P}}_t^{\varepsilon}$ defined by (7.15) and then denote $\tilde{\mathbf{x}}_t^{\varepsilon}$ and $\underset{\sim}{\mathbf{P}}_t^{\varepsilon}$, respectively, by $\overline{\mathbf{x}}_t^{\varepsilon}(\alpha)$ and $\overline{\underset{\sim}{\mathbf{P}}}_t^{\varepsilon}(\alpha)$, and, correspondingly, denote by $\overline{\underset{\sim}{\mathbf{P}}}_s^{\varepsilon}(t, \alpha)$, $\overline{\mathbf{x}}_s^{\varepsilon}(t, \alpha)$, $0 \le s \le t$, $\underset{\sim}{\mathbf{P}}_s^{\varepsilon}(t)$ and $\tilde{\mathbf{x}}_s^{\varepsilon}(t)$ defined by (7.41) and (7.42) with $\tilde{\mathbf{x}}_t^{\varepsilon}$ and $\underset{\sim}{\mathbf{P}}_t^{\varepsilon}$ replaced by $\overline{\mathbf{x}}_t^{\varepsilon}(\alpha)$ and $\overline{\underset{\sim}{\mathbf{P}}}_t^{\varepsilon}(\alpha)$, respectively. Clearly, $\overline{\underset{\sim}{\mathbf{P}}}_s^{\varepsilon}(t, \alpha)$ so obtained is the same as that given by (8.6). But now $\bar{\mathbf{x}}_s^\varepsilon(t,\alpha)$ is no longer an unbiased estimate for $\mathbf{x}_s$ and $\bar{\mathbf{P}}_s^\varepsilon(t,\alpha)$ has nothing to do with the estimation error covariance matrix.

If as $\varepsilon \to 0$ and $\alpha \to \infty$ $\bar{\mathbf{x}}_s^\varepsilon(t,\alpha)$ tends to a limit $\bar{\mathbf{x}}_s(t)$, then it is called the EWKIV of $\mathbf{x}_s$ based on $\{\mathbf{y}_\lambda, 0 \le \lambda \le t\}$. It will be shown later that this coincides with the Gauss–Markov estimate.

From (6.70), we know that

$$\bar{\mathbf{P}}_t^\varepsilon(\alpha) = \mathbf{P}_t^{\varepsilon s} + \mathbf{\Psi}_{t,s}^{\varepsilon s}\bar{\mathbf{P}}_s^\varepsilon(t,\alpha)\mathbf{\Psi}_{t,s}^{\varepsilon s\tau}, \tag{8.9}$$

where $\mathbf{P}_t^{\varepsilon s}$ is defined by (7.37).

Lemma 8.1. (1) *As* $\varepsilon \to 0$, $\bar{\mathbf{P}}_s^\varepsilon(t,\alpha)$ *nonincreasingly tends to a limit denoted by* $\bar{\mathbf{P}}_s(t,\alpha)$. (2) *As* $\varepsilon \to 0$, $\mathbf{\Gamma}_t^\varepsilon$ *is nondecreasing, and for any* $\varepsilon > 0$,

$$\mathbf{\Psi}_{t,0}^{\varepsilon 0\tau}\mathbf{C}_t^\tau(\mathbf{F}_t\mathbf{F}_t^\tau + \varepsilon\mathbf{I})^{-1} = \mathbf{\Gamma}_t^\varepsilon\mathbf{\Gamma}_t^{\varepsilon+}\mathbf{\Psi}_{t,0}^{\varepsilon 0\tau}\mathbf{C}_t^\tau(\mathbf{F}_t\mathbf{F}_t^\tau + \varepsilon\mathbf{I})^{-1} \text{ a.e.,} \quad t \in [0,T]. \tag{8.10}$$

Proof. Set

$$\mathbf{D}_\lambda^1 = \begin{cases} \mathbf{D}_\lambda, & 0 \le \lambda \le s \\ \mathbf{0}, & s < \lambda \end{cases} \qquad \mathbf{D}_\lambda^2 = \begin{cases} \mathbf{0}, & 0 \le \lambda \le s \\ \mathbf{D}_\lambda, & s < \lambda \end{cases}$$

$$\mathbf{z}_t^1 = \boldsymbol{\xi}_0 + \int_0^t \mathbf{\Phi}_{0,\lambda}\mathbf{D}_\lambda^1\,d\mathbf{w}_\lambda, \qquad \mathbf{z}_t^2 = \int_0^t \mathbf{\Phi}_{0,\lambda}\mathbf{D}_\lambda^2\,d\mathbf{w}_\lambda,$$

$$\mathbf{z}_t^\tau = \left[\mathbf{z}_t^{1\tau}\mathbf{z}_t^{2\tau}\right].$$

Then

$$\boldsymbol{\xi}_t = \mathbf{\Phi}_{t,0}\left(\mathbf{z}_t^1 + \mathbf{z}_t^2\right), \qquad \mathbf{z}_0 = \left[\boldsymbol{\xi}_0^\tau\mathbf{0}\right]^\tau,$$

$$d\mathbf{z}_t = \begin{bmatrix} \mathbf{\Phi}_{0,t}\mathbf{D}_t^1 \\ \mathbf{\Phi}_{0,t}\mathbf{D}_t^2 \end{bmatrix} d\mathbf{w}_t,$$

$$d\boldsymbol{\eta}_t = \mathbf{C}_t\mathbf{\Phi}_{t,0}\left[\mathbf{I} \,\vdots\, \mathbf{I}\right]\mathbf{z}_t\,dt + \mathbf{F}_t\,d\mathbf{w}_t.$$

By Lemma 7.7, as $\varepsilon \to 0$, the estimation error covariance matrix of the LUMVE for $\mathbf{z}_t$, based on $\{\boldsymbol{\eta}_\lambda^\varepsilon, 0 \le \lambda \le t\}$, is nonincreasing. Hence for $\boldsymbol{\xi}_s$, the same is true since

$$\boldsymbol{\xi}_s = \boldsymbol{\Phi}_{s,0}\mathbf{z}_t^1,$$

that is, as $\varepsilon \to 0$ $\overline{\mathbf{P}}_s^\varepsilon(t, \alpha)$ is nonincreasing. In particular, for $s = 0$, by (7.41) we know that

$$\overline{\mathbf{P}}_0^\varepsilon(t, \alpha) = \left[\frac{1}{\alpha}\mathbf{I} + \boldsymbol{\Gamma}_t^\varepsilon\right]^{-1} \tag{8.11}$$

as $\varepsilon \to 0$ is nonincreasing. Hence $\boldsymbol{\Gamma}_t^\varepsilon$ is nondecreasing.

Let $\mathbf{a}$ be any n-dimensional vector. We now show that for almost all $t \in [0, T]$

$$\mathbf{a}^\tau\boldsymbol{\Gamma}_t^\varepsilon\mathbf{a} = 0$$

implies

$$f(t) \triangleq \mathbf{a}^\tau\boldsymbol{\Psi}_{t,0}^{\varepsilon 0\tau}\mathbf{C}_t^\tau(\mathbf{F}_t\mathbf{F}_t^\tau + \varepsilon\mathbf{I})^{-1}\mathbf{C}_t\boldsymbol{\Psi}_{t,0}^{\varepsilon 0}\mathbf{a} = 0. \tag{8.12}$$

Suppose the opposite were true, that is, $f(t) > 0$. $f(t)$ is integrable on $[0, T]$. Hence from real analysis, it is known that almost all points on $[0, T]$ are Lebesgue points of $f(t)$, that is,

$$\lim_{h\to 0}\frac{1}{h}\int_{t-h}^{t}|f(t) - f(s)|\, ds = 0.$$

The fact that $\mathbf{a}^\tau\boldsymbol{\Gamma}_t^\varepsilon\mathbf{a} = 0$ implies

$$\int_{t-h}^{t} f(s)\, ds = 0, \qquad \forall h > 0.$$

Hence

$$0 < f(t) = \frac{1}{h}\int_{t-h}^{t} f(t)\, ds \le \frac{1}{h}\int_{t-h}^{t}|f(t) - f(s)|\, ds + \frac{1}{h}\int_{t-h}^{t} f(s)\, ds$$
$$= \frac{1}{h}\int_{t-h}^{t}|f(t) - f(s)|\, ds \underset{h\to 0}{\to} 0.$$

This contradiction shows that (8.12) holds.

Since

$$(\mathbf{I} - \boldsymbol{\Gamma}_t^\varepsilon\boldsymbol{\Gamma}_t^{\varepsilon +})\boldsymbol{\Gamma}_t^\varepsilon(\mathbf{I} - \boldsymbol{\Gamma}_t^\varepsilon\boldsymbol{\Gamma}_t^{\varepsilon +}) = \mathbf{0},$$

from (8.11), it follows that

$$(\mathbf{I} - \boldsymbol{\Gamma}_t^\varepsilon\boldsymbol{\Gamma}_t^{\varepsilon +})\boldsymbol{\Psi}_{t,0}^{\varepsilon 0\tau}\mathbf{C}_t^\tau(\mathbf{F}_t\mathbf{F}_t^\tau + \varepsilon\mathbf{I})^{-1/2} = \mathbf{0}$$

and hence (8.10) holds. □

Let us put

$$\zeta_t^\varepsilon(\alpha) = \Psi_{t,0}^{\varepsilon 0}\overline{P}_0(t,\alpha)\Big\{\int_0^t \overline{P}_0^{\varepsilon-1}(s,\alpha)\Psi_{0,s}^{\varepsilon 0}(D_s F_s^\tau + \overline{P}_s^\varepsilon(\alpha)C_s^\tau)$$

$$\cdot (F_s F_s^\tau + \varepsilon I)^{-1} F_s \, dw_s$$

$$- \int_0^t\int_0^s \overline{P}_0^{\varepsilon-1}(u,\alpha)\Psi_{0,\mu}^{\varepsilon 0}(D_\mu F_\mu^\tau + \overline{P}_\mu^\varepsilon(\alpha)C_\mu^\tau)(F_\mu F_\mu^\tau + \varepsilon I)^{-1} C_\mu \Phi_{\mu,0}\, d\mu$$

$$\times \cdot \Phi_{0,s} D_s \, dw_s\Big\}, \tag{8.13}$$

where we recall that $\overline{P}_0^{\varepsilon-1}$ denotes $(\overline{P}_0^\varepsilon)^{-1}$.

Lemma 8.2. (1) *For $s \le t$, as $\varepsilon \to 0$, $\Psi_{s,t}^{\varepsilon s}\Psi_{t,0}^{\varepsilon 0}\overline{P}_0^\varepsilon(t,\alpha)$ tends to a finite limit. If $\lim_{\varepsilon\to 0}\det \Gamma_t^\varepsilon \neq 0$, then there are $\alpha > 0$, $\varepsilon_1 > 0$ such that $\Psi_{s,t}^{\varepsilon s}\Psi_{t,0}^{\varepsilon 0}\overline{P}_0^\varepsilon(t,\alpha)$ is bounded on $\alpha \ge \alpha_1$ and $\varepsilon \in (0,\varepsilon_1]$ and as $\varepsilon \to 0$ $\Psi_{s,t}^{\varepsilon s}\Psi_{t,0}^{\varepsilon 0}\Gamma_t^{\varepsilon-1}$ converges to a finite limit.* (2) *As $\varepsilon \to 0$, $\overline{x}_s^\varepsilon(t,\alpha)$ tends in the mean square sense to a limit denoted by $\overline{x}_s(t,\alpha)$.*

Proof. From (7.39), (8.9), and (8.11), it is known that

$$\frac{d}{dt}\Psi_{t,0}^{\varepsilon 0}\overline{P}_0^\varepsilon(t,\alpha)\overline{P}_0^{\varepsilon-1}(s,\alpha)\Psi_{0,s}^{\varepsilon 0}$$

$$= (A_t - K_t^{\varepsilon 0}C_t)\Psi_{t,0}^{\varepsilon 0}\overline{P}_0^\varepsilon(t,\alpha)\overline{P}_0^{\varepsilon-1}(s,\alpha)\Psi_{0,s}^{\varepsilon 0}$$

$$- \Psi_{t,0}^{\varepsilon 0}\overline{P}_0^\varepsilon(t,\alpha)\Psi_{t,0}^{\varepsilon 0\tau}C_t^\tau(F_t F_t^\tau + \varepsilon I)^{-1}C_t\Psi_{t,0}^{\varepsilon 0}\overline{P}_0^\varepsilon(t,\alpha)\overline{P}^{\varepsilon-1}(s,\alpha)\Psi_{0,s}^{\varepsilon 0}$$

$$= \big\{A_t - \big[D_t F_t^\tau + \big(\Psi_{t,0}^{\varepsilon 0}\overline{P}_0^\varepsilon(t,\alpha)\Psi_{t,0}^{\varepsilon 0\tau} + P_t^{\varepsilon 0}\big)C_t^\tau\big](F_t F_t^\tau + \varepsilon I)^{-1}C_t\big\}$$

$$\cdot \Psi_{t,0}^{\varepsilon 0}\overline{P}_0^\varepsilon(t,\alpha)\overline{P}_0^{\varepsilon-1}(s,\alpha)\Psi_{0,s}^{\varepsilon 0}$$

$$= \big[A_t - (D_t F_t^\tau + \overline{P}_t^\varepsilon(\alpha)C_t^\tau)(F_t F_t^\tau + \varepsilon I)^{-1}C_t\big]$$

$$\cdot \Psi_{t,0}^{\varepsilon 0}\overline{P}_0^\varepsilon(t,\alpha)\overline{P}^{\varepsilon-1}(s,\alpha)\Psi_{0,s}^{\varepsilon 0}, \tag{8.14}$$

which shows that

$$\Psi_{t,0}^{\varepsilon 0}\overline{P}_0^\varepsilon(t,\alpha)\overline{P}_0^{\varepsilon-1}(s,\alpha)\Psi_{0,s}^{\varepsilon 0}$$

is a fundamental solution matrix. Hence the solution of (8.7) takes the form

$$\tilde{\xi}_t^\varepsilon = \frac{1}{\alpha}\Psi_{t,0}^{\varepsilon 0}\bar{P}_0^\varepsilon(t,\alpha)\mathbf{x} + \Psi_{t,0}^{\varepsilon 0}\bar{P}_0^\varepsilon(t,\alpha)\int_0^t \bar{P}_0^{\varepsilon-1}(s,\alpha)\Psi_{0,s}^{\varepsilon 0}(D_s F_s^T + \bar{P}_s^\varepsilon(\alpha)C_s^T)$$
$$\cdot (F_s F_s^T + \varepsilon I)^{-1}\, d\eta_s. \tag{8.15}$$

Substitute for η_s by its formula and then use the Ito formula. Then we get

$$\tilde{\xi}_t^\varepsilon = \frac{1}{\alpha}\Psi_{t,0}^{\varepsilon 0}\bar{P}_0^\varepsilon(t,\alpha)\mathbf{x} + \Psi_{t,0}^{\varepsilon 0}\bar{P}_0^\varepsilon(t,\alpha)\int_0^t \bar{P}_0^{\varepsilon-1}(s,\alpha)\Psi_{0,s}^{\varepsilon 0}(D_s F_s^T + \bar{P}_s^\varepsilon(\alpha)C_s^T)$$
$$\cdot (F_s F_s^T + \varepsilon I)^{-1} C_s \Phi_{s,0}\, ds\, \xi_0 + \Psi_{t,0}^{\varepsilon 0}\bar{P}_0^\varepsilon(t,\alpha)$$
$$\cdot \int_0^t \bar{P}_0^{\varepsilon-1}(s,\alpha)\Psi_{0,s}^{\varepsilon 0}(D_s F_s^T + \bar{P}_s^\varepsilon(\alpha)C_s^T)(F_s F_s^T + \varepsilon I)^{-1} C_s \Phi_{s,0}\, ds$$
$$\cdot \int_0^t \Phi_{0,s} D_s\, d\mathbf{w}_s + \zeta_t^\varepsilon(\alpha). \tag{8.16}$$

By Theorem 7.1, $E\tilde{\xi}_t^\varepsilon = E\xi_t$. Then taking the expected values of both sides of (8.15) yields

$$\Phi_{t,0} = \frac{1}{\alpha}\Psi_{t,0}^{\varepsilon 0}\bar{P}_0^\varepsilon(t,\alpha) + \Psi_{t,0}^{\varepsilon 0}\bar{P}_0^\varepsilon(t,\alpha)\int_0^t \bar{P}_0^{\varepsilon-1}(s,\alpha)\Psi_{0,s}^{\varepsilon 0}(D_s F_s^T + \bar{P}_s^\varepsilon(\alpha)C_s^T)$$
$$\cdot (F_s F_s^T + \varepsilon I)^{-1} C_s \Phi_{s,0}\, ds \tag{8.17}$$

since $\mathbf{x}$ in (8.15) is an arbitrary vector.

From (8.16) and (8.17), it follows that

$$\tilde{\xi}_t^\varepsilon = \xi_t - \frac{1}{\alpha}\Psi_{t,0}^{\varepsilon 0}\bar{P}_0^\varepsilon(t,\alpha)\left(\xi_0 - \mathbf{x} + \int_0^t \Phi_{0,\lambda} D_\lambda\, d\mathbf{w}_\lambda\right) + \zeta_t^\varepsilon(\alpha). \tag{8.18}$$

In (7.27), instead of P_t^ε, put $\bar{P}_t^\varepsilon(\alpha)$ and denote $\delta_t^\varepsilon(\alpha)$. Then by (7.28)

$$\hat{\xi}_t^\varepsilon = \tilde{\xi}_t^\varepsilon + \delta_t^\varepsilon(\alpha). \tag{8.19}$$

Notice $\hat{\xi}_t^\varepsilon$ is the LUMVE of ξ_t based on $\{\eta_\lambda^\varepsilon, 0 \le \lambda \le t\}$ hence (6.67) becomes

$$\hat{\xi}_t^\varepsilon = \Psi_{t,s}^{\varepsilon s}\hat{\xi}_s^\varepsilon(t) + \Psi_{t,s}^{\varepsilon s}\int_s^t \Psi_{s,\lambda}^{\varepsilon s} K_\lambda^{\varepsilon s}\, d\eta_\lambda^\varepsilon, \tag{8.20}$$

which, after taking expectations, yields

$$\Phi_{t,0} = \Psi_{t,s}^{\varepsilon s}\Phi_{s,0} + \Psi_{t,s}^{\varepsilon s}\int_s^t \Psi_{s,\lambda}^{\varepsilon s} K_\lambda^{\varepsilon s} C_\lambda \Phi_{\lambda,0}\, d\lambda,$$

because $\boldsymbol{\xi}_t^\varepsilon$ is unbiased for $\boldsymbol{\xi}_t$ and $\mathbf{x}$ is arbitrary. The last formula is equivalent to

$$\int_s^t \boldsymbol{\Psi}_{s,\lambda}^{\varepsilon s}\mathbf{K}_\lambda^{\varepsilon s}\mathbf{C}_\lambda\boldsymbol{\Phi}_{\lambda,0}\,d\lambda = \boldsymbol{\Psi}_{s,t}^{\varepsilon s}\left(\boldsymbol{\Phi}_{t,0} - \boldsymbol{\Psi}_{t,s}^{\varepsilon s}\boldsymbol{\Phi}_{s,0}\right). \tag{8.21}$$

By the Ito formula and (8.21), this leads to

$$\begin{aligned}
&\int_s^t \boldsymbol{\Psi}_{s,\lambda}^{\varepsilon s}\mathbf{K}_\lambda^{\varepsilon s}\mathbf{C}_\lambda\boldsymbol{\Phi}_{\lambda,0}\,d\lambda \int_s^t \boldsymbol{\Phi}_{0,\tau}\mathbf{D}_\tau\,d\mathbf{w}_\tau \\
&\quad= \int_s^t d_\mu\left(\int_s^\mu \boldsymbol{\Psi}_{s,\lambda}^{\varepsilon s}\mathbf{K}_\lambda^{\varepsilon s}\mathbf{C}_\lambda\boldsymbol{\Phi}_{\lambda,0}\,d\lambda \int_s^\mu \boldsymbol{\Phi}_{0,\tau}\mathbf{D}_\tau\,d\mathbf{w}_\tau\right) \\
&\quad= \int_s^t\int_s^\mu \boldsymbol{\Psi}_{s,\lambda}^{\varepsilon s}\mathbf{K}_\lambda^{\varepsilon s}\mathbf{C}_\lambda\boldsymbol{\Phi}_{\lambda,0}\,d\lambda\,\boldsymbol{\Phi}_{0,\mu}\mathbf{D}_\mu\,d\mathbf{w}_\mu \\
&\qquad+ \int_s^t \boldsymbol{\Psi}_{s\mu}^{\varepsilon s}\mathbf{K}_\mu^{\varepsilon s}\mathbf{C}_\mu\boldsymbol{\Phi}_{\mu,0}\int_s^\mu \boldsymbol{\Phi}_{0,\tau}\mathbf{D}_\tau\,d\mathbf{w}_\tau\,d\mu
\end{aligned}$$

or, equivalently,

$$\begin{aligned}
&\int_s^t \boldsymbol{\Psi}_{s,\mu}^{\varepsilon s}\mathbf{K}_\mu^{\varepsilon s}\mathbf{C}_\mu\boldsymbol{\Phi}_{\mu,0}\int_s^\mu \boldsymbol{\Phi}_{0,\tau}\mathbf{D}_\tau\,d\mathbf{w}_\tau\,d\mu \\
&\quad= \int_s^t \boldsymbol{\Psi}_{s,\lambda}^{\varepsilon s}\mathbf{K}_\lambda^{\varepsilon s}\mathbf{C}_\lambda\boldsymbol{\Phi}_{\lambda,0}\,d\lambda \int_s^t \boldsymbol{\Phi}_{0,\tau}\mathbf{D}_\tau\,d\mathbf{w}_\tau \\
&\qquad- \int_s^t\int_s^\mu \boldsymbol{\Psi}_{s,\lambda}^{\varepsilon s}\mathbf{K}_\lambda^{\varepsilon s}\mathbf{C}_\lambda\boldsymbol{\Phi}_{\lambda,0}\,d\lambda\,\boldsymbol{\Phi}_{0,\mu}\mathbf{D}_\mu\,d\mathbf{w}_\mu \\
&\quad= \boldsymbol{\Psi}_{s,t}^{\varepsilon s}\left(\boldsymbol{\Phi}_{t,0} - \boldsymbol{\Psi}_{t,s}^{\varepsilon s}\boldsymbol{\Phi}_{s,0}\right)\int_s^t \boldsymbol{\Phi}_{0,\tau}\mathbf{D}_\tau\,d\mathbf{w}_\tau \\
&\qquad- \int_s^t \boldsymbol{\Psi}_{s,\mu}^{\varepsilon s}\left(\boldsymbol{\Phi}_{\mu,0} - \boldsymbol{\Psi}_{\mu s}^{\varepsilon s}\boldsymbol{\Phi}_{s,0}\right)\boldsymbol{\Phi}_{0\mu}\mathbf{D}_\mu\,d\mathbf{w}_\mu \\
&\quad= \boldsymbol{\Psi}_{s,t}^{\varepsilon s}\int_s^t \boldsymbol{\Phi}_{t,\lambda}\mathbf{D}_\lambda\,d\mathbf{w}_\lambda - \int_s^t \boldsymbol{\Phi}_{s,\lambda}\mathbf{D}_\lambda\,d\mathbf{w}_\lambda \\
&\qquad- \int_s^t \boldsymbol{\Psi}_{s,\mu}^{\varepsilon s}\mathbf{D}_\mu\,d\mathbf{w}_\mu + \int_s^t \boldsymbol{\Phi}_{s,\mu}\mathbf{D}_\mu\,d\mathbf{w}_\mu \\
&\quad= \boldsymbol{\Psi}_{s,t}^{\varepsilon s}\int_s^t \boldsymbol{\Phi}_{t,\lambda}\mathbf{D}_\lambda\,d\mathbf{w}_\lambda - \int_s^t \boldsymbol{\Psi}_{s,\lambda}^{\varepsilon s}\mathbf{D}_\lambda\,d\mathbf{w}_\lambda.
\end{aligned} \tag{8.22}$$

Then from (8.21) and (8.22), we obtain

$$\int_s^t \underset{\sim}{\Psi}_{s,\mu}^{\varepsilon s} \underset{\sim}{\mathbf{K}}_\mu^{\varepsilon s} \underset{\sim}{\mathbf{C}}_\mu \int_s^\mu \underset{\sim}{\Phi}_{\mu,\lambda} \underset{\sim}{\mathbf{D}}_\lambda \, d\mathbf{w}_\lambda \, d\mu + \int_s^t \underset{\sim}{\Psi}_{s,\lambda}^{\varepsilon s} \underset{\sim}{\mathbf{K}}_\lambda^{\varepsilon s} \underset{\sim}{\mathbf{F}}_\lambda \, d\mathbf{w}_\lambda$$

$$= \underset{\sim}{\Psi}_{s,t}^{\varepsilon s} (\Phi_{t,0} - \underset{\sim}{\Psi}_{t,s}^{\varepsilon s} \underset{\sim}{\Phi}_{s,0}) \Phi_{0,s} \xi_s + \underset{\sim}{\Psi}_{s,t}^{\varepsilon s} \int_s^t \underset{\sim}{\Phi}_{t,\lambda} \underset{\sim}{\mathbf{D}}_\lambda \, d\mathbf{w}_\lambda$$

$$- \int_s^t \underset{\sim}{\Psi}_{s,\lambda}^{\varepsilon s} \underset{\sim}{\mathbf{D}}_\lambda \, d\mathbf{w}_\lambda + \int_s^t \underset{\sim}{\Psi}_{s,\lambda}^{\varepsilon s} \underset{\sim}{\mathbf{K}}_\lambda^{\varepsilon s} \underset{\sim}{\mathbf{F}}_\lambda \, d\mathbf{w}_\lambda$$

$$= \underset{\sim}{\Psi}_{s,t}^{\varepsilon s} \xi_t - \xi_s - \int_s^t \underset{\sim}{\Psi}_{s,\lambda}^{\varepsilon s} \underset{\sim}{\mathbf{D}}_\lambda \, d\mathbf{w}_\lambda + \int_s^t \underset{\sim}{\Psi}_{s,\lambda}^{\varepsilon s} \underset{\sim}{\mathbf{K}}_\lambda^{\varepsilon s} \underset{\sim}{\mathbf{F}}_\lambda \, d\mathbf{w}_\lambda, \qquad (8.23)$$

and from this and (8.18), (8.19), we have

$$\xi_s^\varepsilon(t) = \underset{\sim}{\Psi}_{s,t}^{\varepsilon s} \xi_t^\varepsilon - \int_s^t \underset{\sim}{\Psi}_{s,\lambda}^{\varepsilon s} \underset{\sim}{\mathbf{K}}_\lambda^{\varepsilon s} \, d\eta_\lambda - \sqrt{\varepsilon} \int_s^t \underset{\sim}{\Psi}_{s,\lambda}^{\varepsilon s} \underset{\sim}{\mathbf{K}}_\lambda^{\varepsilon s} \, d\overline{\mathbf{w}}_\lambda$$

$$= \underset{\sim}{\Psi}_{s,t}^{\varepsilon s} \left[\xi_t - \frac{1}{\alpha} \underset{\sim}{\Psi}_{t,0}^{\varepsilon 0} \underset{\sim}{\overline{\mathbf{P}}}_0^\varepsilon(t,\alpha) \left(\xi_0 - \mathbf{x} + \int_0^t \Phi_{0,\lambda} \underset{\sim}{\mathbf{D}}_\lambda \, d\mathbf{w}_\lambda \right) + \zeta_t^\varepsilon(\alpha) + \delta_t^\varepsilon(\alpha) \right]$$

$$- \underset{\sim}{\Psi}_{s,t}^{\varepsilon s} \xi_s + \xi_s + \int_s^t \underset{\sim}{\Psi}_{s,\lambda}^{\varepsilon s} \underset{\sim}{\mathbf{D}}_\lambda \, d\mathbf{w}_\lambda$$

$$- \int_s^t \underset{\sim}{\Psi}_{s,\lambda}^{\varepsilon s} \underset{\sim}{\mathbf{K}}_\lambda^{\varepsilon s} \underset{\sim}{\mathbf{F}}_\lambda \, d\mathbf{w}_\lambda - \sqrt{\varepsilon} \int_s^t \underset{\sim}{\Psi}_{s,\lambda}^{\varepsilon s} \underset{\sim}{\mathbf{K}}_\lambda^{\varepsilon s} \, d\overline{\mathbf{w}}_\lambda$$

or, equivalently,

$$\xi_s^\varepsilon(t) - \xi_s = -\frac{1}{\alpha} \underset{\sim}{\Psi}_{s,t}^{\varepsilon s} \underset{\sim}{\Psi}_{t,0}^{\varepsilon 0} \underset{\sim}{\overline{\mathbf{P}}}_0^\varepsilon(t,\alpha)(\xi_0 - \mathbf{x})$$

$$- \frac{1}{\alpha} \underset{\sim}{\Psi}_{s,t}^{\varepsilon s} \underset{\sim}{\Psi}_{t,0}^{\varepsilon 0} \underset{\sim}{\overline{\mathbf{P}}}_0^\varepsilon(t,\alpha) \int_0^t \underset{\sim}{\Phi}_{0,\lambda} \underset{\sim}{\mathbf{D}}_\lambda \, d\mathbf{w}_\lambda + \underset{\sim}{\Psi}_{s,t}^{\varepsilon s} \zeta_t^\varepsilon(\alpha)$$

$$- \int_s^t \underset{\sim}{\Psi}_{s,\lambda}^{\varepsilon s} (\underset{\sim}{\mathbf{K}}_\lambda^{\varepsilon s} \underset{\sim}{\mathbf{F}}_\lambda - \underset{\sim}{\mathbf{D}}_\lambda) \, d\mathbf{w}_\lambda + \underset{\sim}{\Psi}_{s,t}^{\varepsilon s} \delta_t^\varepsilon(\alpha) - \sqrt{\varepsilon} \int_s^t \underset{\sim}{\Psi}_{s,\lambda}^{\varepsilon s} \underset{\sim}{\mathbf{K}}_\lambda^{\varepsilon s} \, d\overline{\mathbf{w}}_\lambda.$$

$$(8.24)$$

If on the right-hand side of (8.24) we delete the last two terms, then (8.24) gives the expression for $\tilde{\xi}_s^\varepsilon(t) - \xi_s$. By Theorem 7.2, as $\varepsilon \to 0$, $\tilde{\xi}_s^\varepsilon(t)$ converges in the mean square sense, so by noticing that on the right-hand side of (8.24) the first term is independent of the rest of the terms, we obtain

$$\lim_{\varepsilon_1, \varepsilon_2 \to 0} E \| (\underset{\sim}{\Psi}_{s,t}^{\varepsilon_1 s} \underset{\sim}{\Psi}_{t,0}^{\varepsilon_1 0} \underset{\sim}{\overline{\mathbf{P}}}^{\varepsilon_1}(t,\alpha) - \underset{\sim}{\Psi}_{s,t}^{\varepsilon_2 s} \underset{\sim}{\Psi}_{t,0}^{\varepsilon_2 0} \underset{\sim}{\overline{\mathbf{P}}}_0^{\varepsilon_2}(t,\alpha))(\xi_0 - \mathbf{x}_0) \|^2 = 0.$$

But

$$E(\boldsymbol{\xi}_0 - \mathbf{x})(\boldsymbol{\xi}_0 - \mathbf{x})^{\tau} = \alpha \underset{\sim}{\mathbf{I}},$$

hence the preceding convergence to zero mens that as $\varepsilon \to 0$, $\underset{\sim}{\boldsymbol{\Psi}}_{s,t}^{\varepsilon s} \underset{\sim}{\boldsymbol{\Psi}}_{t,0}^{\varepsilon 0} \underset{\sim}{\overline{\mathbf{P}}}_0^{\varepsilon}(t, \alpha)$ converges to a finite limit.

By Lemma 8.1, as $\varepsilon \to 0$, $\underset{\sim}{\boldsymbol{\Gamma}}_t^{\varepsilon}$ is nondecreasing, hence, if $\lim_{\varepsilon \to 0} \det \underset{\sim}{\boldsymbol{\Gamma}}_t^{\varepsilon} \neq 0$, there is $\varepsilon_1 > 0$ such that for any $\beta > 0$ one can take α_1 sufficiently large so that

$$\beta \underset{\sim}{\boldsymbol{\Gamma}}_t^{\varepsilon} \geq \frac{1}{\alpha} \underset{\sim}{\mathbf{I}}, \qquad \forall \alpha > \alpha_1, \quad \forall \varepsilon \in (0, \varepsilon_1].$$

It then turns out that

$$\begin{aligned}
\frac{1}{(1+\beta)^2} \underset{\sim}{\boldsymbol{\Psi}}_{s,t}^{\varepsilon s} \underset{\sim}{\boldsymbol{\Psi}}_{t,0}^{\varepsilon 0} \underset{\sim}{\boldsymbol{\Gamma}}_t^{\varepsilon - 2} \underset{\sim}{\boldsymbol{\Psi}}_{t,0}^{\varepsilon 0 \tau} \underset{\sim}{\boldsymbol{\Psi}}_{s,t}^{\varepsilon s \tau} &= \underset{\sim}{\boldsymbol{\Psi}}_{s,t}^{\varepsilon s} \underset{\sim}{\boldsymbol{\Psi}}_{t,0}^{\varepsilon 0} \left(\beta \underset{\sim}{\boldsymbol{\Gamma}}_t^{\varepsilon} + \underset{\sim}{\boldsymbol{\Gamma}}_t^{\varepsilon} \right)^{-2} \underset{\sim}{\boldsymbol{\Psi}}_{t,0}^{\varepsilon 0 \tau} \underset{\sim}{\boldsymbol{\Psi}}_{s,t}^{\varepsilon s \tau} \\
&\leq \underset{\sim}{\boldsymbol{\Psi}}_{s,t}^{\varepsilon s} \underset{\sim}{\boldsymbol{\Psi}}_{t,0}^{\varepsilon 0} \left[\underset{\sim}{\overline{\mathbf{P}}}_0^{\varepsilon}(t, \alpha) \right]^2 \underset{\sim}{\boldsymbol{\Psi}}_{t,0}^{\varepsilon 0 \tau} \underset{\sim}{\boldsymbol{\Psi}}_{s,t}^{\varepsilon s \tau} \\
&\leq \underset{\sim}{\boldsymbol{\psi}}_{s,t}^{\varepsilon s} \underset{\sim}{\boldsymbol{\Psi}}_{t,0}^{\varepsilon 0} \underset{\sim}{\boldsymbol{\Gamma}}_t^{\varepsilon - 2} \underset{\sim}{\boldsymbol{\Psi}}_{t,0}^{\varepsilon 0 \tau} \underset{\sim}{\boldsymbol{\Psi}}_{s,t}^{\varepsilon s \tau},
\end{aligned}$$

which leads to

$$\begin{aligned}
\frac{1}{(1+\beta)^2} \overline{\lim_{\varepsilon \to 0}}\, \mathbf{a}^{\tau} \underset{\sim}{\boldsymbol{\Psi}}_{s,t}^{\varepsilon s} \underset{\sim}{\boldsymbol{\Psi}}_{t,0}^{\varepsilon 0} \underset{\sim}{\boldsymbol{\Gamma}}_t^{\varepsilon - 2} \underset{\sim}{\boldsymbol{\Psi}}_{t,0}^{\varepsilon 0 \tau} \underset{\sim}{\boldsymbol{\Psi}}_{s,t}^{\varepsilon s \tau} \mathbf{a} \\
\leq \lim_{\varepsilon \to 0} \mathbf{a}^{\tau} \underset{\sim}{\boldsymbol{\Psi}}_{s,t}^{\varepsilon s} \underset{\sim}{\boldsymbol{\Psi}}_{t,0}^{\varepsilon 0} \left[\underset{\sim}{\overline{\mathbf{P}}}_0^{\varepsilon}(t, \alpha) \right]^2 \underset{\sim}{\boldsymbol{\Psi}}_{t,0}^{\varepsilon 0} \underset{\sim}{\boldsymbol{\Psi}}_{s,t}^{\varepsilon s \tau} \mathbf{a} \\
\leq \underline{\lim_{\varepsilon \to 0}}\, \mathbf{a}^{\tau} \underset{\sim}{\boldsymbol{\Psi}}_{s,t}^{\varepsilon s} \underset{\sim}{\boldsymbol{\Psi}}_{t,0}^{\varepsilon 0} \underset{\sim}{\boldsymbol{\Gamma}}_t^{\varepsilon - 2} \underset{\sim}{\boldsymbol{\Psi}}_{t,0}^{\varepsilon 0 \tau} \underset{\sim}{\boldsymbol{\Psi}}_{s,t}^{\varepsilon s \tau} \mathbf{a},
\end{aligned} \tag{8.25}$$

where $\mathbf{a}$ is any n-dimensional vector, and recall $\underset{\sim}{\boldsymbol{\Gamma}}_t^{\varepsilon - 2} = (\underset{\sim}{\boldsymbol{\Gamma}}_t^{\varepsilon})^{-2}$.

Letting $\beta \to 0$, the arbitrary nature of $\mathbf{a}$ and the preceding inequalities show that as $\varepsilon \to 0$, $\underset{\sim}{\boldsymbol{\Psi}}_{s,t}^{\varepsilon s} \underset{\sim}{\boldsymbol{\Psi}}_{t,0}^{\varepsilon 0} \underset{\sim}{\boldsymbol{\Gamma}}_t^{\varepsilon - 1}$ converges to a finite limit, and from the second inequality of (8.25), it is clear that $\underset{\sim}{\boldsymbol{\Psi}}_{s,t}^{\varepsilon s} \underset{\sim}{\boldsymbol{\Psi}}_{t,0}^{\varepsilon 0} \underset{\sim}{\overline{\mathbf{P}}}_0^{\varepsilon}(t, \alpha)$ is bounded on $\alpha > \alpha_1$, $\varepsilon \in (0, \varepsilon_1]$.

In (8.15), replace $\boldsymbol{\eta}_s$ by $\mathbf{y}_s$ and put $\mathbf{x} = \mathbf{0}$. By definition, it turns into

$$\begin{aligned}
\overline{\mathbf{x}}_t^{\varepsilon}(\alpha) = \underset{\sim}{\boldsymbol{\Psi}}_{t,0}^{\varepsilon 0} \underset{\sim}{\overline{\mathbf{P}}}_0^{\varepsilon}(t, \alpha) \int_0^t \underset{\sim}{\overline{\mathbf{P}}}_0^{\varepsilon - 1}(s, \alpha) \underset{\sim}{\boldsymbol{\Psi}}_{0,s}^{\varepsilon 0} \left(\underset{\sim}{\mathbf{D}}_s \underset{\sim}{\mathbf{F}}_s^{\tau} + \underset{\sim}{\overline{\mathbf{P}}}_s^{\varepsilon}(\alpha) \underset{\sim}{\mathbf{C}}_s^{\tau} \right) \\
\cdot \left(\underset{\sim}{\mathbf{F}}_s \underset{\sim}{\mathbf{F}}_s^{\tau} + \varepsilon \underset{\sim}{\mathbf{I}} \right)^{-1} d\mathbf{y}_s.
\end{aligned} \tag{8.26}$$

Then from here and from the derivative of (8.24), it is easy to see that if in (8.24) we remove the last two terms, take $\mathbf{x} = \mathbf{0}$ and replace $\boldsymbol{\xi}_s$, $\boldsymbol{\xi}_0$ by $\mathbf{x}_s$ and $\mathbf{x}_0$, respectively, we obtain the following formula for $\bar{\mathbf{x}}_s^\varepsilon(t,\alpha)$:

$$\bar{\mathbf{x}}_s^\varepsilon(t,\alpha) = \mathbf{x}_s - \frac{1}{\alpha}\boldsymbol{\Psi}_{s,t}^{\varepsilon 0}\boldsymbol{\Psi}_{t,0}^{\varepsilon 0}\bar{\mathbf{P}}_0^\varepsilon(t,\alpha)\left(\mathbf{x}_0 + \int_0^t \boldsymbol{\Phi}_{0,\lambda}\mathbf{D}_\lambda\, d\mathbf{w}_\lambda\right)$$

$$+\boldsymbol{\Psi}_{s,t}^{\varepsilon s}\boldsymbol{\zeta}_t^\varepsilon(\alpha) - \int_s^t \boldsymbol{\Psi}_{s,t}^{\varepsilon s}(\mathbf{K}_\lambda^{\varepsilon s}\mathbf{F}_\lambda - \mathbf{D}_\lambda)\, d\mathbf{w}_\lambda. \tag{8.27}$$

According to Theorem 7.2, $\boldsymbol{\xi}_s^\varepsilon(t)$ converges in the mean square sense as $\varepsilon \to 0$. Hence so does

$$\boldsymbol{\Psi}_{s,t}^{\varepsilon s}\boldsymbol{\zeta}_t^\varepsilon(\alpha) - \int_s^t \boldsymbol{\Psi}_{s,\lambda}^{\varepsilon s}(\mathbf{K}_\lambda^{\varepsilon s}\mathbf{F}_\lambda - \mathbf{D}_\lambda)\, d\mathbf{w}_\lambda + \boldsymbol{\Psi}_{s,t}^{\varepsilon s}\boldsymbol{\delta}_t^\varepsilon(\alpha)$$

since $\{\bar{\mathbf{w}}_\lambda\}$, $\{\mathbf{w}_\lambda\}$, and $\{\boldsymbol{\xi}_0\}$ are mutually independent. Then, from (8.27), it follows that as $\varepsilon \to 0$, $\bar{\mathbf{x}}_s^\varepsilon(t,\alpha)$ converges in the mean square sense. □

Remark. In this lemma, the condition

$$\lim_{\varepsilon\to 0} \det \boldsymbol{\Gamma}_t^\varepsilon \neq 0 \tag{8.28}$$

is an important property of the system which will be discussed in detail later on.

For sufficiently small $\varepsilon > 0$, denote

$$\boldsymbol{\delta}_t^\varepsilon = \sqrt{\varepsilon}\,\boldsymbol{\Psi}_{t,0}^{\varepsilon 0}\boldsymbol{\Gamma}_t^{\varepsilon -1}\int_0^t \boldsymbol{\Gamma}_\lambda^\varepsilon\boldsymbol{\Psi}_{0,\lambda}^{\varepsilon 0}\left[\mathbf{D}_\lambda\mathbf{F}_\lambda^\tau + (\mathbf{P}_\lambda^{\varepsilon 0} + \boldsymbol{\Psi}_{\lambda,0}^{\varepsilon 0}\boldsymbol{\Gamma}_\lambda^{\varepsilon +}\boldsymbol{\Psi}_{\lambda,0}^{\varepsilon 0\tau})\mathbf{C}_\lambda^\tau\right]$$

$$\cdot(\mathbf{F}_\lambda\mathbf{F}_\lambda^\tau + \varepsilon\mathbf{I})^{-1}\, d\bar{\mathbf{w}}_\lambda, \tag{8.29}$$

$$\mathbf{h}_s^\varepsilon(t) = \boldsymbol{\Psi}_{s,t}^{\varepsilon s}\boldsymbol{\delta}_t^\varepsilon - \sqrt{\varepsilon}\int_s^t \boldsymbol{\Psi}_{s,\lambda}^{\varepsilon s}\mathbf{K}_\lambda^{\varepsilon s}\, d\bar{\mathbf{w}}_\lambda. \tag{8.30}$$

Lemma 8.3. *Assume* (8.28) *holds. Then*

$$\boldsymbol{\delta}_t^\varepsilon(\alpha) \underset{\alpha\to\infty}{\rightarrow} \boldsymbol{\delta}_t^\varepsilon \tag{8.31}$$

in the mean square sense and almost surely and

$$\underset{\varepsilon\to 0}{\text{l.i.m.}}\, \mathbf{h}_s^{\varepsilon}(t) = \mathbf{0}, \qquad s \in [0, t], \tag{8.32}$$

where $\boldsymbol{\delta}_t^{\varepsilon}(\alpha)$ *is given in* (8.19).

Proof. From (7.27) and (8.14), we have

$$\boldsymbol{\delta}_t^{\varepsilon}(\alpha) = \sqrt{\varepsilon}\, \underset{\sim}{\boldsymbol{\Psi}}_{t,0}^{\varepsilon 0} \underset{\sim}{\overline{\mathbf{P}}}_0^{\varepsilon}(t,\alpha) \int_0^t \overline{\mathbf{P}}_0^{\varepsilon -1}(\lambda,\alpha) \underset{\sim}{\boldsymbol{\Psi}}_{0,\lambda}^{\varepsilon 0}$$
$$\times \left[\underset{\sim}{\mathbf{D}}_\lambda \underset{\sim}{\mathbf{F}}_\lambda^{\tau} + \left(\underset{\sim}{\mathbf{P}}_\lambda^{\varepsilon 0} + \underset{\sim}{\boldsymbol{\Psi}}_{\lambda,0}^{\varepsilon 0} \underset{\sim}{\overline{\mathbf{P}}}_0^{\varepsilon}(\lambda,\alpha) \underset{\sim}{\boldsymbol{\Psi}}_{\lambda,0}^{\varepsilon 0\tau} \right) \underset{\sim}{\mathbf{C}}_\lambda^{\tau} \right] \left(\underset{\sim}{\mathbf{F}}_\lambda \underset{\sim}{\mathbf{F}}_\lambda^{\tau} + \varepsilon \underset{\sim}{\mathbf{I}} \right)^{-1} d\overline{\mathbf{w}}_\lambda,$$

and by (8.10) and (8.11),

$$\boldsymbol{\delta}_t^{\varepsilon}(\alpha) = \frac{\sqrt{\varepsilon}}{\alpha} \underset{\sim}{\boldsymbol{\Psi}}_{t,0}^{\varepsilon 0} \underset{\sim}{\overline{\mathbf{P}}}_0^{\varepsilon}(t,\alpha) \int_0^t \underset{\sim}{\boldsymbol{\Psi}}_{0,\lambda}^{\varepsilon 0} \underset{\sim}{\mathbf{K}}_\lambda^{\varepsilon 0}\, d\overline{\mathbf{w}}_\lambda + \sqrt{\varepsilon}\, \underset{\sim}{\boldsymbol{\Psi}}_{t,0}^{\varepsilon 0} \underset{\sim}{\overline{\mathbf{P}}}_0^{\varepsilon}(t,\alpha) \int_0^t \underset{\sim}{\boldsymbol{\Gamma}}_\lambda^{\varepsilon} \underset{\sim}{\boldsymbol{\Psi}}_{0,\lambda}^{\varepsilon 0} \underset{\sim}{\mathbf{K}}_\lambda^{\varepsilon 0}\, d\overline{\mathbf{w}}_\lambda$$
$$+ \sqrt{\varepsilon}\, \underset{\sim}{\boldsymbol{\Psi}}_{t,0}^{\varepsilon 0} \underset{\sim}{\overline{\mathbf{P}}}_0^{\varepsilon}(t,\alpha) \int_0^t \underset{\sim}{\boldsymbol{\Psi}}_{\lambda,0}^{\varepsilon 0} \underset{\sim}{\mathbf{C}}_\lambda^{\tau} \left(\underset{\sim}{\mathbf{F}}_\lambda \underset{\sim}{\mathbf{F}}_\lambda^{\tau} + \varepsilon \underset{\sim}{\mathbf{I}} \right)^{-1} d\overline{\mathbf{w}}_\lambda$$
$$= \frac{\sqrt{\varepsilon}}{\alpha} \underset{\sim}{\boldsymbol{\Psi}}_{t,0}^{\varepsilon 0} \underset{\sim}{\overline{\mathbf{P}}}_0^{\varepsilon}(t,\alpha) \int_0^t \underset{\sim}{\boldsymbol{\Psi}}_{0,\lambda}^{\varepsilon 0} \underset{\sim}{\mathbf{K}}_\lambda^{\varepsilon 0}\, d\overline{\mathbf{w}}_\lambda$$
$$+ \sqrt{\varepsilon}\, \underset{\sim}{\boldsymbol{\Psi}}_{t,0}^{\varepsilon 0} \underset{\sim}{\overline{\mathbf{P}}}_0^{\varepsilon}(t,\alpha) \int_0^t \underset{\sim}{\boldsymbol{\Gamma}}_\lambda^{\varepsilon} \underset{\sim}{\boldsymbol{\Psi}}_{0,\lambda}^{\varepsilon 0} \left[\underset{\sim}{\mathbf{D}}_\lambda \underset{\sim}{\mathbf{F}}_\lambda^{\tau} + \left(\underset{\sim}{\mathbf{P}}_\lambda^{\varepsilon 0} + \underset{\sim}{\boldsymbol{\Psi}}_{\lambda,0}^{\varepsilon 0} \underset{\sim}{\boldsymbol{\Gamma}}_\lambda^{\varepsilon +} \underset{\sim}{\boldsymbol{\Psi}}_{\lambda,0}^{\varepsilon 0\tau} \right) \underset{\sim}{\mathbf{C}}_\lambda^{\tau} \right]$$
$$\cdot \left(\underset{\sim}{\mathbf{F}}_\lambda \underset{\sim}{\mathbf{F}}_\lambda^{\tau} + \varepsilon \underset{\sim}{\mathbf{I}} \right)^{-1} d\overline{\mathbf{w}}_\lambda \underset{\alpha\to\infty}{\longrightarrow} \boldsymbol{\delta}_t^{\varepsilon}. \tag{8.33}$$

By use of the expression for $\boldsymbol{\delta}_t^{\varepsilon}(\alpha)$ in (8.33), it is easy to see that

$$E\left\| \boldsymbol{\delta}_t^{\varepsilon}(\alpha) - \delta_t^{\varepsilon} \right\|^2 \underset{\alpha\to\infty}{\longrightarrow} 0.$$

Thus (8.31) has been proved.

Denote

$$\mathbf{h}_s^{\varepsilon}(t,\alpha) = \underset{\sim}{\boldsymbol{\Psi}}_{s,t}^{\varepsilon s} \delta_t^{\varepsilon}(\alpha) - \sqrt{\varepsilon} \int_s^t \underset{\sim}{\boldsymbol{\Psi}}_{s,\lambda}^{\varepsilon s} \underset{\sim}{\mathbf{K}}_\lambda^{\varepsilon s}\, d\overline{\mathbf{w}}_\lambda. \tag{8.34}$$

From (8.24) and Theorem 7.2, we have

$$\underset{\varepsilon\to 0}{\text{l.i.m.}}\, \mathbf{h}_s^{\varepsilon}(t,\alpha) = \underset{\varepsilon\to 0}{\text{l.i.m.}} \left(\boldsymbol{\xi}_s^{\varepsilon}(t) - \underset{\sim}{\boldsymbol{\xi}}_s^{\varepsilon}(t) \right) = \mathbf{0}, \qquad \forall s \in [0,t], \tag{8.35}$$

and from (8.29) and (8.33)

$$\boldsymbol{\delta}_t^\varepsilon = \underset{\sim}{\boldsymbol{\Psi}}_{t,0}^{\varepsilon 0}\boldsymbol{\Gamma}_t^{\varepsilon-1}\underset{\sim}{\overline{\mathbf{P}}}_0^{\varepsilon-1}(t,\alpha)\underset{\sim}{\boldsymbol{\Psi}}_{t,0}^{\varepsilon 0}\boldsymbol{\delta}_t^\varepsilon(\alpha) - \frac{\sqrt{\varepsilon}}{\alpha}\underset{\sim}{\boldsymbol{\Psi}}_{t,0}^{\varepsilon 0}\underset{\sim}{\boldsymbol{\Gamma}}_t^{\varepsilon-1}\int_0^t \underset{\sim}{\boldsymbol{\Psi}}_{0,\lambda}^{\varepsilon 0}\underset{\sim}{\mathbf{K}}_\lambda^{\varepsilon 0}\,d\overline{\mathbf{w}}_\lambda$$

$$= \boldsymbol{\delta}_t^\varepsilon(\alpha) + \frac{1}{\alpha}\underset{\sim}{\boldsymbol{\Psi}}_{t,0}^{\varepsilon 0}\underset{\sim}{\boldsymbol{\Gamma}}_t^{\varepsilon-1}\underset{\sim}{\boldsymbol{\Psi}}_{0,t}^{\varepsilon 0}\boldsymbol{\delta}_t^\varepsilon(\alpha) - \frac{\sqrt{\varepsilon}}{\alpha}\underset{\sim}{\boldsymbol{\Psi}}_{t,0}^{\varepsilon 0}\underset{\sim}{\boldsymbol{\Gamma}}_t^{\varepsilon-1}\int_0^t \underset{\sim}{\boldsymbol{\Psi}}_{0,\lambda}^{\varepsilon 0}\underset{\sim}{\mathbf{K}}_\lambda^{\varepsilon 0}\,d\overline{\mathbf{w}}_\lambda$$

$$= \boldsymbol{\delta}_t^\varepsilon(\alpha) + \frac{1}{\alpha}\underset{\sim}{\boldsymbol{\Psi}}_{t,0}^{\varepsilon 0}\underset{\sim}{\boldsymbol{\Gamma}}_t^{\varepsilon-1}\left[\underset{\sim}{\boldsymbol{\Psi}}_{0,t}^{\varepsilon 0}\boldsymbol{\delta}_t^\varepsilon(\alpha) - \sqrt{\varepsilon}\int_0^t \underset{\sim}{\boldsymbol{\Psi}}_{0,\lambda}^{\varepsilon 0}\underset{\sim}{\mathbf{K}}_\lambda^{\varepsilon 0}\,d\overline{\mathbf{w}}_\lambda\right]$$

$$= \boldsymbol{\delta}_t^\varepsilon(\alpha) + \frac{1}{\alpha}\underset{\sim}{\boldsymbol{\Psi}}_{t,0}^{\varepsilon 0}\underset{\sim}{\boldsymbol{\Gamma}}_t^{\varepsilon-1}\mathbf{h}_0^\varepsilon(t,\alpha), \tag{8.36}$$

hence by (8.30), (8.34), and (8.36), we obtain

$$\mathbf{h}_s^\varepsilon(t) = \underset{\sim}{\boldsymbol{\Psi}}_{s,t}^{\varepsilon s}\boldsymbol{\delta}_t^\varepsilon(\alpha) + \frac{1}{\alpha}\underset{\sim}{\boldsymbol{\Psi}}_{s,t}^{\varepsilon s}\underset{\sim}{\boldsymbol{\Psi}}_{t,0}^{\varepsilon 0}\underset{\sim}{\boldsymbol{\Gamma}}_t^{\varepsilon-1}\mathbf{h}_0^\varepsilon(t,\alpha)$$

$$-\sqrt{\varepsilon}\int_s^t \underset{\sim}{\boldsymbol{\Psi}}_{s,\lambda}^{\varepsilon s}\underset{\sim}{\mathbf{K}}_\lambda^{\varepsilon s}\,d\overline{\mathbf{w}}_\lambda = \mathbf{h}_s^\varepsilon(t,\alpha) + \frac{1}{\alpha}\underset{\sim}{\boldsymbol{\Psi}}_{s,t}^{\varepsilon s}\underset{\sim}{\boldsymbol{\Psi}}_{t,0}^{\varepsilon 0}\underset{\sim}{\boldsymbol{\Gamma}}_t^{\varepsilon-1}\mathbf{h}_0^\varepsilon(t,\alpha), \tag{8.37}$$

which yields (8.32) by (8.35) since, by Lemma 8.2, for any $s \in [0, t]$ as $\varepsilon \to 0$ $\underset{\sim}{\boldsymbol{\Psi}}_{s,t}^{\varepsilon s}\underset{\sim}{\boldsymbol{\Psi}}_{t,0}^{\varepsilon 0}\underset{\sim}{\boldsymbol{\Gamma}}_t^{\varepsilon-1}$ converges to a finite limit. □

Lemma 8.4. *Assume* (8.28) *holds. Then*

$$\boldsymbol{\zeta}_t^\varepsilon(\alpha) \underset{\alpha\to\infty}{\to} \boldsymbol{\zeta}_t^\varepsilon \tag{8.38}$$

in the mean square sense and almost surely. As $\varepsilon \to 0$, $\underset{\sim}{\boldsymbol{\Psi}}_{s,t}^{\varepsilon s}(\boldsymbol{\zeta}_t^\varepsilon(\alpha) - \boldsymbol{\zeta}_t^\varepsilon)$ *converges in the mean square sense and*

$$\lim_{\alpha\to\infty}\ \underset{\varepsilon\to 0}{\text{l.i.m.}}\ \underset{\sim}{\boldsymbol{\Psi}}_{s,t}^{\varepsilon s}(\boldsymbol{\zeta}_t^\varepsilon(\alpha) - \boldsymbol{\zeta}_t^\varepsilon) = \mathbf{0}, \tag{8.39}$$

where $\boldsymbol{\zeta}_t^\varepsilon(\alpha)$ *is defined by* (8.13) *and*

$$\boldsymbol{\zeta}_t^\varepsilon = \underset{\sim}{\boldsymbol{\Psi}}_{t,0}^{\varepsilon 0}\underset{\sim}{\boldsymbol{\Gamma}}_t^{\varepsilon-1}\int_0^t\Big\{\underset{\sim}{\boldsymbol{\Gamma}}_\lambda^\varepsilon\underset{\sim}{\boldsymbol{\Psi}}_{0,\lambda}^{\varepsilon 0}\big[\mathbf{D}_\lambda\underset{\sim}{\mathbf{F}}_\lambda^T + \big(\underset{\sim}{\mathbf{P}}_\lambda^{\varepsilon 0} + \underset{\sim}{\boldsymbol{\Psi}}_{\lambda,0}^{\varepsilon 0}\underset{\sim}{\boldsymbol{\Gamma}}_\lambda^{\varepsilon +}\underset{\sim}{\boldsymbol{\Psi}}_{\lambda,0}^{\varepsilon 0 T}\big)\underset{\sim}{\mathbf{C}}_\lambda^T\big]$$

$$\cdot\big(\mathbf{F}_\lambda\underset{\sim}{\mathbf{F}}_\lambda^T + \varepsilon\underset{\sim}{\mathbf{I}}\big)^{-1}\underset{\sim}{\mathbf{F}}_\lambda - \underset{\sim}{\boldsymbol{\Gamma}}_\lambda^\varepsilon\underset{\sim}{\boldsymbol{\Psi}}_{0,\lambda}^{\varepsilon 0}\underset{\sim}{\mathbf{D}}_\lambda\Big\}\,d\mathbf{w}_\lambda \tag{8.40}$$

Proof. By (8.9)–(8.11) it is shown that for almost all $s \in [0, T]$

$$\bar{\mathbf{P}}_0^{\varepsilon-1}(\lambda, \alpha)\boldsymbol{\Psi}_{0,\lambda}^{\varepsilon 0}(\mathbf{D}_\lambda \mathbf{F}_\lambda^\tau + \bar{\mathbf{P}}_\lambda^\varepsilon(\alpha)\mathbf{C}_\lambda^\tau)(\mathbf{F}_\lambda \mathbf{F}_\lambda^\tau + \varepsilon \mathbf{I})^{-1}$$

$$= \bar{\mathbf{P}}_0^{\varepsilon-1}(\lambda, \alpha)\boldsymbol{\Psi}_{0,\lambda}^{\varepsilon 0}$$

$$\times\left[\mathbf{D}_\lambda \mathbf{F}_\lambda^\tau + \left(\mathbf{P}_\lambda^{\varepsilon 0} + \boldsymbol{\Psi}_{\lambda,0}^{\varepsilon 0}\bar{\mathbf{P}}_0^\varepsilon(\lambda, \alpha)\boldsymbol{\Psi}_{\lambda,0}^{\varepsilon 0 \tau}\right)\mathbf{C}_\lambda^\tau\right](\mathbf{F}_\lambda \mathbf{F}_\lambda^\tau + \varepsilon \mathbf{I})^{-1}$$

$$= \bar{\mathbf{P}}_0^{\varepsilon-1}(\lambda, \alpha)\boldsymbol{\Psi}_{0,\lambda}^{\varepsilon 0}(\mathbf{D}_\lambda \mathbf{F}_\lambda^\tau + \mathbf{P}_\lambda^{\varepsilon 0}\mathbf{C}_\lambda^\tau)(\mathbf{F}_\lambda \mathbf{F}_\lambda^\tau + \varepsilon \mathbf{I})^{-1}$$

$$+ \boldsymbol{\Psi}_{\lambda,0}^{\varepsilon 0 \tau}\mathbf{C}_\lambda^\tau(\mathbf{F}_\lambda \mathbf{F}_\lambda^\tau + \varepsilon \mathbf{I})^{-1}$$

$$= \frac{1}{\alpha}\boldsymbol{\Psi}_{0,\lambda}^{\varepsilon 0}\mathbf{K}_\lambda^{\varepsilon 0} + \boldsymbol{\Gamma}_\lambda^\varepsilon \boldsymbol{\Psi}_{0,\lambda}^{\varepsilon 0}\mathbf{K}_\lambda^{\varepsilon 0} + \boldsymbol{\Gamma}_\lambda^\varepsilon \boldsymbol{\Gamma}_\lambda^{\varepsilon +}\boldsymbol{\Psi}_{\lambda,0}^{\varepsilon 0 \tau}\mathbf{C}_\lambda^\tau(\mathbf{F}_\lambda \mathbf{F}_\lambda^\tau + \varepsilon \mathbf{I})^{-1}$$

$$= \frac{1}{\alpha}\boldsymbol{\Psi}_{0,\lambda}^{\varepsilon 0}\mathbf{K}_\lambda^{\varepsilon 0} + \boldsymbol{\Gamma}_\lambda^\varepsilon \boldsymbol{\Psi}_{0,\lambda}^{\varepsilon 0}\left[\mathbf{K}_\lambda^{\varepsilon 0} + \boldsymbol{\Psi}_{\lambda,0}^{\varepsilon 0}\boldsymbol{\Gamma}_\lambda^{\varepsilon +}\boldsymbol{\Psi}_{\lambda,0}^{\varepsilon 0 \tau}\mathbf{C}_\lambda^\tau(\mathbf{F}_\lambda \mathbf{F}_\lambda^\tau + \varepsilon \mathbf{I})^{-1}\right]$$

$$= \frac{1}{\alpha}\boldsymbol{\Psi}_{0,\lambda}^{\varepsilon 0}\mathbf{K}_\lambda^{\varepsilon 0} + \boldsymbol{\Gamma}_\lambda^\varepsilon \boldsymbol{\Psi}_{0,\lambda}^{\varepsilon 0}\left[\mathbf{D}_\lambda \mathbf{F}_\lambda^\tau + \left(\mathbf{P}_\lambda^{\varepsilon 0} + \boldsymbol{\Psi}_{\lambda,0}^{\varepsilon 0}\boldsymbol{\Gamma}_\lambda^{\varepsilon +}\boldsymbol{\Psi}_{\lambda,0}^{\varepsilon 0 \tau}\right)\mathbf{C}_\lambda^\tau\right]$$

$$\cdot(\mathbf{F}_\lambda \mathbf{F}_\lambda^\tau + \varepsilon \mathbf{I})^{-1}. \tag{8.41}$$

Substituting (8.17) into (8.13) yields

$$\boldsymbol{\zeta}_t^\varepsilon(\alpha) = \boldsymbol{\Psi}_{t,0}^{\varepsilon 0}\bar{\mathbf{P}}_0^\varepsilon(t, \alpha)\int_0^t \bar{\mathbf{P}}_0^{\varepsilon-1}(s, \alpha)\boldsymbol{\Psi}_{0,s}^{\varepsilon 0}(\mathbf{D}_s \mathbf{F}_s^\tau + \bar{\mathbf{P}}_s^\varepsilon(\alpha)\mathbf{C}_s^\tau)(\mathbf{F}_s \mathbf{F}_s^\tau + \varepsilon \mathbf{I})^{-1}\mathbf{F}_s\, d\mathbf{w}_s$$

$$- \boldsymbol{\Psi}_{t,0}^{\varepsilon 0}\bar{\mathbf{P}}_0^\varepsilon(t, \alpha)\left[\int_0^t \bar{\mathbf{P}}_0^{\varepsilon-1}(s, \alpha)\boldsymbol{\Psi}_{0,s}^{\varepsilon 0}\left(\boldsymbol{\Phi}_{s,0} - \frac{1}{\alpha}\boldsymbol{\Psi}_{s,0}^{\varepsilon 0}\bar{\mathbf{P}}_0^\varepsilon(s, \alpha)\right)\boldsymbol{\Phi}_{0,s}\mathbf{D}_s\, d\mathbf{w}_s\right].$$

By (8.41), this becomes

$$\boldsymbol{\zeta}_t^\varepsilon(\alpha) = \boldsymbol{\Psi}_{t,0}^{\varepsilon 0}\bar{\mathbf{P}}_0^\varepsilon(t, \alpha)\int_0^t \boldsymbol{\Gamma}_\lambda^\varepsilon \boldsymbol{\Psi}_{0,\lambda}^{\varepsilon 0}\left[\mathbf{D}_\lambda \mathbf{F}_\lambda^\tau + \left(\mathbf{P}_\lambda^{\varepsilon 0} + \boldsymbol{\Psi}_{\lambda,0}^{\varepsilon 0}\boldsymbol{\Gamma}_\lambda^{\varepsilon +}\boldsymbol{\Psi}_{\lambda,0}^{\varepsilon 0 \tau}\right)\mathbf{C}_\lambda^\tau\right]$$

$$\cdot(\mathbf{F}_\lambda \mathbf{F}_\lambda^\tau + \varepsilon \mathbf{I})^{-1}\mathbf{F}_\lambda\, d\mathbf{w}_\lambda + \frac{1}{\alpha}\boldsymbol{\Psi}_{t,0}^{\varepsilon 0}\bar{\mathbf{P}}_0^\varepsilon(t, \alpha)\int_0^t \boldsymbol{\Psi}_{0,\lambda}^{\varepsilon 0}\mathbf{K}_\lambda^{\varepsilon 0}\mathbf{F}_\lambda\, d\mathbf{w}_\lambda$$

$$- \boldsymbol{\Psi}_{t,0}^{\varepsilon 0}\bar{\mathbf{P}}_0^\varepsilon(t, \alpha)\int_0^t \boldsymbol{\Gamma}_s^\varepsilon \boldsymbol{\Psi}_{0,s}^{\varepsilon 0}\mathbf{D}_s\, d\mathbf{w}_s - \frac{1}{\alpha}\boldsymbol{\Psi}_{t,0}^{\varepsilon 0}\bar{\mathbf{P}}_0^\varepsilon(t, \alpha)\int_0^t \boldsymbol{\Psi}_{0,s}^{\varepsilon 0}\mathbf{D}_s\, d\mathbf{w}_s$$

$$+ \frac{1}{\alpha}\boldsymbol{\Psi}_{t,0}^{\varepsilon 0}\bar{\mathbf{P}}_0^\varepsilon(t, \alpha)\int_0^t \boldsymbol{\Phi}_{0,s}\mathbf{D}_s\, d\mathbf{w}_s. \tag{8.42}$$

The a.s. convergence of (8.38) directly follows from (8.42) by letting $\alpha \to \infty$ and by using the fact that $\overline{\mathbf{P}}_0^{\varepsilon}(t,\alpha) \underset{\alpha\to\infty}{\to} \boldsymbol{\Gamma}_t^{\varepsilon-1}$.

From (8.40) and (8.42), one can calculate that

$$\boldsymbol{\zeta}_t^{\varepsilon} = \boldsymbol{\Psi}_{t,0}^{\varepsilon 0}\boldsymbol{\Gamma}_t^{\varepsilon-1}\overline{\mathbf{P}}_0^{\varepsilon-1}(t,\alpha)\boldsymbol{\Psi}_{0,t}^{\varepsilon 0}\boldsymbol{\zeta}_t^{\varepsilon}(\alpha)$$

$$-\frac{1}{\alpha}\boldsymbol{\Psi}_{t,0}^{\varepsilon 0}\boldsymbol{\Gamma}_t^{\varepsilon-1}\int_0^t\left(\boldsymbol{\Phi}_{0,\lambda}\mathbf{D}_\lambda - \boldsymbol{\Psi}_{0,\lambda}^{\varepsilon 0}\mathbf{D}_\lambda + \boldsymbol{\Psi}_{0,\lambda}^{\varepsilon 0}\mathbf{K}_\lambda^{\varepsilon 0}\mathbf{F}_\lambda\right)d\mathbf{w}_\lambda.$$

Hence

$$\boldsymbol{\zeta}_t^{\varepsilon}(\alpha) - \boldsymbol{\zeta}_t^{\varepsilon} = \frac{1}{\alpha}\Big[-\boldsymbol{\Psi}_{t,0}^{\varepsilon 0}\boldsymbol{\Gamma}_t^{\varepsilon-1}\boldsymbol{\Psi}_{0,t}^{\varepsilon 0}\boldsymbol{\zeta}_t^{\varepsilon}(\alpha) + \boldsymbol{\Psi}_{t,0}^{\varepsilon 0}\boldsymbol{\Gamma}_t^{\varepsilon-1}$$

$$\times\int_0^t\left(\boldsymbol{\Phi}_{0,\lambda}\mathbf{D}_\lambda - \boldsymbol{\Psi}_{0,\lambda}^{\varepsilon 0}\mathbf{D}_\lambda + \boldsymbol{\Psi}_{0,\lambda}^{\varepsilon 0}\mathbf{K}_\lambda^{\varepsilon 0}\mathbf{F}_\lambda\right)d\mathbf{w}_\lambda\Big]. \quad (8.43)$$

From this, it follows that

$$E\left\|\boldsymbol{\zeta}_t^{\varepsilon}(\alpha) - \boldsymbol{\zeta}_t^{\varepsilon}\right\|^2 \underset{\alpha\to\infty}{\to} 0,$$

since, by Lemma 8.2, $\boldsymbol{\Psi}_{t,0}^{\varepsilon 0}\overline{\mathbf{P}}_0^{\varepsilon}(t,\alpha)$ is bounded on $\alpha \geq \alpha_1$, $\varepsilon \in (0, \varepsilon_1]$ and by (8.42), $E\|\boldsymbol{\zeta}_t^{\varepsilon}(\alpha)\|^2$ is also bounded in this region.

From (8.24), (8.34), and (8.43),

$$-\boldsymbol{\Psi}_{t,0}^{\varepsilon 0}\boldsymbol{\Gamma}_t^{\varepsilon-1}\left(\hat{\boldsymbol{\xi}}_0^{\varepsilon}(t) - \boldsymbol{\xi}_0\right)$$

$$= \boldsymbol{\Psi}_{t,0}^{\varepsilon 0}\boldsymbol{\Gamma}_t^{\varepsilon-1}\Big\{\frac{1}{\alpha}\mathbf{P}_0^{\varepsilon}(t,\alpha)(\boldsymbol{\xi}_0 - \mathbf{x}) + \frac{1}{\alpha}\mathbf{P}_0^{\varepsilon}(t,\alpha)\int_0^t\boldsymbol{\Phi}_{0,\lambda}\mathbf{D}_\lambda\,d\mathbf{w}_\lambda$$

$$-\boldsymbol{\Psi}_{0,t}^{\varepsilon 0}\boldsymbol{\zeta}_t^{\varepsilon}(\alpha) + \int_0^t\boldsymbol{\Psi}_{0,\lambda}^{\varepsilon 0}\left(\mathbf{K}_\lambda^{\varepsilon 0}\mathbf{F}_\lambda - \mathbf{D}_\lambda\right)d\mathbf{w}_\lambda - \mathbf{h}_0^{\varepsilon}(t,\alpha)\Big\}$$

$$= \frac{1}{\alpha}\boldsymbol{\Psi}_{t,0}^{\varepsilon 0}\boldsymbol{\Gamma}_t^{\varepsilon-1}\overline{\mathbf{P}}_0^{\varepsilon}(t,\alpha)\left(\boldsymbol{\xi}_0 - \mathbf{x} + \int_0^t\boldsymbol{\Phi}_{0,\lambda}\mathbf{D}_\lambda\,d\mathbf{w}_\lambda\right)$$

$$-\boldsymbol{\Psi}_{t,0}^{\varepsilon 0}\boldsymbol{\Gamma}_t^{\varepsilon-1}\mathbf{h}_0^{\varepsilon}(t,\alpha) + \alpha\left(\boldsymbol{\zeta}_t^{\varepsilon}(\alpha) - \boldsymbol{\zeta}_t^{\varepsilon}\right) - \boldsymbol{\Psi}_{t,0}^{\varepsilon 0}\boldsymbol{\Gamma}_t^{\varepsilon-1}\int_0^t\boldsymbol{\Phi}_{0,s}\mathbf{D}_s\,d\mathbf{w}_s,$$

from which it follows that

$$\underset{\sim}{\Psi}^{\varepsilon s}_{s,t}(\zeta^{\varepsilon}_t(\alpha) - \zeta^{\varepsilon}_t)$$

$$= \frac{1}{\alpha}\underset{\sim}{\Psi}^{\varepsilon s}_{s,t}\underset{\sim}{\Psi}^{\varepsilon 0}_{t,0}\underset{\sim}{\Gamma}^{\varepsilon-1}_t\Bigg\{\boldsymbol{\xi}_0 - \hat{\boldsymbol{\xi}}^{\varepsilon}_0(t) + \int_0^t \underset{\sim}{\Phi}_{0,s}\underset{\sim}{\mathbf{D}}_s\,d\mathbf{w}_s + \mathbf{h}^{\varepsilon}_0(t,\alpha)$$

$$- \frac{1}{\alpha}\underset{\sim}{\overline{\mathbf{P}}}^{\varepsilon}_0(t,\alpha)\left(\boldsymbol{\xi}_0 - \mathbf{x} + \int_0^t \underset{\sim}{\Phi}_{0,\lambda}\underset{\sim}{\mathbf{D}}_\lambda\,d\mathbf{w}_\lambda\right)\Bigg\}, \tag{8.44}$$

which converges in the mean square sense as $\varepsilon \to 0$. This is because (1) by Lemma 8.2, as $\varepsilon \to 0$, $\underset{\sim}{\Psi}^{\varepsilon s}_{s,t}\underset{\sim}{\Psi}^{\varepsilon 0}_{t,0}\underset{\sim}{\Gamma}^{\varepsilon-1}_t$ converges, (2) by Theorem 7.2, as $\varepsilon \to 0$, $\hat{\boldsymbol{\xi}}^{\varepsilon}_0(t)$ converges in the mean square sense and $\underset{\sim}{\overline{\mathbf{P}}}^{\varepsilon}_0(t,\alpha)$ converges, and, finally, (3) by (8.35), $\text{l.i.m.}_{\varepsilon\to 0}\mathbf{h}^{\varepsilon}_0(t,\alpha) = 0$. Hence as $\varepsilon \to 0$, $\underset{\sim}{\Psi}^{\varepsilon s}_{s,t}(\zeta^{\varepsilon t}(\alpha) - \zeta^{\varepsilon}_t)$ has a limit in the mean square sense for any $s \in [0, t]$.

From (8.37), we also know that

$$\mathbf{h}^{\varepsilon}_s(t) = \left(\underset{\sim}{\mathbf{I}} + \frac{1}{\alpha}\underset{\sim}{\Gamma}^{\varepsilon-1}_t\right)\mathbf{h}^{\varepsilon}_0(t,\alpha)$$

or

$$\mathbf{h}^{\varepsilon}_0(t,\alpha) = \left(\underset{\sim}{\mathbf{I}} + \frac{1}{\alpha}\underset{\sim}{\Gamma}^{\varepsilon-1}_t\right)^{-1}\mathbf{h}^{\varepsilon}_s(t). \tag{8.45}$$

Hence

$$E\|\mathbf{h}^{\varepsilon}_0(t,\alpha)\|^2 \le E\|\mathbf{h}^{\varepsilon}_s(t)\|^2,$$

which tends to zero as $\varepsilon \to 0$ by (8.32). Consequently, $E\|\mathbf{h}^{\varepsilon}_0(t,\alpha)\|^2$ is bounded on $\alpha > 0$, $\varepsilon \in (0, \varepsilon_1]$.

Notice that

$$E\|\boldsymbol{\xi}_0 - \hat{\boldsymbol{\xi}}^{\varepsilon}_0(t)\|^2 \le \operatorname{tr}\underset{\sim}{\overline{\mathbf{P}}}^{\varepsilon}_0(t,\alpha) \le \operatorname{tr}\underset{\sim}{\Gamma}^{\varepsilon-1}_t$$

and that as $\varepsilon \to 0$, $\operatorname{tr}\underset{\sim}{\overline{\mathbf{P}}}^{\varepsilon}_0(t,\alpha)$ is nonincreasing by Lemma 8.1. Then

$$E\Bigg\|\boldsymbol{\xi}_0 - \hat{\boldsymbol{\xi}}^{\varepsilon}_0(t) + \int_0^t \underset{\sim}{\Phi}_{0,s}\underset{\sim}{\mathbf{D}}_s\,d\mathbf{w}_s + \mathbf{h}^{\varepsilon}_0(t,\alpha)$$

$$- \frac{1}{\alpha}\underset{\sim}{\overline{\mathbf{P}}}^{\varepsilon}_0(t,\alpha)\left\|\left(\boldsymbol{\xi}_0 - \mathbf{x} + \int_0^t \underset{\sim}{\Phi}_{0,\lambda}\underset{\sim}{\mathbf{D}}_\lambda\,d\mathbf{w}_\lambda\right)\right\|^2$$

is bounded in the region consisting of $\alpha > 0$ and $\varepsilon \in (0, \varepsilon_1]$. As $\varepsilon \to 0$, $\Psi_{s,t}^{\varepsilon s}\Psi_{t,0}^{\varepsilon 0}\Gamma_t^{\varepsilon -1}$ converges by Lemma 8.2. Hence from (8.44), it can be seen that (8.39) takes place. □

Theorem 8.1. *Under Condition* (8.28), *for any* $s \in [0, t]$, *there exists the* EWKIV $\bar{\mathbf{x}}_s(t)$ *independent of the order in which the limits are taken*:

$$\bar{\mathbf{x}}_s(t) = \underset{\varepsilon \to 0}{\text{l.i.m.}}\ \underset{\alpha \to \infty}{\text{l.i.m.}}\ \bar{\mathbf{x}}_s^{\varepsilon}(t, \alpha) = \underset{\alpha \to \infty}{\text{l.i.m.}}\ \underset{\varepsilon \to 0}{\text{l.i.m.}}\ \bar{\mathbf{x}}_s^{\varepsilon}(t, \alpha). \tag{8.46}$$

The associated estimation error covariance matrix

$$\bar{\mathbf{P}}_s(t) \triangleq E(\mathbf{x}_s - \bar{\mathbf{x}}_s(t))(\mathbf{x}_s - \bar{\mathbf{x}}_s(t))^{\tau}$$

is finite and equal to

$$\begin{aligned} \bar{\mathbf{P}}_s(t) &= \lim_{\varepsilon \to 0} \lim_{\alpha \to \infty} \bar{\mathbf{P}}_s^{\varepsilon}(t, \alpha) = \lim_{\alpha \to \infty} \lim_{\varepsilon \to 0} \bar{\mathbf{P}}_s^{\varepsilon}(t, \alpha) \\ &= \lim_{\varepsilon \to 0} E \Psi_{s,t}^{\varepsilon s} \left[\zeta_t^{\varepsilon} - \int_s^t \Psi_{t,\lambda}^{\varepsilon s} (\mathbf{K}_{\lambda}^{\varepsilon s} \mathbf{F}_{\lambda} - \mathbf{D}_{\lambda})\, d\mathbf{w}_{\lambda} \right] \\ &\quad \cdot \left[\zeta_t^{\varepsilon} - \int_s^t \Psi_{t,\lambda}^{\varepsilon s} (\mathbf{K}_{\lambda}^{\varepsilon s} \mathbf{F}_{\lambda} - \mathbf{D}_{\lambda})\, d\mathbf{w}_{\lambda} \right]^{\tau} \Psi_{s,t}^{\varepsilon s \tau}. \end{aligned} \tag{8.47}$$

Proof. By (8.27) and (8.38), we have

$$\begin{aligned} \underset{\alpha \to \infty}{\text{l.i.m.}}\ \bar{\mathbf{x}}_s^{\varepsilon}(t, \alpha) &= \mathbf{x}_s + \Psi_{s,\lambda}^{\varepsilon s} \zeta_t^{\varepsilon} \\ &\quad - \int_s^t \Psi_{s,\lambda}^{\varepsilon s} (\mathbf{K}_{\lambda}^{\varepsilon s} \mathbf{F}_{\lambda} - \mathbf{D}_{\lambda})\, d\mathbf{w}_{\lambda} \triangleq \bar{\mathbf{x}}_s^{\varepsilon}(t), \end{aligned} \tag{8.48}$$

and from (8.27) and (8.46),

$$\begin{aligned} \bar{\mathbf{x}}_s^{\varepsilon}(t, \alpha) &= \bar{\mathbf{x}}_s^{\varepsilon}(t) - \frac{1}{\alpha} \Psi_{s,t}^{\varepsilon s} \Psi_{t,0}^{\varepsilon 0} \bar{\mathbf{P}}_0^{\varepsilon}(t, \alpha) \left(\mathbf{x}_0 + \int_0^t \Phi_{0,\lambda} \mathbf{D}_{\lambda}\, d\mathbf{w}_{\lambda} \right) \\ &\quad + \Psi_{s,t}^{\varepsilon s} (\zeta_t^{\varepsilon}(\alpha) - \zeta_t^{\varepsilon}). \end{aligned} \tag{8.49}$$

From Lemmas 8.2 and 8.4, it is known that as $\varepsilon \to 0$, the left-hand side of (8.49), as well as the second and the third terms on the right-hand side of (8.49), converge in the mean square sense. Hence the remaining term $\bar{\mathbf{x}}_s^{\varepsilon}(t)$ in (8.49) also converges in the mean square sense. Now in (8.49) pass to the limit first by taking $\text{l.i.m.}_{\varepsilon\to 0}$ and then $\text{l.i.m.}_{\alpha\to\infty}$, then (8.46) follows immediately by Lemma 8.2 and (8.39).

From (8.24) and (8.34),

$$\hat{\xi}_s^\varepsilon(t) - \xi_s = -\frac{1}{\alpha}\Psi_{s,t}^{\varepsilon s}\Psi_{t,0}^{\varepsilon 0}\bar{P}_0^\varepsilon(t,\alpha)(\xi_0 - x)$$

$$-\frac{1}{\alpha}\Psi_{s,t}^{\varepsilon s}\Psi_{t,0}^{\varepsilon 0}\bar{P}_0^\varepsilon(t,\alpha)\int_0^t \Phi_{0,\lambda}D_\lambda\,dw_\lambda + \Psi_{s,t}^{\varepsilon s}\zeta_t^\varepsilon(\alpha)$$

$$-\int_0^t \Psi_{s,\lambda}^{\varepsilon s}(K_\lambda^{\varepsilon s}F_\lambda - D_\lambda)\,dw_\lambda + h_s^\varepsilon(t,\alpha), \tag{8.50}$$

on the right-hand side of which, as $\varepsilon\to 0$, $\alpha\to\infty$, the first and second terms, by Conclusion 1 of Lemma 8.2, converge to zero in the mean square sense independently of the order in which the limits are taken. From (8.35), we know that

$$\lim_{\alpha\to\infty}\ \text{l.i.m.}_{\varepsilon\to 0}\, h_s^\varepsilon(t,\alpha) = 0. \tag{8.51}$$

Hence by (8.31), (8.34), and (8.30)

$$\text{l.i.m.}_{\alpha\to\infty}\, h_s^\varepsilon(t,\alpha) = h_s^\varepsilon(t),$$

and then, by (8.32), it follows that

$$\text{l.i.m.}_{\varepsilon\to 0}\ \text{l.i.m.}_{\alpha\to\infty}\, h_s^\varepsilon(t,\alpha) = 0. \tag{8.52}$$

Hence from (8.39) and (8.50), it is known that

$$\text{l.i.m.}_{\alpha\to\infty}\ \text{l.i.m.}_{\varepsilon\to 0}\,(\hat{\xi}_s^\varepsilon(t) - \xi_s) = \text{l.i.m.}_{\varepsilon\to 0}\ \text{l.i.m.}_{\alpha\to\infty}\,(\hat{\xi}_s^\varepsilon(t) - \xi_s)$$

$$= \text{l.i.m.}_{\varepsilon\to 0}\,\Psi_{s,t}^{\varepsilon s}\zeta_t^\varepsilon - \int_s^t \Psi_{s,\lambda}^{\varepsilon s}(K_\lambda^{\varepsilon s}F_\lambda - D_\lambda)\,dw_\lambda, \tag{8.53}$$

which is just the quantity $\bar{x}_s(t) - x_s$ as seen from (8.46) and (8.48).

Notice that

$$E(\hat{\xi}_s^\varepsilon(t) - \xi_s)(\hat{\xi}_s^\varepsilon(t) - \xi_s)^\tau = \bar{P}_s^\varepsilon(t,\alpha),$$

then (8.47) follows from (8.53) and since

$$E\|\bar{\mathbf{x}}_s(t) - \mathbf{x}_s\|^2 \leq 2E\|\bar{\mathbf{x}}_s(t)\|^2 + 2E\|\mathbf{x}_s\|^2 < \infty,$$

$\bar{\mathbf{P}}_s(t)$ is necessarily finite. □

8.2. STOCHASTIC OBSERVABILITY

In Section 8.1, Condition (8.28) plays a crucial role. In this section, this condition will be considered in detail.

According to Lemma 8.1, as $\varepsilon \to 0$, $\bar{\mathbf{P}}_0^\varepsilon(t, \alpha)$ is nonincreasing, and we shall denote its limit by $\bar{\mathbf{P}}_0(t, \alpha)$.

DEFINITION. If there is $t \geq 0$ such that

$$\lim_{\alpha \to \infty} \lim_{\varepsilon \to 0} \bar{\mathbf{P}}_0^\varepsilon(t, \alpha) = \lim_{\alpha \to \infty} \bar{\mathbf{P}}_0(t, \alpha) < \infty, \tag{8.54}$$

then the system (7.9) and (7.10) is called stochastically observable.

The significance of this definition will be clarified after we have established some equivalent conditions.

Lemma 8.5. *Let $\mathbf{G}_t$ be an $n \times m$ matrix satisfying*

$$\int_0^T \|\mathbf{G}_t \mathbf{C}_t\| \, dt < \infty, \quad \int_0^T \|\mathbf{G}_t \mathbf{F}_t\|^2 \, dt < \infty, \quad \int_0^t \mathbf{G}_\lambda \mathbf{C}_\lambda \mathbf{\Phi}_{\lambda,0} \, d\lambda = \mathbf{\Phi}_{t,0},$$

$$\forall t \in [0, T] \tag{8.55}$$

and let

$$\mathbf{x}_t^G = \int_0^t \mathbf{G}_\lambda \, d\mathbf{y}_\lambda^\varepsilon. \tag{8.56}$$

Then

$$\bar{\mathbf{P}}_t^\varepsilon(\alpha) \leq E\left(\int_0^t \mathbf{G}_\lambda \, d\mathbf{y}_\lambda^\varepsilon - \mathbf{x}_t\right)\left(\int_0^t \mathbf{G}_\lambda \, d\mathbf{y}_\lambda^\varepsilon - \mathbf{x}_t\right)^\tau. \tag{8.57}$$

Proof. Denote

$$\mathbf{F}_t' = \left[\mathbf{F}_t \sqrt{\varepsilon}\,\mathbf{I}\right], \qquad \mathbf{D}_t' = [\mathbf{D}_t \ \mathbf{0}], \qquad \mathbf{w}_t' = \left[\mathbf{w}_t^\tau \ \bar{\mathbf{w}}_t^\tau\right]^\tau.$$

Then

$$d\mathbf{y}_t^\varepsilon = \mathbf{C}_t \mathbf{x}_t \, dt + \mathbf{F}_t' \, d\mathbf{w}_t',$$

and by (8.55),

$$\int_0^t \mathbf{G}_s\, d\mathbf{y}_s^\varepsilon = \int_0^t \mathbf{G}_s\mathbf{C}_s\mathbf{\Phi}_{s,0}\left(\mathbf{x}_0 + \int_0^s \mathbf{\Phi}_{0,\lambda}\mathbf{D}'_\lambda\, d\mathbf{w}'_\lambda\right) ds + \int_0^t \mathbf{G}_\lambda\mathbf{F}'_\lambda\, d\mathbf{w}'_\lambda$$

$$= \mathbf{\Phi}_{t,0}\mathbf{x}_0 + \mathbf{\Phi}_{t,0}\int_0^t \mathbf{\Phi}_{0,\lambda}\mathbf{D}'_\lambda\, d\mathbf{w}'_\lambda - \int_0^t\int_0^\lambda \mathbf{G}_s\mathbf{C}_s\mathbf{\Phi}_{s,0}\, ds\, \mathbf{\Phi}_{0,\lambda}\mathbf{D}'_\lambda\, d\mathbf{w}'_\lambda$$

$$+ \int_0^t \mathbf{G}_\lambda\mathbf{F}'_\lambda\, d\mathbf{w}'_\lambda.$$

Hence

$$\mathbf{x}_t - \mathbf{x}_t^G = -\int_0^t \mathbf{G}_\lambda\mathbf{F}'_\lambda\, d\mathbf{w}'_\lambda + \int_0^t\int_0^\lambda \mathbf{G}_s\mathbf{C}_s\mathbf{\Phi}_{s,0}\, ds\, \mathbf{\Phi}_{0,\lambda}\mathbf{D}'_\lambda\, d\mathbf{w}'_\lambda$$

$$= \int_0^t (\mathbf{L}_\lambda^\tau\mathbf{D}'_\lambda - \mathbf{G}_\lambda\mathbf{F}'_\lambda)\, d\mathbf{w}'_\lambda, \tag{8.58}$$

where

$$\mathbf{L}_s^\tau = \int_0^s \mathbf{G}_\lambda\mathbf{C}_\lambda\mathbf{\Phi}_{\lambda,0}\, d\lambda\, \mathbf{\Phi}_{0,s}.$$

Comparing (8.58) with (6.35), we find it is a special case of (6.35) obtained by setting $\mathbf{H}_t = \mathbf{0}$, $\mathbf{c}_t = \mathbf{0}$, $\mathbf{G}_0(t) = \mathbf{0}$, and $\mathbf{G}_1(t) = \mathbf{I}$ in (6.35) and, replacing $\mathbf{D}_s$, $\mathbf{F}_s$, $\mathbf{w}_s$, and $\mathbf{G}_2(s)$ with $\mathbf{D}'_s$, $\mathbf{F}'_s$, $\mathbf{w}'_s$, and $\mathbf{G}_s$, respectively. Notice that the solution of (7.15) with initial condition $\alpha\mathbf{I}$ is $\overline{\mathbf{P}}_t^\varepsilon(\alpha)$, and it can be written as

$$\dot{\overline{\mathbf{P}}}_t^\varepsilon(\alpha) = \overline{\mathbf{P}}_t^\varepsilon(\alpha)\mathbf{A}_t^\tau + \mathbf{A}_t\overline{\mathbf{P}}_t^\varepsilon(\alpha) + \mathbf{D}'_t\mathbf{D}_t'^\tau - \left(\mathbf{D}'_t\mathbf{F}_t'^\tau + \overline{\mathbf{P}}_t^\varepsilon(\alpha)\mathbf{C}_t^\tau\right)$$

$$\cdot\left(\mathbf{F}'_t\mathbf{F}_t'^\tau\right)^{-1}\left(\mathbf{D}'_t\mathbf{F}'^\tau + \overline{\mathbf{P}}_t^\varepsilon(\alpha)\mathbf{C}_t^\tau\right)^\tau, \qquad \overline{\mathbf{P}}_0^\varepsilon(\alpha) = \alpha\mathbf{I}.$$

Replacing $\mathbf{P}_t$ by $\overline{\mathbf{P}}_t(\alpha)$ in (6.39) and (6.40) yields

$$\overline{\mathbf{P}}_t^\varepsilon(\alpha) = E\left(\mathbf{x}_t - \mathbf{x}_t^G\right)\left(\mathbf{x}_t - \mathbf{x}_t^G\right)^\tau$$

$$-\int_0^t \left[\mathbf{G}_s - \mathbf{L}_s^\tau\left(\mathbf{D}'_s\mathbf{F}_s'^\tau + \overline{\mathbf{P}}_s^\varepsilon(\alpha)\mathbf{C}_s^\tau\right)\cdot\left(\mathbf{F}'_s\mathbf{F}_s'^\tau\right)^{-1}\right]\mathbf{F}'_s\mathbf{F}_s'^\tau$$

$$\times\left[\mathbf{G}_s - \mathbf{L}_s^\tau\left(\mathbf{D}'_s\mathbf{F}_s'^\tau + \overline{\mathbf{P}}_s^\varepsilon(\alpha)\mathbf{C}_s^\tau\right)\left(\mathbf{F}'_s\mathbf{F}_s'^\tau\right)^{-1}\right]^\tau ds,$$

which yields (8.57). □

Lemma 8.6. *Suppose that there is an $m \times m$ matrix $\underset{\sim}{\mathbf{H}}_t$ such that*

$$\int_0^T \left(\|\underset{\sim}{\mathbf{H}}_t \underset{\sim}{\mathbf{C}}_t\| + \|\underset{\sim}{\mathbf{H}}_t \underset{\sim}{\mathbf{F}}_t\|^2 \right) dt < \infty$$

and such that for some fixed $s \in [0, t]$

$$\int_0^t \underset{\sim}{\mathbf{H}}_\lambda \underset{\sim}{\mathbf{C}}_\lambda \underset{\sim}{\mathbf{\Phi}}_{\lambda,0}\, d\lambda = \underset{\sim}{\mathbf{\Phi}}_{s,0}. \tag{8.59}$$

Then

$$\begin{aligned} \underset{\sim}{\mathbf{\Psi}}^{\varepsilon s}_{s,t} \underset{\sim}{\overline{\mathbf{P}}}^{\varepsilon}_t(\alpha) \underset{\sim}{\mathbf{\Psi}}^{\varepsilon s T}_{s,t} \le{} & E\left(\int_0^t \underset{\sim}{\mathbf{H}}_\lambda\, d\mathbf{y}_\lambda - \mathbf{x}_s \right)\left(\int_0^t \underset{\sim}{\mathbf{H}}_\lambda\, d\mathbf{y}_\lambda - \mathbf{x}_s \right)^T \\ & + \int_s^t \underset{\sim}{\mathbf{\Psi}}^{\varepsilon s}_{s,\lambda} (\underset{\sim}{\mathbf{K}}^{\varepsilon s}_\lambda \underset{\sim}{\mathbf{F}}_\lambda - \underset{\sim}{\mathbf{D}}_\lambda)(\underset{\sim}{\mathbf{K}}^{\varepsilon s}_\lambda \underset{\sim}{\mathbf{F}}_\lambda - \underset{\sim}{\mathbf{D}}_\lambda)^T \underset{\sim}{\mathbf{\Psi}}^{\varepsilon s T}_{s,\lambda}\, d\lambda \\ & + \varepsilon \int_0^t \underset{\sim}{\mathbf{H}}_\lambda \underset{\sim}{\mathbf{H}}^T_\lambda\, d\lambda + \varepsilon \int_s^t \underset{\sim}{\mathbf{\Psi}}^{\varepsilon s}_{s,\lambda} \underset{\sim}{\mathbf{K}}^{\varepsilon s}_\lambda \underset{\sim}{\mathbf{K}}^{\varepsilon s T}_\lambda \underset{\sim}{\mathbf{\Psi}}^{\varepsilon s \lambda}_{s,\lambda}\, d\lambda. \end{aligned} \tag{8.60}$$

Proof. To begin with we prove that

$$E\left(\int_0^t \underset{\sim}{\mathbf{H}}_\lambda\, d\mathbf{y}^\varepsilon_\lambda - \mathbf{x}_s \right)\left[\int_s^t \underset{\sim}{\mathbf{\Psi}}^{\varepsilon s}_{s,\lambda} \left(\underset{\sim}{\mathbf{K}}^{\varepsilon s}_\lambda \underset{\sim}{\mathbf{F}}'_\lambda - \underset{\sim}{\mathbf{D}}'_\lambda \right) d\mathbf{w}'_\lambda \right]^T = \underset{\sim}{\mathbf{0}} \tag{8.61}$$

where $\underset{\sim}{\mathbf{F}}'_t$, $\underset{\sim}{\mathbf{D}}'_t$, and $\mathbf{w}'_t$ are the same as in Lemma 8.5. By (8.59), we have

$$\begin{aligned} & \int_0^t \underset{\sim}{\mathbf{H}}_\lambda\, d\mathbf{y}^\varepsilon_\lambda - \mathbf{x}_s \\ &= \int_0^t \underset{\sim}{\mathbf{H}}_\lambda \underset{\sim}{\mathbf{C}}_\lambda \underset{\sim}{\mathbf{\Phi}}_{\lambda,0} \left(\mathbf{x}_0 + \int_0^\lambda \underset{\sim}{\mathbf{\Phi}}_{0,\mu} \underset{\sim}{\mathbf{D}}'_\mu\, d\mathbf{w}'_\mu \right) d\lambda + \int_0^t \underset{\sim}{\mathbf{H}}_\lambda \underset{\sim}{\mathbf{F}}'_\lambda\, d\mathbf{w}'_\lambda - \mathbf{x}_s \\ &= \underset{\sim}{\mathbf{\Phi}}_{s,0} \mathbf{x}_0 + \int_0^t \int_\mu^t \underset{\sim}{\mathbf{H}}_\lambda \underset{\sim}{\mathbf{C}}_\lambda \underset{\sim}{\mathbf{\Phi}}_{\lambda,0}\, d\lambda\, \underset{\sim}{\mathbf{\Phi}}_{0,\mu} \underset{\sim}{\mathbf{D}}'_\mu\, d\mathbf{w}'_\mu + \int_0^t \underset{\sim}{\mathbf{H}}_\lambda \underset{\sim}{\mathbf{F}}'_\lambda\, d\mathbf{w}'_\lambda - \mathbf{x}_s \\ &= \int_s^t \underset{\sim}{\mathbf{\Phi}}_{s,\lambda} \underset{\sim}{\mathbf{D}}'_\lambda\, d\mathbf{w}'_\lambda + \int_0^t \underset{\sim}{\mathbf{H}}_\lambda \underset{\sim}{\mathbf{F}}'_\lambda\, d\mathbf{w}'_\lambda - \int_0^t \int_0^\mu \underset{\sim}{\mathbf{H}}_\lambda \underset{\sim}{\mathbf{C}}_\lambda \underset{\sim}{\mathbf{\Phi}}_{\lambda,0}\, d\lambda\, \underset{\sim}{\mathbf{\Phi}}_{0,\mu} \underset{\sim}{\mathbf{D}}'_\mu\, d\mathbf{w}'_\mu. \end{aligned} \tag{8.62}$$

Hence, in order to prove (8.61), it can be assumed without loss of generality that $\{\mathbf{w}'_\lambda\}$ and $[\mathbf{x}^T_0 \mathbf{y}^T_0]^T$ are independent and normally distributed.

From (6.58), it is seen that

$$E\left(\mathbf{x}_t/\mathscr{F}_t^{\mathbf{y}^\varepsilon}, \mathbf{x}_s\right) = \underset{\sim}{\mathbf{\Psi}}_{t,s}^{\varepsilon s}\mathbf{x}_s + \underset{\sim}{\mathbf{\Psi}}_{t,s}^{\varepsilon s}\int_s^t \underset{\sim}{\mathbf{\Psi}}_{s,\lambda}^{\varepsilon s}\underset{\sim}{\mathbf{K}}_\lambda^{\varepsilon s}\, d\mathbf{y}_\lambda^\varepsilon, \tag{8.63}$$

but from (8.23),

$$\int_s^t \underset{\sim}{\mathbf{\Psi}}_{s,\mu}^{\varepsilon s}\underset{\sim}{\mathbf{K}}_\mu^{\varepsilon s}\, d\mathbf{y}_\mu = \underset{\sim}{\mathbf{\Psi}}_{s,t}^{\varepsilon s}\mathbf{x}_t - \mathbf{x}_s - \int_s^t \underset{\sim}{\mathbf{\Psi}}_{s,\lambda}^{\varepsilon s}\underset{\sim}{\mathbf{D}}_\lambda\, d\mathbf{w}_\lambda + \int_s^t \underset{\sim}{\mathbf{\Psi}}_{s,\lambda}^{\varepsilon s}\underset{\sim}{\mathbf{K}}_\lambda^{\varepsilon s}\underset{\sim}{\mathbf{F}}_\lambda\, d\mathbf{w}_\lambda.$$

Hence

$$\begin{aligned}\underset{\sim}{\mathbf{\Psi}}_{t,s}^{\varepsilon s}\int_s^t \underset{\sim}{\mathbf{\Psi}}_{s,\mu}^{\varepsilon s}\underset{\sim}{\mathbf{K}}_\mu^{\varepsilon s}\, d\mathbf{y}_\mu^\varepsilon &= \mathbf{x}_t - \underset{\sim}{\mathbf{\Psi}}_{t,s}^{\varepsilon s}\mathbf{x}_s - \underset{\sim}{\mathbf{\Psi}}_{t,s}^{\varepsilon s}\int_s^t \underset{\sim}{\mathbf{\Psi}}_{s,\lambda}^{\varepsilon s}\underset{\sim}{\mathbf{D}}_\lambda\, d\mathbf{w}_\lambda \\ &\quad + \underset{\sim}{\mathbf{\Psi}}_{t,s}^{\varepsilon s}\int_s^t \underset{\sim}{\mathbf{\Psi}}_{s,\lambda}^{\varepsilon s}\underset{\sim}{\mathbf{K}}_\lambda^{\varepsilon s}\underset{\sim}{\mathbf{F}}_\lambda\, d\mathbf{w}_\lambda + \sqrt{\varepsilon}\,\underset{\sim}{\mathbf{\Psi}}_{t,s}^{\varepsilon s}\int_s^t \underset{\sim}{\mathbf{\Psi}}_{s,\mu}^{\varepsilon s}\underset{\sim}{\mathbf{K}}_\mu^{\varepsilon s}\, d\overline{\mathbf{w}}_\mu \\ &= \mathbf{x}_t - \underset{\sim}{\mathbf{\Psi}}_{t,s}^{\varepsilon s}\mathbf{x}_s + \underset{\sim}{\mathbf{\Psi}}_{t,s}^{\varepsilon s}\int_s^t \underset{\sim}{\mathbf{\Psi}}_{s,\lambda}^{\varepsilon s}\left(\underset{\sim}{\mathbf{K}}_\lambda^{\varepsilon s}\underset{\sim}{\mathbf{F}}_\lambda' - \underset{\sim}{\mathbf{D}}_\lambda'\right) d\mathbf{w}_\lambda'.\end{aligned} \tag{8.64}$$

Substituting (8.64) into (8.63) yields

$$E\left(\mathbf{x}_t/\mathscr{F}_t^{\mathbf{y}^\varepsilon}, \mathbf{x}_s\right) - \mathbf{x}_t = \underset{\sim}{\mathbf{\Psi}}_{t,s}^{\varepsilon s}\int_s^t \underset{\sim}{\mathbf{\Psi}}_{s,\lambda}^{\varepsilon s}\left(\underset{\sim}{\mathbf{K}}_\lambda^{\varepsilon s}\underset{\sim}{\mathbf{F}}_\lambda' - \underset{\sim}{\mathbf{D}}_\lambda'\right) d\mathbf{w}_\lambda'. \tag{8.65}$$

Further,

$$\begin{aligned}&E\left[E\left(\mathbf{x}_t/\mathscr{F}_t^{\mathbf{y}^\varepsilon}, \mathbf{x}_s\right) - \mathbf{x}_t\right]\left[\int_0^t \underset{\sim}{\mathbf{H}}_\lambda\, d\mathbf{y}_\lambda^\varepsilon - \mathbf{x}_s\right]^T \\ &\quad = EE\left\{\left[E\left(\mathbf{x}_t/\mathscr{F}_t^{\mathbf{y}^\varepsilon}, \mathbf{x}_s\right) - \mathbf{x}_t\right]\left[\int_0^t \underset{\sim}{\mathbf{H}}_\lambda\, d\mathbf{y}_\lambda^\varepsilon - \mathbf{x}_s\right]^T \Big/ \mathscr{F}_t^{\mathbf{y}^\varepsilon}, \mathbf{x}_s\right\} = \underset{\sim}{\mathbf{0}}\end{aligned} \tag{8.66}$$

which, together with (8.65), implies (8.61).

Taking the expectation for both sides of (8.63) gives

$$\underset{\sim}{\mathbf{\Phi}}_{t,0}E\mathbf{x}_0 = \underset{\sim}{\mathbf{\Psi}}_{t,s}^{\varepsilon s}\underset{\sim}{\mathbf{\Phi}}_{t,0}E\mathbf{x}_0 + \underset{\sim}{\mathbf{\Psi}}_{t,s}^{\varepsilon s}\int_s^t \underset{\sim}{\mathbf{\Psi}}_{s,\lambda}^{\varepsilon s}\underset{\sim}{\mathbf{K}}_\lambda^{\varepsilon s}\underset{\sim}{\mathbf{C}}_\lambda\underset{\sim}{\mathbf{\Phi}}_{\lambda,0}\, d\lambda\, E\mathbf{x}_0$$

which, since $E\mathbf{x}_0$ is arbitrary, leads to

$$\underset{\sim}{\mathbf{\Phi}}_{t,0} = \underset{\sim}{\mathbf{\Psi}}^{\varepsilon s}_{t,s}\underset{\sim}{\mathbf{\Phi}}_{t,0} + \underset{\sim}{\mathbf{\Psi}}^{\varepsilon s}_{t,s}\int_s^t \underset{\sim}{\mathbf{\Psi}}^{\varepsilon s}_{s,\lambda}\underset{\sim}{\mathbf{K}}^{\varepsilon s}_{\lambda}\underset{\sim}{\mathbf{C}}_{\lambda}\underset{\sim}{\mathbf{\Phi}}_{\lambda,0}\, d\lambda. \tag{8.67}$$

Let us put

$$\underset{\sim}{\mathbf{G}}_{\lambda} = \underset{\sim}{\mathbf{\Psi}}^{\varepsilon s}_{t,s}\underset{\sim}{\mathbf{H}}_{\lambda} + \underset{\sim}{\mathbf{\Psi}}^{\varepsilon s}_{t,s}\underset{\sim}{\mathbf{\Psi}}^{\varepsilon s}_{s,\lambda}\underset{\sim}{\mathbf{K}}^{\varepsilon s}_{\lambda} I_{[s \le \lambda]}. \tag{8.68}$$

By (8.59) and (8.67), it is known that

$$\int_0^t \underset{\sim}{\mathbf{G}}_{\lambda}\underset{\sim}{\mathbf{C}}_{\lambda}\underset{\sim}{\mathbf{\Phi}}_{\lambda,0}\, d\lambda = \underset{\sim}{\mathbf{\Phi}}_{t,0}, \tag{8.69}$$

and it is easy to see that $\underset{\sim}{\mathbf{G}}_{\lambda}$ given by (8.68) satisfies the other conditions of Lemma 8.5.

Now let

$$\mathbf{z} = \int_0^t \underset{\sim}{\mathbf{G}}_{\lambda}\, d\mathbf{y}^{\varepsilon}_{\lambda} = \underset{\sim}{\mathbf{\Psi}}^{\varepsilon s}_{t,s}\int_0^t \underset{\sim}{\mathbf{H}}_{\lambda}\, d\mathbf{y}^{\varepsilon}_{\lambda} + \underset{\sim}{\mathbf{\Psi}}^{\varepsilon s}_{t,s}\int_s^t \underset{\sim}{\mathbf{\Psi}}^{\varepsilon s}_{s,\lambda}\underset{\sim}{\mathbf{K}}^{\varepsilon s}_{\lambda}\, d\mathbf{y}^{\varepsilon}_{\lambda}. \tag{8.70}$$

Then by (8.64)

$$\mathbf{z} - \mathbf{x}_t = \underset{\sim}{\mathbf{\Psi}}^{\varepsilon s}_{t,s}\left(\int_0^t \underset{\sim}{\mathbf{H}}_{\lambda}\, d\mathbf{y}^{\varepsilon}_{\lambda} - \mathbf{x}_s\right) + \underset{\sim}{\mathbf{\Psi}}^{\varepsilon s}_{t,s}\int_s^t \underset{\sim}{\mathbf{\Psi}}^{\varepsilon s}_{s,\lambda}\left(\underset{\sim}{\mathbf{K}}^{\varepsilon s}_{\lambda}\underset{\sim}{\mathbf{F}}'_{\lambda} - \underset{\sim}{\mathbf{D}}'_{\lambda}\right) d\mathbf{w}'_{\lambda} \tag{8.71}$$

and hence, by Lemma 8.5, it follows that

$$\begin{aligned}
\underset{\sim}{\overline{\mathbf{P}}}{}^{\varepsilon}_t(\alpha) &\le E\left(\int_0^t \underset{\sim}{\mathbf{G}}_{\lambda}\, d\mathbf{y}^{\varepsilon}_{\lambda} - \mathbf{x}_t\right)\left(\int_0^t \underset{\sim}{\mathbf{G}}_{\lambda}\, d\mathbf{y}^{\varepsilon}_{\lambda} - \mathbf{x}_t\right)^{\tau} \\
&= E\left[\underset{\sim}{\mathbf{\Psi}}^{\varepsilon s}_{t,s}\left(\int_0^t \underset{\sim}{\mathbf{H}}_{\lambda}\, d\mathbf{y}^{\varepsilon}_{\lambda} - \mathbf{x}_s\right) + \underset{\sim}{\mathbf{\Psi}}^{\varepsilon s}_{t,s}\int_0^t \underset{\sim}{\mathbf{\Psi}}^{\varepsilon s}_{s,\lambda}\left(\underset{\sim}{\mathbf{K}}^{\varepsilon s}_{\lambda}\underset{\sim}{\mathbf{F}}'_{\lambda} - \underset{\sim}{\mathbf{D}}'_{\lambda}\right) d\mathbf{w}'_{\lambda}\right] \\
&\quad \cdot \left[\underset{\sim}{\mathbf{\Psi}}^{\varepsilon s}_{t,s}\left(\int_0^t \underset{\sim}{\mathbf{H}}_{\lambda}\, d\mathbf{y}^{\varepsilon}_{\lambda} - \mathbf{x}_s\right) + \underset{\sim}{\mathbf{\Psi}}^{\varepsilon s}_{t,s}\int_s^t \underset{\sim}{\mathbf{\Psi}}^{\varepsilon s}_{s,\lambda}\left(\underset{\sim}{\mathbf{K}}^{\varepsilon s}_{\lambda}\underset{\sim}{\mathbf{F}}'_{\lambda} - \underset{\sim}{\mathbf{D}}'_{\lambda}\right) d\mathbf{w}'_{\lambda}\right]^{\tau}.
\end{aligned}$$

By use of (8.61), it turns out from this that

$$\begin{aligned}
\underset{\sim}{\overline{\mathbf{P}}}{}^{\varepsilon}_t(\alpha) &\le \underset{\sim}{\mathbf{\Psi}}^{\varepsilon s}_{t,s} E\left(\int_0^t \underset{\sim}{\mathbf{H}}_{\lambda}\, d\mathbf{y}^{\varepsilon}_{\lambda} - \mathbf{x}_s\right)\left(\int_0^t \underset{\sim}{\mathbf{H}}_{\lambda}\, d\mathbf{y}^{\varepsilon}_{\lambda} - \mathbf{x}_s\right)^{\tau}\underset{\sim}{\mathbf{\Psi}}^{\varepsilon s \tau}_{t,s} \\
&\quad + \underset{\sim}{\mathbf{\Psi}}^{\varepsilon s}_{t,s}\int_s^t \underset{\sim}{\mathbf{\Psi}}^{\varepsilon s}_{s,\lambda}\left(\underset{\sim}{\mathbf{K}}^{\varepsilon s}_{\lambda}\underset{\sim}{\mathbf{F}}'_{\lambda} - \underset{\sim}{\mathbf{D}}'_{\lambda}\right)\left(\underset{\sim}{\mathbf{K}}^{\varepsilon s}_{\lambda}\underset{\sim}{\mathbf{F}}'_{\lambda} - \underset{\sim}{\mathbf{D}}'_{\lambda}\right)^{\tau}\underset{\sim}{\mathbf{\Psi}}^{\varepsilon s \tau}_{s,\lambda}\, d\lambda\, \underset{\sim}{\mathbf{\Psi}}^{\varepsilon s \tau}_{t,s}
\end{aligned}$$

and hence that

$$\bar{\mathbf{P}}_t^\varepsilon(\alpha) \le \boldsymbol{\Psi}_{t,s}^{\varepsilon s} E\left(\int_0^t \mathbf{H}_\lambda \, d\mathbf{y}_\lambda - \mathbf{x}_s\right)\left(\int_0^t \mathbf{H}_\lambda \, d\mathbf{y}_\lambda - \mathbf{x}_s\right)^\tau \boldsymbol{\Psi}_{t,s}^{\varepsilon s \tau}$$

$$+ \varepsilon \boldsymbol{\Psi}_{t,s}^{\varepsilon s} \int_0^t \mathbf{H}_\lambda \mathbf{H}_\lambda^\tau \, d\lambda \boldsymbol{\Psi}_{t,s}^{\varepsilon s \tau} + \boldsymbol{\Psi}_{t,s}^{\varepsilon s} \int_s^t \boldsymbol{\Psi}_{s,\lambda}^{\varepsilon s} (\mathbf{K}_\lambda^{\varepsilon s} \mathbf{F}_\lambda - \mathbf{D}_\lambda)(\mathbf{K}_\lambda^{\varepsilon s} \mathbf{F}_\lambda - \mathbf{D}_\lambda)^\tau$$

$$\cdot \boldsymbol{\Psi}_{s,\lambda}^{\varepsilon s \tau} \, d\lambda \boldsymbol{\Psi}_{t,s}^{\varepsilon s \tau} + \varepsilon \boldsymbol{\Psi}_{t,s}^{\varepsilon s} \int_s^t \boldsymbol{\Psi}_{s,\lambda}^{\varepsilon s} \mathbf{K}_\lambda^{\varepsilon s} \mathbf{K}_\lambda^{\varepsilon s \tau} \boldsymbol{\Psi}_{s,\lambda}^{\varepsilon s \tau} \, d\lambda \boldsymbol{\Psi}_{t,s}^{\varepsilon s \tau},$$

where the mutual independence of $\{\mathbf{w}_t\}$, $\{\bar{\mathbf{w}}_t\}$, and $[\mathbf{x}_0^\tau \mathbf{y}_0^\tau]^\tau$ has been invoked. From this (8.60) follows. □

Denote by $\mathscr{L}_s(t)$ the totality of processes $\mathbf{x}_s^H$ given by

$$\mathbf{x}_s^H = \int_0^t \mathbf{H}_\lambda \, d\mathbf{y}_\lambda, \qquad s \in [0, t] \tag{8.72}$$

together with their limits in the mean square sense, where $\mathbf{H}_\lambda$ satisfies conditions of Lemma 8.6.

Clearly, $E\mathbf{z} = E\mathbf{x}_s$, for all $\mathbf{z} \in \mathscr{L}_s(t)$, and $\mathbf{z}$ is an unbiased estimate of $\mathbf{x}_s$ based on $\{\mathbf{y}_\lambda, 0 \le \lambda \le t\}$ using no initial statistical characteristics of $\mathbf{x}_0$ and $\mathbf{y}_0$.

Theorem 8.2. *System (7.9), (7.10) is stochastically observable if and only if for some $t \in [0, T]$ one of the following equivalent conditions holds*:

1. $\lim_{\alpha \to \infty} \lim_{\varepsilon \to 0} \bar{\mathbf{P}}_0^\varepsilon(t, \alpha) < \infty$.
2. $\lim_{\varepsilon \to 0} \det \boldsymbol{\Gamma}_t^\varepsilon \ne 0$.
3. $\lim_{\alpha \to \infty} \lim_{\varepsilon \to 0} \bar{\mathbf{P}}_s^\varepsilon(t, \alpha) < \infty$, $\forall s \in [0, t]$.
4. $\lim_{\varepsilon \to 0} \lim_{\alpha \to \infty} \bar{\mathbf{P}}_\lambda^\varepsilon(t, \alpha) < \infty$, $\forall \lambda \in [0, t]$.
5. $\mathscr{L}_s(t)$ *is nonempty*, $\forall s \in [0, t]$.
6. $\int_0^t \boldsymbol{\Phi}_{\lambda,0}^\tau \mathbf{C}_\lambda^\tau \mathbf{C}_\lambda \boldsymbol{\Phi}_{\lambda,0} \, d\lambda > 0$.

Proof. Condition 1 is the definition of stochastic observability. Now assume Condition 1 holds. If Condition 2 were not true, that is to say, if

$$\lim_{\varepsilon \to 0} \det \boldsymbol{\Gamma}_t^\varepsilon = 0,$$

then there would be an $\varepsilon_1 > 0$ such that $\boldsymbol{\Gamma}_t^\varepsilon$ would be degenerate for all $\varepsilon \in (0, \varepsilon_1]$, since by Lemma 8.1, $\det \boldsymbol{\Gamma}_t^\varepsilon$ is nondecreasing as $\varepsilon \to 0$. Let λ_i be the eigenvalues of $\boldsymbol{\Gamma}_t^\varepsilon$. Then using an orthogonal matrix to diagonalize the matrix $\bar{\mathbf{P}}_0^\varepsilon(t,\alpha)$ given by (8.11) it is easy to see that the eigenvalues of $\bar{\mathbf{P}}_0^\varepsilon(t,\alpha)$ are

$$\left(\frac{1}{\alpha}+\lambda_i\right)^{-1}=\frac{\alpha}{1+\alpha\lambda_i}. \tag{8.73}$$

Since the minimum eigenvalue of $\mathbf{\Gamma}_t^\varepsilon$ is assumed to be zero for $\varepsilon \in (0, \varepsilon_1]$, it follows from (8.73) that

$$\|\bar{\mathbf{P}}_0^\varepsilon(t,\alpha)\| = \alpha, \qquad \forall \varepsilon \in (0,\varepsilon_1]$$

and

$$\lim_{\varepsilon\to 0}\|\bar{\mathbf{P}}_0^\varepsilon(t,\alpha)\| = \alpha \underset{\alpha\to\infty}{\to} \infty,$$

which contradicts Condition 1. Hence Condition 1 implies Condition 2.

Let Condition 2 hold. According to Theorem 8.1, Conditions 3 and 4 are satisfied.

Now assume Condition 3 holds. Clearly, Condition 3 is stronger than Condition 1. Hence Conditions 1 and then 2 are satisfied.

On the one hand, from (8.48), it is seen that $\bar{\mathbf{x}}_t^\varepsilon$ is unbiased for $\mathbf{x}_t$ for any $E\mathbf{x}_0$. On the other hand, by (8.26) and (8.41), we know that

$$\bar{\mathbf{x}}_t^\varepsilon(\alpha) = \mathbf{\Psi}_{t,0}^{\varepsilon 0}\bar{\mathbf{P}}_0^\varepsilon(t,\alpha)\int_0^t\left\{\frac{1}{\alpha}\mathbf{\Psi}_{0,\lambda}^{\varepsilon 0}\mathbf{K}_\lambda^{\varepsilon 0}+\mathbf{\Gamma}_\lambda^\varepsilon\mathbf{\Psi}_{0,\lambda}^{\varepsilon 0}\right.$$

$$\left.\times\left[\mathbf{D}_\lambda\mathbf{F}_\lambda^\tau+\left(\bar{\mathbf{P}}_\lambda^{\varepsilon 0}+\mathbf{\Psi}_{\lambda,0}^{\varepsilon 0}\mathbf{\Gamma}_\lambda^{\varepsilon+}\mathbf{\Psi}_{\lambda,0}^{\varepsilon 0\tau}\right)\mathbf{C}_\lambda^\tau\right]\left(\mathbf{F}_\lambda\mathbf{F}_\lambda^\tau+\varepsilon\mathbf{I}\right)^{-1}\right\}d\mathbf{y}_\lambda$$

converges to $\bar{\mathbf{x}}_t^\varepsilon$ in the mean square sense. Hence

$$\bar{\mathbf{x}}_t^\varepsilon = \int_0^t\mathbf{G}_\lambda\,d\mathbf{y}_\lambda, \tag{8.74}$$

where

$$\mathbf{G}_\lambda = \mathbf{\Psi}_{t,0}^{\varepsilon 0}\mathbf{\Gamma}_t^{\varepsilon-1}\mathbf{\Gamma}_\lambda^\varepsilon\mathbf{\Psi}_{0,\lambda}^{\varepsilon 0}\left[\mathbf{D}_\lambda\mathbf{F}_\lambda^\tau+\left(\mathbf{P}_\lambda^{\varepsilon 0}+\mathbf{\Psi}_{\lambda,0}^{\varepsilon 0}\mathbf{\Gamma}_\lambda^{\varepsilon+}\mathbf{\Psi}_{\lambda,0}^{\varepsilon 0\tau}\right)\mathbf{C}_\lambda^\tau\right]\left(\mathbf{F}_\lambda\mathbf{F}_\lambda^\tau+\varepsilon\mathbf{I}\right)^{-1}. \tag{8.75}$$

From the unbiasedness of $\bar{\mathbf{x}}_t^\varepsilon$ for any $E\mathbf{x}_0$, it follows that

$$\int_0^t\mathbf{G}_\lambda\mathbf{C}_\lambda\mathbf{\Phi}_{\lambda,0}\,d\lambda = \mathbf{\Phi}_{t,0}.$$

Now $\underset{\sim}{\mathbf{F}}_\lambda^\tau(\underset{\sim}{\mathbf{F}}_\lambda \underset{\sim}{\mathbf{F}}_\lambda^\tau + \varepsilon \underset{\sim}{\mathbf{I}})^{-2} \underset{\sim}{\mathbf{F}}_\lambda \leq (1/\varepsilon)\underset{\sim}{\mathbf{I}}$ and since $\underset{\sim}{\mathbf{\Psi}}_{t,0}^{\varepsilon 0}$, $\underset{\sim}{\mathbf{\Gamma}}_t^\varepsilon$, and $\underset{\sim}{\mathbf{P}}_t^{\varepsilon 0}$ are continuous in t, it is clear, by (8.10), that

$$\int_0^T \|\underset{\sim}{\mathbf{G}}_\lambda\|^2 \, d\lambda < \infty$$

and that $\underset{\sim}{\mathbf{G}}_\lambda$ satisfies all the conditions of Lemma 8.5.

Hence

$$\overline{\underset{\sim}{\mathbf{P}}}_t^\varepsilon(\alpha) \leq E\left(\int_0^t \underset{\sim}{\mathbf{G}}_\lambda \, d\mathbf{y}_\lambda - \mathbf{x}_t\right)\left(\int_0^t \underset{\sim}{\mathbf{G}}_\lambda \, d\mathbf{y}_\lambda - \mathbf{x}_t\right)^\tau + \varepsilon \int_0^t \underset{\sim}{\mathbf{G}}_\lambda \underset{\sim}{\mathbf{G}}_\lambda^\tau \, d\lambda.$$

Consequently,

$$\lim_{\varepsilon \to 0} \lim_{\alpha \to \infty} \overline{\underset{\sim}{\mathbf{P}}}_t^\varepsilon(\alpha) < \infty,$$

and by (8.9)

$$\lim_{\alpha \to \infty} \overline{\underset{\sim}{\mathbf{P}}}_\lambda^\varepsilon(t, \alpha) = \lim_{\alpha \to \infty} \underset{\sim}{\mathbf{\Psi}}_{\lambda,t}^{\varepsilon\lambda}\left(\overline{\underset{\sim}{\mathbf{P}}}_t^\varepsilon(\alpha) - \underset{\sim}{\mathbf{P}}_t^{\varepsilon\lambda}\right)\underset{\sim}{\mathbf{\Psi}}_{\lambda,t}^{\varepsilon\lambda\tau} < \infty.$$

But by Lemma 8.1, as $\varepsilon \to 0$, $\overline{\underset{\sim}{\mathbf{P}}}_\lambda^\varepsilon(t, \alpha)$ is nonincreasing, that is, for $0 < \varepsilon_1 < \varepsilon_2$

$$\overline{\underset{\sim}{\mathbf{P}}}_\lambda^{\varepsilon_1}(t, \alpha) \leq \overline{\underset{\sim}{\mathbf{P}}}_\lambda^{\varepsilon_2}(t, \alpha).$$

Hence, as $\varepsilon \to 0$, $\lim_{\alpha \to \infty} \overline{\underset{\sim}{\mathbf{P}}}_\lambda^\varepsilon(t, \alpha)$ is nonincreasing, and

$$\lim_{\varepsilon \to 0} \lim_{\alpha \to \infty} \overline{\underset{\sim}{\mathbf{P}}}_\lambda^\varepsilon(t, \alpha) < \infty.$$

That is, Condition 4 holds.

Now assume Condition 4 holds.

For $\lambda = 0$, there is an $\varepsilon > 0$ such that

$$\lim_{\alpha \to \infty} \overline{\underset{\sim}{\mathbf{P}}}_0^\varepsilon(t, \alpha) < \infty \quad \text{or} \quad \lim_{\alpha \to \infty} \left(\frac{1}{\alpha} I + \underset{\sim}{\mathbf{\Gamma}}_t^\varepsilon\right)^{-1} < \infty.$$

Analogously to (8.73), it can be shown that $\underset{\sim}{\mathbf{\Gamma}}_t^\varepsilon > 0$, and by Theorem 8.1, Condition 3 holds. Hence $\underset{\sim}{\mathbf{G}}_\lambda$ defined by (8.75) satisfies the conditions of Lemma 8.5 and $\underset{\sim}{\mathbf{H}}_\lambda \triangleq \underset{\sim}{\mathbf{\Phi}}_{s,t} \underset{\sim}{\mathbf{G}}_\lambda$ satisfies the conditions of Lemma 8.6. This means that Condition 5 is true.

Now suppose that Condition 5 holds. In particular, for $s = t$, there is a function $\underset{\sim}{\mathbf{G}}_\lambda$ with

$$\int_0^T \|\underset{\sim}{\mathbf{G}}_\lambda\|^2 \, d\lambda < \infty,$$

such that

$$\int_0^t \Phi^T_{\lambda,0} C^T_\lambda G^T_\lambda \, d\lambda = \Phi^T_{t,0}. \tag{8.76}$$

If the converse were true, then there would be a nonzero n-dimensional vector $\mathbf{x} \neq \mathbf{0}$ such that

$$\mathbf{x}^T \int_0^t \Phi^T_{\lambda,0} C^T_\lambda C_\lambda \Phi_{\lambda,0} \, d\lambda \; \mathbf{x} = 0.$$

That is, $\int_0^t \| C_\lambda \Phi_{\lambda,0} \mathbf{x} \|^2 \, d\lambda = 0$. Hence

$$C_\lambda \Phi_{\lambda,0} \mathbf{x} = \mathbf{0}, \qquad \text{a.s. } \lambda \in [0, t],$$

and so

$$\mathbf{x}^T \int_0^t \Phi^T_{\lambda,0} C^T_\lambda G^T_\lambda \, d\lambda = \mathbf{0}. \tag{8.77}$$

On the other hand, by (8.76),

$$\mathbf{x}^T \int_0^T \Phi^T_{\lambda,0} C^T_\lambda G^T_\lambda \, d\lambda = \mathbf{x}^T \Phi^T_{t,0} \neq \mathbf{0}$$

since $\Phi_{t,0}$ is of full rank. This latter fact contradicts (8.77) and hence Condition 6 holds.

Finally, assume that Condition 6 is true.

Denote

$$G_\lambda = \Phi_{t,0} \left(\int_0^t \Phi^T_{s,0} C^T_s C_s \Phi_{s,0} \, ds \right)^{-1} \Phi^T_{\lambda,0} C^T_\lambda.$$

Clearly, G_λ satisfies conditions of Lemma 8.5, and by (8.57)

$$\lim_{\alpha \to \infty} \bar{P}^\varepsilon_t(\alpha) < \infty.$$

But by (8.9)

$$\bar{P}^\varepsilon_0(t, \alpha) = \Psi^{\varepsilon 0}_{0,t} \left(\bar{P}^\varepsilon_t(\alpha) - P^{\varepsilon 0}_t \right) \Psi^{\varepsilon 0 T}_{0,t}.$$

Hence

$$\lim_{\alpha \to \infty} \bar{P}^\varepsilon_0(t, \alpha) < \infty \quad \text{or} \quad \Gamma^\varepsilon_t > 0.$$

That is, Condition 2 holds. Then, by Theorem 8.1, Condition 3 holds, and hence Condition 1 holds.

Thus we have proved the chain of implications:

$$\text{Condition } 1 \Rightarrow 2 \Rightarrow 3 \Rightarrow 4 \Rightarrow 5 \Rightarrow 6 \Rightarrow 1,$$

which establishes the validity of the theorem. □

Remark. $\overline{\mathbf{P}}_0^{\varepsilon}(t, \alpha)$ is defined independently of whether $\mathbf{D}_t \mathbf{D}_t^{\tau}$ and $\mathbf{F}_t \mathbf{F}_t^{\tau}$ are nondegenerate or not. Hence the definition of stochastic observability given here is applicable to both stochastic and deterministic systems. From Condition 6 of Theorem 8.2 we see that for deterministic systems, stochastic observability coincides with complete observability.

8.3. GAUSS–MARKOV FILTERING, INTERPOLATION, AND PREDICTION

Let $\mathscr{L}_s(t)$ be the set of unbiased estimates for $\mathbf{x}_s$ based on $\{\mathbf{y}_\lambda, 0 \le \lambda \le t\}$ defined in Section 8.2. If it is nonempty, let

$$E(\mathbf{x}_s^* - \mathbf{x}_s)(\mathbf{x}_s^* - \mathbf{x}_s)^{\tau} = \inf_{\mathbf{z} \in \mathscr{L}_s(t)} E(\mathbf{z} - \mathbf{x}_s)(\mathbf{z} - \mathbf{x}_s)^{\tau}. \tag{8.78}$$

$\mathbf{x}_s^*$ is called the Gauss–Markov estimate for $\mathbf{x}_s$ based on $\{\mathbf{y}_\lambda, 0 \le \lambda \le t\}$. If $s \in [0, t)$, then it is called the Gauss–Markov interpolation and if $s = t$, the Gauss–Markov filtered estimate.

We now give explicit expressions for them and, in case $\mathbf{F}_\lambda \mathbf{F}_\lambda^{\tau} > \mathbf{0}$, we provide recursive formulas for these estimates.

Theorem 8.3. *Suppose that for some $t \in [0, T]$, one of the conditions presented in Theorem 8.2 is fulfilled, then the Gauss–Markov estimate coincides with the EWKIV. That is, $\mathbf{x}_s^* = \overline{\mathbf{x}}_s(t)$, where $\overline{\mathbf{x}}_s(t)$ and its estimation error covariance matrix are given in Theorem 8.1.*

Proof. Since $\mathscr{L}_s(t)$ is nonempty, there is a function $\mathbf{H}_\lambda$ satisfying the conditions of Lemma 8.6.

By (8.65), we have

$$\begin{aligned}
\mathbf{P}_t^{\varepsilon s} &= E\left[E\left(\mathbf{x}_t / \mathscr{F}_t^{\mathbf{y}^\varepsilon}, \mathbf{x}_s\right) - \mathbf{x}_t\right]\left[E\left(\mathbf{x}_t / \mathscr{F}_t^{\mathbf{y}^\varepsilon}, \mathbf{x}_s\right) - \mathbf{x}_t\right]^{\tau} \\
&= \boldsymbol{\Psi}_{t,s}^{\varepsilon s} \int_s^t \boldsymbol{\Psi}_{s,\lambda}^{\varepsilon s} \left(\mathbf{K}_\lambda^{\varepsilon s} \mathbf{F}_\lambda' - \mathbf{D}_\lambda'\right)\left(\mathbf{K}_\lambda^{\varepsilon s} \mathbf{F}_\lambda' - \mathbf{D}_\lambda'\right)^{\tau} \boldsymbol{\Psi}_{s,\lambda}^{\varepsilon s \tau} \, d\lambda \, \boldsymbol{\Psi}_{t,s}^{\varepsilon s \tau} \\
&= \boldsymbol{\Psi}_{t,s}^{\varepsilon s} \int_s^t \Big[\boldsymbol{\Psi}_{s,\lambda}^{\varepsilon s} (\mathbf{K}_\lambda^{\varepsilon s} \mathbf{F}_\lambda - \mathbf{D}_\lambda)(\mathbf{K}_\lambda^{\varepsilon s} \mathbf{F}_\lambda - \mathbf{D}_\lambda)^{\tau} \boldsymbol{\Psi}_{s,\lambda}^{\varepsilon s \tau} \, d\lambda \\
&\quad + \varepsilon \boldsymbol{\Psi}_{s,\lambda}^{\varepsilon s} \mathbf{K}_\lambda^{\varepsilon s} \mathbf{K}_\lambda^{\varepsilon s \tau} \boldsymbol{\Psi}_{s,\lambda}^{\varepsilon s \tau}\Big] \, d\lambda \, \boldsymbol{\Psi}_{t,s}^{\varepsilon s \tau},
\end{aligned} \tag{8.79}$$

and from (8.9)

$$\overline{\mathbf{P}}_s^{\varepsilon}(t,\alpha) = \boldsymbol{\Psi}_{s,t}^{\varepsilon s}(\overline{\mathbf{P}}_t^{\varepsilon}(\alpha) - \mathbf{P}_t^{\varepsilon s})\boldsymbol{\Psi}_{s,t}^{\varepsilon s \tau}. \tag{8.80}$$

Putting (8.79) and (8.80) into (8.60) leads to the conclusion that for any $\mathbf{H}_\lambda$ satisfying the conditions of Lemma 8.6 the following inequality holds:

$$\overline{\mathbf{P}}_s^{\varepsilon}(t,\alpha) \le E\left(\int_0^t \mathbf{H}_\lambda\, d\mathbf{y}_\lambda - \mathbf{x}_s\right)\left(\int_0^t \mathbf{H}_\lambda\, d\mathbf{y}_\lambda - \mathbf{x}_s\right)^{\tau} + \varepsilon\int_0^t \mathbf{H}_\lambda \mathbf{H}_\lambda^{\tau}\, d\lambda. \tag{8.81}$$

From Theorem 8.2, $\lim_{\varepsilon\to 0}\det \boldsymbol{\Gamma}_t^{\varepsilon} \neq 0$, and hence Theorem 8.1 can be applied. By passing to the limit in (8.81), it then turns out that

$$\lim_{\alpha\to\infty}\ \lim_{\varepsilon\to 0} \overline{\mathbf{P}}_s^{\varepsilon}(t,\alpha) = \overline{\mathbf{P}}_s(t) \le E\left(\int_0^t \mathbf{H}_\lambda\, d\mathbf{y}_\lambda - \mathbf{x}_s\right)\left(\int_0^t \mathbf{H}_\lambda\, d\mathbf{y}_\lambda - \mathbf{x}_s\right)^{\tau}. \tag{8.82}$$

Now from (8.23), we have

$$\int_0^t \boldsymbol{\Psi}_{s,\mu}^{\varepsilon s}\mathbf{K}_\mu^{\varepsilon s}\, d\mathbf{y}_\mu - \boldsymbol{\Psi}_{s,t}^{\varepsilon s}\mathbf{x}_t = -\mathbf{x}_s + \int_s^t \boldsymbol{\Psi}_{s,\lambda}^{\varepsilon s}(\mathbf{K}_\lambda^{\varepsilon s}\mathbf{F}_\lambda - \mathbf{D}_\lambda)\, d\mathbf{w}_\lambda, \tag{8.83}$$

which, by substitution into (8.48), leads to

$$\overline{\mathbf{x}}_s^{\varepsilon}(t) = \boldsymbol{\Psi}_{s,t}^{\varepsilon s}(\boldsymbol{\zeta}_t^{\varepsilon} + \mathbf{x}_t) - \int_s^t \boldsymbol{\Psi}_{s,\mu}^{\varepsilon s}\mathbf{K}_\mu^{\varepsilon s}\, d\mathbf{y}_\mu, \tag{8.84}$$

and from this

$$\overline{\mathbf{x}}_t^{\varepsilon} = \boldsymbol{\zeta}_t^{\varepsilon} + \mathbf{x}_t.$$

Hence

$$\overline{\mathbf{x}}_s^{\varepsilon}(t) = \boldsymbol{\Psi}_{s,t}^{\varepsilon s}\overline{\mathbf{x}}_t^{\varepsilon} - \int_s^t \boldsymbol{\Psi}_{s,\mu}^{\varepsilon s}\mathbf{K}_\mu^{\varepsilon s}\, d\mathbf{y}_\mu,$$

and by (8.74)

$$\overline{\mathbf{x}}_s^{\varepsilon}(t) = \boldsymbol{\Psi}_{s,t}^{\varepsilon s}\boldsymbol{\Psi}_{t,0}^{\varepsilon 0}\boldsymbol{\Gamma}_t^{\varepsilon -1}\int_0^t \boldsymbol{\Gamma}_\lambda^{\varepsilon}\boldsymbol{\Psi}_{0,\lambda}^{\varepsilon 0}\left[\mathbf{D}_\lambda\mathbf{F}_\lambda^{\tau} + \left(\mathbf{P}_\lambda^{\varepsilon 0} + \boldsymbol{\Psi}_{\lambda,0}^{\varepsilon 0}\boldsymbol{\Gamma}_\lambda^{\varepsilon +}\boldsymbol{\Psi}_{\lambda,0}^{\varepsilon 0 \tau}\right)\mathbf{C}_\lambda^{\tau}\right]$$

$$\cdot(\mathbf{F}_\lambda\mathbf{F}_\lambda^{\tau} + \varepsilon\mathbf{I})^{-1}\, d\mathbf{y}_\lambda - \int_s^t \boldsymbol{\Psi}_{s,\mu}^{\varepsilon s}\mathbf{K}_\mu^{\varepsilon s}\, d\mathbf{y}_\mu. \tag{8.85}$$

As can be seen from (8.48), $\bar{\mathbf{x}}_s^\varepsilon(t)$ is an unbiased estimator for $\mathbf{x}_s$ for any $E\mathbf{x}_0$, hence

$$\int_0^t \underset{\sim}{\mathbf{H}}_\lambda \underset{\sim}{\mathbf{C}}_\lambda \underset{\sim}{\mathbf{\Phi}}_{\lambda,0}\, d\lambda = \underset{\sim}{\mathbf{\Phi}}_{s,0},$$

where

$$\underset{\sim}{\mathbf{H}}_\lambda = \underset{\sim}{\mathbf{\Psi}}_{s,t}^{\varepsilon s} \underset{\sim}{\mathbf{\Psi}}_{t,0}^{\varepsilon 0} \underset{\sim}{\mathbf{\Gamma}}_t^{\varepsilon -1} \underset{\sim}{\mathbf{\Gamma}}_\lambda^{\varepsilon} \underset{\sim}{\mathbf{\Psi}}_{0,\lambda}^{\varepsilon 0} \left[\underset{\sim}{\mathbf{D}}_\lambda \underset{\sim}{\mathbf{F}}_\lambda^\tau + \left(\underset{\sim}{\mathbf{P}}_\lambda^{\varepsilon 0} + \underset{\sim}{\mathbf{\Psi}}_{\lambda,0}^{\varepsilon 0} \underset{\sim}{\mathbf{\Gamma}}_\lambda^{\varepsilon +} \underset{\sim}{\mathbf{\Psi}}_{\lambda,0}^{\varepsilon 0 \tau}\right) \underset{\sim}{\mathbf{C}}_\lambda^\tau\right]$$

$$\cdot \left(\underset{\sim}{\mathbf{F}}_\lambda \underset{\sim}{\mathbf{F}}_\lambda^\tau + \varepsilon \underset{\sim}{\mathbf{I}}\right)^{-1} - \underset{\sim}{\mathbf{\Psi}}_{s,\lambda}^{\varepsilon s} \underset{\sim}{\mathbf{K}}_\lambda^{\varepsilon s} I_{[s \le \lambda \le t]}. \tag{8.86}$$

It is easy to verify that $\underset{\sim}{\mathbf{H}}_\lambda$ also satisfies the other conditions of Lemma 8.6. Hence

$$\bar{\mathbf{x}}_s^\varepsilon(t) = \int_0^t \underset{\sim}{\mathbf{H}}_\lambda\, d\mathbf{y}_\lambda \in \mathscr{L}_s(t),$$

and

$$\underset{\varepsilon \to 0}{\text{l.i.m.}}\, \bar{\mathbf{x}}_s^\varepsilon(t) = \bar{\mathbf{x}}_s(t) \in \mathscr{L}_s(t).$$

By Theorem 8.1, the estimation error covariance matrix of the estimate $\bar{\mathbf{x}}_s(t)$ for $\mathbf{x}_s$ is just $\bar{\underset{\sim}{\mathbf{P}}}_s(t)$, which is the infimum of the right-hand side of (8.82). Thus $\bar{\mathbf{x}}_s(t)$ coincides with the Gauss–Markov estimate. □

Corollary. *From the proof of the theorem it can be seen that if* $\underset{\sim}{\mathbf{H}}_\lambda$ *is given by* (8.86), *then*

$$\bar{\mathbf{x}}_s^\varepsilon(t) = \int_0^t \underset{\sim}{\mathbf{H}}_\lambda\, d\mathbf{y}_\lambda$$

is the suboptimal Gauss–Markov interpolation ($s < t$) *and filtering* ($s = t$) *formula for* $\mathbf{x}_s$ *based on* $\{\mathbf{y}_\lambda, 0 \le \lambda \le t\}$. *We shall write*

$$\underset{\sim}{\mathbf{R}}_\lambda = \underset{\sim}{\psi}_{t,0}^{\varepsilon 0} \underset{\sim}{\mathbf{\Gamma}}_t^{\varepsilon -1} \underset{\sim}{\mathbf{\Gamma}}_\lambda^{\varepsilon} \underset{\sim}{\mathbf{\Psi}}_{0,\lambda}^{\varepsilon 0} \left[\underset{\sim}{\mathbf{D}}_\lambda \underset{\sim}{\mathbf{F}}_\lambda^\tau + \left(\underset{\sim}{\mathbf{P}}_\lambda^{\varepsilon 0} + \underset{\sim}{\mathbf{\Psi}}_{\lambda,0}^{\varepsilon 0} \underset{\sim}{\mathbf{\Gamma}}_\lambda^{\varepsilon +} \underset{\sim}{\mathbf{\Psi}}_{\lambda,0}^{\varepsilon 0 \tau}\right) \underset{\sim}{\mathbf{C}}_\lambda^\tau\right]$$

$$\cdot \left(\underset{\sim}{\mathbf{F}}_\lambda \underset{\sim}{\mathbf{F}}_\lambda^\tau + \varepsilon \underset{\sim}{\mathbf{I}}\right)^{-1} \underset{\sim}{\mathbf{F}}_\lambda - \underset{\sim}{\mathbf{\Gamma}}_\lambda^{\varepsilon} \underset{\sim}{\mathbf{\Psi}}_{0,\lambda}^{\varepsilon 0} \underset{\sim}{\mathbf{D}}_\lambda - \underset{\sim}{\mathbf{\Psi}}_{0,\lambda}^{\varepsilon s} \left(\underset{\sim}{\mathbf{K}}_\lambda^{\varepsilon s} \underset{\sim}{\mathbf{F}}_\lambda - \underset{\sim}{\mathbf{D}}_\lambda\right) I_{[s \le \lambda \le t]}.$$

Then from (8.40) *and* (8.48), *it is easy to see that*

$$E\left(\bar{\mathbf{x}}_s^\varepsilon(t) - \mathbf{x}_s\right)\left(\bar{\mathbf{x}}_s^\varepsilon(t) - \mathbf{x}_s\right)^\tau = \underset{\sim}{\mathbf{\Psi}}_{s,t}^{\varepsilon s} \int_0^t \underset{\sim}{\mathbf{R}}_\lambda \underset{\sim}{\mathbf{R}}_\lambda^\tau\, d\lambda\, \underset{\sim}{\mathbf{\Psi}}_{s,t}^{\varepsilon s \tau}. \tag{8.87}$$

We shall now use the notation introduced in (6.50)–(6.59).

Theorem 8.4. *For the system* (7.9), (7.10), *in addition to the conditions imposed on the coefficient matrices, assume that* $\mathbf{F}_\lambda \mathbf{F}_\lambda^\tau > \mathbf{0}$ *and*

$$\int_0^T \mathbf{C}_\lambda^\tau (\mathbf{F}_\lambda \mathbf{F}_\lambda^\tau)^{-1} \mathbf{C}_\lambda \, d\lambda < \infty .$$

Then for

$$t > t_0 \triangleq \inf\left\{ t \colon \det \int_0^t \mathbf{\Phi}_{\lambda,0}^\tau \mathbf{C}_\lambda^\tau \mathbf{C}_\lambda \mathbf{\Phi}_{\lambda,0} \, d\lambda \neq 0 \right\},$$

the Gauss–Markov estimate of $\mathbf{x}_s$ *based on* $\{\mathbf{y}_\lambda,\ 0 \le \lambda \le t\}$ *satisfies the following stochastic integral equation*:

$$\bar{\mathbf{x}}_s(t) = \bar{\mathbf{x}}_s + \int_s^t \bar{\mathbf{P}}_s(\lambda) \mathbf{\Psi}_{\lambda,s}^{s\tau} \mathbf{C}_\lambda^\tau (\mathbf{F}_\lambda \mathbf{F}_\lambda^\tau)^{-1} (d\mathbf{y}_\lambda - \mathbf{C}_\lambda \bar{\mathbf{x}}_\lambda \, d\lambda) \qquad (8.88)$$

$$\bar{\mathbf{P}}_s(t) = \left[\mathbf{I} + \bar{\mathbf{P}}_s \int_s^t \mathbf{\Psi}_{\lambda,s}^{s\tau} \mathbf{C}_\lambda^\tau (\mathbf{F}_\lambda \mathbf{F}_\lambda^\tau)^{-1} \mathbf{C}_\lambda \mathbf{\Psi}_{\lambda,s}^{s} \, d\lambda \right]^{-1} \bar{\mathbf{P}}_s . \qquad (8.89)$$

Similarly, replacing $\mathbf{P}_t$ *by* $\bar{\mathbf{P}}_t$ *in* (6.31) *gives the recursive formula of Gauss–Markov filtering and the equation for* $\bar{\mathbf{P}}_t$ *is given by* (6.8) *with initial values* $\bar{\mathbf{x}}_\lambda, \bar{\mathbf{P}}_\lambda$, $\lambda > t_0$. *Finally,* (6.75) *and* (6.76) *become the backwards Gauss–Markov interpolation equations if the quantities* $\mathbf{P}_s(t)$ *and* $\mathbf{P}_s$ *in them are replaced by* $\bar{\mathbf{P}}_s(t)$ *and* $\bar{\mathbf{P}}_s$, *respectively, for* $t > t_0$.

Proof. According to the definition of $\tilde{\boldsymbol{\xi}}_s^\varepsilon(t)$, if in (8.24) we remove the last two terms on the right-hand side, we obtain $\tilde{\boldsymbol{\xi}}_s^\varepsilon(t)$. That is,

$$\begin{aligned}
\tilde{\boldsymbol{\xi}}_s^\varepsilon(t) = \boldsymbol{\xi}_s &- \frac{1}{\alpha} \mathbf{\Psi}_{s,t}^{\varepsilon s} \mathbf{\Psi}_{t,0}^{\varepsilon 0} \bar{\mathbf{P}}_0^\varepsilon(t,\alpha)(\boldsymbol{\xi}_0 - \mathbf{x}) \\
&- \frac{1}{\alpha} \mathbf{\Psi}_{s,t}^{\varepsilon s} \mathbf{\Psi}_{t,0}^{\varepsilon 0} \bar{\mathbf{P}}_0^\varepsilon(t,\alpha) \int_0^t \mathbf{\Phi}_{0,\lambda} \mathbf{D}_\lambda \, d\mathbf{w}_\lambda + \mathbf{\Psi}_{s,t}^{\varepsilon s} \boldsymbol{\zeta}_t^\varepsilon(\alpha) \\
&- \int_s^t \mathbf{\Psi}_{s,\lambda}^{\varepsilon s} (\mathbf{K}_\lambda^{\varepsilon s} \mathbf{F}_\lambda - \mathbf{D}_\lambda) \, d\mathbf{w}_\lambda .
\end{aligned} \qquad (8.90)$$

On the other hand, since $\mathbf{F}_t \mathbf{F}_t^\tau > \mathbf{0}$ and $\mathbf{C}_t^\tau (\mathbf{F}_t \mathbf{F}_t^\tau)^{-1} \mathbf{C}_t$ are integrable on $[0, T]$, in analogy with (8.20)–(8.24), it can be shown that

$$\begin{aligned}
\boldsymbol{\xi}_s(t) = \boldsymbol{\xi}_s &- \frac{1}{\alpha} \mathbf{\Psi}_{s,t}^{s} \mathbf{\Psi}_{t,0}^{0} \bar{\mathbf{P}}_0(t,\alpha)(\boldsymbol{\xi}_0 - \mathbf{x}) - \frac{1}{\alpha} \mathbf{\Psi}_{s,t}^{s} \mathbf{\Psi}_{t,0}^{0} \bar{\mathbf{P}}_0(t,\alpha) \int_0^t \mathbf{\Phi}_{0,\lambda} \mathbf{D}_\lambda \, d\mathbf{w}_\lambda \\
&+ \mathbf{\Psi}_{s,t}^{s} \boldsymbol{\zeta}_t(\alpha) - \int_s^t \mathbf{\Psi}_{s,\lambda}^{s} (\mathbf{K}_\lambda^{s} \mathbf{F}_\lambda - \mathbf{D}_\lambda) \, d\mathbf{w}_\lambda ,
\end{aligned} \qquad (8.91)$$

where the notation of (6.50)–(6.52) is invoked and $\zeta_t(\alpha)$ is defined by (8.13) with ε removed. By Lemma 8.2, as $\varepsilon \to 0$, $\underset{\sim}{\Psi}^{\varepsilon s}_{s,t}\underset{\sim}{\Psi}^{\varepsilon 0}_{t,0}\overline{\underset{\sim}{\mathbf{P}}}{}^{\varepsilon}_{0}(t,\alpha)$ converges, and by Theorem 7.2, as $\varepsilon \to 0$, $\tilde{\boldsymbol{\xi}}^{\varepsilon}_s t)$ tends to $\hat{\boldsymbol{\xi}}_s(t)$ in the mean square sense. But by the definition in (8.90), if we replace $\boldsymbol{\xi}_0$ by $\mathbf{x}_0$, $\boldsymbol{\xi}_s$ by $\mathbf{x}_s$ and $\mathbf{x}$ by $\mathbf{0}$, then we obtain $\overline{\mathbf{x}}^{\varepsilon}_s(t,\alpha)$ [see (8.27)]. Hence

$$\underset{\varepsilon\to 0}{\text{l.i.m.}}\, \overline{\mathbf{x}}^{\varepsilon}_s(t,\alpha) = -\frac{1}{\alpha}\underset{\sim}{\Psi}^{s}_{s,t}\underset{\sim}{\Psi}^{0}_{t,0}\overline{\underset{\sim}{\mathbf{P}}}_{0}(t,\alpha)\mathbf{x}_0 - \frac{1}{\alpha}\underset{\sim}{\Psi}^{s}_{s,t}\underset{\sim}{\Psi}^{0}_{t,0}\overline{\underset{\sim}{\mathbf{P}}}_{0}(t,\alpha)$$

$$\cdot\int_0^t \underset{\sim}{\Phi}_{0,\lambda}\underset{\sim}{\mathbf{D}}_\lambda\, d\mathbf{w}_\lambda + \mathbf{x}_s + \underset{\sim}{\Psi}^{s}_{s,t}\zeta_t(\alpha) - \int_s^t \underset{\sim}{\Psi}^{s}_{s,\lambda}(\underset{\sim}{\mathbf{K}}^{s}_{\lambda}\underset{\sim}{\mathbf{F}}_\lambda - \underset{\sim}{\mathbf{D}}_\lambda)\, d\mathbf{w}_\lambda. \tag{8.92}$$

Now from (8.91), it is easy to see that if we directly replace $E\mathbf{x}_0$ by $\mathbf{0}$ and $\underset{\sim}{\mathbf{R}}_{\mathbf{x}_0}$ by $\alpha\underset{\sim}{\mathbf{I}}$ in the expression of $\hat{\mathbf{x}}_s(t)$, we obtain exactly the formula (8.92). Hence for the present case, the Gauss–Markov interpolation is given by

$$\overline{\mathbf{x}}_s(t) = \mathbf{x}_s + \underset{\sim}{\Psi}^{s}_{s,t}\zeta_t - \int_s^t \underset{\sim}{\Psi}^{s}_{s,\lambda}(\underset{\sim}{\mathbf{K}}^{s}_{\lambda}\underset{\sim}{\mathbf{F}}_\lambda - \underset{\sim}{\mathbf{D}}_\lambda)\, d\mathbf{w}_\lambda, \tag{8.93}$$

where

$$\zeta_t = \underset{\sim}{\Psi}^{0}_{t,0}\underset{\sim}{\Gamma}_t^{-1}\int_0^t \Big\{\underset{\sim}{\Gamma}_\lambda\underset{\sim}{\Psi}^{0}_{0,\lambda}\big[\underset{\sim}{\mathbf{D}}_\lambda\underset{\sim}{\mathbf{F}}^{\tau}_\lambda + (\underset{\sim}{\mathbf{P}}^{0}_\lambda + \underset{\sim}{\Psi}^{0}_{\lambda,0}\underset{\sim}{\Gamma}^{+}_\lambda\underset{\sim}{\Psi}^{0\tau}_{\lambda,0})\underset{\sim}{\mathbf{C}}^{\tau}_\lambda\big](\underset{\sim}{\mathbf{F}}_\lambda\underset{\sim}{\mathbf{F}}^{\tau}_\lambda)^{-1}\underset{\sim}{\mathbf{F}}_\lambda$$

$$-\underset{\sim}{\Gamma}_\lambda\underset{\sim}{\Psi}^{0}_{0,\lambda}\underset{\sim}{\mathbf{D}}_\lambda\Big\}\, d\mathbf{w}_\lambda, \tag{8.94}$$

and

$$\underset{\sim}{\Gamma}_t = \int_0^t \underset{\sim}{\Psi}^{0\tau}_{\lambda,0}\underset{\sim}{\mathbf{C}}^{\tau}_\lambda(\underset{\sim}{\mathbf{F}}_\lambda\underset{\sim}{\mathbf{F}}^{\tau}_\lambda)^{-1}\underset{\sim}{\mathbf{C}}_\lambda\underset{\sim}{\Psi}^{0}_{\lambda,0}\, d\lambda. \tag{8.95}$$

From (8.83),

$$\int_s^t \underset{\sim}{\Psi}^{s}_{s,\mu}\underset{\sim}{\mathbf{K}}^{s}_{\mu}\, d\mathbf{y}_\mu - \underset{\sim}{\Psi}^{s}_{s,t}\mathbf{x}_t = -\mathbf{x}_s + \int_s^t \underset{\sim}{\Psi}^{s}_{s,\lambda}(\underset{\sim}{\mathbf{K}}^{s}_{\lambda}\underset{\sim}{\mathbf{F}}_\lambda - \underset{\sim}{\mathbf{D}}_\lambda)\, d\mathbf{w}_\lambda.$$

Hence by (8.93), it follows that

$$\overline{\mathbf{x}}_s(t) = \underset{\sim}{\Psi}^{s}_{s,t}\zeta_t + \underset{\sim}{\Psi}^{s}_{s,t}\mathbf{x}_t - \int_s^t \underset{\sim}{\Psi}^{s}_{s,\lambda}\underset{\sim}{\mathbf{K}}^{s}_{\lambda}\, d\mathbf{y}_\lambda,$$

and

$$\overline{\mathbf{x}}_s(t) = \underset{\sim}{\Psi}^{s}_{s,t}\overline{\mathbf{x}}_t - \int_s^t \underset{\sim}{\Psi}^{s}_{s,\lambda}\underset{\sim}{\mathbf{K}}^{s}_{\lambda}\, d\mathbf{y}_\lambda. \tag{8.96}$$

In the equations for $\hat{\mathbf{x}}_t$, if we replace $E\mathbf{x}_0$ by $\mathbf{0}$ and $\mathbf{P}_t$ by $\overline{\mathbf{P}}_t$ it turns out by (8.74) that

$$\overline{\mathbf{x}}_t = \mathbf{\Psi}_{t,0}^0 \mathbf{\Gamma}_t^{-1} \int_0^t \mathbf{\Gamma}_\lambda \mathbf{\Psi}_{0,\lambda}^0 \left[\mathbf{D}_\lambda \mathbf{F}_\lambda^\tau + \left(\mathbf{P}_\lambda^0 + \mathbf{\Psi}_{\lambda,0}^0 \mathbf{\Gamma}_\lambda^+ \mathbf{\Psi}_{\lambda,0}^{0\tau}\right)\mathbf{C}_\lambda^\tau\right]\left(\mathbf{F}_\lambda \mathbf{F}_\lambda^\tau\right)^{-1} d\mathbf{y}_\lambda, \tag{8.97}$$

and from (6.71) that

$$\overline{\mathbf{P}}_t = \mathbf{P}_t^s + \mathbf{\Psi}_{t,0}^0 \mathbf{\Gamma}_t^{-1} \mathbf{\Psi}_{t,0}^{0\tau}. \tag{8.98}$$

Taking the stochastic differential of $\overline{\mathbf{x}}_t$ gives

$$d\overline{\mathbf{x}}_t = \mathbf{A}_t \overline{\mathbf{x}}_t\, dt + \left(\mathbf{D}_t \mathbf{F}_t^\tau + \overline{\mathbf{P}}_t \mathbf{C}_t^\tau\right)\left(\mathbf{F}_t \mathbf{F}_t^\tau\right)^{-1}\left(d\mathbf{y}_t - \mathbf{C}_t \overline{\mathbf{x}}_t\, dt\right), \qquad t > t_0,$$

and, for $t > t_0$, $\overline{\mathbf{P}}_t$ clearly satisfies (6.8). Hence, for $t > t_0$, $\overline{\mathbf{x}}_t$ satisfies (6.8) and (6.31). The initial conditions $\overline{\mathbf{x}}_\lambda, \overline{\mathbf{P}}_\lambda$ are defined at $\lambda > t_0$ by the Gauss–Markov estimate for $\mathbf{x}_\lambda$ and its associated estimation error covariance matrix based on $\{\mathbf{y}_s, 0 \le s \le \lambda\}$, respectively.

Taking the stochastic differential of (8.94) and using (8.97) and (8.98) leads to (8.88), while (8.89) can be directly obtained from (6.66). The rest of the assertions are proved analogously. □

Remark 1. Let $t \ge s > 0$, $s > t_0$. It is easy to see that the Gauss–Markov prediction of $\mathbf{x}_t$ based on $\{\mathbf{y}_\lambda, 0 \le \lambda \le s\}$ is

$$\overline{\mathbf{x}}_t(s) = \mathbf{\Phi}_{t,s} \overline{\mathbf{x}}_s.$$

Remark 2. As $\alpha \to \infty$, $\overline{\mathbf{x}}_s^\varepsilon(t,\alpha)$ tends to the Gauss–Markov estimate $\overline{\mathbf{x}}_s(t)$. Notice that $\overline{\mathbf{x}}_s^\varepsilon(t,\alpha)$ can be computed in real time, hence it is a suitable practical approximation to $\overline{\mathbf{x}}_s(t)$. In particular, if $\mathbf{F}_t \mathbf{F}_t^\tau > \mathbf{0}$, Theorem 8.3 gives a theoretical basis for the engineering practice of taking $E\mathbf{x}_0 = \mathbf{0}$ and $\mathbf{R}_{\mathbf{x}_0} = \alpha \mathbf{I}$ when $E\mathbf{x}_0$ and $\mathbf{R}_{\mathbf{x}_0}$ are unknown.

8.4. STOCHASTIC CONTROL WITH A MINIMAX INDEX OF QUADRATIC TYPE

We shall consider system (6.1), (6.2) on $[0, T]$, that is,

$$d\mathbf{x}_t = \mathbf{A}_t \mathbf{x}_t\, dt + \mathbf{B}_t \mathbf{u}_t\, dt + \mathbf{D}_t\, d\mathbf{w}_t, \tag{8.99}$$

$$d\mathbf{y}_t = \mathbf{C}_t \mathbf{x}_t\, dt + \mathbf{F}_t\, d\mathbf{w}_t, \tag{8.100}$$

where the coefficient matrices are the same as in Chapter 6, but now we allow $\underset{\sim}{\mathbf{F}}_s\underset{\sim}{\mathbf{F}}_s^{\tau}$ to be degenerate.

The set $\mathscr{U}$ of admissible controls is still defined by (7.52)–(7.54).

Initial values $E\mathbf{x}_0$ and $\underset{\sim}{\mathbf{R}}_{\mathbf{x}_0}$ will be unknown, but will be assumed to be finite.

Without loss of generality, we can assume $\{\mathbf{w}_t\}$ and $[\mathbf{x}_0^{\tau}, \mathbf{y}_0^{\tau}]^{\tau}$ are normally distributed and mutually independent.

Let us put

$$J(t_1, \mathbf{u}, E\mathbf{x}_0 \underset{\sim}{\mathbf{R}}_{\mathbf{x}_0}) = E\left[\mathbf{x}_T^{\tau}\underset{\sim}{\mathbf{Q}}_0\mathbf{x}_T + \int_{t_1}^{T}\left(\mathbf{x}_t^{\tau}\underset{\sim}{\mathbf{Q}}_1(t)\mathbf{x}_t + \mathbf{u}_t^{\tau}\underset{\sim}{\mathbf{Q}}_2(t)\mathbf{u}_t\right) dt\right], \tag{8.101}$$

where $\underset{\sim}{\mathbf{Q}}_0 \geq \underset{\sim}{\mathbf{0}}$, $\underset{\sim}{\mathbf{Q}}_1(t) \geq \underset{\sim}{\mathbf{0}}$, $\underset{\sim}{\mathbf{Q}}_2(t) > \underset{\sim}{\mathbf{0}}$.

Since $E\mathbf{x}_0$ and $\underset{\sim}{\mathbf{R}}_{\mathbf{x}_0}$ are unknown, we shall use a worst case or minimax analysis; hence we shall consider the performance index:

$$\inf_{\mathbf{u}\in\mathscr{U}} \sup_{E\mathbf{x}_0, \underset{\sim}{\mathbf{R}}_{\mathbf{x}_0}} J(t_1, \mathbf{u}, E\mathbf{x}_0, \underset{\sim}{\mathbf{R}}_{\mathbf{x}_0}). \tag{8.102}$$

In (6.102), instead of 0 put t_1 and take the expected value, then

$$J(t_1, \mathbf{u}, E\mathbf{x}_0, \underset{\sim}{\mathbf{R}}_{\mathbf{x}_0}) = E\mathbf{x}_{t_1}^{\tau}\underset{\sim}{\mathbf{S}}_{t_1}\mathbf{x}_{t_1} + \int_{t_1}^{T}\operatorname{tr}\underset{\sim}{\mathbf{D}}_t\underset{\sim}{\mathbf{D}}_t^{\tau}\underset{\sim}{\mathbf{S}}_t\, dt$$
$$+E\int_{t_1}^{T}(\mathbf{u}_t - \underset{\sim}{\mathbf{L}}_t\mathbf{x}_t)^{\tau}\underset{\sim}{\mathbf{Q}}_2(t)(\mathbf{u}_t - \underset{\sim}{\mathbf{L}}_t\mathbf{x}_t)\, dt, \tag{8.103}$$

where $\underset{\sim}{\mathbf{S}}_t$ and $\underset{\sim}{\mathbf{L}}_t$ are given by (6.97) and (6.99).

Together with (8.99) and (8.100), we shall also consider the system (7.55), (7.56), which is assumed to be stochastically observable.

Suppose

$$t_1 > t_0 \triangleq \inf\left\{t: \det\int_0^t \underset{\sim}{\mathbf{\Phi}}_{\lambda,0}^{\tau}\underset{\sim}{\mathbf{C}}_{\lambda}^{\tau}\underset{\sim}{\mathbf{C}}_{\lambda}\underset{\sim}{\mathbf{\Phi}}_{\lambda,0}\, d\lambda \neq 0\right\}, \tag{8.104}$$

and denote by $\tilde{\boldsymbol{\xi}}_t$ the Gauss–Markov filtered estimate of $\boldsymbol{\xi}_t$ based on $\{\boldsymbol{\eta}_\lambda, 0 \leq \lambda \leq t\}$. Let $\bar{\boldsymbol{\xi}}_t^{\varepsilon}$ be defined by (8.74) with $\mathbf{y}_\lambda$ replaced by $\boldsymbol{\eta}_\lambda$.

Lemma 8.7. *Assume that*

$$\int_0^T\left(\|\underset{\sim}{\mathbf{A}}_t\| + \|\underset{\sim}{\mathbf{B}}_t\|^2 + \|\underset{\sim}{\mathbf{C}}_t\|^2 + \|\underset{\sim}{\mathbf{F}}_t\|^2 + \|\underset{\sim}{\mathbf{D}}_t\|^2\right) dt < \infty$$

and that $\underset{\sim}{\mathbf{M}}_s$ *is an* $r \times n$ *deterministic matrix with*

$$\int_0^T \|\underset{\sim}{\mathbf{M}}_s\|^2 \, ds < \infty .$$

Then $\{\mathbf{u}_t\} \in \mathscr{U}$, *if*

$$\mathbf{u}_t = \underset{\sim}{\mathbf{M}}_t \boldsymbol{\zeta}_t^{\varepsilon}, \tag{8.105}$$

where

$$\boldsymbol{\zeta}_t^{\varepsilon} = \bar{\boldsymbol{\xi}}_t^{\varepsilon} I_{[t_1 \le t]} + \int_0^t \underset{\sim}{\boldsymbol{\Phi}}_{t,s} \underset{\sim}{\mathbf{B}}_s \underset{\sim}{\mathbf{M}}_s \boldsymbol{\zeta}_s^{\varepsilon} \, ds. \tag{8.106}$$

Proof. For $t < t_1$, $\boldsymbol{\zeta}_t^{\varepsilon}$ is a deterministic vector and the assertion of the lemma is trivial.

By differentiating (8.74), we have

$$d\bar{\boldsymbol{\xi}}_t^{\varepsilon} = \underset{\sim}{\mathbf{A}}_t \bar{\boldsymbol{\xi}}_t^{\varepsilon} \, dt = (\underset{\sim}{\mathbf{D}}_t \underset{\sim}{\mathbf{F}}_t^{\tau} + \overline{\underset{\sim}{\mathbf{P}}}_t^{\varepsilon} \underset{\sim}{\mathbf{C}}_t^{\tau})(\underset{\sim}{\mathbf{F}}_t \underset{\sim}{\mathbf{F}}_t^{\tau} + \varepsilon \underset{\sim}{\mathbf{I}})^{-1} (d\boldsymbol{\eta}_t - \underset{\sim}{\mathbf{C}}_t \bar{\boldsymbol{\xi}}_t^{\varepsilon} \, dt),$$

and now the proof can be completed along the lines of the proof of Lemma 7.4. □

Lemma 8.8. *If*

$$\int_0^{t_1} \underset{\sim}{\boldsymbol{\Phi}}_{0,\lambda} \underset{\sim}{\mathbf{B}}_{\lambda} \underset{\sim}{\mathbf{B}}_{\lambda}^{\tau} \underset{\sim}{\boldsymbol{\Phi}}_{0,\lambda}^{\tau} \, d\lambda > \underset{\sim}{\mathbf{0}}, \tag{8.107}$$

then there is an admissible control $\{\mathbf{u}_t\} \in \mathscr{U}$ *on* $[0, t_1)$ *such that*

$$\sup_{E\mathbf{x}_0, \underset{\sim}{\mathbf{R}}_{\mathbf{x}_0}} E\mathbf{x}_{t_1}^{\tau} \underset{\sim}{\mathbf{S}}_{t_1} \mathbf{x}_{t_1} < \infty .$$

Proof. Take

$$\mathbf{u}_{\lambda} = -\underset{\sim}{\mathbf{B}}_{\lambda}^{\tau} \underset{\sim}{\boldsymbol{\Phi}}_{0,\lambda}^{\tau} \left(\int_0^{t_1} \underset{\sim}{\boldsymbol{\Phi}}_{0,s} \underset{\sim}{\mathbf{B}}_s \underset{\sim}{\mathbf{B}}_s^{\tau} \underset{\sim}{\boldsymbol{\Phi}}_{0,s}^{\tau} \, ds \right)^{-1} \mathbf{x}_0, \qquad \lambda \in [0, t_1],$$

then

$$\mathbf{x}_{t_1} = \underset{\sim}{\boldsymbol{\Phi}}_{t_1,0} \mathbf{x}_0 - \underset{\sim}{\boldsymbol{\Phi}}_{t_1,0} \int_0^{t_1} \underset{\sim}{\boldsymbol{\Phi}}_{0,\lambda} \underset{\sim}{\mathbf{B}}_{\lambda} \underset{\sim}{\mathbf{B}}_{\lambda}^{\tau} \underset{\sim}{\boldsymbol{\Phi}}_{0,\lambda}^{\tau} \, d\lambda \left(\int_0^{t_1} \underset{\sim}{\boldsymbol{\Phi}}_{0,s} \underset{\sim}{\mathbf{B}}_s \underset{\sim}{\mathbf{B}}_s^{\tau} \underset{\sim}{\boldsymbol{\Phi}}_{0,s}^{\tau} \, ds \right)^{-1} \mathbf{x}_0$$

$$+ \int_0^{t_1} \underset{\sim}{\boldsymbol{\Phi}}_{t_1,\lambda} \underset{\sim}{\mathbf{D}}_{\lambda} \, d\mathbf{w}_{\lambda} = \int_0^{t_1} \underset{\sim}{\boldsymbol{\Phi}}_{t_1,\lambda} \underset{\sim}{\mathbf{D}}_{\lambda} \, d\mathbf{w}_{\lambda}$$

and

$$E\mathbf{x}_{t_1}^{\tau}\underset{\sim}{\mathbf{S}}_{t_1}\mathbf{x}_{t_1} = \operatorname{tr}\underset{\sim}{\mathbf{S}}_{t_1}\int_0^{t_1}\underset{\sim}{\mathbf{\Phi}}_{t_1,\lambda}\underset{\sim}{\mathbf{D}}_{\lambda}\underset{\sim}{\mathbf{D}}_{\lambda}^{\tau}\underset{\sim}{\mathbf{\Phi}}_{t_1,\lambda}\,d\lambda,$$

which is independent of $E\mathbf{x}_0$ and $\underset{\sim}{\mathbf{R}}_{\mathbf{x}_0}$. □

Theorem 8.5. *Suppose that $\underset{\sim}{\mathbf{A}}_t$, $\underset{\sim}{\mathbf{B}}_t$, $\underset{\sim}{\mathbf{C}}_t$, $\underset{\sim}{\mathbf{F}}_t$, and $\underset{\sim}{\mathbf{D}}_t$ satisfy the conditions of Lemma* 8.7,

$$\int_0^T \|\underset{\sim}{\mathbf{B}}_t\underset{\sim}{\mathbf{Q}}_2^{-1}(t)\|^2\,dt < \infty,$$

that (8.104) *and* (8.107) *hold, and further suppose that $\mathbf{u}_t^1$, $t \in [0, t_1)$ has been chosen such that*

$$\sup_{E\mathbf{x}_0,\,\underset{\sim}{\mathbf{R}}_0} E\|\mathbf{x}_{t_1}\|^2 < \infty. \tag{8.108}$$

Then

$$\mathbf{u}_t^{\varepsilon} \triangleq \begin{cases} \mathbf{u}_t^1, & t < t_1 \\ \underset{\sim}{\mathbf{L}}_t\boldsymbol{\zeta}_t^{\varepsilon}, & t \geq t_1 \end{cases}$$

is an optimal sequence of controls. That is,

$$J\big(t_1, \mathbf{u}^{\varepsilon}, E\mathbf{x}_0, \underset{\sim}{\mathbf{R}}_{\mathbf{x}_0}\big) \underset{\varepsilon\to 0}{\to} \inf_{\mathbf{u}\in\mathscr{U}}\ \sup_{E\mathbf{x}_0,\,\underset{\sim}{\mathbf{R}}_{\mathbf{x}_0}} J\big(t_1, \mathbf{u}, E\mathbf{x}_0, \mathbf{R}_{\mathbf{x}_0}\big)$$

$$= E\mathbf{x}_{t_1}^{\tau}\underset{\sim}{\mathbf{S}}_{t_1}\mathbf{x}_{t_1} + \int_{t_1}^{T}\operatorname{tr}\underset{\sim}{\mathbf{D}}_t\underset{\sim}{\mathbf{D}}_t^{\tau}\underset{\sim}{\mathbf{S}}_t\,dt + \int_{t_1}^{T}\operatorname{tr}\underset{\sim}{\mathbf{L}}_t^{\tau}\underset{\sim}{\mathbf{Q}}_2(t)\underset{\sim}{\mathbf{L}}_t\overline{\underset{\sim}{\mathbf{P}}}_t\,dt, \tag{8.109}$$

where $\overline{\underset{\sim}{\mathbf{P}}}_t$ is $\overline{\underset{\sim}{\mathbf{P}}}_t(t)$ given by Theorem 8.1.

Proof. Denote by $\underset{\sim}{\mathbf{P}}_t$ the filtering error covariance matrix of $\boldsymbol{\xi}_t$ by using $\{\boldsymbol{\eta}_s,\ 0 \leq s \leq t\}$; in order to emphasize its dependence on $\underset{\sim}{\mathbf{R}}_{\mathbf{x}_0}$, we shall write it as $\underset{\sim}{\mathbf{P}}_t(\underset{\sim}{\mathbf{R}}_{\mathbf{x}_0})$.

By Theorem 7.1,

$$\lim_{\varepsilon\to 0}\underset{\sim}{\mathbf{P}}_t^{\varepsilon}(\alpha\underset{\sim}{\mathbf{I}}) = \underset{\sim}{\mathbf{P}}_t(\alpha\underset{\sim}{\mathbf{I}}),$$

and since (8.104) holds, by Theorem 8.2, we have

$$\lim_{\alpha\to\infty}\lim_{\varepsilon\to 0}\underset{\sim}{\mathbf{P}}_t^{\varepsilon}(\alpha\underset{\sim}{\mathbf{I}}) = \lim_{\alpha\to\infty}\underset{\sim}{\mathbf{P}}_t(\alpha\underset{\sim}{\mathbf{I}}) < \infty, \qquad t \geq t_1,$$

or, in the notation of Section 8.3,

$$\lim_{\varepsilon\to 0}\lim_{\alpha\to\infty}\overline{\underset{\sim}{\mathbf{P}}}_t^{\varepsilon}(\alpha) = \lim_{\varepsilon\to 0}\overline{\underset{\sim}{\mathbf{P}}}_t^{\varepsilon} = \lim_{\alpha\to\infty}\lim_{\varepsilon\to 0}\overline{\underset{\sim}{\mathbf{P}}}_t^{\varepsilon}(\alpha) = \overline{\underset{\sim}{\mathbf{P}}}_t < \infty, \qquad \forall t \geq t_1.$$

From Lemma 7.7, $\underset{\sim}{\mathbf{P}}_t(\underset{\sim}{\mathbf{R}}_{\mathbf{x}_0})$ is nondecreasing as $\underset{\sim}{\mathbf{R}}_{\mathbf{x}_0}$ increases. Hence

$$\sup_{\underset{\sim}{\mathbf{R}}_{\mathbf{x}_0}}\underset{\sim}{\mathbf{P}}_t(\underset{\sim}{\mathbf{R}}_{\mathbf{x}_0}) = \lim_{\alpha\to\infty}\underset{\sim}{\mathbf{P}}_t(\alpha\underset{\sim}{\mathbf{I}}) = \lim_{\alpha\to\infty}\lim_{\varepsilon\to 0}\overline{\underset{\sim}{\mathbf{P}}}_t^{\varepsilon}(\alpha) = \overline{\underset{\sim}{\mathbf{P}}}_t < \infty.$$

Notice that $\overline{\underset{\sim}{\mathbf{P}}}_t$ is independent of $E\mathbf{x}_0$. Hence from (8.103), it follows that

$$\begin{aligned}
&\inf_{\mathbf{u}_t, t\geq t_1}\ \sup_{E\mathbf{x}_0, \mathbf{R}_{\mathbf{x}_0}} J(t_1, \mathbf{u}, E\mathbf{x}_0, \mathbf{R}_{\mathbf{x}_0})\\
&= E\mathbf{x}_{t_1}^{\tau}\underset{\sim}{\mathbf{S}}_{t_1}\mathbf{x}_{t_1} + \int_{t_1}^{T}\operatorname{tr}\underset{\sim}{\mathbf{D}}_t\underset{\sim}{\mathbf{D}}_t^{\tau}\underset{\sim}{\mathbf{S}}_t\,dt + \inf_{\mathbf{u}_t, t\geq t_1}\ \sup_{E\mathbf{x}_0, \underset{\sim}{\mathbf{R}}_{\mathbf{x}_0}}\\
&\quad\times\left\{E\int_{t_1}^{T}(\mathbf{u}_t - \underset{\sim}{\mathbf{L}}_t\hat{\mathbf{x}}_t)^{\tau}\underset{\sim}{\mathbf{Q}}_2(t)(\mathbf{u}_t - \underset{\sim}{\mathbf{L}}_t\hat{\mathbf{x}}_t)\,dt + \int_{t_1}^{T}\operatorname{tr}\underset{\sim}{\mathbf{L}}_t^{\tau}\underset{\sim}{\mathbf{Q}}_2(t)\underset{\sim}{\mathbf{L}}_t\underset{\sim}{\mathbf{P}}_t(\underset{\sim}{\mathbf{R}}_{\mathbf{x}_0})\,dt\right\}\\
&\geq E\mathbf{x}_{t_1}^{\tau}\underset{\sim}{\mathbf{S}}_{t_1}\mathbf{x}_{t_1} + \int_{t_1}^{T}\operatorname{tr}\underset{\sim}{\mathbf{D}}_t\underset{\sim}{\mathbf{D}}_t^{\tau}\underset{\sim}{\mathbf{S}}_t\,dt + \int_{t_1}^{T}\operatorname{tr}\underset{\sim}{\mathbf{L}}_t^{\tau}\underset{\sim}{\mathbf{Q}}_2(t)\underset{\sim}{\mathbf{L}}_t\overline{\underset{\sim}{\mathbf{P}}}_t\,dt. \qquad (8.110)
\end{aligned}$$

By Lemma 8.7, $\{\mathbf{u}_t^{\varepsilon}\} \in \mathscr{U}$, hence by (8.103), (8.47), and (8.48), it follows that

$$\begin{aligned}
J(t_1, \mathbf{u}^{\varepsilon}, E\mathbf{x}_0, \underset{\sim}{\mathbf{R}}_{\mathbf{x}_0}) &= E\mathbf{x}_{t_1}^{\tau}\underset{\sim}{\mathbf{S}}_{t_1}\mathbf{x}_{t_1} + \int_{t_1}^{T}\operatorname{tr}\underset{\sim}{\mathbf{D}}_t\underset{\sim}{\mathbf{D}}_t^{\tau}\underset{\sim}{\mathbf{S}}_t\,dt + \int_{t_1}^{T}\operatorname{tr}\underset{\sim}{\mathbf{L}}_t^{\tau}\underset{\sim}{\mathbf{Q}}_2(t)\underset{\sim}{\mathbf{L}}_t\overline{\underset{\sim}{\mathbf{P}}}_t^{\varepsilon}\,dt\\
&\underset{\varepsilon\to 0}{\to} E\mathbf{x}_{t_1}^{\tau}\underset{\sim}{\mathbf{S}}_{t_1}\mathbf{x}_{t_1} + \int_{t_1}^{T}\operatorname{tr}\underset{\sim}{\mathbf{D}}_t\underset{\sim}{\mathbf{D}}_t^{\tau}\underset{\sim}{\mathbf{S}}_t\,dt + \int_{t_1}^{T}\operatorname{tr}\underset{\sim}{\mathbf{L}}_t^{\tau}\underset{\sim}{\mathbf{Q}}_2(t)\underset{\sim}{\mathbf{L}}_t\overline{\underset{\sim}{\mathbf{P}}}_t\,dt.
\end{aligned}$$

□

Remark. As in Section 7.4, we can consider the case where $\mathbf{Q}_2(t)$ is singular and the stochastic differential game problem when $E\mathbf{x}_0$ and $\underset{\sim}{\mathbf{R}}_{\mathbf{x}_0}$ are unknown. Corresponding results in these cases are not difficult to obtain. We can also obtain the discrete-time analogue of Theorem 8.5.

References

[1] Doob, J. L., *Stochastic Process*, John Wiley & Sons, New York, 1953.

[2] Liptser, R. S. and Shiryayev, A. N., *Statistics of Random Processes*, Springer, New York, 1977.

[3] Loeve, M., *Probability Theory*, Van Nostrand Reinhold, 1960.

[4] Wang, Z. K., *Theory of Random Processes* (in Chinese), Science Press, Brijing, 1965.

[5] Chow, Y. C., Local convergence of martingales and the law of large numbers, *The Annals of Mathematical Statistics*, Vol. 36, 1965, 552–558.

[6] Hall, P. and Heyde, C. C., *Martingale Limit Theory and Its Applications*, Academic Press, New York, 1980.

[7] Kiefer, E. and Wolfowitz, J., Stochastic estimation of the maximum of a regression function, *The Annals of Mathematical Statistics*, Vol. 23, 1952, 462–466.

[8] Robbins, H., and Monro, S., A stochastic approximation method, *The Annals of Mathematical Statistics*, Vol. 22, 1951, 400–407.

[9] Dvoretsky, A., On stochastic approximation, *Proceedings of the Third Berkeley Symposium on Mathematical Statistics and Probability*, Vol. I, University of California Press, Berkeley and Los Angeles, 1956, pp. 39–55.

[10] Gladyshev, E. T., On stochastic approximation (in Russian), *Probability Theory and Its Applications*, Vol. 10, No. 2, 1965, 297–300.

[11] Lai, T. L. and Robbins, H., Local convergence theorems for adaptive stochastic approximation schemes, *Proceedings of the National Academy of Sciences of the United States of America*, Vol. 76, No. 7, 1979, 3065–3067.

[12] Tsypkin, Ya. Z., *Self-Adaptive and Learning Systems* (in Russian), Nauka, Moscow, 1968.

[13] Wasan, M. T., *Stochastic Approximation*, Cambridge University Press, Cambridge, 1969.

[14] Nevelson, M. B. and Khasminsky, R. Z. *Stochastic Approximation and Recursive Estimation* (in Russian), Nauka, Moscow, 1972.

[15] Ljung, L., Analysis of recursive stochastic algorithms, *IEEE Transactions on Automatic Control*, AC-22, No. 4, 1977, 551–575.

[16] Ljung, L., On positive real transfer functions and the convergence of some recursive schemes, *IEEE Transactions on Automatic Control*, AC-22, No. 4, 1977, 539–550.

[17] Kushner, H. K. and Clark, D. S., *Stochastic Approximation Methods for Constrained and Unconstrained Systems*, Springer, New York, 1978.

[18] Chen, H. F., Stochastic approximation under correlated measurement errors, *Scientia Sinica* (Series A) (in Chinese), No. 3, 1983, 264–274.

[19] Chen, H. F., Stochastic approximation with ARMA measurement errors, *Journal of Systems Science and Mathematical Sciences*, Vol. 2, No. 3, 1982, 227–239.

[20] Chen, H. F., On continuous-time stochastic approximation, in A. Bensoussan and J. L. Lions (Eds.), *Analysis and Optimization of Systems*, Lecture Notes in Control and Information Science, Vol. 44, Springer, Berlin, 1982, pp. 203–214.

[21] Chen, H. F., Recursive algorithms for adaptive beam-formers, *Kexue Tongbao* (A Monthly Journal of Science), Vol. 26, No. 6, 1981, 10–13.

[22] Lai, T. L., Robbins, H., and Wei, C. Z., Strong consistency of least squares estimate in multiple regression, *Proceedings of the National Academy of Sciences of the United States of America*, Vol. 75, 1978, 3034–3036.

[23] Lai, T. L. and Wei, C. A., Least squares estimates in stochastic regression models with applications to identification and control of dynamic systems, *The Annals of Statistics*, Vol. 10, No. 1, 1982, 154–166.

[24] Chen, X. R., Consistency of least squares estimate in linear models, *Scientia Sinica, Special Issue (II) on Mathematics*, 1979, 162–176.

[25] Chen, H. F., Least squares identification for continuous-time systems, in A. Bensoussan and J. L. Lions (Eds.), *Analysis and Optimization of Systems*, Lecture Notes in Control and Information Science, Vol. 28, 264–277, Springer, Berlin, 1978.

[26] Ljung, L., Consistency of the least squares identification method, *IEEE Transactions on Automatic Control*, AC-21, No. 5, 1976, 779–781.

[27] Moore, J. B., On strong consistency of least squares identification algorithms, *Automatica*, Vol. 14, 1978, 505–509.

[28] Chen, H. F., Strong consistency and convergence rate of least squares indentification, *Scientia Sinica* (Series A), Vol. 25, No. 7, 1982, 771–784.

[29] Akaike, H., A new look at the statistical model identification, *IEEE Transactions on Automatic Control*, AC-19, No. 6, 1974, 716–722.

[30] Eykhoff, P., *System Identification*, John Wiley & Sons, London, 1974.

[31] Goodwin, G. C. and Payne, R. L., *Dynamic System Identification*, Academic Press, New York, 1977.

[32] Chen, H. F., Quasi-least-squares identification and its strong consistency, *International Journal of Control*, Vol. 34, No. 5, 1981, 921–936.

[33] Chen, H. F., Strong consistency in system identification under correlated noise, *Identification and System Parameter Estimation* (Proceedings of the 6th IFAC Symposium, Washington DC, June 1982), Pergamon Press, Oxford, 1983.

[34] Sin, K. S. and Goodwin, G. C., Stochastic adaptive control using a modified least squares algorithm, *Automatica*, Vol. 18, No. 3, 1982, 315–321.

[35] Solo, V., The convergence of AML, *IEEE Transactions on Automatic Control*, AC-24, No. 6, 1979, 958–962.

[36] Kushner, H. J., A projected stochastic approximation method for adaptive filters and identifiers; *IEEE Transactions on Automatic Control*, AC-25, 1980, 836–838.

[37] Popov, V. M., *Hyperstability of Automatic Control Systems*, Springer, New York, 1973.

[38] Landau, Y. D., *Adaptive Control, The Model Reference Approach*, Mercel Dekker, New York, 1979.

[39] Chen, H. F., Strong consistency of recursive identification under correlated noise. *Journal of Systems Science and Mathematical Sciences*, Vol. 1, No. 1, 1981, 34–52.

[40] Chen, H. F., Recursive system identification and adaptive control by use of the modified least squares algorithm, *SIAM J. Control and Optimization* Vol. 22, No. 5, 1984.

[41] Aström, K. J. and Wittenmark, B., On self-tuning regulators, *Automatica*, Vol. 9, 1973, 185–199.

[42] Clarke, D. W. and Gawthrop, P. J., Self-tuning controller, *Proceedings of IEE*, Vol. 122, 929–934, 1975.

[43] Lu, G. Z. and Yuan, Z. Z., Self-tuning controller of MIMO discrete-time systems, *Digital Computer Applications to Process Control* (Proceedings of the 6th IFAC/IFIP Conference, Dusseldorf, FRG, October 1980), Pergamon Press, Oxford, 1981.

[44] Yuan, Z. Z. and Lu, G. Z., Self-tuning controller with application to a distillation column, *Control Science and Technology for the Progress of Society* (Proceedings of the 8th Triennial World Congress of IFAC, Kyoto, Japan, August 1981), Vol. 22, Pergamon, Press, Oxford, 1982.

[45] Caines, P. E. and Lafortune, S., Adaptive control with recursive identification for stochastic linear systems, *IEEE Transactions on Automatic Control*, AC-28, No. 2, April 1984.

[46] Goodwin, G. C., Ramadge, P. T., and Caines, P. E., Discrete-time multi-variable adaptive control, *IEEE Transactions on Automatic Control*, AC-25, No. 3, 1980, 449–456.

[47] Goodwin, G. C., Ramadge, P. T., and Caines, P. E., Discrete-time stochastic adaptive control, *SIAM Journal on Control and Optimization*, Vol. 19, No. 6, 1981, 829–853.

[48] Chen, H. F., Self-tuning controller and its convergence under correlated noise, *International Journal of Control*, Vol. 35, No. 6, 1982, 1051–1059.

[49] Chen, H. F. and Caines, P. E., Adaptive linear quadratic control for stochastic discrete time systems, *Proceedings of the 9th IFAC Congress*, Budapest, 1984.

[50] Anderson, O. D., *Time Series*, North-Holland, Amsterdam, 1980.

[51] Box, G. E. P. and Jenkins, G. M., *Time Series Analysis*, *Forecasting and Control*, Holden-Day, San Francisco, 1970.

[52] Chen, H. F., Optimality of estimates without the knowledge of initial values, *Scientia Sinica*, Vol. 22, No. 6, 1979, 615–627.

[53] Chen, H. F., On stochastic observability, *Scientia Sinica*, Vol. 20, No. 3, 1977, 305–325.

[54] Chen, H. F., Linear unbiased minimum covariance interpolation and extrapolation (in Chinese), *Acta Mathematica Sinica*, Vol. 23, No. 1, 1980, 88–97.

[55] Chen, H. F., On stochastic observability and controllability, *Automatica*, Vol. 16, No. 2, 1980, 179–190.

[56] Aoki, M., *Optimization of Stochastic Systems*, Academic Press, New York, 1967.

[57] Sunahara, Y., Aikara, S., Kishino, K., On the stochastic observability and controllability for nonlinear systems, *International Journal of Control*, Vol. 21, No. 1, 1975, 65–82.

[58] Sunahara, Y., Aikara, S., and Shiraiwa, M., The stochastic observability for noisy nonlinear stochastic systems, *International Journal of Control*, Vol. 22, No. 4, 1975, 461–480.

[59] Aström, K. J., *Introduction to Stochastic Control Theory*, Academic Press, New York, 1970.

[60] Balakrishnan, A. V., *Stochastic Differential Systems I*, Lecture Notes, Springer, Berlin, 1971.

[61] Fujisaki, M., Kallianpur, G., and Kunita, H., Stochastic differential equation for the nonlinear filtering problems, *Osaka Journal of Mathematics*, Vol. 9, No. 1, 1972, 19–40.

[62] Kalman, R. E. and Bucy, R. S., New results in linear filtering and the prediction theory, *Journal of Basic Engineering*, March 1961, 95–108.

[63] Sun, F. K. and Ho, Y. C., Role of information in the stochastic zero-sum differential game, in G. Leitmann (Ed.), *Multicriteria Decision Making*, Plenum, New York, 1976.

[64] Wonham, W. M., On the separation theorems of stochastic control, *SIAM Journal of Control*, Vol. 6, 1968, 312–326.

[65] Ho, Y. C., Linear stochastic singular control problems, *Journal of Optimization Theory and Application*, Vol. 9, No. 1, 1972, 24–31.

[66] Chen, H. F., Stochastic control problem with the quadratic performance index (in Chinese), *Acta Mathematica Sinica*, Vol. 22, No. 4, 1979, 438–447.

[67] Chen, H. F., Stochastic control under the minimax performance index of the quadratic type (in Chinese), *Scientia Sinica*, *Special Issue (I) on Mathematics*, 1979, 165–177.

[68] Chen, H. F., Singular games for the linear quadratic Gaussian system, *Scientia Sinica*, Vol. 24, No. 6, 1981, 182–193.

[69] Chen, H. F., Unified controls and strategies applicable to both singular and nonsingular cases, *Control Science and Technology for the Progress of Society* (Proceedings of the 8th Triennial World Congress of IFAC, Kyoto, Japan, August 1981), Vol. 6, Pergamon Press, Oxford, 1982.

[70] Chen, H. F., Unified control laws under the quadratic performance index (in Chinese), *Acta Mathematicae Applicatae Sinica*, Vol. 5, No. 1, 1982, 45–52.

[71] Chen, H. F., The stochastic observability and the estimates without the knowledge of initial values for the continuous-time systems, *Scientia Sinica*, *Special Issue (II) on Mathematics*, 1979, 281–293.

[72] Dunford, N. and Schwatz, J. T., *Linear Operators*, Part 1, John Wiley & Sons, New York, 1966.

[73] Hahn, W., *Stability of Motion*, Springer, Berlin, 1967.

[74] Hardy, G. H., Polya, G., and Littlewood, J. E., *Inequalities*, Cambridge University Press, Cambridge, England, 1934.

[75] Anderson, B. D. O. and Moore, J. B., *Linear Optimal Control*, Prentice-Hall, Englewood Cliffs, New Jersey, 1971.

[76] Wonham, W. M., *Linear Multivariable Control: a Geometric Approach* Springer, New York, 1979.

[77] Anderson, B. D. O. and Moore, J. B., *Optimal Filtering*, Prentice-Hall, Englewood Cliffs, New Jersey, 1979.

Index

Applied Probability and Statistics (Continued)

DILLON and GOLDSTEIN • Multivariate Analysis: Methods and Applications
DODGE and ROMIG • Sampling Inspection Tables, *Second Edition*
DOWDY and WEARDEN • Statistics for Research
DRAPER and SMITH • Applied Regression Analysis, *Second Edition*
DUNN • Basic Statistics: A Primer for the Biomedical Sciences, *Second Edition*
DUNN and CLARK • Applied Statistics: Analysis of Variance and Regression
ELANDT-JOHNSON and JOHNSON • Survival Models and Data Analysis
FLEISS • Statistical Methods for Rates and Proportions, *Second Edition*
FOX • Linear Statistical Models and Related Methods
FRANKEN, KÖNIG, ARNDT, and SCHMIDT • Queues and Point Processes
GALAMBOS • The Asymptotic Theory of Extreme Order Statistics
GIBBONS, OLKIN, and SOBEL • Selecting and Ordering Populations: A New Statistical Methodology
GNANADESIKAN • Methods for Statistical Data Analysis of Multivariate Observations
GOLDBERGER • Econometric Theory
GOLDSTEIN and DILLON • Discrete Discriminant Analysis
GREENBERG and WEBSTER • Advanced Econometrics: A Bridge to the Literature
GROSS and CLARK • Survival Distributions: Reliability Applications in the Biomedical Sciences
GROSS and HARRIS • Fundamentals of Queueing Theory, *Second Edition*
GUPTA and PANCHAPAKESAN • Multiple Decision Procedures: Theory and Methodology of Selecting and Ranking Populations
GUTTMAN, WILKS, and HUNTER • Introductory Engineering Statistics, *Third Edition*
HAHN and SHAPIRO • Statistical Models in Engineering
HALD • Statistical Tables and Formulas
HALD • Statistical Theory with Engineering Applications
HAND • Discrimination and Classification
HILDEBRAND, LAING, and ROSENTHAL • Prediction Analysis of Cross Classifications
HOAGLIN, MOSTELLER, and TUKEY • Understanding Robust and Exploratory Data Analysis
HOEL • Elementary Statistics, *Fourth Edition*
HOEL and JESSEN • Basic Statistics for Business and Economics, *Third Edition*
HOGG and KLUGMAN • Loss Distributions
HOLLANDER and WOLFE • Nonparametric Statistical Methods
IMAN and CONOVER • Modern Business Statistics
JAGERS • Branching Processes with Biological Applications
JESSEN • Statistical Survey Techniques
JOHNSON and KOTZ • Distributions in Statistics
Discrete Distributions
Continuous Univariate Distributions—1
Continuous Univariate Distributions—2
Continuous Multivariate Distributions
JOHNSON and KOTZ • Urn Models and Their Application: An Approach to Modern Discrete Probability Theory
JOHNSON and LEONE • Statistics and Experimental Design in Engineering and the Physical Sciences, Volumes I and II, *Second Edition*
JUDGE, HILL, GRIFFITHS, LÜTKEPOHL and LEE • Introduction to the Theory and Practice of Econometrics
JUDGE, GRIFFITHS, HILL, LÜTKEPOHL and LEE • The Theory and Practice of Econometrics, *Second Edition*
KALBFLEISCH and PRENTICE • The Statistical Analysis of Failure Time Data
KUH, NEESE, and HOLLINGER • Structural Sensitivity in Econometric Models